ANALYTICAL CHEMISTRY SYMPOSIA SERIES — volume 11

stable isotopes

Proceedings of the 4th International Conference, Jülich, March 23–26, 1981

ANALYTICAL CHEMISTRY SYMPOSIA SERIES

Volume 1 Recent Developments in Chromatography and Electrophoresis. Proceedings of the 9th International Symposium on Chromatography and Electrophoresis, Riva del Garda, May 15–17, 1978
edited by A. Frigerio and L. Renoz

Volume 2 Electroanalysis in Hygiene, Environmental, Clinical and Pharmaceutical Chemistry, Proceedings of a Conference, organised by the Electroanalytical Group of the Chemical Society, London, held at Chelsea College, University of London, April 17–20, 1979
edited by W.F. Smyth

Volume 3 Recent Developments in Chromatography and Electrophoresis, 10. Proceedings of the 10th International Symposium on Chromatography and Electrophoresis, Venice, June 19–20, 1979
edited by A. Frigerio and M. McCamish

Volume 4 Recent Developments in Mass Spectrometry in Biochemistry and Medicine, 6. Proceedings of the 6th International Symposium on Mass Spectrometry in Biochemistry and Medicine, Venice, June 21–22, 1979
edited by A. Frigerio and M. McCamish

Volume 5 Biochemical and Biological Applications of Isotachophoresis. Proceedings of the First International Symposium, Baconfoy, May 4–5, 1979
edited by A. Adam and C. Schots

Volume 6 Analytical Isotachophoresis. Proceedings of the 2nd International Symposium on Isotachophoresis, Eindhoven, September 9–11, 1980
edited by F.M. Everaerts

Volume 7 Recent Developments in Mass Spectrometry in Biochemistry, Medicine and Environmental Research, 7. Proceedings of the 7th International Symposium on Mass Spectrometry in Biochemistry, Medicine and Environmental Research, Milan, June 16–18, 1980
edited by A. Frigerio

Volume 8 Ion-selective Electrodes. Proceedings of the Third Symposium, Mátrafüred, October 13–15, 1980
edited by E. Pungor

Volume 9 Affinity Chromatography and Related Techniques. Theoretical Aspects/ Industrial and Biomedical Applications. Proceedings of the 4th International Symposium, Veldhoven, The Netherlands, June 22–26, 1981
edited by T.C.J. Gribnau, J. Visser and R.J.F. Nivard

Volume 10 Advances in Steroid Analysis. Proceedings of the Symposium on the Analysis of Steroids, Eger, May 20–22, 1981
edited by S. Görög

Volume 11 Stable Isotopes. Proceedings of the 4th International Conference, Jülich, March 23–26, 1981
edited by H.-L. Schmidt, H. Förstel and K. Heinzinger

ANALYTICAL CHEMISTRY SYMPOSIA SERIES — volume 11

stable isotopes

Proceedings of the 4th International Conference, Jülich, March 23—26, 1981

edited by

H.-L. Schmidt,
Technische Universität München, Lehrstuhl für Allgemeine Chemie und Biochemie, Freising-Weihenstephan, F.R.G.

H. Förstel
Kernforschungsanlage Jülich GmbH, Jülich, F.R.G.

and

K. Heinzinger
Max-Planck-Institut für Chemie (Otto-Hahn-Institut), Mainz, F.R.G.

ELSEVIER SCIENTIFIC PUBLISHING COMPANY
Amsterdam — Oxford — New York   1982

ELSEVIER SCIENTIFIC PUBLISHING COMPANY
Molenwerf 1
P.O. Box 211, 1000 AE Amsterdam, The Netherlands

Distributors for the United States and Canada:

ELSEVIER SCIENCE PUBLISHING COMPANY, INC.
52, Vanderbilt Avenue
New York, N.Y. 10164

Library of Congress Cataloging in Publication Data
Main entry under title:

Stable isotopes.

    (Analytical chemistry symposia series ; v. 11)
    Includes index.
    1. Biological chemistry--Technique--Congresses.
2. Isotopes--Congresses. 3. Tracers (Biology)--Congresses. 4. Tracers (Chemistry)--Congresses.
I. Schmidt, H.-L. (Hanns-Ludwig), 1930- .
II. Förstel, H. (Hilmar), 1941- . III. Heinzinger,
K. (Karl), 1930- . IV. Series.
QP519.7.S8   574.19'285   82-2463
ISBN 0-444-42076-2   AACR2

ISBN 0-444-42076-2 (Vol. 11)
ISBN 0-444-41786-9 (Series)

© Elsevier Scientific Publishing Company, 1982

All rights reserved. No part of this publication may be reproduced, stored in a retrieval system or transmitted in any form or by any means, electronic, mechanical, photocopying, recording or otherwise, without the prior written permission of the publisher, Elsevier Scientific Publishing Company, P.O. Box 330, 1000 AH Amsterdam, The Netherlands

Printed in The Netherlands

CONTENTS

EDITORS PREFACE   XVII

I. ISOTOPE EFFECTS, THEORY AND CONSEQUENCES

   THE THEORETICAL ANALYSIS OF ISOTOPE EFFECTS   3
   M. Wolfsberg

   VAPOUR PRESSURE ISOTOPE EFFECT OF ACETONITRILE   15
   Gy. Jákli, G. Jancsó and T. Koritsánszky

   THERMODYNAMIC PROPERTIES OF SOLUTIONS OF UREA, DEUTEROUREA
   AND TETRAMETHYLUREA IN H_2O AND D_2O 23
   Gy. Jákli and W.A. van Hook

   HYDROGEN ISOTOPE EFFECTS IN THE DEHYDRATION OF POLYSTYRENE-
   DIVINYL-BENZENE(PS-DVB) TYPE ION-EXCHANGERS   29
   A.R. Gupta

   INTERPRETATION OF ISOTOPE FRACTIONATION IN BIOLOGICAL SYSTEMS   35
   E.M. Galimov

   OXYGEN KINETIC ISOTOPE EFFECT IN THE THERMAL ISOMERIZATION OF
   BIS-(5,5-DIMETHYL-2-OXO-1,3,2-DIOXAPHOSPHORINANIL) SULFIDE   49
   W. Reimschüssel and P. Paneth

   KINETIC STUDY OF DEUTERIUM EXCHANGE IN METHYL GROUPS OF
   PYRIMIDINE DERIVATIVES   55
   P. Dizabo and E. Picquenard

   DEUTERIUM SOLVENT ISOTOPE EFFECTS ON REACTIONS OF
   PENTAAQUAORGANOCHROMIUM (III) IONS WITH ELECTROPHILES   61
   V. Gold and D.L. Wood

HEAVY-ATOM ISOTOPE EFFECTS ON ENZYME-CATALYZED REACTIONS 67
M.H. O'Leary

ISOTOPE EFFECTS ON EACH, C- AND N-ATOMS, AS A TOOL FOR THE
ELUCIDATION OF ENZYME CATALYZED AMIDE HYDROLYSES 77
R. Medina, T. Olleros and H.-L. Schmidt

PARAMETERS AFFECTING THE $^{13}CO_2/^{12}CO_2$ ISOTOPE DISCRIMINATION
OF THE RIBULOSE-1,5-BISPHOSPHATE CARBOXYLASE REACTION 83
F.J. Winkler, H. Kexel, C. Kranz and H.-L. Schmidt

II. GEOCHEMISTRY AND COSMOCHEMISTRY

STABLE ISOTOPES AND THE ORIGIN OF THE SOLAR SYSTEM 93
F. Begemann

STABLE ISOTOPES AND THE EVOLUTION OF LIFE: AN OVERVIEW 95
M. Schidlowski

ISOTOPE GEOCHEMISTRY OF CARBON 103
J. Hoefs

THE $^{13}C/^{12}C$ ISOTOPE RATIOS IN A NORTH-GERMAN PODZOL 115
H.-G. Bertram and G.H. Schleser

STABLE ISOTOPE INVESTIGATIONS IN ANTARCTICA 121
H. Schütze, G. Strauch and K. Wetzel

CARBON ISOTOPE FRACTIONATION FACTORS OF THE CARBON DIOXIDE-
CARBONATE SYSTEM AND THEIR GEOCHEMICAL IMPLICATIONS 127
L.E. Maxwell and Z. Sofer

ISOTOPE GEOCHEMICAL MODELS OF THE EVOLUTION OF THE EARTH'S CRUST 135
K. Wetzel

THE ISOTOPIC FRACTIONATION DURING THE OXIDATION OF CARBON
MONOXIDE BY HYDROXYLRADICALS AND ITS IMPLICATION FOR THE
ATMOSPHERIC CO-CYCLE   147
H.G.J. Smit, A. Volz, D.H. Ehhalt and H. Knappe

MODELLING OF STABLE ISOTOPE COMPOSITION OF ATMOSPHERIC WATER
VAPOUR AND PRECIPITATION   153
K. Rozanski, K.O. Münnich and C. Sonntag

THE STABLE ISOTOPIC COMPOSITION OF INFILTRATION MOISTURE IN
THE UNSATURATED ZONE OF THE ENGLISH CHALK   161
A.H. Bath, W.G. Darling and A.P. Brunsdon

THE USE OF DEUTERIUM OXIDE (^2H_2O) AS A WATER TRACER IN A STUDY
OF THE HYDROLOGY OF THE UNSATURATED ZONE OF THE ENGLISH CHALK   167
S.R. Wellings

$^{18}O/^{16}O$-RATIO OF GROUNDWATER AT THE FEDERAL REPUBLIK OF GERMANY   173
H. Förstel and H. Hützen

OXYGEN-18 STUDY OF NON-LIQUID MECHANISMS OF ATMOSPHERIC
SULFATE FORMATION   179
B.D. Holt, P.T. Cunningham, A.G. Engelkemeir, D.G. Graczyk
and R. Kumar

III. BIOMEDICAL APPLICATIONS
    A. PHARMACOLOGY AND DRUG METABOLISM

APPLICATIONS OF STABLE ISOTOPES IN PHARMACOLOGICAL RESEARCH   187
T.A. Baillie, H. Hughes and D.S. Davies

APPLICATION OF ^{13}C-LABELLING IN THE BIOAVAILABILITY ASSESSMENT
OF EXPERIMENTAL CLOVOXAMINE FORMULATIONS   203
H. de Bree, D.J.K. van der Stel, J.H.M.A. Kaal and
J.B. van der Schoot

MEASUREMENT OF THE PHARMACOKINETICS OF DI-($[15,15,16,16-D_4]$-
LINOLEOYL)-3-sn-GLYCEROPHOSPHOCHOLINE AFTER ORAL ADMINISTRATION
TO RATS 211
A. Brekle, F. Wirtz-Peitz and O. Zierenberg

DEUTERIUM ISOTOPE EFFECTS IN THE METABOLISM OF THE
ANTI-CANCER AGENT 6-MERCAPTOPURINE 217
M. Jarman, J.H. Kiburis, G.B. Elion, V.C. Knick, G. Lambe,
D.J. Nelson and R.L. Tuttle

DEUTERIUM ISOTOPE EFFECTS IN THE METABOLISM OF DIPHENYLHYDANTOIN 223
J.A. Hoskins and P.B. Farmer

INVESTIGATIONS WITH ^{18}O-LABELLED COMPOUNDS OF OXYGEN TRANSFER
BY CYTOCHROME P-450 AND METHEMOGLOBIN 229
H. Kexel, B. Limbach, M. Möller and H.-L. Schmidt

ANALYTICAL TECHNIQUES FOR USING MULTIPLE, SIMULTANEOUS STABLE
ISOTOPIC TRACERS 235
D.L. Hachey, K. Nakamura, M.J. Kreek and P.D. Klein

DETERMINATION OF URAPIDIL IN HUMAN SERUM USING SOLID PROBE CHEMICAL
IONIZATION MASS SPECTROMETRY AND STABLE ISOTOPE LABELING 241
E. Sturm and K. Zech

DETERMINATION OF CATECHOL-O-METHYLTRANSFERASE (COMT) ACTIVITY BY
GAS CHROMATOGRAPHY-MASS SPECTROMETRY USING A MIXTURE OF
DEUTERATED CATECHOLAMINE AS MULTI-SUBSTRATE SYSTEM 247
H. Miyazaki and Y. Hashimoto

III. BIOMEDICAL APPLICATIONS
   B. CLINICAL DIAGNOSIS

THE APPLICATION OF THE STABLE ISOTOPES OF OXYGEN TO
BIOMEDICAL RESEARCH AND NEUROLOGY 255
D. Samuel

^{13}C-LABELLED VALPROIC ACID PULSE DOSING STEADY STATE
ANTIEPILEPTIC THERAPY FOR PHARMACOKINETIC STUDIES DURING
PREGNANCY 265
W. Wittfoht, H. Nau, D. Rating and H. Helge

THE APPLICATION OF DEUTERATED VALPROIC ACID (VPA) IN CHRONICALLY
TREATED EPILEPTIC PATIENTS UNDER MONOTHERAPY AND POLYPHARMACY 271
W. Kochen, B. Tauscher, M. Klemens and E. Depene

IN VIVO MEASUREMENT OF ENZYMES WITH DEUTERATED PRECURSORS
AND GC/SIM 277
H.-Ch. Curtius

PRENATAL DIAGNOSIS OF PROPIONIC AND METHYLMALONIC ACIDEMIA BY
STABLE ISOTOPE DILUTION ANALYSIS OF METHYLCITRIC AND
METHYLMALONIC ACIDS IN AMNIOTIC FLUIDS 287
L. Sweetman, G. Naylor, T. Ladner, J. Holm, W.L. Nyhan,
C. Hornbeck, J. Griffiths, L. Morch, S. Brandange, L. Gruenke
and J.C. Craig

MASS FRAGMENTOGRAPHIC DETERMINATION OF METHYLMALONIC ACID IN
MATERNAL URINE AND AMNIOTIC FLUID USING ^{2}H$_3$-METHYLMALONIC
ACID AS INTERNAL STANDARD 295
F.K. Trefz, B. Tauscher and W. Kochen

THE FATE OF ORALLY ADMINISTERED TESTOSTERONE-19-d$_3$ AND ITS
INFLUENCE ON THE PLASMA LEVELS OF ENDOGENOUS TESTOSTERONE IN HUMANS 301
S. Baba, Y. Shinohara and Y. Kasuya

IN VIVO STUDIES OF STEROID-METABOLISM USING DEUTERATED
PREGNENOLONE 307
Th. Kuster, M. Zachmann and B. Zagalak

STUDIES OF RENAL AND EXTRARENAL URIC ACID EXCRETION DURING
A DIETARY PURINE LOAD USING ^{15}N-URIC ACID 313
W. Löffler, W. Gröbner, R. Medina and N. Zöllner

ESTIMATION OF PROTEIN TURNOVER IN PATIENTS WITH LIVER
DISEASES USING ^{15}N-LABELLED GLYCINE   319
H. Faust, P. Junghans, R. Matkowitz, W. Hartig and K. Jung

ISOLATION OF ^{15}N LABELLED HUMAN SERUM PROTEINS AND NON-PROTEIN
NITROGEN COMPOUNDS OF SERUM AND URINE   325
H. Faust, H. Bornhak and K. Hirschberg

DYNAMIC ASPECT OF AMINO ACIDS IN NEONATES, PROBED BY
NITROGEN-15 AND GCMS   331
A. Lapidot, J. Amir and S.H. Reisner

USE OF WATER LABELLED WITH DEUTERIUM FOR MEDICAL APPLICATIONS   337
E. Roth, G. Basset, J. Sutton, M. Apfelbaum and J. Marsac

III. BIOMEDICAL APPLICATIONS
C. BREATH TESTS AND LUNG FUNCTION TESTS

THE COMMERCIAL FEASIBILITY OF ^{13}C BREATH TESTS   347
P.D. Klein and E.R. Klein

DECONVOLUTION OF ^{13}C-BREATH TEST DATA BY COMPARTMENTAL
MODELING   353
C.S. Irving, D.A. Schoeller, K. Nakamura and P.D. Klein

THE EFFECT OF ACARBOSE ON SUCROSE ABSORPTION MEASURED BY
THE ^{13}C-SUCROSE BREATH TEST   359
D. Rating, N. Gryzewski, W. Burger, C. Jakobs, B. Weber
and H. Helge

APPLICATION OF ^{13}C-FATTY ACIDS BREATH TESTS IN MYOCARDIAL
METABOLIC STUDIES   367
M. Suehiro, K. Ueda, M. Iio, J. Morikawa, M. Nakajima and R. Ohsawa

^{13}C-GLYCOCHOLATE BREATH TESTS IN CHILDREN WITH CYSTIC FIBROSIS   373
K. Siafarikas, D. Glaubitt and H. Steinhauer

BREATH TEST USING ^{13}C-PHENYLALANINE AS SUBSTRATE 379
D. Glaubitt and K. Siafarikas

METABOLIC RATE OF ^{13}C LABELLED AMINOPYRINE APPLIED TO THE
EXPLORATION OF HEPATIC FUNCTIONAL ACTIVITY 385
F. Botter, M. Drifford, J. Sutton, F. Degos and J.P. Benhamou

COMPARISON OF NATURALLY AND ARTIFICIALLY LABELLED GLUCOSE
UTILIZATION TO STUDY GLUCOSE OXIDATION BY MEANS OF $^{13}C/^{12}C$
BREATH TEST 393
M. Lacroix, N. Pallikarakis and F. Mosora

USE OF THE $^{13}C/^{12}C$ BREATH TEST TO STUDY SUGAR METABOLISM IN
ANIMALS AND MEN 399
J. Duchesne, M. Lacroix and F. Mosora

ASSESSMENT OF PULMONARY FUNCTION: APPLICATION OF STABLE ISOTOPES
TO MEASUREMENT OF LUNG DIFFUSING CAPACITY 409
M. Meyer, P. Scheid and J. Piiper

A CONSTANT-FLOW SINGLE-EXHALATION MEASUREMENT OF REGIONAL
LUNG FUNCTION 415
C. Hook and M. Meyer

IV. LIFE SCIENCES, AGRICULTURE AND ENVIRONMENTAL RESEARCH

STABLE ISOTOPES IN AGRICULTURE 421
H. Faust

BALANCE OF ^{15}N FERTILIZER IN SOIL/PLANT SYSTEM 433
N. Sotiriou and F. Korte

NITROGEN-15 FOR THE CONTROL OF NITROGEN UTILIZATION IN THE
ACTIVATED SLUDGE PROCESS 439
H. Noguchi, K. Muraoka, S. Noda, T. Matsuzaki and T. Morishita

^{15}N IN THE STUDY OF NITROGEN TRANSFORMATIONS AND AMMONIUM
EXCHANGE REACTIONS IN SOIL 445
Marie Králová, K. Drazdák, J. Kubát and M. Ebeid

ISOTOPIC FRACTIONATION IN SOYBEAN NODULES 451
D.H. Kohl, Barbara A. Bryan, Lori Feldman, P.H. Brown
and Georgia Shearer

EXPERIMENTAL DETERMINATION OF KINETIC ISOTOPE FRACTIONATION
OF NITROGEN ISOTOPES DURING DENITRIFICATION 459
A. Mariotti, J.C. Germon, A. Leclerc, G. Catroux and R. Letolle

NITROGEN ISOTOPE RATIO VARIATIONS IN BIOLOGICAL MATERIAL -
INDICATOR FOR METABOLIC CORRELATIONS 465
R. Medina and H.-L. Schmidt

NATURAL VARIATIONS OF ^{15}N-CONTENT OF NITRATE IN GROUND AND
SURFACE WATERS AND TOTAL NITROGEN OF SOIL IN THE WADI EL-NATRUN
AREA IN EGYPT 475
A.I.M. Aly, M.A. Mohamed and E. Hallaba

POSSIBILITIES OF STABLE ISOTOPE ANALYSIS IN THE CONTROL OF
FOOD PRODUCTS 483
J. Bricout

DETECTION OF ADDED WATER AND SUGAR IN NEW ZEALAND COMMERCIAL WINES 495
J. Dunbar

^{18}O/^{16}O-RATIO OF WATER IN PLANTS AND IN THEIR ENVIRONMENT
(RESULTS FROM FED. REP. GERMANY) 503
H. Förstel

USE OF WATER WITH DIFFERENT ^{18}O-CONTENT TO STUDY TRANSPORT
PROCESSES IN PLANTS 511
H. Förstel and H. Hützen

TRACER STUDIES ON THE OXYGEN EVOLUTION IN PHOTOSYNTHESIS 517
H. Metzner, K. Fischer and O. Bazlen

QUANTIFICATION OF INDOLE-3-ACETIC ACID BY GC/MS USING DEUTERIUM
LABELLED INTERNAL STANDARDS   529
J.R.F. Allen, J.-E. Rebeaud, L. Rivier and P.-E. Pilet

QUANTIFICATION OF ABSCISIC ACID AND ITS 2-TRANS ISOMER IN
PLANT TISSUES USING A STABLE ISOTOPE DILUTION TECHNIQUE   535
L. Rivier and P.E. Pilet

THE BIOSYNTHETIC ORIGIN OF THE SULFUR ATOMS IN LIPOIC ACID   543
R.H. White

PLANT UPTAKE OF ISOTOPICALLY ENRICHED SO_2 and NO_2 PRESENT AT
SUBNECROTIC CONCENTRATIONS IN THE ATMOSPHERE   549
M. Dubois, F. Botter, M. Drifford, J. Sutton and J. Guenot

DYNAMIC STUDY ON ANIMAL EXPERIMENTS USING ^{15}N-LABELED
NITROGEN DIOXIDE   557
Y. Otha, M. Yamada, Y. Yoneyama, A. Suzuki and I. Wakisaka

GC-MS ANALYSIS OF ^{2}H- AND ^{15}N-LABELED ANALOGUES OF N-NITROSO-
2(3',7'-DIMETHYL-2',6'-OCTADIENYL)AMINOETHANOLS IN ORGANIC
TISSUES   563
S.L. Abidi and A. Idelson

V. METHODS
A. ANALYTICAL DEVELOPMENTS

RECENT APPLICATIONS OF ^{13}C NMR SPECTROSCOPY TO BIOLOGICAL SYSTEMS   573
N.A. Matwiyoff

INTER- AND INTRAMOLECULAR ISOTOPIC HETEROGENEITY IN
BIOSYNTHETIC ^{13}C-ENRICHED AMINO ACIDS   587
E. Bengsch, J.-Ph. Grivet and H.-R. Schulten

QUANTITATIVE MASS SPECTROMETRY WITH STABLE ISOTOPE LABELLED
INTERNAL STANDARD AS A REFERENCE TECHNIQUE 593
I. Björkhem

GC/MS ASSAY OF MYO-INOSITOL (AS HEXAACETATE) IN HUMANS UTILIZING
A DEUTERATED INTERNAL STANDARD 605
E. Larsen, J.R. Andersen, H. Harbo, B. Bertelsen, J.E. Christensen
and G. Gregersen

ANALYSIS OF I-METHYL-I,2,3,4-TETRAHYDRO-β-CARBOLINE BY GC/MS
USING DEUTERIUM LABELLED INTERNAL STANDARDS 611
J.R.F. Allen and O. Beck

THE CARRIER EFFECT IN CAPILLARY GAS CHROMATOGRAPHY-MASS SPECTRO-
METRY AND ITS IMPLICATION ON THE ACCURATE MEASUREMENT OF ION
ABUNDANCE RATIOS 617
L. Dehennin, A. Reiffsteck and R. Scholler

APPLICATION AND MEASUREMENT OF METAL ISOTOPES 623
K. Habfast

STABLE ISOTOPES IN BIOMEDICAL AND ENVIRONMENTAL ANALYSIS
BY FIELD DESORPTION MASS SPECTROMETRY 635
W.D. Lehmann and H.-R. Schulten

CALCIUM ABSORPTION STUDIES IN MAN BY STABLE ISOTOPE DILUTION
AND FIELD DESORPTION MASS SPECTROMETRY 649
W.D. Lehmann and M. Kessler

AN ORGANIC PREPARATION SYSTEM FOR ^{13}C STABLE ISOTOPE RATIO
ANALYSIS 655
A. Barrie, M.C. Clarke and R.A. Cokayne

A NEW METHOD FOR THE DETERMINATION OF THE $^{18}O/^{16}O$ RATIO IN
ORGANIC COMPOUNDS 661
C.A.M. Brenninkmeijer and W.G. Mook

ANCILLARY TECHNIQUES FOR $^{13}CO_2$ GENERATION IN BIOMEDICAL STUDIES 667
W.W. Wong, D.A. Schoeller and P.D. Klein

A METHOD OF MASS SPECTROMETRIC ISOTOPIC ANALYSIS OF NITROGEN
AND OXYGEN IN DOUBLY LABELLED NITRIC OXIDE 673
M.N. Kerner, K.G. Ordzhonikidze, L.P. Parulava and G.A. Tevzadze

APPLICATION OF THE MICROWAVE PLASMA DETECTOR IN METABOLIC
STUDIES WITH THE ANTIARRHYTHMIC DRUG PROPAFENONE 679
H.G. Hege and J. Weymann

AUTORADIOGRAPHIC DETERMINATION OF ^{18}O BY PROTON ACTIVATION
ANALYSIS 685
Y. Ohta, T. Inada, A. Maruhashi, T. Kanai, K. Kouchi and M. Aihara

V. METHODS
  B. ISOTOPE SEPARATION AND SYNTHESIS OF LABELLED COMPOUNDS

THE PRODUCTION OF STABLE ISOTOPES OF OXYGEN 693
I. Dostrovsky and M. Epstein

SEPARATION OF THE STABLE ISOTOPES OF CHLORINE, SULFUR,
AND CALCIUM 703
W.M. Rutherford

ENRICHMENT OF HEAVY CALCIUM ISOTOPES BY ION EXCHANGER RESINS
WITH CYCLOPOLYETHERS AS ANCHOR GROUPS 711
K.G. Heumann, H.-P. Schiefer and W. Spiess

THE SYNTHESIS OF MONO- AND OLIGOSACCHARIDES ENRICHED WITH
ISOTOPES OF CARBON, HYDROGEN, AND OXYGEN 719
R. Barker, E.L. Clark, H.A. Nunez, J. Pierce, P.R. Rosevear
and A.S. Serianni

THE PREPARATION OF SOME C-DEUTERATED L-ASCORBIC ACIDS 731
H.J. Koch and R.S. Stuart

SPECIFIC DEUTERIUM LABELLING OF 1,4-BENZODIAZEPINES 735
A.A. Liebman, G.J. Bader, W. Burger, J. Cupano, C.M. Delaney,
Y.-Y. Liu, R.R. Muccino, C.W. Perry and E. Thom

SYNTHESIS OF ^{2}H-LABELLED PROSTAGLANDINS 743
C.O. Meese and J.C. Frölich

NEW APPROACHES IN THE PREPARATION OF ^{15}N LABELED AMINO ACIDS 747
Zvi E. Kahana and Aviva Lapidot

SYNTHESIS OF CARBON-13 AND OXYGEN-18 LABELLED COMPOUNDS 753
M.B. Chkhaidze, Z.N. Morchiladze, Ts.I. Obolashvili and
E.D. Oziashvili

AUTHOR INDEX 759

SUBJECT INDEX 763

EDITORS' PREFACE

The International Conferences on Stable Isotopes aim at the presentation of recent scientific results, originating from the determination of variations in natural isotope abundances, and from the application of stable isotopes as tracers. Thus, although basing on common analytical methodology, preferably quantitative mass spectrometry, the scope of these meetings is quite interdisciplinary.

The first three conferences, organized by Peter Klein at the Argonne National Laboratory, emphasized and demonstrated the importance of the stable isotopes of the main bioelements in biochemical, environmental, clinical and pharmacological research. The fourth conference, organized by the "Arbeitsgemeinschaft Stabile Isotope" for the first time in Europe, and mainly sponsored by the "Kernforschungs-anlage Jülich", continued this tradition, however, added as a second crucial point topics dealing with the theory of isotope effects, and with the reasons and consequences of isotope abundance' variations in the fields of geochemistry, cosmochemistry and agriculture.

The idea was to bring together scientists from rather different disciplines, in order to give them the possibility of exchanging experiences and discussing methods and results from quite different points of view. With the aim to make participants and readers familiar with the aspects of many disciplines, a large part of the meeting was dedicated to reviews, given by invited experts, who summarized methods, results and aspects of their proper field, thus demonstrating the possibilities and the methodology of isotope application from different points of view. The organizers strived to complete this framework by a number of selected original contributions of high quality, showing the present state in special topics, and stimulating and promoting future investigations.

When we succeeded in attaining, at least in part, this goal, we are largely indepted to all participants contributing a manuscript, and to the members of the International Scientific Committee

T.A. Baillie, Seattle WA	E. Galimov, Moscow	M.H. O'Leary, Madison
H.Ch. Curtius, Zürich	J.M. Hayes, Bloomington	D. Samuel, Rehovot
J. Duchesne, Liège	P.D. Klein, Houston	J. Sjövall, Stockholm
H. Faust, Leipzig	N. Matwiyoff, Los Alamos	K. Wetzel, Leipzig
A. Frigerio, Milan	H. Miyazaki, Tokio	M. Wolfsberg, Irvine

for their invaluable help in preparing the meeting and the proceedings.
Finally we want to thank the KFA for its really generous support and organisation, to the IAEA and many commercial companies for their sponsorship, to J. Dunbar for help in revising the English, to H. Kexel for preparing the subject index, and last not least to our secretaries for many hours of typing and correcting the manuscripts.

      H.-L. Schmidt   H. Förstel   K. Heinzinger

I. ISOTOPE EFFECTS, THEORY AND CONSEQUENCES

THE THEORETICAL ANALYSIS OF ISOTOPE EFFECTS

M. WOLFSBERG
University of California, Irvine, California 92717 (U.S.A.)

ABSTRACT

The theory of isotope effects on equilibrium constants and on chemical rate constants is briefly reviewed. Some current topics in isotope effect calculations are discussed: exact calculations of equilibrium constant isotope effects, information about condensed phases and transition states from vapor pressure isotope effects and rate isotope effects respectively, the dissection of isotope effects.

REVIEW OF THE THEORY OF ISOTOPE EFFECTS
Introduction

Most discussions of isotope effects on a variety of properties including equilibrium constants and rate constants are carried out within the framework of the Born-Oppenheimer approximation. Within this framework, the potential function for nuclear motion depends on the charges of the nuclei in a molecular system and on the number of electrons in the system but is independent of the isotopic masses of the nuclei. Isotope effects result then from the effect of nuclear mass on motion on the same potential energy surface. The potential energy surface of a molecule is usually described in terms of the so-called force constants of the molecule; these force constants are therefore independent of isotopic substitution and, for instance, the species $H_2^{16}O$, $H_2^{18}O$, $HD^{16}O$, $D_2^{16}O$, etc. all have the same force constants. Use is made of this mass independence of force constants in deducing force constants from observed vibrational spectra. Thus, observed data on $H_2^{16}O$ is not sufficient to deduce the force constants of the water molecule, but spectral data on a number of isotopic water molecules, which all move on the same potential energy surface, permits one to deduce these force constants (ref. 1).

Isotope effects on equilibrium constants

Consider the equilibria

AH + B = CH + E	K_1	(1)
AD + B = CD + E	K_2	(2)
AH + CD = AD + CH	K_3	(3)

Here a hydrogen in molecule AH (CH) has been replaced by a deuterium in AD (CD). The ratio K_1/K_2 (or more precisely the deviation of this ratio from unity) is referred to as an isotope effect; this isotope effect is clearly also equal to the equilibrium constant K_3 of the exchange reaction of Eq. (3).

By the methods of statistical mechanics, equilibrium constants may be expressed in terms of relevant molecular partition functions Q, and the isotope effect on an equilibrium constant may be expressed in terms of a ratio of isotopic partition function ratios. One obtains

$$K_3 = \frac{K_1}{K_2} = \frac{Q_{AD}/Q_{AH}}{Q_{CD}/Q_{CH}} \tag{4}$$

It can be shown that, in classical mechanics, isotope effects vanish if symmetry number contributions are ignored (ref. 2); symmetry number contributions merely reflect the effects arising from counting the number of ways in which a molecule may be formed and these effects will be ignored here. One can thus divide the right hand side of Eq. (4) by the corresponding partition function ratios evaluated classically. After some rearrangement, one obtains the isotope effect in terms of reduced isotopic partition function ratios $(s_2/s_1)f$,

$$K_3 = \frac{(s_2/s_1)f(AD/AH)}{(s_1/s_2)f(CD/CH)} \tag{5}$$

$$(s_2/s_1)f(AD/AH) = \frac{Q_{AD}/Q_{AH}}{(Q_{AD}/A_{AH})_{classical}} \tag{6}$$

By convention, the reduced isotopic partition function ratio is given as heavy molecule/light molecule.

With the approximations that the rotational motions are

classical, that there is no rotational-vibrational interaction, and that the vibrational motion is harmonic, Bigeleisen and Mayer (ref. 2) showed that

$$(s_2/s_1)f[\text{molecule 2/molecule 1}] = \prod_i \frac{u_{2i}}{u_{1i}} \frac{1-e^{-u_{1i}}}{1-e^{-u_{2i}}} e^{(u_{1i}-u_{2i})/2} . \quad (7)$$

Here $u = h\nu/kT$, where ν refers to a normal mode vibrational frequency, h is Planck's constant, k is Boltzmann's constant, and T is the absolute temperature. The product is over the 3N-6 (3N-5 for a linear molecule) normal mode vibrational frequencies of the N-atomic molecule.

Isotope effects on rate constants

If one considers isotope effects on rates within a transition state theory framework

$$AH + B \rightleftharpoons (AHB)^{\neq} \rightarrow \text{products} \quad k_H \quad (8)$$
$$AD + B \rightleftharpoons (ADB)^{\neq} \rightarrow \text{products} \quad k_D , \quad (9)$$

where $\neq$ designates transition state, one obtains (refs. 3-4) for the isotope effect on the rate constant

$$\frac{k_H}{k_D} = \frac{\nu_{1L}^{\neq}}{\nu_{2L}^{\neq}} \frac{(s_2/s_1)f(AD/AH)}{(s_2/s_1)f'(ADB^{\neq}/AHB^{\neq})} \quad (10)$$

The transition state here differs from a normal molecule in that one of its normal mode vibrational frequencies, the one corresponding to motion along "the reaction path" which is designated $\nu_{1L}^{\neq}$ in the H transition state here and $\nu_{2L}^{\neq}$ in the D transition state, is an imaginary number. The reduced isotopic partition function ratio $(s_2/s_1)f'$ is evaluated by omitting this frequency when the product is formed according to Eq. (7). Often a quantum mechanical tunnelling correction is added to Eq. (10). Except for the tunnelling correction and for the factor $\nu_{1L}^{\neq}/\nu_{2L}^{\neq}$, this theoretical framework for isotope effects on rate constants is quite similar to that for isotope effects on equilibrium constants.

DISCUSSION

Consider briefly the individual factors

$$b_i = \frac{u_{2i}}{u_{1i}} \frac{1-e^{-u_{1i}}}{1-e^{-u_{2i}}} e^{(u_{1i}-u_{2i})/2} \tag{11}$$

the product of which goes to make up the reduced isotopic partition function ratio. If the subscript 1 is used to refer to the isotopically lighter molecule, these factors will be equal to or larger than unity. The frequency ν_{1i} of the lighter molecule will always be equal to or larger than the frequency ν_{2i} of the heavier molecule, and consequently u_{1i} will be larger than u_{2i}. At low temperature, where $u = h\nu/kT$ is large, the term $\exp[(u_{1i} - u_{2i})/2]$ in Eq. (11) will dominate in making b_i larger than unity. This exponential term is the so-called zero-point energy factor. At high temperature, where u tends to be small and b_i tends to be close to unity, b_i takes the form

$$b_i = 1 + (1/24)(u_{1i}^2 - u_{2i}^2) \quad . \tag{12}$$

The reduced isotopic partition function ratio monotonically decreases from large values at low temperatures to unity at very high temperatures. The absolute magnitude of a reduced isotopic partition function ratio depends on the relative difference between the masses of the isotopic atoms and on the magnitude of the force constants at or near the position of isotopic substitution.

When a ratio of reduced isotopic partition function ratios is taken to calculate an isotope effect on an equilibrium constant (Eq. (5)) or to calculate an isotope effect on a rate constant (Eq. 10)), the heavy isotope (deuterium in the example here) will concentrate in that species which has the larger $(s_2/s_1)f$ value (i.e. the higher force constants at the position of isotopic substitution). Thus, in a calculation of an isotope effect on a rate constant, in a typical case one would expect $(s_2/s_1)f$ for the isotopically substituted reactant to be larger than $(s_2/s_1)f'$ for the corresponding transition state so that the heavy isotope concentrates in the reactant. Since the light isotope concentrates in the transition state, the light isotope (H here) corresponds to the higher rate constant. The word "typical" has been used here because one expects the force constants in transition states to be smaller than those in reactants. Rate constant isotope effects are therefore often referred to as "normal" if the light isotope reacts more rapidly than the heavy isotope (i.e. $k_H/k_D > 1$).

In practice, one also observes inverse isotope effects. Thus, for reactions in which the isotopically substituted reactant is an atomic species, the heavy atomic species often reacts more rapidly than the light atomic species. From the point of view of Eq. (10) this result comes about since $(s_2/s_1)f$ for an atomic reactant equals unity. Of course, both $\nu^{\neq}_{1L}/\nu^{\neq}_{2L}$ and tunnelling would favor a larger rate constant for the light atom. For the rates

$$H + H_2 \longrightarrow H_2 + H \qquad k_H \qquad (13)$$
$$D + H_2 \longrightarrow DH + H \qquad k_D \qquad (14)$$

k_D is found experimentally to be larger than k_H (ref. 5).

While it has been noted that, in an equilibrium, the heavy isotope will tend to concentrate in that species where the force constants are larger, the situation becomes complicated if, when comparison is made between the two species of the isotopic equilibrium, it is found that certain force constants involving the position of isotopic substitution are larger in the first species than in the second while others are smaller in the first species. Indeed, Stern, Spindel, and Monse (ref. 6) have shown that temperature anomalies often exist in equilibrium constant isotope effects, e.g. the isotope effect corresponds to an equilibrium constant for the corresponding isotopic exchange reaction which is larger than unity in one temperature range and smaller than unity in another temperature range. As an aside, it is noted that Skaron and this author (ref. 7) have recently proposed that these temperature anomalies might be responsible for the anomalous $^{18}O/^{17}O/^{16}O$ ratios observed by Clayton et al. (ref. 8) in meteoritic samples.

SOME CURRENT TOPICS IN ISOTOPE EFFECT CALCULATIONS
Force constants

Calculations of isotope effects on equilibrium constants require the normal mode vibrational frequencies of the relevant isotopically substituted species for the evaluation of $(s_2/s_1)f$ values. Even if all the vibrational frequencies of both isotopic species have been obtained experimentally, it is best not to employ these frequencies directly to calculate $(s_2/s_1)f$ from Eq. (7) since a small error in observed frequency can lead to a large error in $(s_2/s_1)f$. Rather, it is preferrable to start from isotope independent force constants, equilibrium geometries of molecules, and atomic masses to calculate normal mode vibrational frequencies and then reduced isotopic partition functions ratios. Computer programs exist to carry out such calculations readily (ref. 9). The force constants have usually been obtained in the past by fitting observed vibrational spectra for a molecule and its isotopic variants to isotope independent force fields (ref. 1). Another method which is just now beginning to find some application is the a priori calculation of force constants by the methods of quantum chemistry (ref. 10).

Accurate calculation of isotope effects on equilibrium constants

The approximations leading to the Bigeleisen-Mayer formula Eq. (7) for the reduced isotopic partition function ratio are lumped together as the harmonic approximation. If great accuracy is desired (say, $\pm$ 1% for hydrogen/deuterium isotope effects) in the calculations, corrections must be introduced. Such accurate calculations become necessary if one wants to test the completeness of the theory of isotope effects by comparing theoretical calculations with very precise experimental determinations. Also, for some isotope considerations (e.g. geochemistry), it may be important to have accurate values of isotope effects (fractionation factors) for systems on which it is difficult to make precise measurements directly.

The corrections which must be introduced into the reduced isotopic partition function ratios for gas phase molecules in order to achieve the previously stated accuracy around room temperature include the correction for the effect of anharmonicity on the zero-point factor and the correction for quantum effects on rotational motion (ref. 11). It has been found by considering a number of diatomic molecules and albeit only two polyatomic molecules (water and ammonia) that the effect of quantum mechanical rotation on calculated reduced isotopic partition function ratios for D/H isotopic substitution at 300K may be as large as 2% while the corresponding anharmonicity effect may be as large as 10% (refs. 11-12). This last correction generally decreases the reduced isotopic partition function ratio so that there may be considerable cancellation between the two ratios in the evaluation of an equilibrium constant isotope effect. For heavy atom isotope effects (effects other than hydrogen isotope effects), calculations at 300K have shown that quantum rotation leads to corrections of the order of 0.01% while anharmonicity can lead to corrections of a few tenths of one percent (ref. 11-12).

For accurate calculations, corrections to the Born-Oppenheimer approximation must also be taken into account. This so-called electronic isotope effect generally increases the reduced isotopic partition function ratio; for HD/H_2 the correction to the Born-Oppenheimer approximation requires that the partition function ratio must be multiplied by 1.13 at 300K while for DF/HF the factor is 1.09 (ref. 13). Note that in the calculation of an equilibrium constant, there will often be a good deal of cancellation of these correction factors when the ratio of $(s_2/s_1)f$ values is taken.

For the gas phase isotopic exchange equilibrium H_2O + HD = HDO + H_2, accurate experimental results are available (ref. 14). Bardo and this author (ref. 15) have compared these results with accurate theoretical calculation. Over the experimental temperature range 280K-475K, theory and experiment differ by at most 0.7%

Information about the liquid phase from isotope effects on the vapor pressure of liquids

The vapor pressure of a liquid can be considered the equilibrium constant between the liquid phase and the gas phase. The isotope effect on vapor pressure has been formulated by Bigeleisen (ref. 16) in terms of the reduced isotopic partition function ratio of the liquid phase molecule $(s_2/s_1)f_\ell$ and of the gas phase molecule $(s_2/s_1)f_g$,

$$\frac{P_1}{P_2} = \frac{(s_2/s_1)f_\ell(2/1)}{(s_2/s_1)f_g(2/1)} \tag{15}$$

where 2 and 1 refer to heavy and light molecules respectively. Some small correction terms which should be included in Eq. (15) will not be discussed here. Bigeleisen proposed the calculation of $(s_2/s_1)f_\ell$ in terms of 3N-6 (3N-5 for a linear molecule) internal normal mode vibrational frequencies of the molecule in the liquid, which are similar to the vibrational frequencies of a gas phase molecule, and 6 (5 for a linear molecule) additional vibrational frequencies which correspond to the unbound (zero frequency) translational and rotational motions of the molecule in the gas phase. Since these six translational-rotational motions obviously correspond to larger force constants in the liquid phase than in the gas phase, the contribution of these degrees of freedom will tend to make $(s_2/s_1)f_\ell$ larger than $(s_2/s_1)f_g$, i.e. to make p_1/p_2 larger than unity. Moreover, since the translational-rotational frequencies lead to relatively small values of $u = h\nu/kT$, their contributions will be of the form of Eq. (12) so that for small isotope effects (if one remembers that $\ln(1 + x) = x$ for small x) the contributions of these motions to the $\ln(p_1/p_2)$ will be of the form A/T^2 with A positive. The 3N-6 internal vibrational motions usually correspond to high frequencies so that their contributions to $(s_2/s_1)f$ in both the liquid and gas phase will be dominated by the zero-point energy term $\exp[(u_{1i} - u_{2i})/2]$. Thus, the contribution of the internal vibrational frequencies to $\ln(p_1/p_2)$ will be of the form B/T where B is a constant. The overall expected form of the vapor pressure isotope effect is

$$\ln(p_1/p_2) = A/T^2 + B/T \ . \tag{16}$$

Here B is positive if the force constants for internal vibrational motion at the position of isotopic substitution are larger in the liquid phase than in the gas while B is negative if the converse is true. If B is negative, as it is for many liquids, then a temperature dependence may occur in which $\ln(p_1/p_2)$ goes from a positive to a negative value at the so-called cross-over temperature. Information about B exists in many cases from spectroscopic data on gases and liquids although, as has been pointed out by Van Hook and Jancsó (ref. 17), care must be taken in interpreting the liquid data.

The study of vapor pressure isotope effects over a temperature range will give information about motions and forces in the liquid phase if one starts with a knowledge of $(s_2/s_1)f_g$ (ref. 18). In actual practice, one does not use Eq. (16) but uses the full expressions for $(s_2/s_1)f_g$ and $(s_2/s_1)f_\ell$ in conjunction with Eq. (15). The liquid phase model is the one outlined before or some more sophisticated variants of it which have been proposed recently (ref. 19).

Information about transition states from isotope effects on rate constants

Measurements of isotope effects on rate constants give information about the transition state (in particular information about force constants) if the reduced isotopic partition function ratio for the isotopically substituted reactant is known. While model calculations on transition states were carried out with hand calculators for reactions involving small molecules (and atoms), such as the calculations on the H_2 + Cl reactions by Bigeleisen et al. (ref. 20), calculations on larger molecular systems awaited the availability of large digital computers for the evaluation of reduced isotopic partition function ratios from force constant and geometrical information. Stern and Wolfsberg (ref. 21) then showed that the observation of an experimental rate isotope effect implies a change in force constant at the position of isotopic substitution between the isotopically substituted reactant and the transition state.

Willi (ref. 22) was among those who pioneered in obtaining information about transition states from rate isotope effects. Shiner and his co-workers (ref. 23) and also others (ref. 24) have been active in this area in recent years.

Dissection of isotope effects

Methods have been developed for the evaluation of $(s_2/s_1)f$ values which do not require, as does Eq. (7), a knowledge of normal mode vibrational frequencies. These methods are the orthogonal polynomial expansion methods of Bigeleisen, Ishida, and their co-workers (ref. 25) and the perturbation theory method developed by Singh and this author (ref. 26). The importance of these methods is that they permit one to see directly the various factors which determine $(s_2/s_1)f$ values and which then also determine isotope effects on chemical equilibria. This short discussion will be restricted to the perturbation theory approach.

In the perturbation theory approach, the unperturbed problem corresponds to independent coordinate oscillators while the perturbation corresponds to the interaction between these oscillators, both in the kinetic energy and potential energy. Thus, for H_2O, the unperturbed problem corresponds to two O-H stretching oscillators and one H-O-H bending oscillator. The frequencies of these oscillators can be calculated with a hand calculator. Thus, the frequencies ν' are given by

$$\nu'_i = (f_{ii} g_{ii})^{1/2} / 2\pi \tag{17}$$

where f_{ii} is the diagonal element of the F matrix and g_{ii} refers to the diagonal G matrix element of Wilson (ref. 27). For an O-H stretching oscillator, f_{ii} is the O-H stretching force constant in water and g_{ii} is $(m_O^{-1} + m_H^{-1})$, where m_O refers to oxygen atomic mass. The perturbation arises from the off-diagonal elements of the F matrix and the G matrix.

The reduced isotopic partition function ratio is evaluated by statistical mechanical perturbation theory. The first non-vanishing contribution from the perturbation theory arises in second order. The reduced isotopic partition function ratio corresponding to the unperturbed problem is

$$(s_2/s_1)f_0 = \prod_i \frac{\nu_{2i}}{\nu_{1i}} \frac{1-e^{-\nu_{1i}}}{1-e^{-\nu_{2i}}} e^{(\nu_{1i} - \nu_{2i})/2} \tag{18}$$

$$= \prod_i c(i)$$

with $\nu = h\nu'/kT$. This expression is the same as Eq. (7) except that normal mode frequencies ν_i have been replaced by coordinate frequencies ν'_i. The $c(i)$ factors, one for each coordinate of the molecule, are the analogues of the b_i factors of Eq. (11). It should be noted that $c(i)$ factors for coordinates not isotopically substituted are unity. Thus

$$(s_2/s_1)f_0 \text{ (HDO/H}_2\text{O)} = c(O-D) \times c(H-O-D) \tag{19}$$
$$(s_2/s_1)f_0 \text{ (NH}_2\text{D/NH}_3\text{)} = c(N-D) \times c(N-H-D)^2 \tag{20}$$

In the perturbation development

$$(s_2/s_1)f = (s_2/s_1)f_0 \times \text{CORR} \tag{21}$$

where CORR is the perturbation theory correction. Singh and Wolfsberg have demonstrated for a variety of molecules that, with CORR evaluated by second order perturbation theory, values of $(s_2/s_1)f$ are obtained which closely approximate those obtained by the use of Eq. (7).

It is the purpose to emphasize here the finding of Skaron and Wolfsberg (ref. 28) that zeroeth order perturbation theory and the $c(i)$ factors are sufficient to explain many isotope effects. Thus, consider the equilibrium $NH_3 + HDO = NH_2D + H_2O$ with

$$K = (s_2/s_1)f(NH_2D/NH_3)/(s_2/s_1)f(HDO/H_2O) \tag{22}$$

With use of Eq. (7), one finds at 298K from force field and geometry data for water and ammonia,

$$K = 13.92/13.48 = 1.033 \tag{23}$$

From the unperturbed problems, one finds with Eqs. (19) and (20),

$$K = \frac{c(N-D)c^2(H-N-D)}{c(O-D)c(H-O-D)} = \frac{7.422(1.368)^2}{9.350(1.435)} = 1.032 \tag{24}$$

Thus, the fact that the equilibrium constant is quite close to unity is a consequence of two competing factors: the contribution of the stretches tends to give a value smaller than unity while the two isotopically substituted bends in NH_2D, compared to the one substituted bend in HDO, give an effect in the opposite direction. The work of Skaron and Wolfsberg gives many more examples of the use of the $c(i)$ factor approach and of perturbation theory in dissecting isotope effects.

AKNOWLEDGEMENT

This research was supported by the U.S. Department of Energy under Contract No. DE-AT03-76ER70188.

REFERENCES

1. G. Hertzberg, Infrared and Raman Spectra of Polyatomic Molecules, Van Nostrand, New York, 1945.
2. J. Bigeleisen and M.G. Mayer, J. Chem. Phys., 15(1947)261-267.
3. J. Bigeleisen, J. Chem. Phys., 17(1949)675-678.
4. J. Bigeleisen and M. Wolfsberg, Adv. Chem. Phys., 1(1958)15-76.
5. D.J. Le Roy, B.A. Ridley, and K.A. Quickert, Disc. Faraday Soc., 44(1967)92-107.
6. M.J. Stern, W. Spindel, and E.U. Monse, J. Chem. Phys., 48(1968) 2908-2919; W. Spindel, M.J. Stern, and E.U. Monse, J. Chem Phys., 52(1970)2022-2035.
7. S. Skaron and M. Wolfsberg, J. Chem. Phys., 72(1980)6810-6811.
8. R.N. Clayton, L. Grossman, and K.T. Mayeda, Science, 182(1973) 485-488.
9. M. Wolfsberg and M.J. Stern, Pure Appl. Chem., 8(1964)225-242, 325-338.
10. R.F. Hout, M. Wolfsberg, and W.J. Hehre, J. Am. Chem. Soc., 102 (1980)3296-3298.
11. C.F. Chang, J. Bron, and M. Wolfsberg, Z. Naturforsch, 28a(1973) 129-136.
12. D. Goodson, S.K. Sarpal, P. Bopp, and M. Wolfsberg, to be published.
13. M. Wolfsberg and L.I Kleinman in P.A. Rock (Ed.), Isotopes and Chemical Principles, ACS Symposium Series 11(1975)64-76.
14. J.H. Rolston, J. den Hartog, and J.P. Butler, J. Phys. Chem., 80 (1976)1064-1067.
15. R.D. Bardo and M. Wolfsberg, J. Phys. Chem., 80(1976)1068-1070.
16. J. Bigeleisen, J. Chem. Phys., 34(1961)1485-1493.
17. G. Jancsó and W.A. Van Hook, Chem. Phys, Lett., 48(1977)481-482.
18. G. Jancsó and W.A. Van Hook, Chem. Rev., 74(1974)689-750.
19. Y. Yato, M.W. Lee, and J. Bigeleisen, J. Chem. Phys., 63(1975) 1555-1563; J.S. Pollino and T. Ishida, J. Chem. Phys., 66(1977) 4433-4444.
20. J. Bigeleisen, F.S. Klein, R.E. Weston, and M. Wolfsberg, J. Chem. Phys., 30(1959)1340-1351.
21. M.J. Stern and M. Wolfsberg, J. Chem. Phys., 45(1966)2618-2629.
22. A.V. Willi, Can. J. Chem., 44(1966)1889-1897.
23. S.R. Hartshorn and V.J. Shiner, Jr., J. Am. Chem. Soc., 94(1972) 9002-9012.
24. For a review, see L. Melander and W.H. Saunders, Jr., Reaction Rates of Isotopic Molecules, Wiley, New York, 1980.
25. J. Bigeleisen and T. Ishida, J. Chem. Phys., 48(1968)1311-1330; T. Ishida and J. Bigeleisen, J. Chem. Phys., 64(1976)4775-4789, and references therein. Recently these expansion methods have been combined with the unperturbed portion of the perturbation scheme of ref 26: J. Bigeleisen, T. Ishida, and M.W. Lee, J. Chem. Phys., 74(1981)1799-1816.
26. G. Singh and M. Wolfsberg, J. Chem. Phys., 62(1975)4165-4180.
27. E.B. Wilson, Jr., J.C. Decius, and P.C. Cross, Molecular Vibrations, McGraw-Hill, New York, 1955.
28. S. Skaron and M. Wolfsberg, J. Am. Chem. Soc., 99(1977)5252-5261.

VAPOUR PRESSURE ISOTOPE EFFECT OF ACETONITRILE

GY. JÅKLI and G. JANCSÓ
Central Research Institute for Physics,
H-1525 Budapest, P.O.B. 49 (Hungary)

T. KORITSÅNSZKY
Central Research Institute for Chemistry,
H-1525 Budapest, P.O.B. 17 (Hungary)

ABSTRACT

The vapour pressure difference between CH_3CN and CD_3CN was measured by differential capacitance manometry between -40 and +80 °C and the results can be expressed by the equation: $\ln(p_H/p_D) = 871.761/T^2 - 13.577/T + 0.006874$. The experimental data were interpreted within the framework of the statistical theory of isotope effects in condensed systems. The largest contribution to be VPIE arises from the shifts in the CH stretching vibrations on condensation which were found to be temperature (density) dependent in accordance with the available spectroscopic information. The results also indicate that the rotation of molecules about the 3-fold symmetry axis can be considered quasi-free whereas the rotation about the axes perpendicular to the top axis is restricted.

INTRODUCTION

The experimental data on the vapour pressure isotope effect (VPIE) can be related - in terms of the statistical theory of isotope effects in the condensed phase (ref. 1) - to the motions of molecules and to the intermolecular forces acting between them in this phase. The literature in this field has been critically reviewed by Jancsó and Van Hook (ref. 2). The interpretation of recent experimental VPIE data on simple molecular systems, such as carbon disulfide (ref. 3) and benzene (ref. 4), has indicated the importance of the use of vibrational frequencies corrected for dielectric effect (refs. 5, 6) and has also necessitated the introduction of a temperature dependent internal force constant (ref. 7).

Acetonitrile, an important organic solvent with a large dipole moment (3.94 D (ref. 8)), represents the class of symmetric top molecules. In the liquid phase the short-range structure - judging by X-ray and neutron scattering investigations (refs. 9-11) - can be described by the presence of clusters consisting of five molecules. A number of other experimental techniques (refs. 8,12,13) lend support to the hypothesis on the antiparallel orientation of the neighbouring dipole molecules in the liquid phase.

We present here the results from determining the vapour pressure differences between CH_3CN and CD_3CN in the temperature range -40 to +80 °C and we attempt to correlate the experimental findings with the internal vibrations and external motions of acetonitrile molecules by a model calculation of the VPIE in terms of the harmonic cell approximation.

EXPERIMENTAL

Reagent grade acetonitrile and deuterated acetonitrile (99 at% D) (Isocommerz) were purified by preparative scale gas chromatography. Before measurement the protio and deutero compounds were dried in a vacuum apparatus over phosphorus pentoxide and distilled to the degassing bulb, and they then underwent many freeze-pump-thaw-freeze cycles.

The vapour pressure difference measurements were made using the differential capacitance manometer apparatus described elsewhere (refs. 14,15). The pressure transducer was calibrated against the vapour pressure of water (refs. 16,17). The temperature of the equilibrium vessel was measured by a platinum resistance thermometer to ±0.03 °C.

The absolute vapour pressure of acetonitrile up to a pressure of 100 mmHg was also determined by comparing the pressure of the protiated sample against vacuum.

RESULTS

The vapour pressures of acetonitrile between the melting point (-43.835 °C (ref. 18)) and 30 °C (altogether 130 data were taken)[+] can be adequately described by the equation

$$\ln(P/mmHg) = 16.3147 - 2969.71/T - 166508/T^2 \qquad (1)$$

For the vapour pressure differences between the pure CH_3CN and CD_3CN samples we took 151 data in three series of measurements.[+] To calculate the VPIE values from the isotopic pressure differences below 100 mmHg absolute vapour pressure Eq.(1) was used; above this value we used the equation in Jordan's compilation (ref. 19). The data were fitted by least squares giving the equation

$$\ln \frac{P_H}{P_D} = \frac{(871.761 \pm 213)}{T^2} - \frac{(13.577 \pm 1.6)}{T} + (0.00687 \pm 0.0029) \qquad (2)$$

[+]Lists of experimental data are available from the authors on request.

The standard deviation of the calculated values is about ± 2×10^{-4}. The vapour pressure of deuterated acetonitrile is higher than that of acetonitrile and our data are in good agreement (0.5-3%) with those of Miljević, Dokić and Pupezin (ref. 20) who carried out VPIE measurements in the solid phase and in the liquid phase below room temperature.

The theoretically important quantity is the ratio of the reduced partition function ratios of the liquid and gas phases (f_c/f_g) obtainable from the experimentally observed VPIE using the equation (refs. 1,2)

$$\ln(f_c/f_g) = \ln \frac{P_H}{P_D} - \frac{P_H V_H - P_D V_D}{RT} - (B_o P + \frac{1}{2} C_o P^2 + \ldots)_D + (B_o P + \frac{1}{2} C_o P^2 + \ldots)_H + \ln f_{rot} \tag{3}$$

where P_H and P_D are the vapour pressures of the isotopic species H and D; f_c and f_g the reduced partition function ratios for the condensed and gaseous phase; V_H and V_D molar volumes in the liquid phase of CH_3CN and CD_3CN; B_o and C_o are virial coefficients; f_{rot} is a correction term for the nonclassical rotation of the molecules in the gaseous phase. The magnitude of the corrections turned out to be 0.1% and 8% of the logarithm of the vapour pressure ratios at -44 °C and 80 °C, respectively. The experimental reduced partition function ratio for acetonitrile is plotted vs. temperature in Fig. 1.

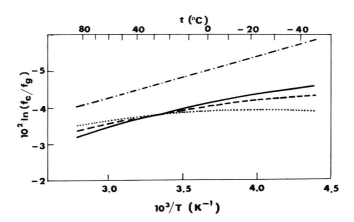

Fig. 1. Comparison of calculated and experimental reduced partition function ratios (f_c/f_g) for CH_3CN/CD_3CN. ——— experimental; -.-.- VPIE from the internal frequencies; ----- calculated from temperature independent liquid force field; ..... calculated with temperature dependent external force constants.

DISCUSSION

In the harmonic oscillator approximation, assuming a cell model for the liquid phase, the theory of isotope effects gives for f_c/f_g the expression (refs. 2,21)

$$\frac{f_c}{f_g} = \prod_{\text{int.freq.}}^{3N-6} \frac{(u_i/u_i')_c}{(u_i/u_i')_g} \frac{\exp((u_i'-u_i)_c/2)}{\exp((u_i'-u_i)_g/2)} \frac{(1-\exp(-u_i')_c)/(1-\exp(-u_i)_c)}{(1-\exp(-u_i')_g)/(1-\exp(-u_i)_g)} \times$$

$$\times \prod_{\text{ext.freq.}}^{6} \frac{u_i}{u_i'} \left[\exp\left(\frac{u_i'-u_i}{2}\right)\right] \left[\frac{1-\exp(-u_i')}{1-\exp(-u_i)}\right], \quad (4)$$

where $u_i = hc\nu_i/kT$; c and g refer to the condensed and gaseous phases; N is the number of atoms in the molecule; ν_i is the i-th normal mode frequency in cm^{-1}.

The gas phase force constants were first determined by least-squares fitting to the observed vibrational frequencies (refs. 22-28), this force field was then employed to calculate the frequencies of CH_3CN and CD_3CN (see Table 1). All calculational details will be given elsewhere (ref. 29). The liquid phase force field was obtained by changing only the diagonal gas phase force constants and by using the observed frequencies of Kakimoto and Fujiyama (ref. 30). These CH_3CN frequencies were observed at 25 °C and corrected for "dielectric effect" (refs. 31,32), thus the entire shifts of the internal frequencies on condensation are due to the effect of the intermolecular interactions in the liquid phase.

TABLE 1

Calculated gas and liquid phase frequencies for CH_3CN and CD_3CN in cm^{-1} (25 °C)

Vibrational mode	CH_3CN		CD_3CN	
	gas	liquid	gas	liquid
ν_1 sym. C-H stretch	2954.246	2943.752	2125.756	2117.891
ν_2 sym. C≡N stretch	2270.303	2256.289	2274.713	2260.952
ν_3 sym. CH_3 def.	1388.887	1376.729	1108.922	1100.890
ν_4 sym. C-C stretch	921.101	919.860	829.640	827.449
ν_5 deg. C-H stretch	3009.793	3006.704	2256.528	2254.229
ν_6 deg. CH_3 def.	1455.164	1443.703	1045.250	1036.986
ν_7 deg. CH_3 rocking	1041.886	1040.014	844.764	846.347
ν_8 deg. C-C≡N bend	364.428	381.870	331.073	345.856
ν_{T_x}, ν_{T_y} translation		64.381		61.188, 62.762[a]
ν_{T_z} translation		107.564		103.792
ν_{R_x}, ν_{R_y} rotation		75.373		72.263, 70.724[a]
ν_{R_z} rotation		119.342		84.420

[a]The splitting in these frequencies is due to the external-internal coupling elements in G.

At 25 °C the difference between the VPIE calculated by Eq.(4) from the internal vibrational frequencies (Table 1) and the experimental value is 0.0085 (see Fig. 1). This amounts to 30% of the measured VPIE and can be assumed to be due to the contribution from the different external modes to the isotopic ratios. To evaluate the contributions from the hindered translational and rotational motions, selected observed lattice frequencies of the high temperature (β) crystalline phase (ref. 33) were transformed by assuming that the ratio of the external force constants to each other remains the same during the transition from the solid phase into the liquid phase at 25 °C. The external frequencies thus obtained are listed in Table 1 and are in good agreement with the observed FIR frequencies (refs. 8,12,34-37).

The values for the different hindered rotational force constants $F_{R_x} = F_{R_y} = 0.183$ and $F_{R_z} = 0.027$ mdyne·Å lend support to the conclusion reached by different experimental techniques (refs. 38-41) according to which the rotation of the acetonitrile molecules in the liquid phase is much less hindered about the 3-fold symmetry axis than about the axis perpendicular to the top axis. The relatively free rotation about the symmetry axis seems to be in accordance with the hypothesis that the molecules form dipole-dipole complexes in the liquid phase (e.g. refs. 8,35,42,43).

The external frequencies decrease with increasing temperature (ref. 37) and this effect can properly be described by volume dependent external force constants. The same volume dependence for the different translational and rotational force constants was assumed and an average, temperature independent value of 1.5 for the Grüneisen parameter ($\gamma_i = -d\ln\nu_i/d\ln V$) (ref. 44) was used.

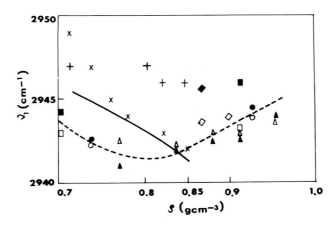

Fig. 2. Density (ρ) dependence of ν_1 stretching frequency in liquid acetonitrile. +: (ref. 46); x: (ref. 45); Δ,o,□,◊: $\Delta\nu_{VV}$ at 30, 60, 90 and 120 °C, respectively, at different pressures; filled symbols denote $\Delta\nu_{VH}$: (ref. 47) ----- corresponds to the average of the above Raman frequency shifts; ——— from VPIE measurements: present work.

Fig. 1 shows comparisons between the calculated and experimental VPIE's. It can be seen that the difference between the calculated and experimental points is significant, eventually it became even larger on the introduction of temperature dependent lattice modes. Since the largest contribution to the VPIE arises from the CH stretching vibrations (ν_1, ν_5) and there is also spectroscopic evidence on the temperature dependence of these frequencies (refs. 45-47), the introduction of the temperature dependent CH stretching force constant seems to be warranted just as in the case of benzene (ref. 7). In Fig. 1 no distinction is made between calculation and experiment (solid line) since the temperature dependence of the CH stretching force constant was chosen to reproduce the experiment. The frequency shifts are to the blue with increasing temperature ($\Delta\nu_1/\Delta t \approx 0.03$ cm^{-1}/°C) in agreement with the spectroscopically observed temperature dependences (refs. 45-47).

It is of interest to plot the available data on the ν_1 CH stretching vibration as a function of liquid density. (The Raman line shifts observed by Jonas (refs. 47,48) have been measured as a function of pressure up to 4 kbar within the temperature interval of 30-120 °C). It can be seen in Fig. 2 that the characteristic of the density dependence of ν_1 vibrational frequency is similar to that observed spectroscopically, almost up to the density at the melting point ($\rho \approx 0.85$ gcm^{-3}). At the liquid-solid phase transition the density increases

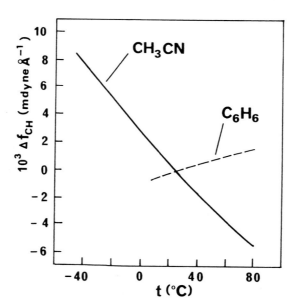

Fig. 3. Temperature dependence of CH stretching force constant (mdyne/Å) in liquid acetonitrile (———) and benzene (-----) as derived from VPIE data. $\Delta f_{CH} = f_{CH}(t_o °C) - f_{CH}(t °C)$, where t_o is equal to 25 °C.

by 10.1% and there is a decrease of about 1 cm^{-1} in the ν_1 vibrational frequency, whereas according to the ν_1-density curve corresponding to Raman investigations ref. 47 this density increase should go together with a frequency increase in the liquid phase. This strongly suggests that the structure of the solid phase differs greatly from that of the liquid phase (refs. 12,42).

In Fig. 3 the temperature dependence of the CH stretching force constant of acetonitrile is compared with that of benzene (ref. 7). The difference in the characteristics of the temperature dependences can qualitatively be understood in terms of structural differences in the two liquids. In contrast with acetonitrile, in liquid benzene the average molecular arrangement is surprisingly similar to that of the solid, and the solid-like somewhat disordered (quasicrystalline) structure persists (refs. 34,49). For acetonitrile the repulsive part of the intermolecular potential seems to play an important role (refs. 48,50) and apparently is strengthened with increasing temperature which goes together with a blue shift in the ν_1, ν_5 frequencies (ref. 51).

ACKNOWLEDGEMENT

We wish to thank to Professor J. Jonas of the University of Illinois for making available to us their Raman frequency shifts.

REFERENCES

1 J. Bigeleisen, J. Chem. Phys. 34(1961)1485.
2 G. Jancsó and W.A. Van Hook, Chem. Rev. 74(1974)689.
3 R.H. Betts and W.D. Buchannon, Can. J. Chem. 54(1976)3007.
4 Gy. Jákli, P. Tzias and W.A. Van Hook, J. Chem. Phys. 68(1978)3177.
5 G. Jancsó and W.A. Van Hook, Can. J. Chem. 55(1977)3371.
6 G. Jancsó and W.A. Van Hook, Chem. Phys. Lett. 48(1977)481.
7 G. Jancsó and W.A. Van Hook, J. Chem. Phys. 68(1978)3191.
8 E. Knözinger, D. Leutloff and R. Wittenbeck, J. Mol. Struct. 60(1980)115.
9 A. Kratochwill, J.U. Weidner and H. Zimmermann, Ber. Bunsenges., Phys. Chem. 77(1973)408.
10 H. Bertagnolli, P. Chieux and M.D. Zeider, Mol. Phys. 32,(1976)759,1731.
11 H.W. Zimmermann, in A.D. Buckingham, E. Lippert and S. Bratos (Eds.), Organic Liquids, John Wiley, New York, 1978, p.1.
12 H. Michel and E. Lippert, ibid. p. 293 and references therein.
13 K.R. Srinivasan and R.L. Kay, J. Sol. Chem. 6(1977)357.
14 Gy. Jákli and H. Illy, KFKI Report 80-15 (1980).
15 G. Jancsó and Gy. Jákli, Aust. J. Chem. 33(1980)2357.
16 G. Jancsó, J. Pupezin and W.A. Van Hook, J. Phys. Chem. 74(1970)2984.
17 J.A. Goff, Humidity and Moisture, Ed. Wexler A., Rheinhold, New York, 1963, Vol.3, p.289.
18 W. Putnam, D.M. Mc Eachern, Jr. and J.E. Kilpatrick, J. Chem. Phys. 42(1965)749.
19 E.T. Jordan, Vapor Pressure of Organic Compounds, Interscience Publ., New York - London, 1954, p.175.
20 N.R. Miljević, V.R. Dokić and J.D. Pupezin, Bull. Soc. Chim. Beograd, 46(1981)87.
21 M.J. Stern, W.A. Van Hook and M. Wolfsberg, J. Chem. Phys. 39(1963)3179.
22 K. Venkateswarlu, J. Chem. Phys. 19(1951)293.

23 S. Kondo and W.B. Person, J. Mol. Spectrosc. 52(1974)287.
24 I. Nakagawa and T. Shimanouchi, Spectrochim. Acta 18(1962)513.
25 F.W. Parker, A.H. Nielsen and W.H. Fletcher, J. Mol. Spectrosc. 1(1957)107.
26 W.H. Fletcher and C.S. Shoup, J. Mol. Spectrosc. 10(1963)300.
27 F.N. Masri, J.L. Duncan and G.K. Speirs, J. Mol. Spectrosc. 47(1973)163.
28 J.C. Evans and H.J. Bernstein, Can. J. Chem. 33(1955)1746.
29 Gy. Jákli, T. Koritsánszky and G. Jancsó (to be published).
30 M. Kakimoto and T. Fujiyama, Bull. Chem. Soc. Japan 45(1972)3021.
31 S.Kh. Akopyan, O.P. Girin and N.G. Bakshiev, Opt. Spectrosc. 36(1974)185 and references therein.
32 A.A. Clifford and B. Crawford Jr., J. Phys. Chem. 70(1966)1536.
33 M.P. Marzocchi and M.G. Migliorini, Spectrochim. Acta 29A(1973)1643.
34 L.A. Blatz, J. Chem. Phys. 47(1967)841.
35 R.J. Jakobsen and J.W. Brasch, J. Am. Chem. Soc. 86(1964)3571.
36 B.J. Bulkin, Helv. Chim. Acta 52(1969)1348.
37 S.G. Kroon and J. Van der Elsken, Chem. Phys. Lett. 1(1967)285.
38 W.J. Jones and N. Sheppard, Trans. Far. Soc. 56(1960)625.
39 W.G. Rotschild, J. Chem. Phys. 57(1972)991.
40 T.E. Bull and J. Jonas, J. Chem. Phys. 53(1970)3315.
41 D.E. Woessner, B.S. Snowden, Jr. and E.T. Strom, Mol. Phys. 14(1968)265.
42 R.J. Jakobsen and Y. Mikawa, Appl. Opt. 9(1970)17.
43 J.E. Griffiths, J. Chem. Phys. 59(1973)751.
44 J.C. Slater, Introduction to Chemical Physics, McGraw-Hill, New York, 1939.
45 A.Z. Gadzhiev, Opt. Spectrosc. 23(1967)722.
46 G.I. Baranov and A.V. Setskarev, Vopr. Mol. Spectrosc. 1974, 89.
47 J. Jonas (personal communication)
48 J. Schroeder, V.H. Schiemann, P.T. Sharko and J. Jonas, J. Chem. Phys. 66(1977)3215.
49 A.H. Narten, J. Chem. Phys. 48(1968)1630.
50 J. Yarwood, R. Ackroyd, K.E. Arnold, G. Döge and R. Arndt, Chem. Phys. Lett. 77(1981)239.
51 A.M. Benson, Jr., H.G. Drickamer, J. Chem. Phys. 27(1957)1164.

THERMODYNAMIC PROPERTIES OF SOLUTIONS OF UREA, DEUTEROUREA AND
TETRAMETHYLUREA IN H_2O AND D_2O

GY. JÁKLI[#] and W. ALEXANDER VAN HOOK
Chemistry Department, University of Tennessee, Knoxville, TN 37916 (USA)

ABSTRACT

Thermodynamic data on the osmotic coefficients, activity coefficients, solubilities, and apparent molar volumes, and their isotope effects are presented. The data are discussed in the context of the theory of isotope effects in condensed phase systems.

INTRODUCTION

Aqueous urea solutions comprise a system of considerable importance. Such solutions not only serve as the final sink in the protein nitrogen metabolic cycle in mammals but are also of marked biological interest as denaturing agents in protein chemistry (ref. 1). In addition vast quantities of urea are consumed in agriculture both as fertilizer and as a feed additive. Additional tonnage amounts of this material are employed in the polymer industry. In spite of the industrial and scientific interest in this system, thermodynamic data on urea water solutions were available only over modest temperature and concentration ranges in H_2O as solvent (ref. 2), and were almost nonexistent in D_2O as solvent (refs. 3,4). Comparisons of the thermodynamic properties in the two solvent systems is of interest because such data can be used in a reasonably straightforward fashion to test models of the solution structure. Detailed measurements (to saturation, 10 to 60°C) of osmotic coefficients and apparent molar volumes of $CO(NH_2)_2/H_2O$ and $CO(ND_2)_2/D_2O$, together with information on the solubilities and related properties have been completed and are reported in detail in another place (ref. 4). More recently we have obtained similar data on the TetraMethylUrea (TMU)/H_2O and TMU/D_2O systems and are making a preliminary report of these data together with comparisons with the Urea-h_4(Uh)/H_2O and Ud/D_2O system in this paper. These and other studies (ref. 5) of isotope effects in nonelectrolyte aqueous solutions complement our earlier studies of isotope effects in electrolyte containing solutions (refs. 6,7).

[#]Permanent address: Central Research Institute for Physics, Budapest.

EXPERIMENTAL

Vapor pressures and vapor pressure differences leading to osmotic coefficients and osmotic coefficient isotope effects were obtained on the University of Tennessee apparatus as previously described (refs. 4,7). Densities were measured with a Mettler-Paar densitometer using well established techniques (ref. 8), solubilities were determined gravimetrically. Concentrations are expressed in aqua-molality (moles solute per 55.5082 moles solvent).

RESULTS

Osmotic coefficients of urea solutions in H_2O (4, 8, 12, 16, and 20 m, 10<t<60) are described with an RMS deviation of 2×10^{-3} unit by Eq. 1. Above 60° there is considerable decomposition and measurements are not possible.

$$\phi_H(m,t) = \phi_H^*(m,25) + D\tau + E\tau^2 + F\tau^2 m^2 \quad ; 10 \leq t \leq 60 \quad (1)$$

In Eq. 1, $\phi_H^*(m,25)$ is the fit reported by Stokes (ref. 2) to isopiestic data at 25°C, $\tau = (t-25)$, $D = 1.47 \times 10^{-3}$, $E = -1.77 \times 10^{-5}$, $F = -7.5 \times 10^{-9}$ and

$$\phi_H^*(m,25) = 1 + Ax + Bx^2; \quad x = m/(1 + bm); \quad 0 \leq m \leq 20 \quad (2)$$

with $A = -0.042783$, $B = -0.0004198$ and $b = 0.15$. Experiments on saturated solutions yield the following results for the solubility ($X_{H,s}$ is the mole fraction of Uh in saturated H_2O solution),

$$\ln X_{H,s} = (-2070 \pm 117)/T + (17.496 \pm 2.56) + (-2.086 \pm 0.380) \ln T; \quad 273<T<343 \quad (3)$$

solubility isotope effect ($\Delta \ln X_s = \ln X_{H,s} - \ln X_{D,s}$),

$$\Delta \ln X_s = (95482 \pm 15446)/T^2 - (565.3 \pm 101.0)/T + (0.8607 \pm 0.1645); \quad \sigma^2 = 9.7 \times 10^{-7} \quad (4)$$

and osmotic coefficient (at saturation) in H_2O

$$\phi_H(m_s) = (0.7737 \pm 0.0026) + ((1.166 \pm 0.211) \times 10^{-3}) m_s - ((5.55 \pm 0.40) \times 10^{-5}) m_s^2 \quad (5)$$

and in D_2O

$$\phi_D(m_s) = (0.7566 \pm 0.0037) + ((1.505 \pm 0.295) \times 10^{-3}) m_s - ((5.60 \pm 0.55) \times 10^{-5}) m_s^2 \quad (6)$$

In Fig. 1 the osmotic coefficient data as described by Eqs. 1 through 5 are plotted. Several isotherms below saturation are shown. Since the information extends to saturation it is possible with a Gibbs-Duhem integration to calculate the standard state (infinite dilution) thermodynamic properties. Values for several temperatures are reported in Table I. Previously no information on the free energy of transfer from the crystal to the infinitely dilute solution, $(\mu^\bullet - \mu^\circ)_H$, was available at any temperature, while the standard enthalpy (ref. 9) and heat capacity (ref. 10) were only available at 25°C. The agreement between the standard enthalpy and heat capacity derived from the vapor pressure measurements with those determined calorimetrically (refs. 9,10) is excellent and lends confidence to our experimental and data handling techniques.

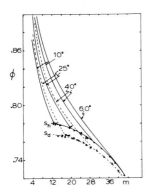

Fig. 1. Osmotic Coefficients at several temperatures. Osmotic Cfs. of Saturated Solutions. Solid lines refer to Uh/H$_2$O and dotted lines to Ud/D$_2$O.

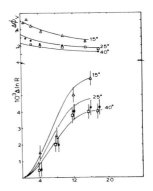

Fig. 2. Isotope Effects on Solvent Activity and Apparent Molar Volume of Urea Solutions. The lines are calculated from the eq. in Table 2.
△ 15°C, ● 25°C, □ 40°C.

TABLE I

Standard Thermodynamic Properties of Solution and their Isotope Effects for Urea-y$_4$/Y$_2$O Solutions; Y=H or D.

Temp., °C	$-(\mu^o-\mu^\bullet)$ [IE]	$(H^o-H^\bullet)$ [IE]	$Cp^\bullet$	Cp^o [IE]	V^o [IE]
15	5.14 [0.29]	15.33 [-2.16]	90.3	119	43.46 [0.50]
25	5.85 [0.24]	15.35* [-1.29]	93.1	82* [4.4]	44.20 [0.34]
40	6.90 [0.20]	14.68 [-0.66]	97.4	26	45.19 [0.24]

*(Ref. 9) $H^o-H^\bullet$ = 15.30; (Ref.10) Cp^o = 87.5. Superscript o's refer to infinitely dilute solution. Superscript ●'s refer to the crystal standard state.

Isotope effects on the excess free energies are most conveniently discussed in terms of the isotopic vapor pressure differences, Δln R, which are defined in terms of the measured pressures and the isotope effects on osmotic coefficients in Eq. 7

$$\Delta \ln R = [\ln(P_H^o(t)/P_D^o(t)) - \ln(P_H(m,t)/P_D(m,t))] = m\Delta\phi/55.5 \qquad (7)$$

The effects observed for the Uh/Ud system are positive and very small. At a given temperature the IE's can be described with a function which rises rapidly from zero, then saturates at or above a concentration around 10m. In fact within experimental error no difference in the IE was observed for runs at 12, 16 and 18 molal. A thermodynamic rationalization of these isotope effects via the measured volumetric properties of the solutions is possible. In Fig. 2 $\Delta\phi_v$'s and $\Delta\ln R$ are compared for several isotherms. The lines are calculated from the parameters reported in Table II. The procedure involves a single fit of combined $\Delta\ln R$ and $\Delta\phi_v$ data and is consistent with the rationalization of the excess thermodynamics of the solutions in terms of a volume fraction statistical model which emphasizes the importance of the volume dependence of the excess free energy and its isotope effect (refs. 3,11).

TABLE II
Parameters of Fit of Volume and Free Energy Data to Relations A and B.

(A) $\phi_v = \phi_v^o (1 + \alpha m + \beta m^{3/2})$ (B) $\ln R = -C(\alpha\Delta\phi_v^o m + 1.5\beta\Delta\phi_v^o m^{3/2})m$

where $\Delta\phi_v^o = 0.50 - 0.020\tau + 3.8 \times 10^{-4}\tau^2$; $\tau = t-15$

$C = C_o + C_1/T^2$

and $m \leq 16$ for $m>16$ then $\Delta\ln R = (\Delta\ln R)_{16}$.

Finally $\alpha = -0.064$ $\beta = 8.8 \times 10^{-3}$ $C_o = 0.0149$ $C_1 = -888$

We turn now to a comparison of the properties of $TMU/H_2O//TMU/D_2O$ solutions with the corresponding urea solutions. These comparisons are given in Figs. 3, 4, and 5. Both the solvent partial molar free energy (as measured by the osmotic coefficient) and the excess volumetric effects (as measured by the apparent molar volume) show dramatic differences from the unsubstituted urea solutions. We have not yet succeeded in deducing a functional representation for these data which preserves an economy in number of parameters and therefore in this report restrict attention to a qualitative description of some of the features of the data. TMU is infinitely soluble in water and the present data for the solutions extend to 70m. Pronounced minima in ϕ or ϕ_v vs. m curves are found in the region of 2 to 10m (Fig. 3). The dilute solutions are highly nonideal (Fig. 4) and osmotic coefficients of as large as -5 have been observed at 1 and 2m at 15°. The nonideality is a sharp function of the temperature and at 80° has been essentially damped out, but even at that temperature the solutions are more nonideal than are urea-h_4/d_4 solutions at any of the temperatures investigated (compare the upper part of figure 4 with the rest of that figure). The minima in the excess free energy or volume curves are functions of both temperature and isotopic label. This results in large isotope effects which show sensitive temperature dependences including pronounced maxima at the lower temperatures investigated (Fig. 5). It is reassuring that similar effects are demonstrated for the isotope effects on the excess volumes when these are plotted appropriately. The thermodynamic consistency thus demonstrated by two widely different kinds of experimental measurements lends confidence to our procedures. Again, the effects for urea itself are much smaller and show no evidence of unusual behavior in the low concentration range (compare the lower parts of Figs. 3 and 5 with the rest of the data in those figures). We have very recently finished measurements of the osmotic coefficients and osmotic coefficient isotope effects of 1,3 dimethyl urea-h_2 (DMUh) and DMUd solutions in H_2O and D_2O, and are presently engaged in determining the apparent molar volumes of these solutions. The single most interesting features of the osmotic coefficient isotope effect measurements is that $\Delta\phi$ is very small, even smaller than $\Delta\phi$ for urea itself. In other words, the effect of methyl substitution on the partial molar free energies of the solutions is not proportional to the number of methyl groups added. This important observation will be of interest in testing models of hydrophobic hydration in aqueous solutions of substituted ureas.

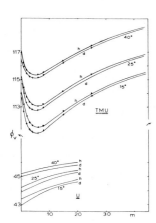
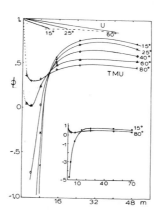
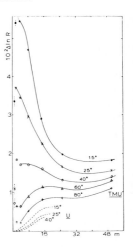

Fig. 3. Apparent Molar Volumes of Urea and Tetramethylurea solutions in H_2O and D_2O Note the pronounced minuma.

Fig. 4. Osmotic Cfs. of Urea and TMU solutions in H_2O. Note the sensitive dependence of TMU on concentration and temperature. The steep minimum in solvent activity is reflected by an azeotrope.

Fig. 5. Isotope Effects on solvent activity for U and TMU solutions.

ACKNOWLEDGEMENTS

Support from the National Institutes of Health, Program in General Medical Sciences, and from the National Science Foundation, Chemical Thermodynamics Program is appreciated.

REFERENCES

1. F. Franks, in F. Franks, Ed., Water; A Comprehensive Treatise, Vol. 2, pp 355, 370, Plenum, New York, 1973; D.B. Wetlaufer, S.K. Malik, L. Stroller, and R.J. Coffin, J. Am. Chem. Soc. 86(1964)509; Y. Nozaki and C. Tanford, J. Biol. Chem., 238(1963)4074; J.P. Simko, and W. Kauzmann, Biochemistry 1(1962)1005.
2. R.H. Stokes, Aust. J. Chem. 20(1967)2087.
3. G. Jancso and W.A. Van Hook, Chem. Reviews 74(1974)689.
4. Gy. Jakli and W.A. Van Hook, J. Phys. Chem. (submitted).
5. Gy. Jakli and W.A. Van Hook, J. Chem. Thermodynamics 4(1972)857; T.C. Chan and W.A. Van Hook, J. Chem. Soc., Far. Trans. I. 72(1976)583.
6. Gy. Jakli, T.C. Chan and W.A. Van Hook, J. Soln. Chem. 4(1975)71; Q.D. Craft and W.A. Van Hook, J. Soln. Chem. 5(1975)901,923; G. Dessauges and W.A. Van Hook, J. Inorg. Nucl. Chem. (in press).
7. J. Pupezin, G. Jakli, G. Jancso, and W.A. Van Hook, J. Phys. Chem. 76(1972)743.
8. M.K. Dutta-Choudhury and W.A. Van Hook, J. Phys. Chem. 84(1980)2735.
9. E.P. Egan and B.B. Luff, J. Chem. Eng. Data, 11(1966)192.
10. P.R. Philip, G. Perron and J.E. Desnoyers, Can. J. Chem. 52(1974)1709.
11. G. Jancso and W.A. Van Hook, Physica, 91A(1978)619; Acta Chem. Sci. Hung 98(1978)183; G. Jakli and W.A. Van Hook, Geochemical J. (in press).

HYDROGEN ISOTOPE EFFECTS IN THE DEHYDRATION OF POLYSTYRENE-DIVINYL-BENZENE(PS-DVB) TYPE ION-EXCHANGERS

A.R. GUPTA
Chemistry Division, Bhabha Atomic Research Centre, Bombay 400085 (INDIA)

ABSTRACT

Hydrogen isotope effects in the dehydration of PS-DVB type ion exchangers have been studied. A modified Rayleigh distillation technique was used for the purpose. Fully swollen strong acid cation exchanger Dowex 50WX8 in Li^+ and Cs^+ forms and a strong base anion exchanger, Dowex 1X8 in Cl^- and NO_3^- forms have been studied. The distillations were performed at 353K in a vacuum line. Distillate samples were analysed for their H/D ratios in an isotope ratio mass spectrometer. The higher values for the separation factor, α, for Li^+ (1.068) and NO_3^- (1.054) forms of resin than α_{water} (1.034) indicate that the water in these resins is more structured than normal water. On the other hand, Cs^+ (1.029) and Cl^- (1.039) forms of resins have α values comparable to water, though in the former a less structured state of water is indicated in the light of generally accepted ideas about the structure of aqueous electrolyte solutions.

INTRODUCTION

It is fairly well-established that the ions inside the moderately and low cross-linked polystyrene-divinylbenzene type ion exchangers are fully hydrated (ref. 1). The excess water in these exchangers, which is not part of the hydration sphere of the ions, has been described as 'free water' or 'osmotically active water' (refs. 2, 3). The existence of high concentrations of the ions as well as swelling pressures inside the exchanger phase makes the water inside the swollen resinates a very interesting substance from the point of view of electrolyte solutions. In a recent thermal study (ref. 4) of the dehydration process of fully swollen resins of PS-DVB type, it has been shown that the state (structure) of the water inside the resinates is primarily governed by the properties of counter ions. As the hydrogen isotope effects in the distillation of free water arise because of the hydrogen bonded structure of liquid water (ref. 5), a study of the hydrogen isotope effects in the dehydration of the resinates in different ionic forms, which would contain water with varying degree of structure depending upon the ion-water interactions, would provide valuable information about ion-water, and hydration sphere water-water interactions inside the exchanger phase. In the present paper, the results of such studies on fully swollen Li^+ - and Cs^+- forms of Dowex 50WX8 and Cl^- and NO_3^- forms of Dowex 1X8 resins are being presented.

MATERIALS AND METHODS

Strong base anion exchanger Dowex 1X8 and strong acid cation exchanger Dowex 50WX8 (J.T. Baker and Co., Pennsylvania, USA) were conditioned in the usual way by recycling through Na^+/H^+ cycle for cation exchangers and OH^-/Cl^- cycle for the anion exchanger. The resins were finally converted to Li^+- and Cs^+- forms (cation exchanger) and Cl^- and NO_3^- forms (anion exchanger). The maximum water uptake of the resins are given in Table I.

ISOTOPE EFFECT STUDIES

A Rayleigh distillation type technique was used for these studies. An aliquot of fully swollen resin was taken in a flask which was connected to a vacuum system. The system was slowly evacuated and the temperature of the flask was raised to 353 K and the distillate water samples were collected in weighed sample tubes immersed in liquid nitrogen. Usually four to five samples were collected in this way. The last traces of water were distilled from the resin sample by raising the temperature of the flask to ~ 410 K. The water samples were weighed and analysed in an isotope ratio H/D mass spectrometer, specially fabricated by the Technical Physics Division of BARC. In the 100-200 ppm deuterium range the accuracy of the isotopic analysis was better than ± 0.5 ppm.

RESULTS

The single stage separation factor, α, was calculated from the measured isotopic ratios in these distillations by the well known Rayleigh formula, $(N/N_o)^{\alpha/\alpha-1} = W_o/W$ where N and N_o (both assumed to be $\ll 1$) refer to the mole fraction of the rare isotope in the final sample and initial water composition in the resin sample, respectively; W and W_o are respectively the weight of the final sample and the total water content in the resin sample used. The separation factors and the relevant data are recorded in Table I.

TABLE I

Single stage separation factors (α_D^H) in the dehydration of fully swollen PS-DVB type ion exchangers

Resin	Total water in resin (g)	Water recovered (g)	Weight of last sample (g)	D content (ppm)	α_D^H
Dowex 50WX8 Li^+-form	0.7633	0.7579	0.2294	166.5	1.068
Dowex 50WX8 Cs^+-form	1.0982	1.1079	0.1570	164.7	1.029
Dowex 1X8 Cl^--form	0.7010	0.7049	0.0842	171.1	1.039
Dowex 1X8 NO_3^--form	0.8376	0.8350	0.1273	170.1	1.054
water	0.9039	0.8967	0.0262	171.5	1.034

The hydrogen isotope separation factor for pure water (1.034 ± 0.002), determined by this technique is in excellent agreement with the literature value of 1.035 for H_2O/HDO separation factor at 353 K (ref. 6) in water distillation. This emphasizes the reliability of the technique for the determination of isotope separation factors in the present systems.

DISCUSSION

The separation factors in Table I follow the sequence $Li^+ > NO_3^- > Cl^- \geqslant$ water $> Cs^+$.

In a recent thermal study of the kinetics of dehydration process of the various ionic forms of PS-DVB type ion exchangers, i.e., Li^+ form of Dowex 50WX8 and Cl^- and NO_3^- forms of Dowex 1X8 resins (ref. 4) and Cs^+ form of Dowex 50WX8 (ref. 7), it was observed that the activation energies for the loss of the osmotically imbibed water follow the same sequence (ref. 4) i.e., $Li^+ > NO_3^- >$ water $\geqslant Cl^- \geqslant Cs^+$. This similar behaviour of the activation energies and hydrogen isotope effects in the dehydration of ion exchange resins strongly suggests that these quantities depend upon the same properties of the water inside the resins i.e., its structure.

These results can be understood in the light of known general features of aqueous electrolyte solutions. Currently accepted ideas about the structure of aqueous electrolyte solutions visualize the water as existing in the following three forms (ref. 8) :- (i) water in the hydration sphere of the ions, (ii) water in between the hydration sphere and bulk water, and (iii) bulk water. The high ionic concentrations inside the exchanger phase practically rules out the possibility of normal bulk water existing in the exchanger phase. In the light of these general considerations, the data on hydrogen isotope effects and activation energies in the dehydration of ion exchangers will now be discussed systemwise.

Cl^- form of Dowex 1X8

In the dehydration of chloride form of Dowex 1X8, the activation energy as well as the hydrogen isotope separation factor is very close (or nearly the same) as for pure water. Kristoff and Inczedy (ref. 9) have also recently investigated the dehydration of Cl^- form of Dowex AG 1X8 resin by thermoanalytical techniques. They also observed as in the study reported above that the whole of the water was lost in one step and consequently concluded that this resin contains water either in the 'free' state or in a loosely bound state, i.e., the normal hydrogen bonded structure of free water is not very much disturbed in this resin. This obviously would lead to H/D separation factor of the same magnitude as for free water.

Li^+ form of Dowex 50 WX8

Lithium ions are strongly hydrated in the aqueous solutions. Coordination/hydration number for lithium ion has been reported as ~ 4.0 based on X-ray and neutron diffraction studies (ref. 10). In this connection it is relevant that Li^+ form of Dowex 50WX8 in fully swollen form contains 11.4 water molecules per lithium ion (ref. 11). Therefore, besides primary solvation of lithium ion, secondary solvation of lithium ions can also be expected. The larger activation energy as well as the larger isotope effect in the dehydration of Li^+ form of resin indicates that water is present in a more structured form than normal water. Hence, the existence of second category water (see above), i.e. water having more broken structure, is ruled out. These data then strongly suggest that excess water in Li^+ form of resin (over and above the primary solvation water) is part of a secondary solvation shell around the hydrated lithium ions, in which water is more strongly bound than in ordinary bulk water. The evidence of infra-red spectra of electrolyte solutions shows that the OH stretching modes of water in primary hydration shell of cations usually shift to lower frequencies than in normal water, indicating a loosening of OH bonds i.e., $M^+O\!\!\diagup\!\!\!^H_H$ (ref. 12). This would increase the positive charge on the protons in the water molecules in primary hydration shell and thus they interact more strongly with the water molecules around the hydrated cation.

NO_3^- form of Dowex 1X8

The activation energy as well as hydrogen isotope effects data again suggest that water in this resin is present in a more ordered state than ordinary water. Infrared studies of nitrate ions in aqueous solutions (ref. 13) have indicated the presence of two strong hydrogen bonds between the nitrate ions and water molecules. In a neutron diffraction study of the hydration of nitrate ions by Neilson and Enders (ref. 14) a coordination number of 1.4 ± 0.2 has been reported. These authors have interpreted their data as demonstrating that the potential energy surface in the region of the nitrate anion is relatively flat and will therefore allow a variable degree of hydration. X-ray studies of the hydration of nitrate anion (ref. 15) have indicated the absence of any strong interaction with one or two water molecules. Instead the data showed that nitrate anions interact weakly with a large number of molecules. The present value of the activation energy for the dehydration of NO_3^- form of resin also support the X-ray and neutron diffraction studies.

Cs^+ form of resin

The activation energy of the dehydration process of Cs^+ form of Dowex 50WX8 along with the hydrogen isotope effect data indicate that ion-water interactions in this system are weaker than water-water interactions in ordinary water. The hydration number for this cation in 2.5-10 molar solution by X-ray diffraction (ref. 16), neutron diffraction (ref. 17) and molecular dynamic (ref. 18) studies have been reported as 2.0-6.0, 8 and around 7 respectively. The structure breaking character of this large cation has been emphasized by Samoilov (ref. 19), in particular the difference between the activation energy for a water jumping from the hydration shell of this cation and from bulk water is negative. A similar behaviour has been observed for the dehydration of Cs^+ form of resin. Therefore, the water molecules in the coordination sphere of caesium cations are in a more disordered state, having weaker hydrogen bond interaction amongst them than bulk water. This leads to smaller hydrogen isotope effects in the dehydration of Cs^+ form of resin as observed in the present study.

In conclusion, the one point which emerges from the present study needs to be emphasized. Both activation energy and hydrogen isotope effect data of the dehydration of ion-exchange resins could be interpreted solely in terms of the counterion-water interactions. In contrast, hydrogen isotope effects in the distillation of aqueous electrolyte solutions are generally very small (even for Li salts) because of the presence of both anions and cations. In that sense, these studies on ion exchangers provide a means for studying the properties of single ions - particularly single ion-water interactions.

ACKNOWLEDGEMENT

The author wishes to express his sincere thanks to Technical Physics Division, BARC, for the isotopic analysis of water samples, to Dr. K. N. Rao for his interest and encouragement during the course of this investigation and to Drs S. K. Sarpal and Deoki Nandan for their help in the experimental work.

REFERENCES

1. F. Helfferich, Ion Exchange, McGraw Hill, New York, 1962.
2. G. E. Boyd and B. A. Soldano, Z. Electrochem., 57 (1953) 162.
3. D. Nandan and A. R. Gupta, Indian J. Chem., 12 (1974) 808.
4. A. R. Gupta, J. Chromatogr.,205 (1981) 263-270.
5. F. T. John, in H. London (Ed.), Separation of Isotopes, George Newnes Ltd., London, 1961, p 41-94.
6. Ref. 5, p 79.
7. A. R. Gupta, (unpublished work).
8. H. L. Friedman and C. V. Krishnan, in F. Franks (Ed.), Water, A Comprehensive Treatise, Vol. 3, Plenum Press, London, 1973, p 1-118.
9. J. Kristoff and J. Inczedy, presented at IV International Symposium on Ion Exchange, held at Lake Balaton, Hungary, May 27-30, 1980.
10. A. H. Narten, F. Vaslow and H. A. Levy, J. Chem. Phys., 58 (1973) 5017.
11. D. Nandan, A. R. Gupta and J. Shankar, Indian J. Chem., 10 (1972) 83.
12. W. C. McCabe and H. F. Fischer, J. Phys. Chem., 74 (1970) 2990.
13. T. E. Chang and D. E. Irish, J. Phys. Chem., 77 (1973) 52; ibid. J. Solution Chem., 3 (1974) 175; D. E. Irish in W. A. P. Luck (Ed.), Structure of Water and Aqueous Solutions, Verlag Chemie, Weinheim , 1974, p 333.
14. C. W. Neilson and J. E. Enderby, unpublished work.
15. R. Caminiti, G. Licheri, G. Piccaluga and G. Pinna, J. Chem. Phys., 68 (1978) 1967.
16. R. M. Lawrence and R. F. Kruh, J. Chem. Phys., 37 (1967) 4758.
17. N. Ohtomi and K. Arakawa, Bull. Chem. Soc. Japan, 52 (1979) 2755.
18. K. Heinzinger and P. Vogel, Z. Naturforsch.,31a (1976) 463.
19. O. Ya. Samoilov, in R. A. Horne (Ed.), Water and Aqueous Solutions, John. Wiley, New York, London, Sydney, Toronto, 1971.

INTERPRETATION OF ISOTOPE FRACTIONATION IN BIOLOGICAL SYSTEMS

E.M.GALIMOV
V.I.Vernadski Institute of Geochemistry and Analytical Chemistry
of the Academy of Sciences of the USSR, Moscow (USSR)

ABSTRACT

As distinguished from previous views which assumed kinetic type of isotope fractionation on some few biochemical pathways, in this paper the idea is maintained that the isotope fractionation is characteristic of all enzyme controlled reactions with the isotope effects being often of thermodynamic nature.

Theoretical models of the isotope fractionation both in a single enzyme reaction and metabolic system as a whole are developed. The expressions obtained describe quantitively the isotope fractionation in biological systems and allow a conclusion regarding the conditions under which thermodynamic or kinetic type of isotope effects could be expected.

INTRODUCTION

For a long time it was generally accepted that the biological fractionation of isotopes was a process which mainly occured on the initial step of photosynthetic fixation of CO_2. After profound intermolecular and intramolecular isotope heterogeneity of biological systems has been found, it became clear that many barriers of isotope fractionation other than that occuring on the photosynthetic stage should be taken into account. One of the most intriguing facts which was recently discoverd is that the carbon isotope compositions of various biomolecules in an organism are often in such correlations as if there had been isotope exchange equilibrium between them. Such an exchange is absolutely inconceivable for carbon which forms the structural skeleton of biomolecules. Under normal temperatures the carbon isotope exchange doesn't reach an equilibrium even for the simple organic compounds.

It was clear from the outset that the so called thermodynamically ordered isotope distribution in certain biological systems was in someways related to the fact that in living organisms the chemical reactions occur under control of enzymes (ref. 1). This understanding has not, however, helped so far to find a satisfactory isotope fractionation mechanism to account for the observed data. In this paper a theoretical model which, I believe, allows to describe the basic facts will be considered.

EXPERIMENTAL OBSERVATIONS

The experimentally observed facts may be summerized as follows.

1. In many biological systems a <u>thermodynamically ordered</u> distribution of isotopes is observed. It manifests itself through a correlation between $\delta^{13}C$ and $\beta^{13}C$ of biomolecules

$$\delta^{13}C_k = \delta^{13}C + \overline{\mathcal{H}}(\overline{\beta} - \beta_k) \cdot 10^3 \tag{1}$$

where the β-factor is a partition function ratio of the isotopic species of a compound reduced with respect to the symmetry numbers (see for example Fig. 1).

2. The observed isotope effects are always smaller than the equilibrium values determined by the ratio of the β-factors. The factor $\mathcal{H}$ is a measure of the <u>reduction</u>. Its value ranges between zero and one.

3. The thermodynamically ordered distribution of isotopes occurs not only between biomolecules but also on the <u>intramolecular level</u> (see for example Fig. 2).

4. The correlation between $\delta^{13}C$ and $\beta^{13}C$ is quite <u>universal</u>. It is observed in compounds of various structures and compositions, organisms of various ecological environment and taxonomy.

5. However in certain systems an isotope distribution of this kind may not be valid. The fact that the thermodynamically ordered distribution of isotopes manifests itself through a correlation, the correlation being of various strength (at times no correlation can be observed at all) means that realization of a thermodynamically ordered distribution depends on some prerequisites.

THEORETICAL MODEL

The theoretical model must explain these facts. First of all the question arises how can a thermodynamic isotope effect (which is

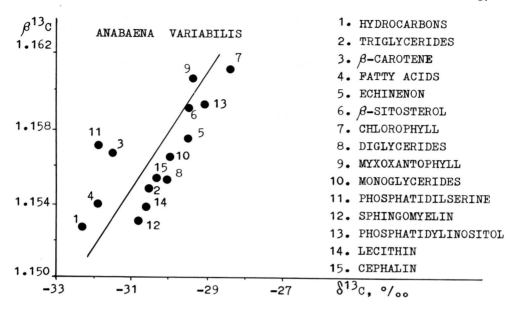

Fig. 1. An example of the correlation between $\delta^{13}C$ and $\beta^{13}C$ values of biomolecules. Components of the lipid fraction of a fresh water blue-green alga (ref. 2).

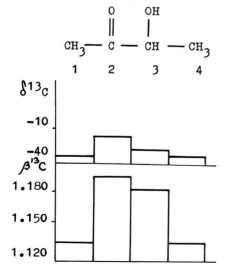

Fig. 2. Intramolecular carbon isotope effects in biologically produced acetoin (ref. 3) correlate with $\beta^{13}C$-values for corresponding positions of carbon atoms in the molecule.

of equilibrium nature) ever occur at all in biological systems. We believe that this can be explained if we refer to the concept of reversibilly of substrate-product conversion with in an enzyme-substrate complex. The enzymatic reaction as a whole may well be irreversible.

If the lifetime of the enzyme-substrate complex is much longer than the characteristic time of the bond rearrangement there is reason to expect that the process on the enzyme's active center will begin with the transition of substrate from an initial state to the state of the product, which may be followed by the reversed process. And this cycle may recur many times during the lifetime of a given enzyme-substrate complex.

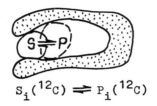
$S_i({}^{12}C) \rightleftharpoons P_i({}^{12}C)$

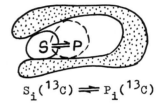
$S_i({}^{13}C) \rightleftharpoons P_i({}^{13}C)$

What distinguishes such an interpretation is that the concept of reversibility is here applied to the transformation of a single molecule.

It implies that the energy released in the course of product formation does not dissipate into the environment but can be immediately reutilized to return the product to the initial state; that is, the enzyme behaves like a "spring" or an elastically reflecting wall.

Processes of this type can be symbolized as follows.

$E + S_o \rightleftharpoons E(S \rightleftharpoons P) \rightarrow P + E$

The process of transformation of a compound from a state S to a state P and back is stopped at some moment in time either due to dissociation of the complex into a free enzyme and substrate (from the ES state) or into a free enzyme and the product (from the EP state).

It is obvious that dissociation of the enzyme-substrate complex will more frequently find a molecule in the relatively more stable isotope form. And this is controlled by the value of the β-factor. Therefore the relative probabilites of the presence of the molecule in a certain isotopic form of the substrate or the product state at the moment of the complex dissociation will be in proportion to their β-factor values

$$\frac{S_i^*}{S_i} : \frac{P_i^*}{P_i} = \beta_{S_i} : \beta_{P_i}$$

If we sum up the processes occurring in a single enzyme-substrate complex for all such complexes available in the cells, we will come up with an equation which is formally quite similar to the conventional isotope exchange equation:

$$S_i(^{13}C) + P_i(^{12}C) \rightleftharpoons S_i(^{12}C) + P_i(^{13}C)$$

and

$$\alpha = \frac{R_{P_i}}{R_{S_i}} = \frac{\beta_{P_i}}{\beta_{S_i}}$$

But actually there is no isotope exchange, it is rather selection of the isotopic species according to energy that leads to the same ultimate result, i.e., to the thermodynamic isotope effect.

Following this model one can write an enzymatic reaction in the following form:

$$E + S_o \underset{k_{-1}}{\overset{k_1}{\rightleftharpoons}} (ES \underset{k_b}{\overset{k_a}{\rightleftharpoons}} EP) \overset{k_2}{\longrightarrow} P + E \qquad (2)$$

Under steady state conditions the following kinetic relationships hold true:

$$[ES](k_a - k_{-1}) = ([E_T] - [ES] - [EP])[S_o]k_1 + [EP]k_b$$

$$[ES]k_a = [EP](k_2 + k_b)$$

$$v = [EP]k_2$$

Combination of them yields an equation for the reaction rate

$$\frac{v(k_2 + k_b)(k_a + k_{-1})}{k_2 k_a} = [E_T][S_o]k_1 - \frac{v(k_2 + k_b)}{k_a k_2}[S_o]k_1$$

This equation may be transformed to look like the Michaelis-Menten equation

$$v = \frac{\dfrac{k_2 k_a [E_T]}{k_2 + k_b + k_a}[S_o]}{\dfrac{k_2 k_{-1}\left(1 + \dfrac{k_b}{k_2} + \dfrac{k_a}{k_{-1}}\right)}{k_1(k_2 + k_b + k_a)} + [S_o]}$$

In a general case it is rather difficult to investigate the isotope effect of this reaction. It would therefore be convenient to consider two extreme cases: those with a high and a low substrate concentration.

When $K_m \gg [S_o]$, then $v = \dfrac{V_{max}}{K_m}[S_o]$. Hence

$$-\frac{d[S_o]}{dt} = \frac{k_a k_1 [E_T]}{k_{-1}\left(1 + \dfrac{k_b}{k_2} + \dfrac{k_a}{k_{-1}}\right)}[S_o]$$

The fraction before $[S_o]$ is the rate constant of the reaction. Let us denote it as k_e. Correspondingly for the isotope-substituted form

$$-\frac{d[S_o^*]}{dt} = k_e^* [S_o]$$

and it is the ratio of the rate constants for the basic and isotope substituted forms which defines the isotope effect of reaction

$$\alpha_e = \frac{k_e}{k_e^*}$$

A number of transformations give the following expression for the isotope effect:

$$\alpha_e = \left[1 - \frac{\dfrac{k_b}{k_2}}{1 + \dfrac{k_b}{k_2} + \dfrac{k_a}{k_{-1}}}\left(\frac{\beta_P}{\beta_S} - 1\right)\right]\left[1 + \frac{1}{1 + \dfrac{k_b}{k_2} + \dfrac{k_a}{k_{-1}}}\left(\frac{k_a}{k_a^*} - 1\right)\right] \quad (3)$$

The first square bracket contains the β-factor ratio. It therefore describes the thermodynamic isotope effect. The second square bracket containes the ratio of the rate constants of transformation of isotopic forms of the substrate, i.e. it describes the kinetic isotope effect.

By use of the corresponding denotations the expression (3) can be written in a more compact form

$$\alpha_e = \left[1 - \varkappa\left(\frac{\beta_P}{\beta_S} - 1\right)\right]\left[1 + \lambda(\alpha_{kin} - 1)\right] \qquad (4)$$

Another extreme case, when $K_m \ll [S_o]$, brings about an expression that is similar to (3). Therefore it may be concluded that an expression of the form (4) has a rather general applicability to isotope fractionation effects in an enzymatic process (if one accepts the model proposed).

The thermodynamic component may be isolated and the equation may be rewritten in terms of $\delta^{13}C$ values in the following final form.

$$\delta^{13}C_P = \delta^{13}C_{S_o} + \varkappa\left(\frac{\beta_P}{\beta_S} - 1\right)10^3 + \Delta_{kin} \qquad (5)$$

Obviously, the magnitude of the enzymatic isotope effect and its nature (thermodynamic or kinetic) depends on the values of the factors $\varkappa$ and λ which, in turn, depend on the ratio of reaction rate constants.

The following limiting cases are possible.

When $k_b \gg k_2$, and $k_a \gg k_{-1}$, both λ and $\varkappa$ are close to zero. This situation corresponds to a sequence of irreversible processes in which isotope effects cannot manifest themselves:

$$E + S_o \longrightarrow (ES \longrightarrow EP) \longrightarrow P + E$$

When $k_b \ll k_2$ and $k_a \gg k_{-1}$ then $\lambda \simeq 1$ and $\varkappa \simeq 0$. Here, the process is reversible at the substrate binding step; but conversion of the substrate to product is irreversible. This latter case corresponds to the conventional conception of an irreversibility of a enzymatic catalysis step:

$$E + S_o \rightleftharpoons (ES \longrightarrow EP) \longrightarrow P + E$$

In this case the isotope effect is of purely kinetic nature.

Finally, when $k_b \gg k_2$ and $k_a \ll k_{-1}$ then $\lambda \simeq 0$ and
$$\mathcal{H} = \frac{1}{1 + \frac{k_a}{k_{-1}} + \frac{k_b}{k_2}}$$
i.e. the thermodynamic isotope effect takes place. This is the case which falls under our assumption of microscopic reversibility of the intra-complex substrate-product transition:

$$E + S_o \rightleftharpoons (ES \rightleftharpoons EP) \rightarrow P + E$$

Thus, depending on the particular content of the enzymatic process, the net isotope effect may be either kinetic or thermodynamic, or both at once in a certain combination.

The following example illustrates how the formula (4) works. Recently DeNiro and Epstein (ref. 4) reported their investigation of carbon isotope fractionation in the process of enzymatic oxidation of pyruvate. Table 1 shows their experimental data.

TABLE 1

	CH_3	CO	COOH
Pyruvate	-20.1	-19.9	-22.3
	CH_3	CHO	CO_2
Acetoaldehyde + CO_2	-21.1	-34.0	-28.8

The enzymatic reaction in this case may be presented in the following form:

$$E + \begin{array}{c} CH_3 \\ | \\ CO \\ | \\ COOH \end{array} \rightleftharpoons \begin{bmatrix} CH_3 \\ | \\ CO \\ | \\ COOH \end{bmatrix} \rightleftharpoons \begin{array}{c} CH_3 \\ | \\ CHO \\ \\ CO_2 \end{array} \rightarrow \begin{array}{c} CH_3 \\ | \\ CHO \end{array} + CO_2 + E$$

Isotope fractionation in each carbon position is assertained by the formula (5).

Thus we have got a system:

$$-28.8 = -22.3 + \mathcal{H}\left(\frac{1.1916}{1.2019} - 1\right)10^3 + \Delta_{kin}$$

$$-34.0 = -19.9 + \mathcal{H}\left(\frac{1.1706}{1.1933} - 1\right)10^3 + \Delta_{kin}$$

$$-21.2 = -20.1 + \mathcal{H}\left(\frac{1.1344}{1.1357} - 1\right)10^3$$

From the two first equations we are able to calculate both variables: $\Delta_{kin} = -0.3°/_{oo}$ and $\mathcal{H} = 0.73$. The Kinetic isotope effect is found to be small when it is compared with the total isotope effect. The magnitude of $\mathcal{H}$ is within reasonable theoretical limits between zero and unity and what is important is that it independently satisfies the third of the equations.

The formula of the isotope effect of an enzymatic reaction accounts for two essential features of isotope distribution observed in natural biological materials. First, it points to the principal possibility of the occurrence of thermodynamic isotope effects in a biological system under certain conditions. Second, it explains why the observed isotope effects are always smaller than the equilibrium ones determined immediately by the β-factor ratio. The reason is that the factor $\mathcal{H}$ is always within the range from zero to unity. The factor $\mathcal{H}$ which we have introduced empirically as a measure showing how much the observed isotope effect is smaller than the equilibrium one hereby acquires the physical meaning of a value depending on the rate constant ratio of an enzymatic process.

However, the formula is only valid for the isotope effect between the product and its immediate precursor. It cannot describe the $\delta^{13}C - \beta^{13}C$ correlation between random biomolecules. It cannot account, either, for the possibility of the intramolecular thermodynamically ordered distribution of isotopes.

To explain these aspects of isotope fractionation one has to have recourse to a more general model capable of covering a system of enzymatic reactions.

Biosynthesis involves a succession of enzymatic reactions, so that the product of one serves as the substrate of the next. Under steady conditions the concentrations of the intermediate metabolites may, within certain limits, be considered as constant. Besides, the metabolic paths are interrelated and anabolic processes are balanced with katabolic ones.

On these grounds it is possible to idealize the metabolic system by the following model which, apart from a straightforward path of the reactions, implies branching and back pathway at every stage:

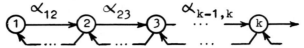

These are not necessarily reversible reactions, just any back paths. While the forward reaction is characterized by the isotope effect, the back path may in the general case not show the isotope effect or have the zero net isotope effect. It is easy to see a principal similarity between this model of metabolism and the process occurring in an isotope fractionation column (Fig. 3).

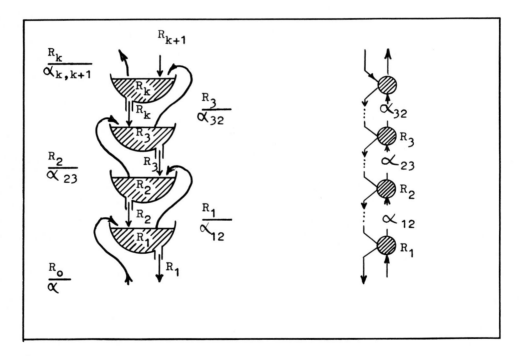

Fig. 3. To the similarity between an idealized model of metabolism and a process occuring in an isotope fractionation column.

Under steady state conditions, the amount of matter on each plate of the fractionation column is constant and what is important is that it operates under a "reflux" that is analogous of cathabolic process in a methabolic system. It is known that under steady state conditions the ratio of isotopic compositions between any column plates is defined as the product of the separation factors of all the separation stages between the two plates in question. Accordingly, in our model of a system of enzymatic reactions the isotopic compositions of any two metabolites (say, the 1-st and k-th) occurring in one and the same metabolic path is defined as a product of the separation factors of all the intermediate enzymatic reactions:

$$\frac{R_k}{R_1} = \frac{1}{\alpha_{12}} \frac{1}{\alpha_{23}} \cdots \frac{1}{\alpha_{k-1,k}} = \prod_{i=1}^{k-1} \alpha_{i,i+1}^{-1} \qquad (6)$$

Here α is defined by the formula (4) for an isotope fractionation factor in an isolated enzymatic reaction. The $\mathcal{H}$ value at each particular stage should be represented as a mean value for a given biosynthesis path and the deviation from it: $\mathcal{H}_{i,i+1} = \overline{\mathcal{H}} + \Delta\mathcal{H}_{i,i+1}$. After a number of rather cumbersome, yet simple, transformations we obtain the following final expression.

$$\frac{R_k}{R_1} = \mathcal{H}\left(\frac{\beta_1}{\beta_k} - 1\right) + 1 + \frac{\beta_1}{\beta_k}\sum_{i=1}^{k-1}\left\{\Delta\mathcal{H}_{i,i+1}\left(\frac{\beta_{i+1}}{\beta_i} - 1\right) + \left(\alpha\lambda_{i,i+1}-\right)\right\} \quad (7)$$

Its meaning is as follows. The ratio of isotope compositions of any compounds related to one and the same pathway may be expressed as a sum of two terms. The first term is the reduced thermodynamic isotope effect expressed through the ratio of the thermodynamic isotope factors of the components in question and the averaged value of $\overline{\mathcal{H}}$. The second term (let denote it as Δ) includes an algebraic sum of the deviations due to the difference of the real $\mathcal{H}_{i,i+1}$ values from the mean $\overline{\mathcal{H}}$ value, as well as an algebraic sum of the kinetic isotope effects along the path between the biosyntheses of the 1-st and k-th carbon.

In terms of $\delta^{13}C$ this relationship can be rewritten in the following form:

$$\delta^{13}C_1 - \delta^{13}C_k = \mathcal{H}\left(\frac{\beta_1}{\beta_k} - 1\right) 10^3 \pm |\Delta| \qquad (8)$$

On the basis of this formula one is now able to account for the observed cases of $\delta^{13}C - \beta^{13}C$ correlations between random biomolecules as well as for the intramolecular isotopic distribution.

Indeed the resultant formula can be easily transformed to characterize the correlation between $\delta^{13}C$ and $\beta^{13}C$:

$$\delta^{13}C_k = \overline{\delta^{13}C} + \overline{\mathcal{H}}(\beta_k - \overline{\beta}) \cdot 10^3 \pm |\Delta| \tag{9}$$

This equation is none other but the regression equation in that the very form we obtain it by processing the experimental data with Δ accounting for the dispersion.

The same model can be used to clarify the possibility of the intramolecular thermodynamically ordered isotope distribution.

If there is a biochemical path involving two atoms of one and the same molecule, then the thermodynamic isotope effect may occur between them as well as between any 1-st and k-th atom of a conversion chain:

$$\delta^{13}C_{P_i} - \delta^{13}C_{P_{i+1}} = \mathcal{H}\left(\frac{\beta_{P_i}}{\beta_{P_{i+1}}} - 1\right) \cdot 10^3 \pm |\Delta| \tag{10}$$

Consequently the intramolecular thermodynamic ordering of isotopes is a property of molecules generated not merely in an enzymatic reaction but rather in the course of a metabolic process taking place in an interrelated system of enzyme reactions, i.e., practically in an organism.

Thus now we are in a position to explain the facts mentioned in the beginning of this paper including the $\delta^{13}C - \beta^{13}C$ correlation between random biomolecules and intramolecular ordering as well as the cases of the lack of the correlation. The latter is determined by the limitations inherent in the theoretical model.

The first limitation is connected with the requirement that the mechanism of the enzymatic catalysis be such as to assure the thermodynamic isotope effect of an enzymatic reaction. Obviously, this is not always so.

Second, it is required that there be branching and back conversion paths at every stage of the biosynthetic process. This is a rather strong idealization too.

Thirdly, the biochemical processes are assumed to be steady. In addition, for realization of the intramolecular ordered isotope distribution presence of such a distribution in the precursor is essential. This condition may not be met in the case of heterotrophic assimilation of organic compounds.

It is clear that such properties as steadiness of the biochemical processes, reversibility of biosynthesis pathways and the like are inherent in methabolism of an organism as a whole rather then individual biosynthesis path. Therefore the phenomenon of thermodynamically ordered isotope distribution has a statistical meaning. Its manifistation is the more apparent the greater is the number of the processes or the experimental facts we are taking into consideration.

Since organic matter in sediments and rocks integrates biomass of many individual components of many organisms, thermodynamically ordered isotope distribution should have a geochemical significance. One could suggest the following geological applications of this phenomenon.

First, being an inherent feature of biological systems, the phenomenon of thermodynamically ordered isotope distribution may be used as a means of identification of biogenic or abiogenic substances both under terrestrial and extraterrestrial conditions.

Second, since in biogenic compounds that have undergone a considerable post-biogenic evolution (fragmentation and so on) the originally thermodynamically ordered isotope distribution is gradually eliminated, the pattern of $\delta^{13}C$ versus $\beta^{13}C$ relationships may reflect a history of the geochemical maturation of organic matter.

Thirdly, knowing the intramolecular isotopic heterogeneity pattern in organic matter it is possible to predict the isotopic composition of its derivatives such as hydrocarbons, CO_2 and so on or, vice versa, by solving the reverse problem to identify the biological precursors, and indicate the sources of organic carbon, of these derivatives what is of great importance in oil and gas geology.

REFERENCES

1 E.M. Galimov, Carbon isotopes in oil-gas geology, Nedra Press, Moscow 1973, NASA TT F-682, Washington D.C., 1975, p.395.
2 E.M. Galimov and V.G.Shirinski, Geochimia (in Russian), (1975) 503-528.
3 G. Rinaldi, W.J. Meinschein and J.M. Hayes, Biomedical Mass Spectrometry, 1(1974) 415-417.
4 M.DeNiro and S. Epstein, Science, 197(1977) 261.

H.-L. Schmidt, H. Förstel and K. Heinzinger (Editors), *Stable Isotopes* 49

OXYGEN KINETIC ISOTOPE EFFECT IN THE THERMAL ISOMERIZATION OF BIS-(5,5-DIMETHYL-2-OXO-1,3,2-DIOXAPHOSPHORINANIL) SULFIDE

W. REIMSCHÜSSEL and P. PANETH
Institute of Applied Radiation Chemistry, Technical University, Łódź (Poland)

ABSTRACT

The oxygen kinetic isotope effect in the rearrangement of symmetrical monothiopyrophosphate 1 to the unsymmetrical isomer 2 was studied. Theoretical values predicted for three alternative mechanisms were compared with the experimental results. The dissociative mechanism was suggested.

INTRODUCTION

The rearrangement of organic monothiopyrophosphates with a sulfur atom in the central position into the unsymmetrical isomer (ref. 1) was first observed by Michalski (refs. 2,3). At high temperatures this process is spontaneous. Kinetics of this isomerization were studied in our laboratory (refs. 1,4) on the example bis-(5,5-dimethyl-2-oxo-1,3,2-dioxaphosphorinanil)-sulfide 1.

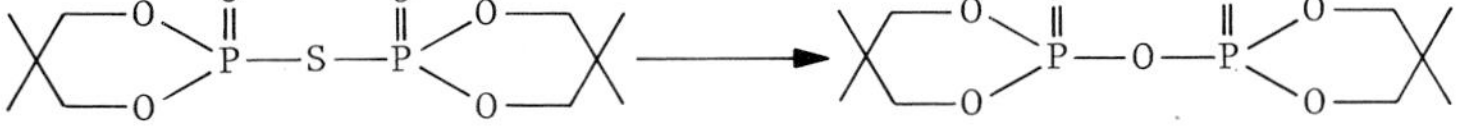

1 2

In dilute solutions of 1-methylnaphtalene or benzonitrile the above reaction is monomolecular and the only product is 2-oxo-2'-thioxo-bis-(5,5-dimethyl-1,3,2-dioxaphosphorinanil)-oxide 2.

In this paper theoretical and experimental values of the oxygen kinetic isotope effect are presented. Theoretical values were predicted for three alternative mechanisms: concerted and nonconcerted cyclic intramolecular rearrangement and dissociative rearrangement. Models of transition state structures for these mechanisms are shown in Table 2.

EXPERIMENTAL

Materials

Substrat 1, selectively labelled with ^{18}O in exocyclic positions, was synthesized from ^{18}O-enriched water by modified procedures (refs. 5,6) outlined in the following scheme:

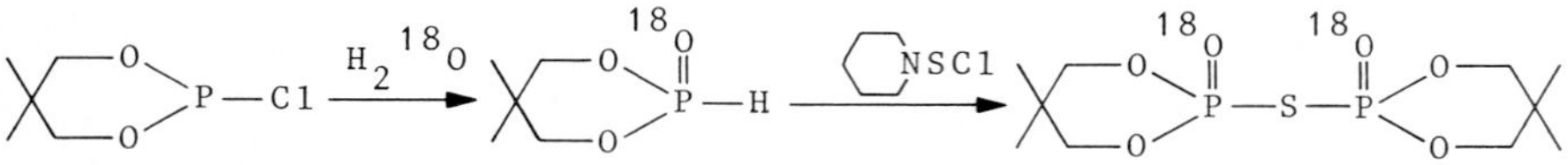

^{18}O-labelled compound 1 was purified by a two fold crystallization from benzene-dichloromethane (1:1).

Kinetic runs

Samples of about 15 mg of substrate 1 in 5 ml of solvent were heated at 120 ± 0.1 °C in sealed ampuls to obtain a conversion of about 5%. From the reaction mixture the solvent was evaporated at reduced pressure and the residual solid was dissolved in dichloromethane. Product 2 was isolated by TLC technique. Benzene-chloroform-acetone (4:2:1) as the eluent and Kieselgel G (Merck) were used. Compound 2 (R_F 0.65) was extracted from the gel with dichloromethane, which was then evaporated off.

Determination of $k_{16,16}/k_{18,18}$

Isotopic ratios $^{18}O/^{16}O$ of compounds 1 and 2 were measured by the mass spectrometric method (ref. 7). Mass spectra were obtained with a Finnigan 4000 mass spectrometer. The background and mass fragmentograms of peaks used in calculations were investigated to prevent errors caused by impurities. The values of the isotopic ratio R were calculated as:

$$R = \frac{^{18}O}{^{16}O} = \frac{I_{M+2} + 2I_{M+4}}{2I_M + I_{M+2}} \tag{1}$$

where I are the intensities of the peaks m/e 330(M), 332(M+2) and 334(M+4).

Kinetic isotope effects for molecules containing one ^{18}O atom - $k_{16,16}/k_{16,18}$ and two ^{18}O atoms - $k_{16,16}/k_{18,18}$, assuming the statistical distribution of isotopes and low fraction of reaction f, were calculated from:

$$k_{16,16}/k_{16,18} = R_o/R_f \quad \text{and} \quad k_{16,16}/k_{18,18} = (R_o/R_f)^2 \qquad (2)$$

where R_o and R_f are isotopic ratios of substrate 1 and product 2, respectively.

The experimental values of isotopic ratios and kinetic isotope effects are given in Table 1. Statistical errors were computed at a 0.95 confidence level.

TABLE 1
Experimental values of isotopic ratios and kinetic isotope effects

	1-methylnaphtalene	benzonitrile
R_o	0.7006 ± 0.0022	0.7007 ± 0.0016
f	0.050	0.056
R_f	0.6993 ± 0.0021	0.6999 ± 0.0025
$k_{16,16}/k_{16,18}$	1.002 ± 0.004	1.001 ± 0.004
$k_{16,16}/k_{18,18}$	1.004 ± 0.009	1.002 ± 0.009

Theoretical calculations of $k_{16,16}/k_{18,18}$

To simplify the calculations only the molecules containing two ^{18}O atoms were considered. Predicted values were based on the Bigeleisen-Mayer equation for heavy atoms (ref. 8):

$$k_{16,16}/k_{18,18} = SN \cdot TIF \cdot TDF \qquad (3)$$

The ratio of symmetry numbers SN was assumed to be unity since the isotopic substitution of two oxygen atoms does not alter the symmetry of the molecule.

The temperature independent factor TIF for model 3 (see Table 2) was computed from the reduced masses of fragments separated by the breaking bond (ref. 9). The cut-off method (ref. 10) was employed for the calculation of these masses in case of model 4. In structure 5 of the transition state corresponding to cyclic concerted rearrangement, four bonds are changing simultaneously. Therefore TIF was obtained on the basis of molecular masses of isotopomers.

The temperature dependent factor TDF for the dissociative mechanism was taken as unity because there are no bonding changes of phosphoryl oxygens in going from substrat 1 to transition state 3.

For the nonconcerted cyclic intramolecular mechanism (model $\underline{4}$) TDF was calculated in accordance with:

$$TDF = 1 + 2G/u_{P=^{18}O}/\Delta u_{P=O} - G/u^{\neq}_{P-^{18}O^-}/\Delta u^{\neq}_{P-O^-} - G/u^{\neq}_{P-^{18}O-P}/\Delta u^{\neq}_{P-O-P} \quad (4)$$

where $u_i = hc\nu_i/k_B T$; $i = P=O$, $P-O^-$ or $P-O-P$.

Values of wave number $\nu_{P=^{18}O} = 1248.0$ cm^{-1} and $\Delta\nu_{P=O} = 37.5$ cm^{-1} for the labelled substrate $\underline{1}$ were obtained from an IR spectrum (Perkin Elmer 580 spectrophotometer). Wave number $\nu_{P-^{18}O^-}$ and $\Delta\nu_{P-O^-}$ were estimated, assuming harmonic oscillation (ref. 11), from equations:

$$\nu_{P-O^-} = \nu_{P=O}(k_{P-O^-}/k_{P=O})^{1/2} \quad (5)$$

$$k_i = 1.67 N_i (X_P X_O/d_i^2)^{3/4} + 0.3 \quad (6)$$

The degrees of bonding N_i, were calculated from the Cruickshank relation (ref. 12) as equal to 1.8 for P=O and to 1.4 for P-O$^-$. The lengths of bonds d_i (1.45 for P=O and 1.60 for P-O$^-$) from crystallographic data (refs. 13,14) and the Pauling electronegativities X (2.1 for P and 3.5 for O) were used. As a result values $\nu_{P-^{18}O^-} = 1031.7$ cm^{-1} and $\Delta\nu_{P-O^-} = 31.0$ cm^{-1} were obtained. The force constant k_{P-O-P} was calculated from relationship (6) and the wave number for the P-O-P vibration was estimated, assuming angle $\alpha = 105°$ at the oxygen atom, using the equation (ref. 15):

$$\nu^{as}_{P-O-P} = 1303 \left[k_{P-O-P}(1/M_P + (1-\cos\alpha)/M_O) \right]^{1/2} \quad (7)$$

The values obtained were $\nu_{P-^{18}O-P} = 827.8$ cm^{-1} and $\Delta\nu_{P-O-P} = 32.1$ cm^{-1}. Substituting G(u) values, found in the tables of (ref. 16) for the corresponding wave number into equation (4), one obtains the value of TDF for the nonconcerted mechanism (Table 2).

The cyclic concerted intramolecular mechanism is intermediate between the dissociative one, in which there are no bonding changes of phosphoryl oxygens, and the cyclic nonconcerted mechanism, in which the full bond formation was assumed. Therefore TDF was taken as the average (ref. 17) of values for the above two mechanisms. The TDF value in the latter mechanism was corrected for bonding changes of only one phosphoryl oxygen.

The estimated kinetic isotope effect $k_{16,16}/k_{18,18}$ for the considered three alternative mechanisms is presented in Table 2.

TABLE 2

Theoretical values of kinetic isotope effect

Alternative mechanism	Transition state structure	Effective mass model	TIF	TDF	$\dfrac{k_{16,16}}{k_{18,18}}$
dissociative	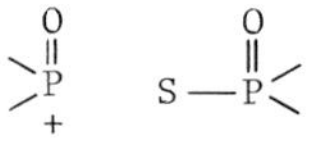 >P(=O)-S ... P<(=O) (+) **3**	133 (+16/+18) ... 165 (+16/+18)	1.0060	1	1.0060
cyclic nonconcerted	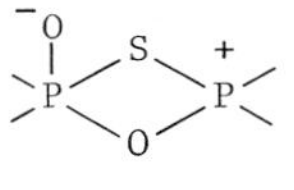 >P(−O)(S)(+)P< cyclic **4**	150 (+16/+18) ... 95 (+16/+18)	1.0078	1.0247	1.0327
cyclic concerted	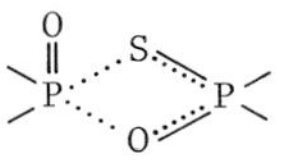 >P(=O)...S...P< cyclic **5**	298 (+2 16/+2 18)	1.0060	1.0070	1.0131

CONCLUSIONS

Comparing the theoretical kinetic isotope effect for three alternative structures of the transition state (Table 2) with the experimental values (Table 1), it can be seen that the best agreement is obtained for model 3. Consequently, the most probable mechanism of the reaction in both solvents is the dissociative one with cleavage of one of the P-S bonds. Because of the relatively large experimental errors, however, the concerted mechanism cannot be excluded. We believe that further studies involving measurements of sulfur kinetic isotope effect will allow us to determine the real mechanism.

REFERENCES

1 J. Michalski, W. Reimschüssel and R. Kamiński, Usp. Khim., 47(1978) 1528.
2 J. Michalski, Ann. N.Y. Acad. Sci., 192(1972)90.
3 J. Michalski, B. Mlotkowska and A. Skowrońska, J. Chem. Soc., Perkin Trans. I, (1974)319.
4 R. Kamiński, Ph.D. Thesis, Centre of Molecular and Macromolecular Studies, Polish Academy of Sciences, Łódź, 1978.

5 A. Zwierzak, Can. J. Chem., 45(1967)2501.
6 J. Michalski, M. Mikolajczyk and B. Mlotkowska, Chem. Ber., 102 (1969)90.
7 W. Reimschüssel and P. Paneth, Org. Mass Spectrom., 15(1980)302.
8 J. Bigeleisen, J. Chem. Phys., 17(1949)675.
9 L. Melander, Isotope Effects on Reaction Rates, Ronald Press Co., New York, 1960, p.38.
10 M.J. Stern and M. Wolfsberg, J. Chem. Phys., 45(1966)4105.
11 D.E.C. Corbridge, in M. Grayson and E.J. Griffith(Eds.), Topics in Phosphorus Chemistry, Vol. 6, The Infrared Spectra of Phosphorus Compounds, J. Wiley and Sons, New York, 1969, p.240.
12 D.W. Cruickshank, J. Chem. Soc., (1961)5486.
13 M. Bukowska-Strzyzewska and W. Dobrowolska, Acta Cryst., B36(1980) 3169.
14 D.E.C. Corbridge, The Structural Chemistry of Phosphorus, Elsevier, Amsterdam, 1974, p.176.
15 see: ref. 11, p.283.
16 J. Bigeleisen and M.G. Mayer, J. Chem. Phys., 15(1947)261.
17 D. Benko, Diss. Abstr., Int. B, 40(1979)411.

KINETIC STUDY OF DEUTERIUM EXCHANGE IN METHYL GROUPS OF PYRIMIDINE DERIVATIVES

P. DIZABO and E. PICQUENARD[+]

Laboratoire de Spectrochimie Moléculaire, Université Pierre et Marie Curie, 4, Place Jussieur, F 75230 Paris-Cédex, France

ABSTRACT

The isotopic deuteration kinetics of C-methyl protons is investigated in acidic medium in a series of pyrimidines : methylpyrimidines, amino-methylpyrimidines and methyl-2-pyrimidones. The reactions are followed by means of ^{1}H-n.m.r. spectroscopy. For the methyl and aminomethyl derivatives the study enables us to determine that deuteration proceeds via a stepwise reaction. The influence of an amino group substituent attached to the ring is discussed. For 2-pyrimidones, the reaction rate passes through a maximum when the pD of the solution is equal to the pKA of the conjugate acid. This maximum is due to the neutral molecule acting as a base and removing a proton from one of the methyl groups of its conjugate acid.

INTRODUCTION

Protons in the α-position of alkyl substituents of pyrimidines can undergo exchange reactions. In the case of alkyl- and aminoalkylpyrimidines these exchanges take place only in a strongly acidic medium [1,2], while certain alkylpyrimidones exchange these protons in acid, basic and neutral medium. As a contribution to the elucidation of the mechanism of the exchange reaction the proton exchange of a series of pyrimidine derivatives in D_2O was pursued by means of nmr spectroscopy, and the signals of the 5-methyl group protons, which do not exchange, were used as internal standards.

RESULTS AND DISCUSSION

Methyl- and aminomethylpyrimidines

The proton exchange of methyl- and aminomethylpyrimidines is only observed in a strongly acid medium. Therefore a doubly protonated intermediate has to be postulated. With this intermediate two mechanisms have to be discussed :

[+]Present address : Laboratoire de Spectrochimie Infrarouge et Raman, C.N.R.S., 2, rue Henri Dunant, 94320 Thiais, France.

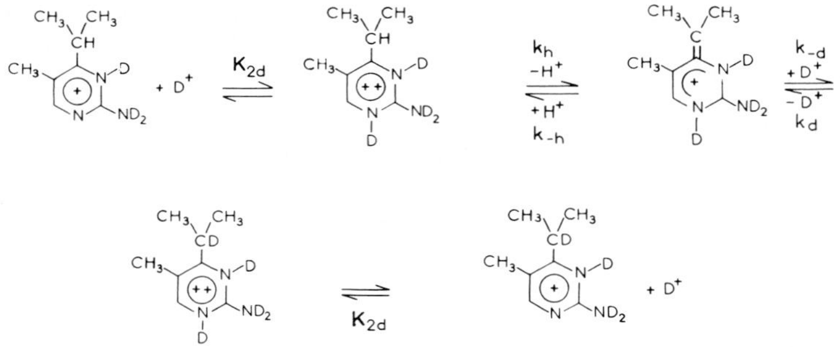

[Scheme 1 diagram showing stepwise mechanism (A) and concerted mechanism (B) with chemical structures of pyrimidine derivatives]

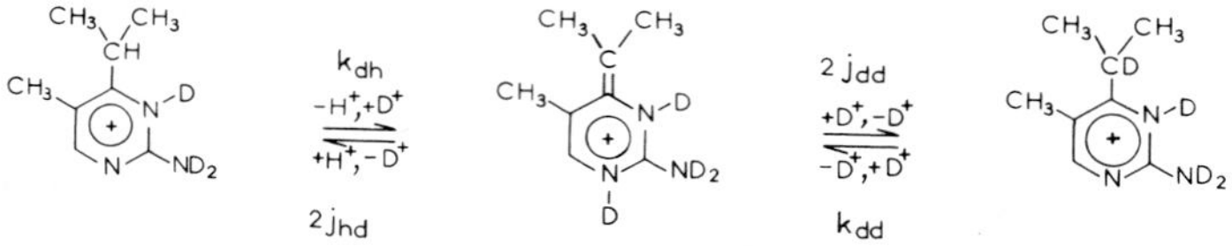

Scheme 1 : Possible mechanisms

A) a stepwise reaction in which the protonation is fast, and the proton loss is slow
B) a concerted mechanism in which the second protonation is slow and immediately followed by a rapid loss of another proton.

If we make the following approximations : we disregard the secondary isotope effects when exchange occurs for a $-CH_3$, $-CH_2D$ or $-CHD_2$ group on the one hand and a H^+ before a D^+ owing to the high content of deuterium in the medium on the other hand, both mechanisms led to the equation [3] :

$$Y(t) = u \log \frac{u(w+\xi)}{w(v+\xi)} = C.t.$$

the parameters u, v, w are obtained from initial composition. The reaction rate is defined by

$$\xi = \frac{\text{peak area at time 0} - \text{peak area at time t}}{\text{peak area at time 0}}$$

C constant depends on mechanism:

- stepwise mechanism $C = \dfrac{k_h}{K_{2D} + (D^+)}$

with K_{2D} second acidity constant.
- concerted mechanism $C = k_{dh}$

A discrimination between these two mechanisms is possible owing to Arrhenius law. For a concerted mechanism log C versus the inverse of the temperature is a line and log C ($K_{2D} + (D^+)$) versus I/T for a stepwise mechanism. In this case K_{2D} is unknown and is used as a parameter for the least square fit. We can, then, calculate a value of K_{2D}.

Experimental data (Table 1) and their mathematical treatment show that for two compounds only 2 and 4 exchange occurs owing to a stepwise mechanism. For the other compounds, it is not possible to make a choice between the two mechanisms. However, comparing the activation energies and the frequency factor values let us assume that the mechanism for all compounds is the same.

TABLE 1

Calculated values for methyl- and aminomethylpyrimidines

Pyrimidine	Mechanism	Activation Energy kJ Mol^{-1}	Frequency factor Observed L s^{-1} Mol^{-1}	pK_{2D}
(1) 4,5,6-Trimethyl	A, B	79 $\pm$ 2	2 . 10^8	$-$ 1
(2) 2-Amino 4,5,6-Trimethyl	A	81 $\pm$ 2	1.5 . 10^{10}	0.0 $\pm$ 0.3
(3) 2-Amino 4-Ethyl-5-Methyl	A, B	73 $\pm$ 6	5 . 10^8	$-$ 1
(4) 4-Amino 2,6-Dimethyl	A	80 $\pm$ 3	2 . 10^8	1 $\pm$ 1

The comparison of our results (Table 1) shows that the amino group does not interfere in the exchange mechanism of a methyl group ; it only modifies the electronic distribution in the ring.

An amino group attached to C-2 of the pyrimidine ring increases the ring nitrogen atoms basicity. Its electron donating power has not much effect on a C-4 or C-6 atom. Likewise in the 4 position, the amino group electron donating power does not interfere with the C-2 atom. By contrast, the amino group strongly loads the C-6 atom and then the methyl group attached to it is inactive.

The ring-nitrogen atoms seem therefore to form a barrier to the substituent electrodonating effects.

Comparing the activation energy of the compound 2 and compound 4 shows that our approximations are justified.

Methyl-2-pyrimidones

As opposed to methyl and aminomethylpyrimidines, for 4,5,6-trimethyl-2pyrimidone [4] the exchange of the 4- and 6-methyl protons occurs in pure D_2O, acidic D_2O or basic D_2O. We just study the exchange for acidic medium (40°C).

As the concentration of DCl increases, the exchange rate increases, passes through a maximum for (DCl) = 1/2 (pyrimidone) and then decreases (Fig. 1).

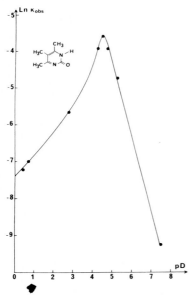

Fig. 1. Acidity effect on the rate of exchange of 4,5,6-trimethyl-2-pyrimidone

The previous mathematical treatment is not suitable in this case. The exchange reaction is a superposition of two mechanisms in which each mechanism is predominant according to the medium used.

When (DCl) > (pyrimidone), the pyrimidone exists mostly as a conjugate acid, and we can consider that the reaction is mainly directed by the attack of the conjugate acid by a base i.e. water. The decrease in the rate, as the acidity further increases, is due to the decrease of the activity of the water.

When (DCl) < (pyrimidone), the exchange rate is maximum when the pD of the medium is equal to the pK of the conjugate acid. The proposed mechanism is the reaction of the neutral pyrimidone acting as a base and removing a proton from the methyl group of its conjugate acid, then reaction with D_3O^+ giving the deuterated compound.

All exchange reactions occur pseudo first order. The kinetic equations in the second case, having made previous approximations, lead to the following equation, identical to that established by Stewart et al. for the proton exchange in creatinine [5].

$$-\frac{d\omega}{dt} = k_1 \frac{(D^+) K_d}{((D^+) + K_d)^2} X_0 \omega = k_{obs} \omega$$

with (D^+) measured at the glass electrode, K_d acidity constant of the conjugate acid, X_0 initial concentration of pyrimidone and ω reaction relative rate defined by

$$\omega = \frac{\text{Area of 4,6-methyl peak}}{2 \times (\text{area of 5-methyl peak})}$$

The condition k_{obs} maximum is obtained for $(D^+) = K_d$.
When plotting k_{obs} versus the expression $\frac{(D^+) K_d}{((D^+) + K_d)^2}$ a satisfactory linear relationship is obtained (fig. 2).

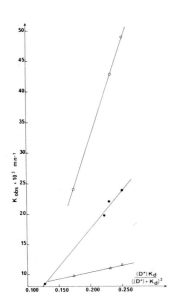

Fig. 2. Plot of k_{obs} vs $\frac{(D^+) K_d}{((D^+) + K_d)^2}$ for

4,5,6-trimethyl-2-pyrimidone (—●—)
1,4,5,6-tetramethyl-2-pyrimidone (Me-6 =—O—, Me-4 =—△—)

The exchange study of the methyl protons in 1,4,5,6-tetramethyl-2-pyrimidone corroborates this mechanism. For this compound, the rate of exchange is different for the 4-methyl and the 6-methyl group. The rate for the 6-methyl is greater than for the 4-methyl, this fact is due to the inhibiting effect of the 6-methyl substituent on the deuteration of the 4-methyl group as our results for 4-amino-2,6 dimethylpyrimidine show. However, the two methyl groups both react like those of 4,5,6-trimethyl-2-pyrimidone and a linear relationship is also obtained (Fig. 2).

The slopes of the lines enable us to obtain the rate constants k_1 (table 2).

TABLE 2

Rate constant k_1 of 2-pyrimidones

2-Pyrimidone	k_1^* L Mol^{-1} S^{-1}
4,5,6-Trimethyl	$3.92 \cdot 10^{-3}$
1,4,5,6-Tetramethyl	
4-Me	$7.67 \cdot 10^{-4}$
6-Me	$10.82 \cdot 20^{-3}$

*For $3 < pD < 7$

As we can see, the isotopic kinetic exchange study, owing to measurable parameters, shows some properties of pyrimidine, the electronic distribution in the ring and a particular type of reactivity for 2-pyrimidones when the acidity of the medium is near the pK_a of the conjugate acid.

REFERENCES

1 T.J. Batterham, D.J. Brown and M.N. Paddon-Row, J. Chem. Soc. (B) 2(1967)171-173.
2 P. Dizabo, J.C. Monier and A. Pompon, J. Label. Compounds 7(1971)399-408.
3 A. Pompon, J. Chim. Phys. 72(1975)505-508.
4 E. Picquenard and P. Dizabo, Can. J. Chem. 59(1981)1270-1276.
5 R. Srinivasan and R. Stewart, Can. J. Chem. 53(1975)224-231.

DEUTERIUM SOLVENT ISOTOPE EFFECTS ON REACTIONS OF PENTAAQUAORGANOCHROMIUM (III) IONS WITH ELECTROPHILES

V. GOLD and D. L. WOOD

Department of Chemistry, King's College London.

ABSTRACT

The reaction rates for reactions of pentaaquaorganochromium (III) ions with electrophiles (resulting in scission of the carbon-chromium bond) have been measured in ordinary water, deuterium oxide and in mixtures of the two as solvent. In all cases the reaction rate decreases as the deuterium content of the solvent increases, but the effect is particularly marked for reactions in which the aqueous hydrogen ion is the electrophile. It is shown that the effects are satisfactorily accounted for in terms of fractionation factor theory, provided that isotopic fractionation of the water of aquation of the ions is explicitly considered.

The measurement of kinetic solvent isotope effects, involving the comparison of rate constants (k) for a reaction in ordinary water, deuterium oxide, and in isotopically mixed water, has been widely applied to the elucidation of organic reaction mechanisms [1]. There is a generally accepted theoretical basis for the interpretation of the effects in terms of fractionation theory, expressible to a good approximation by equation (1) [2],

$$\frac{k_n}{k_{H_2O}} = \frac{\Pi(1-n + n\phi_i^\ddagger)}{\Pi(1-n + n\phi_j)} \tag{1}$$

where n is the atom fraction of deuterium in the solvent,

$$n = ([HOD] + 2[D_2O])/2([HOD] + [H_2O] + [D_2O]) \tag{2}$$

and ϕ_i represents the fractionation factor for the site i in a solute, i.e. the ratio

$$\phi_i = \frac{\left(\frac{D}{H}\right) \text{ in site i}}{\left(\frac{D}{H}\right) \text{ in solvent}} \tag{3}$$

The limiting case n→1 corresponds to solution in D_2O, i.e.

$$\frac{k_{D_2O}}{k_{H_2O}} = \frac{\Pi \phi_i^{\ddagger}}{\Pi \phi_j} \qquad (4)$$

In equations (1) and (4) the products are taken over all exchangeable hydrogen nuclei (general chemical symbol L) in the reactants (denominator) and in the transition state (numerator). (The symbol Π represents such a product operator.)

Examples of the successful application of equation (1) are some reactions in which there is a rate-limiting proton transfer from aqueous hydrogen ions (fractionation factor $\phi_{H_3O} = \ell = 0.69$) to an olefinic carbon atom [3]. The other relevant fractionation factors in such reactions are indicated in equation (5).

$$L_3O^+ + CR_2=CX_2 \longrightarrow \begin{bmatrix} R_2C \cdots CX_2 \\ \vdots \\ L \quad \phi_1^{\ddagger} \\ \vdots \\ O \\ / \backslash \\ L \quad L \\ \phi_2^{\ddagger} \quad \phi_2^{\ddagger} \end{bmatrix}^{\ddagger +} \qquad (5)$$

ℓ

Their insertion in equation (1) leads to equation (6)

$$\frac{k_n}{k_{H_2O}} = \frac{(1-n+n\phi_1^{\ddagger})(1-n+n\phi_2^{\ddagger})^2}{(1-n+n\ell)^3} \qquad (6)$$

The unknown parameters $\phi_1^{\ddagger}$ and $\phi_2^{\ddagger}$ can be evaluated by measurement of k_n/k_{H_2O} at two (or preferably, more than two) values of n. Alternatively, it can be found by measurement of k_{D_2O}/k_{H_2O} and of the isotopic composition of the product for at least one value of n, since the D/H ratio for the hydrogen nucleus transferred (and retained in the product) is entirely governed by $\phi_1^{\ddagger}$.

By contrast with the established status of solvent isotope effects in physical organic chemistry, the interpretation of such measurements for reactions of transition metal complexes is not well understood. We have tried to extend the methodology of physical organic chemistry to such reactions by selecting examples in which the scission of a metal-carbon bond is synchronous with a rate-limiting proton trarsfer to carbon, as for the reaction of equation (5) [4]. Pentaaquaorganochromium (III) complexes react with hydrogen ions as in equation (7).

$$L_3O^+_{,,\ell} + (L_2O)_5 CrCH_2R^{2+} \longrightarrow \begin{bmatrix} (L_2O)_5 Cr\cdots CH_2R \\ \phi^\ddagger_{Cr} \quad\quad L \;\phi^\ddagger_1 \\ \vdots \\ O \\ L \diagup \;\diagdown L \\ \phi^\ddagger_2 \quad \phi^\ddagger_2 \end{bmatrix}^{3+} \quad (7)$$

(with formation of CH_3R and $(H_2O)_5 Cr^{3+}$ as products, the latter being rapidly transformed into $[(H_2O)_6 Cr]^{3+}$) [5]. Substituting the fractionation parameters indicated in equation (7) in equation (1), we obtain equation (8)

$$\frac{k_n}{k_{H_2O}} = \frac{(1-n + n\phi^\ddagger_{Cr})^{10} (1-n + n\phi^\ddagger_1)^{10} (1-n + n\phi^\ddagger_2)^2}{(1-n + n\phi_{Cr})^{10} (1-n + n\ell)^3} \quad (8)$$

Unambiguous solution of equation (8) for $\phi^\ddagger_1$ and $\phi^\ddagger_2$ is in this case frustrated by the occurrence of the fractionation factors ϕ_{Cr} and $\phi^\ddagger_{Cr}$ relating to the ten hydrogen atoms of the water of aquation. Since the values of ϕ_{Cr} and $\phi^\ddagger_{Cr}$ are close to unity, the approximation (9) reduces the number of unknown factors to the single parameter ϕ_Δ.

$$\frac{(1-n + n\phi^\ddagger_{Cr})^{10}}{(1-n + n\phi_{Cr})^{10}} = \frac{(\phi^\ddagger_{Cr})^{10n}}{(\phi_{Cr})^{10n}} = \phi_\Delta^n \quad (9)$$

An estimate of ϕ_Δ is obtainable by studying solvent isotope effects on reactions of organochromium (III) complexes with electrophiles not containing exchangeable hydrogen, e.g. CH_3Hg^+ (equation (10)) and Br_2 [5]. Subject to small correction factors (to allow for solvation of CH_3Hg^+ and of the

$$MeHg^+ + (L_2O)_5CrCH_2R^{2+} \longrightarrow \begin{bmatrix} (L_2O)_5Cr \cdots CH_2R \\ \vdots \\ HgCH_3 \end{bmatrix}^{3+} \quad (10)$$

incipient bromide ion in the transition state for the reaction with bromine) the solvent isotope effects (k_n/k_{H_2O}) of such reactions are thought to be given only by Φ_Δ^n. Because of the remarkable parallelism of substituent effects on reactions of organochromium (III) complexes with a range of electrophiles, it seems reasonable to assume that the same values of Φ_Δ apply to all these reactions. Substitution in equation (9), and hence (8), of the value obtained with methylchromium (0.62) then allows $\phi_1^\ddagger$ and $\phi_2^\ddagger$ to be evaluated by fitting equation (8) to experimental values of k_n/k_{H_2O} at different n-values, as shown in (Fig. 1) for reaction (7) (R = H). The parameter $\phi_1^\ddagger$ for this reaction is thus found to have the value 0.16 ± 0.03, and $\phi_2^\ddagger = 0.73 \pm 0.04$. These values are similar to those derived for proton transfer reactions to

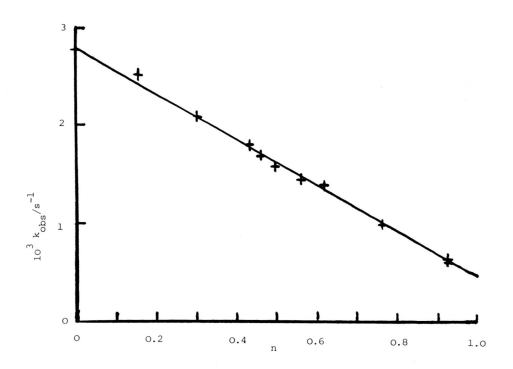

Fig. 1. Solvent isotope effect for reaction of pentaaquamethylchromium (III) with aqueous hydrogen ions (I = 1.5M). (The curve is calculated from equation (8) with $\Phi_\Delta = 0.62$, $\phi_1^\ddagger = 0.16$, $\phi_2^\ddagger = 0.74$, and $k_{H_2O} = 2.79 \times 10^{-3}$ s^{-1}.)

organic substrates [3] and lend support to the correctness of the analysis presented. It should be pointed out that the apparent linear dependence of k_n/k_{H_2O} upon n arises (Fig. 1) from the accidental compensation of opposed curvatures in the two factors 0.62^n and $(1-n + 0.16 n)(1-n + 0.73 n)^2/(1-n + 0.69 n)^3$ occurring in the simplified equation (8).

The method of analysis has also successfully been applied to a reaction in which H_2O is the proton donor, viz. the aquation of $[(H_2O)_5 CrCH_2OH]^{2+}$ with resultant formation of methanol [4]. The fit of the calculated curve to the experimental data points is shown in Fig. 2.

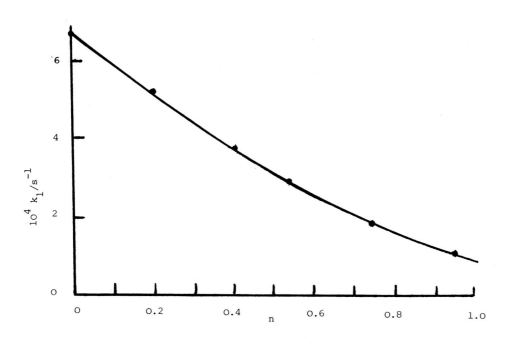

Fig. 2. Solvent isotope effect for reaction of pentaaquahydroxymethyl-chromium (III) with water (uncatalysed component): experimental points and curve calculated from theory.

REFERENCES

1. For a survey, see P. M. Laughton and R. E. Robertson, in "Solute-Solvent Interactions", J. F. Coetzee and C. D. Ritchie (Eds.), Marcel Dekker, New York, 1969, Ch. 6.
2. V. Gold, Adv. Phys. Org. Chem., 7 (1969), 259-331. (review)
3. See, for example, V. Gold and M. A. Kessick, J. Chem. Soc., (1965), 6718-6729; V. Gold and D. C. A. Waterman, J. Chem. Soc. (B), (1968) 839-855.
4. V. Gold and D. L. Wood, J. Chem. Soc., Dalton Trans. Submitted.
5. For a review, see M. D. Johnson, Acc. Chem. Res., 11 (1978), 57-65.

HEAVY-ATOM ISOTOPE EFFECTS ON ENZYME-CATALYZED REACTIONS*

MARION H. O'LEARY

Departments of Chemistry and Biochemistry, University of Wisconsin, Madison, WI 53706 (U.S.A.)

ABSTRACT

Heavy-atom isotope effects are receiving increasing attention for studying mechanisms of enzymatic reactions. We review the methods which can be used for determining heavy-atom isotope effects and present a variety of examples of applications of isotope effects to enzymatic reaction mechanisms.

INTRODUCTION

Heavy-atom isotope effects have long been used for the investigation of mechanisms of organic reactions (1-4). Interest in the application of heavy-atom isotope effects in enzymatic systems has developed more slowly; only within the last ten years have an appreciable number of publications appeared dealing with such applications. The reasons for this later development are twofold. In the first place, heavy-atom isotope effects are small (almost always <10%) and high-precision techniques are needed for their measurement. Second, enzymatic reactions generally involve a large number of reaction steps and the relationship between the observed isotope effect and the rates of the various steps is complex. Several reviews dealing with heavy-atom isotope effects on enzymatic reactions have appeared (5-8).

MEASUREMENT OF HEAVY-ATOM ISOTOPE EFFECTS

Because heavy-atom isotope effects are so small, ordinary kinetic techniques are not very useful for their measurement. Instead, the commonly-used procedure is the competitive method, in which the isotope effect is calculated from the change in isotopic composition of starting material or product over the course of the reaction (8). It is common to use the small natural abundance of carbon, nitrogen, or oxygen in these measurements, although enriched materials are occasionally

---

*Work in the author's laboratory was supported by the National Science Foundation and the National Institutes of Health.

used. The advantages of using natural abundance substrates are (1) there is no need to synthesize isotopically enriched compounds; (2) contaminants in the samples cause less error in the measurements when both substrate and contaminant contain only the natural abundance of the desired isotope than when enriched materials are used. The application of the competitive method to enzymatic reactions has recently been described in detail (9).

The isotopic composition measurements used in the calculation of isotope effects must be of very high precision. This precision can be achieved only by use of a specially designed double-collector mass spectrometer which makes simultaneous measurement of the ion intensities for the labelled and unlabelled materials and then compares this measurement with the corresponding measurement for a standard sample of the same substance. In addition to the initial cost of the instrument, this procedure suffers from the disadvantage that only very volatile substances can be studied. Commonly used materials include CO_2, CO, N_2, and CH_3Cl. The usefulness of the competitive technique has been somewhat restricted because of this problem.

Carbon isotope effects on decarboxylations have been particularly widely studied both in enzymatic and in non-enzymatic systems because of the ease with which the carbon dioxide produced in the reaction can be isolated and measured. Quite high precision can be obtained by this method. For example, arginine decarboxylase catalyzes the reaction

$$(H_2N)_2\overset{+}{C}NH(CH_2)_3\underset{\underset{NH_3^+}{|}}{C}HCO_2^- \longrightarrow CO_2 + (H_2N)_2\overset{+}{C}NH(CH_2)_4\overset{+}{N}H_3 \quad (1)$$

In a recent study of carbon isotope effects on this reaction, nine measurements of the isotope effect under a single set of conditions gave $k^{12}/k^{13} = 1.0147 \pm 0.0004$ (9).

If the necessary volatile product is not obtained directly in the reaction, it may be necessary to subject a substrate or product to chemical conversion prior to isotopic analysis. For example, phosphoenolpyruvate carboxylase catalyzes the reaction

$$CH_2{=}\underset{\underset{}{|}}{\overset{OPO_3^{2-}}{C}}{-}CO_2^- + HCO_3^- \longrightarrow {}^-O_2C{-}CH_2{-}\overset{O}{\overset{\|}{C}}{-}CO_2^- + P_i \quad (2)$$

The carbon isotope effect for the carbon atom of bicarbonate is of interest because of carbon isotope fractionations occurring during photosynthesis in C_4 plants (10). This effect was measured by reduction of the product oxalacetate to malate using malate dehydrogenase and NADH. This malate was purified and subjected to decarboxylation using malic enzyme and NADP. Comparison of the isotopic composition of the CO_2 thus produced with that of the HCO_3^- starting material gave a carbon isotope effect k^{12}/k^{13} = 1.002 at pH 8.0, 25°, for the enzyme from maize (11).

An alternate approach to this same isotope effect was taken by Schmidt et al. (12) who conducted the carboxylation of phosphoenolpyruvate in the presence of an excess of phosphoenolpyruvate and a limiting amount of CO_2. They measured the change in the isotopic composition of the residual CO_2 as the reaction proceeded and obtained a carbon isotope effect k^{12}/k^{13} = 1.002 at 25°, pH 8, in excellent agreement with the value obtained by the other method.

In the case of carbon isotope effects, the radioactive isotope ^{14}C can be used rather than the stable ^{13}C, in which case the isotopic measurements are done by scintillation counting. This method is sometimes simpler, but it is not generally as accurate as the stable isotope method and is less often used.

A modification of the competitive method has recently been described by O'Leary and Marlier (13) which extends its usefulness to cases in which the isotopic site is inaccessible to direct analysis by mass spectrometry. In this technique, an isotopic label at a site accessible to measurement is used as a tracer for the label at the site desired. A mixture of doubly-labelled and unlabelled materials is prepared. Measurement of the "isotope effect" using this material provides (after appropriate corrections) the product of the isotope effects at the two sites. The procedure is difficult and requires multiple isotopic syntheses, but it may be useful in cases where there is no other convenient way to measure the isotope effect. This method has been used, for example, to measure carbonyl carbon and oxygen isotope effects on the hydrolysis of methyl benzoate (13,14) and oxygen isotope effects on enzymatic decarboxylations (15).

A new method for determination of isotope effects in reversible reactions has recently been introduced by Cleland and collaborators (16,17). The "equilibrium perturbation" method depends on the fact that if a reversible reaction is at equilibrium except that one component of the equilibrium contains isotopic label whereas the other contains none, addition of enzyme will result in a temporary excursion of the system away from equilibrium as a result of the fact that the labelled and unlabelled materials scramble through the system at different rates. If one of the substrates or products is chromophoric, then the excursion away from equilibrium can be observed and the isotope effect can be calculated. For heavy-atom isotope effects, the departure from equilibrium is small and considerable care is needed to avoid experimental artifacts (17). Furthermore, highly-labelled substrate is needed in this method. Nonetheless, this method can be used for the determination of heavy-atom isotope effects in favorable cases.

In the one case where comparison has been possible, the equilibrium perturbation method gave results in excellent agreement with the competitive method. Malic enzyme catalyzes the reversible oxidative decarboxylation of malic acid:

$$^-O_2C-CH_2-\underset{\underset{H}{|}}{\overset{\overset{OH}{|}}{C}}-CO_2^- + NADP \longrightarrow NADPH + CO_2 + CH_3-\overset{\overset{O}{\|}}{C}-CO_2^- \qquad (3)$$

When the carbon isotope effect on this reaction was measured by the equilibrium perturbation method, a value $k^{12}/k^{13} = 1.031$ was obtained at pH 8.0, 25° (16). Measurements of the isotope effect under the same conditions by the competitive method gave a value of $k^{12}/k^{13} = 1.030$ (18).

It is also possible to obtain heavy-atom isotope effects by direct comparison of rate constants for labelled and unlabelled substrates. However, the rate difference is expected to be a few percent at most, and extreme care is needed to obtain data of the required precision. Nonetheless, this method has been used in a few cases (19-21). It is unlikely that precision approaching that of the competitive method can ever be achieved by this procedure, and the procedure is also difficult because of the need to synthesize highly-labelled substrates. Nonetheless, this method is useful in a number of cases where mass spectrometry of labelled substrates or products would not be convenient. In addition, this is the only method by which it is possible to obtain heavy-atom isotope effects on both V_{max} and V_{max}/K_m (both the competitive method and the equilibrium perturbation method give only the isotope effect on V_{max}/K_m).

INTERPRETATION OF HEAVY-ATOM ISOTOPE EFFECTS

In the case of enzymatic reactions, isotope effects measured by either the competitive method or the equilibrium perturbation method are invariably isotope effects on V_{max}/K_m. As such, they do not necessarily reflect the "rate-determining step" in the reaction. Instead, they reflect the relative rates of all steps through the first irreversible step (often the release of the first product). For example, the chymotrypsin catalyzed hydrolysis of N-acetyl-L-tryptophan ethyl ester shows an oxygen isotope effect $k^{16}/k^{18} = 1.018$, in spite of the fact that the overall rate-determining step is hydrolysis of the acyl enzyme (22).

When the reaction under study consists of a single step, then the isotope effect reflects directly the structure of the transition state for that step. For example, formate dehydrogenase catalyzes the reaction

$$HCO_2^- + NAD \longrightarrow NADH + CO_2 \qquad (4)$$

At pH 7.8, 25°, the reaction shows a carbon isotope effect $k^{12}/k^{13} = 1.042$ and a hydrogen isotope effect on V_{max}/K_m $k^H/k^D = 2.8$ (23). Both of these effects are in the range expected for a reaction involving a single chemical step. If acetylpyridine-NAD is used in place of NAD, the carbon isotope effect decreases to 1.036 and the hydrogen isotope effect becomes 6.9. Both these changes reflect the occurrence of an "earlier" transition state in the case of acetylpyridine-NAD (23).

Most enzymatic reactions do not consist of a single step, but rather involve a number of steps. In such a case, the observed isotope effect reflects the structure of the transition state for the isotopically-sensitive step and the rate of this step relative to rates of other steps in the sequence. For a reaction having two kinetically significant steps

$$\rightleftharpoons E \cdot S \underset{k_2}{\overset{k_1}{\rightleftharpoons}} E \cdot S' \overset{k_3}{\longrightarrow} E \cdot P \rightleftharpoons \tag{5}$$

in which we assume that all preceding and following steps are relatively fast, the observed isotope effect k/k^* (in which the asterisk indicates the heavier isotope) is given by

$$\text{observed } \frac{k}{k^*} = \frac{k_3/k_3^* + k_3/k_2}{1 + k_3/k_2} \tag{6}$$

provided that k_3 is not reversible under the reaction conditions and provided that k_1 and k_2 show no isotope effects. Unstarred rate constants are for the normal isotopic species (^{12}C, ^{16}O, or ^{14}N). Thus, the observed isotope effect depends on the actual isotope effect on the isotopically-sensitive step (called the "intrinsic isotope effect") and on the ratio k_3/k_2 (called the "partition factor"), which reflects the relative rates of various steps in the sequence. In more complex reaction schemes, the factor k_3/k_2 is replaced by a more complex ratio of rate constants (7).

The effect of a multistep mechanism such as this is to make the observed isotope effect smaller than the intrinsic isotope effect. For example, the enzymatic decarboxylation of acetoacetic acid proceeds by way of a mechanism involving a covalent enzyme-substrate Schiff base between the carbonyl group of the substrate and an amino group of the enzyme (here represented as $E-NH_2$):

$$E-NH_2 + CH_3\overset{O}{\overset{\|}{C}}CH_2CO_2^- \underset{k_2}{\overset{k_1}{\rightleftharpoons}} \overset{E-NH^+}{\overset{\|}{CH_3\overset{O}{\overset{\|}{C}}CH_2CO_2^-}} \downarrow k_3 \qquad (7)$$

$$E-NH_2 + CH_3\overset{O}{\overset{\|}{C}}CH_3 + CO_2 \leftarrow \underset{CH_3C=CH_2}{\overset{E-NH}{|}} + CO_2$$

The expected carboxyl carbon isotope effect on the decarboxylation step is in the range k^{12}/k^{13} = 1.04-1.06. The observed effect is 1.018 (24), indicating that decarboxylation is not entirely rate-determining. The partitioning factor k_3/k_2 is in the range 1-2.

In some cases it is not adequate to assume that only a single step shows a significant isotope effect. For example, the hydrolysis of N-acetyl-L-tryptophanamide by chymotrypsin appears to proceed by way of a tetrahedral intermediate:

$$E-OH + R-\overset{O}{\overset{\|}{C}}-NH_2 \underset{k_2}{\overset{k_1}{\rightleftharpoons}} E-O-\underset{R}{\overset{O^-}{\underset{|}{\overset{|}{C}}}}-NH_2 \overset{k_3}{\longrightarrow} E-O-\overset{O}{\overset{\|}{C}}-R + NH_3 \qquad (8)$$

Steps following the release of ammonia are not important because the release of ammonia is irreversible under these conditions (25). The second step in this sequence is expected to show an appreciable nitrogen isotope effect because the carbon-nitrogen bond is broken in this step. However, the first step may also show a significant nitrogen isotope effect because of the fact that the carbon-nitrogen bond in the starting amide is appreciably stronger than that in the intermediate because of resonance. In this case the observed isotope effect is given by

$$\text{observed } \frac{k^{12}}{k^{13}} = (k_1/k_1^*)\frac{(k_3/k_3^*)/(k_2/k_2^*) + k_3/k_2}{1 + k_3/k_2} \qquad (9)$$

The nitrogen isotope effect is k^{14}/k^{15} = 1.010 at pH 8.0, 25° (26). In this case it is not currently possible to predict the magnitudes of the isotope effects on k_1, k_2, and k_3. Thus it is not possible to derive a value for k_3/k_2. However, the fact that the isotope effect varies with pH probably indicates that k_3/k_2 is not too different from unity (26).

MORE COMPLEX EXAMPLES

In some cases it is possible to measure an isotope effect and compare this effect with the expected intrinsic isotope effect using equation 6. However, in many cases, such intrinsic isotope effects are not readily available and it is more useful to approach the subject by studying the variation of the observed isotope effect with different reaction conditions. Possible variations include pH, metal ion, substrate, temperature, coenzyme, and others.

Several investigations have been made of isotope effects in decarboxylations catalyzed by the pyridoxal-5'-phosphate dependent enzymes glutamate decarboxylase and arginine decarboxylase (27-30). The reactions proceed through a Schiff base between substrate and pyridoxal phosphate, forming a quinone-like intermediate (Q) which eventually gives rise to product:

$$E + S \underset{k_2}{\overset{k_1}{\rightleftharpoons}} \text{Schiff base} \xrightarrow{k_3} CO_2 + Q \longrightarrow \text{products} \qquad (10)$$

Again, isotope effects can be interpreted within the framework of equation 6. In most cases the isotope effects observed are near $k^{12}/k^{13} = 1.02$, whereas the intrinsic isotope effect on the decarboxylation step is expected to be 1.04-1.07. This indicates that the partitioning factor k_3/k_2 is not too different from unity and neither the Schiff base interchange step nor the decarboxylation step is entirely rate-determining.

In the case of glutamate decarboxylase, the carbon isotope effect increases with increasing pH (27). This reflects not a change in the intrinsic isotope effect with pH, but rather the fact that k_2 and k_3 have different pH dependences; thus, the partition factor changes with pH.

In the case of arginine decarboxylase, $k^{12}/k^{13} = 1.015$ at pH 5.25, 25°, for arginine (28). Homoarginine is a very poor substrate for this enzyme, being decarboxylated about 100x more slowly than arginine, but with no appreciable change in K_m. Homoarginine shows a carbon isotope effect $k^{12}/k^{13} = 1.057$ at 25°, pH 5.25. Thus, the decarboxylation step is entirely rate-limiting for homoarginine. This indicates that the specificity of this enzyme is manifested in the decarboxylation step, rather than in substrate binding or in Schiff-base interchange (28).

Many pyridoxal phosphate dependent enzymes are active with coenzyme analogs which differ slightly from the natural coenzyme. For example, substitution of other alkyl groups for the methyl group which ordinarily occurs at the 2-position of the cofactor results in coenzyme analogs which have a substantial degree of catalytic activity. Replacement of the 2-methyl group of the cofactor by a hydrogen results in the compound 2-nor-pyridoxal 5'-phosphate, which shows about 1.6 times more activity with arginine decarboxylase than does pyridoxal phosphate itself. The question of why this rate increase occurs can easily be answered by use of carbon isotope effects: at pH 5.25, 25°, arginine shows a carbon isotope effect k^{12}/k^{13} = 1.015 with pyridoxal phosphate, whereas the isotope effect is 1.016 with the nor-methyl compound. This indicates that the rates of both the Schiff-base interchange step and the decarboxylation step have been increased by nearly the same amount as a result of coenzyme modification (31).

Solvent isotope effects can sometimes be used in conjunction with carbon isotope effects to provide a complete picture of the intrinsic isotope effects and partition factors in an enzymatic reaction (29). Glutamate decarboxylase, for example, shows solvent isotope effects $V_{max}^{H_2O}/V_{max}^{D_2O}$ = 5.0 and $(V_{max}/K_m)^{H_2O}/(V_{max}/K_m)^{D_2O}$ = 2.6. The carbon isotope effect depends on the isotopic nature of the solvent: k^{12}/k^{13} = 1.018 in H_2O at pH 4.7, 37°, whereas in D_2O under otherwise identical conditions the isotope effect is 1.009. The intrinsic carbon isotope effect on the decarboxylation step should be the same in H_2O as in D_2O; the change in isotope effect occurs as a result of a change in the partitioning factor. In this case it is possible to derive a complete set of isotope effects on the various steps. The substrate binding step shows a large solvent isotope effect: substrate binding is about twofold weaker in H_2O than in D_2O, probably because of the desolvation which accompanies binding of the substrate to the enzyme (a similar phenomenon has been seen in other carbon isotope effects (30)). The Schiff-base interchange step shows a solvent isotope effect of about 7, and a proton inventory analysis reveals that this large effect is due to small contributions from at least four different protonic sites. The decarboxylation step shows a solvent isotope effect of about 2 and a carbon isotope effect of about 1.06. Schiff-base interchange and decarboxylation are both partially rate-determining.

Sometimes the dependence of carbon isotope effects on other reaction conditions can reveal unexpected aspects of reaction mechanism. For example, the carbon isotope effect on the enzymatic decarboxylation of arginine is k^{12}/k^{13} = 1.027 at 5°, pH 5.25. Addition of ethylene glycol to the solvent results in little change in decarboxylation rate but a large change in the carbon isotope effect: in 16 mole % ethylene glycol the isotope effect is only 1.003 (30). This change occurs because of a change in the relative rates of the decarboxylation and Schiff-base interchange steps. As the solvent becomes less polar, the rate of the decarboxylation step increases and the rate of the Schiff-base interchange step decreases. In the limit of high ethylene glycol concentration, the isotope effect approaches 1.002. This residual effect is believed to be due to a carbon isotope effect on the binding of substrate to the enzyme which results from desolvation of the substrate carboxyl group. This is consistent with the results cited above for kinetics in D_2O.

CONCLUSION

Heavy-atom isotope effects can provide useful information about the details of enzymatic reaction mechanisms. The preferable approach is to use variations in the observed isotope effect as a function of some other reaction parameter, rather than relying solely on the magnitude of the observed isotope effect.

REFERENCES

1. L. Melander and W. H. Saunders, Jr., Reaction Rates of Isotopic Molecules, Wiley, New York, 1980.
2. A. MacColl, Annu. Rep. Chem. Soc., 71B (1974) 77.
3. A. Fry, in C. J. Collins and N. S. Bowman (Eds.), Isotope Effects in Chemical Reactions, Van Nostrand-Reinhold, Princeton, NJ, 1970, p. 364.
4. E. Buncel and C. C. Lee (Eds.), Isotopes in Organic Chemistry, Vol. 3, Elsevier, Amsterdam, 1977.
5. M. H. O'Leary, in E. E. van Tamelen (Ed.), Bioorganic Chemistry, Vol. 1, Academic Press, New York, 1977, p. 259.
6. M. H. O'Leary, in W. W. Cleland, M. H. O'Leary and D. B. Northrop (Eds.), Isotope Effects on Enzyme-Catalyzed Reactions, University Park Press, Baltimore, Maryland, 1977, p. 233.
7. M. H. O'Leary, in R. Gandour and R. L. Schowen (Eds.), Transition States of Biochemical Processes, Plenum, New York, 1978, p. 285.
8. J. Bigeleisen and M. Wolfsberg, Adv. Chem. Phys., 1(1958) 15.
9. M. H. O'Leary, Methods in Enzymology, 64 (1980) 83.
10. M. H. O'Leary, Phytochemistry, (1981) in press.
11. M. H. O'Leary, J. E. Rife and J. D. Slater, unpublished work.
12. H.-L. Schmidt, F. J. Winkler, E. Latzko, and E. Wirth, Isr. J. Chem., 17 (1978) 223.
13. M. H. O'Leary and J. F. Marlier, J. Am. Chem. Soc., 101 (1979) 3300.
14. J. F. Marlier and M. H. O'Leary, J. Org. Chem., (1981) in press.
15. J. Hermes, W. W. Cleland and M. H. O'Leary, unpublished work.
16. M. I. Schimerlik, J. E. Rife and W. W. Cleland, Biochemistry, 14 (1975) 5347.
17. W. W. Cleland, Methods in Enzymology, 64 (1980) 104.
18. C. Roeske and M. H. O'Leary, unpublished work.
19. D. G. Gorenstein, J. Am. Chem. Soc., 94 (1972) 2523.
20. M. F. Hegazi, R. T. Borchardt and R. L. Schowen, J. Am. Chem. Soc., 101 (1979) 4359.
21. S. Rosenberg and J. F. Kirsch, Biochemistry, (1981) in press.
22. C. B. Sawyer and J. F. Krisch, J. Am. Chem. Soc., 97 (1975) 1963.
23. J. S. Blanchard and W. W. Cleland, Biochemistry, 19 (1980) 3543.
24. M. H. O'Leary and R. L. Baughn, J. Am. Chem. Soc., 94 (1972) 626.
25. M. H. O'Leary and M. D. Kluetz, J. Am. Chem. Soc., 93 (1971) 7341.
26. M. H. O'Leary and M. D. Kluetz, J. Am. Chem. Soc., 94 (1972) 665.
27. M. H. O'Leary, D. T. Richards and D. W. Hendrickson, J. Am. Chem. Soc., 92 (1970) 4435.
28. M. H. O'Leary and G. J. Piazza, J. Am. Chem. Soc., 100 (1978) 6632.
29. M. H. O'Leary, H. Yamada and C. J. Yapp, Biochemistry, (1981) in press.
30. M. H. O'Leary and G. J. Piazza, Biochemistry, (1981) in press.
31. G. J. Piazza and M. H. O'Leary, unpublished work.

ISOTOPE EFFECTS ON EACH, C- AND N-ATOMS, AS A TOOL FOR THE ELUCIDATION OF
ENZYME CATALYZED AMIDE HYDROLYSES

R. MEDINA, T. OLLEROS and H.-L. SCHMIDT
Institut für Chemie Weihenstephan, Technische Universität München,
D-8050 Freising (F.R.G.)

ABSTRACT

The usefulness of a double isotope effect determination for the elucidation of reaction mechanisms was evaluated, using enzyme catalyzed amide hydrolyses as an example. In the urease reaction the nitrogen isotope effect found was half the carbon isotope effect. From the result the existence of an enzyme-carbamoyl-intermediate is derived. On the other hand, the identical isotope effects found on the C and the N atoms in the side chain of arginine proved, that in the arginase reaction not the bond splitting between the N-atom in δ-position and the adjacent C-atom is rate limiting, but the coordination of the guanido group nitrogen atoms with the manganese ion of arginase. Temperature and pH dependences of the isotope effects found were interpreted as changes in the rate limiting steps or as conformational changes of the enzymes.

Thus the double isotope effect determination yielded more detailed information on the mechanisms of the two seemingly identical enzyme catalyzed reactions, and it is therefore a valuable tool for the elucidation of reaction mechanisms.

INTRODUCTION

Among the methods for the elucidation of the enzyme reaction mechanisms, the determination of kinetic isotope effects is often used for the detection of the rate limiting step, in many cases a bond fission. In order to get more information on reaction mechanisms we tested the method of "double isotope effect determination in which the discrimination of both atoms of the bond cleaved is determined. We studied the urease and the arginase reactions as models for the amide hydrolysis catalyzed by metalloenzymes.

METHODS AND RESULTS

Urease reaction

The incubation was performed in a buffer free medium in an autotitrator combined to a mass spectrometer (Fig. 1). After a defined turnover, CO_2 and NH_3 were collected for the mass spectrometric isotope ratio analysis. The isotope effects of the reaction were calculated from the δ-values of the substrate and of the products after a given turnover. The experiments were made with two different preparations of urease (jack bean urease "Zerner" and urease from Canavalia ensiformis "Boehringer").

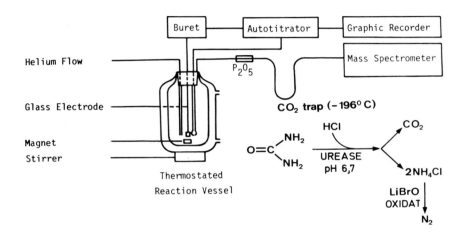

Fig. 1. Device for the isotope effect determination of the urease reaction

The experimental kinetic isotope effect of the urease reaction for nitrogen was less than half the isotope effect found for the carbon atom (Tab. 1). This can be interpreted by the assumption, that only the first bond splitting between N and C is rate limiting, while the intermediate carbamate is hydrolyzed spontaneously in a rapid step, yielding a second molecule of NH_3. Therefore, for a comparison of the two isotope effects, the experimental value for nitrogen must be corrected in regard to the dilution by the second N-atom, and by taking into account the difference between the maximum theoretical isotope effects of nitrogen and carbon, related to the reduced masses of these atoms. The identity of the two isotope effects found after these corrections (Tab. 1) justifies our supposition of a two step mechanism.

After Barth and Michel the active site of urease is containing histidine [1]. Studies by Zerner's group prove Ni^{2+}, a carboxyl- and an essential SH-group

TABLE 1

Kinetic N- and C-isotope effects of the urea hydrolysis on two urease preparations at 40°C and pH 6.7. $f = (14 \cdot 13/15 \cdot 12)^{1/2}$

	k^{14}/k^{15} experim.	$2 \cdot \dfrac{k^{14}}{k^{15}} - 1$	$2 \cdot \dfrac{k^{14}}{k^{15}} - 1 \cdot f$	k^{12}/k^{13} experim.	Quotient of N-/C-isotope effects
urease Zerner	1.0086 ± 0.0005	1.0172	1.0228	1.0216 ± 0.0003	1.0012
urease Boehringer	1.0075 ± 0.0005	1.0150	1.0206	1.0183 ± 0.0003	1.0023

to be present in the active site [2,3]. Our results fit well with the mechanism proposed by this group [4], and prove the existence of an intermediate enzyme-carbamoyl-complex.

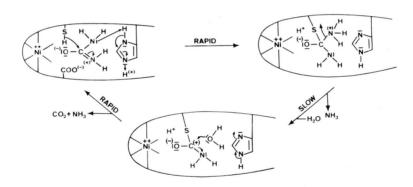

Fig. 2. Proposed mechanism of the urease reaction.

Considering the facts known so far, the following mechanism can be proposed. Ni^{2+} as a Lewis acid favors the tautomeric form of urea, the $=\overset{+}{N}H_2$ group of which interacts with the adjacent carboxylate ion [1,4]. An electron transfer from the sulfide group supports the substrate protonation by the imidazolyl group. The so-formed ammonium ion corresponds to the tetrahedral intermediate as proposed for the papain catalyzed hydrolysis of amides [5]. The complex decomposes, releasing NH_3 in a slow rate limiting step, leading to a carbamoyl-urease-intermediate, which has already been proved by other facts [4]. The hydrolysis of the carbamoyl-urease-complex completes the reaction in a rapid step, liberating CO_2, NH_3 and urease.

The isotope effect of the urease reaction increased with temperature, however, the ratio of the carbon and nitrogen isotope effects remained constant. While the isotope effect for nitrogen did not change remarkably with the pH, a dramatic dependence of the carbon isotope effect on the pH-value was observed. These results seem to indicate a change of the rate limiting step.

By partial heat denaturation of the enzyme the isotope effects (Tab. 1) were shifted to 1.012 for C and 1.002 for N respectively. This could be in agreement with the existence of two forms of the enzyme proposed by Barth and Michel [1].

Arginase reaction

Arginine was incubated with arginase in a buffer free medium, and the reaction was pursued in aliquots by a spectrophotometric method [6]. After a defined turnover the reaction was stopped by addition of HCl, then the pH was adjusted to 6.7, and the urea formed was hydrolyzed by an excess of urease, and analyzed as before. The C- and N-atoms of the side chain and the N-atom in the δ-position were isolated and analyzed as shown in the following scheme (Fig. 3). From the δ-values of these atoms after 10 % and 100 % turnover the isotope effects were calculated.

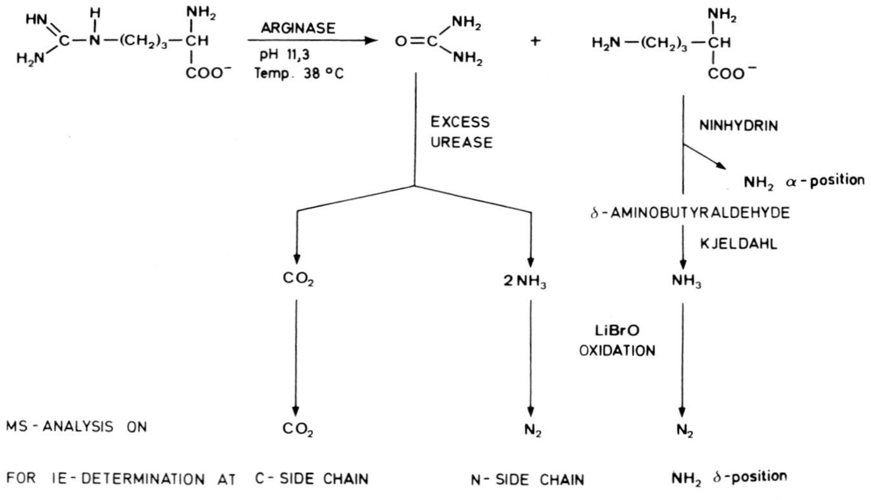

Fig. 3. Analytical procedures for the determination of the isotope effects of the arginase reaction.

The arginase reaction proceeded with identical corrected isotope effects for the C- and N-atoms of the side chain, however, no isotope effect was found for the nitrogen in δ-position of the main chain (Table 2), which is directly involved in the bond that is split.

TABLE 2

Kinetic isotope effects of the arginine hydrolysis on the N- and C-atoms of the side chain at 37°C and pH 11.3. $f=(14\cdot13/15\cdot12)^{1/2}$

k^{14}/k^{15} experim.	$\frac{k^{14}}{k^{15}} \cdot f$	k^{12}/k^{13} experim.	Quotient of correct. N/C- isotope effects
1.0104 ± 0.0010	1.0160	1.0150 ± 0.0007	1.0010

This result can be interpreted by the following mechanism:

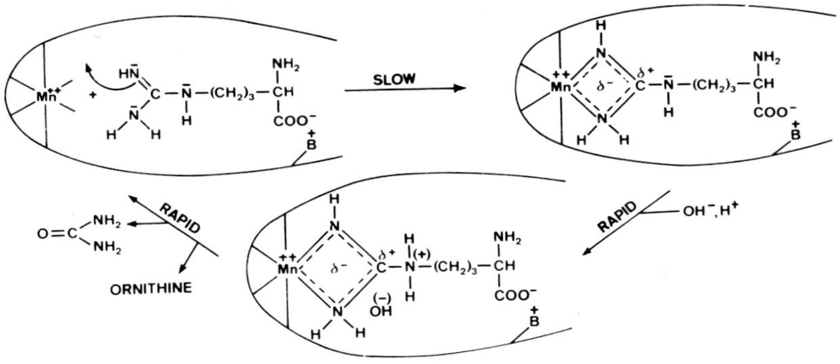

Fig. 4. Mechanism proposed for the arginase reaction.

Arginase is a Mn^{2+}-containing enzyme. After our results and in agreement with data from the literature [7] we propose, that the manganese ion coordinates with the two free nitrogen atoms of the guanido group. A delocalization of the π-bond electrons takes place due to the electrophilic character of the cation. The C-N bond is then attacked by a dissociated water molecule leading to the subsequent hydrolysis of the substrate to urea and ornithine. In the proposed mechanism the rate limiting step would be the electrophilic attack of the Mn^{2+} to the side chain of the nitrogen atoms of the arginine, which has as a consequence the polarization of the double bond and the formation of a complex, in which both nitrogen atoms of the guanidine group are equivalent. Therefore, we observe an isotope effect with the carbon and with the two nitrogen atoms of the guanido group of arginine and not with the nitrogen atom in δ-position.

The carbon isotope effect of the arginase reaction increased with temperature. Corresponding observations with alcohol dehydrogenase have been interpreted as a change of the rate limiting step [8]. However, at the present we are not yet able to give a similar interpretation. The isotope effect of the arginase reaction does not show a dependence on the pH-value. Therefore, in spite of the participation of a proton in the reaction, the concentration of H^+ seems not to interfer with the rate limiting step.

CONCLUSIONS

Both hydrolyses investigated involve a bond splitting between a carbon and a nitrogen atom, catalyzed by a metal containing enzyme. In both cases the carbon isotope effect by itself indicates a carbon bond splitting as the rate limiting step, and hence one would postulate a similar mechanism for both reactions. However, the additional determination of the isotope effect on the N-atom of the bond in question gives more information, from which different mechanisms for the two reactions can be derived. This proves the value of the "double effect determination", which we are going to apply to the elucidation of other reactions.

ACKNOWLEDGEMENTS

We thank Prof. B. Zerner from the University of Queensland (Australia) for kindly providing a highly purified urease preparation and for his discussion of the draft. This contribution was supported by a grant of the "Deutsche Forschungsgemeinschaft".

REFERENCES

1. A.Barth and H.J. Michel, Biochem. Physiol. Pflanz., 163(1972)103-109.
2. R.L. Blakeley, J.A. Hinds, H.E. Kunze, E.C. Webb and B. Zerner, Biochemistry, 8(1969)1991-2000.
3. N.E. Dixon, C. Gazzola, R.L. Blakeley and B. Zerner, Science, 191(1976)1144-1150.
4. N.E. Dixon, P.W. Riddles, C. Gazzola, R.L. Blakeley and B. Zerner, Can. J. Biochem., 58(1980)1335-1344.
5. M.H. O'Leary, M. Urberg and P. Young, Biochemistry, 13(1974)2077-2081.
6. J.J. Coulombe and L.Favreau, Clin. Chem., 9(1963)102.
7. H. Hirsch-Kolb, H.J. Kolb and D.M. Greenberg, J. Biol. Chem., 246(1971)395-401.
8. D. Palm, Z. Naturforsch., 21b(1966)540.

PARAMETERS AFFECTING THE $^{13}CO_2/^{12}CO_2$ ISOTOPE DISCRIMINATION OF THE RIBULOSE-1,5-BISPHOSPHATE CARBOXYLASE REACTION

F.J. WINKLER, H. KEXEL, C. KRANZ, and H.-L. SCHMIDT

Institut für Chemie Weihenstephan, Technische Universität München, D-8050 Freising (F.R.G.)

ABSTRACT

The $^{13}CO_2/^{12}CO_2$ isotope fractionation of the ribulose-1,5-bisphosphate carboxylase reaction has been determined as a function of temperature, pH value, Mg^{2+} concentration, ionic strength, and enzyme source. The practicability of this work has much profited by the speed and accuracy of a new experimental procedure applied, the determination of the isotope-ratio change in the substrate CO_2 after 40 % turnover. By the results the large ^{13}C-isotope effect variations reported in the literature can generally be confirmed and explained. A strong pH dependence of the isotope discrimination is observed (pH 7.5: 18 ‰; pH 8.0: 26 ‰; pH 8.5: 35 ‰). The cofactor Mg^{2+} also shows a substantial influence on the isotope effect (values at pH 8.0, 8 mM: 21 ‰; 100 mM: 36 ‰). The results obtained are in line with a random-order mechanism of the ribulosebisphosphate carboxylase reaction, and contribute to the interpretation of the carbon isotope fractionation pattern in plants.

INTRODUCTION

Since the first investigation of Park and Epstein [1] in 1960 dealing with the key role of RuBP* carboxylase/oxygenase in the carbon isotope fractionation during plant photosynthesis, a series of further studies on the subject has been published [2-7]. The $^{13}CO_2/^{12}CO_2$ isotope discrimination values reported from in vitro experiments with RuBP carboxylase (see Eq. 1) range from about 8 to 90 ‰. However, the factors which really influence the magnitude of this isotope effect are not yet well understood.

$$CO_2 + RuBP \xrightarrow[k_E]{RuBP\ carboxylase} 2\ PGA \qquad (1)$$

---

*Abbreviations used: RuBP(-C) = ribulose-1,5-bisphosphate(-carboxylase), PEP(-C) = phosphoenolpyruvate(-carboxylase), PGA = 3-phosphoglycerate, CAM = crassulacean acid metabolism, PDH = pyruvate dehydrogenase, decarboxylating.

Referring to reaction (1) the relevant <u>isotope effect</u>, the ratio of the reaction-rate constants for $^{12}CO_2$ and $^{13}CO_2$, is designated as k_E^{12}/k_E^{13} (the subscript E indicates that the isotope effect of the enzyme-catalyzed reaction is considered, see below). It is used as a base term within our data evaluation. In addition, the results are expressed - as commonly done in plant physiology - in terms of the <u>isotope discrimination</u> D*, the relative depletion of the heavy isotopic species notated in per mil (‰). The interconversion of the two terms is given by Eq. (2).

$$D = (1 - k_E^{13}/k_E^{12}) \times 1000 \ [‰] \quad (2)$$

In the experimental standard technique for the isotope effect determination used so far, the isotopic compositions of the substrates CO_2 and RuBP and of the product PGA are compared. The procedure includes isolation and combustion steps, it is very susceptible to certain errors and time consuming. Therefore we applied an alternative experimental approach [9-12]. The change of the isotope ratio of the substrate CO_2 was determined, when the carboxylation had proceeded to approximately 40 %. The evaluation of the rate-constant ratio from the experimental data in this type of competitive measurement [13,14] is basically calculated by means of Eq. (3) (F = fraction of reaction; R_o and R = initial and final $^{13}C/^{12}C$ ratios of the CO_2 + HCO_3^- pool).

$$k^{12}/k^{13} = \frac{\ln[(1-F)(1+R_o)/(1+R)]}{\ln[(1-F)(1+1/R_o)/(1+1/R)]} \quad (3)$$

k^{12}/k^{13} represents the isotope effect of the overall reaction which includes a HCO_3^-/CO_2 prequilibrium with a thermodynamic isotope discrimination. The relevant isotope effect on the enzyme-satalyzed RuBP carboxylation, k_E^{12}/k_E^{13}, can be computed by means of Eq. (4), which is appropriately obtained from Eq. (3), and which takes into account the pH-dependent contribution of the HCO_3^-/CO_2 isotope fractionation.

$$k_E^{12}/k_E^{13} = \frac{1 + 10^{(6.46 - pH)}}{1.009 + 10^{(6.46 - pH)}} \times k^{12}/k^{13} \quad (4)$$

This new procedure demands strict control in handling the CO_2 samples during incubation and transfer steps. However, the precision of the results is improved, and the time per measurement is reduced so that larger series of experiments can be performed more readily.

---

*D is linked to the $\delta^{13}C$ <u>value</u> which is used as an expression of the isotopic composition. The relevant definitions are:

$$D = \frac{\delta^{13}C_{initial} - \delta^{13}C_{final}}{1 + \delta^{13}C_{initial}/1000} \ ; \quad \delta^{13}C\ [‰] = (\frac{(^{13}C/^{12}C)_{sample}}{(^{13}C/^{12}C)_{PDB}} - 1) \times 1000$$

PDB is a carbonate standard (Pee Dee Belemnite) with an isotope ratio of ($^{13}C/^{12}C$) = 0.0112372 in the CO_2 released [8].

In this paper we briefly summarize our results [9-12] on the temperature, pH, CO_2 concentration, and enzyme source dependence of the $^{13}CO_2/^{12}CO_2$ discrimination by the RuBP carboxylase reaction, and we add new data - with full details demonstrating the principle of the method in Table 1 further below - on the influence of the concentration of the cofactor Mg^{2+} and other cofactors, and of the medium ionic strength. The importance and the consequences of these results on the isotope effect are discussed with regard to the interpretation of the RuBP carboxylase enzyme mechanism, and to the carbon isotope fractionation model of C_3 plants.

RESULTS AND DISCUSSION

Our present results [9-12] on the $^{13}CO_2/^{12}CO_2$ isotope discrimination of the RuBP carboxylase reaction cover values from 18 ‰ to about 110 ‰. The discrimination values above 40 ‰ have been obtained under extreme incubation conditions, and their physiological validity and interpretation is still somewhat doubtful, but similar high figures of up to 90 ‰ have also been observed [3] with the standard technique. Apparently both methods independently reveal a full primary carbon isotope effect for $^{13}CO_2/^{12}CO_2$ in the RuBP carboxylase reaction under optimal enzyme test conditions, and indicate considerable additional isotope effect contributions under modified conditions.

Parameters affecting the in-vitro isotope effect

We found that - above CO_2-concentrations for enzyme saturation - the ^{13}C isotope effect on the RuBP carboxylation was independent of the CO_2 concentration in the medium. Our measurements show a very small, negative temperature effect between 20 and 45 °C below pH 8.0, and indicate steeper decreases at pH 8.3 and in the 5 to 15 °C interval. These observations are in line with theoretical predictions for the temperature effect [13-14].

The most interesting outcome is the influence of the pH value on the ^{13}C isotope discrimination. A very steep increase is observed above pH 7.8 which is within the physiological H^+ ion concentration range in plants. Mean values at pH 7.5 are 18 ‰; at pH 8.0: 26 ‰; at pH 8.5: 35 ‰. The importance of the observed pH effect will be discussed in the final section.

The influence of the enzyme source on the $^{13}CO_2/^{12}CO_2$ isotope effect of the RuBP carboxylation has also been studied [12], and comparison of C_3, C_4, and CAM plant RuBP carboxylases showed a close agreement of the isotope effect, measured at pH 7.6 as well as at pH 8.3. These data underline the identity of the enzymes from different sources and thus combine different results [2-7] obtained by the standard technique. Differences of $\delta^{13}C$ values within a certain photosynthetic group of plants must be therefore related to other, endogenous and exogenous, factors and not to varying RuBP carboxylase types, as already known from other enzyme data.

Magnesium ions (Mg^{2+}) are essential as a cofactor for the activation of RuBP carboxylase [15]. Two series of in vitro experiments to measure the Mg^{2+} effect have been performed (complete raw data in Table 1), which include the physiological pH value. The experimentally accessible Mg^{2+} concentrations are limited by insufficient enzyme activity at low values, or otherwise by pending Mg^{2+} precipitation.

TABLE 1
Application of the substrate procedure to measurement of $^{13}CO_2/^{12}CO_2$ discrimination by RuBP carboxylase. Influence of Mg^{2+} concentration, cofactor type, and ionic strength (25 C, various pH).[a]

exptl. condit.		substrate (CO_2 + HCO_3^-) in incub.				observed effects	
pH	cofactor or salt [mM]	start[b]		end		k_E^{12}/k_E^{13}	discriminat. D [‰]
		c_0 [mM]	$R_0 \cdot 10^6$	F	$R \cdot 10^6$		
8.0	Mg^{2+}						
	8			0.391	11285	1.0208	20.6 ± 0.2
	8	2.908	11123	0.342	11261	1.0212	
	12.5	(2.891)	(11120)	0.389	11302	1.0244	23.8
	20	(2.925)	(11126)	0.349	11281	1.0246	
	20			0.349	11285	1.0255	25.5 ± 1.5
	20	-----------------------		0.357	11329	1.0276	
	20			0.320	11306	1.0271	
	50	3.263	11154	0.340	11312	1.0261	26.0 ± 0.6
	50	(3.252)	(11152)	0.349	11322	1.0273	
	100	(3.274)	(11157)	0.274	11309	1.0366	
	100			0.289	11323	1.0371	35.6 ± 0.2
	100			0.284	11318	1.0369	
8.5	2			0.083	11167	1.0235	
	2			0.082	11162	1.0190	23.0 ± 4.5
	2			0.118	11187	1.0281	
	5	2.917	11137	0.177	11224	1.0321	
	5	(2.932)	(11137)	0.257	11268	1.0320	31.3 ± 0.4
	5	(2.902)	(11134)	0.362	11338	1.0328	
	20		(11139)	0.289	11303	1.0361	34.6 ± 0.3
	20			0.340	11335	1.0355	
	50			0.298	11315	1.0377	37.6 ± 1.4
	50			0.296	11321	1.0404	
7.6	Mn^{2+};[c]						
	20	2.609	11168	0.310	11277	1.0181	18.9 ± 1.2
	20	(2.628)	(11169)	0.409	11338	1.0205	
		(2.590)	(11167)				
8.0	NaCl[d]						
	90			0.303	11299	1.0282	27.4
	240	3.263	11154	0.354	11324	1.0269	
	240	(3.252)	(11152)	0.304	11310	1.0315	29.0 ± 2.2
	240	(3.274)	(11157)	0.324	11324	1.0312	

[a] Data evaluation according to Eq. (2) and (4).
[b] Initial concentrations c_0 and isotope ratios R_0 used within a series of experiments, are mean values of multiple determinations given in ( ).
[c] Experiment with 5 mM Ni^{2+} (pH 8.5; F = 0.043): D = 31.3 ‰.
[d] Cofactor 20 mM Mg^{2+}.

A substantial increase of the $^{13}CO_2/^{12}CO_2$ isotope effect on the RuBP carboxylation was observed as a function of the Mg^{2+} concentration. At pH 8.0, mean discrimination values are 21 ‰ (8 mM Mg^{2+}) and 36 ‰ (100 mM Mg^{2+}). At pH 8.5 the increase is from 23 ‰ (2 mM Mg^{2+}) to 38 ‰ (50 mM Mg^{2+}). This outcome will be discussed later together with the above mentioned pH dependence.

As an extension of the Mg^{2+} studies, the influence of other metal cofactors and of high ionic strength conditions was examined. Experiments with $\underline{Mn^{2+}}$ and $\underline{Ni^{2+}}$ as cofactors showed $^{13}C/^{12}C$ isotope effects of 19 ‰ and 31 ‰ (pH 7.6 and pH 8.5). These values are similar to the results with Mg^{2+} obtained under the same conditions, but they differ from earlier data [6]. Experimental difficulties arise with these unphysiological cofactors as the enzyme activity is 10-20 times lower compared to the Mg^{2+} system.

High <u>ionic strength</u> of the incubation medium, simulating salt stress conditions of plants in certain environments, also leads to a change of the carbon isotope effect on the RuBP carboxylation (Table 1). 240 mM NaCl (pH 8.0; 20 mM Mg^{2+}) gives a minor, but distinct increase of the isotope discrimination from 26.5 to 29.0 ‰. This salt effect is small in comparison to the influence of Mg^{2+}; e.g. 100 mM Mg^{2+} increases the isotope effect up to 36 ‰. Therefore the Mg^{2+} influence obviously reflects the metal cofactor specifity, which could merely be corrected for by an unspecific salt contribution.

Finally, among the influences observed on the RuBP carboxylase isotope effect, the pH and Mg^{2+} dependences are particularly important, and this is in line with the importance of these factors for the enzyme mechanism. The generally accepted mechanism [15] includes a base-catalyzed enediol formation of the enzyme-bound RuBP and a random order for the addition of the substrates. Thus, the observed pH effect on the $^{13}CO_2/^{12}CO_2$ discrimination indicates a controlling role of the enediol formation and pH-depending contributions of the possible sequences. As the Mg^{2+} ion is part of the active site of the enzyme, the influence of the Mg^{2+} concentration on the isotope effect might be interpreted as a change of the substrate addition behaviour.

Consequences upon the in vivo ^{13}C discrimination

Within the photosynthetic group of C_3 plants the range of $\delta^{13}C$ variations of the assimilation products [11,16-20] is determined by three main parameters (Scheme 1). The exact $\delta^{13}C$ value of the <u>source CO_2</u> is time and location dependent. Incomplete <u>isotopic equilibration</u> between atmospheric and leaf-absorbed CO_2 (due to stomatal diffusion resistance) reduces the expression of the isotope effect of the RuBP carboxylation. Finally, the fundamental value of the <u>enzyme isotope effect</u> is influenced by various factors which have been the subject of our above investigation.

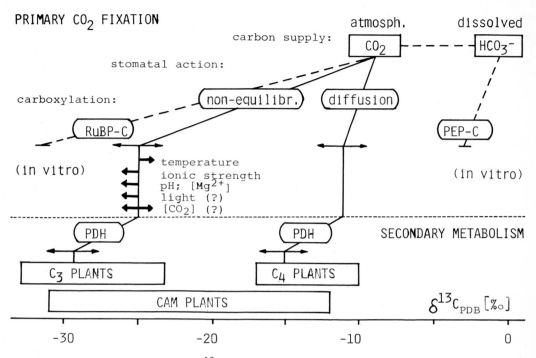

Scheme 1. Parameters affecting $\delta^{13}C$ values of plants (revised from [11]).

Some, but not all, of these factors investigated with respect to the enzyme isotope effect should show an impact on plant $\delta^{13}C$ values. Thus, temperature and ionic strength effects on k_E^{12}/k_E^{13} are too small for any detectable manifestation in vivo. Therefore corresponding temperature and salt effects reported in ecological studies as quoted in [11,16] must occur at the levels of source CO_2 or stomatal action.

However, the strong pH and Mg^{2+} dependences of k_E^{12}/k_E^{13} can be linked to ecologically controlled $\delta^{13}C$ variations in C_3-plants, at least tentatively [11]. The light induced enzyme activation in photosynthesizing chloroplasts is accompanied by pH value and Mg^{2+} concentration increases in the stroma [21]. Increases of the pH value and Mg^{2+} concentration lead to higher enzymatic $^{13}CO_2/^{12}CO_2$ isotope effects as shown above. Therefore higher light intensities in areas with hot climates could possibly cause enhanced H^+ and Mg^{2+} gradients in the chloroplasts and, thus, explain the stronger ^{13}C isotope discrimination observed for corresponding plants.

An additional factor with respect to the fundamental value of the RuBP carboxylase isotope effect, not yet investigated, is likely to be of importance for model calculations of $\delta^{13}C$ values in plants. Whereas the physiological H^+ and Mg^{2+}

concentration ranges of the RuBP carboxylation can easily be realized in vitro, the physiological stroma CO_2 level of C_3-plants (∼6 μM) has not yet been studied in in-vitro experiments. Different discrimination values for RuBP carboxylase might well be found at very low, rate limiting CO_2 concentrations, which better reflect the in vivo conditions.

ACKNOWLEDGEMENT

This work was supported by the Deutsche Forschungsgemeinschaft.

REFERENCES

1 R. Park and S. Epstein, Geochim. Cosmochim. Acta, 21(1960)110-126.
2 T. Whelan, W.M. Sackett and C.R. Benedict, Plant Physiol., 51(1973)1051-1054.
3 E. Deleens, J.C. Lerman, A. Nato and A. Moyse, in M. Avron (Ed.), Proc. 3rd Int. Congress on Photosynthesis, Rehovot/Israel, 1974, Elsevier, Amsterdam, 1975, pp. 1267-1276.
4 J.T. Christeller, W.A. Laing and J.H. Troughton, Plant. Physiol., 57(1976) 580-582.
5 W.W.L. Wong, C.R. Benedict, T. McGrath and R.J. Kohel, Plant Physiol., 59(1977), Suppl., p. 42.
6 M.F. Estep, F.R. Tabita, P.L. Parker and C. Van Baalen, Plant Physiol., 61(1978)680-687.
7 W.W. Wong, C.R. Benedict and R.J. Kohel, Plant Physiol., 63(1979)852-856.
8 H. Craig, Geochim. Cosmochim. Acta, 3(1953)53.
9 H.-L. Schmidt, F.J. Winkler, E. Latzko and E. Wirth, Israel J. Chem., 17(1978)223-224.
10 H.-L. Schmidt and F.J. Winkler, in E.R. Klein and P.D. Klein (Eds.), Stable Isotopes: Proc. 3rd Int. Conference, Oak Brook, Illinois, 1978, Academic Press, New York, 1979, pp. 295-298.
11 H.-L. Schmidt and F.J. Winkler, Ber. Deutsch. Bot. Ges., 92(1979)185-191.
12 B. Lenhart, F.J. Winkler, H.-L. Schmidt and H. Ziegler, (to be publ.).
13 J. Bigeleisen and M. Wolfsberg, Adv. Chem. Phys., 1(1958)15-76.
14 L. Melander and W.H. Saunders Jr., Reaction Rates of Isotopic Molecules, John Wiley a. Sons, New York, 1980.
15 G.H. Lorimer, Ann. Rev. Plant Physiol., 32(1981)349-383.
16 M.H. O'Leary, Phytochemistry, 20(1981)553-567.
17 J.C. Vogel, in Sitzungsberichte der Heidelberger Akademie der Wissenschaften, Mathematisch-naturwissenschaftliche Klasse, 1980, 3. Abhandlung, Springer-Verlag, Berlin, 1980, pp. 111-135.
18 Cf. G.D. Faequhar, S. von Caemmerer, J.A. Berry, Planta, 149(1980)78-90.
19 Cf. M. Peisker, Kulturpflanze, 26(1978)81-98.
20 Cf. E.M. Galimov, V.G. Shirinskiy, Geochim. Int., 12(1975)157-180.
21 R.G. Jensen and J.T. Bahr, Ann. Rev. Plant Physiol., 28(1977)379-400.

II. GEOCHEMISTRY AND COSMOCHEMISTRY

STABLE ISOTOPES AND THE ORIGIN OF THE SOLAR SYSTEM

F. BEGEMANN

Max-Planck-Institut für Chemie (Otto-Hahn-Institut), 6500 Mainz (G.F.R.)

Until several years ago it was an apparently well-established fact of cosmochemistry that the planetary objects of our solar system formed from a well-mixed primordial nebula of chemically and isotopically uniform composition. Recent measurements have shown this conception to be erroneous, however. Anomalies have been discovered in the isotopic composition of a number of elements which cannot be explained by processes known to be going on within the solar system at present. Rather, they appear to reflect primordial heterogeneities, testifying to variations in space and/or time of the isotopic composition of these elements within the proto-solar nebula.

Since there are a number of recent review articles on the data and their interpretations the interested reader is referred to (in chronological order)

R.N. CLAYTON: "Isotopic anomalies in the early solar system", Ann. Rev. Nucl. Part. Sci.,28 (1978) 501-22.

D.D. CLAYTON: "Supernovae and the origin of the solar system", Space Sc. Revs., 24 (1979) 147-226.

G.J. WASSERBURG, D.A. PAPANASTASSIOU, and TYPHOON LEE: "Isotopic heterogeneities in the solar system", Proc. XXII. Colloq. Intern.d'Astrophysiq., Liege, 1978, p. 203-255.

TYPHOON LEE: "New isotopic clues to solar system formation", Rev.Geophys. and Space Phys., 17 (1979) 1591-1611.

G.J. WASSERBURG, D.A. PAPANASTASSIOU, and T. LEE: "Isotopic heterogeneities in the solar system", Early Solar System Processes and the Pres. Solar System, 1980, LXXIII Corso Soc.Ital.di Fisica, Bologna.

F. BEGEMANN: "Isotopic anomalies in meteorites", Rep.Prog.Phys., 43 (1980) 1309-1356.

STABLE ISOTOPES AND THE EVOLUTION OF LIFE: AN OVERVIEW

M. SCHIDLOWSKI

Max-Planck-Institut für Chemie, D-6500 Mainz (F.R.G.)

ABSTRACT

Biologically mediated fractionations of the stable isotopes of carbon and sulfur can be traced back over most of the sedimentary record, their first appearance placing limits on the antiquity of (i) biological (autotrophic) carbon fixation and (ii) dissimilatory sulfate reduction which processes must have emerged prior to 3.5 and 2.7 x 10^9 years ago, respectively. Hence, biological control of the terrestrial carbon and sulfur cycles has been established very early in the Earth's history.

INTRODUCTION

With the exception of phosphorus, the principal elements on which life processes are based (C, O, H, N, S, P) are mixtures of isotopes. It is now well established that incorporation into living systems, and/or biochemical processing in the widest sense, of these key elements of life entail sizable isotope fractionations as a result of both thermodynamic and kinetic effects imposed on the principal metabolic pathways. With biochemical reactions largely enzyme-controlled, and living systems generally constituting dynamic states undergoing rapid cycles of anabolism and catabolism, it has come to be widely accepted that most biological fractionations are due to kinetic rather than equilibrium effects, although the latter have been claimed to be important in inter- and intramolecular isotope exchange among, and within, individual metabolites (ref. 1; see also paper by GALIMOV, this volume).

Since biological isotope fractionations are basically retained when organic matter is incorporated in sediments, the above effects are propagated into the rock section of the geochemical cycle where they have left their signatures on the crustal inventories of the respective elements as from almost the beginning of the geological record. This is particularly evident in the case of carbon and sulfur. Continuous biological processing in the exogenic exchange reservoir (atmosphere, ocean) of these two elements is responsible for the characteristic dichotomy of their fluxes into the crust caused by (i) photosynthetic carbon fixation, and (ii) bacterial sulfate reduction. This has consequently given rise to crustal depositories of organic carbon and bacteriogenic sulfide (Fig. 1).

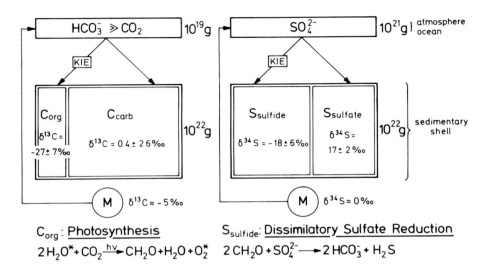

Fig. 1. Box models showing splitting of mantle (M)-derived primordial carbon and sulfur between biogenic and non-biogenic crustal reservoirs. The biogenic reservoirs store the <u>reduced</u> forms of the elements (organic carbon, bacteriogenic sulfide) while the inorganic depositories are made up of <u>oxidized</u> phases (carbonate, sulfate). Note that the kinetic isotope effects ($\overline{KIE}$) inherent in the processes of photosynthetic carbon fixation and bacterial (dissimilatory) sulfate reduction are propagated from the exogenic exchange reservoir into the crust, this having consequently caused an isotopic disproportionation of terrestrial carbon and sulfur into "light" and "heavy" partial reservoirs.

As a result, the sedimentary reservoirs of carbon and sulfur are typically bipartite, storing a biologically-derived (reduced) and a non-biological (oxidized) element species. An important corollary of this partitioning is a large-scale isotopic disproportionation of terrestrial carbon and sulfur into a "light" and a "heavy" fraction since the biogenic phases preferentially concentrate the light isotopes (^{12}C, ^{32}S), leaving the heavy species (^{13}C, ^{34}S) to accumulate in the residual inorganic phase. Accordingly, the geochemical cycles of carbon and sulfur are outstanding examples of element cycles governed by the terrestrial biota (ref. 2).

BIOLOGICAL FRACTIONATION OF CARBON ISOTOPES

In the case of carbon, the geochemically most important fractionations are linked to processes of photosynthetic carbon fixation, most notably "Calvin cycle" or C3 photosynthesis operated by the majority of higher plants, algae and autotrophic bacteria. The principal isotope-discriminating steps responsible for the gross isotope composition of plants (Fig. 2) are (i) the diffusion of

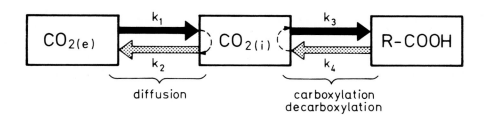

Fig. 2. Synopsis of principal isotope discriminating reactions determining the isotopic composition of plants (black: assimilatory pathway; stippled: dissimilatory and other reverse reactions). $CO_{2(e)}$ and $CO_{2(i)}$ represent, respectively, external and internal CO_2, and R-COOH stands for the product of the first CO_2-fixing carboxylation; k_{1-4} are the rate constants assigned to the individual reactions. The net effect of these reactions is a considerable enrichment of ^{12}C in biosynthesized materials as compared to the inorganic substrate (external CO_2).

CO_2 into, and out of, the plant tissue, (ii) the initial fixation of CO_2 in the carboxyl group of an organic acid (carboxylation), and (iii) decarboxylation and related dissimilatory processes (notably photorespiration). Accordingly, the overall fractionation between the inorganic substrate (CO_2) and biosynthesized materials will depend on an interplay of these processes. The quantitatively most important contribution towards total fractionation is provided by the first carbon-fixing carboxylation reaction, notably the one catalyzed by the enzyme ribulose-1.5-bisphosphate (RuBP) carboxylase by which CO_2 enters the reductive pentose phosphate or "Calvin" cycle. The kinetic isotope effect inherent in this reaction discriminates very effectively against ^{13}C, the wide range of values observed (-17 to -42 °/oo) reflecting variations of temperature, pH, metal ion availability, and other variables which bear on the magnitude of isotope fractionations in enzymatic reactions (ref. 3). The corresponding average enrichment of ^{12}C in organic substances (living and fossil) by about 20 to 30 °/oo as compared to the Earth's inorganic carbon pool (carbonate) results, therefore, principally from the isotope selection properties of this particular enzyme (refs. 2, 3, 4). Consequently, primordial terrestrial carbon (with $\delta^{13}C_{prim}$ = -5 °/oo [PDB]) has been partitioned, through the ages, between a "light" organic species ($\delta^{13}C_{org}$ = -27 ± 7 °/oo) and a "heavy" carbonate species ($\delta^{13}C_{carb}$ = + 0.4 ± 2.6 °/oo).

BIOLOGIAL FRACTIONATION OF SULFUR ISOTOPES

The decisive control of the geochemistry of sulfur is exercised by a few genera of sulfate reducing bacteria utilizing an energy-yielding reaction that couples the reduction of sulfate to the oxidation of organic substrates. This process of dissimilatory sulfate reduction (see Fig. 1) brings about a large-scale low-temperature conversion of sulfate to sulfide in the exogenic reservoir which provides a pivotal link in the terrestrial sulfur cycle, but has never been demonstrated to proceed at $t \lesssim 150°$ C unless biologically mediated. Substantial parts of the hydrogen sulfide thus generated end up as sedimentary sulfide.

The biochemical pathway of sulfide reduction comprises a number of component reactions characterized by kinetic isotope effects of variable magnitude. Discriminations against ^{34}S over the total pathway have been shown to sum up to -46 $°/oo$ in culture experiments conducted with single bacterial species and to about -60 $°/oo$ in the natural environment, with average fractionations lying in the range -30 to -40 $°/oo$ (refs. 2, 5). Accordingly, the $\delta^{34}S$ values of biogenic sulfide are markedly shifted to the negative side of the scale as compared to the parent (marine) sulfate pool. Since bacteriogenic H_2S is largely preserved as sulfide minerals (mostly pyrite, FeS_2), the $\delta^{34}S$ patterns of sedimentary sulfides reflect the primary values, allowing the identification of bacteriogenic sulfide in ancient sedimentary environments. Due to continuous processing in the geochemical cycle, primordial sulfur (with $\delta^{34}S_{prim} = 0$ $°/oo$ [CDT]) has been subjected to an isotopic disproportionation into "light" biogenic ($\delta^{34}S = -18 \pm 6$ $°/oo$) and "heavy" sulfate sulfur ($\delta^{34}S = +17 \pm 2$ $°/oo$).

ANTIQUITY OF BIOLOGICALLY MEDIATED ISOTOPE FRACTIONATIONS

With both the biogenic and non-biogenic species of carbon and sulfur preserved in sediments, it is possible to trace the biological fractionations of these elements back into the geologic past, thereby constraining the time of emergence of the underlying biochemical processes (refs. 2, 6, 7, 8, 9). The respective isotope records are summarized in Fig. 3.

Biological carbon fixation

As can be deduced from the carbon isotope age curves, biological (viz. autotrophic) carbon fixation must be a very ancient process, since the fractionation of about 25 $°/oo$ typically observed between organic and inorganic (carbonate) carbon is a characteristic feature of the record over the last 3.5×10^9 years. Since this fractionation is, for the most part, the geochemical manifestation of the activities of one key enzyme of the assimilatory pathway (RuBP carboxylase), the sedimentary carbon isotope record gives a remarkably consistent

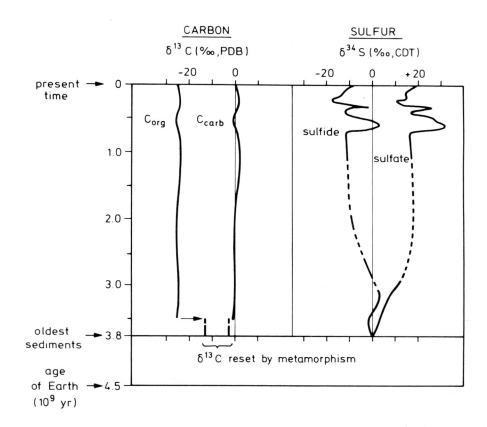

Fig. 3. Average isotope fractionation between organic carbon (C_{org}) and carbonate (C_{carb}) and between sulfide and sulfate through geologic time. Note that these biologically mediated fractionations date back to at least 3.5×10^9 yr in the case of carbon and to about 2.7×10^9 yr in the case of sulfur. Accordingly, biological control of the terrestrial carbon and sulfur cycles had been established early in the Earth's history, with photosynthesis apparently predating the emergence of bacterial sulfate reduction.

signal of biological activity as from 3.5×10^9 years ago. Since the isotopic shifts observed for both sedimentary carbon species in the 3.8×10^9 year-old Isua sediments of West Greenland can be accounted for by metamorphism, there is little doubt that biological control of the terrestrial carbon cycle had already been established by Isua times (refs. 2, 4, 9, 10). For such interpretation of the record to be invalidated, we would have to postulate an inorganic process that had mimicked fractionations in biological carbon fixation with a remarkable degree of precision.

Bacterial sulfate reduction

As for the antiquity of bacterial sulfate reduction, the record suggests that incipient partitioning of terrestrial sulfur into isotopically light (biogenic) sulfide and heavy sulfate might have occurred around 3×10^9 years ago. The oldest isotope distribution patterns of sedimentary sulfides presumably attributable to the activities of microbial sulfate reducers are about 2.7×10^9 years old (refs. 11, 12). On the other hand, coeval sulfide and sulfate from sediments older than about 3.2×10^9 years (refs. 13-15) show virtually no fractionation indicative of bacterial activity, the averages for both sulfur species staying close to the value of primordial sulfur ($\delta^{34}S$ between 0 and $+1\ ^o/oo$). Accordingly, dissimilatory sulfate reduction seems to be a relatively late achievement in the evolution of bioenergetic processes, being possibly bracketed by the time limits 2.8 and 3.2×10^9 years (refs. 7, 8). It should be noted, however, that a conclusive interpretation notably of the oldest record ($t > 2 \times 10^9$ years) is subject to the severe limitations imposed by the scanty data base hitherto available (ref. 2).

CONCLUSIONS

In sum, stable isotope studies of sedimentary carbon and sulfur have proved to be of considerable potential for imposing temporal constraints on the earliest involvement of life processes in the geochemical cycles of these elements and thus, by inference, on some major quantum steps of early organic evolution. Unfortunately, studies of the most ancient (Precambrian) record lag well behind those of younger rocks. Apart from the poor data base for sulfur, virtually no record of $\delta^{15}N$ and δD is as yet available for Precambrian organic matter. Much additional data will be required before such studies can provide the considerable degree of insight into ancient biochemistries of which they are potentially capable.

ACKNOWLEDGEMENTS

Research leading to the views expressed in this paper was carried out as part of the program of the Sonderforschungsbereich No. 73 (Atmospheric Trace Components) of the Deutsche Forschungsgemeinschaft and of the Precambrian Paleobiology Research Group, University of California, Los Angeles, funded by NASA Grant NSG 7489 and the NSF Waterman Foundation Award to J.W. Schopf. All support thus received is gratefully acknowledged.

REFERENCES

1. E.M. Galimov, in B. Durand (Ed.), Kerogen, Editions Technip, Paris, 1980 Ch. 9, p. 271-299.
2. M. Schidlowski, J.M. Hayes and I.R. Kaplan, in J.W. Schopf (Ed.), Origin and Evolution of Earth's Earliest Biosphere, Princeton Univ. Press, Princeton, N.J., 1981 (in press).
3. M.H. O'Leary, Phytochemistry, 20 (1981) 553-767.
4. M. Schidlowski, in H.D. Holland and M. Schidlowski (Eds.), Mineral Deposits and the Evolution of the Biosphere, Springer, Berlin, 1981, p. 103-122.
5. L.A. Chambers and P.A. Trudinger, Geomicrobiol. J., 1 (1979) 249-299.
6. R. Eichmann and M. Schidlowski, Geochim. Cosmochim. Acta, 39 (1975) 585-595.
7. J. Monster, P.W.U. Appel, H.G. Thode, M. Schidlowski, C.M. Carmichael and D. Bridgwater, Geochim. Cosmochim. Acta, 43 (1979) 405-413.
8. M. Schidlowski, Origins of Life, 9 (1979) 299-311.
9. M. Schidlowski, in B.J. Ralph, P.A. Trudinger and M.R. Walter (Eds.), Biogeochemistry of Ancient and Modern Environments, Springer, Berlin, 1980, p. 47-54.
10. M. Schidlowski, P.W.U. Appel, R. Eichmann and C.E. Junge, Geochim. Cosmochim. Acta, 43 (1979) 189-199.
11. A.M. Goodwin, J. Monster and H.G. Thode, Econ. Geol., 71 (1976) 870-891.
12. E.M. Ripley and D.L. Nicol, Geochim. Cosmochim. Acta, 45 (1981) 839-846.
13. E.C. Perry, J. Monster and T. Reimer, Science, 171 (1971) 1015-1016.
14. V.I. Vinogradov, T.O. Reimer, A.M. Leites and S.B. Smelov, Lithology & Mineral Res., 11 (1976) 407-420.
15. I.B. Lambert, T.H. Donnelly, J.S.R. Dunlop and D.I. Groves, Nature, 276 (1978) 808-811.

ISOTOPE GEOCHEMISTRY OF CARBON

J. HOEFS

Geochemisches Institut der Universität Göttingen, Göttingen (F.R.G.)

ABSTRACT

In this review three topics are discussed:
1) Are there any changes in the isotopic composition of the carbonate carbon and organic carbon reservoir with geologic time?
   As more and more carbon isotope data became available this seems to be the case. A tentative "age curve" is given and some implications are discussed.
2) What happens to the isotopic composition of carbonate carbon and organic carbon with increasing temperatures?
   The possible changes in the isotopic compositions are described in a very generalized way. In more detail the fractionation mechanisms occurring during the graphitization of carbonaceous matter are demonstrated in rocks from a metamorphic profile from New Caledonia.
3) What is the isotopic composition of "juvenile" carbon?
   A simple balance calculation gives a mean value of -5.5 ‰, which is in agreement with the isotopic composition of kimberlites, carbonatites and most diamonds. However, there are some other findings - showing low $\delta^{13}C$ values around -25 ‰ - which do not fit into such a simple model. Some implications of this dual distribution are discussed.

INTRODUCTION

This paper does not discuss all aspects of carbon isotope geochemistry, but centers around three different topics, which is, of course, a very subjective selection.

I will not discuss the isotopic composition of atmospheric CO_2, nor the isotopic composition of the oceanic reservoir, nor questions of the living biosphere and oil and gas. To start with, Fig. 1 summarizes some geologically important carbon reservoirs in which the absolute carbon content and their mean $\delta^{13}C$ isotopic composition are given. Out of the five reservoirs only three are important concerning their masses: a) sedimentary carbonates; b) sedimentary organic matter and c) so-called "juvenile" carbon.

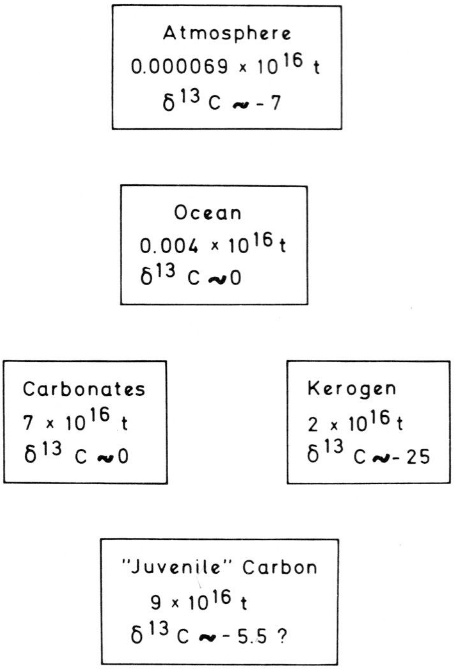

```
          ┌─────────────────────────┐
          │       Atmosphere        │
          │  0.000069 × 10^16 t     │
          │    δ^13 C ~ -7          │
          └─────────────────────────┘

          ┌─────────────────────────┐
          │         Ocean           │
          │   0.004 × 10^16 t       │
          │    δ^13 C ~ 0           │
          └─────────────────────────┘

  ┌──────────────────┐      ┌──────────────────┐
  │   Carbonates     │      │     Kerogen      │
  │   7 × 10^16 t    │      │   2 × 10^16 t    │
  │   δ^13 C ~ 0     │      │  δ^13 C ~ -25    │
  └──────────────────┘      └──────────────────┘

          ┌─────────────────────────┐
          │    "Juvenile" Carbon    │
          │      9 × 10^16 t        │
          │    δ^13 C ~ -5.5 ?      │
          └─────────────────────────┘
```

Fig. 1. Some geologically important reservoirs of carbon. (Carbon contents are from ref. [3], $\delta^{13}C$ values are in ‰ relative to PDB.)

RESULTS AND DISCUSSION

1) <u>Are there any changes in the isotopic composition of carbonate carbon and organic carbon with geologic time?</u>

Since the very first measurements [1,2] we know that carbonates concentrate ^{13}C relative to organic compounds. While marine carbonates have $\delta^{13}C$ values around 0 (zero), the organic matter in sediments (kerogen) has $\delta^{13}C$ values around -25 ‰. Although there have been some speculations about systematic changes in the isotopic composition

of carbonates during the earth's history [4-6], most workers agreed
until a few years ago that carbonates do not show any definite age
trend. The arguments for this assumption have been that carbonates
are very susceptible to diagenetic changes and therefore have not
preserved their primary composition. This is, however, true only for
oxygen, not for carbon, because the diagenetic solutions obviously
do not contain enough carbon to modify on a large scale the carbon
isotopic composition. During the last few years - as more and more
carbon isotope analyses become available and as the stratigraphic
derivation of samples has become refined - a pattern of ^{13}C variation
with time is beginning to emerge (see Fig. 2).

In this figure the variation of carbonate carbon and organic carbon
with time is shown. The different curves show a very similar trend,
although there are also some dissimilarities. It is especially
noteworthy that there are two time periods where we obviously have
definite changes in the carbon isotope composition; that is during
the Devonian-Carboniferous-Permian time and during the Cretaceous.

This secular variation in carbon isotope ratios can be explained
in the following way: If we assume that the total exogenic carbon
reservoirs remain constant, growth of one reservoir must be compensated
for by shrinking of the other with a resulting shift in the ^{13}C
concentration, specifically this means that the amount of carbon fixed
in the oxidized and reduced reservoirs has changed with time. These
changes might be due to fluctuations in the deposition of carbonates
and organic carbon into the sediments. Variations in the withdrawal of
photosynthetically fixed carbon might be due to an increase or decrease
in the rate of carbon burial related to changes in the oxidizing
potential of the oceanic system. This is postulated by Fischer and
Arthur [8] and Scholle and Arthur [12] for the Cretaceous. An
alternative interpretation relates these variations to an increase in
the photosynthetic activity at constant carbon burial rates, which is
geologically plausible for the Devonian-Carboniferous time period [13].

As shown in Fig. 1 the ratio of the carbonate carbon reservoir to
the organic carbon reservoir is about 4:1, which implies that a change
of 1 ‰ in the carbonate reservoir is accompanied by a change of 4 ‰
in the organic carbon reservoir. Therefore the organic carbon reservoir
should be much more sensitive for isotopic variations than the
carbonate carbon reservoir. However, as shown in Fig. 2, this is not
reflected by the data which may mean that during the diagenesis of
organic carbon the primary isotopic imprint more or less disappears.

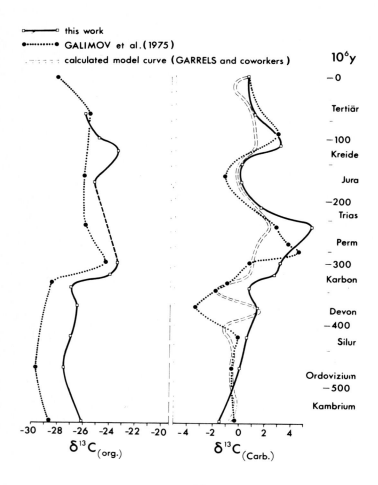

Fig. 2. "Age curve" of $\delta^{13}C$ values of carbonate-carbon and organic carbon. Data base: Refs. [7-13]. The calculated $\delta^{13}C$ curve for carbonates is after Garrels and Lerman [14] and depends on the $\delta^{34}S$ age curve for oceanic sulfate. (For further details see also ref. [14].)

$^{13}C/^{12}C$ variations with time cannot be seen as an isolated phenomenon, but have important implications for other geological parameters such as the atmospheric oxygen and carbon dioxide content. They imply for instance that the atmospheric CO_2 content in the geological past must have fluctuated. It has been argued [15] that throughout the last 500 million years the CO_2 content has fluctuated within the limits of 0.1 to 0.4 %. By the way, I am not absolutely convinced that the increase in the atmospheric CO_2 content which we observe today is entirely due to anthropogenic effects, but could also be due to some natural fluctuations.

2) <u>What happens to the carbonate carbon and organic carbon isotopic composition with increasing temperatures?</u>

To illustrate the pathways the following schematic Fig. 3 is given:

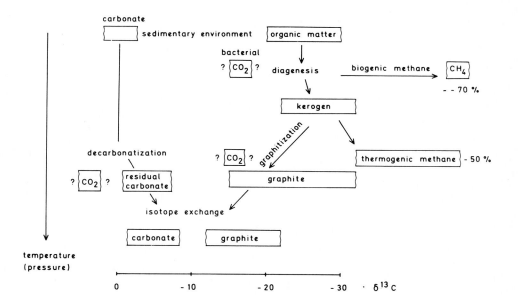

Fig. 3. Schematic diagram of the evolution of $\delta^{13}C$(carb) and $\delta^{13}C$(org) with increasing temperatures (geological depths).

On the left side the behavior of carbonate carbon is demonstrated: the carbon isotopic composition of the carbonate in carbonate-rich, organic carbon-poor sediments will remain essentially unchanged until decarbonatization reactions with silicate minerals occur, which will not happen much below 400° C. The isotopic composition of the liberated CO_2 is about 2 to 3 ‰ heavier than the carbonate [16]. However, recent experiments by Maxwell and Sofer [17] demonstrate that during decarbonatization the CO_2 is depleted in ^{13}C relative to the carbonate. With increasing temperatures isotope exchange reactions with graphitic material take place and may shift the isotopic composition of the carbonate towards light values depending upon the proportions of carbonate carbon and organic carbon.

Looking at the behavior of the organic carbon, the situation seems to be more complicated. Immediately after burial of the biological organic material into the sediments, complex diagenetic changes occur in the organic matter. On the one hand mainly bacterial action results in methane very much enriched in ^{12}C. On the other hand, microbiological activity also leads to CO_2 whose isotopic composition is not very well known, but could be equal to or a little bit heavier than the organic matter. Furthermore the preferential elimination of ^{13}C-rich compounds such as proteins and carbohydrates could also lead to a ^{12}C-enrichment. All in all, recent marine sediments show a mean $\delta^{13}C$ value of -25 ‰ [18]. With further transformation to kerogen some ^{13}C loss obviously occurs. A careful tabulation of the literature data [19] provides an average value of -26.8 ‰, which has been modified to -27.5 ‰ by Hayes et al. [19]. If we accept this last value as representative of kerogen then a depletion of 2.5 ‰ is indicated. This isotope effect might be best explained by the large losses of CO_2 that occur during the synthesis of kerogen and which are especially pronounced during the decarboxylation of some ^{13}C-rich carboxyl groups [20,21].

I will now discuss more detailed the question what happens to the kerogen with further temperature increase until fully ordered graphite is formed. During the graphitization process, which is mainly characterized by the loss of volatiles, isotopic fractionations could accompany the chemical changes of the carbonaceous matter. As we know from the literature, in some cases [22-24] a shift towards higher ^{13}C contents with increasing grade of metamorphism has been observed. However, in others it was not [25,26].

To elucidate in further detail the relationship between graphitization and isotopic composition a metamorphic profile from the Ouégoa District, New Caledonia was analyzed in cooperation with Phillippa Black from Auckland, New Zealand. In this profile Dr. Black was able to calculate the composition of the fluid phase which is in equilibrium with the mineral paragenesis. The composition of the fluid phase (see Fig. 4) was calculated from the chemical equilibrium data of certain mineral reactions such as

1) paragonite + calcite + quartz $\rightleftharpoons$ plagioclase + CO_2 + H_2O
2) lawsonite + calcite + quartz $\rightleftharpoons$ grossulare + CO_2 + H_2O

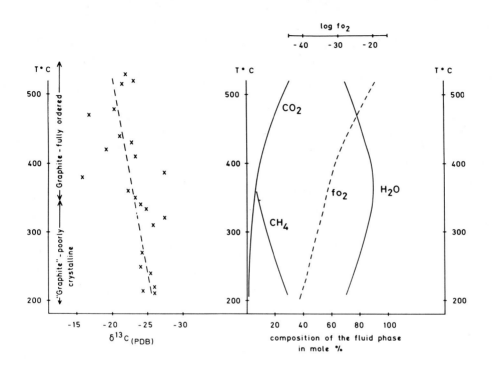

Fig. 4. Isotopic composition of carbonaceous matter (graphite) in a metamorphic profile from Ouégoa District, New Caledonia in relationship to the composition of the fluid phase.

At low temperatures (up to ~350° C) methane is the most abundant carbon-containing species in the gas phase, at high temperatures (~500° C) CO_2 is much more abundant than CH_4. At temperatures around 350° C CO_2 and CH_4 are about equally abundant. Furthermore, Fig. 4 demonstrates that there is a small but continuous shift towards larger ^{13}C values in the graphite, although there is considerable scattering in the data. This shift towards larger $\delta^{13}C$ values was interpreted earlier [24] as mainly being due to the liberation of relatively large amounts of CH_4. However, due to the predominance of CO_2 at high temperatures this liberation of methane cannot be the only process being responsible for the observed shift. Therefore it is postulated that a further mechanism leads to a ^{13}C enrichment in the graphite, most probably isotope exchange with carbonates [27,28].

3) **What is the isotopic composition of so-called "juvenile" carbon?**

Already Goldschmidt [29] showed that the sedimentary mass originates from the weathering of igneous rocks. However, this balance does not fit for the volatiles, which was first pointed out by Goldschmidt [29] and later on by Rubey [30]. The carbon now present in the sedimentary shell has to be derived from the interior of the earth by degassing. With this assumption we can calculate the $\delta^{13}C$ values of primary magmatic carbon by a $\delta^{13}C$ mass balance of sediments for which the following relation holds [31]:

$\delta^{13}C$ of primary magmatic carbon = $(1-f)(\delta^{13}C$ of carbonate$)$
$\qquad\qquad\qquad\qquad\qquad\qquad + f(\delta^{13}C$ of organic carbon$)$

where f = the fraction of primary magmatic CO_2 ended up as organic carbon

$\delta^{13}C$ of carbonate = 0
$\delta^{13}C$ of organic carbon = -25

The values for f given in the literature vary widely, for example 0.27 [30], 0.23 [32], 0.18 [33] and 0.13 [34]. I have adopted an f value of 0.22 according to Holland [3] and solving the equation for $\delta^{13}C(prim)$:

$\delta^{13}C = 0.78 \times 0 + 0.22 \times -25 = -5,5$

As is demonstrated in Fig. 5, this calculated mean value corresponds to measured $\delta^{13}C$ values from kimberlites, carbonatites, diamonds - materials where a lot of petrological and chemical arguments prove a deep-seated mantle origin. However, as is also shown in Fig. 5, several arguments do not agree with such a simple isotope distribution: some diamonds have very low $\delta^{13}C$ values, down to -30 %o [35]. The CO_2 from fluid inclusions in olivine nodules, products of the mantle, show also very low $\delta^{13}C$ values. The "reduced" carbon - reduced in the sense of being not oxidized (neither carbonates nor CO_2) - found in all igneous rocks in trace amounts around 100 to 200 ppm is also very light [36]. A dual distribution of $\delta^{13}C$ values - one group has values around -5, another around -25 - is also found in extra-terrestrial materials, in meteorites and lunar rocks where biological activity can be excluded.

This dual distribution can be explained by at least two different models:

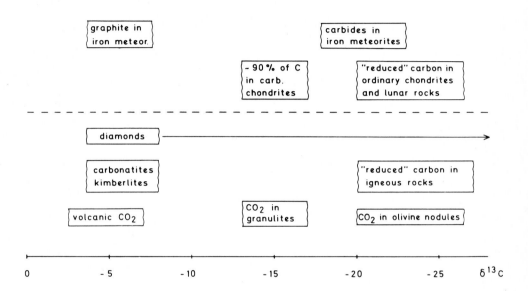

Fig. 5. Carbon isotopic composition of deep-seated carbon on earth in comparison with extraterrestrial materials.

a) Galimov [37] argued that two different classes of carbon compounds existed from the very beginning of the history of the earth, one around -5 and a second one around -25 ‰.

b) According to Pineau et al. [38], magmas - derived from the mantle - are able to dissolve about 1 % CO_2. When the magmas move towards the surface, the CO_2 solubility decreases and CO_2 exsolves. As shown experimentally [39], the fractionation factor between this exsolved CO_2 and magma carbon is 1.004 at temperatures between 1,100 and 1,300° C. The initially exsolved CO_2 has a δ-value of around -5 ‰. When the magma has reached the surface more than 95 % of the initial carbon content has exsolved. The remaining ~5 % of magma carbon will then have very low $\delta^{13}C$ values as observed in nature.

Which of the two models is the more probable one, must remain an open question. Furthermore the isotopic composition of the earth as a whole is unknown. However, it is tempting to make a comparison with carbonaceous chondrites type I - the most primitive matter from our solar system. In these meteorites carbon occurs in at least three different forms, as carbonates, as extractable organic matter and as insoluble highly condensed and polymerized carbonaceous matter, the latter being present to ~90 % of the total carbon. This nonextractable carbon has a relatively constant carbon isotopic composition of around -15 ‰ [40]. Very similar $\delta^{13}C$ values have been measured in granulites [41-43], rocks typically found in the deep crust and which contain abundant probably mantle-derived CO_2. Assuming that the carbon in these rocks is of juvenile origin, we might speculate that these values around -15 ‰ reflect a third reservoir which might represent the primordial value of the earth as a whole.

ACKNOWLEDGEMENT

I would like to thank Dr. P. Black (Auckland, N.Z.) for providing me with rock samples and data about the composition of the fluid phase.

REFERENCES

1 A.O. Nier, E.A. Gulbransen, J.Am.Chem.Soc., 61(1939)697.
2 B.F. Murphey, A.O. Nier, Phys.Rev., 59(1941)771.
3 H.D. Holland, The Chemistry of the Atmosphere and Oceans, John Wiley & Sons, New York, 1978.
4 P.M. Jeffery, W. Compston, D. Greenhalgh, J. de Laeter, Geochim. Cosmochim.Acta,7(1955)255.
5 W. Compston, Geochim.Cosmochim.Acta,18(1960)1.
6 J.N. Weber, Geochim.Cosmochim.Acta, 31(1967)2343.
7 J.D. Arneth, Unpublished measurements, manuscript in preparation.
8 A.G. Fischer, M.A. Arthur, Soc.Econ.Paleont.Mineral., Spec.Publ.,25 (1977)19.
9 E.M. Galimov, A.A. Migdisov, A.B. Ronov, Geochemistry Intern.,12 (1975)1.
10 M. Magaritz, K.H. Schulze, Contr.Sedimentology,9(1980)269.
11 R. Renner, Diplomarbeit Universität Göttingen, 1978.
12 P.A. Scholle, M.A. Arthur, Am.Ass.Petrol.Geol., Bull., 64(1980)67.
13 D.H. Welte, W. Kalkreuth, J. Hoefs, Naturwissenschaften, 62(1975) 482.
14 R.M. Garrels, A. Lerman, Proc.US.Nat.Acad.Sci. (in press).
15 M.I. Budyko, A.B. Ronov, Geochemistry Intern., 16, No. 3(1979)1.
16 Y. Bottinga, Geochim.Cosmochim.Acta, 33(1969)49.
17 L.E. Maxwell, Z. Sofer, This volume, in press (1981).
18 P. Deines, in P. Fritz and J.Ch. Fontes (Eds.), Handbook of Environmental Geochemistry, Vol. 1, Elsevier, Amsterdam, 1980, p.329.
19 J.M. Hayes, I.R. Kaplan, K.W. Wedeking, in J.W. Schopf (Ed.), Origin and Evolution of Earth's Earliest Biosphere (unpublished manuscript).
20 P.H. Abelson, T.C. Hoering, Proc.Nat.Acad.Sci.USA, 47(1961)623.

21. E.M. Galimov, Carbon isotopes in oil-gas geology (Russ.), Nedwa, Moscow, 1973.
22. F. Barker, I. Friedman, Bull.Geol.Soc.Am., 80(1969)1403.
23. D.M. McKirdy, T.G. Powell, Geology, 2(1974)591.
24. J. Hoefs, M. Frey, Geochim.Cosmochim.Acta, 40(1976)945.
25. S. Gavelin, Geochim.Cosmochim.Acta, 12(1957)297.
26. M.O. Andreae, Contrib.Mineral.Petrol., 47(1974)299.
27. H. Wada, Geochem.J., 11(1977)183.
28. J.W. Valley, J.R. O'Neil, Geochim.Cosmochim.Acta, 45(1981)411.
29. V.M. Goldschmidt, Fortschr.Mineralogie,Kristallographie und Petrographie, 17(1933)112.
30. W.W. Rubey, Geol.Soc.Am.,Bull., 62(1951)1111.
31. Y.H. Li, Am.J.Sci., 272(1972)119.
32. H. Borchert, Geochim.Cosmochim.Acta, 2(1951)62.
33. A.B. Ronov, Sedimentology, 10(1968)25.
34. A.B. Ronov, A.A. Yaroshevsky, Am.Geophys.Union Geophys.Mon.,13 (1969)37.
35. E.M. Galimov, in VII. Nat.Symp.Isotope Geochem., Moscow, 1978, Abstr., p. 13.
36. J. Hoefs, Contrib.Mineral.Petrol., 41(1973)277.
37. E.M. Galimov, Geokhimiya, 5(1967)530.
38. F. Pineau, M. Javoy, Y. Bottinga, Earth Planet.Sci.Lett., 29 (1976)413.
39. M. Javoy, F. Pineau, I. Iiyama, Contrib.Mineral.Petrol., 67 (1978)35.
40. J.W. Smith, I.R. Kaplan, Science, 167(1970)1367.
41. J. Hoefs, J. Touret, Contrib.Mineral.Petrol., 52(1975)165.
42. J. Hoefs, J. Touret, Abstracts of the Spring Meeting of the "Sektion für Geochemie, Deutsche Mineralogische Gesellschaft", 1981, p. 22.
43. F. Pineau, M. Javoy, F. Behar, J. Touret, Bull.Soc.fr.Min., Crist. (in press)(1981).

THE $^{13}C/^{12}C$ ISOTOPE RATIOS IN A NORTH-GERMAN PODZOL

H.-G. BERTRAM and G.H. SCHLESER
Abt. Biophysikalische Chemie - ICH -
Kernforschungsanlage Jülich GmbH, D-5170 Jülich

ABSTRACT

In conjunction with transport and decomposition problems of organic carbon compounds in soils, a north german humus podzol has been investigated. Small sampling intervals have made it possible to detect a certain fine structure in the organic carbon content, the corresponding $\delta^{13}C$-values and C/N ratios not previously observed in such details. An attempt is made to use these results for an interpretation of organic carbon transport and decomposition within this soil.

INTRODUCTION

The ever increasing demand in energy has led to the combustion of tremendous amounts of fossil fuels. As a result yearly increasing amounts of CO_2 liberated into the atmosphere have been recorded, which may lead to an increase in temperature (greenhouse effect) and to changes in plant response.

Predictions for the future development necessitate a thorough understanding of the global carbon cycle with its sources and sinks. This implies that the various carbon compartments and their carbon flux exchange rates have to be known. At present however the correct world circulation of carbon is elusive, due to the paucity and uncertainty of experimental data.

The compartments involved which exchange their carbon in terms of a few years are atmosphere, biosphere and soil organic matter. We have chosen soil organic matter for more detailed analysis because
- it is a poorly known compartment and
- it represents by far the largest carbon reservoir.

The latter point is of importance because small changes within this pool may have large effects on the other reservoirs. In order to receive more detailed information about the role of carbon in soils i.e. of carbon transport and decomposition, an activity presently centers on the analysis of variations of the stable carbon isotopes with soil depth. One of the first soils selected for investigation has been a podzol simply because of two reasons:

- Podzols show very distinctive layers with marked characteristics promising pronounced isotope effects and
- they are found in large areas representing an important type of soil on a global scale.

Podzols are mainly covered by boreal forests and their habitat comprises the humid climate of the northern hemisphere.

MATERIAL AND METHODS

The soil samples originate from a Humus Podzol at Brobergen near Hamburg (9° 10' E, 53° 36' N), which is located near the North Sea Coast of Germany. This soil is now covered by grass.

Sampling of soil has been restricted to soil layers of 2 cm width in the upper horizons, and 5 cm width in the lower horizons. This procedure minimizes uncertainties, which may exist due to variations in a horizon itself.

The cut samples were dried for 14 hours at 60^oC and afterwards thoroughly homogenized. pH-values were determined by dissolving 10 g of soil in 25 ml of a 0.1 molar solution of KCl. The organic carbon content of each sample was analyzed by its combustion to CO_2 and by coulometrical determination of this CO_2. Application of Kjeldahl's method yielded the nitrogen content of each sample, and atomic absorption spectrometry was used for the analysis of manganese. $^{13}C/^{12}C$ determinations were performed by converting the organic carbon content of soil samples quantitatively into CO_2 /1/ and then performing a mass spectrometric analysis. The carbon isotope results are stated as $\delta^{13}C$-values relative to the PDB standard.

RESULTS

Significant variations of organic carbon concentration have been found, reflecting the intense optical differences in the course of the soil profile as indicated by the soil layer description (Fig. 1). In the upper region of the A_h-horizon (0-35 cm), the carbon concentration remains nearly unaltered at values of about 3 to 4 %. However at the bottom of the A_h-horizon, which comprises the first 43 cm, the carbon concentration rises to almost 14 %.

Chemistry, Life Sciences, Biochemistry, Molecular Biology, Pharmacology, Pharmaceutical Sciences, Earth Sciences, Environmental Research

Stable Isotopes

Proceedings of the Fourth International Conference held in Julich, March 23-26, 1981

edited by H.-L. SCHMIDT, H. FÖRSTEL *and* K. HEINZINGER

ANALYTICAL CHEMISTRY SYMPOSIA SERIES, 11

**1982 xvii + 758 pages
Price: US $127.75 / Dfl. 275.00
ISBN 0-444-42076-2**

The 85 contributions and 15 reviews in this volume make it a comprehensive overview of the state of investigations on stable isotopes of the main bioelements as they occur in nature. The conference emphasized stable tracer applications in medicine, pharmacology, agriculture and biochemistry, and added the dimension of the theory and consequences of isotope effects, and the importance of stable isotopes in geochemistry, cosmochemistry and environmental research.

The papers compare recent results and methodology in the different disciplines obtained on the basis of isotope measurements. Interaction between the disciplines was facilitated by a common methodology which revealed that a particular problem could be approached from various angles, and stimulus given for further investigations. Examples of new possibilities are the use of NMR in biological research and the introduction of "naturally labelled" compounds in nutrition physiology. The book is essential to all those working on stable isotopes but even scientists not familiar with isotope research may benefit from the demonstration of the wide possibilities isotope application can have as a common tool in most biosciences from agriculture to zoology.

A SELECTION OF THE CONTENTS: **I. Isotope Effects, Theory and Consequences.** The Theoretical Analysis of Isotope Effects *(M. Wolfsberg).* Vapour Pressure Isotope Effects of Acetonitrile *(Gy. Jakli et al.).* Heavy-Atom Isotope Effects on Enzyme-Catalyzed Reactions *(M.H. O'Leary).* Isotope Effects on Each, C- and N-Atoms, as a Tool for the Elucidation of Enzyme Catalyzed Amide Hydrolyses *(R. Medina et al.).* **II. Geochemistry and Cosmochemistry.** Stable Isotopes

ELSEVIER SCIENTIFIC PUBLISHING COMPANY

Amsterdam *and* New York

and the Evolution of Life: An Overview *(M. Schidlowski)*. Isotope Geochemistry of Carbon *(J. Hoefs)*. Carbon Isotope Fractionation Factors of the Carbon Dioxide-Carbonate System and their Geochemical Implications *(L.E. Maxwell and Z. Sofer)*. The Isotopic Fractionation During the Oxidation of Carbon Monoxide by Hydroxylradicals and its Implication for the Atmospheric Co-Cycle *(H.G.J. Smit et al.)*. **III. Biomedical Applications. A. Pharmacology and Drug Metabolism.** Applications of Stable Isotopes in Pharmacological Research *(T.A. Baillie et al.)*. Application of ^{13}C-Labelling in the Bioavailability Assessment of Experimental Clovoxamine Formulations *(H. de Bree et al.)*. Measurement of the Pharmacokinetics of DI-([15,15,16,16-D4]-Linoleoyl)-3-sn-Glycerophosphocholine After Oral Administration to Rats *(A. Brekle et al.)*. **B. Clinical Diagnosis.** The Application of the Stable Isotopes of Oxygen to Biomedical Research and Neurology *(D. Samuel)*. In Vivo Measurement of Enzymes with Deuterated Precursors and GC/SIM *(H.-Ch. Curtius)*. Prenatal Diagnosis of Propionic and Methylmalonic Acidemia by Stable Isotope Dilution Analysis of Methylcitric and Methylmalonic Acids in Amniotic Fluids *(L. Sweetman et al.)*. Use of Water Labelled with Deuterium for Medical Applications *(E. Roth et al.)*. **C. Breath Tests and Lung Function Tests.** The Commercial Feasibility of ^{13}C Breath Tests *(P.D. Klein and E.R. Klein)*. Application of ^{13}C-Fatty Acids Breath Tests in Myocardial Metabolic Studies *(M. Suehiro et al.)*. Breath Test Using ^{13}C Phenylalanine as Substrate *(D. Glaubitt and K. Siafarikas)*. Use of the $^{13}C/^{12}C$ Breath Test to Study Sugar Metabolism in Animals and Men *(J. Duchesne et al.)*. **IV. Life Sciences, Agriculture, and Environmental Research.** Stable Isotopes in Agriculture *(H. Faust)*. Balance of ^{15}N Fertilizer in Soil/Plant System *(N. Sotiriou and F. Korte)*. Nitrogen Isotope Ratio Variations in Biological Material - Indicator for Metabolic Correlations *(R. Medina and H.-L. Schmidt)*. Possibilities of Stable Isotope Analysis in the Control of Food Products *(J. Bricout)*. **V. Methods. A. Analytical Developments.** Recent Applications of ^{13}C NMR Spectroscopy to Biological Systems *(N.A. Matwiyoff)*. Quantitative Mass Spectrometry with Stable Isotope Labelled Internal Standard as a Reference Technique *(I. Björkem)*. Application and Measurement of Metal Isotopes *(K. Habfast)*. Stable Isotopes in Biomedical and Environmental Analysis by Field Desorption Mass Spectrometry *(W.D. Lehmann and H.-R. Schulten)*. **B. Isotope Separation and Synthesis of Labelled Compounds.** The Production of Stable Isotopes of Oxygen *(I. Dostrovsky and M. Epstein)*. Separation of the Stable Isotopes of Chlorine, Sulfur and Calcium *(W.M. Rutherford)*. The Synthesis of Mono-and Oligosaccharides Enriched with Isotopes of Carbon, Hydrogen and Oxygen *(R. Barker et al.)*. New Approaches in the Preparation of ^{15}N Labeled Amino Acids *(Z.E. Kahana and A. Lapidot)*.

Send your order to **your bookseller** or
ELSEVIER SCIENCE PUBLISHERS
P.O. Box 211, 1000 AE Amsterdam, The Netherlands

Distributor in the U.S.A. and Canada:
ELSEVIER SCIENCE PUBLISHING CO., INC.
52 Vanderbilt Ave., New York, N.Y. 10017

Continuation orders for series are accepted.

Orders from individuals must be accompanied by a remittance, following which books will be supplied postfree.

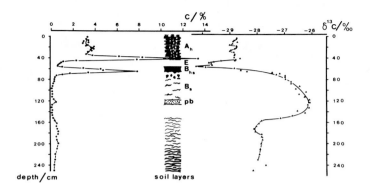

Fig. 1. Organic carbon content and the corresponding $\delta^{13}C$-values in a north german podzol. (A_h = humus layer, E = bleached layer, B_{hs} = humus enriched layer, B_s = sesquioxid enriched layer, pb = pebble bed layer).

Below the E-horizon, which represents a rather strongly bleached layer, exhibiting about 0.5 % in carbon concentration, a further carbon increase is observed as shown in Fig. 1 leading to 3 % at the lower part of the B_{hs}-horizon (55-65 cm). It seems to be remarkable that both peaks appear at the bottom of the corresponding horizons. Below the B_{hs}-horizon the carbon content decreases to almost zero, but exhibits a further increase starting from about 150 cm, which can optically be seen from a darkening of this layer.

pH-values (Fig. 2) show strong variations and display their lowest values within those horizons which contain the highest carbon concentrations. Values between 3.6 and 5.8 have been measured in the course of the whole soil profile, indicating the acidic character of this soil.

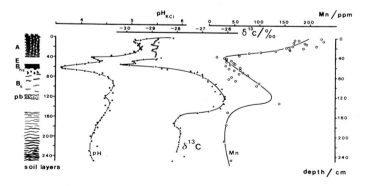

Fig. 2. pH-values and Mn-concentration in relation to the $\delta^{13}C$-values in a north german podzol (soil layer explanation see Fig. 1).

The Mn-concentration exhibits values of about 200 ppm in the upper part of the A_h-horizon. This concentration however decreases rapidly to a fifth of the original amount at the bottom of this horizon. Within the B_s-horizon an increase of the Mn-concentration has been found, reaching almost 150 ppm directly below the pebble bed layer (pb). A further decrease is to be seen below the pebble bed layer and the Mn-concentration finally reaches between 30 and 40 ppm at 150 cm depth.

In spite of the strong carbon variations within the upper soil layers only moderate changes are observed in the C/N-ratio (Fig. 3). Below 70 cm depth a distinct decrease is to be seen, ranging from about 40 to 12, while at 150 cm depth the C/N-ratio rapidly increases to values of about 37. In the lower horizon no further changes have been found.

The overall pattern of $\delta^{13}C$-data exhibits rather large variations which partly coincide with the corresponding variations of the carbon content (Fig. 1). Within the A_h-horizon $\delta^{13}C$-data show only small variations with a mean value of -28.80 o/oo. The low carbon values between 45-55 cm depth coincide with the lowest $\delta^{13}C$-values, displaying a minimum of -30.25 o/oo at 55 cm. Within the following 15 cm a strong enrichment in $\delta^{13}C$ is found, ranging from about -30 to -27 o/oo. This enrichment continues somewhat less dramatically down to 130 cm ending at a value of about -26 o/oo. The transition into the next lower horizon is accompanied by a strong depletion in $\delta^{13}C$ covering about 2 o/oo across 20 cm (-26 to -28 o/oo). From about 160 cm downward to the lowest investigated soil horizons (250 cm) the $\delta^{13}C$-data remain almost constant.

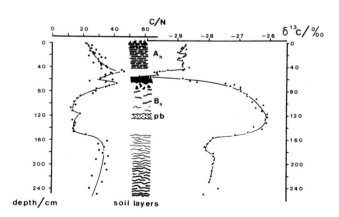

Fig. 3. The C/N-ratio and the $\delta^{13}C$-values in a north german podzol (soil layer explanation see Fig. 1).

DISCUSSION

Organic material originating from plant material is vertically transported by water and at the time undergoes complex reactions and polymerisations, the latter being promoted by the presence of Fe and Mn. As a result pH-values and Mn content are decreasing strongly within the A_h-horizon (Fig. 2), while the C/N-ratio increases (Fig. 3).

Gradual immobilization of displaced organic substances due to sorption and mechanical filtration leads to an accumulation of organic material within the very lowest part of the A_h-horizon (Fig. 1). Transportation and accumulation of carbon do not lead to an isotope shift, therefore leaving the $\delta^{13}C$-data of this horizon unaffected. Organic carbon compounds released from the bottom A_h-horizon, are being retained only in very small amounts within the bleached eluvial horizon (Fig. 1). They are adsorbed to sandy particles with a preferential adherence to the isotopically lighter soil fraction, thus leading to the very depleted values. This in turn leads to an enrichment of ^{13}C in the adjacent B_{hs}-horizon. Mineralisation effects which normally would lead to an enrichment of the remaining solid fraction [2, 3] have to be ruled out, as there is practically no air oxygen available.

The enrichment in ^{13}C across the B_{hs}-horizon which is continuing down the following soil layers may partly be understood by referring to the C/N-ratio (Fig. 3). The strong ^{13}C enrichment across B_{hs} is accompanied by an increase in the C/N-ratio, indicating that a decomposition of organic compounds may take place, retaining isotopically lighter soil fractions, while more enriched fractions are being removed to lower layers. The less pronounced ^{13}C increase across the B_s-horizon which is accompanied by a sharp decrease of the C/N-ratio, again indicates a disintegration of organic compounds, by which isotopically enriched ^{13}C-fractions are preferentially being removed downwards.

No separate soil organic fractions such as humic and fulvic acids have been analysed so far. Nissenbaum and Schallinger [4] however have shown, that for example fulvic acids of peat organic matter change by -1.6 o/oo with soil depth. Similar changes have been reported for polysaccharides [4]. The sharp depletion of the carbon isotopes starting at about 1.50 m depth once more coincides with a drastic change of the C/N-ratio (Fig. 3). This again indicates a change in the composition of soil organic fractions, whereby enriched carbon ^{13}C-compounds are being preferentially retained in the lower B_s and pebble bed (pb) layers.

The constant ^{13}C-values below about 1.70 m are probably due to groundwater dynamics by which the formation of a further differentiation is prevented.

REFERENCES

1 G. H. Schleser and R. Pohling, Int. J. Appl. Rad. Isot., 31 (1980) 769-773
2 I. D. Stout and T. A. Rafter, Dep. Sci. Ind. Res. Bull., 220 (1978) 75-83
3 G. H. Schleser, R. Pohling and W. Kerpen, Z. Pflanzenernaehr. Bodenk., 144 (1981) 23-29
4 A. Nissenbaum and K. M. Schallinger, Geoderma, 11 (1974) 137-145

STABLE ISOTOPE INVESTIGATIONS IN ANTARCTICA

H. SCHÜTZE, G. STRAUCH and K. WETZEL
Zentralinstitut für Isotopen- und Strahlenforschung der Akademie der Wissenschaften der DDR, Leipzig (DDR)

ABSTRACT

First results on the stable isotope content of biological and hydrological material from antarctic regions are given. δD-, $\delta^{13}C$-, and $\delta^{15}N$-measurements were made on algae, mosses, and lichens. The δ-values of these plants show the influences of the environmental conditions. A remarkably high hydrogen isotope enrichment could be found in the case of all lichens. The degree of evaporation of a small water pool and the seasonal variations of the inner dynamics of an antarctic lake were estimated by isotopic measurements. The results of δD-measurements on air moisture during seatrip suggest that antarctic precipitations originate from sources which may be situated far away.

INTRODUCTION

Since 1978 scientists of the Central Institute of Isotope and Radiation Research have been taking part in the Soviet Antarctic Expeditions (S.A.E.). In Antarctica, a region which is relatively free from anthropogenic influences, it was the aim of our work to further the knowledge on the geochemical cycles of the light elements hydrogen, oxygen, carbon, and nitrogen. On the basis of isotope measurements it should be possible to draw conclusions about hydrological, meteorological, glaciological, and biological processes in Antarctica.

EXPERIMENTAL

Biological samples such as algae, lichens, and mosses have been collected around the Schirmacher-Oasis (antarctic station Novolasarevskaya). The abundance ratios of stable isotopes were measured mass-spectrometrically relative to the abundance ratio in a standard. The deviations are given in the form of δ-values in o/oo (e. g. for hydrogen):

$$\delta D = [(D/H)_{sample} / (D/H)_{standard} - 1] \cdot 1000 \text{ o/oo}.$$

We have been able to directly process certain samples in the antarctic station whereas the remaining preparations and the mass-spectrometrical measurements had to be carried out in the Leipzig institute.

RESULTS

Biosphere
- The $\delta^{13}C$-values of lichens and mosses are in the range of other C_3-plants [1,2]. Any similarity in $\delta^{13}C$-values to C_4-plants could not be found. Only when lichens assimilated CO_2 under very hard conditions [3] an enrichment of ^{13}C had been observed [4].
- The algae are subdivided into two groups: algae from the bottom of lakes (type Pomornika), and algae from soils of dried out pools. The latter group obviously use carbonates as their carbon source [5].

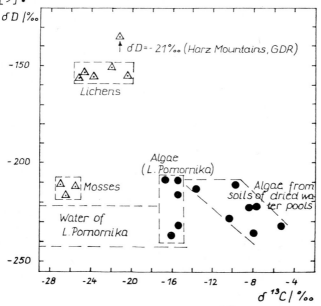

Fig. 1. Correlation of δD- and $\delta^{13}C$-values of antarctic algae, mosses, and lichens (Omphalodiscus antarcticus, Omphalodina exsulans, Acarospora cf. emergens, Xanthoria elegans var. pulvinata).

- The δD-values of water flowing into the Schirmacher-Oasis range from -210 to -300 o/oo. As shown in Fig. 1 the δD-values of algae and mosses lie within this range. However, all lichen samples clearly differ by about 60 o/oo from the other species

and by about 100 o/oo from the mean precipitation. Obviously there exists in the assimilation process a high hydrogen isotope effect because in comparison with lichens from the Harz Mountains (GDR) a similar difference from the mean precipitation was found.
- From the results of δ^{15}N-measurements (Fig. 2) it can be assumed, that atmospheric nitrogen is the nitrogen source of lichens, mosses, and of one part of algae. In contrast to this the very positive δ^{15}N-values of algae from Lake Pomornika indicate, that the nitrogen originates from lake sediments as in the case of lakes in Victoria Land [6].

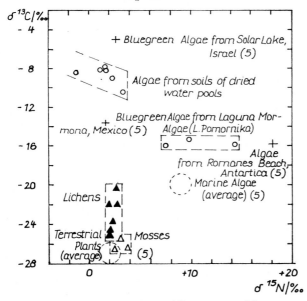

Fig. 2. Correlation of δ^{13}C- and δ^{15}N-values of antarctic algae, mosses, and lichens. (5) in the Figure means ref. [5].

Lake dynamics

Investigations on evaporation

By means of the relation

$$\ln V_o/V = \frac{\ln \delta'/\delta'_o}{\varepsilon/\alpha} \quad (1)$$

the degree of evaporation

$$\eta = 1 - V/V_o \quad (2)$$

can be calculated using isotopic measurements (in this example isotope ratios of hydrogen).

$$\delta' = \frac{\delta D}{1000} + 1; \quad \delta'_o = \frac{\delta D_o}{1000} + 1$$

V_o = volume of the lake with the isotopic composition δD_o
(start of observations)

V = volume of the lake with the isotopic composition δD
(end of observations)

$\varepsilon/\alpha = \frac{\ln \alpha}{\alpha}$ = separation of hydrogen isotopes in the system $H_2O_{liquid}/H_2O_{vapour}$.

The expression ε/α consists of two terms: a temperature dependent thermodynamic term [7] $\ln \alpha_{th} = 0.0972$ ($\vartheta = +7\ °C$) and an, in a first approximation, temperature independent kinetic term [8] $\ln \alpha_{kin} = 0.0083$.

How the two terms participate in the separation is determined by the relative humidity h

$$\varepsilon/\alpha = h\,(\varepsilon/\alpha)_{th} + (1-h)\,(\varepsilon/\alpha)_{kin}. \tag{3}$$

Using the measured values
$\delta D_o = -195\ o/oo;\ \delta D = -140\ o/oo$ (30.XI. - 25.XII. 1978)
$\overline{h} = 0.30;\ \overline{\vartheta} = +7\ °C;\ \varepsilon/\alpha = 0.0322$
we obtain, according to the equations 1 and 2, the degree of evaporation $\eta = 0.87$ in accordance with the direct determination of V/V_o.

Studies of seasonal deuterium variations in Lake Pomornika

The results of these studies, summarized in Fig. 3, form a complex of four distribution patterns. These patterns characterize typical states of Lake Pomornika. In the main melting period there occurs an input of glacier water with δD-values about $-260\ o/oo$. In this case the 5 m-level enrichment decreased, and the lake water body is underflown at the 7 m-level. After the melting period the state of Lake Pomornika is characterized by an approximatly straight lined D-distribution. Further investigations should however be carried out by also measuring $\delta^{18}O$-values.

Origin of antarctic precipitations

Registration of air humidity during sea trip [9] showed D-variations caused not only by condensation processes [10, 11]

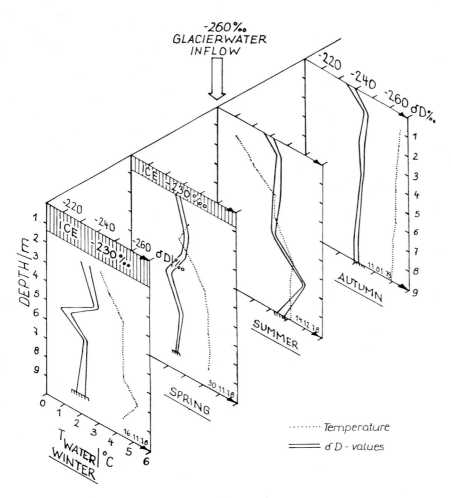

Fig. 3 Seasonal vertical distribution of deuterium within Lake Pomornika

but also by transport effects. Obviously antarctic precipitations originate substantially from areas of lower geographical latitude and went through longer transport distances (Fig. 4).

CONCLUSION

We intend to continue isotopic investigations of biological and hydrological material to get further information about relations between δ-values and environmental conditions. Long-range observations of δD- and $\delta^{18}O$-values of humidity in antarctica should be carried out.

ACKNOWLEDGEMENTS

The authors wish to acknowledge the preparation of biological materials by I. Maaß and A. Runge. Thanks are due to W. Richter for helpful discussions as well as to E. Kirschner, Ch. Schachtschneider, and H. Birkenfeld for the mass-spectrometrical measurements.

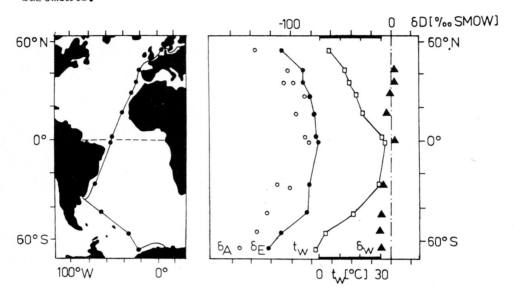

Fig. 4. Points of sampling with results (t_w: surface ocean temperature; δ_W: δD-values of surface ocean water; δ_A: δD-values of atmospheric moisture; δ_E: calculated δD-values of atmospheric moisture being in equilibrium with surface ocean water).

REFERENCES

1 R. Park and S. Epstein, Plant Physiol., 36 (1961) 133-138.
2 B.N. Smith and S. Epstein, Plant Physiol., 47 (1971) 380-384.
3 O.L. Lange, L. Kappen and E.-D. Schulze (Eds.). Water and Plant Life, Springer Verlag Berlin, Heidelberg, New York, 1976.
4 A. Shomer-Ilan, A. Nissenbaum, M. Galun and Y. Waisel, Plant Physiol., 63 (1979) 201-205.
5 D.H. Stuermer, K.E. Peters and I.R. Kaplan, Geochim. Cosmochim. Acta, 42 (1978) 989-997.
6 T. Torü, N. Yamagata, S. Nakaya, S. Murata, T. Hashimoto, O. Matsubaya and H. Sakai, Memoirs of National Inst. Pol. Res. Spec. Issue No. 4 (1975) 5-29.
7 M. Majzoub, J. Chim. Phys., 68 (1971) 1423-1436.
8 H. Schütze and M. Mohnke, Isotopenpraxis, 17 (1981) 159-163.
9 P. Kowski, G. Strauch and H. Schütze, ZfI-Mitt. 29 (1980) 309-318.
10 H. Craig, Science, Washington, 133 (1961) 1702-1703.
11 K. Fröhlich and H. Schütze, Geol. Geophys. Veröff. Reihe I, Heft 7 (1980) 122-144.

Chemistry, Life Sciences, Biochemistry, Molecular Biology, Pharmacology, Pharmaceutical Sciences, Earth Sciences, Environmental Research

Stable Isotopes

Proceedings of the Fourth International Conference held in Julich, March 23-26, 1981

edited by H.-L. SCHMIDT, H. FÖRSTEL *and* K. HEINZINGER

ANALYTICAL CHEMISTRY SYMPOSIA SERIES, 11

1982 xvii + 758 pages
Price: US $127.75 / Dfl. 275.00
ISBN 0-444-42076-2

APPROXIMATE MONTH OF PUBLICATION: APRIL

The 85 contributions and 15 reviews in this volume make it a comprehensive overview of the state of investigations on stable isotopes of the main bioelements as they occur in nature. The conference emphasized stable tracer applications in medicine, pharmacology, agriculture and biochemistry, and added the dimension of the theory and consequences of isotope effects, and the importance of stable isotopes in geochemistry, cosmochemistry and environmental research.

The papers compare recent results and methodology in the different disciplines obtained on the basis of isotope measurements. Interaction between the disciplines was facilitated by a common methodology which revealed that a particular problem could be approached from various angles, and stimulus given for further investigations. Examples of new possibilities are the use of NMR in biological research and the introduction of "naturally labelled" compounds in nutrition physiology. The book is essential to all those working on stable isotopes but even scientists not familiar with isotope research may benefit from the demonstration of the wide possibilities isotope application can have as a common tool in most biosciences from agriculture to zoology.

A SELECTION OF THE CONTENTS: **I. Isotope Effects, Theory and Consequences.** The Theoretical Analysis of Isotope Effects *(M. Wolfsberg)*. Vapour Pressure Isotope Effects of Acetonitrile *(Gy. Jakli et al.)*. Heavy-Atom Isotope Effects on Enzyme-Catalyzed Reactions *(M.H. O'Leary)*. Isotope Effects on Each, C- and N-Atoms, as a Tool for the Elucidation of Enzyme Catalyzed Amide Hydrolyses *(R. Medina et al.)*. **II. Geochemistry and Cosmochemistry.** Stable Isotopes

ELSEVIER SCIENTIFIC PUBLISHING COMPANY

Amsterdam *and* New York

and the Evolution of Life: An Overview *(M. Schidlowski)*. Isotope Geochemistry of Carbon *(J. Hoefs)*. Carbon Isotope Fractionation Factors of the Carbon Dioxide-Carbonate System and their Geochemical Implications *(L.E. Maxwell and Z. Sofer)*. The Isotopic Fractionation During the Oxidation of Carbon Monoxide by Hydroxylradicals and its Implication for the Atmospheric Co-Cycle *(H.G.J. Smit et al.)*. **III. Biomedical Applications. A. Pharmacology and Drug Metabolism.** Applications of Stable Isotopes in Pharmacological Research *(T.A. Baillie et al.)*. Application of ^{13}C-Labelling in the Bioavailability Assessment of Experimental Clovoxamine Formulations *(H. de Bree et al.)*. Measurement of the Pharmacokinetics of DI-([15,15,16,16-D$_4$]-Linoleoyl)-3-sn-Glycerophosphocholine After Oral Administration to Rats *(A. Brekle et al.)*. **B. Clinical Diagnosis.** The Application of the Stable Isotopes of Oxygen to Biomedical Research and Neurology *(D. Samuel)*. In Vivo Measurement of Enzymes with Deuterated Precursors and GC/SIM *(H.-Ch. Curtius)*. Prenatal Diagnosis of Propionic and Methylmalonic Acidemia by Stable Isotope Dilution Analysis of Methylcitric and Methylmalonic Acids in Amniotic Fluids *(L. Sweetman et al.)*. Use of Water Labelled with Deuterium for Medical Applications *(E. Roth et al.)*. **C. Breath Tests and Lung Function Tests.** The Commercial Feasibility of ^{13}C Breath Tests *(P.D. Klein and E.R. Klein)*. Application of ^{13}C-Fatty Acids Breath Tests in Myocardial Metabolic Studies *(M. Suehiro et al.)*. Breath Test Using ^{13}C Phenylalanine as Substrate *(D. Glaubitt and K. Siafarikas)*. Use of the ^{13}C/^{12}C Breath Test to Study Sugar Metabolism in Animals and Men *(J. Duchesne et al.)*. **IV. Life Sciences, Agriculture, and Environmental Research.** Stable Isotopes in Agriculture *(H. Faust)*. Balance of ^{15}N Fertilizer in Soil/Plant System *(N. Sotiriou and F. Korte)*. Nitrogen Isotope Ratio Variations in Biological Material - Indicator for Metabolic Correlations *(R. Medina and H.-L. Schmidt)*. Possibilities of Stable Isotope Analysis in the Control of Food Products *(J. Bricout)*. **V. Methods. A. Analytical Developments.** Recent Applications of ^{13}C NMR Spectroscopy to Biological Systems *(N.A. Matwiyoff)*. Quantitative Mass Spectrometry with Stable Isotope Labelled Internal Standard as a Reference Technique *(I. Björkem)*. Application and Measurement of Metal Isotopes *(K. Habfast)*. Stable Isotopes in Biomedical and Environmental Analysis by Field Desorption Mass Spectrometry *(W.D. Lehmann and H.-R. Schulten)*. **B. Isotope Separation and Synthesis of Labelled Compounds.** The Production of Stable Isotopes of Oxygen *(I. Dostrovsky and M. Epstein)*. Separation of the Stable Isotopes of Chlorine, Sulfur and Calcium *(W.M. Rutherford)*. The Synthesis of Mono-and Oligosaccharides Enriched with Isotopes of Carbon, Hydrogen and Oxygen *(R. Barker et al.)*. New Approaches in the Preparation of ^{15}N Labeled Amino Acids *(Z.E. Kahana and A. Lapidot)*.

Send your order to **your bookseller** or
ELSEVIER SCIENCE PUBLISHERS
P.O. Box 211, 1000 AE Amsterdam, The Netherlands

Distributor in the U.S.A. and Canada:
ELSEVIER SCIENCE PUBLISHING CO., INC.
52 Vanderbilt Ave., New York, N.Y. 10017

Continuation orders for series are accepted.

Orders from individuals must be accompanied by a remittance, following which books will be supplied postfree.

The Dutch guilder price is definitive. US$ prices are subject to exchange rate fluctuations. Prijzen zijn excl. B.T.W.

CARBON ISOTOPE FRACTIONATION FACTORS OF THE CARBON DIOXIDE-CARBONATE SYSTEM AND THEIR GEOCHEMICAL IMPLICATIONS

LESLIE E. MAXWELL and ZVI SOFER
Cities Service Company, P. O. Box 3908, Tulsa, Oklahoma 74102 (USA)

ABSTRACT

Carbon isotope fractionation factors were determined experimentally for the CO_2-carbonate system at various temperatures in the range 330°C to 850°C. Carbonates studied were natural rocks containing calcite, dolomite, and silicate minerals, as well as laboratory-prepared calcite controls.

The results show that the carbon isotopic composition of CO_2 in natural gases containing high concentrations of CO_2 can be explained by a low temperature, low yield metamorphism of fine-grain, silicate-containing carbonates.

INTRODUCTION

The merits of carbon dioxide (CO_2) flooding in enhanced oil recovery has caused the value of this gas to increase greatly. In particular, it is a definite asset when it is available in high concentrations from natural gas wells in the vicinity of enhanced oil recovery operations. Until recently, the discovery of such gases was generally an undesirable result of exploration for liquid petroleum and/or hydrocarbon gases. Better understanding of the origin of the gases containing high percentages of carbon dioxide (high CO_2 gases) could either greatly improve exploration efforts for this gas or help avoid its discovery when the gas is undesirable.

The theory of the origin of the high CO_2 gases as being the result of thermal metamorphism of carbonate rocks was advocated by several authors (refs. 1-5). These authors show a strong correlation between gases containing high concentrations of CO_2 and past thermal events. In contrast, gases containing low concentrations of CO_2 have been explained by diverse theories of origin, all of which are related to maturation or oxydation of organic matter (refs. 4, 6).

The overall known range of carbon isotopes ($\delta^{13}C$, PDB) for carbon dioxide found in the subsurface is rather large: from -42°/oo to +12°/oo (ref. 6). However, the range is much narrower for high CO_2 gases (>20 mole % CO_2): from -9°/oo to 0°/oo, and these gases are always associated with methane which, based on stable carbon isotopes, seems to be thermally mature (refs. 2-4). According to theoretical fractionation factors calculated by Bottinga (ref. 7), carbon dioxide generated under isotopic equilibrium conditions from marine calcites (limestones, which have $\delta^{13}C$ values close to 0°/oo) should be enriched in $\delta^{13}C$ relative to the limestone for temperatures above 200°C. This is in contradiction with the values observed in nature. Evidence from stable carbon isotope studies, therefore, do not generally support an origin of high CO_2 gases by thermal decarbonation of marine carbonates.

Few experimental studies of carbon isotope fractionations in the CO_2-carbonate system have been conducted at ambiant and intermediate (200-400°C) temperatures (refs. 8-11). Although the observed fractionation factors agree quite well with the theoretical values in this temperature range, decarbonation reactions are kinetically unfavorable and substantial amounts of CO_2 cannot be generated. No experimental values were available for temperatures above 400°C. Since theory often fails to explain the complex solid-gas interactions, this study attempts to determine the equilibrium fractionation factors at metamorphic temperatures between CO_2 and natural carbonate rocks and obtain experimental confirmation to Bottinga's theoretical calculations.

EXPERIMENTAL

Five different carbonate materials were used for the carbon isotope fractionation experiments: rock limestone, rock dolomite, and a siliceous calcite were chosen as representative of common natural carbonate lithologies. Since natural carbonates often contain impurities that alter the chemical properties of the CO_2-carbonate system, and since Bottinga (ref. 7) calculated fractionation factors for pure calcite, a laboratory-prepared synthetic calcite was chosen to serve as a pure control. The siliceous calcite contained some organic carbon. Therefore, to eliminate possible effects of that carbon, a mixture of fine-grain synthetic calcite and organic carbon-free montmorillonite was prepared by mixing the two in an aqueous suspension. Montmorillonite was chosen over quartz because of its fine-grain size. Table 1 summarizes the mineralogical composition of the five carbonates as determined by X-ray diffraction. As seen in Table 1, the limestone and dolomite, as well as the siliceous calcite, are not pure carbonate minerals, but each contains subordinate amounts of dolomite and calcite, respectively. With the exception of the synthetic materials, all samples were pulverized and extracted with a chloroform-methanol solution followed by boiling to dryness in a 30 percent hydrogen peroxide solution to oxidize as much organic matter as possible.

TABLE 1. Mineralogy and Isotopic Composition of Samples

	Mineral	%	$\delta^{13}C$ PPB
Synthetic Calcite	Calcite	100	-7.36
Limestone	Calcite	60.2	
	Dolomite	38.5	+2.60
	Quartz	2.3	
Dolomite	Dolomite	67.9	
	Calcite	20.1	-0.04
	Quartz	12.0	
Siliceous Calcite	Calcite	61.5	
	Quartz	20.2	
	Dolomite	13.6	
	Illite	2.3	+2.66
	Plagioclase	1.8	
	Microcline	0.6	
	Org. Carbon	0.3	
Synthetic Calcite-montmorillanite mixture	Calcite	69.0	
	Montmorillonite	27.9	-7.50
	Quartz	3.1	

Most experiments were conducted with approximately 100 milligrams of sample sealed in fused quartz tubes (7 mm I.D., 16 cm long) under initial vacuum. Borosilicate glass tubes were used at temperatures below 580°C. Below 400°C, experiments using synthetic calcite, limestone, and the synthetic calcite-montmorillonite mixture necessitated larger volume tubes (20 mm I.D., 15 cm long reduced on one end to a 9 mm O.D. by 20 cm long tube) in order to obtain an adequate volume of carbon dioxide for isotope measurement (due to the low partial pressure of CO_2 generated by decarbonation of carbonates at these lower temperatures). In order to determine the effect of water, distilled water was added to some systems in amounts which would create vapor saturation; however, it is realized that none of the other experiments can be regarded as being conducted in an absolutely water-free atmosphere because water absorbed chemically or on grain surfaces was not removed before sealing the tubes.

The tubes and their contents were heated in a resistance furnace to release carbon dioxide from the carbonate material. Experiments were conducted in the temperature range of 330°C to 850°C for varying durations of time. Reaction tubes were immediately quenched after heating by immersion in liquid nitrogen in order to prevent lower temperature reequilibration of the CO_2 and carbonate. Tubes remained at liquid nitrogen temperature until the evolved CO_2 gas was transferred to a vacuum line for cryogenic separation of the carbon dioxide from any water present. After separation of CO_2 from the water, the volume of carbon dioxide was measured in a manometer and then transferred to glass tubes for mass spectrometric analyses. The solid residue from each experiment was reacted with 100% phosphoric acid, and the evolved carbon dioxide was collected and purified for mass spectrometer analysis as described by McCrea (ref. 12).

During the heating and subsequent residue acidification of the siliceous calcite and the dolomite, some H_2S was evolved which was carried through the collection and purification steps along with the CO_2. This could cause an interfering reaction in the mass spectrometer whereby hydrogen ionized from the H_2S reacted with CO_2 giving rise to an abnormal large mass 45:

$$H^+ + {}^{12}C^{16}O_2 \rightleftarrows H^{12}C^{16}O_2{}^+$$

The resulting large mass 45 ions can lead to a false apparent enrichment in ^{13}C (i.e., the CO_2 gas appears isotopically more positive (heavier) than in reality). This interference can be rectified by sealing cupric oxide in the collection tubes of both thermally-evolved and acid-evolved gas and reacting in a furnace at 550°C for two hours, converting the H_2S and cupric oxide to water and copper sulfate. The produced water can than be cryogenically separated from the carbon dioxide. Some H_2S was also detected and was removed similarily from the gas evolved by heating limestone at 390°C.

The carbon isotope data is reported as the usual $\delta^{13}C$ value:

$$\delta^{13}C = [\frac{R_{sample}}{R_{standard}} - 1] \times 1000 ,$$

where R is the atom ratio $^{13}C/^{12}C$ and the standard is carbon dioxide prepared from the Pee Dee Belemnite (P.D.B.) (ref. 13). All $\delta^{13}C$ values were corrected for ^{17}O contribution. Standard deviation for δ measurements is $\pm 0.05°/oo$ (1σ).

The fractionation factor, α, between carbon dioxide and carbonate is defined as

$$\alpha = \frac{R_{CO_2}}{R_{CARB.}} = \frac{\delta^{13}CO_2 + 1000}{\delta^{13}CARB. + 1000} .$$

RESULTS

As expected, impurities alter the chemical properties of the CO_2-carbonate system, and, hence, different carbonates generate various amounts of CO_2 at given temperatures. In general, for any given temperature, the amount of CO_2 generated in an increasing order is: synthetic calcite, limestone, dolomite, siliceous calcite, synthetic calcite-montmorillonite mixture. Results of the fractionation factors are summerized in Table 2 and Figure 1. Results show that the fractionation factors calculated from the experimental data of synthetic calcite, limestone, and dolomite are all more positive than the theoretical values calculated by Bottinga (1969).

Low temperature data were obtained for limestone (390°C). The limestone data are relatively precise. The fractionation factors here are smaller than the theoretical values and extrapolate to crossover to negative fractionations at a temperature of about 320°C (Figure 1).

Results also show that below 800°C the carbonate-CO_2 fractionations occurring in the siliceous calcite system were markedly more negative than those for the limestone, dolomite, and synthetic calcite systems. As can be seen on Figure 1, the trend of the data is linear and reproducible with the exception of the lowest temperature experiments. Because of the great differences observed for this system, and the pure carbonate systems, the possibility of an apparent fractionation due to CO_2 evolved from any residual organic carbon was raised. Therefore, a control clay-carbonate mixture was prepared in the laboratory. This system was free of any organic carbon, but contained montmorillonite instead of fine-grain quartz. Despite this difference, the observed fractionation factors are similar to those of the siliceous calcite system (see Figure 1).

It is not clear whether the differences encountered between the 330°C experiments for the siliceous calcite and the 350°C experiments for the synthetic system are real. Although data for the siliceous calcite represent different durations of heating, the data do not reveal a time dependence. The direction of equilibrium is, therefore, uncertain. Also, analytical problems encountered when handling such small samples ($\approx$0.06 cc STP CO_2) greatly increase the analytical error.

Figure 1 proposes an alternative curve to Bottinga's theoretical curve (ref. 7) for the temperature dependance of $10^3 \ln\alpha$ in a CO_2-pure carbonate system and a linear relation for the temperature-$10^3 \ln\alpha$ dependance of the fine-grain silicate-calcite mixtures.

TABLE 2. Average values of $10^3 \ln\alpha$

Sample	\multicolumn{11}{c}{Temperature (°C)}
	850	800	750	700	650	600	500	400	390	350	330
Synthetic calcite	3.2	3.4	3.5		2.9						
Limestone	3.1	3.2	3.4		3.4				1.3		
Dolomite	3.4	3.6	3.2		2.9						
Siliceous calcite	3.1	3.1	2.4		0.3		-2.9	-6.4			-6.4
Synthetic calcite-montmorillonite				1.5		-1.6	-3.4			-4.4	

Based on chemical theory, the difference in the fractionation factors between the synthetic calcite, limestone and dolomite, and the siliceous calcite and synthetic calcite-montmorillonite cannot be predicted. The amount of noncarbonate materials does not seem to affect the fractionation factor either, as is evident from the dolomite which contains as much as 12% quartz.

A possible explanation for the difference in the fractionation factors involves the grain size of the carbonate and noncarbonate minerals and different chemical reactions that might be involved in CO_2 generation. The synthetic calcite and the limestone generate CO_2 via a dissociation process:

$$CaCO_3 \underset{}{\overset{heat}{\rightleftarrows}} CaO + CO_2$$

the synthetic calcite-montmorillonite mixture on the other hand generate CO_2 via a

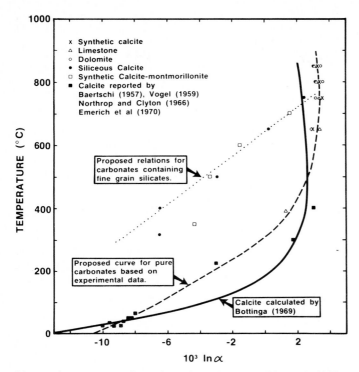

Figure 1. Average fractionations between CO_2 and different carbonates.

solid phase acid dissociation process whereby hydroxyl groups in the montmorillonite liberate protons which act like an acid:

$CaCO_3 + 2H^+ \rightleftarrows Ca^{++} + H_2O + CO_2$.

In the above reaction calcium (or magnesium) ions presumably interact immediately with the montmorillonite to form a different mineral. By the nature of the fractionation factors obtained for the siliceous calcite, it appears that a similar reaction holds; however, here the quartz acts as the acid (ref. 15):

$CaCO_3 + SiO_2 \overset{H_2O}{\rightleftarrows} CaSiO_3 + CO_2$

The reason why dolomite, which also has high amounts of quartz, acts more like the relatively pure limestone may be related to its grain size and the intimacy of contact between the quartz and the carbonate. Microscopic examination reveals that the grains of the minerals in the dolomite rock are relatively uniform in size and are much larger than the size of the quartz grains in the siliceous calcite. Virtually no grain size can be recognized in the carbonate minerals of siliceous calcite. Since the acid dissociation occurs in a solid phase, the contact area between the acid and the carbonate will determine which process--acid dissociation or heat dissociation--is dominant. There is a very intimate contact between the montmorillonite and the synthetic calcite and between the different minerals of the siliceous calcite; therefore, acid dominated fractionation occurs. In the dolomite rock the contact area is limited and, hence, heat dissociation dominates.

GEOCHEMICAL IMPLICATIONS

As was mentioned previously, the carbon isotopic composition of CO_2 in high CO_2 gases cannot be explained by a simple metamorphism of pure carbonates. The variations in the isotopic composition of pure limestone (or dolomite) and the CO_2 under high temperature metamorphic conditions can, however, be described by a set of Rayleigh distillation equations for a situation where CO_2 is generated from limestone under isotope equilibrium conditions. Figure 2 is a graphical solution of the Rayleigh equations for a marine limestone with an isotope composition of 0.0°/°° giving off CO_2 through a fractionation factor of +3°/°°. The three solid curves on Figure 2 describe the evolution of the isotopic composition of the limestone and the incremental and cumulative composition of the CO_2 as a function of the carbonate fraction metamorphosed. It can be seen that the limestone composition becomes progressively more negative as metamorphism proceeds and the composition of the incremental CO_2 follows this curve with a composition 3°/°° more positive. The isotopic composition of the cumulative CO_2 varies with the fraction from +3.0 to 0.0°/°°, provided that all of the CO_2 generated is also accumulated.

When, for some reason, the first fractions of the generated CO_2 are lost and the the carbon dioxide subsequently generated accumulates, the isotopic composition of the accumulating CO_2 will change (depending on the initial fraction lost) along the dashed curves on Figure 2. Situations, in which the initial carbon dioxide generated is lost, can be described by processes occurring between the migrating front of the CO_2 and the rock matrix forming the migration path. Carbon dioxide is a relatively reactive gas and several reactions can cause its fixation as carbonate minerals. Such reactions may occur with calcium and magnesium ions absorbed on clay minerals, or as Bray and Foster (ref. 16) maintain, minerals such as smectite and feldspar can rapidly react with CO_2 and water to form kaolinite, silica, and bicarbonate. In summary, this model requires high temperatures and high yields of CO_2 generation in the metamorphic process as well as substantial loss of CO_2 during migration. The new data reported in this paper suggest a somewhat different model for CO_2 generation and accumulation.

The suggested model allows an assemblage of fine-grain silicate-carbonate minerals to generate large quantities of CO_2 at relatively low metamorphic temperatures (~300°C). Evolution of the isotopic composition in this system also follows a Rayleigh distillation process. Achievement of isotopic equilibrium is the most important assumption in Rayleigh distillation processes, and this work has demonstrated that it is achieved in a short time. It should also be noted that the experiments in this study were not aimed at achieving chemical equilibrium between carbonate and CO_2 but rather isotopic equilibrium between the two. The rate at which chemical equilibrium is achieved is controlled by a slow solid state diffusion (ref.15). This is particularly true for the silicate containing carbonate. On the other hand, isotopic equilibrium is achieved much faster because it does not involve a solid state diffusion but rather a gas-solid interaction.

The evolution of the carbon isotopic composition of CO_2 and the residual carbonate in low temperature siliceous carbonate metamorphism is described in Figure 3. This figure describes the carbon isotopic composition of CO_2 and a siliceous carbonate of marine origin (i.e., the initial isotopic composition of the carbonate is 0.0°/°°). Based on Figure 1, a fractionation of -9°/°° was chosen (i.e., the CO_2 generated is 9°/°° more negative than the carbonate). Because of the negative fractionation, the residual carbonate becomes isotopically more positive (as metamorphism progresses), and the incremental CO_2 follows this trend with a constant difference of 9°/°°. The cumulative CO_2, however, changes from -9°°° at the beginning of the process and ends at 0.0°/°° when all the carbonate has been metamorphosed. When the process is interrupted in the middle, intermediate isotope values are observed as described on Figure 3. In contrast to the previous model for pure carbonate metamorphism, this model requires insignificant losses of the CO_2 during migration and does not require high yields of carbonate metamorphism in order to produce carbon isotope ratios of CO_2 similar to those often observed in high CO_2 gases.

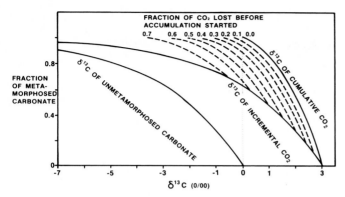

Figure 2. Evolution of the isotopic composition of calcite and CO_2 in a high temperature metamorphic system.

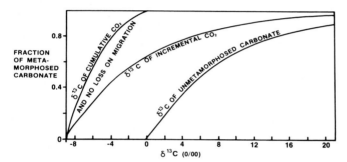

Figure 3. Evolution of the isotopic composition of carbonate and CO_2 in a low temperature (300°C) siliceous carbonate matamorphic system.

CONCLUSIONS

This work has demonstrated that the carbon isotope fractionation factors between CO_2 and carbonates greatly depend on the presence of quartz and clay minerals. The data also suggest that the grain size of the noncarbonate minerals incorporated with the carbonate is important in determining the fractionation factor: pure carbonates and carbonates containing coarse-grain quartz show fractionation factors with a trend similar to the theoretical values; fine-grain carbonates mixed with fine-grain silicates show more negative fractionations and exhibit a linear relation with temperature.

Based on the experimentally observed fractionation factors, natural high CO_2 gases can be genetically related to low temperature metamorphism of fine-grain carbonates containing fine-grain silicates and clay minerals. They can also be related to high temperature pure carbonate metamorphism with CO_2 loss on migration.

ACKNOWLEDGMENTS

The authors would like to thank Cities Service Company, Energy Resources Group, for supporting this investigation; Drs. C. Sutton and J. E. Zumberge for helpful comments and suggestions; Mr. C. F. Schiefelbein for isotope analyses, Miss A. Kinsaul for drafting, and Mrs. M. Draughon for typing the text.

REFERENCES

1. W. B. Land, J. Geophys. Research, 64 (1959) 127-131.
2. R. E. Zartman, G. J. Wasserburg and J. H. Reynolds, 1961, Contri. Mineralogy and Petrology, 26 (1961) 61-198.
3. U. Colombo., F. Gazarrini, R. Gonfiantini, G. Sironi, and E. Tongiorgi, in G. D. Hobson and M. C. Louis Advances in Organic Geochemistry, 1964, Pergamon Press, Oxford, 1966, pp. 279-292.
4. C. Panichi and E. Tongiorgi, in 2nd U.S. Symposium on the Development and Use of Geothermal resources, 1966, pp. 815-815.
5. K. A. Kvenvolden, K. Weliky, and H. Nelson, Science, 205 (1979) 1264-1266.
6. R. G. Pankina, V. L. Mekhtiyeva, S. M. Guriyeva and Y. N. Shkutnik, Int. Geology Review, 21 (1979) 535-539.
7. Y. Bottinga, Geochim. et Cosmochim. Acta, 33 (1969) 49-64.
8. P. Baertschi, Schweizer. Mineralog. u. Petrog. Mitt., 37 (1957) 73-152.
9. J. C. Vogel, Heidelberg Univ. Ph.D thesis, 1959, 196 p.
10. D. A. Northrop and R. N. Clayton, J. Geology, 74 (1966) 174-196.
11. K. Emrich, D. H. Enhalt, and J. C. Vogel, Earth and Planetary Science Letters, 8 (1970) 363-371.
12. J. M. McCrea, Jour. Chem. Physics, 18 (1950) 849-857.
13. H.,Craig, Geochim. et Cosmochim. Acta, 12 (1957) 133-149.
14. S. M. F. Sheppard, and H. P. Schwarez, Contri. Mineralogy and Petrology, 26 (1970) 161-198.
15. S. J. Kridelbaugh, Amer. Jour. of Science, 273 (1973) 757-777.
16. E. E. Bray and W. R. Foster, Am. Assoc. of Pet. Geol. Bull., 64 (1980) 107-114.

ISOTOPE GEOCHEMICAL MODELS OF THE EVOLUTION OF THE EARTH'S CRUST

K. WETZEL

Central Institute of Isotope and Radiation Research of the Academy of Sciences of the German Democratic Republic, Leipzig (DDR)

ABSTRACT

Investigations of the isotopic composition of oxygen, silicon, strontium and lead in their natural abundances provide evidence for a model of the evolution of the earth's crust with the following main features: Apart from periodical fluctuations the rate of plate tectonics has been constant during the earth's history. Plate tectonics probably has not only been constant during the earth's history with respect to its rate but also according to its chemism. The successive accumulation of silicic rocks on the continents is mainly due to the fact that the oceanic crust descending in the trench zones is somewhat less silicic than the crust ascending in the mid-ocean ridges. The most important primary products of this interchange between ascending and descending oceanic crust probably are the volcanic rocks of the subduction zones.

INTRODUCTION

The evolution of the earth's crust can be understood on the basis of the modern concept of plate tectonics. Following this concept we have to consider the outer shell of the earth, the lithosphere, to consist of about a dozen rigid plates that move with respect to each other. The edges of these plates are mainly either mid-ocean ridges, where new lithosphere is created by the up-welling and crystallization of magma from the earth's interior, or subduction zones, where old lithosphere is pushed down into the earth's mantle. Such continuous creation and destruction of oceanic crust gives rise to sea-floor spreading and continental drift as well as to continental growth. As far as continental growth is concerned, we should particularly look for an explanation for the transformation of an originally basaltic crust into the recent crust consisting of gabbro-basaltic (lower crust) and granitic (upper crust) rocks.

In spite of the great progress in understanding the earth's history since the late 1960's we have to keep in mind that there are a lot of open or controversial questions concerning the dynamics and the mechanisms of crustal evolution.

In the first part of the present paper we will analyse the situation of plate tectonics from a geochemical point of view. In the second part we will show how isotope geochemistry can contribute to answering some of the open questions.

As far as the dynamics of crustal growth is concerned, we have to decide between the following alternatives:
- fast growth during the early history of the earth's crust
  (type I (ref. 1))
- continuous growth during the entire history of the earth's crust
  (type II (refs.2-6))
- maximum growth during a period between 3×10^9 and $2,5 \times 10^9$ years ago
  (type III (refs. 7-10))
- growth accelerating in the course of the earth's history (type IV
  (ref. 11)).

Evaluating the different types of crustal growth we have to take into consideration that all kinds of transistions between these types are possible. We will assume that continental growth follows type II which will be superimposed by some periodical fluctuations of orogenic activity.

As far as the mechanism of crustal evolution is concerned most geologists agree with the concept that continental growth takes place mainly in the subduction regions, where the oceanic crust is trenched by the mantle. But there are controversial ideas about the source of the material accreted to the continents (Fig. 1).

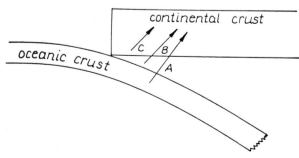

Fig. 1. Mechanism of the evolution of the earth's crust
(type A (refs. 12-13) type B (refs. 2-3) type C (refs. 8,14))

Independent of any particular assumption about the source material of continental growth there is one severe problem: Unlike basaltic rocks granitic rocks which are typical of the continental crust do not enter roots extending into the mantle. If we assume mantle material to be the source material for continental growth (type B in Fig. 1), two additional problems arise: We cannot explain why continental growth is limited to subduction zones and we can hardly answer the question why silicic rocks are increasingly accumulated on the continents instead of a decreasing accumulation of such rocks as a consequence of a depletion of the mantle wedge with respect to silicic material. If we try to avoid answering the question about the source material (type C in Fig. 1), we are unable to understand why continental growth is limited to subduction zones and we are faced with an additional severe problem: The stoichiometry of magmatic differentiation demands a very much higher excess of simatic rocks than is observed in the earth's crust.

Faced with this situation we consider the oceanic crust descending into the mantle, the Benioff zone, to be the source of continental growth (type A in Fig. 1). Partial melting of this material gives rise to the ascension of silicic or intermediate magmas into the subduction zones of the continents and leaves behind large amounts of basaltic material with a lower content of SiO_2 as compared with the tholeiitic basalts ascending at the mid-ocean ridges, thus fulfilling the stoichiometric condition mentioned above.

During the last two or three years we have made some efforts in developing isotope geochemical models to get a deeper insight into the history of the earth's crust on the basis of the above mentioned considerations. Within the framework of these investigations we have obtained interesting results on the isotope geochemistry of oxygen, silicon, strontium and lead. Further investigations of this kind include the isotope geochemistry of elements such as nitrogen, hydrogen and sulphur.

Consequences of the isotopic composition of oxygen in rocks and minerals

The isotopic composition of oxygen in magmatic and metamorphic rocks and minerals depends mainly on the chemical composition and on the temperature of the formation of the rocks and minerals (refs. 15-18). Particularly on the basis of Schütze's incremental method (ref. 19) we were able to calculate the equilibria of oxygen isotope exchange between magmas and their solidification products even at temperatures occuring at the Mohorovičič discontinuity between the earth's crust and the mantle which are about 900 K in the area of the oceanic crust and about 1100 K in the area of the continental crust.

We assume that the original ancient crust had the same chemical and isotopic composition as the tholeiitic basalts ascending at the recent mid-ocean ridges which can be described by the formula $Al_{0.58}[Si_{3.42}Al_{0.58}O_{11.33}]_K$ (ref. 20). The recent crust is represented by the formula $Al_{0.42}[Si_{3.31}Al_{0.69}O_{10.24}]$. (The Al within the bracket is at tetrahedron centres, the Al outside the brackets at octahedron centres.)

Thus we can describe the geochemical evolution of the earth's crust by the chemical equation

$$\text{original crust} \qquad\qquad \text{ascending crust}$$
$$Al_{0.58}[Si_{3.42}Al_{0.58}O_{11.33}] + r\, Al_{0.58}[Si_{3.42}Al_{0.58}O_{11.33}]$$

$$\text{recent crust} \qquad\qquad \text{descending crust}$$
$$\rightarrow \frac{Q_{UK} + O_{OK}}{Q_{UK}} Al_{0.42}[Si_{3.31}Al_{0.69}O_{10.24}] + s\, Al_z[Si_xAl_{4-x}O_y] \qquad (1)$$

Q_{UK} and Q_{OK} are the amounts of oxygen in the original and in the recent upper crust, respectively (ref. 20). The coefficients r and s describe the rate of crustal evolution: x, y and z characterize the (unknown) chemical composition of the descending crust.

As we have shown element geochemistry supplies 4 equations for the determination of 5 unknown values (r, s, x, y and z): the aluminium balance, the silicon balance, the oxygen balance and the tetrahedron balance, the last one following from the postulate that the number of SiO_4- and AlO_4-tetrahedra has been constant during the earth's history.

The missing fifth equation has to be supplied by isotope geochemistry. Using the global mean values of $\delta^{18}O$ in magmatic, parametamorphic and sedimentary rocks and in the hydrosphere (ref. 20), we can establish an ^{18}O-balance and are able to calculate the coefficients r and s in the chemical equation (1) describing the geochemical evolution of the earth's crust and the indices x, y and z in the chemical formula of the descending crust:

$$\text{Al}_{0.58}\underset{\text{original crust}}{\left[\text{Si}_{3.42}\text{Al}_{0.58}\text{O}_{11.33}\right]} +14.3\ \text{Al}_{0.58}\underset{\text{ascending crust}}{\left[\text{Si}_{3.42}\text{Al}_{0.58}\text{O}_{11.33}\right]}$$

$$\rightarrow 1.54\ \text{Al}_{0.42}\underset{\text{recent crust}}{\left[\text{Si}_{3.31}\text{Al}_{0.69}\text{O}_{10.24}\right]} +13.8\ \text{Al}_{0.60}\underset{\text{descending crust}}{\left[\text{Si}_{3.43}\text{Al}_{0.57}\text{O}_{11.45}\right]} \quad (2)$$

Equation (2) shows the descending crust to be only a little less silicic than the ascending one. Just this small difference between the chemical composition of the ascending and the descending crust, however, is the driving force of the evolution of the earth's crust, in the course of which the upper continental crust with its silicic rocks as well as the ^{18}O excess of the earth's crust over the ^{18}O of the mantle was created.

If we assume that the replacement of the crust refers to the oceanic and the lower continental crust as a whole, we calculate 250 million years for one of these cycles (ref. 20). In the (more probable) case of a replacement of the oceanic crust only we get about 150 million years. This is in rather good agreement with the results on recent seafloor spreading.

Thus we conclude that the evolution of the earth's crust has proceeded in the same way as we observe now at the mid-ocean ridges and in the subduction zones both with respect to the chemism and with respect to the rate.

Consequences of the isotopic composition of silicon in rocks and minerals

In order to confirm our conclusions we tried to apply our model to the isotope geochemistry of silicon (ref. 20). The following empirical equation allows us to calculate the δ^{30}Si values of the ascending (δ^{30}Si = -0.777) and the descending crust (δ^{30}Si = -0.813):

$$\delta^{30}\text{Si} = 0.85\ (n + 0.5\ m)\ \frac{2\ [4\ (n+m) - \Sigma]}{n\ (n+m)} - 2.93 \quad (3)$$

In this equation n, m and Σ denote the numbers of Si, Al (at the tetrahedron centres) and O atoms in the chemical formula of the silicate. The δ^{30}Si value of the recent crust calculated on the basis of our model using equation (3) agrees rather well with experimental results of Epstein and Taylor and of Tilles on the isotopic composition of silicon in granitic and in basaltic rocks (refs. 21-23).

Consequences of the isotopic composition of strontium in rocks and minerals

The isotopic composition of strontium in rocks and minerals depends upon the ratio of the concentrations of Rb and Sr, the age t and the origin of the rock or the mineral. If the initial isotopic ratio of strontium is denoted by $(^{87}Sr/^{86}Sr)_o$, the recent isotopic ratio of strontium is described by

$$(^{87}Sr/^{86}Sr)_r = (^{87}Rb/^{86}Sr)_r (e^{\lambda t} - 1) + (^{87}Sr/^{86}Sr)_o \qquad (4)$$

where $(^{87}Rb/^{86}Sr)_r$ denotes the recent ratio of ^{87}Rb and ^{86}Sr in the rock or the mineral, $\lambda = 1.4 \times 10^{-11}$ $[a^{-1}]$ is the decay constant and t the age of the sample. Because of $\lambda \cdot t \ll 1$ equation (4) can be simplified:

$$(^{87}Sr/^{86}Sr)_r = (^{87}Rb/^{86}Sr)_r \cdot \lambda \cdot t + (^{87}Sr/^{86}Sr)_o \qquad (5)$$

An analysis of the isotopic composition of Sr in granitic rocks of different ages from all over the world should lead to a deeper understanding of continental growth. In accordance with our picture of the dynamics of crustal growth we will assume h_o to be the constant amount of silicic rocks annually ascending from the lower crust or from the mantle into the upper crust. Because of equation (5) the initial isotopic ratio $(^{87}Sr/^{86}Sr)^j_\tau$ of these granitic rocks is determined by the time τ which has elapsed since the formation of the earth's crust about 3.8×10^9 years ago and by the $^{87}Rb/^{86}Sr$ ratio in the mantle $(^{87}Rb/^{86}Sr)_M = 0.0706$ (ref. 7):

$$(^{87}Sr/^{86}Sr)^j_\tau = (^{87}Rb/^{86}Sr)_M \cdot \lambda \cdot \tau + 0.699 \qquad (6)$$

We emphasize that equation (6) is valid independent of the particular mechanism of the formation of these granitic rocks. Validity of equation (6) means either crystallization from a magma which stems directly from the mantle or palingenesis of a material which ascended from the mantle less than 100 or 150 million years ago.

In the course of crustal evolution a second process giving rise to the formation of granitic rocks has gained importance, namely, the conversion of silicic rocks or their weathering products into younger granitic rocks which can be distinguished by their elevated initial $^{87}Rb/^{86}Rb$ ratios. This process is usually called reworking. We are probably right in assuming the rate of this reworking to be proportional to the amount of granitic rocks which has already been accumulated during crustal evolution. Therefore t years after the formation of the amount h_o the amount h

$$h = h_o \cdot e^{-\eta \cdot t} \qquad (7)$$

is left over, where η denotes a conversion coefficient.

With these two assumptions only (constancy of h_o and proportionality between reworking and amount of material to be reworked) we get the following connection between the isotopic ratio $(^{87}Sr/^{86}Sr)_\tau$ of granitic rocks and the time τ after the formation of the earth's crust:

$$(^{87}Sr/^{86}Sr)_\tau = \frac{1}{1+\eta\tau} \left[(^{87}Rb/^{86}Sr)_M \cdot \lambda \cdot \tau + 0.699 \right]$$
$$+ \frac{\eta\tau}{1+\eta\tau} \left[(^{87}Rb/^{86}Sr)_M \cdot \lambda \cdot \frac{\tau}{2} + (^{87}Rb/^{86}Sr)_G \cdot \lambda \cdot \frac{\tau}{2} + 0.699 \right]$$
$$= \frac{1}{1+\eta\tau} (0.988 \cdot 10^{-12} \tau + 0.699)$$
$$+ \frac{\eta\tau}{1+\eta\tau} (4.05 \cdot 10^{-12} \tau + 0.699) \qquad (8)$$

where $(^{87}Rb/^{86}Sr)_G = 0.508$ denotes the ratio of ^{87}Rb and ^{86}Sr in granitic rocks.

There are two independent ways to calculate the conversion coefficient η in equation (8). Equation (7) describes the distribution of the global abundances of granitic rocks as a function of their age t. Thus we can calculate η by fitting equation (7) to the experimentally determined age distribution resulting in $\eta = 4.3 \times 10^{-10}$. A second way can be founded on the hypothesis that all granitic rocks are palingenetic. Provided this is true and taking into account that the global reservoir of sediments has been approximately constant during the last 2×10^9 or 3×10^9 years η should be equal to the ratio of the sedimentation rate ($km^3 \, a^{-1}$) and the total amount of silicic rocks including sediments (km^3):

$$\eta = \frac{0.733 \, km^3 \, a^{-1}}{2 \cdot 10^9 \, km^3} = 3.7 \cdot 10^{-10} \, a^{-1},$$

using Jaroshevskij's value for the mean sedimentation rate during the last 1.5×10^9 years (ref. 24).

Fig. 2 shows the global mean values of the initial isotopic ratios $(^{87}Sr/^{86}Sr)_\tau$ of 208 granitic rocks from all continents as a function their age (refs. 2,3,7,11,12) in comparision with the curve calculated from equation 8 (using $\eta = 3.7 \times 10^{-10}\ a^{-1}$). The figures indicate the number of samples investigated by the different authors. Fig. 2 shows satisfactory agreement between the experimental results on the isotopic composition of strontium in granitic rocks and our model of the evolution of the earth's crust. Particularly there is no systematic deviation from the calculated curve which could give rise to any doubt about the continuous dynamics of continental growth.

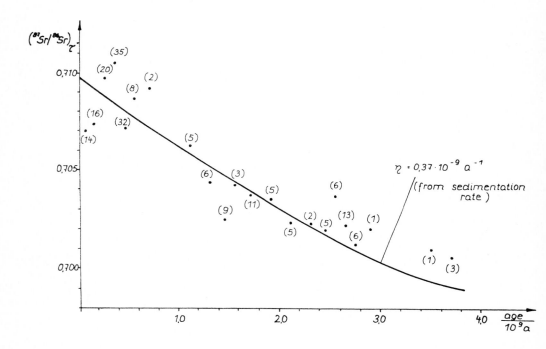

Fig. 2. Comparision of experimentally determined and calculated initial isotopic ratios $(^{87}Sr/^{86}Sr)$ of granitic rocks as a function of their age

But the isotope geochemistry of strontium can also help to answer the question about the source material of continental growth. It is evident that this source material should show low initial $^{87}Sr/^{86}Sr$

ratios as they are typical for material derived from the mantle. It is also evident that this source material should be characterized by approximately identical initial $^{87}Sr/^{86}Sr$ ratios in subduction zones all over the world. Indeed the majority of all volcanic rocks typical of trench zones (andesites and dacites in Andes-type trench zones or andesites, rhyolites and tonalites in island arc trench zones) show just this behaviour.

Our model of crustal evolution does give an unconstrained explanation of the isotopic composition of strontium in andesites, dacites and rhyolites. The mean value of the $^{87}Sr/^{86}Sr$ ratios in tholeiitic basalts of mid-ocean ridges was determined to be $(^{87}Sr/^{86}Sr)_B$ = 0.7026. If we assume a residence time of 200 million years for the oceanic crust, we can calculate the $^{87}Sr/^{86}Sr$ ratio of the mentioned volcanic rocks $(^{87}Sr/^{86}Sr)_v$ using the following equation (compare equation (5)):

$$(^{87}Sr/^{86}Sr)_v = (^{87}Rb/^{86}Sr)_o \cdot \lambda \cdot 200 \cdot 10^6 + (^{87}Sr/^{86}Sr)_B = 0.7034 \quad (9)$$

$(^{87}Rb/^{86}Sr)_o = 0.286$ denotes the $^{87}Rb/^{86}Sr$ ratio of basalts. This is in satisfactory agreement with the experimentally determined mean $^{87}Sr/^{86}Sr$ ratio of 0.7037.

Consequences of the isotopic composition of lead in rocks and minerals

Considering the fundamental character of the question about the mechanism of crustal growth we tried to utilize data on the isotopic composition of lead in volcanic rocks of subduction zones for verifying our picture of the evolution of the earth's crust, using equation of the same type as equation 9.

Except $^{206}Pb/^{204}Pb$ in andesites and dacites of Andes-type regions there is satisfactory agreement between theory and experiment giving further support to the evidence of our picture of the mechanism of continental growth.

The results are summarized in table 1.

TABLE 1

Experimentally determined lead isotope ratio of tholeiitic basalts and volcanic rocks of subduction zones in comparision with values calculated from our model of continental growth. (The figures in brackets indicate the number of experimentally investigated samples (refs. 25-26).)

	tholeiitic basalts	andesites, dacites and rhyolites of island arc regions	andesites and dacites of Andes-type regions	volcanics of subduction zones
	(experimental values)	(experimental values)	(experimental values)	(calculated values)
$^{206}Pb/^{204}Pb$	18.34 (6)	18.48 (31)	18.87 (25)	18.47
$^{207}Pb/^{204}Pb$	15.57 (6)	15.68 (31)	15.59 (25)	18.58
$^{208}Pb/^{204}Pb$	38.08 (6)	38.69 (31)	38.54 (25)	38.33

CONCLUSIONS

Investigations of the isotopic composition of selected elements in magmatic, metamorphic and sedimentary rocks and in the hydrosphere (oxygen, silicon), in granitic rocks (strontium) and in volcanic rocks of subduction zones (strontium, lead) give a good deal of evidence for a new model of the evolution of the earth's crust with the following main features:

1. The earth's crust is formed by differentiation of material from the mantle.

2. Plate tectonics already began in the Archaean, probably about 3.8×10^9 years ago.

3. Apart from periodical fluctuations the rate of plate tectonics has been constant during the earth's history.

4. In the course of plate tectonics the original continental crust, mainly consisting of basaltic rocks, was transformed into the recent crust, consisting of basaltic as well as of silicic rocks.

5. In the course of plate tectonics the crust has grown more laterally than vertically. So far our investigations have confirmed the widely, but not commonly accepted results obtained by a great number of geologists, geochemists and petrologists during the last

10 or 15 years (ref. 3). Apart from these statements our investigations lead to some more or less new conclusions:

6. The successive accumulation of silicic rocks on the continents is mainly due to the fact that the oceanic crust descending in the trench zones is somewhat less silicic than the crust ascending in the mid-ocean ridges.

7. The most important primary products of this interchange between ascending and descending oceanic crust probably are the volcanic rocks of the subduction zones.

8. Plate tectonics probably has not only been constant during the earth's history with respect to its rate but also concerning its chemism.

REFERENCES

1 R.L. Armstrong, Rev. Geophys., 6 (1968) 175-200.
2 C.I. Allègre and D.B. Othman, Nature, 286 (1980) 335-341.
3 S. Moorbath, Chemical Geology, 20 (1977) 151-187.
4 E.W. Pawlowskij, Iswestija Akad. Nauk SSSR, Ser. geologiceskaja, 1, (1970) 23-39.
5 R.K. O'Nions, R.J. Hamilton and N.M. Evensen, Scientific American, 242 (1980) 91-101.
6 A.I. Tugarinow and E.W. Bibikowa, in:Die Trennung der Elemente und Isotope bei geochemischen Prozessen, Verlag Nauka, Moskau, 1979, 5-14.
7 G. Faure and J.L. Powell, Strontium Isotope Geology, Springer Verlag, Berlin - Heidelberg - New York, 1972.
8 A.I. Tugarinow, Z. Angew. Geol., 14 (1968) 568-578.
9 S.M. McLennan and S.R. Taylor, Nature, 285 (1980) 621-624.
10 J. Veizer and W. Compton, Geochim. Cosmochim. Acta, 40 (1976) 905-914.
11 M. Hurley and I.R. Rand, Science, 164 (1969) 1229-1242.
12 W.B. Dickinson, Rev. Geophys. Space Phys., 8 (1970) 813-860.
13 D. Bridgewater, V.R. McGregor and J.S. Meyers, Precambrian Research, 1 (1974) 179-197.
14 E.C. Kraus, Die Entwicklungsgeschichte der Kontinente und Ozeane, Akademie-Verlag Berlin, 1971, 27-29.
15 G.D. Garlick, Earth Planet. Sci. Lett., 1 (1966) 361-368.
16 H.P. Taylor and S. Epstein, Geol. Soc. Am. Bull., 73 (1962) 675-694.
17 K. Wetzel, D. Mißbach and K. Mühle, ZfI-Mitteilungen, 30 (1980) 168-180.
18 K. Wetzel, ZfI-Mitteilungen, 29 (1980) 13-47.
19 H. Schütze, Chem. Erde, 39 (1980) 321-334.
20 K. Wetzel and H. Schütze, Chem. Erde, 40 (1981) 58-67.
21 S. Epstein and H.P. Taylor jr., Science, 167 (1970) 533-536.
22 D. Tilles, J. Geophys. Res., 66 (1961) 3003-3013.
23 D. Tilles, J. Geophys. Res., 66 (1961) 3015-3020.
24 A.A. Jaroshevskij, in Die Trennung der Elemente und Isotope bei geochemischen Prozessen, Verlag Nauka, Moskau, 1979, 15-34.
25 S.E. Church and G.R. Tilton, Geol. Soc. Am. Bull., 84 (1973) 431-454.

THE ISOTOPIC FRACTIONATION DURING THE OXIDATION OF CARBON MONOXIDE BY HYDROXYL-RADICALS AND ITS IMPLICATION FOR THE ATMOSPHERIC CO-CYCLE

H.G.J. SMIT, A. VOLZ, D.H. EHHALT, and H. KNAPPE
Institut für Chemie 3: Atmosphärische Chemie, Kernforschungsanlage Jülich GmbH, Postfach 1913, 5170 Jülich (F.R.G.)

ABSTRACT

The fractionation of $^{13}C/^{12}C$ for the reaction of CO and OH was investigated over a pressure range of 100-780 Torr. At high pressures the reaction was found to favour ^{12}CO by 6 $^{o}/oo$, while at low pressures the inverse effect was observed.

From the results, the relative contribution of the CH_4 oxidation to the global sources of CO is estimated to be about 30 % in excellent agreement with results from the global budget of $^{14}CO/^{12}CO$.

INTRODUCTION

Investigations by Stevens et al. [1] and Ehhalt and Volz [2] indicated, that atmospheric CO is enriched in ^{13}C compared to its known sources. It was postulated, that this enrichment must be due to a fractionation during the oxidation of CO with OH radicals, which is the major sink for atmospheric CO.

In this paper, measurements of this fractionation are presented. The results are used to uptake the ^{13}C balance in atmospheric CO.

EXPERIMENTAL

The fractionation of ^{13}C during the reaction of CO with OH radicals was investigated in a stationary photolysis experiment. A schematic diagram of the apparatus used is given in Fig. 1.

The OH radicals were produced in a stainless steel reaction chamber by photolysis of H_2O with radiation from a discharge through a mixture of J_2 and He. This light source emitts strong resonance lines of the J-atoms in the VUV region, which are absorbed by H_2O leading to the dissociation reaction (1).

$$H_2O + h\nu \ (\lambda < 1860 \ \text{Å}) \rightarrow OH + H \qquad (1)$$

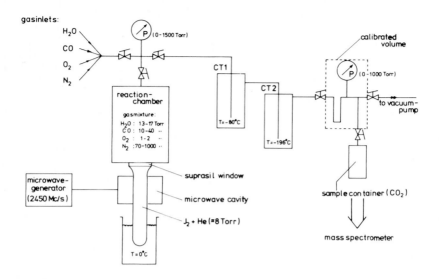

Fig. 1. Schematic diagram of the photolysis apparatus.

The experiment was performed in the presence of 17 Torr H_2O, 1 Torr O_2, and 10-40 Torr CO. The total pressure in the photolysis chamber was varied in several experiments by addition of N_2 between 100 and 780 Torr.

Under these conditions the OH radicals produced react almost exclusively with CO through reactions (2) and (3).

$$CO + OH \rightarrow CO_2 + H \qquad (2)$$
$$CO + OH + O_2 \xrightarrow{M} CO_2 + HO_2 \qquad (3)$$

O_2 was added to scavenge the H-atoms produced and to simulate atmospheric conditions for the reactions (2) and (3). The reaction of CO + OH was found to be pressure dependent only in the presence of O_2 [3].

After an irradiation time of 3-5 hours, a few percent of the CO were oxidized to CO_2, which was then separated from the mixture by pumping it slowly through two cold traps (CT 1 and CT 2). CT 1 was chilled with dry ice and served to retain the H_2O while CT 2 was chilled with liquid N_2 to collect the CO_2. The total amount of CO_2 produced was determined in a calibrated volume. It was eventually transferred to a mass spectrometer for the determination of its ^{13}C content. Prior to the photolysis experiments the ^{13}C content in the CO used had been determined after quantitative oxidation on J_2O_5 at 150°C.

RESULTS

The fractionation, ε_{CO}, is described by equation (4), where ^{12}k and ^{13}k, respectively, are the different rate coefficients for the reaction of ^{12}CO and ^{13}CO with OH.

$$\varepsilon_{CO} = \left(\frac{^{12}k}{^{13}k} - 1\right) \qquad (4)$$

It was calculated according to equation (5) from the ^{13}C-contents in the initial CO, δ_{CO}, and in the CO_2 formed, δ_{CO_2}, as a function of the fraction of the CO, F, which was oxidized to CO_2.

$$\varepsilon_{CO} = \frac{\ln(1 - F)}{\ln(1 - (1 + \delta_{CO_2} - \delta_{CO})F)} - 1 \qquad (5)$$

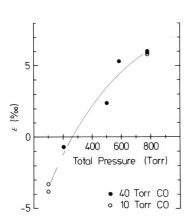

Fig. 2. Observed pressure dependence of the $^{12}C/^{13}C$ fractionation ε_{CO} during the reaction of CO with OH.

The results for the fractionation ε_{CO} found in several experiments at different total pressures are plotted in Fig. 2. As can be seen, a well defined pressure dependence is observed, namely at high pressure ^{12}CO is found to react faster by 6 °/oo while at low pressure ^{13}CO reacts faster by 3 °/oo. In addition, the partial pressure of CO was varied in several experiments. However, no significant influence on the fractionation was observed. The reproducibility of the whole analytical procedure was ± 0.1 °/oo.

The results presented here are in excellent agreement with a recent independent study [4] with respect to both, magnitude and pressure dependence of the observed $^{12}CO/^{13}CO$ fractionation. These authors used a different photochemical system to produce the OH radicals as well as slightly different experimental conditions.

DISCUSSION

With the results presented here for the $^{12}C/^{13}C$ fractionation during the reaction of CO with OH, which provides the main sink for atmospheric CO, we shall try to reinvestigate the global ^{13}CO balance.

The global cycle of CO is shown in Figure 3 as far as it is of interest for the $^{13}C/^{12}C$ budget. The magnitudes of the various sources and sinks were taken from a recent investigation of the $^{14}CO/^{12}CO$ budget [5]. They refer to the respective annual fluxes of ^{12}CO. Also listed are the ^{13}C-content of the different reservoirs and the measured or estimated fractionation for the various processes involved [1,2,6,7].

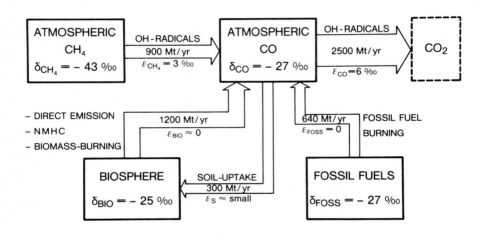

Fig. 3. The global budget of atmospheric ^{12}CO and ^{13}CO. The numbers in the arrows are the annual ^{12}CO fluxes [5]. The $^{13}\delta$ values are from [1,2,6]. The fractionation ε_{CH_4} is from [7], ε_{CO} is the high pressure value from this paper, the others are estimated to be negligible.

As can be seen, the ^{13}C-content in atmospheric CO is with -27 $^o/oo$ close to that of the majority of its sources (fossil fuel and biomass burning and biological sources). However, through the oxidation of atmospheric CH_4 a much lighter fraction of CO is formed (-46 $^o/oo$ if the fractionation of 3 $^o/oo$ is taken into account). Thus, from its combined sources atmospheric CO should exhibit a somewhat lower ^{13}C-content than actually found [1].

CO is removed from the atmosphere almost exclusively (~ 90 %) by reaction with OH [5]. This reaction is most effective in the lowest 2 km of the troposphere, as a simple argument shows:

The destruction term, D_{OH}, is given by equation (6).

$$D_{OH} = [CO] \cdot [OH] \cdot k_{CO} \qquad (6)$$

All three terms on the right hand side of (6), namely the concentrations of CO and OH as well as the rate coefficient for the reaction of CO with OH decrease with altitude. The product, D_{OH}, therefore, decreases even more rapidly: the scale-height for the destruction term is calculated to be on the order of 2 km. Hence, for the fractionation during the destruction, ε_{CO}, we have to adopt the value found here for high pressure (600-760 Torr), namely $\varepsilon_{CO} \sim 6\ ^o/oo$ (Fig. 2). This corresponds to a faster removal of ^{12}CO from the atmosphere, thus leading to an enrichment of ^{13}C in atmospheric CO.

We can use the ^{13}CO balance to estimate the relative contribution of CH_4, P_{CH_4}, to the total sources of CO. For this estimate, we lump all these sources, which exhibit about the same ^{13}C-content, namely the contributions from fossil fuels and from the biosphere, P_{com}, together to a common source. Furthermore, we can neglect the small sink provided by soil uptake. The global balance equation of atmospheric CO then becomes

$$P_{CH_4} + P_{com} \simeq D_{OH} \qquad (7)$$

For ^{13}CO we have to multiply each term in equation (7) with its respective ^{13}C-content, in addition we have to account for fractionation. In the δ-notation eq. (7) then becomes for ^{13}CO:

$$P_{CH_4} \cdot (\delta_{CH_4} - \varepsilon_{CH_4}) + P_{com} \cdot \delta_{com} = D_{OH} \cdot (\delta_{CO} - \varepsilon_{CO}) \qquad (8)$$

Elimination of D_{OH} gives:

$$\frac{P_{CH_4}}{P_{CH_4} + P_{com}} \equiv \frac{P_{CH_4}}{\Sigma P} = \frac{\delta_{CO} - \varepsilon_{CO} - \delta_{com}}{\delta_{CH_4} - \varepsilon_{CH_4} - \delta_{com}} \qquad (9)$$

With the numbers given in Figure 3 and adopting $\delta_{com} = -26\ ^o/oo$ for a mean value, the relative contribution of P_{CH_4} required to balance the ^{13}CO cycle is calculated to amount to 30 %.

As can be calculated from the fluxes given in Figure 3, this result is in excellent agreement with independent investigations of the global CO-budget from ^{14}CO measurements [5]. Due to the observed variability in the ^{13}C-content of the sources involved, our estimate cannot be seen as an independent reinvestigation of the global cycle of atmospheric CO. However, it clearly demonstrates that the ^{13}C-content of atmospheric CO is in agreement with our present knowledge of its sources and sinks.

REFERENCES

1 C.M. Stevens, L. Krout, D. Walling, A. Venters, A. Engelkemeir and L.E. Ross, Earth Planet. Sci. Lett., 16(1972)147-165.
2 D.H. Ehhalt and A. Volz, Proc. Symp. on Microbial Production and Utilization of Gases, Akademie d. Wiss. Göttingen, 1975, Goltze Druck, Göttingen, 1976, pp.23-33.
3 H.W. Biermann, C. Zetsch, F. Stuhl, Ber. Bunsenges. Phys. Chem., 82(1978) 633-638.
4 C.M. Stevens, L. Kaplan, R. Gorse, S. Durkee, M. Compton, S. Cohen and K. Bielling, Int. J. Chem. Kin. 12(1980)935-948.
5 A. Volz, D.H. Ehhalt and R.G. Derwent, J. Geophys. Res., in press (paper 1C0083) (1981).
6 J. Levin, Diplomarbeit, Universität Heidelberg (1978).
7 C.M. Stevens, F. Rust, Int. J. Chem. Kin. 12(1980)371-377.

MODELLING OF STABLE ISOTOPE COMPOSITION OF ATMOSPHERIC WATER VAPOUR AND PRECIPITATION

K. ROZANSKI[+], K.O. MÜNNICH and C. SONNTAG
Institut für Umweltphysik, Universität Heidelberg, Heidelberg /F.R.G./

ABSTRACT

The results of numerical modelling of the stable isotope composition of atmospheric waters /water vapour and precipitation/ are presented. A multibox cloud model is proposed to explain the vertical distribution of deuterium content in atmospheric water vapour. A simple model explaining seasonal and spatial variations of the stable isotope ratios in present day European precipitation is also presented. The calculations show that isotopic composition of local precipitation is primarily controlled by regional scale processes, i.e. by the water vapour transport patterns into the continent and by the average precipitation-evapotranspiration history of the air masses precipitating at a given place.

INTRODUCTION

The stable isotope composition of atmospheric waters /water vapour and precipitation/ is a subject of interest for over twenty years. Numerous applications of stable isotopes to hydrology, climatology and glaciology require thorough understanding of the processes primarily responsible for the observed variations in the stable isotope composition of atmospheric water vapour and precipitation.

Contrary to relatively good evidence of the isotopic composition of precipitation /1/, the relevant data for atmospheric water vapour are scarce till now. Systematic studies of vertical deuterium distribution in atmospheric vapour have been carried out by Taylor /2/ and Ehhalt /3/. Recently Hübner et al. /4/ have undertaken systematic measurements of deuterium content in the atmospheric moisture near the ground.

+permanent address: Institute of Physics and Nuclear Techniques, 30-059 Kraków, Poland.

The aim of the present study was to gain better understanding of the processes controlling the stable isotope composition of atmospheric moisture and precipitation. With simple numerical models we try to explain the observed variations in stable isotope composition of atmospheric waters.

DEUTERIUM IN ATMOSPHERIC WATER VAPOUR

Fig. 1 summarizes the results of measurements of deuterium content in tropospheric water vapour, published by Taylor [2] and Ehhalt [3]. The Fig. 1a shows the average vertical deuterium profiles evaluated on the basis of the data taken from the papers cited above. These profiles are considered to be representative for the average climatological conditions over a given sampling region. Fig. 1b illustrates the time variation observed in deuterium isotopic composition of atmospheric vapour close to the average cloud base level, as derived from the vertical deuterium profiles published by Taylor [2] and Ehhalt [3]. Despite the considerable scatter of the data points, the seasonal variation /isotopically more depleted water vapour in winter/ is obvious.

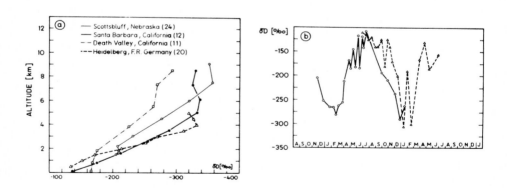

Fig. 1. (a) - average vertical deuterium distribution in the tropospheric water vapour over mid-latitude regions. Numbers in parenthesis indicate the number of sampling flights over the given area. (b) - seasonal variations in deuterium isotopic composition of atmospheric moisture at the 1500 m altitude. The experimental points derived from the vertical deuterium profiles measured by Taylor [2] and Ehhalt [3].

For describing the isotopic evolution of the water vapour in an isolated ascending air mass several different models can be used. Neither the simple Rayleigh distillation model /which is basically one-phase model/, nor two-phase models considering the formation of a liquid phase in the system, do account for the observed vertical slope of deuterium distribution in tropospheric moisture /cf. Fig. 3/. Also the isotope model of Eriksson [5], which is based on a general circulation model of atmospheric vapour, provides considerably less steep deuterium profiles than observed ones. Isotope exchange between falling raindrops as well as ascending cloud droplets and water vapour in developing cloud systems has been postulated by Ehhalt and Östlund [6] to explain this discrepancy. For better discussion of this process, we have developed a simple numerical model.

Fig. 2 illustrates the basic features of this multibox cloud model. The cloud is represented by a vertical multibox system operating as a counterflux distillation column. Air with given initial parameters enters the bottom part of the system. The isotope exchange process between vapour, cloud droplets and raindrops acts within each box /see the right side of Fig. 2/. The isotopic composition of the vapour and the liquid phase in each box is calculated according to the two-phase model for adiabatically ascending separate air mass with rainout, proposed recently by Merlivat and Jouzel [7]. This model is an extended version of the known Craig-Gordon model [8]. Details of our multibox model will be published elsewhere [9].

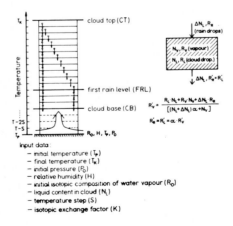

Fig. 2. Schematic diagram of the multibox cloud model /see text for details/.

Reliable comparison between model prediction and experimental data has only been possible for the average vertical deuterium profile representing Taylor`s data /Fig. 3/. Only in that case the input parameters needed to run the model can be assessed with sufficient accuracy. Having defined the input parameters, we compared the predictions of different models used previously with the response of the proposed multibox model. This comparison includes the Rayleigh condensation model /10/, the two-phase model without rainout /11/, the two-phase model with rainout /7/ and the multibox cloud model briefly presented above. As can be seen from Fig. 3, the multibox model reproduces sufficiently well the experimental data, whereas all other models do not give acceptable agreement. This seems to confirm the role of isotope exchange in producing the observed vertical gradient of deuterium in atmospheric water vapour.

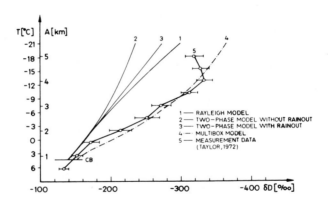

Fig. 3. Comparison between average vertical deuterium distribution in the tropospheric water vapour over continental Europe /2/ and the predictions of different models. CB - cloud base level.

STABLE ISOTOPE COMPOSITION OF MODERN EUROPEAN PRECIPITATION

An attempt to model the observed variations of deuterium content in monthly precipitation collected by the European stations of the IAEA/WMO network has been undertaken with the aim to understand in more detail the well established empirical relations between isotopic composition of precipitation and various environmental parameters. Two distinct regularities become obvious from the meteorological and isotope data:
a/ the correlation between deuterium content in monthly precipitation and local surface air temperature /seasonal temperature effect/, becoming increasingly stronger towards the center of the continent,

b/ the continuous depletion of deuterium in monthly precipitation averages with increasing distance from the Atlantic coast /continental effect/.

The continental effect, as observed in summer and winter precipitation in Europe, is shown in Fig. 4. The points are long-term δD mean values for ten selected stations with the longest deuterium record. The resulting inland gradient of deuterium content amounts to −3.3‰/100 km in winter and −1.3‰/100 km in summer /linear fit/. A significantly lower inland gradient for the winter season /−2.4‰/100 km/ if compared with summer data was already suggested by a previous estimate [12,13], based on smaller data set.

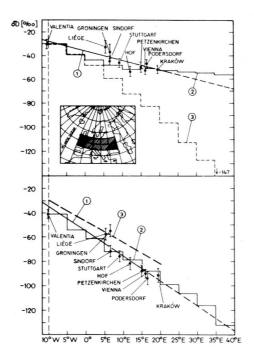

Fig. 4. Upper part: continental effect observed in deuterium isotopic composition of European summer precipitation /April-September/. ①- best linear fit of the data points, ②- model prediction obtained with 35% contribution of winter precipitation to summer evapotranspiration, ③- model prediction without evapotranspiration. Lower part: continental effect observed in deuterium isotopic composition of European winter precipitation /October-March/. ①- best linear fit of the data points, ②- model prediction for the estimated mean value of the winter water vapour flux entering the continent of 88 kg/m·s, ③- continental effect observed in shallow European groundwaters [16].

The numerical approach briefly presented here is based on the fact that atmospheric circulation over European continent has essentially zonal character. Available data for the global water vapour transport in the atmosphere clearly show that the meridional component of this transport over Europe, is practically absent in winter and negligibly small in summer [14]. One, therefore, needs to consider only the water vapour flux entering Europe across its western boundary. In the model the part of the European continent in question has been aproximated by a rectangular field with corners at: $55°N, 10°W$; $55°N, 40°E$; $45°N, 10°W$; $45°N, 40°E$ /see Fig. 4/. This field has been subdivided into 10 boxes, each 300 x 900 km in size /$5°$ longitude, $10°$ latitude, respectively/. The mean values of precipitation and of actual evapotranspiration rates as well as of the surface air temperature in each box have been evaluated on the basis of the published data [14,15].

The basic concept in the here attempted model reproduction of isotope variations observed in precipitation is that the marine water vapour entering the continent undergoes a Rayleigh process [5,13,17], in which the increasing loss of vapour from the air mass by precipitation /minus evapotranspiration/ makes the vapour increasingly depleted in the heavier isotope species. Details of this numerical approach will be published elsewhere [18]. Model predictions of the average continental effect for the summer and winter season, respectively, are presented in Fig. 4. The same calculations performed for each individual month separately further yields a prediction of the seasonal temperature effect at a given place. The comparison between the observed and calculated temperature effect for Vienna Station is shown in the Fig.5.

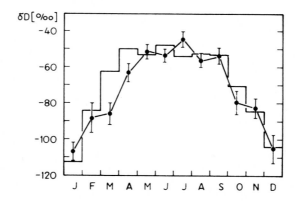

Fig. 5. Observed and predicted seasonal temperature effect in deuterium isotopic composition of Vienna precipitation.

The model provides, despite its rough character, an acceptable prediction of the observed variations in isotopic composition of precipitation over the main part of the European continent. The histograms of the continental effect produced by the model generally nicely reproduce the experimental data. Model predictions for the summer months reveal an important role of evapotranspiration for the observed inland gradient of deuterium content, especially for the more continental areas.

Line 3 in the lower part of Fig. 4 represents the continental effect of deuterium content in European shallow groundwaters as measured by Sonntag et al. $\sqrt{16\,\_\!\!/}$. The position of this line suggests that both summer and winter precipitation contribute to groundwater formation in Europe. A rough estimate based on Fig. 4 shows, for example, that for the region 0° - 10°E the contribution of summer precipitation to the total yearly infiltration reaches some 40%. Long-term isotopic lysimeter studies in this area $\sqrt{19\,\_\!\!/}$ yield also a figure of about 40%. The role of summer precipitation in groundwater formation seems, however, to be smaller for more continental areas. This is probably due to the higher air temperatures and lower precipitation rates in these regions.

CONCLUSIONS

The numerical models briefly presented in previous sections provide a better understanding of the processes primarily responsible for the observed variations of stable isotope ratios in atmospheric moisture and in precipitation.

Despite a very complex character of the general atmospheric circulation, the vertical distribution of heavy isotopes in atmospheric water vapour seems to be primarily controlled by convection-type processes like cloud formation. Turbulent vertical mixing apparently does not significantly disturb this distribution. The predictions of the multi-box model seem to suggest that the isotope exchange process between gaseous and liquid phases within developing cloud systems plays an important role in establishing the observed vertical isotope gradient in atmospheric moisture.

The horizontal transport of water masses over the continents results in the rainout process, which further causes a continuous depletion in heavy isotope content of both atmospheric water vapour and precipitation. This depletion is significantly diminished during the summer months due to evapotranspiration. From the model calculations carried out for the Europe region it becomes obvious that stable isotope composition of local precipitation is controlled by regional scale processes like the water vapour flux variations in the atmosphere, the evaporation conditions from the ocean, and the precipitation-evapotranspiration rates over the whole distance to the sampling point. If one, therefore, tries to predict the isotope composition of local precipitation, the whole history of precipitating air masses should be known. This conclusion is of importance for the application of stable isotopes as a paleoclimatic indicator preserved in ice-cores and ancient groundwaters. If major climatic fluctuations in the past were associated with dramatic changes of atmospheric circulation, it may be that isotopic variations preserved in ice-cores and groundwaters will have a quite different relation to the local temperature than extrapolated from the present situation. Therefore, any attempts in quantitative interpretation of stable isotope ratios in ancient precipitation should be preceded by careful examination of possible changes of atmospheric circulation in the past.

REFERENCES
1. IAEA, World survey of isotope concentrations in precipitation. Technical Report Series,/1969,1970,1971,1973,1975,1979/.
2. C.B. Taylor, Report INS-R-107, Lower Hutt, New Zealand,/1972/.
3. D.H. Ehhalt, NCAR Technical Note, NCAR-TN/STR-100,/1974/.
4. H. Hübner, P. Kowski, W.D. Hermichen, W. Richter and H. Schütze, Proc. Int. Symp. Isotope Hydrology 1978, Neuherberg, June 19-23, IAEA, Vienna,/1979/, pp.289-307.
5. E. Eriksson, Tellus,17/1965/, pp.498-512.
6. D.H. Ehhalt and G.G. Östlund, J.Geophys.Res.,75/1970/, pp.2323-2327.
7. L. Merlivat and J. Jouzel, J.Geophys.Res.,84/1979/, pp.5029-5033.
8. H. Craig and L.I. Gordon, Int. Conf. Stable Isotopes in Oceanographic Studies and Paleotemperatures, Spoleto,/1965/, pp.9-129.
9. K. Rozanski and C. Sonntag, submitted for publication.
10. W. Dansgaard, Tellus,16/1964/, pp.436-467.
11. J. Jouzel,L. Merlivat and E. Roth,J.Geophys.Res.,80/1975/,pp.5015-30.
12. C. Sonntag, P. Neureuther, Chr. Kalinke, K.O. Münnich, E. Klitzsch and K. Weistroffer, Naturwissenschaften,63/1976/, pp.479.
13. C. Sonntag, E. Klitzsch, E.M. Shazly, Chr. Kalinke and K.O. Münnich, Geolog.Rundsch.,67/1978/, pp.413-423.
14. UNESCO, World water balance and water resources of the earth. Studies and reports in hydrology, UNESCO Press,/1978/.
15. CLINO, Climatological normals for climat and climat ship stations for the period 1931-1960, WMO/OMM Report No.117,/1962/.
16. C. Sonntag, U. Thorweihe, J. Rudolph, E.P. Löhnert, Chr. Junghans, K.O. Münnich, E. Klitzsch, E.M. Shazly and F.M. Swailem, 2nd Arbeitstagung Isotope in der Natur, Leipzig, November 5-9,/1979/.
17. E. Salati, A. Dall`Olio, E. Matsui and J.R. Gat, Water Res.Research 15/1979/, pp.1250-1258.
18. K. Rozanski, C. Sonntag, and K.O. Münnich, submitted for publication.
19. B. Blavoux, Thesis, Université Pierre et Marie Currie, Paris,/1978/.

THE STABLE ISOTOPIC COMPOSITION OF INFILTRATION MOISTURE IN THE UNSATURATED ZONE OF THE ENGLISH CHALK

A.H. BATH, W.G. DARLING and A.P. BRUNSDON*

Institute of Geological Sciences (Hydrogeology Unit) and *Institute of Hydrology, Maclean Building, Wallingford, Oxon, OX10 8BB, U.K.

ABSTRACT

The oxygen and hydrogen isotopic compositions of interstitial moisture removed from cored profiles of unsaturated chalk at two sites are presented. For one site these are compared with the compositions of drainage from an undisturbed lysimeter and of daily collections of rainfall. The variations of isotopic compositions with depth in the unsaturated zone are discussed in the context of mechanisms proposed for unsaturated zone moisture movement.

INTRODUCTION

The mechanism by which water traverses the unsaturated zone to recharge the underlying chalk aquifer has been a subject of debate: both rapid movement through fissures and very slow interstitial transfer have been proposed. The nature of this mechanism is fundamental to the interpretation of concentration profiles of thermonuclear tritium through the unsaturated zone in terms of net infiltration rates [1-3]. More recently, concern over the accumulation of pollutants, principally agricultural nitrate, in the unsaturated zone and the rate of their downwards migration has placed further significance in the understanding of infiltration mechanisms [4]. Some early measurements of stable isotopic composition of soil water were combined with deuterated and tritiated water tracing experiments. These led to the establishment of the hypothesis of 'piston-displacement' with dispersion and diffusive mixing for soil moisture movement [5-6]. The variations in natural oxygen and hydrogen isotopic compositions of unsaturated zone moisture are being investigated to obtain further information on the infiltration mechanism in chalk.

LOCATIONS

The investigation comprises 3 sites on chalk from which unsaturated zone moisture samples were recovered. These are located in Hampshire and Dorset in southern England, and Cambridgeshire in eastern England. The results from the

Hampshire site are described in a separate paper at this symposium [7] and are not discussed here.

The Cambridgeshire chalk was sampled at the Fleam Dyke research site in 1979 where 2 cored boreholes were drilled below arable land (FD prefix) as well as a further borehole below undisturbed grass 6 km away (GM prefix). The soft, weathered chalk is overlain by 25-35 cm of brown soil; a hard band of chalk at about 1.5-2 m depth is thought to have an important role in the infiltration characteristics of the profile (D.Cooper, personal communication). The moisture content of the chalk was typically 20-30% by weight at the time of sampling, and the minimum depth to saturated water table conditions at the FD site is about 15-16 m, and considerably more at the GM site.

An undisturbed lysimeter has been in operation for several years under permanent grass at the FD site. This is artificially sealed at the sides and base, the latter at about 5 m depth, such that drainage through it should represent natural conditions of infiltration [8]. Samples of drainage water were recovered from the base of the lysimeter by means of trench access. Infiltration, mostly during winter months, is 150-200 mm per year. Daily collections of rainfall at this site (average precipitation 550-600 mm/year) were also available for isotopic analysis.

The Dorset chalk was sampled at Gussage (prefix OG) in July 1979 at 2 sites less than one kilometre apart but with very different depths to saturated water table (about 15 m at OG 9A and more than double this at OG 10A). Both sites are on permanent rough grassland with about 30-50 cm of soil overlying chalk. The range of moisture contents is similar to that in Cambridgeshire chalk. This site has been sampled previously for tritium measurements which have suggested an average rate of downward movement of interstitial moisture of about 1 m/year. The problems of deriving quantitative estimates of infiltration from these tritium data have been discussed by Foster and Smith-Carington [3]. Estimates based on meteorological data suggest 400-450 mm infiltration per year out of an average 880 mm rainfall.

SAMPLING AND ANALYTICAL METHODS

Samples of unsaturated chalk were obtained by driving tubes and subsequently extruding the chalk core with a hydraulic ram. This dry sampling method avoids the risk of contamination by drilling fluid invasion. The interstitial water was extracted by centrifuging the crushed sample after removal of the outer rim of the core [9]; this was carried out mostly at a field laboratory as soon as possible after drilling. The collection of samples for stable isotope analysis was complementary to sampling for the study of the movement of nitrate and other solute species and tritium through the unsaturated zone. 7 ml aliquot samples of extracted solution were preserved in sealed containers for stable isotope analysis at the Wallingford laboratory. $^{18}O/^{16}O$ and $^{2}H/^{1}H$ determinations were made on CO_2 and H_2 gases respectively in a VG Micromass 602E mass spectrometer after conventional preparation techniques. The standard used in all δ^2H and $\delta^{18}O$ values in this study is the international reference sample SMOW (Standard Mean Ocean Water).

RESULTS

The $^{18}O/^{16}O$ and $^{2}H/^{1}H$ analyses of moisture extracted from 20 cm-spaced samples of the upper part of the unsaturated zone of Cambridgeshire chalk are shown in Fig. 1.

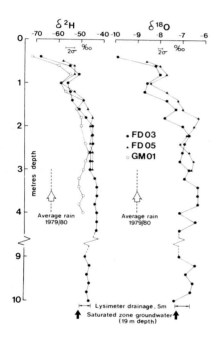

Fig. 1. $^{2}H/^{1}H$ and $^{18}O/^{16}O$ of moisture extracted from the unsaturated zone of Cambridgeshire chalk. The composition of drainage from a lysimeter at 5 m depth and of saturated zone groundwater at 19 m depth are also shown. The average composition of rain at the FD site for the period Nov. 1979-Oct. 1980 is from data shown in Fig. 2.

The average isotopic composition of rain at this site over the period November 1979-October 1980, calculated by taking a weighted mean from analyses of daily rainfall collections, is also shown on Fig. 1. The variation found in weighted weekly and monthly means of $^{2}H/^{1}H$ from the daily samples is shown in Fig. 2. This illustrates the large fluctuations in short-term averages and also the seasonal trend in rainfall isotopic composition from lighter in winter to

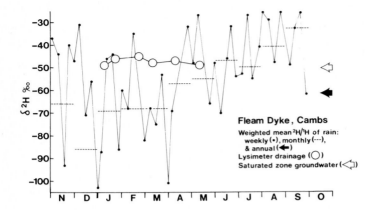

Fig. 2. ^{2}H/^{1}H of rainfall in Cambridgeshire; weighted mean values from analyses of daily collections. ^{18}O/^{16}O data shows correlated variations according to δ^2H = 8.1 δ^{18}O + 11.2.

heavier in summer. The compositions of lysimeter drainage (5 m depth) and saturated zone groundwater (approx 19 m depth) at the FD site are shown in Figs 1 and 2, and resemble each other and the composition of extracted moisture from the lower part of the sampled profile.

The isotopic compositions of unsaturated zone moisture in the two profiles of Dorset chalk are shown in Fig. 3.

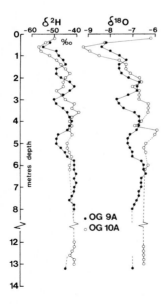

Fig. 3. ^{2}H/^{1}H and ^{18}O/^{16}O of unsaturated zone moisture of Dorset chalk.

COMMENT

The first feature to be noted from Figs. 1 and 3 is the close similarity between isotope profiles in each case - particularly in Fig. 1 where the FD and GM sites are 6 km apart, in different stratigraphic horizons of chalk, and have different vegetation characteristics. This suggests that these profile shapes may be generally representative of unsaturated zone moisture in chalk at these locations.

The isotopic composition of moisture in the **Cambridgeshire** chalk varies through the top 2 m, below which it is relatively uniform down to the water table with a composition significantly heavier than any moisture found in the top 2 m at the time of sampling (May, 1979). Therefore moisture in the shallow sub-surface must undergo mixing or modification resulting in a heavier isotopic composition before moving downwards below 2 m. Isotopic modification of moisture in the shallow sub-surface might occur as a result of evapotranspiration, although plant transpiration has been reported to cause no isotopic fractionation effects in residual soil moisture [5]. Any isotopic enrichment due to bare soil evaporation should be seen in this profile as a result of the antecedent winter; this also seems to be negligible in the sampled profile below 0.4 m. Mixing within a shallow sub-surface 'buffer zone' prior to downwards displacement has been proposed to explain the behaviour of tritium in the unsaturated zone of chalk [3]. A substantial input of isotopically-heavier water would be required to shift the overall composition of moisture within the upper zone of Fig. 1 to that found below 2 m. For example, the top 2 m at FD contained about 700 mm moisture when sampled in May 1979; this would require mixing with 480 mm of water with -30 $^o/oo$ $\delta^2 H$. Thus, the amount and composition (Fig. 2) of summer rain is insufficient to cause this shift <u>throughout</u> the upper zone. The mechanism by which the isotopic composition of soil moisture between, say, 1.5 and 2 m could undergo enrichment prior to downward movement is not yet understood; soil physics studies indicate that summer rainfall does not infiltrate directly to that depth because of the high soil moisture deficit developed very close to the surface. It is hoped to make progress towards resolving this problem by analysing profiles taken at different stages of the annual infiltration cycle. The average isotopic composition of rain at the FD site measured for 1979/80 is substantially lighter than unsaturated and saturated zone groundwater (Fig. 1); however this single annual cycle of rain so far measured may not be representative of long-term average composition and it is therefore uncertain how significant is this observation to the overall isotopic relationship between rain and groundwater. Continued monitoring of rainfall composition in addition to the further profiles described above will clarify this and its relevance to processes occurring in the upper unsaturated zone.

The profiles through Dorset chalk (Fig. 3) exhibit a cyclicity which probably reflects the seasonal variations in isotopic compositions of infiltrating rain; unfortunately no direct measurements of rain composition are available. The period of cylces - approximately 1 m - is in general agreement with estimates of annual infiltration from tritium displacement.

The different shapes of the profiles at the Cambridgeshire and the Dorset sites must reflect the contrasting rates of net infiltration (150-200 mm and 400-450 mm per year respectively, estimated from meteorological data). Thus the equivalent annual 'piston' displacements would be about 0.5 m and 1-1.5 m in the two locations. Diffusive mixing and consequent loss of seasonal cyclicity would be operative over a distance of about 0.2 m in the first year, 0.3 m in the second year and so on ($\alpha\ t^{\frac{1}{2}}$). Therefore the cyclic pattern would inevitably be lost by diffusive mixing more rapidly where infiltration rate is lowest. Conversely the preservation of seasonal isotopic cyclicity in the Dorset chalk moisture suggests that the vertical scale of the mixing zone or infiltration 'buffer', as inferred by Foster and Smith-Carington (1980) is considerably less than the annual displacement, i.e. less than 1 m.

In summary, the stable isotope profiles are consistent with piston-displacement movement occurring in unsaturated zone interstitial moisture of Dorset chalk, but indicate a process of mixing and possibly modification of isotopic composition in the shallow sub-surface of Cambridgeshire chalk. Further work aims to investigate seasonal changes in the isotopic composition of the upper zone and the significance of 'by-pass' water movement under varying conditions.

ACKNOWLEDGEMENTS

Colleagues at IGS and IH provided samples and discussed results - A. Smith-Carington, L.R. Bridge (IGS), S. Wellings and D. Cooper (IH). This paper is published with permission of the Directors of the Institute of Geological Sciences and of the Institute of Hydrology, Natural Environment Research Council, UK.

REFERENCES

1 D.B. Smith, P.L. Wearn, H.J. Richards and P.C. Rowe, in Isotope Hydrology, Proc. Symp. Vienna 1970. IAEA Vienna, 1970, pp. 73-86.
2 S.S.D. Foster, J. Hydrol., 25 (1975) 159-165.
3 S.S.D. Foster and A. Smith-Carington, J. Hydrol., 46 (1980) 343-364.
4 S.S.D. Foster and C.P. Young, Bull. Bur. Rech. Géol. Min., 3. (1979) 245-256
5 U.Zimmerman, D Ehhalt and K.O. Münnich, in Isotopes in Hydrology, Proc. Symp. Vienna 1966, IAEA Vienna, 1967, 567-584.
6 U. Zimmerman, K.O. Münnich and W. Roether, in Isotope Techniques in the Hydrologic Cycle, Geophys. Monograph Series 11, Amer. Geophys. Union, Washington, 1967, 28-36.
7 S.R. Wellings, this symposium, 1981.
8 R. Kitching and T.R. Shearer, in preparation, 1981.
9 W.M. Edmunds and A.H. Bath, Environ. Sci. Technol., 10 (1976) 467-472.

THE USE OF DEUTERIUM OXIDE (^2H_2O) AS A WATER TRACER IN A STUDY OF THE HYDROLOGY OF THE UNSATURATED ZONE OF THE ENGLISH CHALK

S.R. WELLINGS

Institute of Hydrology, Maclean Building, Wallingford, Oxon OX10 8BB, U.K.

ABSTRACT

This paper describes a field experiment in which soil water and solute fluxes are measured beneath grass and arable plots. Deuterium oxide is being used as a water tracer. The emphasis of the study is on the physics of water movement in the unsaturated zone, to understand the mechanism of water movement and to measure the downward velocity of solutes.

INTRODUCTION

The chalk is a soft Cretaceous limestone, varying in purity between 65 and 99 % $CaCO_3$. It is the major aquifer in England, supplying London and the surrounding region. In the last decade, the concentration of some solutes in the groundwater has been increasing; nitrate in particular has caused much concern, as it is a potential health hazard in domestic water at low concentrations [1]. Nitrate nitrogen concentrations in groundwater at some sites have exceeded the WHO advisory limit of 11.3 mg l^{-1}, considered the maximum for domestic supply. Higher concentrations than this have been found in the unsaturated zone of the Chalk at many sites in south and east England [2]. It is generally thought that some of this extra nitrate is derived from agricultural practices.

The mechanism of water movement through the unsaturated zone of the Chalk has been a subject for debate for a long time. The rock consists of microfossil fragments, the pores between being of a very uniform size distribution, with a mean diameter of between 1.0 and 1.5 μm. The rock is traversed by much wider (100 to 1000 μm) fissures. The spacing between fissures varies stratigraphically but is typically 0.05 to 0.2 m. Because of the fine pore size of the matrix it has been assumed that the hydraulic conductivity would be very low, and the water in the matrix would be essentially static. Recharge would then have to occur by water flow through the fissures. There is some evidence from the response of water table

recorders to rainfall [3] that this flow occurs under certain conditions. Lysimeter studies on undisturbed monoliths of soil and chalk [4] have also shown rapid leaching of solutes that could only occur by fissure flow. The nitrate content of the chalk unsaturated zone has been surveyed by core sampling and analysing the interstitial water [2]. Attempts to explain the observed vertical distribution of solutes in terms of agricultural management and climate, have so far been empirical. An understanding of the physics of water movement is fundamental to understanding solute movement, and this paper describes a field experiment designed to do this. The aims of the experiment are:
1. To measure unsaturated water fluxes in the upper 3m of the chalk.
2. To combine these data with nitrate concentration data in the interstitial water, to determine nitrate flux and velocity.
3. To measure the relationships between water potential, water content and unsaturated hydraulic conductivity to understand the physics of water movement.

METHODS

The experimental site is at Bridgets Experimental Husbandry Farm, owned by the Ministry of Agriculture Fisheries and Food, near Winchester, Hampshire. It is on Upper Chalk, the purest of the horizons, and the most extensive in outcrop in southern England. The soil is a rendzina consisting of 0.3 m of a silt loam overlying 1.0 m of cryoturbated chalk, with undisturbed chalk below 1.2-1.4 m depth. The water table varies between 38 and 42 m depth, and the mean annual rainfall is 790 mm.

A neutron probe was used to measure soil water content, and mercury manometer tensiometers and gypsum resistance blocks were used to measure soil water potential. Meteorological data from an adjacent site were used to calculate potential evaporation [5]. Samples of soil and chalk were removed every 8 weeks from beneath the plots with manual percussion coring tools, taking samples every 0.2 m from 0.1 to 2.9 m depth. Samples were centrifuged to extract interstitial water [6], and analysed for nitrate nitrogen and chloride. The relative deuterium abundance was measured by mass spectrometry. All abundances are quoted with respect to Standard Mean Ocean Water (SMOW). The experiment has been conducted in two phases. The first phase (1976-8) studied the leaching of nitrate from grass plots treated with animal slurry [7]. The second phase (1978 to date) studied spring barley treated with inorganic fertilizers. A tracer solution was applied to one arable plot in November 1979, containing chloride (500 mg l^{-1}) and deuterium oxide (+ 1000 D^o/oo) in the equivalent of 10 mm water in one hour. Core samples were taken within a 4 x 8 metre area from each plot, each profile being one metre laterally from the previous one. The sampling area was 3 metres away from the instrumented area. The relationship between water potential and unsaturated hydraulic conductivity was measured for 12 horizons between 0.2 and 3.0 m depth on one of the grass plots, using the steady-state infiltration and instantaneous profile methods [8].

RESULTS AND DISCUSSION

Wellings and Bell [7] concluded from phase I results that the predominant flow mechanism in the Upper Chalk at Bridgets EHF is through the fine pores of the matrix. There were two reasons for reaching this conclusion:
1. During winter when the predominant water movement in the unsaturated zone is downwards, matric water potentials below 1.4 m depth are normally between -0.3 and -0.5 bars. Fissure flow would be very unlikely to occur at such potentials, as higher potentials, between -0.03 and -0.003 bars, would have to occur for fissures of 0.1 to 1.0 mm wide to hold, and therefore conduct water.
2. Chloride concentration profiles beneath slurry-treated grass could be explained by piston displacement of the chloride derived from the slurry, which was applied from November to March. A cyclic profile could be explained by the downward movement of discrete peaks of chloride, which remained separate in the upper 3 metres.

Data from other Upper Chalk sites [9, 10], suggest that the Upper Chalk at Bridgets is representative of that elsewhere.

Results from phase II so far support the mechanism of water and solute flow through the fine pores of the matrix. In fig. 1 the relationship between the unsaturated hydraulic conductivity and matric water potential is shown for 2 horizons at Britgets. The conductivity varies between 6 and 2.5 mm day^{-1} at the matric potentials which normally occur during winter drainage (-0.3 to -0.5 bars) when fissures could not contain water. The mean winter daily drainage rate for the 2 winters of 1976-78 was 2.7 mm day^{-1} [7]. Although unsaturated soil water flow is an unsteady process, the mean daily rate could be conducted through the matrix. Larger inputs could be temporarily stored in the upper metre, which has a much wider pore size distribution and large water holding capacity at high potentials.

Darling and Bath [11] measured the $\delta^{18}O$ and $\delta^{2}H$-values of pore water from beneath the unfertilized grass plot of the phase I experiment, sampled on 6 December 1978. They found a cyclic pattern in the δ-values with depth; this has since been found at other chalk sites [12]. Further profiles sampled during that winter showed a downward movement of the whole $\delta^{2}H$-profile. Zimmermann et al. [13] also observed a cyclic variation in $\delta^{2}H$ of soil water, which indicated layered movement, or piston displacement, of water. Some preliminary $\delta^{2}H$-values from the arable plot which received the tracer solution are shown in fig. 2. The whole of the profile moves down during the drainage period, with $\delta^{2}H$-values in the top 0.2 m closely following the δ-values in rainfall. Nitrate nitrogen and chloride data show a similar downward movement for the same period.

The data from both Bridgets EHF and another of the Institute's sites on Middle Chalk near Cambridge show that both fissure and matrix flow can occur in the chalk. What needs to be known for water quality prediction is the dominant mode in a particular area. This emphasises the important role of the physics of water movement in understanding aquifer behaviour.

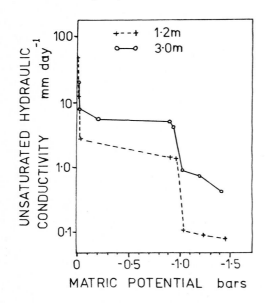

Fig. 1. Relationships between unsaturated hydraulic conductivity and matric potential for 2 depths of the Bridgets EHF Upper Chalk.

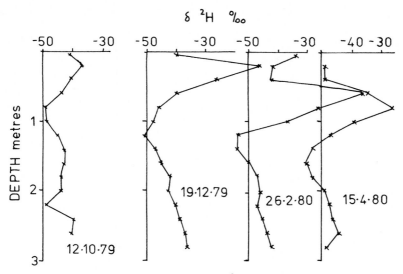

Fig. 2. Downward movement of the ^{2}H profile during the winter of 1979-80 beneath the arable plot to which tracer ^2H_2O was applied on 29th November 1979.

REFERENCES

1. WHO (Ed.), European Standards for drinking water. WHO, Geneva 1970, 2nd ed.
2. S.S.D. Foster and C.P. Young, Bull. BRGM 1979 No. 3, p.222.
3. H.G. Headworth, J. Inst. W. Eng., 26(1972)107-121.
4. E.R. Mercer, J.R. Burford, R.J. Dowdell, D. Hill, C.P. Webster and R.J. Thompson, Rep. ARC Letcombe Lab., 1977(1978)69-70.
5. H.L. Penman, Proc. Roy. Soc. A, 193(1948)120-146.
6. W.M. Edmunds, A.H. Bath, Env. Sci. Techn., 10(1976)468-472.
7. S.R. Wellings and J.P. Bell, J. Hydrol., 48(1980)119-136.
8. A. Poulovassilis, V.D. Krentos, Y. Stylianou and Ch. Metochis, in Proc. IAEA Symp. Isotope and Radiation Techniques in Soil Physics and Irrigation Studies, Vienna, 1973, pp.205-222.
9. W.O. Binns, W.H. Hinson, R. Kitching and D.F. Fourt, Rep. Forest Res., Forest Commission, 1964, pp.48-51.
10. A.H. Bunting and J.F. Elston, Sci. Hort., 18(1966)116-120.
11. W.G. Darling and A.H. Bath, IGS Stab. Is. Techn. Rep., 1(1979).
12. A.H. Bath, W.G. Darling and A.P. Brunsdon, in H.-L. Schmidt, H. Förstel and K. Heinzinger (Eds.), Proc. 4th Int. Conf. Stable Isotopes, Elsevier, Amsterdam, 1982.
13. V. Zimmermann, D. Ehhalt and K.O. Münnich, in Proc. IAEA-IUGG Symp., Vienna, 1967, pp.567-586.

$^{18}O/^{16}O$-RATIO OF GROUNDWATER AT THE FEDERAL REPUBLIC OF GERMANY

H. FÖRSTEL, and H. HÜTZEN

Institute of Radioagronomy, Nuclear Research Center (KFA),
P.O. Box 1913, 5170 Jülich, (F.R.G.)

ABSTRACT

About 900 well-distributed samples of the municipal water supply allowed us to construct a geographical pattern of the $^{18}O/^{16}O$-ratio of the groundwater and consequently of the precipitation at the area of the Federal Republic of Germany. From north to south a distinct decrease of the $^{18}O/^{16}O$-ratio may be explained by the consecutive precipitation and evaporation of rain and snow. Generally one observes a decrease of -3 °/oo per 1000 km distance from sea and of -0.44 °/oo per 100 m altitude above sea level.

INTRODUCTION

The natural variation of the stable isotope ratio of an element can be used as a tracer of the origin and the turnover of material in its natural cycle. Especially, the water circulation at the surface of the earth is a very rapid process, and it is accompanied by very large variations of the oxygen isotope ratio. The isotopic composition of the oceanic water does not vary markedly. Therefore, a sample of sea water is generally accepted as the international standard SMOW [1]. The water vapour, escaping from the surface of the ocean into the atmosphere, has a depleted $^{18}O/^{16}O$-ratio. Consequently, the rain has a depleted oxygen isotope ratio, too.

The local $^{18}O/^{16}O$-ratio of precipitation mainly depends on the geographical circumstances at the observation site. Its mean local value determines the oxygen isotopic composition of soil water, plant biomass and even air humidity, in summary the oxygen isotope ratio of the whole ecosystem at one place. Consequently the ^{18}O-content of any product from fresh plant biomass depends on the isotopic composition of rain and snow. Therefore the origin of vegetable products can be tested by the help of the isotopic composition of their water content. If one uses water of different origin to prepare the nutrient solution, one can use it as a tracer of transport processes within the plant also. Another idea arises from the correlation between the climate (and/or temperature) and the $^{18}O/^{16}O$-ratio of local precipitation [2]. The isotopic content of precipitation should be stored in the cellulose of the tree rings.

Average values of the oxygen isotope ratio of local precipitation usually are calculated from single samples, which have been collected during longer periods, at least over one year. The result of this procedure is sensitive to a variety of disturbancies. An international network of stations has been organized by the IAEA, which publishes the results regularly [3]. A global map of the results [4] shows the world-wide tendency: the $^{18}O/^{16}O$-ratio of precipitation decreases from the equator towards the poles. The network is not dense enough to inform about local differences at a single territory. Therefore another procedure was tested: the groundwater samples from the whole area of a country should reflect the average oxygen isotope ratio of precipitation at each location. Disturbancies from local effects should be eleminated by the large number of samples. The origin and the quality of the samples are ensured by the co-operation with the municipial waterworks. They have sent us only samples directly taken from a fountain or other parts of the factory itself.

METHODS
Collection of samples

About thousand public companies are responsible for the water supply of the Federal Republic of Germany. We asked all of them to collect samples for us, half of them answered. Each factory sent us usually two, sometimes more samples. Each sample was taken from a separate fountain or other water source. As a result we got samples of 900 well-distributed stations.

The sample bottles are usual plastic vials for liquid scintillation counting (25 ml, polypropylen, aluminium foil covered cap). The vessels are mechanically stable, easy to handle and tight enough. After the end of the measurements one sample of each series was choosen randomly and measured again. The mean of the first and the second measurement agreed within -0.09 ^{o}/oo (±0.32 ^{o}/oo standard deviation). No instrumental or other methodological shift between the two series was demonstrated. To avoid a systematical error, the sampleholders were mailed by an alphabetical list of stations, and measured randomly.

Sample preparation and measurement

The large number of samples necessitates a semi-automatic preparation. Twenty six samples and two laboratory standards were prepared simultaneously. One milliliter of sample water was equilibrated with gaseous carbon dioxide overnight at a constant temperature. Next morning the water was frozen out and the remaining gas was fed into the double inlet and double collector mass spectrometer micromass 602 (VG Isogas, Winsford, U.K.). Each sample was compared to an internal standard. Within each series two samples of a wellknown water standard (so-called JÜL I) indicated variations of the whole procedure. The results are reported as δ-values in permil, refered to the international standard Vienna-SMOW [1]:

$$\delta [°/oo] = \left(\frac{R_{sample}}{R_{standard}} - 1 \right) \cdot 10^3.$$

RESULTS

General pattern

The δ-values ($^{18}O/^{16}O$) in groundwater samples of the Fed. Rep. Germany show a decrease from the coast of Northern Germany towards the mountain regions of the South (fig. 1). The water of various regions can be distinguished by its different δ-values. At our station in Jülich we got the experience that the $^{18}O/^{16}O$-ratio of groundwater is a very time-constant value (table 1), and represents the mean oxygen isotope ratio of precipitation.

TABLE 1

δ-values of groundwater and precipitation from Nuclear Research Center Jülich

water source	year	δ-value
tap water	1974	- 7.65
groundwater 7 m fountain	1980	- 7.66
110 m		- 7.62
precipitation (weighted mean) [5]	1974-1977	- 7.42

Distance from sea

The decrease of the $^{18}O/^{16}O$-ratio from the coast towards the continental region can be seen in fig. 2. The whole area of Sleswick-Holstein and Northern Lower Saxonia is considered as a region with

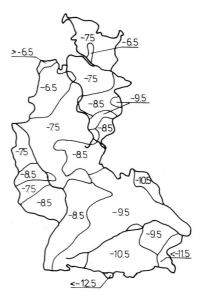

Fig. 1. $^{18}O/^{16}O$-ratio of about 900 groundwater samples, collected at the territory of the Federal Republic of Germany. The δ-values are related to Vienna-SMOW [1].

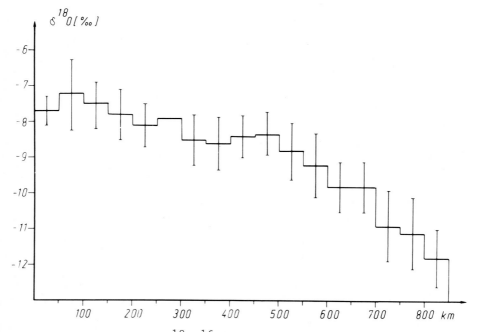

Fig. 2. Decrease of the $^{18}O/^{16}O$-ratio (δ-values) from coastal towards continental stations alongside a line from Emden (NW) to Berchtesgaden (SE). The standard deviation is indicated by a vertical line.

marine climate. The distance from the coast inside the land has been taken alongside a line from Emden (Eastern Frisia) to Berchtesgaden (Southern Bavaria). Its direction (NW to SE) represents the major influence of weather conditions from the North Atlantic Ocean into Central Europe. Our results indicate a decrease of the δ-value by the so-called "continental effect" of -3 °/oo per 1000 kilometers.

Effect of Altitude

Fig. 3 shows the decrease of the ^{18}O-content with increasing altitude. Single results are reported to demonstrate the scattering under field conditions. The large variation may be caused by the relief of the landscape. Going from North to South: The northern third or our country is covered by a low plain, the middle by a hilly region, and the southern third by an elevated plain. The alpine mountains influence only a small part alongside the southern border (km 700-800 in fig. 2).

Generally an increase of 100 m above sea level is accompanied by a decrease of the δ-value in groundwater: -0.44 °/oo per 100 m altitude, but recently we have not yet made a multicorrelation analysis for our data to eliminate local effects and the influence of the distance from the coast.

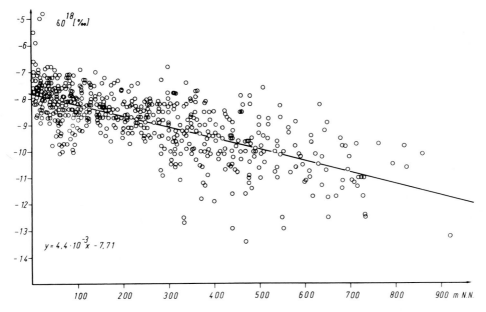

Fig. 3. Dependence of δ-value on altitude (-0.44 °/oo per 100 m above sea level) without correction of the distance from sea.

Special local effects

The results of two groups of stations show an aberration: stations close to the coast in Eastern Frisia (NW) and some stations along the river Rhine. The low δ-values of East Frisia probably indicate the influence of seawater. The second group of data represents water, which was produced as a filtrate at the banks of the river Rhine and originates mainly from the southern part of Germany with low δ-values (Bonn: -9.41, Koblenz-Oberwerth -9.73 and Koblenz-Metternich -7.85 $^o/oo$). The two values of Koblenz represent the mean ^{18}O-content of the rivers Rhine (Oberwerth) and Moselle (Metternich). This remark supports our idea, that a combination of a large number of stations and detailed informations about each source of water enable one to construct a general pattern of the $^{18}O/^{16}O$-ratio in groundwater and consequently of precipitation.

DISCUSSION

The decrease of the $^{18}O/^{16}O$-ratio from the northern towards the southern part of the Federal Republic of Germany can be explained as successive steps of precipitation and evaporation. The mean δ-value of precipitation at Jülich from 1974 to 1977 agrees well with the result of groundwater samples [5]. The δ-values of precipitation in Jülich as well as in Stuttgart (1962-1975: -8.2 $^o/oo$ [3]) seem to be slightly more positive. This small difference could be caused by the complete evaporation of rainfalls during the summer months. Instead of this small difference under our climatic conditions of Central Europe samples of groundwater may be considered as a longtime average of local precipitation, more representative than measurements of precipitations during short periods.

REFERENCES

1 H. Craig, Science, 133(1961)1833-1834.
2 W. Dansgaard, Tellus, 16(1964)436-467.
3 IAEA (Ed.), Technical Reports Series, 96(1969), 117(1970), 129(1971), 147(1973), 165(1975), 192(1979).
4 H. Förstel, A. Putral, G. Schleser and H. Lieth, in FAO/IAEA-Symposium on Isotope Ratios as Pollutant Source and Behaviour Indicators, IAEA Symp., 191(1975)3-20.
5 H. Förstel and H. Hützen, JÜL-Bericht 1524(1978), p. 26.

ACKNOWLEDGEMENT

The publication was supported by many colleagues and the Association of German Gas and Water Factories (Bonn).

OXYGEN-18 STUDY OF NON-LIQUID MECHANISMS OF ATMOSPHERIC SULFATE FORMATION

B. D. HOLT, P. T. CUNNINGHAM, A. G. ENGELKEMEIR, D. G. GRACZYK and R. KUMAR

Chemical Engineering Division, Argonne National Laboratory, 9700 South Cass Avenue, Argonne, Illinois, 60439, (U.S.A.)

ABSTRACT

Oxygen isotope ratios were measured in sulfates and in the SO_2 and water vapors from which they were formed to study the mechanisms of formation. The SO_2 and water vapors of different ^{18}O contents were rapidly isotopically equilibrated in a 3-liter glass chamber, followed by oxidation of the SO_2 to sulfate by four different methods. Oxidation was induced by a high-voltage electric discharge in the SO_2-air-water vapor mixture, by NO_2 addition, by gamma irradiation, and by adsorption on activated charcoal.

The $\delta^{18}O(SO_4^{2-})$-vs.-$\delta^{18}O(H_2O)$ relationships observed for these non-liquid transformations of SO_2 to sulfate, as well as those in aqueous-phase transformations previously reported (Fe^{3+}-catalyzed air oxidation, charcoal-catalyzed air oxidation, and H_2O_2 oxidation), were compared to sulfate in rain and snow collected at Argonne, Illinois. The $\delta^{18}O$ of sulfate in precipitation water was higher than might be expected by any of the investigated transformation mechanisms.

INTRODUCTION

In addition to aqueous-phase mechanisms of formation of sulfates from SO_2 which may occur in the atmosphere and on which we have reported isotopic results (ref. 1), we have pursued isotopic examinations of non-liquid, laboratory reactions which also may occur in the atmosphere. These reactions are (1) isotopic equilibration of SO_2 and water vapor (molar ratio, $H_2O/SO_2 \approx 31$) in air, in the absence of liquid water; (2) oxidation of the SO_2 in the SO_2-air-$H_2O(g)$ mixture to H_2SO_4 induced by a high-voltage spark, simulating atmospheric lightning (ref. 2); (3) oxidation induced by addition of NO_2, simulating atmospheric NO_x (ref. 3); (4) oxidation induced by gamma irradiation (ref. 4), simulating atmospheric production of OH radicals (ref. 5); and (5) oxidation on activated charcoal, simulating soot particles in the atmosphere (ref. 6).

The isotopic equilibration reaction between SO_2 and water in air and the oxidation reactions (except for the oxidation on charcoal) were carried out in a

3-liter glass chamber. In the oxidation reactions, the SO_2 was quantitatively converted to H_2SO_4. The equilibration reaction, the rate of which was not measured, went to completion in <0.5 h at 22°C; the oxidation reactions were essentially complete within 16 h at 22°C. The details of the experimental procedures are to be reported elsewhere.

RESULTS

Isotopic Equilibration of SO_2 and Water Vapor

In Fig. 1, the $\delta^{18}O$ of SO_2 is plotted versus the $\delta^{18}O$ of the water vapor to which it was exposed in two sets of experiments. The data in Curve A were obtained from experiments in which the water vapor was cryogenically separated at -79°C from the SO_2-air-$H_2O(g)$ mixture; in Curve B, the water was separated by adsorption on a desiccant (P_2O_5) at 22°C. The average $\delta^{18}O$ for five samples of the tank SO_2 used in these experiments was 13.7 ± 1.2‰; the $\delta^{18}O$ values obtained from two control runs (same procedure but with dry air) also averaged 13.7‰.

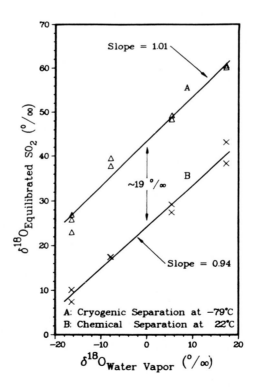

Fig. 1. Isotopic equilibration between SO_2 and water vapor in air.

These results strongly indicate very rapid equilibration between SO_2 and water vapor in air under the conditions imposed by the experiment. The slopes of approximately 1 indicate that the $\delta^{18}O$ of the oxygens in the equilibrated SO_2 was effectively established by the $\delta^{18}O$ of the water vapor to which it was exposed. The difference of $\sim 19\%_{oo}$ between the y-intercepts (43.4‰ and 24.2‰) is in approximate agreement with the difference in the thermodynamic fractionation factors that are calculable from spectroscopic data for the gaseous species, SO_2 and $H_2O(g)$, at the temperatures $\sim -79°C$ and $22°C$. This agreement could only be realized if the isotopic exchange between SO_2 and $H_2O(g)$ had been very rapid within the inlet zones of the dry-ice cold trap ($-79°C$) through which the gas mixture was drawn during evacuation of the 3.1-liter reaction chamber in the first set of experiments, Curve A.

Implications of these results are (1) that SO_2 in the atmosphere may undergo rapid equilibration with ambient water vapor, resulting in the dynamic control of the $\delta^{18}O$ of the SO_2 by the $\delta^{18}O$ of ambient water vapor, regardless of the $\delta^{18}O$ of the SO_2 at its point of origin; (2) that measurement of the $\delta^{18}O$ of SO_2 cannot, therefore, be used to determine its source of emission; and (3) that the $\delta^{18}O$ of secondary sulfates, formed in the atmosphere from SO_2 and water vapor, is affected primarily by the $\delta^{18}O$ of atmospheric water.

Oxidation of SO_2 in Air with High-Voltage Spark

Kinetic data, obtained by monitoring the temporal decline of SO_2 concentration in the SO_2-air-water vapor mixture in the 3.1-liter chamber after a relatively brief exposure to a high-voltage Tessla-coil spark, showed that the spark generated an oxidant which caused oxidation of the SO_2 to continue for several hours after the spark was terminated. In an experiment in which the air ($\sim 80\%$ N_2-20% O_2) was replaced by a gas mixture of 80% Ar-20% O_2, the oxidation of the SO_2 did not continue after termination of the high-voltage electric discharge. These kinetic data (to be reported elsewhere) strongly indicate that the oxidant produced in air was NO_2, probably participating in the oxidation of the SO_2 to H_2SO_4 through a chain reaction mechanism. The presence of NO_2 in the gas mixture after a 10-min exposure to the Tessla-coil spark was confirmed by mass spectrometric analysis and by the appearance of a slightly brownish color in the reaction chamber. Neither O_3 nor gaseous H_2O_2 were detected in the mass spectrum of the air mixture. The free radicals OH and HO_2 were not detectable by the mass spectrometer used. The depletion of SO_2 in each of the experiments was accompanied by a corresponding appearance of very finely divided droplets of H_2SO_4 on the inner walls of the chamber.

In Fig. 2, Curve C shows the $\delta^{18}O$ of sulfate, plotted versus the $\delta^{18}O$ of prevaporized liquid water. The slope, when corrected for an increase of $\sim 6\%$, caused by change in the $\delta^{18}O$ of each water vapor during equilibration with the SO_2, approximates the ratio of 3/4, indicating that the isotopy of three of the four oxygen atoms in the sulfate were controlled by the $\delta^{18}O$ of the water vapor. A mechanism of transformation of SO_2 to sulfate that is satisfied by this condition comprises isotopic equilibration between the SO_2 and $H_2O(g)$ (in which, as shown above, the two oxygens of SO_2 are isotopically controlled by the water vapor), followed by oxidation that incorporates one oxygen from water vapor, two oxygens from the water-vapor controlled SO_2, and one oxygen from the oxidant.

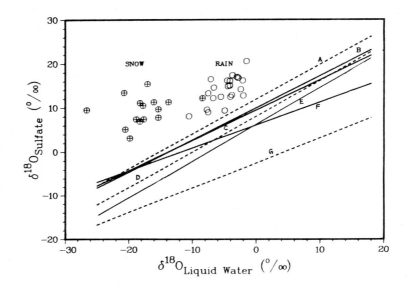

Fig. 2. Sulfates in precipitation water and sulfates formed by laboratory methods, represented by Curves A through G, as follows: A, aqueous air oxidation (Fe^{3+}-catalyst); B, γ-irradiation in humidified air; C, high-voltage spark in humidified air; D, aqueous air oxidation (charcoal catalyst); E, NO_2 addition in humidified air; F, non-liquid oxidation on charcoal; G, aqueous H_2O_2 oxidation.

Oxidation with NO_2

Because of the apparent involvement of NO_2 in the oxidation of SO_2 in the high-voltage spark experiments, NO_2 was added to the SO_2-air-water vapor mixture in the absence of a spark. The kinetic behavior of the reactants was essentially indistinguishable from that of the spark experiments.

To test for the possibility of gas-phase air oxidation of the SO_2 in high-humidity air without initiation by either electrical spark or NO_2 addition, the SO_2-air-water vapor mixture in one experiment was allowed to stand untreated at room temperature for three days. Daily analyses showed no decline in the SO_2 concentration and, therefore, no oxidation. On the third day, the chamber was chilled to produce condensation of water on the walls of the chamber. A slight decline in concentration occurred which may have indicated a small amount of aqueous air oxidation. Injection of NO_2 into the system on the fourth day was followed by a relatively rapid decline in SO_2 concentration.

The $\delta^{18}O(SO_4^{2-})$-vs.-$\delta^{18}O(H_2O)$ relationship obtained for the addition of NO_2 to SO_2 in humidified air is represented by Curve E in Fig. 2. The lack of complete agreement in oxygen isotopy for the two procedures (Curves C and E for high-voltage spark and NO_2 addition) may indicate non-identical sets of chain reactions (isotopic and/or chemical) leading to the formation of H_2SO_4.

Oxidation in Air with Gamma Irradiation

Mass spectrometric determinations of SO_2 concentrations before and after γ-irradiation revealed a decrease of ∼8% after a 60-min exposure and a decrease of >95% after a 1000-min (overnight) exposure. The 1000-min exposure was used in the four runs that are reported. The inner walls of the glass chamber were dry before irradiation and coated with a deposit of finely divided H_2SO_4 after irradiation. The $\delta^{18}O(SO_4^{2-})$-vs.-$\delta^{18}O(H_2O)$ relationship obtained in the irradiation experiments is shown as Curve B in Fig. 2.

Oxidation on Charcoal

Curve F, Fig. 2, represents results of experiments in which the SO_2 in a pre-equilibrated mixture of SO_2, air, and water vapor was adsorbed and oxidized on activated charcoal at 22°C.

Intercomparison of Sulfates

The $\delta^{18}O(SO_4^{2-})$-vs.-$\delta^{18}O(H_2O)$ relationships for sulfates in seven methods of preparation from SO_2 are plotted for intercomparison in Fig. 2. The broken lines represent previously reported aqueous-phase preparations (ref. 1); the solid lines represent the non-liquid preparations of this report. Listed in the order of decreasing $\delta^{18}O$ in the sulfates at $\delta^{18}O(H_2O) = 0$ (Curves A through G), the methods of preparation were Fe^{3+}-catalyzed aqueous-phase air oxidation, gamma irradiation in air, high-voltage spark in air, charcoal-catalyzed aqueous-phase air oxidation, NO_2 in air, oxidation on "dry" charcoal, and aqueous H_2O_2 oxidation.

Also plotted in Fig. 2 are the isotopic data obtained for samples of precipitation water (rain or snow) collected at Argonne, Illinois, during the period of October, 1976, to March, 1978 (ref. 7). The $\delta^{18}O$ of the sulfate dissolved in the precipitation water is plotted versus the $\delta^{18}O$ of the water. All of the precipitation samples contained sulfates that were higher in $\delta^{18}O$ than the sulfate formed by any of the mechanisms tested by the seven laboratory methods of preparation. It appears, therefore, that the precipitation sulfates were not formed solely by these mechanisms, either singly or in combination. Apparently, there was a component of highly enriched ^{18}O in the atmosphere which, when mixed with sulfates of lower ^{18}O content, as may be formed by one or more of the mechanisms listed above, gave the observed $\delta^{18}O$ values of field samples of precipitation sulfates.

ACKNOWLEDGMENT

This work was performed under the auspices of the U.S. Department of Energy.

REFERENCES

1 B. D. Holt, R. Kumar and P. T. Cunningham, Atmos. Environ., 15(1981)557-566.
2 W. L. Chameides, D. H. Stedman, R. R. Dickerson, D. W. Rusch and R. J. Cicerone, J. Atmos. Sci., 34(1977)143-149.
3 W. H. Schroeder and P. Urone, Environ. Sci. Technol., 12(1978)545-550.
4 S. Gordon and W. A. Mulac, Int. J. Chem. Kinet. Symp., 1(1975)289-299.
5 W. P. Wood, A. W. Castleman, Jr., and I. N. Tang, J. Aerosol Sci., 6(1975) 367-374.
6 T. Novakov, Lawrence Berkeley Laboratory, Berkeley, California, Report No. LBL-7887, 1978.
7 B. D. Holt, P. T. Cunningham and R. Kumar, Environ. Sci. Technol., in press.

III. BIOMEDICAL APPLICATIONS
A. PHARMACOLOGY AND DRUG METABOLISM

APPLICATIONS OF STABLE ISOTOPES IN PHARMACOLOGICAL RESEARCH

T. A. BAILLIE[1], H. HUGHES[2] and D.S. DAVIES[3]

1. Department of Medicinal Chemistry, University of Washington, Seattle, WA 98195, U.S.A.
2. Institute for Lipid Research, Baylor College of Medicine, Houston, TX 77030, U.S.A.
3. Department of Clinical Pharmacology, Royal Postgraduate Medical School, London W12 OHS, U.K.

ABSTRACT

The use of stable isotopes in pharmacological research has increased dramatically in recent years, principally as a result of the growing popularity of highly sensitive and specific mass spectrometric assay procedures based on the reverse stable isotope dilution approach. In addition to their important role in the preparation of internal standards for such assays, however, compounds labelled at specific positions with the stable isotopes of carbon, hydrogen, nitrogen or oxygen may be utilized effectively to study such diverse topics as drug pharmacokinetics under steady-state conditions, the molecular basis of drug interactions, the effect of drug administration on the biosynthesis and metabolism of endogenous compounds and the biochemical mechanisms of metabolic processes themselves. These and other areas of application may be classified under one of the following broad headings: (1) quantitative applications, (2) qualitative applications, and (3) mechanistic applications, and are discussed accordingly in the text.

In order to illustrate the utility of stable isotope labelling techniques for applications in one important area, viz. mechanistic studies of drug metabolism, examples are presented from an investigation into the metabolic fate of N,N'-dimethylclonidine, a close structural analog of the potent anti-hypertensive agent, clonidine. By the use of substrates labelled in specific positions with deuterium and carbon-13, coupled with sample analysis by combined gas chromatography-mass spectrometry, information was obtained on the sequence of certain multi-step biotransformation pathways and on the biological fate of metabolically-labile carbon atoms. In addition, details of the mechanism by which the heterocyclic ring system of this compound undergoes oxidative cleavage, and information on the nature of the chemically unstable intermediates involved, was obtained effectively by the use of the stable isotope of oxygen, ^{18}O. Finally, the complementary roles of stable and radioactive isotopes in studies of drug metabolism is illustrated and the great value of the combined use of the two types of isotope in such investigations is emphasized.

INTRODUCTION

Applications of the stable isotopes of carbon, hydrogen, nitrogen and oxygen to studies in pharmacology have increased dramatically over the past few years, when stable isotope labelling techniques have made particularly valuable contributions to investigations of drug metabolism and pharmacokinetics [1-5]. This resurgence of interest in the use of stable isotopes has been due in part to their increased commercial availability, both at high isotopic enrichment and in a wide variety of chemical forms, and in part to recent developments in the manufacture of gas chromatography-mass spectrometry (GC-MS) instrumentation suitable for the detection and quantitative measurement of isotopically labelled compounds at the nanogram level and below. Pharmacological applications of stable isotopes may be divided into three broad categories viz (1) quantitative applications, (2) qualitative applications, and (3) mechanistic applications. Salient areas of stable isotope usage in each of these three areas are summarized below.

1. QUANTITATIVE APPLICATIONS

   (a) Internal Standards

The use of stable isotope labelled compounds as internal standards in mass spectrometric assay procedures traditionally has been a major area of application in pharmacology and has been adopted widely for those investigations where a high degree of both sensitivity and specificity of analysis is required. Practical and theoretical considerations in the development of stable isotope dilution assay procedures have been dealt with in the text by Millard [6] and a recent comprehensive review by Garland and Powell [7] serves to illustrate the diversity of compounds amenable to analysis by this approach. While deuterium remains by far the most widely employed nuclide for the preparation of stable isotope labelled internal standards, carbon-13 and nitrogen-15 are gaining in popularity for this purpose, despite their relatively high cost. This trend can be attributed to the chemical "stability" of the isotopes of carbon and nitrogen, in contrast to the situation with deuterium and oxygen-18, where inadvertent chemical back-exchange of label may occur during sample preparation and analysis.

(b) Pharmacokinetics

In addition to the now classical usage of stable isotope dilution assay procedures for drug concentration-effect studies, specialized stable isotope methodology has been developed for certain types of pharmacokinetic studies [4]. An important example is the investigation of bioavailability, either relative [8] or absolute [9], of different formulations of the same drug. Stable isotope methodology may also be conveniently employed to determine the pharmacokinetics of a drug under chronic dosing conditions, when "steady state" pharmacokinetic parameters may be obtained [10]. Despite the advantages of stable isotope labelled compounds for studies such as these, caution should be exercised when deuterium-labelled drugs are administered, since untoward in vivo deuterium isotope effects may give rise to erroneous results. An additional factor which may complicate bioavailability studies performed by stable isotope techniques arises when the drug-of-interest is subject to a saturable "first-pass" effect [11].

(c) Optical Isomers

Enantiomeric differences in the metabolism and disposition of racemic drugs is an area of growing interest and a number of analytical methods have been developed to study the phenomenon. In one of these, a "pseudoracemic" mixture of the drug, in which one of the two enantiomers is labelled with a stable isotope, is employed as metabolic substrate [12]. In this case, the heavy isotope label becomes a stereochemical marker, and the ratio of labelled to unlabelled molecules provides a direct measure of the optical purity of the compound under study.

(d) $^{13}CO_2$ Breath Test

The stable isotope counterpart to the well-established $^{14}CO_2$ breath test has yet to find widespread application in pharmacological research, although preliminary investigations have indicated that the $^{13}CO_2$ breath test may serve as a useful indicator of hepatic mixed function oxidase activity in man [13]. The value of the $^{13}CO_2$ breath test is that it is both non-invasive and lacks the radiation exposure associated with the use of carbon-14. As such, it is ideally suited to investigations in young infants and pregnant women. A major disadvantage, however, is that it is indirect, the metabolic pathway from the ^{13}C-labelled drug to exhaled $^{13}CO_2$ involving a number of steps including absorption of the labelled drug and a variety of metabolic transformations catalyzed by several distinct enzyme systems.

(e) Inter-Individual Differences in Metabolism

In 1976, Baty and Robinson [14] proposed a technique for the study of inter-individual differences in drug metabolism based on the use of two analgesics, acetanilide and phenacetin, as metabolic substrates. Since both drugs are metabolized to a common product, acetaminophen, the co-adminsitration of deuterium-labelled acetanilide and unlabelled phenacetin leads to the formation of a mixture of unlabelled and deuterated forms of acetaminophen. The ratio of unlabelled to labelled acetaminophen thus affords a measure of the relative activities of the respective enzyme systems responsible for the O-demethylation of phenacetin and the para-hydroxylation of acetanilide. This technique for the simultaneous study of two types of metabolic transformations in vivo would appear to hold promise for investigations of enzyme induction and drug interaction.

(f) Metabolism of endogenous compounds

An increasingly important area of application of stable isotope labelling techniques in pharmacological research is the study of the biosynthesis and metabolism of a variety of endogenous compounds in vivo and the effects of physiological or pharmacological factors on the rates of these processes. The group of endogenous compounds which has been most widely investigated by stable isotope techniques comprises the neurotransmitter substances acetylcholine, norepinephrine, dopamine, serotonin and GABA [5]. In addition, the effects of drug administration on the biosynthesis and metabolism of both steroids [15, 16] and prostaglandins [17] has been investigated by stable isotope techniques.

Noteworthy among the stable isotope techniques under this heading is the series of studies first reported by Sedvall and colleagues [18] on the incorporation of ^{18}O into brain catecholamines during exposure of animals to an $^{18}O_2$-containing atmosphere These early studies first demonstrated the feasibility of "pulse-labelling" of brain neurotransmitter pools by inhalation of ^{18}O-enriched molecular oxygen, a potentially attractive, non-invasive procedure by which quantitative determination of the turnover of brain neurotransmitters could be assessed in animals and man.

2. QUALITATIVE APPLICATIONS

(a) Isotope Cluster Technique

The so-called "isotope cluster," "ion doublet" or "twin-ion" technique [19] refers to the procedure in which the substrate under investigation is enriched at a level of approximately 50% with one or more atoms of a suitable heavy isotope. The mass spectrum of the substrate, and any metabolites derived from it, will thus exhibi artificially generated "twin" peaks for the molecular ion and those fragment ions retaining the labelled atoms. The isotope cluster technique has proved to be an invaluable aid to the mass spectrometric recognition and characterization of drug metabolites in complex biological extracts, and has found widespread application in studies of drug biotransformation [3, 4, 20, 21]. However, as with other techniques in which a stable isotope labelled compound is administered to animals or humans, extreme caution should be exercised in selecting the site for heavy isotope incorporation, particularly if the nuclide to be employed is deuterium. In cases where deuterium-labelled drugs are to be used in such studies, the site of deuterium substitution should be remote from known or suspected sites of metabolic attack in order to minimize the risk of loss of label through oxidative attack and also to avoid deuterium isotope effects which may lead to the phenomenon of "metabolic switching" [22].

(b) Deuterium Labelling in Defining Sites of Metabolic Attack

Loss or retention of deuterium substituents at specific positions in a molecule can afford valuable information on sites of metabolic attack [23]. Stereochemical aspects of metabolic reactions may also be investigated by the deuterium labelling technique if one of two stereochemically distinct centers in the substrate is labelled specifically [24]. It should be noted, however, that retention of deuterium in a hydroxylated metabolite need not necessarily exclude the site originally bearing the label as being the center of metabolic attack; deuterium rearrangements may occur in certain situations, particularly with aromatic ring systems which undergo hydroxylation by way of intermediate arene oxides (see 3 (e) below).

(c) Stable Isotope Labelled Derivatizing Reagents

Reaction of an unknown compound with both an unlabelled and corresponding stable isotope labelled derivatizing reagent and comparison of the mass spectra of the products so obtained is a useful aid to structure elucidation in studies of drug metabolism and has been used for this purpose for many years. Thus, derivatization with the labelled, as compared with the normal, reagent indicates directly the number of functional groups which have undergone reaction, while analysis of the mass spectral fragmentation patterns in terms of molecular structure is greatly facilitated by knowledge of which fragment ions retain certain functional groups [5].

(d) Drug Interactions

Stable Isotope labelling techniques have proven to be extremely useful in defining the nature of the molecular interactions which may take place as a result of the co-administration of two or more xenobiotics. Examples of the use of deuterium labelling to reveal the interaction of ethanol co-administered with lidocaine [25] and with chloral hydrate [26] serve to illustrate this type of application.

3. MECHANISTIC APPLICATIONS

(a) Deuterium Labelled Compounds

Deuterium labelled compounds have been employed profitably to define the nature of chemically reactive, and potentially toxic, electrophilic intermediates of metabolism. Examples of such investigations include studies on the nature of the reactive species generated from the carcinogen dimethylnitrosamine [27], from the anticancer nitrosoureas [28] and the antitumor agent cyclophosphamide [29], from the hydrazine drugs isoniazid and iproniazid [30] and from the analgesics phenacetin and acetaminophen [30]. In each of these cases, specifically deuterium labelled analogs of the parent drugs were employed as metabolic substrates in order to provide information on the molecular nature of the reactive metabolites formed.

(b) Carbon-13 and Nitrogen-15 Labelled Compounds

Compounds labelled with these heavy isotopes may be used to define the metabolic fate of selected carbon and nitrogen atoms in a given molecule. Carbon-13 labelling has proved to be particularly useful in studies of the halocarbons chloroform [31] and carbon tetrachloride [32], each of which is metabolized to phosgene. Compounds labelled with ^{15}N may be similarly employed to trace the metabolic fate of selected nitrogen atoms. Examples of the latter application would be to studies on the metabolic fates of dimethylnitrosamine [33], nitrous oxide [34] and the antihypertensive drug hydralazine [35]. Nitrogen-15 is an especially important tracer in the case of nitrogen-containing compounds, since there are no sufficiently long-lived radioactive isotopes of this element which can be used for biological studies.

(c) Oxygen-18 Labeleld Compounds

Oxygen-18 is rapidly becoming the most important stable isotope for mechanistic applications in view of the fact that the vast majority of metabolic transformations are oxidative in nature. As is the case with nitrogen, no suitable long-lived radioactive isotope exists for oxygen, and one is therefore obliged to use the stable isotopes ^{17}O or ^{18}O. For reasons of availability, isotopic enrichment, cost and mass increment over ^{16}O, ^{18}O has been the isotope of choice in virtually all applications. With the growing interest in the mechanism of biological oxidation processes, oxygen-18 has an assured future in pharmacological research. An example of the utility of oxygen-18 for distinguishing between alternative mechanisms of metabolism is presented below.

(d) Deuterium Isotope Effects

Primary kinetic deuterium isotope effects yield information on those metabolic processes in which a carbon-hydrogen bond undergoes cleavage in the rate-determining step, while secondary deuterium isotope effects may be useful in studies of processes in which the bond to deuterium is not broken during the reaction. In the latter case, deuterium substitution in close proximity to the reaction center exerts an effect on the energy of the transition state of the reaction [36].

(e) Molecular Re-arrangements

Hydrogen rearrangements accompany certain metabolic transformations, the best known of which is the so-called "NIH shift," which constituted the first experimental evidence for the now accepted arene oxide mechanism for the aromatic ring hydroxylati [37]. More recently, however, an analogous 1,2-hydrogen atom rearrangement has been shown to occur during the metabolism of certain terminal acetylenes [38]. The recognition of each of these types of rearrangements, which was made possible by the use of specifically deuterium labelled substrates, has contributed greatly to our understanding of the nature of the intermediates involved in biological oxidation processes.

While each of the above areas of stable isotope usage have achieved, or are now gaining, widespread acceptance in pharmacological research, the use of specifically labelled compounds in investigations of a mechanistic nature would appear to offer especially intriguing possibilities. Some recent examples of the use of deuterium, carbon-13 and oxygen-18 in studies of drug metabolism are discussed below in order to illustrate this point; these examples are taken from an investigation of the metabolism of the anti-hypertensive drug clonidine and related imidazoline derivatives [39].

(A) The Use of Oxygen-18 to Investigate Pathways of Oxidative Drug Metabolism.

In a continuation of the studies presented at the previous conference in this series [40], information was sought on the mechanism by which the heterocyclic ring system of clonidine (I) undergoes oxidative cleavage both in vitro and in vivo to yield the substituted guanidine (II) and the two-carbon aldehyde, glyoxal (III) (Fig. 1).

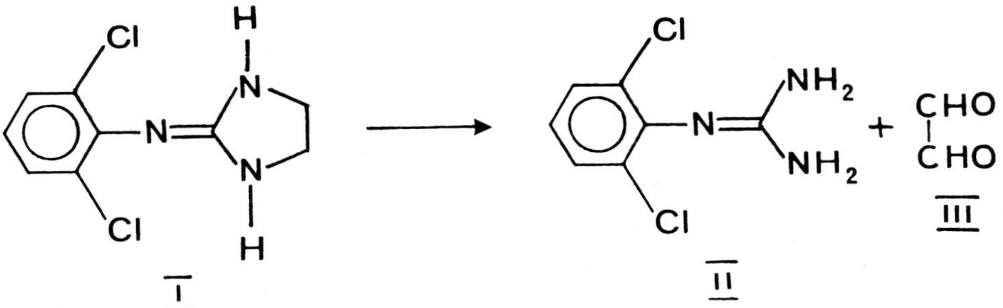

Fig. 1. Metabolic degradation of clonidine (I) to 2,6-dichlorophenylguanidine (II) and glyoxal (III).

Attempts to investigate details of this pathway, however, were hampered by the chemical instability of the presumed intermediate metabolites, bearing oxygen substituents on one or both of the carbon atoms C-4 and C-5. In contrast, preliminary work with the N,N'-dimethyl analog of clonidine (IV) showed that, while this compound followed essentially the same metabolic pathway as clonidine itself, the proposed 4-hydroxy (V) and 5-hydroxy (VIII) metabolites of this compound were sufficiently stable chemically to be characterized by GC-MS as their trimethylsilyl (TMS) ether derivatives. The origin of the diol VIII was of particular interest, since one pathway for its formation could have involved a highly reactive (and potentially toxic) epoxide (VII), formed from the known unsaturated metabolite VI (Fig. 2). On the other hand, an alternative, simpler, mechanism for the formation of VIII would entail two successive hydroxylations at adjacent carbon atoms (Fig. 3), a process not previously known for such heterocyclic systems. In order to distinguish, therefore, between the "epoxide-diol" and "successive hydroxylation" pathways, an in vitro experiment utilizing $^{18}O_2$ was devised, in which dimethyl-

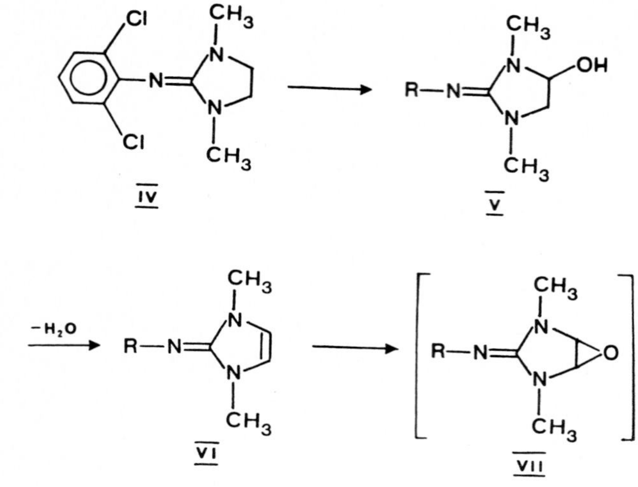

Fig. 2 Proposed "epoxide-diol" pathway for the metabolism of N,N'-dimethylclonidine. (The epoxide VII was not isolated from metabolic experiments).

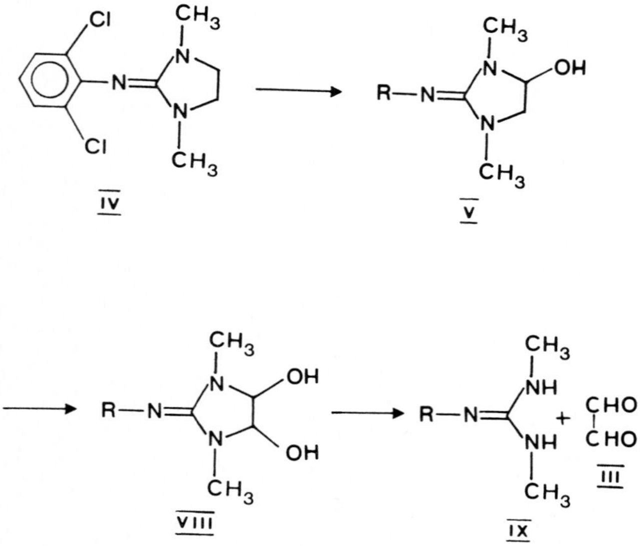

Fig. 3 "Successive hydroxylation" pathway for the metabolism of N,N'-dimethylclonidine.

clonidine was incubated under an atmosphere enriched in this isotope of oxygen. If the principle pathway of metabolism was as depicted in Fig. 2, the dihydroxy compound VIII would be expected to contain one atom of ^{18}O (the label being retained from the putative epoxide VII) and one atom of ^{16}O (introduced during the hydrolytic attack of VII by unlabelled water of the medium). On the other hand, if the mechanism in Fig. 3 operated, the diol would become labeled with two atoms of ^{18}O, each derived from molecular oxygen and introduced, most likely, by the cytochrome P-450 system. In the event, incubation of N,N'-dimethylclonidine under an atmosphere of $N_2:{}^{18}O_2$ (4:1) yielded the labelled diol whose mass spectrum (TMS derivative) is reproduced in Fig 4. Analysis of the isotope cluster representing the molecular ion region of this compound (m/z 433-444) indicated there to be a 47% incorporation of two atoms of ^{18}O, demonstrating that molecular oxygen, and not water, was the source of both oxygens in a high proportion of the metabolite. This finding is not

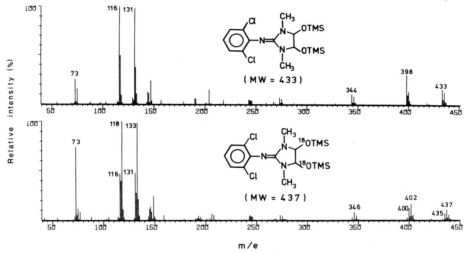

Fig. 4 Mass spectra (70 eV) of the TMS ether derivatives of (upper frame) unlabelled diol VIII and (lower frame) diol isolated from $^{18}O_2$ experiment.

compatible with hydration of an intermediate epoxide, and suggests successive insertion of two atoms of oxygen, as shown in Fig. 3. Failure to incorporate a higher percentage of two atoms of ^{18}O into VIII may have been due to partial loss of the first ^{18}O atom through reversible dehydration of the 4-hydroxy metabolite V; the low ^{18}O content found for the labelled mono-hydroxy species V (16 atom % excess) lends support to this view. Thus, for N,N'-dimethylclonidine, the "epoxide-diol" pathway is not the only route involved in oxidative degradation of the heterocyclic ring system, and may not operate at all. Whether the same applies to the metabolism of clonidine must await further studies.

(B) The NIH Shift in Studies of Aromatic Hydroxylations.

A minor pathway of metabolism of clonidine and its N,N'-dimethyl derivative in the rat is aromatic ring hydroxylation at the para position, a transformation which is presumed to occur via an intermediate arene oxide. In such cases, a deuterium substituent at the para position of the substrate would be expected to be partially retained in the phenolic product as a consequence of the NIH shift accompanying rearrangement of this arene oxide. The availability of para-[^{2}H] analogs of both clonidine and dimethylclonidine allowed us to investigate the magnitude of the NIH shift obtained with compounds of this class during para-hydroxylation in vitro.

When para-[^{2}H]clonidine was incubated with rat liver microsomal preparations, the para-hydroxyclonidine (X) formed was found to have retained only 2% of the deuterium originally present in the substrate (Fig. 5). Such low values for deuterium retention are typically observed in molecules where a substituent para to the site of

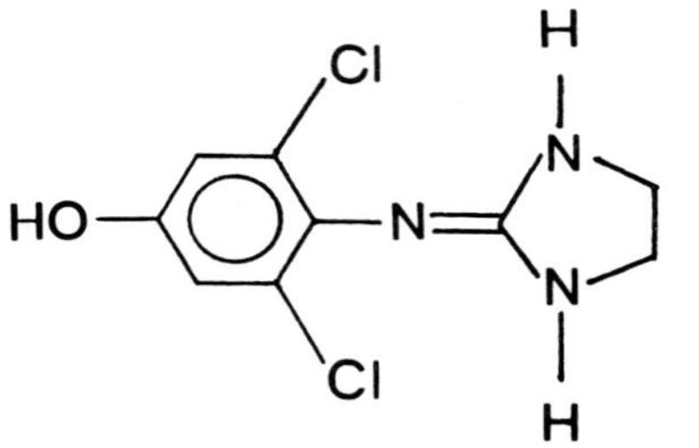

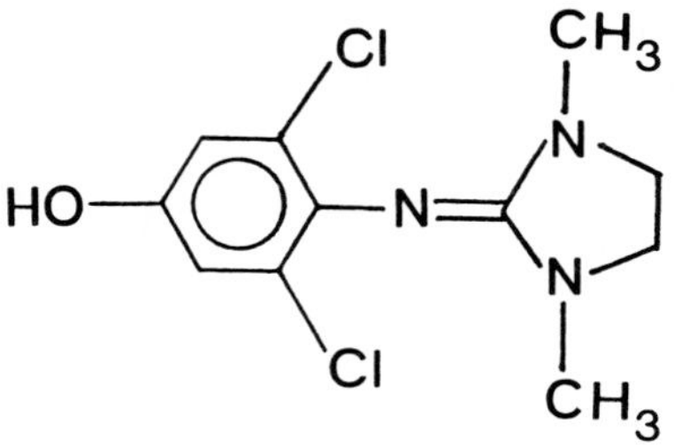

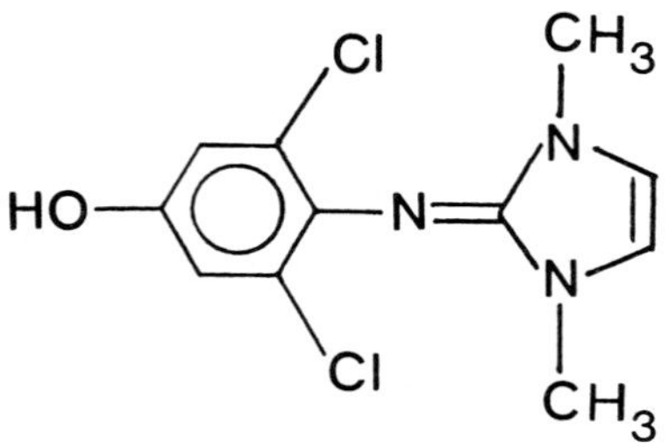

Fig. 5 Magnitudes of the NIH shifts (expressed as % deuterium retention) associated with the para-hydroxylation of clonidine (I), N,N'-dimethyl-clonidine (IV) and metabolite VI.

hydroxylation possesses (or may acquire through tautomerism, as in the case of clonidine) an ionizable hydrogen [41]. Somewhat surprisingly, the corresponding phenolic metabolite XI from para-[^{2}H] N,N'-dimethylclonidine also exhibited a small NIH shift (deuterium retention = 8%), despite the absence of such an ionizable hydrogen atom, a situation which normally leads to relatively high degrees of label retention. In addition, and of some interest mechanistically, was the finding that an additional phenolic metabolite (XII) of para-[^{2}H] dimethylclonidine exhibited a deuterium retention of 13%, thereby indicating that oxidative metabolism of dimethylclonidine in the heterocyclic moiety must precede hydroxylation in the aromatic ring in order to account for the observed NIH shift values. This example thus illustrates how the magnitude of the NIH shift may afford useful information on the sequence of metabolic transformations when one step entails aromatic ring hydroxylation.

(C) Carbon-13 as a Tracer for Metabolically-Labile Carbon Atoms.

In some of our preliminary work on the metabolic fate of clonidine in rat liver, a radioactive analog of the drug, [4,5-^{14}C$_2$]clonidine, was synthesized for use as a tracer (Fig. 6).

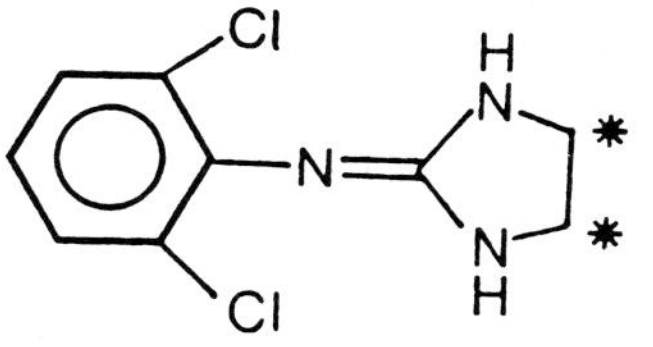

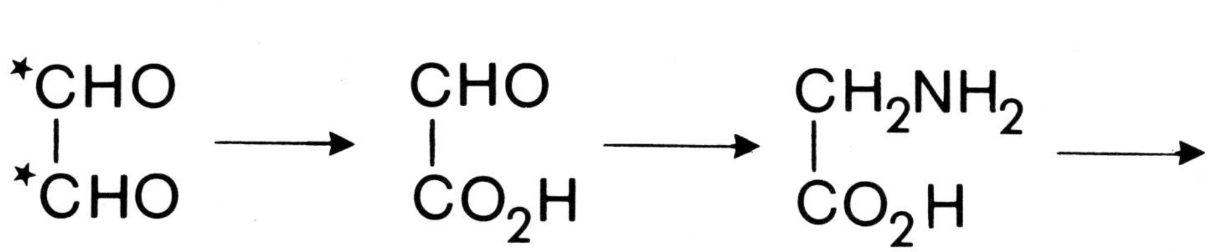

Fig. 6 Structure of [4,5-^{14}C$_2$]clonidine and scheme for the metabolic conversion of labelled glyoxal into glycine.

While it was appreciated at the time these early experiments were carried out that carbons 4 and 5 of the imidazoline ring were metabolically labile, the precise fate of these carbon atoms was not known; indeed, it was hoped that the overall disposition of the ^{14}C label would provide some indication as to the molecular entity (— ies) to which carbons 4 and 5 were converted (e.g., CO_2 or two-carbon compounds).

When $[4,5-^{14}C_2]$clonidine was employed as a tracer in an isolated perfused liver experiment (recycling perfusion system), it was found that an unexpectedly high proportion (29%) of the radioactivity administered to the system became tightly associated with liver tissue; repeated washing of the homogenized liver with a variety of solvents failed to remove the bound radiolabel, which thus appeared to be covalently attached to hepatic protein. This observation, taken in isolation, may have been interpreted (erroneously) to indicate the formation of a reactive electrophilic metabolite of clonidine which had become covalently attached to cellular macromolecules. However, subsequent liver perfusion experiments, performed with an analog of clonidine labelled at the same positions with the corresponding stable isotope of carbon, ^{13}C, led to the identification of the two-carbon species glyoxal (III) and its product of oxidation, glyoxylic acid, in both bile and perfusion medium [40]. Since glyoxylic acid is known to undergo transamination in rat liver cytosol to afford the amino acid glycine [42], and since protein synthesis (utilizing glycine) will occur in the isolated perfused liver, labelled carbons entering the glyoxylate pool will ultimately become incorporated into protein through endogenous pathways of metabolism (Fig. 6). Thus, the observed "covalent binding" of radiolabel from the experiment with $[4,5-^{14}C_2]$clonidine most likely did not reflect the formation of large amounts of a chemically reactive metabolite of the drug; rather, the bound radioactivity provided a valuable indication of the extent to which endogenous amino acid pools became labelled with ^{14}C via the intermediate two-carbon species glyoxal and glyoxylate. This example, while emphasizing the danger of relying solely on data obtained from experiments performed with radioisotopes [43], illustrates effectively the strong complementary nature of radioactive and stable isotope labelled tracers in studies of drug metabolism.

CONCLUSIONS

The examples presented above serve to illustrate how deuterium, carbon-13 and oxygen-18 labelling procedures may be employed to provide invaluable data on mechanistic aspects of drug metabolism, where information may be sought on the nature of unstable intermediates, on the fate of specific carbon atoms and on the sequence of multi-step metabolic processes. The value, as metabolite substrates, of compounds labelled in specific positions with deuterium or ^{13}C is self-evident from the studies discussed under (B) and (C) above, while the use of $^{18}O_2$ for investigations of drug oxidation is an area with great potential in pharmacological research.

The dramatic growth of stable isotope usage in the pharmacological sciences witnessed over the past decade will no doubt continue through the 1980's when many of the specialized stable isotope techniques currently available will enjoy increased application to an even wider variety of problems in experimental and clinical pharmacology. It is important, however, to recognize the complementary features of stable and radioactive isotopes in order to take full advantage of the unique properties of each whenever experimental protocols allow.

REFERENCES

1. T.A. Baillie (Ed.), Stable Isotopes: Applications in Pharmacology, Toxicology and Clinical Research, MacMillans, London, 1978, 314 pp.
2. E.R. Klein and P.D. Klein (Eds.), Stable Isotopes. Proc. 2nd Int. Conf., Academic Press, New York, 1979, 627 pp.
3. D.R. Hawkins, in J.W. Bridges and L.F. Chasseand (Eds.), Progress in Drug Metabolism, Vol. 2, Wiley and Sons, London, 1977, p. 163.
4. P.J. Murphy and H.R. Sullivan, Ann. Rev. Pharmacol. Toxicol., 20(1980) 609-621.
5. T.A. Baillie, Pharmacol. Rev., in press 1981.
6. B.J. Millard, Quantitative Mass Spectrometry, Heyden, London, 1978, 171 pp.
7. W.A. Garland and M.L. Powell, J. Chromarogr., in press 1981.
8. H.d'A Heck, S.E. Buttrill, Jr., N.W. Flynn, R.L. Dyer, M. Anabar, T. Cairns, S. Dighe and B.E. Cabana, J. Pharmacokin. Biopharm., 7(1979) 233-248.
9. J.M. Strong, J.S. Dutcher, W.-K. Kee and A.J. Atkinson, Jr., Clin. Pharm. Ther., 18(1975) 613-622.
10. H.R. Sullivan, P.G. Wood and R.E. McMahon, Biomed. Mass Spectrom. 3(1976) 212-216.
11. J. Schmid, A. Prox, H. Zipp and F.W. Koss, Biomed. Mass Spectrom. 7(1980) 560-564.
12. R.E. McMahon, H.R. Sullivan and S.L. Due, Acta Pharm. Suecica, 11(1974) 639-640.
13. J.F. Schneider, D.A. Scholler, B. Nenchausky, J.L. Boyer and P.D. Klein, Clin. Chim. Acta, 84(1978) 153-162.
14. J.D. Baty and P.R. Robinson, Clin. Pharm. Ther. 21(1976) 177-186.
15. J.J. deRidder and P.C.J.M. Koppens, in T.A. Baillie (Ed.), Stable Isotopes: Applications in Pharmacology, Toxicology and Clinical Research, MacMillans, London, 1978, pp. 157-165.
16. S. Baba, Y. Shinohara and Y. Kasuya, J. Clin. Endocr. Metab. 50(1980) 889-894.
17. A.R. Brash and M.E. Conolly, Prostaglandins, 15(1978) 983-993.
18. A. Mayevsky, B. Sjoquist, C.-G. Fri, D. Samuel and G. Sedvall, Biochem. Biophys. Res. Commun., 51(1973) 746-755.
19. R.F. Morfin, I. Leav, P. Ofner and J.C. Orr, Fed. Proc., 29(1970) 247.
20. S.D. Nelson and L.R. Pohl, Ann. Rep. Med. Chem., 12(1977) 319-329.
21. D. Halliday and I.M. Lockhart, Prog. Med. Chem., 15(1978) 1-86.
22. M.G. Horning, K.D. Haegele, K.R. Sommer, J. Nowlin, M. Stafford and J.-P. Thenot, in E.R. Klein and P.D. Klein (Eds.), Proc. 2nd Int. Conf. on Stable Isotopes, Oak Brook, October 20-23, 1975, National Technical Information Service Document CONF-751027, pp. 41-54.
23. R.E. McMahon, H.R. Sullivan, S.L. Due and F.J. Marshall, Life Sci., 12(1973) 463-473.
24. R.E. McMahon, H.R. Sullivan, J.C. Craig and W.E. Pereira, Jr., Arch. Biochem. Biophys., 132(1969) 575-577.
25. S.D. Nelson, G.D. Breck and W.F. Trager, J. Med. Chem., 16(1973) 1106-1112.
26. L.K. Wong and K. Biemann, Biochem. Pharmacol., 27(1978) 1019-1022.
27. W. Lijinsky, J. Loo and A.E. Ross, Nature 218(1968) 1174-1175.
28. W. Lijinsky, H. Garcia, L. Keefer, J. Loo and A.E. Ross, Cancer Res., 32(1972) 983-987.
29. T.A. Connors, P.J. Cox, P.B. Farmer, A.B. Foster, M. Jarman and J.K. Macleod, Biomed. Mass Spectrom., 1(1974) 130-136.
30. J.R. Mitchell and S.D. Nelson, Adv. Pharmacol. Ther., 7(1978) 203-214.
31. L.R. Pohl, B. Bhoosan, N.R. Whittaker and G. Krishna, Biochem. Biophys. Res. Commun., 79(1977) 684-691.
32. V.L. Kubic and M.W. Anders, Life Sci., 26(1980) 2151-2155.
33. R.C. Cottrell, B.G. Lake, J.C. Phillips and S.D. Gangolli, Biochem. Pharmacol., 26(1977) 809-813.
34. K. Hong, J.R. Trudell, J.R. O'Neil and E.N. Cohen, Anesthesiology, 52(1980) 16-19.
35. J.A. Timbrell and S.J. Harland, Clin. Pharm. Ther., 26(1979) 81-85.

36. K.B. Wiberg, Chem. Rev., 55(1955) 713-743.
37. J.W. Daly, D.M. Jerina and B. Witkop, Experientia, 28(1972) 1129-1264.
38. P. Ortiz de Montellano and K.L. Kunze, J. Amer. Chem. Soc., 102(1980) 7373-7375.
39. H. Hughes, Ph.D. Thesis, University of London, 1980.
40. T.A. Baillie, E. Neill, H. Hughes, D.L. Davies and D.S. Davies, in E.R. Klein and P.D. Klein (Eds.), Stable Isotopes. Proc. 3rd Int. Conf., Academic Press, New York, 1979, pp. 415-425.
41. M. Jarman, P.B. Farmer, A.B. Foster and P.J. Cox, in T.A. Baillie (Ed.), Stable Isotopes: Applications in Pharmacology, Toxicology and Clinical Research, Macmillans, London, 1978, pp. 85-95.
42. T. Noguchi, Y. Takada and K. Kido, Hoppe-Seyler's Z. Physiol. Chem., 358(1977) 1533-1542.
43. S.S. Thorgeirsson and P.J. Wirth, J. Toxicol. Environ. Health, 2(1977) 873-881.

APPLICATION OF ^{13}C-LABELLING IN THE BIOAVAILABILITY ASSESSMENT
OF EXPERIMENTAL CLOVOXAMINE FORMULATIONS

H. DE BREE, D.J.K. VAN DER STEL, J.H.M.A. KAAL and J.B. VAN DER SCHOOT
Duphar B.V. Research Laboratories,
P.O. Box 2, 1380 AA Weesp, The Netherlands.

ABSTRACT

The bioavailability of two experimental oral clovoxamine fumarate formulations was assessed in healthy volunteers using the ^{13}C-labelled drug as the plain standard. Test formulation and plain drug were administered simultaneously. Blood samples were taken and after addition of an internal standard (fluvoxamine maleate), both the labelled and the unlabelled drug were quantitated using gas chromatography - mass fragmentography. The results warrant the conclusion, that, even with drugs showing large inter-individual variations in the area under the curve (AUC), it is possible to obtain useful information about the bioavailability of a test formulation with as few as four subjects.

INTRODUCTION

Clovoxamine (Fig. 1) is currently being evaluated as a non-tricyclic antidepressant drug acting as a noradrenaline- and serotonin-uptake inhibitor (ref. 1). The half life time in humans ranges from 4.6 to 13.3 h (mean 6.4, n=17). For patient convenience it was decided to develop a sustained release formulation. For the comparative testing of drug formulations the most rational method is a cross-over set up in which the performance of both formulations is assessed within individuals; thus variations among individuals are eliminated.

If two dose forms are ingested successively, with a necessary wash-out period in between, each individual varies in the handling of the dose forms. Then to acquire sufficient statistical power as a rule at least 18 subjects are necessary. Accordingly the number of samples to be analysed easily mounts up to several hundred.

The use of mass fragmentography offers the possibility to quantify (stable) isotope ratios. In its turn the use of (stable) isotope labelled drugs makes it feasible to compare the bioavailability of formulations simultaneously in the same subject (refs. 2-6). In that way the number of subjects can be markedly reduced while the quality of the results is improved since the inter- and intra- individual variations are now eliminated. In the present study we measured the bioavailability of two experimental sustained-release formulations of clovoxamine fumarate, each with respect to the plain drug. As the plain dosage form we used [^{13}C] clovoxamine fumarate in capsules. We preferred the use of ^{13}C over D as the stable isotope, in order to eliminate the risk of isotope effects, such as have been encountered if deuterium-labelling is used (ref. 7).

MATERIALS AND METHODS

Clovoxamine and internal standard

The unlabelled clovoxamine fumarate used in the study was of pharmaceutical quality. The ^{13}C-labelled drug was synthesized according to the scheme presented in Figure 2 and met the specifications set for pharmaceutical quality. Fluvoxamine maleate (Fig. 1), the internal standard, was synthesized in our laboratory and was also of pharmaceutical grade.

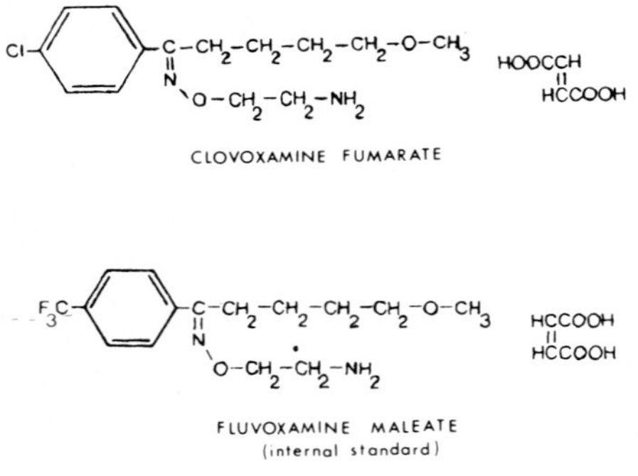

Fig. 1 Structures of the compounds involved

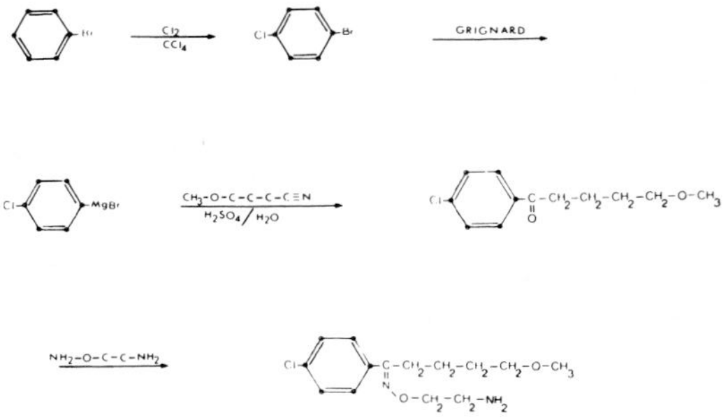

Fig. 2 Synthesis scheme for ^{13}C-clovoxamine starting with 90% uniformly labelled ^{13}C bromobenzene

Human volunteers

Healthy female and male volunteers gave their informed consent. They were between 30 and 36 years of age and of normal weight. They had normal values in the clinical tests performed. For at least one week prior to the studies they had taken no drugs. In the first trial, formulation I, four female subjects were involved, in the second trial, formulation II, four male subjects were added to three participants of the first study.

Protocols

In both trials the volunteers received the plain ^{13}C-drug and the ^{12}C-test formulation simultaneously. The drugs were administered in hard gelatine capsules containing 60 mg of the active compound. The drug was taken after an overnight fast. Half an hour later the subjects were allowed a light breakfast. Lunch and dinner were also light on the day of intake. Blood samples of 10 ml were drawn at 0, 1/2, 1, 1 1/2, 2, 3, 4, 6, 8, 12, 24, 32 and 48 hours after drug intake. From the heparinized blood samples the plasma was separated by centrifuging and stored frozen at -20°C.

Sample preparation

In a separatory funnel 1 to 5 ml of plasma sample were mixed with 500 ng of internal standard (fluvoxamine maleate). After addition of saturated aqueous potassium carbonate solution (0.1 ml per ml of plasma) the mixture was extracted with 100 ml of isooctane (distilled from sodium). The compounds were reextracted from the isooctane with 3.0 ml of 5% v/v aqueous phosphoric acid solution. The acid layer was drawn off into a 3 ml screw capped teflon closed reaction vial and subsequently the oxim ethers were hydrolysed to the parent ketones by heating at 90°C for 1 hr. Finally, after cooling to room temperature, the ketones were extracted into 100 µl of isooctane. Aliquots of the isooctane layer were subjected to GC-MF.

Gas chromatography-mass fragmentography (GC-MF)

GC-MF was performed on a Finnigan 4000 operated with a 2 m x 3 mm silanized glass column packed with 5% SE-30 on silanized chromosorb G-HP. The conditions were: temperatures - 250°C (injector) 230°C (column) 250°C (separator) and 270°C (ion source), the gas flow rate was 25 ml.min^{-1} of helium, ionizing voltage 60 V, current 200 µA. The multi ion detector was focused on m/z 139, 145 and 173. Fig. 3 shows the mass spectra of the parent ketones of the compounds involved.

Quantification

The results were calculated from substance/internal standard peak height ratios with the aid of calibration graphs. For this purpose mixtures of varying amounts of both ^{12}C- and ^{13}C- clovoxamine fumarate (5 - 500 ng) with a constant amount of fluvoxamine maleate (500 ng) were hydrolysed with 3 ml of phosphoric acid solution and further processed according to the procedure used in the sample preparation. The calibration graphs (Fig. 4) were constructed by plotting the substance/internal standard peak height

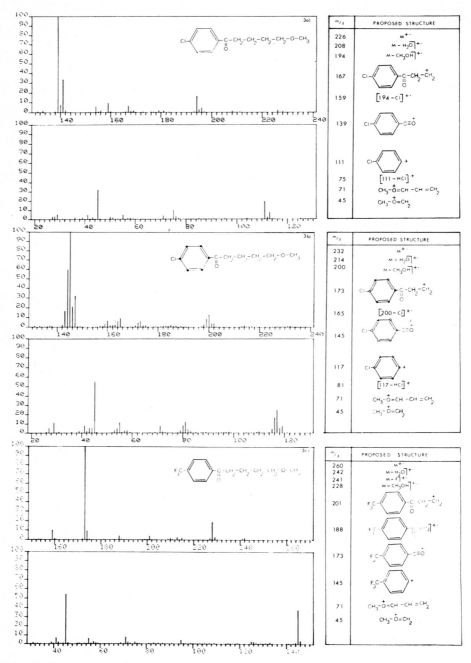

Fig. 3 EI mass spectra of the parent ketones of the compounds involved; 3a) clovoxamine ketone, 3b) ^{13}C clovoxamine ketone, 3c) fluvoxamine ketone.

ratios versus the corresponding amounts of substance in the calibration mixtures. The end results were expressed in terms of the free bases.

Pharmacokinetic evaluation

The area under the plasma level vs. time curve (AUC) was used to calculate the extent of bioavailability (EBA). For the evaluation of the retard-effect we used the half-value duration (HVD), i.e. the time that the plasma concentration exceeded one-half of the maximum plasma concentration (ref. 8).

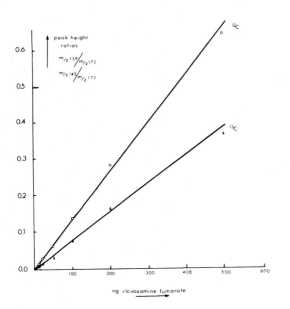

Fig. 4 Calibration curves for the GC-MF determination of ^{12}C and ^{13}C clovoxamine. The calibraton mixtures contained 500 ng of internal standards.

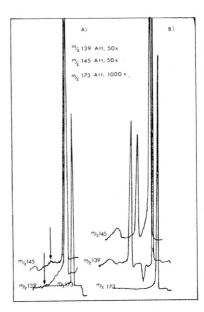

Fig. 5 Mass fragmentograms of; A) blank plasma sample, B) plasma sample cont. 6 resp. 8 ng.ml^{-1} ^{12}C- and ^{13}C-clovoxamine

RESULTS AND DISCUSSION

Sample analysis

The calibration graphs show a linear response in the range of interest (5 - 200 ng). The slope of the ^{13}C-clovoxamine graph differs from the ^{12}C-graph as would be expected from the statistical distribution of the ^{13}C-label in the phenyl moiety (90°/o uniformly labelled benzene will contain approximately 53°/o of completely ^{13}C-labelled benzene). The duplicate analysis of samples yielded an average variation coefficient of 7°/o (n=28, range 2-50 ng.ml^{-1} clovoxamine).

Typical fragmentograms of blank and sample are presented in Figure 5.

Evaluation of the Formulations

The mean plasma level curves of both experimental sustained release formulations are presented in Figure 6, and some salient pharmacokinetic data are tabulated in Table 1.

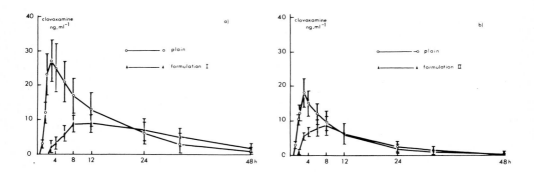

Fig. 6 Mean plasma curves (± SEM); a) plain versus formulation I, 4 subjects, b) plain versus formulation II, 7 subjects involved.

Formulation I, which showed the slower in-vitro dissolution of the two, did increase the HVD by a factor of 2 (range: 1.6–2.5); but its EBA dropped to 70%/o of the plain reference (range: 57–100%/o).

TABLE 1
Values of areas under the curve (AUC), half value duration (HVD) and elimination half lives of clovoxamine in capsules and of two test formulations (retard I and II). In the third column the ratios of AUC-values and of HVD are shown.

volunt	Formulation I							Formulation II						
	plain			retard		retard/plain		plain			retard		retard/plain	
	AUC	HVD	t1/2	AUC	HVD	AUC	HVD	AUC	HVD	t1/2	AUC	HVD	AUC	HVD
A	800	15.4	13.3	470	38.5	0.59	2.5							
B	265	4.4	6.2	168	8.6	0.59	2.0	250	6.0	6.5	150	8.0	0.60	1.3
C	100	7.6	6.1	57	.*	0.57	.*	80	4.8	4.6	38.5	8.5	0.48	1.8
D	350	12.3	9.7	350	19.8	1.00	1.6	550	14.1	10.5	440	16.7	0.80	1.2
E								124	6.2	6.0	77	7.2	0.62	1.2
F								102	7.4	7.4	45.5	9.2	0.45	1.2
G								230	7.1	6.4	172	9.0	0.75	1.3
H								27	4.6	4.6	16.4	5.2	0.61	1.1

* no reliable data available.

In accordance with its quicker in vitro dissolution, formulation II showed a HVD only slightly longer than that of the plain formulation (factor: 1.3, range: 1.1-1.8). The EBA, however, was about the same as that of formulation I: 60°/o (range: 45-80°/o).

Although the number of volunteers who participated in both studies was limited to three, permitting only tentative conclusions, the results seem to indicate that with clovoxamine the intra-individual variation of the AUC is much smaller than the differences between subjects, which can read a factor of 30. The plasma half-lives of each subject between studies also run very much parallel.

The loss of bioavailability, demonstrated by both experimental formulations, is not satisfactory explained. Neither of the two shows a diminished in vitro availability.

This study demonstrates how different formulations can be succesfully compared in a limited number of volunteers, using ^{13}C and non labelled drug. Especially with drugs featuring large inter-individual variations this approach should be considered. The improved quality of the results, the reduced number of subjects required and the reduced assaying costs greatly outweigh the investment in the synthesis of the ^{13}C-labelled compound.

REFERENCES

1. V. Claassen, Th.A.C. Boschman, K.M. Dhasmana, F.C. Hillen, W.J. Vaatstra and J.M.A. Zwagemakers, Pharmacology of clovoxamine, a new non-tricyclic antidepressant. Arzneim-Forsch./Drug Res., 28 (1978) 1756-1766.
2. R. Roncucci, M.J. Simon, G. Jacques and G. Lambelin, Eur. J. Drug Metab. Pharmacol, 9 (1976).
3. T.A. Baillie, in Macmillan (Ed.), Stable Isotopes: Applications in Pharmacology, Toxicology and Clinical Research, London, 1978
4. J.M. Strong, J.S. Dutcher, W.K. Lee and A.J. Atkinson, J. Clin. Pharmacol. Ther., 18 (1975) 613.
5. T.Walle and U.K. Walle, Res. Commun. Chem. Path. Pharmacol., 23 (1979)453.
6. R.L. Wolen, R.H. Carmichael, A.S. Ridolfo, L. Thompkins and E.A. Ziege, Biomed. Mass Spectrom., 6 (1979) 173.
7. J.R. Carlin, R.W. Walker, R.O. Davies, R.T. Ferguson, and W.J.A. Vandenheuvel, J. Pharm. Sci., 10 (1980) 1111.
8. J. Meyer, E. Nuesch, R. Schmidt, Eur. J. Clin. Pharmacol. 7 (1974) 429.

MEASUREMENT OF THE PHARMACOKINETICS OF DI-($[15,15,16,16-D_4]$-LINO-
LEOYL)-3-sn-GLYCEROPHOSPHOCHOLINE AFTER ORAL ADMINISTRATION TO RATS

A. BREKLE, F. WIRTZ-PEITZ, O. ZIERENBERG
A. Nattermann & Cie. GmbH, Abteilung Radiochemie, Nattermannallee 1,
D-5000 Köln 30, FRG

ABSTRACT

The pharmacokinetics of deuterium-labelled di-($[15,15,16,16-d_4]$-linoleoyl)-3-sn-glycerophosphocholine were studied after oral administration to rats. With the method used deuterium-labelled 3-sn-phosphatidylcholine and 3-sn-phosphatidylcholine could be isolated from blood in a highly purified form and measured by gas chromatography/ mass spectrometry. Only 200 pg deuterium-labelled fatty acid and a ratio of one molecule deuterium-labelled stearic acid to 500 molecules stearic acid are necessary for reproducible measurement. Hence the sensitivity of deuterium-labelled 3-sn-phosphatidylcholine measurement was greater than that of radioactive 3-sn-phosphatidylcholine.

6 hours after administration of deuterium-labelled di-($[15,15,-16,16-d_4]$-linoleoyl)-3-sn-glycerophosphocholine the greatest concentration was detected in the total blood volume (0.12 % dose/ml, i.e. 2.1 % of the maximum absorbed dose). These results are comparable with those found after administration of di-$[1-^{14}C]$-linoleoyl-3-sn-glycerophosphocholine.

The procedure described for deuterium-labelled di-($[15,15,16,16-d_4]$-linoleoyl)-3-sn-glycerophosphocholine may thus be used as an example for studying the pharmacokinetics of a natural product in vivo, omitting radioactive tracers.

INTRODUCTION

Dilinoleoyl-phosphatidylcholine (PPC) is the major component of natural polyene-3-sn-phosphatidylcholine isolated from soybean [1].
The antiatherosclerotic properties of the compound correspond to the following observations made after PPC administration:

1) The activity of cholesterol ester hydrolase (EC 3.1.1.13) in the arterial wall is increased
2) Cholesterol is more easily mobilised from cholesterol-overloaded cells of the arterial wall
3) Triglyceride lipase (EC 3.1.1.3) and lecithin:cholesterol acyltransferase (EC 2.1.3.43) activity in serum are increased
4) The high density lipoprotein/low density lipoprotein ratio is increased

Pharmacokinetic studies using PPC labelled with radioactive isotopes have revealed that following oral administration PPC is partly hydrolysed during the absorption process in the mucosa cells. In rats, about half the absorbed PPC is hydrolysed to 1-acyl-lysophosphatidylcholine and reacylated to 3-sn-phosphatidylcholine (PC) in the mucosa cells, while the other half is completely hydrolysed to free fatty acids and glycerophosphocholine or its hydrolysis products [2]. The greatest PPC concentration in blood was measured 6 - 8 hours after dosing.

At the time of the highest concentration, 4 % of the applied dose of PPC labelled with radioactive fatty acids was calculated for the total blood volume.

To date, only one study has been performed in man using radioactively labelled PPC [3]. Hence a method using harmless di-([15,15,16,-16-d_4]-linoleoyl)-3-sn-glycerophosphocholine (d-PPC) was developed with a view to studying the pharmacokinetics of PPC clinically in man.

Di-([15,15,16,16-d_4]-linoleoyl)-3-sn-glycerophosphocholine

MATERIALS AND METHODS

d-PPC was synthesized from d_4-linoleic acid and glycerophosphatidylcholine according to established procedures in our laboratory [2]. $[15,15,16,16-d_4]$-linoleic acid was the reaction product of $[5,5,6,6-d_4]$-1-bromo-2-octyne and 9-decynoic acid, $[5,5,6,6-d_4]$-1-bromo-2-octyne being obtained after reduction of 5-methoxy-pentyne with deuterium and reaction with 3-methoxy-propyne. 9-Decynoic acid was a product of a multistage reaction of 2-propyn-1-ol and 1-bromoheptane.

Before oral administration 500 mg d-PPC was ultrasonicated in 17 ml water 20 min, at 4 °C and 60 W under an argon atmosphere with a Branson Sonifier B.

Quantification was carried out using the GC/MS system in the "Multiple Ion Detection" mode. The relevant ions being monitored were m/z 298 (18:0) and m/z 302 (d_4-18:0). PC was measured according to Rouser [4]. The fatty acid composition was determined using a Silar 10 C-column. The extraction procedure and purification of blood PC is shown on Fig. 1.

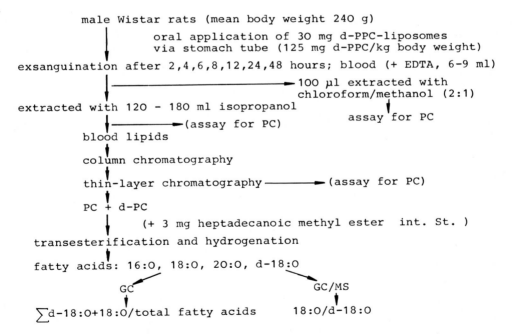

Fig. 1: Flow diagram of PC-purification and reaction to the saturated fatty acid methyl ester.

RESULTS AND DISCUSSION

After purification and derivatisation (Fig. 1) 27 % of the PPC administered could be measured as fatty acid methyl ester (Tab. 1). No specific loss of any group of the different fatty acids by transesterification and hydrogenation could be detected (Tab. 2) using heptadecanoic acid as an internal standard. The greatest concentration in blood was measured 6 hours after oral administration of d-PPC (Fig. 2). At this point 0.12 % of the dose/ml or 2.1 % of the d-PPC administered was detected in the total blood volume.

Tab. 1 Yield of PC during the purification steps

	chloroform-methanol phase	isopropanol phase	column chromat. and TLC	transesterification and hydrogenation
Yield	100 % ⟶	54 % ⟶	34 % ⟶	27 %
SEM (n = 11)		± 4 %	± 4 %	± 6 %

Tab. 2 Comparison of heptadecanoic acid (internal standard) to fatty acid ratio before and after transesterification and hydrogenation

heptadecanoic acid to fatty acids ratio			% loss after transesterification and hydrogenation
hours	before	after	
6	1.36	1.23	− 10 %
8	1.04	1.03	− 1 %
24	0.41	0.44	7 %
48	0.99	1.02	3 %

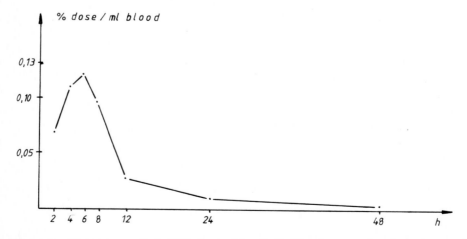

Fig. 2: Kinetics of d-PPC from blood following oral administration to rats (mean of 2 experiments with 2 animals each)

For comparison ^{14}C-labelled PPC with a specific activity of 6 mCi/mmol was administered simultaneously with d-PPC. The results obtained (% dose absorbed/ml blood) using the GC/MS-technique as well as the radiochemical calculations (Tab. 3) were comparable.

Tab. 3 Comparison of the absorption of a mixture of ^{14}C-PPC and d-PPC (dose of the mixture: 82 μCi ^{14}C-PPC (6 mCi/mmol)/kg body weight and 125 mg d-PPC/kg body weight)

hours	^{14}C-PC % dose / ml	d-PC % dose / ml
2	0.087	0.052
6	0.069	0.097
12	0.035	0.030

A minimum quantity of 200 pg deuterium-labelled fatty acid methyl ester (corresponding to 259 pg d_8-PPC) and a deuterium-labelled to non-labelled linoleic methyl ester ratio of 1 : 500 are used for calculation of the 18:0/d-18:0 ratio by mass spectrometry. On the other hand, a minimum amount of 1000 dpm was used for reproducible measurements. This corresponds to 59 ng ^{14}C-PPC. Consequently, with an 18:0/d-18:0 ratio of up to 1 : 500 deuterium-labelling is a more sensitive method for PPC measurement than radiolabelling. When the ratio 18:0/d-18:0 exceeds 1 : 500 radiolabelling will have to be used.

CONCLUSIONS

The results demonstrate that, with the sensitivity of the method used, the metabolic fate of d-labelled PC can be measured in blood.

Pharmacokinetic parameters obtained in experiments with radiolabelled and d-labelled PC are comparable.

These results are an encouraging step towards performing clinical studies in man without exposure to radioactivity.

REFERENCES

1 O. Zierenberg et al., Atherosclerosis 34 (1979) 259 - 276
2 D. LeKim and H. Betzing, Hoppe-Seyler's Z. Physiol. Chem. 357 (1976) 1321 - 1331
3 O. Zierenberg and S.M. Grundy (1981) submitted to J. Lipid Res.
4 G. Rouser et al., Lipids 5 (1970) 494 - 496

DEUTERIUM ISOTOPE EFFECTS IN THE METABOLISM OF THE ANTI-CANCER AGENT 6-MERCAPTOPURINE

M. JARMAN and J. H. KIBURIS[1]
Mass Spectrometry-Drug Metabolism Group, Institute of Cancer Research, Clifton Avenue, Sutton, Surrey, England

G. B. ELION, V. C. KNICK, G. LAMBE, D. J. NELSON and R. L. TUTTLE
The Wellcome Research Laboratories, Burroughs Wellcome Company, 3030 Cornwallis Road, Research Triangle Park, North Carolina 27709, U.S.A.

ABSTRACT

A major detoxification pathway for the antitumour agent 6-mercaptopurine is oxidation, via 8-hydroxy-6-thio-purine, to thiouric acid, mediated by the enzyme xanthine oxidase. Allopurinol, a xanthine oxidase inhibitor, potentiates the antitumour and toxic effects of 6-mercaptopurine by inhibiting its catabolism. Deuterium substitution at C-2 and/or C-8 in 6-mercaptopurine might elicit a similar result by introducing an isotope effect into one or both detoxification steps, hence the metabolism of these deuterated analogues was examined. Large and virtually identical isotope effects (3.5-3.8) were found for the xanthine oxidase catalysed oxidation of 8-[^{2}H]-6-mercaptopurine and its 2,8-[^{2}H]$_2$-analogue in vitro, but none for the 2-[^{2}H]-variant. In two experiments mice given the 2,8-[^{2}H]$_2$-analogue excreted 2.2 and 3.7 times as much unchanged drug and 54 and 70%, respectively, of the thiouric acid produced by the unlabelled drug. This also resulted in an increased antitumour activity (3-5 fold) of the dideuterated analogue against the adenocarcinoma Ca 755 in mice. Neither monodeuterated analogue produced significantly altered levels of drug or metabolite in urine or altered antitumour activity, despite the isotope effect elicited by the 8-[^{2}H]-analogue in vitro. Possibly an alternative metabolic pathway e.g. aldehyde oxidase, contributes to the oxidation of this analogue in vivo.

INTRODUCTION

When a metabolism step for a drug or other xenobiotic occurs by rate-determining cleavage of a C-H bond, it is generally retarded by deuterium substitution. If the isotope effect is large, and occurs in a pathway important for biological activity, then a significant change in activity can result. Thus, deuteration at position 5 in the antitumour agent cyclophosphamide retarded by 5-fold the β-elimination pathway leading to the active metabolite phosphoramide mustard, and produced a corresponding

---

[1]Visiting scientist from the Department of Biological Chemistry, Medical School, University of Athens, Goudi, Athens, Greece.

reduction in potency (ref. 1). Conversely, if a major deactivation pathway is inhibited by deuterium substitution, increased potency could result. This possibility is illustrated in the present report on the deuterated analogues of the antitumour agent 6-mercaptopurine (6-MP).

6-MP is used clinically in the maintenance therapy of acute lymphocytic leukaemia (ref. 2). Like other purine and pyrimidine antimetabolites, it requires metabolic conversion into the mononucleotide, which is the active species (ref. 3). 6-MP is also a substrate for a degradative pathway which results in its inactivation by conversion into 6-thiouric acid by a two-stage hydroxylation (Figure 1) mediated by xanthine oxidase. This detoxification pathway is important since conversion of 6-MP into 6-thiouric acid was reduced 3-5 fold in mice on coadministration of the xanthine oxidase inhibitor allopurinol, and there was a corresponding increase in antitumour activity (ref. 4). Deuterium substitution at positions 2 and 8 is a possible alternative strategy for retarding the catabolism of 6-MP and increasing its potency. Therefore, we have determined the isotope effect for xanthine oxidase-mediated conversion of 2- and 8-monodeuterated and 2,8-dideuterated analogues of 6-MP into thiouric acid in vitro, the urinary outputs of unchanged drug, intermediary metabolites and thiouric acid in mice given 6-MP and each analogue, and activities against the sensitive and resistant sublines of the Ca 755 mammary adenocarcinoma in mice. The effect of deuteration on the rate of oxidation in vitro of the natural substrates hypoxanthine and xanthine (Figure 1) has also been determined.

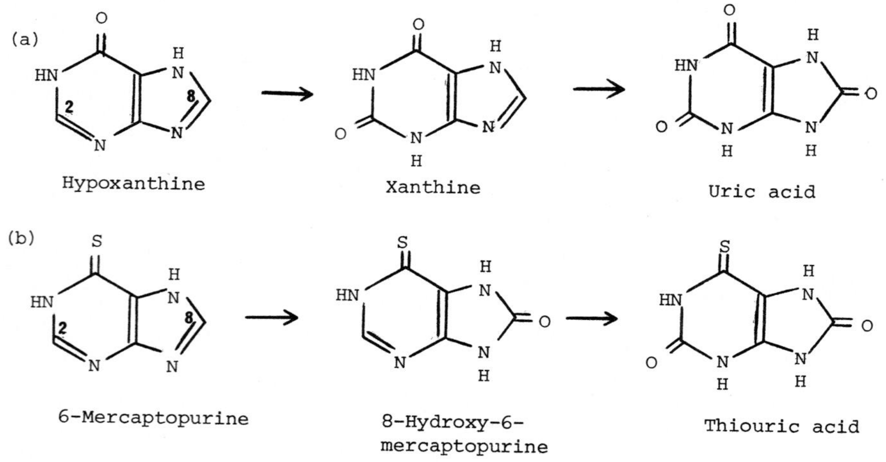

Fig. 1. Xanthine oxidase-mediated catabolism of (a) hypoxanthine (b) 6-mercaptopurine

METHODS

Synthesis of deuterium-labelled analogues

The 8-[^{2}H]-6-MP (2.9 g) was prepared (cf. ref. 5) by boiling for 5 h a solution of 6-MP (3.3 g) in ^{2}H$_2$O (330 ml). 8-[^{2}H]-Xanthine (0.055 g) was similarly prepared from xanthine (0.080 g in 120 ml ^{2}H$_2$O, 16 h reflux). The slower exchange at position 2 of

6-MP in ^2H_2O is preceded by decomposition (ref. 5) but the 2,8-$[^2H]$-analogue (3.7 g) was obtained after boiling 6-MP (5 g) in 0.2M K_2HPO_4 in ^2H_2O (250 ml) during 14 days in the dark. 2,8-$[^2H]_2$-Hypoxanthine (0.31 g) was similarly prepared from hypoxanthine (0.5 g, 1% w/v solution, 2 days reflux). The 2-$[^2H]$-6-MP (2.3 g) was produced by heating under reflux during 16 h an aqueous solution (450 ml) of the 2,8-$[^2H]_2$-6-MP (3 g) whereupon the 8-$[^2H]$-substituent underwent complete exchange with little accompanying loss of the 2-$[^2H]$-substituent.

The deuterated analogues were examined for isotopic purity by ^1H-n.m.r. spectroscopy (cf. ref. 5). The 2,8-$[^2H]_2$- and 8-$[^2H]$-6-MP contained respectively 2% of monodeuterated and nondeuterated congeners, whereas the 2-$[^2H]$-form contained 10% of unlabelled 6-MP. The hypoxanthine analogue was isotopically pure; the xanthine analogue contained 7.5% of nondeuterated material.

Isotope effects in xanthine oxidase-mediated catabolism of deuterated analogues

Large changes in absorbancy accompany the conversion of 6-MP into 6-thiouric acid, those at 315 and 345 nm, near the respective λ_{max} values for substrate and end product respectively being sufficiently great to permit reliable assay of 6-MP with xanthine oxidase (ref. 6). Since the initial rate of oxidation is ca. 100-fold slower for 6-MP than for hypoxanthine (ref. 6), the relative enzyme concentrations were appropriately modified. Due attention was also given to the need to saturate enzyme binding sites. The concentration of 6-MP used was 6.94 x K_i and 7.14 x K_m (ref. 6).

6-MP and its deuterated analogues (1.25 x 10^{-4} M) were each incubated at 37°C with a mixture (2.5 ml) of $Na_2P_4O_7$ (0.055 M) and sodium EDTA (mM), pH 8.2 (made by adding 0.5 ml of the former solution to 40 ml of the latter) and cow milk xanthine oxidase (Boehringer Mannheim GmbH, 25 μl, 10 mg/ml, 0.4 U/mg). Absorbance changes at the stated wavelengths were followed using a Cary Model 16 UV spectrometer and were linear for the duration of the experiments (10-15 min). The procedure used for xanthine and hypoxanthine and the mono- and dideuterated analogues respectively was identical except that the enzyme solution was diluted 100-fold and conversion to uric acid was monitored by following the increase in absorbance at 295 nm (ref. 7).

Metabolism of 6-MP and deuterated analogues in mice

Table 1 compares the amount of unchanged drug with the amount of thiouric acid and other metabolites excreted in the urine following administration of 6-MP and the various deuterated forms (10 mg/kg, i.p.) to CD-1 mice (20-27 g). The drugs were prepared in 0.9% NaCl at 1 mg/ml, and one equivalent of NaOH was added. Animals were housed in metabolism cages. Urine samples were collected at 4 h and each cage was rinsed with 5 ml of water. Dithiothreitol (nM) was added as an antoxidant to each urine specimen. Creatinine was measured colorimetrically. Thiopurines were measured by HPLC, using an ODS-reverse phase column (25 cm x 4.6 mm) and isocratic elution with 1% acetonitrile in 0.05 M potassium phosphate buffer, pH 3.5. Retention times were: thiouric acid 8.4-8.6 min, 8-hydroxy-6-thio-purine 9.7-9.9 min, 6-MP 11.1-11.4 min, and thioxanthine 14.7-15.1 min at a flow rate of 1.0ml/min.

TABLE 1.

Urinary Metabolites of 6-Mercaptopurine and its Deuterated Forms given i.p.(10 mg/kg) to CD-1 Mice

Compound		6-MP	6-thio-xanthine	8-OH-6-SH purine	6-thiouric acid	Ratio 6-MP: thiouric acid	Animals per Test
		0-4 hr Urine Metabolites (nmol/μg creatinine)					
6-MP	(i)[a]	5.5±1.4	0.44±0.05	0.25±0.02	6.6±0.6	0.71±0.16	5
	(ii)[b]	4.4±1.0			5.0±1.4	0.92±0.1	6
2-[^{2}H]-6-MP	(i)	2.2±1.0	0.43±0.18	0.08±0.03	2.3±0.4	0.81±0.32	4
	(ii)	5.0±1.3			2.7±1.4	2.2±0.7	6
8-[^{2}H]-6-MP	(i)	4.0±0.3	0.21±0.06	0.14±0.03	4.3±0.5	0.97±0.08	5
	(ii)	7.7±2.8			6.7±3.2	1.2±0.2	6
2,8-[^{2}H]$_2$-6-MP	(i)	12.1±2.7*	0.50±0.10	0.25±0.01	3.6±0.8*	3.6±0.2*	5
	(ii)	16.2±3.9*			3.5±0.8*	4.8±0.3*	6

* Significant differences from 6-MP; $p < 0.05$
[a] Animals were housed individually. [b] Mice were housed two to a cage.

Antitumour activities

6-MP and its dideuterated analogue were compared using the subline of the Ca 755 mammary adenocarcinoma which is sensitive to 6-MP. Tumour fragments (1 mm) were implanted subcutaneously into B6D2-F$_1$ mice and the drug was administered i.p. 7, 8, 9, 12, 14 and 16 days later at the stated doses. The experiment was terminated 19 days after implant and tumours were resected and weighed. Further tests additionally incorporated 8-[^{2}H]-6-MP. The results are summarized in Table 2.

TABLE 2.

Activity of 6-Mercaptopurine and Deuterated Analogues against the Sensitive Strain of the Ca 755 Mammary Adenocarcinoma in Mice

Compound	dose (mg/kg)	Tumour Weight (control = 1.00)		Tumours remaining	
		(i)	(ii)	(i)	(ii)
6-MP	1	0.81	0.71	6/6	6/6
8-[^{2}H]-6-MP			0.81		6/6
2,8-[^{2}H]$_2$-6-MP		0.35	0.53	6/6	6/6
6-MP	5	0.41	0.38	6/6	6/6
8-[^{2}H]-6-MP			0.59		6/6
2,8-[^{2}H]$_2$-6-MP		0.034	0.10	6/6	5/6
6-MP	25	0.015	0.24	4/6	4/6
8-[^{2}H]-6-MP			0.05		6/6
2,8-[^{2}H]$_2$-6-MP		0.001	0.14	1/6	3/5
6-MP	75	0.001	0.22	1/6	6/6
8-[^{2}H]-6-MP			0.16		5/5
2,8-[^{2}H]$_2$-6-MP		5/6 died	0.02	5/6 died	4/5

RESULTS

Isotope effects

The oxidations of 8-$[^2H]$- and 2,8-$[^2H]_2$-analogues of 6-MP in vitro exhibited virtually identical isotope effects whether measured by decrease in absorbance near the λ_{max} for 6-MP (315 nm; $k[_1H]/k[_2H]$ 3.8 for 8-$[^2H]$-6-MP and 3.5 for the 2,8-$[^2H]$-6-MP) or by increase in absorbance near the λ_{max} for 6-thiouric acid (345 nm; $k[_1H]/k[_2H]$ 3.5 for each analogue). There was no isotope effect for the oxidation of 2-$[^2H]$-6-MP nor for 8-$[^2H]$-xanthine and only a small effect ($k[_1H]/k[_2H]$ 1.16) for 2,8-$[^2H]_2$-hypoxanthine. Mice given 2,8-$[^2H]_2$-6-MP excreted 2.2 [Table 1 (i)] and 3.7 [Table 1 (ii)] times as much unchanged drug and 54 and 70%, respectively, of the thiouric acid found for the normal isotopic 6-MP. This resulted in an increased 6-MP/thiouric acid ratio of ca. 5-fold. In mice given either 8-$[^2H]$-6-MP or 2-$[^2H]$-6-MP there was no significant shift of the ratio of excreted 6-MP/thiouric acid away from normal, indicating no decrease in the rate of oxidation.

Antitumour activities

In the first experiment [Table 2 (i)], in which the activities of 6-MP and its dideuterated analogue against the sensitive subline of the CA 755 mammary carcinoma were compared, from 1 to 25 mg/kg, deuterated 6-MP at 5-fold lower doses appeared as effective as the parent compound in retarding tumour growth. The increased toxicity of deuterated 6-MP became evident only when comparison was made at a higher dose level (75 mg/kg). In the second experiment [Table 2 (ii)], the 8-$[^2H]$-6-MP was also included (2-$[^2H]$-6-MP was not tested because it showed no evidence for an isotope effect in thiouric acid production either *in vitro* or *in vivo*). Again, the dideuterated analogue suppressed tumour growth substantially more than the parent compound at all doses tested. However, 8-$[^2H]$-6-MP showed no advantage over the parent compound.

In a further test against the 6-MP-resistant subline, the dideuterated analogue retarded tumour growth to a greater extent at 25 mg/Kg (tumour weight reduction 69%) than the parent compound (34%), but there was no significant difference at 50 mg/Kg.

DISCUSSION

6-Mercaptopurine deuterated in the 2 and 8 positions, or in the 8 position alone, was oxidized 3.5 times more slowly by xanthine oxidase *in vitro* than was isotopically normal 6-MP. Although no attempt was made in the experiment *in vitro* to follow the formation or degradation of either of the potential intermediates (Figure 1) 8-hydroxy-6-mercaptopurine or 6-thioxanthine, the question as to whether either of these compounds contributed to the absorbance at the wavelength values used in these experiments can be answered by reference to published studies (ref. 8), which show that they are attacked at rates 6-12 times higher than is 6-MP (ref. 8) and therefore are unlikely to accumulate and contribute substantially to the observed absorption. Also, only 6-MP and 6-thiouric acid were detected by paper chromatography during incubation of 6-MP with xanthine oxidase under conditions similar to those used here (6). Moreover, the 8-hydroxy derivative and 6-thioxanthine were much less abundant than unchanged drug and thiouric acid in the urine of mice given 6-MP and its deuterated analogues [Table 1 (i)].

In contrast to the 6-MP analogue, the dideuterated analogue of the naturally occurring substrate for xanthine oxidase, namely hypoxanthine, exhibited a minimal isotope effect for oxidation to uric acid, and 8-deuterated xanthine was oxidized at the same rate as xanthine. The transformation of 6-MP to thiouric acid occurs (ref. 8) via an 8-hydroxylated intermediate (Figure 1), in contrast to the oxidation of hypoxanthine, where 2-hydroxylation occurs first, though it is not clear why this difference should produce the observed result. Only the first step in the oxidation of 6-MP is subject to an isotope effect, since 6-MP deuterated at position 2 only is oxidized at virtually the same rate as 6-MP.

In the metabolism experiments in vivo (Table 1), the extent of decrease in the oxidation of $2,8-[^2H]_2$-6-MP to thiouric acid was comparable to that observed in vitro, 2.2-fold in one experiment [Table 1 (i)] and 3.7-fold in the other [Table 1 (ii)]. The metabolism of $8-[^2H]$-6-MP in mice was not significantly different from that of 6-MP or $2-[^2H]$-6-MP, although there was a large decrease in the rate of oxidation of $8-[^2H]$-6-MP (but not $2-[^2H]$-6-MP) in vitro compared with 6-MP. The lack of effect of deuteration at position 8 upon the fate of the drug in vivo is substantiated by the equivalent antitumour effectiveness of $8-[^2H]$-6-MP and 6-MP [Table 2 (ii)]. These results may reflect an alternative enzyme operating in vivo, e.g. aldehyde oxidase, in addition to xanthine oxidase, to contribute to the oxidation of $8-[^2H]$-6-MP in the mouse.

The change in the metabolism of 6-MP produced by introduction of deuterium in the 2 and 8 positions is reflected in the observed increased anti-tumour effect of $2,8-[^2H]_2$-6-MP against the 6-MP sensitive line of Ca 755. In the first antitumour experiment [Table 2 (i)], the $2,8-[^2H]_2$-6-MP appeared to be ca. 5 times as potent as 6-MP and was also more toxic, as shown by the deaths at 75 mg/kg. An increase in toxicity had also been observed when high doses of 6-MP were given in combination with allopurinol, which also reduced the conversion of 6-MP into thiouric acid ca. 3.5-fold (ref. 4). In the second antitumour experiment [Table 2 (ii)], $2,8-[^2H]_2$-6-MP was somewhat less active than in the first experiment but still showed greater activity than 6-MP at all levels. Only 1/6 animals died at the 75 mg/kg level of $2,8-[^2H]_2$-6MP in this experiment. In summary, the aim of exploiting a deuterium isotope effect to increase the potency of an antitumour agent was realized in the case of 6-mercaptopurine.

ACKNOWLEDGMENTS

This investigation was supported in part by a grant (G973/786) to the Institute of Cancer Research from the Medical Research Council. We thank Professor A. B. Foster for his interest in this work, Dr. A. H. Calvert and Dr. L. I. Hart for helpful discussions and Mr. L. J. Griggs for skilled technical assistance.

REFERENCES

1 P.J. Cox, P.B. Farmer, A.B. Foster, E.D. Gilby and M. Jarman, Cancer Treat. Rep. 60: 1976, 483-491
2 J.V. Simone, R.J.A. Aur, H.O. Hustu and D. Pinkel, Cancer 30: 1972, 1488-1494
3 G.B. Elion, Fed. Proc. 26: 1967, 898-904.
4 G.B. Elion, S. Callahan, H. Nathan, S. Bieber, R.W. Rundles and G.H. Hitchings, Biochem. Pharmacol., 12: 1963, 85-93.
5 J.L. Wong and J.H. Keck, J. Chem. Soc. (Chem. Comm.): 1975, 125-126.
6 H.R. Silberman and J.B. Wyngaarden, Biochem. Biophys. Acta 47: 1961, 178-180.
7 P.G. Avis, F. Bergel and R.C. Bray, J. Chem. Soc.: 1955, 1100-1105.
8 F. Bergmann and H. Ungar, Fed. Proc. 82: 1960, 3957-3960.

DEUTERIUM ISOTOPE EFFECTS IN THE METABOLISM OF DIPHENYLHYDANTOIN

J.A. HOSKINS and P.B. FARMER
MRC Toxicology Unit, Woodmansterne Road, Carshalton, Surrey, U.K.

ABSTRACT

The metabolism of 5-$[^2H]_5$-phenyl-5-phenylhydantoin has been studied in rat liver microsomes and in man to discover any inter- or intramolecular isotope effect.

INTRODUCTION

As part of a study on pregnancy and anti-convulsant drug therapy we have prepared phenytoin (5,5-diphenylhydantoin, DPH) labelled with $[^2H]_5$ in one of the phenyl rings for administration in place of the normal unlabelled material. However before the drug could be used in patients the pharmacological equivalence of the deuterated and non-deuterated forms of the drug had to be established. The comparison of the pharmacokinetic behaviour of the $[^2H]_0$- and $[^2H]_5$-labelled forms of DPH is a study of intermolecular isotope effects. Additionally $[^2H]_5$-DPH is an ideal molecule to investigate if any intramolecular isotope effect occurs during metabolism. Most reported isotope effects in the metabolism of deuterated molecules have been measured intermolecularly: that is, with a mixture of deuterated and non-deuterated analogues. When the rate-determining step leading to the metabolites involves breaking of a carbon-hydrogen bond normally the product formation is enhanced with the non-deuterated species [1]. Intramolecular isotope effects in molecules containing two equivalent groups, one of which is labelled, may be considerable, and larger than the corresponding intermolecular effects found with equivalent molecules. The microsomal demethylation of p-trideuteromethoxyanisole occurs ten times faster on the $C[^1H]_3$ group compared to the $C[^2H]_3$ group, whereas p-dimethoxybenzene is demethylated only 2.1 times faster than p-di(trideuteromethoxy)-

benzene [2]. Similarly an intramolecular isotope effect of 1.6-2.0 could be detected [3] for the cytochrome P450 dependent N-demethylation of N-methyl-N-trideuteromethylphentermine, whereas no intermolecular isotope effect could be seen in the comparison of the N,N-dimethyl and the N,N-di(trideuteromethyl) analogues.

The main initial pathway for the metabolism of phenytoin is the formation of the p-hydroxylated derivative (p-HPPH). The m-hydroxylated isomer (m-HPPH) is also formed to a minor extent in humans and rats although it is a major metabolite in dogs [4]. Metabolism of $[^2H]_5$-DPH will lead to a mixture of $[^2H]_4$ and $[^2H]_5$ hydroxylated derivatives and the ratio of $[^2H]_5:[^2H]_4$ will give the extent of any intramolecular isotope effects in the phenyl ring hydroxylation. p-Substitution of aromatic compounds occurs via rate determining formation of an epoxide intermediate and thus no isotope effect would be expected. Many examples of the lack of intermolecular isotope effects for p-hydroxylation have been reported [5,6,7]. m-Hydroxylation on the other hand has been shown to have an intermolecular deuterium isotope effect for several substrates [6]. It is possible, therefore, that the metabolic formation of these m-hydroxy derivatives may not be via an intermediate arene oxide. For this reason it was of particular interest to determine any intramolecular isotope effect in the formation of m-HPPH from the $[^2H]_5$-labelled DPH. The metabolism was studied using *in vitro* rat liver preparations and humans following oral administration.

METHODS

5-$[^2H]_5$phenyl-5-phenylhydantoin was prepared by a method published [8] for the unlabelled compound from the corresponding labelled benzophenone. The reaction under Friedel-Crafts conditions in the presence of aluminium chloride of benzoyl chloride with $[^2H]_6$-benzene (99.5 atom % D) gave $[^2H]_5$-benzophenone.

Metabolism of $[^2H]_5$-DPH by rat liver microsomes

A microsomal fraction (10,000 g) was prepared from the livers of LACP male rats that had been induced with either phenobarbitone (0.1% in drinking water) or β-naphthoflavone (i.p. in glycerolformal). Metabolism of $[^2H]_5$-DPH by the liver microsomes occurred in the presence of each of two co-factor mixtures [9,10]. The ratio of m- to p-hydroxylation was not significantly affected by the method of induction. A greater proportion of the DPH was metabolised using the co-factors suggested by Kutt and Verebely [10]. The metabolites were extracted from the mixture using the method recommended by Dill [11] for the metabolites in plasma.

Metabolism of $[^2H]_5$-DPH by humans

Two healthy adult volunteers were given phenytoin capsules (200 mg and 400 mg) containing a mixture of equal weights of unlabelled and $[^2H]_5$-DPH. Total urine passed for the succeeding 4 days was collected in timed portions and analysed for metabolites.

Gas chromatography-mass spectrometry (GC-MS)

Samples were chromatographed as their trimethylsilyl (TMS) derivatives on a fused silica capillary column (25 m x 0.3 mm) coated with SE52 phase using helium (2 ml/min) as the carrier gas and the column temperature programmed from 100° to 260°C. This column gave enhanced chromatographic properties for the derivatives compared to a similar column made from pyrex glass (see also Egger [12]). Samples were applied to the column via a falling needle injector. The column was linked directly to a VG 70-70 double focusing mass spectrometer (electron impact (EI) mode: 70 eV) scanning at 1 sec/decade at a resolution (10%) of 1000. The spectra were processed by a VG 2035 Data System.

RESULTS AND DISCUSSION

The EI mass spectra of the compounds all showed weak molecular ions (DPH m/z 396 3%, m-HPPH m/z 484 4.9%, p-HPPH m/z 484 5.3%) and stronger $(M-15)^+$ ions m/z 381 10.7%, m/z 469 12.8%, m/z 469 10.6%. Despite the low intensity of these ions they were chosen for quantification by multiple ion detection as they were of sufficiently high mass to yield uncontaminated analyses.

Intramolecular isotope effects

The hydroxylated metabolites of $[^2H]_5$-DPH from incubation with rat liver microsomal fractions were extracted and converted to their TMS derivatives. The products were monitored at m/z 489 and 488 ($M^{+\cdot}$ for the $[^2H]_5$- and $[^2H]_4$-labelled species) and at m/z 474 and 473 ($(M-15)^+$ for $[^2H]_5$ and $[^2H]_4$-species). A major peak corresponding in retention time to the p-HPPH derivative and a minor peak (ca 1-2% intensity)

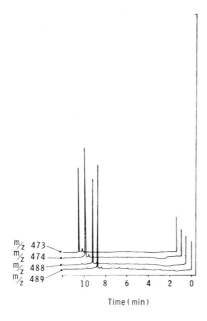

Fig. 1. Multiple ion detection trace for the TMS-derivatives of the metabolic products of $[^2H]_5$-DPH extracted from a rat liver microsomal fraction (m/z 473, 474 $(M-CH_3)^+$; m/z 488, 489 $M^{+\cdot}$).

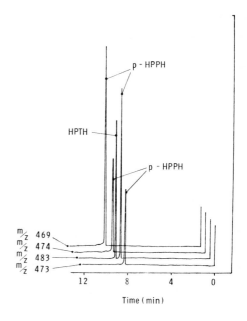

Fig. 2. Multiple ion detection trace for the TMS derivatives of hydroxyphenytoins isolated from human urine following an oral dose of $[^2H]_0$-DPH plus $[^2H]_5$-DPH (m/z 469, 473, 474 $(M-CH_3)^+$). The internal standard was 5-(p-hydroxyphenyl)-5-(p-toluyl)-hydantoin (HPTH) (m/z 483 $(M-CH_3)^+$)

(Fig. 1) at the retention time of the m-HPPH derivative were seen. Mass spectral scanning confirmed the identity of the compounds. The relative intensities of m/z 488:489 and m/z 473:474 were corrected for the inherent $(M-1)^+$, $(M+1)^+$, $(M-14)^+$ and $(M-16)^+$ ions and additionally for the p-$[^2H]_4$- content in the $[^2H]_5$-DPH administered. Average $[^2H]_5:[^2H]_4$ ratios ($[^1H]:[^2H]$ isotope effects) determined from two analyses for p-HPPH were 1.14 (from m/z 488 and 489) and 0.97 (from m/z 473 and 474). Similar determinations on $[^2H]_5/[^2H]_4$ p-HPPH from human urine gave a ratio of 0.83 ± 0.07 (n=27). These results suggest that there is no intramolecular isotope effect for p-hydroxylation as would be expected from the rate determining intermediate formation of an arene oxide.

Determination of the $[^2H]_5:[^2H]_4$ ratio for the m-hydroxy derivative was more problematical owing to the very small amount available.

However measurable peaks were obtained at m/z 488, 489 and at m/z 473, 474. The results from an in vitro rat liver preparation indicate a $[^2H]_5:[^2H]_4$ ratio of 0.90. Results obtained for the m-hydroxy derivative were more variable than those for the p-hydroxylated product owing to the weakness of the signals. No significant isotope effect could be seen in any samples and one may conclude that the rate determining step for the formation of m-HPPH does not involve breakage of a $C-[^2H]$ bond. Rearrangement of an epoxide is therefore a likely possibility for the m-hydroxylation process although other pathways such as dehydration of a dihydrodiol (formed via an arene oxide) could be occurring.

Intermolecular isotope effects

Following administration of an equimolar mixture of $[^2H]_0$- and $[^2H]_5$-DPH to human volunteers, urines were deconjugated with glucuronidase and analysed for p-hydroxylated and m-hydroxylated products using 5-(p-hydroxyphenyl)-5-(p-toluyl)-hydantoin (HPTH) as an internal standard [13].

An example of a urinary analysis is shown in Figure 2. The internal standard is monitored at m/z 483 (M-15)$^+$, $[^2H]_0$-hydroxyphenytoins at m/z 469 (M-15)$^+$, $[^2H]_4$-hydroxyphenytoins at m/z 473 and $[^2H]_5$-

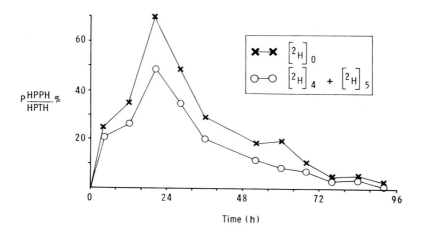

Fig. 3. Elimination curve of p-hydroxylated phenytoins in urine following an oral dose. Internal standard : 77.5 µg/ml urine.

hydroxyphenytoins at m/z 474. The ratio of $[^2H]_0:([^2H]_4 + [^2H]_5)$ gives an indication of the intermolecular isotope effect. The average of 23 determinations of unlabelled : labelled peak heights was 1.22 ± 0.22, compared to a weight ratio of unity for the administered drug. The excretion curve for $[^2H]_0$ and $([^2H]_4 + [^2H]_5)$ p-HPPH is shown in Figure 3. In the absence of data from the spectra of the two pure labelled species we conclude that there is no evidence for an isotope effect in the production of this metabolite.

In conclusion, these results, with others obtained to date, suggest that the pharmacokinetic behaviour of $[^2H]_5$-DPH is such that the administration of this compound in place of the unlabelled material should not upset the therapeutic regimen. No evidence for significant isotope effects in p- and m-hydroxylation has been found.

ACKNOWLEDGEMENTS

We gratefully acknowledge the help of Drs M Landon and A J Cummings of the Division of Perinatal Medicine, CRC, London.

REFERENCES

1. D.R. Hawkins in J.W. Bridges, and L.F. Chasseaud, (Eds.) Progress in Drug Metabolism, Vol. 2, Wiley, 1977, pp.163-218.
2. A.B. Foster, M. Jarman, J.D. Stevens, P. Thomas and J.H. Westwood, Chem.-Biol. Interactions, 9(1974)327-340.
3. G.T. Miwa, W.A. Garland, B.J. Hodshon, A.Y.H. Lu and D.B. Northrop, J. Biol. Chem., 255(1980)6049-6054.
4. A.J. Atkinson, J. Macbee, J. Strong, D. Garteiz and T.E. Gaffney, Biochem. Pharmacol., 19(1970)2483-2491.
5. M. Tanabe, D. Yasuda, J. Tagg and C. Mitoma, Biochem. Pharmacol., 16(1967)2230-2233.
6. J.E. Tomaszewski, D.M. Jerina and J.W. Daly, Biochemistry, 14(1975)2024-2031.
7. P.B. Farmer, A.B. Foster and M. Jarman, Biomed. Mass Spectrom., 2(1975)107-111.
8. H.R. Henze, Method for obtaining hydantoins, U.S. pat. 2,409,754; 1946.
9. P. Mazel in B.N. LaDu, H.G. Mandel and E.L. Way (Eds) Fundamentals of Drug Metabolism and Drug Disposition, Williams and Wilkins, Baltimore, 1971, p.570.
10. H. Kutt and K. Verebely, Biochem. Pharmacol., 19(1970)675-686.
11. W.A. Dill, J. Baukema, T. Chang and A.J. Glazko, Proc. Soc. Exp. Med., 137(1971)674-679.
12. H.J. Egger, W. Wittfoht, H. Nan, G. Dielmann, W. Rapp in A. Quayle, (Ed.) Advances in Mass Spectrometry, Vol. 8, Heyden, 1980, pp.1219-1226.
13. K.M. Witkin, D.L. Bius, B.L. Teague, L.S. Wicse, W.L. Boyles and K.H. Dudley, Therap. Drug Monit., 1(1979)11-34.

INVESTIGATIONS WITH ^{18}O-LABELLED COMPOUNDS OF OXYGEN TRANSFER BY CYTOCHROME P-450 AND METHEMOGLOBIN

H. KEXEL, B. LIMBACH, M. MÖLLER[+] and H.-L. SCHMIDT
Institut für Chemie Weihenstephan,
Technische Universität München, D-8050 Freising (G.F.R.)

[+]Institut für Rechtsmedizin der Universität des Saarlandes,
D-6650 Homburg/Saar (G.F.R.)

ABSTRACT

The "active oxygen" produced by cytochrome P-450 from the reaction O_2 + 2e during the oxygenation of xenobiotics is assumed to have an Fe⋯O structure. Cytochrome P-450 can use organic peroxides in the place of the natural oxidant. In incubations with microsomes the peripheral oxygen atom of cumene [$^{18}O_2$]-hydroperoxide was transferred to several substrates. However, a significant amount of the oxygen introduced into the products originated from water. As neither the oxidant nor the products themselves exchanged oxygen with the water, this oxygen exchange must have taken place at the stage of the active oxygen intermediate.

In a nitrogen atmosphere a mixture of N,N-[$^{14}CH_3$]-dimethylaniline and N,N-dimethylaniline oxide is demethylated by microsomes to N-methylaniline and almost exclusively non-radioactive formaldehyde. In order to elucidate the fate of the oxygen atom, another porphyrin-catalysed isomerization was investigated. N,N-Dimethylaniline [^{18}O]-oxide yielded o-N,N-dimethylamino[^{18}O]-phenol when incubated with methemoglobin. The products obtained from the incubation of a mixture of the oxygen- and the nitrogen-labelled compound proved that the porphyrin-catalysed oxygen shift proceeded is an intramolecular reaction. The results are in line with the concept of a homolytic O-O bond fission in the course of the oxygen activation by cytochrome P-450.

INTRODUCTION

Cytochrome P-450, occurring in the 100,000 g fraction of cell homogenates ("microsomes"), catalyses hydroxylations of aromatic and aliphatic compounds, N-demethylations and oxygenations on sulfur atoms. The reactions are so-called mixed function oxidations because, in addition to the oxidant O_2, two electrons (mainly from NADPH) are needed for its activation:

$$A-H + {}^{**}O_2 + 2\ H^+ + 2\ e \longrightarrow A-{}^*OH + H_2{}^*O \quad (A-H = \text{substrate})$$

In some cases hydroperoxides can replace NADPH + O_2 [1]. A Fe···O-structure, proposed for the normally produced "active oxygen", may therefore also be formed from other oxidants. In this investigation we compared porphyrin-catalysed oxygen transfer reactions from O_2 and other oxidants in order to help elucidate the nature of the active oxygen intermediate.

MATERIALS AND METHODS

Syntheses of labelled compounds

a) Cumene [$^{18}O_2$]-hydroperoxide was prepared according to the method of Armstrong et al. [2]. Cumene, $^{18}O_2$ and an aqueous mixture of sodium carbonate and potassium stearate were shaken at 85°C for 2.5 h in a sealed glass tube. The cumene hydroperoxide formed was extracted with benzene and distilled under reduced pressure.

b) N,N-Dimethylaniline [^{18}O]-oxide and N,N-dimethyl[^{15}N]-aniline were prepared as described earlier [3]; N,N-[$^{14}CH_3$]-dimethylaniline was synthesized from N-methylaniline and $^{14}CH_3I$.

Incubations

a) The incubation mixtures for the oxygenations with cumene hydroperoxide contained the oxidant (0.8 mM), Tris buffer (pH 7.4)(0.1 M), substrate (2 mM) and 10-15 mg/ml of microsomal protein. After the incubation the products were extracted, purified by TLC on silica gel using light petroleum - ethyl acetate (7:1) as developing solvent, and analysed by mass spectrometry. Cyclohexanol was derivatized with trifluoracetic anhydride and analysed by GC/MS using a Tenax column. The unreacted cumene hydroperoxide, after purification by TLC, was distilled under reduced pressure and analysed by the method of Rittenberg and Ponticorvo [4].

b) Incubations of mixtures of labelled N,N-dimethylaniline oxides with methemoglobin were performed according to Kiese et al. [5]. The products were extracted with diethyl ether, and the extract was analysed either by TLC or by GC/MS using a Tenax column. The mass spectrum was evaluated by a subtraction method. Formaldehyde was determined according to Nash [6], and radioactive formaldehyde, after addition of carrier and distillation, by liquid scintillation counting.

RESULTS
Oxygen transfer from cumene hydroperoxide

The extent of the oxygen incorporation from cumene hydroperoxide into the products was similar to that from O_2 (Table 1). No isotope dilution was detected in the remaining oxidant, and the dimethylphenylcarbinol formed in stoichiometric amounts from cumene hydroperoxide had the same specific label as the oxidant itself. Likewise, the products obtained by the oxidation did not exchange oxygen with water under the conditions of the incubation. However, in all incubations with ^{18}O-labelled oxidants, appreciable incorporation of

TABLE 1

Oxygen transfer from ^{18}O-labelled oxidants by incubation of substrates with rat liver microsomes

Reaction[a] type	Oxygenation system	^{18}O-incorporation into product [%] from:		Relative reaction velocity
		oxidant	water	
$\geq$P=S → $\geq$P=O	$^{18}O_2$/NADPH, $H_2^{16}O$	82 ± 4	(18)[b]	–
	[$^{18}O_2$]-CumOOH, $H_2^{16}O$	80 ± 2	(20)	1.0
	[$^{16}O_2$]-CumOOH, $H_2^{18}O$	(80)	20 ± 5	1.0
-SO-CH$_3$ ↓ -SO-CH$_3$	$^{18}O_2$/NADPH, $H_2^{16}O$	80 ± 4	(20)	–
	[$^{18}O_2$]-CumOOH, $H_2^{16}O$	85 ± 2	(15)	1.7
$\geq$C-H ↓ $\geq$C-OH	[$^{18}O_2$]-CumOOH, $H_2^{16}O$	92 ± 2	(8)	2.5
	[$^{16}O_2$]-CumOOH, $H_2^{18}O$	(92)	8[c]	

[a] abbreviations used: $\geq$P=S = parathion; $\geq$P=O = paraoxon; CumOOH = cumene hydroperoxide; -SO-CH$_3$ = p-methylsulfinylaniline; -SO$_2$-CH$_3$ = p-methylsulfonylaniline; $\geq$C-H = cyclohexane; $\geq$C-OH = cyclohexanol.

[b] data in parentheses are calculated as differences from values determined.

[c] after Nordblom et al. [7].

oxygen from water into the product was observed, and the extent of this "exchange" was inversely proportional to the reaction velocity. Therefore, this oxygen exchange must have occurred at the stage of an oxygenated intermediate of the enzyme and must depend on its lifetime during the reaction.

In addition, the fact that the specific label of the oxygen atom bound directly to the carbon atom of the α,α-dimethylbenzyl residue did not change during the reaction indicates, that only the peripheral oxygen atom is involved in the oxygenation.

Like cumene hydroperoxide, thioacetamide sulfoxide could replace O_2 + 2e in microsomal oxidations, at least in N-demethylations and in the oxidation of the sulfur atom of p-aminophenyl methyl sulfide. Among potential oxidants tested without success were nitrosobenzene and N,N-dimethylaniline oxide. Nevertheless, as can be seen in the following chapter, under modified conditions the oxygen of the latter compound can be transferred by porphyrin proteins.

Oxygen shift in amine oxides

Rat liver microsomes dealkylate N,N-dimethylaniline oxide (DMAO) to methylaniline (MA) and formaldehyde without further need for oxygen. This reaction is an intramolecular shift, because during the incubation of a mixture of DMAO and N,N-[$^{14}CH_3$]-dimethylaniline ([$^{14}CH_3$]-DMA) in a nitrogen atmosphere, nearly exclusively non-radioactive formaldehyde was produced. However, the fate of the oxygen atom is not elucidated by this experiment. The demethylation of ^{18}O-labelled DMAO would not be conclusive because of the rapid oxygen exchange of formaldehyde with water.

TABLE 2
Isotope labels in o-hydroxy-N,N-dimethylaniline (HDMA) originating from incubation of a mixture of ^{15}N- and ^{18}O-labelled N,N-dimethyl-aniline oxide (DMAO) with methemoglobin

Substrate	Products expected	MW	Relative mass peak intensities		
			calculated		found
			intermol. reaction	intramol. reaction	
	[^{14}N,^{16}O]-HDMA	137	2	0	0.2
[^{15}N,^{16}O]-DMAO (50 %) + [^{14}N,^{18}O]-DMAO (50 %)	[^{15}N,^{16}O]-HDMA	138	1	3	2.9
	[^{14}N,^{18}O]-HDMA	139	1	3	2.9
	[^{15}N,^{18}O]-HDMA	140	2	0	0

As is the case with cytochrome P-450, methemoglobin can hydroxylate aniline to p-aminophenol [8] with NADPH and O_2 as the oxidizing system and can also dealkylate DMAO [5]. In addition, methemoglobin catalyses the formation of o-hydroxy-N,N-DMA (HDMA) from DMAO. Experiments with [^{18}O]-DMAO showed that the oxygen atom of the phenolic group originates from the N-O group. In order to obtain final proof for the intramolecular oxygen transfer, a mixture of [^{15}N,^{16}O]-DMAO and [^{14}N,^{18}O]-DMAO was incubated with methemoglobin. As can be seen from Table 2, the reaction proceeded without isotope scrambling. Thus, the methemoglobin-catalysed isomerization of the substrate is an intramolecular oxygen shift.

DISCUSSION

Porphyrin-containing oxygenases catalyse the transfer of the oxygen not only from O_2 but also, without additional need for electrons, from organic oxygen carriers with oxygen bound to heteroatoms such as I [9], O [1], S or N. Probably the O atom of these oxidants is taken by the porphyrin iron and then transferred to the substrate. In cumene hydroperoxide-mediated oxidation, the oxidant and the substrate seem to be bound simultaneously to the active centre of cytochrome P-450. On the other hand, as shown by binding spectra, the hydrophile oxidant DMAO has only a low affinity to the enzyme and may even be displaced by more lipophilic substrates. From DMAO bound to the active site at the moment of the loss of the oxygen atom, the more lipophilic substrate DMA is formed, which is tightly bound to the enzyme and immediately becomes oxygenated. Thus, the reaction appears as an intramolecular shift. Nevertheless, the nature of the active oxygen my be similar in both instances.

According to White et al. [10], the active oxygen intermediate originating from cumene hydroperoxide is formed by homolytic bond fission. The following scheme (fig. 1) proposed by these authors can be completed by our own results, especially concerning the observations of the oxygen exchange described before.

After the binding of the oxidant and the substrate to the active site of the enzyme a homolytic scission of the peroxy group yields an iron(III)-porphyrin complex with the peripheral oxygen atom. This complex can exchange its OH$^-$ group with water, and the extent of this exchange is inversely proportional to the velocity of the substrate oxygenation. Our results exclude the reversibility of the peroxide splitting suggested by White et al. [10], because under no conditions could any oxygen exchange between the peroxide and water be observed.

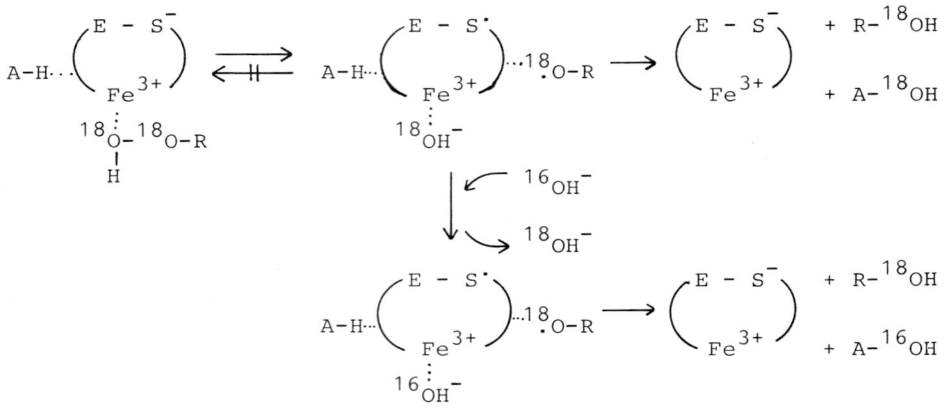

Fig. 1. Proposed mechanism of the oxygen activation by cytochrome P-450 [10].

REFERENCES

1 F.F. Kadlubar, K.C. Morton and D.M. Ziegler, Biochem. Biophys. Res. Commun. 54(1973)1255-1261
2 G.P. Armstrong, R.H. Hall and D.C. Quin, J. Chem. Soc. (1950) 666-670
3 H.-L. Schmidt, N. Weber and M. Halmann, J. Labelled Compds. 7(1971) 171-174
4 D. Rittenberg and L. Ponticorvo, Intern. J. Appl. Radiation Isotopes 1(1956)208-214
5 M. Kiese, G. Renner and R. Schlaeger, Naunyn-Schmiedebergs Arch. Pharmacol. 268(1971)247-263
6 T. Nash, Biochem. J. 55(1953)416
7 G.D. Nordblom, R.E. White and M.J. Coon, Arch. Biochem. Biophys. 175(1976)524-533
8 J.J. Mieyal, R.S. Acherman, J.L. Blumer and L.S. Freeman, J. Biol. Chem. 251(1976)3436-3441
9 F. Lichtenberger, W. Nastainczyk and V. Ullrich, Biochem. Biophys. Res. Commun. 70(1976)939-946
10 R.E. White, S.G. Sligar and M.J. Coon, J. Biol. Chem. 255(1980) 11108-11111.

H.-L. Schmidt, H. Förstel and K. Heinzinger (Editors), *Stable Isotopes*
© 1982 Elsevier Scientific Publishing Company, Amsterdam — Printed in The Netherlands

ANALYTICAL TECHNIQUES FOR USING MULTIPLE, SIMULTANEOUS STABLE ISOTOPIC TRACERS

D.L. HACHEY, K. NAKAMURA*, M.J. KREEK[†], and P.D. KLEIN
Stable Isotope Laboratory, Children's Nutrition Research Center, Department of Pediatrics, Baylor College of Medicine, Houston, Texas 77030, *Sankyo Co. Ltd., Tokyo, Japan, [†]Rockefeller University, New York, New York 10021

ABSTRACT

Analytical techniques have been developed for the simultaneous use of several stable isotopically labeled forms of methadone. These methods permit simultaneous quantitation of both methadone enantiomers down to 5 ng per g plasma and permit pharmacokinetic studies for 7 days following a single tracer dose.

INTRODUCTION

Stable isotopic tracer studies most commonly are executed using only a single enriched isotopic species. Single tracer studies satisfy the majority of experimental needs and offer a number of advantages, including a simplified experimental protocol and simplified isotopic tracer calculations. However, the disadvantages become apparent in studies that require multiple isotopic tracers, for example, to investigate the pharmacokinetic behavior of enantiomeric drugs. Such protocols previously required multiple, sequential studies that are expensive and impractical for routine use in a clinical environment.

Multiple isotopic tracer studies most frequently have been done using simple molecules or elements without a complex isotopic distribution. Elaborate isotopic distribution patterns are the usual result of chemical or biochemical synthesis of multiply-labeled organic molecules (ref. 1). When two or more such complex isotopic species are used simultaneously, the true composition of the mixture can be determined only using mathematical techniques to deconvolute the resulting isotopic pattern in the mass spectrum (ref. 2). We herein describe methods for the simultaneous quantitation of unlabeled methadone and two deuterium labeled species, S-(+)-methadone-$^2H_3$ and R-(-)-methadone-$^2H_5$, in plasma using racemic methadone-$^2H_8$.

EXPERIMENTAL METHODS

Three isotopically labeled forms of methadone, S-(+)-methadone-$^2H_3$, R-(-)-methadone-$^2H_5$, and (±)-methadone-$^2H_8$, were prepared using modifications of established procedures (ref. 3). The isotopic abundances were determined using ammonia chemical ionization mass spectrometry (Table 1). These data were used to construct a set of simultaneous linear equations (Equation 1) that describe completely all isotopic contributors measured at a specific ion. These equations were used to obtain corrected isotopic abundances. A set of calibration curves was prepared that covered a useful linear range of 5-200 ng/g of plasma for the -$^2H_3$ and -$^2H_5$ isotopic species and 20-1200 ng/g of plasma for unlabeled methadone.

TABLE 1

Relative isotopic abundances of the deuterated methadone isotopic species used in the clinical studies.

Mass (MH$^+$)	-$^2H_0$	-$^2H_3$	-$^2H_5$	-$^2H_8$
310	100.00	0.16	0.21	
311	23.41	0.26	NA	
312	2.94	2.64	1.03	
313	0.23	100.00	5.20	
314		23.20	23.75	
315		2.86	100.00	0.09
316		0.22	29.41	0.54
317			10.29	8.58
318			7.03	100.00
319				22.70
320				2.68

ISOTOPIC ABUNDANCE CALCULATIONS

The ion intensity ratios were measured at m/z 310, m/z 313, and m/z 315 and were normalized to the ion intensity of the internal standard at m/z 318. The ion intensities at a specific mass (I_x) were the sum of several terms consisting of the product of an abundance coefficient ($A_{i,j}$) and an unknown mole fraction (X_k), as shown in equation 1.

$$I_{310} = A_{1,1} X_0 + A_{1,2} X_3 + A_{1,3} X_5 + A_{1,4} X_8$$
$$I_{313} = A_{2,1} X_0 + A_{2,2} X_3 + A_{2,3} X_5 + A_{2,4} X_8$$
$$I_{315} = A_{3,1} X_0 + A_{3,2} X_3 + A_{3,3} X_5 + A_{3,4} X_8$$
$$I_{318} = A_{4,1} X_0 + A_{4,2} X_3 + A_{4,3} X_8 + A_{4,4} X_8$$

(1)

These equations could be expressed in matrix notation as equation 2.

$$I_x = A X \qquad (2)$$

The solution to obtain X in the set of equations can be obtained by matrix algebraic methods. The solution is the product of the inverse of A transpose A, A transpose and the normalized ion abundance (I_x), as shown in equation 3.

$$X = (A^T A)^{-1} A^T I_x \qquad (3)$$

This method assumes that one has a homogenous set of ions comprising one unique elemental formula, although the equations can be expanded to include as many terms as necessary to handle ions formed by hydride abstraction or other ion fragmentation processes. The penalty, however, is decreased statistical accuracy due to propagation of experimental errors associated with the empirically determined matrix coefficients and the experimentally determined isotopic ratios (ref. 4).

CLINICAL PROTOCOLS

Methadone maintenance treatment patients were admitted to a metabolic research unit. Unlabeled, racemic methadone (0.5-1.0 mg/kg) was administered to each patient p.o. prior to the tracer dose to establish a kinetic steady-state. A pseudo-racemic mixture of S-(+)-methadone-$^2H_3$ and R-(-)-methadone-$^2H_5$, the active enantiomer, was given to the patient in place of his normal daily dose. Blood was collected hourly for the first 10 h, and then at 24 h intervals for a two week period. Total urine and stool samples were collected daily for 2 wk. Unlabeled methadone-$^2H_0$, S-(+)-methadone-$^2H_3$ and R-(-)-methadone-$^2H_5$ were simultaneously measured in plasma and urine using ($\pm$) methadone-$^2H_8$ as the internal standard.

RESULTS

The methods described above permit quantitation of unlabeled methadone and both S-(+)-methadone-$^2H_3$, and R-(-)-methadone-$^2H_5$ with detection limits in the range of 1-5 ng/g plasma. This sensitivity enables us to follow methadone levels in plasma for up to one week following administration of the labeled tracer. These methods are being used currently in several on-going projects to study methadone metabolism in alcoholism and in liver disease. The data obtained so far support our earlier findings that the inactive enantiomer is cleared from the body more rapidly than the active one. Figure 1 shows the plasma methadone clearance for one patient. In this instance, unlabeled methadone reached a steady-state level of 200 ng/g. S-(+)-methadone-^{2}H$_3$ reached a peak plasma level of 120 ng/g and was cleared with a half-life of 28.1 h. R-(-)-methadone-^{2}H$_5$ reached a peak plasma value of only 70 ng/g and was cleared with a half-life of 37.9 h. The ratio of the two half-lives for these enantiomers (R:S) was 1.35:1.

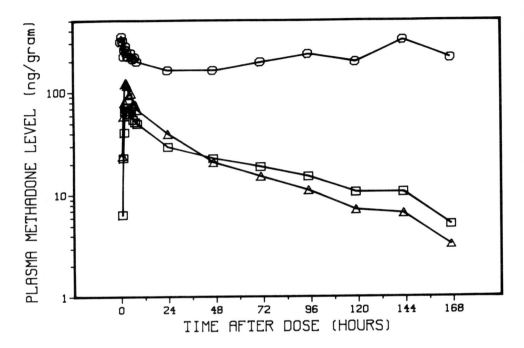

Fig. 1. Plasma clearance of (±) methadone (0), S-(+)-methadone-^{2}H$_3$ (△), and R-(-)-methadone-^{2}H$_5$ (□) in a methadone maintenance patient.

These methods also can be used to study urinary excretion of the two enantiomers. Figure 2 illustrates the cumulative urinary excretion of the two enantiomers over an eight day period. S-(+)-methadone-^{2}H$_3$ reached a peak urinary excretion of about 5% of the administered dose, whereas 10% of the R-(-)-methadone-^{2}H$_5$ was excreted over the same time period. The mechanism underlying the differential renal clearance of the two enantiomers is uncertain.

In summary, the methods presented here for simultaneous isotopic tracer studies are nearly as rapid and convenient as single tracer studies. They offer the unique advantage of permitting multidimensional kinetic and quantitative studies that are impossible using radiolabeled tracers.

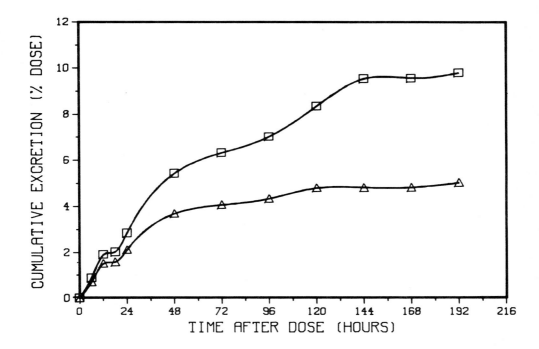

Fig. 2. Cumulative urinary excretion of S-(+)-methadone-$^2H_3$ (Δ) and R-(-)-methadone-$^2H_5$ (□) following a single 70 mg dose of methadone.

ACKNOWLEDGMENTS

Publication of this manuscript was supported by the USDA/SEA Children's Nutrition Research Center in the Department of Pediatrics, Baylor College of Medicine and Texas Children's Hospital. This work was supported by the National Institute on Drug Abuse grant DA-01138, and the National Institute of Health General Clinical Research Center grant RR-00102. Dr. Kreek is the recipient of a Research Scientist Award (DA-00049) from HEW-ADMAHA-NIDA. The authors wish to acknowledge the support given to Dr. Nakamura by D.K. Murayama, Director of Central Research Laboratories, Sankyo Company, Ltd., Tokyo Japan, during his sabbatical visit to the Argonne National Laboratory.

REFERENCES

1 H. Yamamoto and J.A. McCloskey, Anal. Chem., 49(1977)281-283.
2 J.I. Brauman, Anal. Chem., 38(1966)607-611.
3 K. Nakamura, D.L. Hachey, M.J. Kreek, C.S. Irving, and P.D. Klein, J. Pharm. Sci., (1981) In press.
4 D.L. Hachey, J.-C. Blais, and P.D. Klein, Anal. Chem., 52(1980)1131-1135.

DETERMINATION OF URAPIDIL IN HUMAN SERUM USING SOLID PROBE CHEMICAL IONIZATION MASS SPECTROMETRY AND STABLE ISOTOPE LABELING

E. STURM and K. ZECH
From the Research Laboratories of Byk Gulden Lomberg Chemische Fabrik GmbH,
D-7750 Konstanz (F.R.G.)

ABSTRACT

A quantitative analytical procedure for the analysis of the antihypertensive drug 6-(3-(4-(o-methoxyphenyl)-1-piperazinyl)-propylamino)-1,3-dimethyluracil (urapidil, Ebrantil®) in human serum has been developed which involves the addition of a deuterium-labeled internal standard to serum, followed by extraction and direct sample insertion into a mass spectrometer operating under chemical ionization conditions. Peak height ratios of the $(M + H)^+$-ions formed upon ionization with isobutane as reactant, used to calculate serum levels, were obtained from repetitive scan data. In this manner, serum levels from 20 ng to 1 µg per ml serum could be quantitated. Examples are given of serum levels obtained in the course of pharmacokinetic studies after i.v. (10 mg) and oral (30 mg) administration of urapidil.

INTRODUCTION

Urapidil (1a) is a new antihypertensive agent (ref. 1) whose efficacy on intravenous and oral administration has been demonstrated in clinical trials (ref. 2). In order to better adapt the onset and persistence of activity to clinical requirements, a new sustained released form has been developed. The present study was conducted to gain further insight into the pharmacokinetics of urapidil on administration as capsule and on bolus injection. From an earlier study (ref. 3) with ^{14}C-labeled urapidil we know that therapeutic concentrations could be less than 500 ng per ml serum. In this range, application of the HPLC-technique, as described for monitoring urapidil-levels (ref. 4), is difficult, because extraction from serum is incomplete and subject to intra- and interindividual variation. Therefore we tried to develop a new analytical method based on mass spectrometry using a deuterium labeled analog of urapidil as internal standard. As we did not succeed in converting urapidil into a compound volatile enough for detection by GC-MS, we tried to utilize the advantages of direct sample insertion mass spectrometry (refs. 5 - 8).

1a R = ^{1}H Urapidil (MW 387)

1b R = ^{2}H (^{2}H)$_4$-Urapidil (MW 391)

EXPERIMENTAL

Mass spectral analysis

The instrument used in this study was a VG MM 7070F high resolution mass spectrometer equipped with a dual electron impact/chemical ionization source and interfaced with a VG datasystem 2035. Samples were introduced using silica tubes. All CI-spectra were obtained using isobutane as reagent gas at 2×10^{-5} torr source housing pressure and 200°C ion chamber temperature. The instrument was scanned over the mass range of interest and the sample was warmed up by induction from the ion chamber. Serum sample analysis was performed at 3000 resolving power in order to overcome the problem of interference from endogenous substances in the serum. The relevant protonated molecular ions $(M+H)^+$-peak heights of approximately four scans were measured manually and averaged to determine ion abundance.

Synthesis of (^{2}H)$_4$-urapidil (1b)

The compound was prepared by the method of Ludwig et al. (ref. 4) for the synthesis of ^{14}C-urapidil, starting from (^{2}H)$_4$-dibromoethane (E. Merck, Darmstadt) and the corresponding diaminocompound. Mass spectral analysis showed that the product contained less than 0.33% unlabeled urapidil.

Serum sample analysis

During the development of this method, and later on for control purposes, standard samples were produced by adding known amounts of urapidil (0 - 500 ng/ml) and (^{2}H)$_4$-urapidil (500 ng/ml) to 3 blank serum samples (Biotest-Serum-Institut, Frankfurt). After shaking and equilibration for 10 min, 40 µl 1 N NaOH was added and the basic mixture successively extracted with 1x20 ml and 1x10 ml of glass-distilled dichlormethane. The extract was dried with Na$_2$SO$_4$, and 20 ml of the solution was evaporated to dryness. The residue was dissolved in 1 ml of dichloromethane, the solution transferred into a minivial, and the solvent removed by a N$_2$-stream. The residue was dissolved in 20 µl dichloromethane, and 10 µl was placed on the silica sample tube of the direct inlet probe.

Blood samples were obtained from 6 volunteers, who had received a 10 mg intravenous dose and, one week later, a 30 mg oral dose of urapidil. Just before administration, blood was withdrawn to obtain a blank value. Successive blood samples were than taken during the 24 hours following administration. Serum was separated by centrifugation. Internal standard was added to 3 ml serum and each sample extracted as described.

RESULTS AND DISCUSSION

The mass spectra of urapidil upon electron impact (EI) and chemical ionization (CI) are shown in figs. 1a and 1b. Although the base peak in the EI-spectrum is predominant and could be used for quantitation, in principle it seemed to be more expedient to develop a method based on measurements of the protonated molecular ion $(M + H)^+$ obtained upon chemical ionization with isobutane, for several reasons: (1) evaporation of equal amounts of urapidil is faster under CI- than under EI-conditions and increases the sensitivity; (2) CI reduces the chance that background ions contribute essentially to the signal of a particular mass; (3) the $(M+H)^+$-ion appearing at a higher m/e value (388) than the base peak (225) in the EI-spectrum should less likely to suffer interference from endogenous substances; (4) detection of the $(M+H)^+$-ion provides higher flexibility in the synthesis of a suitable internal standard because there are no restrictions concerning labeling positions due to fragmenta-

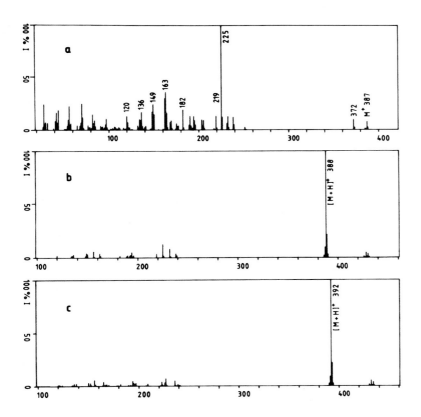

Fig. 1. Mass spectra of urapidil (a: EI, 70 eV; b: isobutane-CI) and of $(^2H)_4$-urapidil (c: isobutane-CI).

tion processes or isotope effects.

To avoid cross contributions, it appeared desirable to use an internal standard containing at least three deuterium atoms. For chemical reasons, it was preferable to synthesise the $(^2H)_4$-analog 1b. Its CI-spectrum is shown in fig. 1c.

The presumption that deuterium-labeling does not cause differences in the evaporation profile of urapidil and internal standard was investigated and confirmed by low resolution selected ion monitoring (SIM) of mixtures containing the two compounds (fig. 2).

SIM-analysis of serum samples, spiked with urapidil and standard, gave erroneous results. High resolution scans showed that two factors were responsible for this: a) serum samples contained dioctylphthalate (DOP), a compound widely used as plasticizer and present in solvents, reagents and other materials (ref. 9), and b) endogenous cholesterol. DOP produces a $(M+H)^+$-ion at mass 391 accompanied by a ^{13}C-isotope-peak at nominal mass 392 of the internal standard. The concentration of DOP could be reduced to an acceptable level by distilling the dichloromethane in an all-glass apparatus, but it was not possible to remove it completely from the serum samples, probably

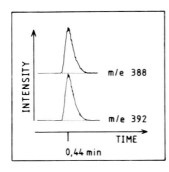

Fig. 2

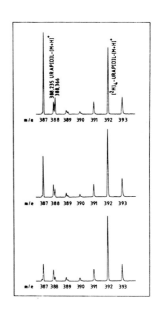

Fig. 2. Computer-reconstructed evaporation profiles of urapidil (m/e 388) and $(^2H)_4$-urapidil.

Fig. 3. Typical sequence of CI-mass spectra obtained during the evaporation of the same serum sample

Fig. 3

because the contaminant had already been present in the serum used. The remaining DOP is evaporated before the signal of urapidil appears, since phthalates are relatively volatile compounds. Control of the time-dependent intensity of the DOP-$(M+H)^+$-signal showed that there was now no contribution by the $^{13}C-(M+H)^+$ to the signal of the internal standard. Urapidil could not be completely separated from cholesterol. Cholesterol produces a ^{13}C-signal from the $(M+H)^+$-ion at nominal mass 388, which contributes to the signal of urapidil. It was, therefore, necessary to increase the resolving power in order to separate the two signals (urapidil $(M+H)^+$: 388.235, cholesterol $^{13}C-(M+H)^+$: 388.366). Fig. 3 shows typical scans obtained at different times during evaporation of the same serum sample. The ratio of the peak heights of the doublet due to urapidil and cholesterol changes as expected, during the time of evaporation; the ratio of the peak heights of urapidil and internal standard, however, remains constant. This observation prompted us to analyze serum samples by scanning over the mass range from 387 to 393 at a resolving power of 3000 and to calculate concentrations from peak height ratios.

This approach was tested using standard serum samples. Plotting observed levels versus added levels, a straight calibration line was obtained (fig. 4).

The ordinate intercept reflects the fact that there is still a trace level of background contamination at the mass of urapidil. The very good correlation between expected and observed values invites the assumption that the background level is nearly constant for all serum samples from the same subject. This assumption was confirmed by the results obtained with blank sera from blood samples drawn from two volunteers at different times. The background level corresponds to 18 ng per ml serum ($\bar{X} \pm SD$: 18 ± 4; n = 8). Therefore, it appeared justified to quantitate urapidil serum levels by interpolation of the calibration line after correction by the blank value. The results from six standard samples containing 50 and 350 ng per ml, respectively, are shown in table 1. Precision of the test system varied between 2.5% and 13%, depending on the range of urapidil concentration (350 ng/ml and 50 ng/ml, resp.). The lower limit of detection is 20 ng per ml serum.

TABLE 1

Precision of the results from serum samples containing 50 an 350 ng ml^{-1} urapidil.

Sample	Urapidil added (ng/ml)	Urapidil observed (ng/ml)	Mean ± SD
1	50	40	
2	50	47	46 ± 6
3	50	51	
4	350	346	
5	350	363	355 ± 9
6	350	356	

The above method was used to measure urapidil in the serum of six healthy volunteers after bolus injection (10 mg) and a single dose of urapidil as capsule (30 mg). Typical results from one subject are shown in fig. 5. Serum levels could be measured up to 8 h after i.v. and 14 h after oral administration. Comprehensive results will be reported elsewhere.

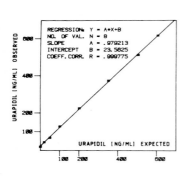
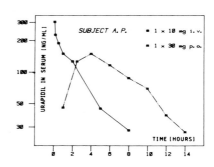

Fig. 4   Fig. 5

Fig. 4. Standard calibration line of urapidil from serum.

Fig. 5. Serum concentrations (log) of urapidil versus time in subject A. P. after i.v. administration (10 mg) of urapidil and after a single oral dose of a urapidil capsule (30 mg).

ACKNOWLEDGEMENTS

We are grateful to Priv.-Doz. Dr. E.G. Bruckschen for providing the blood samples, to Dr. W. Prüsse for the synthesis of $(^2H)_4$-urapidil and to Dr. U. Krüger for his helpful discussions.

REFERENCES

1 K. Klemm, W. Prüsse and U. Krüger, Arzneim.-Forsch./Drug Res. 27 (1977) 1895.
2 E.G. Bruckschen, F. Henze and G. Michael, Arzneim.-Forsch./Drug Res. 28 (1978) 1176.
3 W.R. Kukovetz, G. Ludwig, H. Vergin, K. Zech, V. Steinijans and E.G. Bruckschen, Arzneim.-Forsch./Drug Res. 27 (1977) 2406.
4 G. Ludwig, H. Vergin and K. Zech, Arzneim.-Forsch./Drug Res. 27 (1977) 2077.
5 B.J. Millard, Quantitative Mass Spectrometry, Heyden, London 1978, pp. 91-115, and literature cited therein.
6 R.J. Weinkam, M. Rowland and P.J. Meffin, Biomed. Mass Spectrom. 4, (1977) 42.
7 R.J. Weinkam and Huey-shin Lin, Anal. Chem. 51 (1979) 972.
8 R.W. Walker, V.F. Gruber, A. Rosenburg, F.J. Wolf and W.J.A. Vanden Heuvel, Anal. Biochem. 95 (1979) 579.
9 M. Ishida, K. Suyama and S. Adachi, J. of Chromatography 189 (1980) 421.

DETERMINATION OF CATECHOL-O-METHYLTRANSFERASE (COMT) ACTIVITY BY
GAS CHROMATOGRAPHY-MASS SPECTROMETRY USING A MIXTURE OF DEUTERATED
CATECHOLAMINE AS MULTI-SUBSTRATE SYSTEM

H. MIYAZAKI and Y. HASHIMOTO
Research Laboratories, Pharmaceutical Division, Nippon Kayaku Co.,
3-31, Shimo Kitaku Tokyo 115, Japan.

ABSTRACT

An assay method is described for the determination of COMT activity by selected ion monitoring (SIM) after chemical ionisation (CI) with isobutane as a reagent gas. The assay method uses a mixture of $[^2H_2]$-dopamine, $[^2H_3]$-norepinephrine and $[^2H_3]$-epinephrine as a "multi-substrate system", and a mixture of $[^2H_5]$-3-methoxytyramine, $[^2H_6]$-normetanephrine and $[^2H_6]$-metanephrine as internal standards. This "multisubstrate system" has been applied for the determination of the COMT activity in the brain of rats stressed by restraint and water immersion. The activity was significantly decreased in comparison with that in the corresponding tissue of control rats.

INTRODUCTION

Often enzymes or enzyme systems have several endogenous substrates in vivo. Their competitive individual transformation may play an important role in the metabolic regulation of biologically active substances such as catecholamines.

In conventional assay methods for the determination of enzyme activities, however, a single substrate is exclusively used because of analytical restrictions. Yet if the enzyme activity for the transformation of several competitive substrates could be determined by their simultaneous use, the result might be informative for the description of the competitive turnover of the substrates by the enzyme system in vivo.

Gas chromatography-mass spectrometry permits the use of mixtures of endogenous substrates when they are differently labelled with stable isotopes [1]. In order to investigate the competitive transformation of catecholamines by COMT, we attempted to use a mixture of deuterated catecholamines as a "multisubstrate system". This paper describes the assay of COMT activities in rat brain and liver by selected ion monitoring (SIM) in chemical ionization (CI) mode.

MATERIALS AND METHODS

Preparation of COMT

Freshly prepared rat liver and brain were minced and homogenized by a Potter-Elvehjem type homogenizer in 0.1 M phosphate buffer at pH 7.6. The homogenate was centrifuged at 48,000 x g for 30 min. The supernatant fluid was used as a crude enzyme solution. The protein content of these solutions prepared from rat liver and brain was found to be 46 mg and 15 mg protein/ml, respectively.

Syntheses

Substrates: $[1,1-^2H_2]$-dopamine (DA)·HCl; after the esterification of 3,4-dihydoxyphenylacetic acid with MeOH and HCl, the ester was treated with liquid ammonia in a sealed glass tube at room temperature for 12 hr. The resulting 3,4-dihydroxyphenylacetoamide was reduced with $LiAlD_4$ to $[1,1-^2H_2]$-DA·HCl. $[1,1,2-^2H_3]$-norepinephrine (NE)·HCl and $[1,1,2-^2H_3]$-epinephrine (E)·HCl were purchased from Merck, Sharp and Dohme, Canada.

Internal standards: $[1,2,2',5',6'-^2H_5]$-3-methoxytyramine (3-MT)·HCl, $[1,1,2,2',5',6'-^2H_6]$-normetanephrine (NM)·HCl, $[1,1,2,2',5',6'-^2H_6]$-metanephrine (MN)·HCl were synthesized in our laboratories. $[1,2,2',5',6'-^2H_5]$-3-MT was synthesized as follows; benzylvanilline was condensed with CH_3NO_2 at 110° for 2 hr in an acetate buffer solution and the resulting liquid was evaporated under reduced pressure. The residue was reduced with $LiAlD_4$ and the reaction product was hydrolyzed with DCl/MeOD at 100° for 6 hr in a sealed glass tube. The solvent was evaporated to dryness and the residue was recrystallized from water to give $[1,2,2',5',6'-^2H_5]$-3-MT HCl. Other deuterated internal standards were synthesized according to the procedure described above.

Other reagents: Pentafluoropropionyl-1,2,4-triazole (PFPT) was synthesized according to the method of our previous report [2]. Pentafluoropropionic acid (PFPA) was purchased from Tokyo Kasei Co., Japan. S-adenosyl-L-methionine (SAM) was obtained from Boehringer, Mannheim, West Germany. Controlled pore glass-10 (CPG-10) (120-200 mesh, mean pore diameter, 74.4 Å) was purchased from Electro-Nucleonics, USA.

Assay for COMT activity by a multi-substrate system

In a multi-substrate system an aliquot of the enzyme solution was pipetted into a test tube. To the tube were added 50 µl of 1 M phosphate buffer pH 7.6, 100 µl of 0.1 M $MgCl_2$, 100 µl of a mixture of 0.01 M $[^2H_2]$-DA, $[^2H_3]$-NE and $[^2H_3]$-E, and 50 µl of 3 mM SAM. The reaction mixture was incubated at 37° for 20 min, then 100 µl of PFPA were added to stop the reaction. The solution was lyophilized immediately after the addition of 100 μl of a mixture of 0.25 mM [^{2}H$_5$]-3-MT, [^{2}H$_6$]-NM and [^{2}H$_6$]-MN as the internal standards. The extraction and clean-up procedure were carried out according to the method described in our previous report [1].

Gas chromatography-chemical ionization mass spectrometry: A Shimadzu-LKB 9000 gas chromatography-mass spectrometer equipped with a chemical ionization source (Shimadzu Seisakusho, Kyoto, Japan) was employed. The operating conditions were as follows: Shimadzu MID-PM 9060s multi-ion detector (MID); column, 3 % OV-17 on Chromosorb W-HP (80-100 mesh) 1.5 m x 2.5 mm (Applied Science Labs., USA); column temperature 145°; flash heater temperature 250°; ion source temperature 180-200°; carrier gas (helium) flow rate 30 ml/min; ionization potential 500 eV in CI mode and 20 eV in EI mode; reagent gas, isobutane, accelerating voltage, 3500 V.

RESULTS AND DISCUSSION

Syntheses of stable isotope labelled substrates and internal standards

When a substrate labelled with a stable isotope, especially with deuterium, is used for an enzymatic reaction, it is indispensable to confirm if the labelled substrate exhibits a biological isotope effect on the enzymatic reaction.

Fig. 1 shows EI mass spectra of the PFP-derivatives of [1,1-^{2}H$_2$]-DA and [1,2,2',5',6'-^{2}H$_5$]-3-MT as representative of a substrate and an internal standard. The content of [^{2}H$_0$]-DA in its [^{2}H$_2$] variant and [^{2}H$_2$]-3-MT in its [^{2}H$_5$] variant was less than 0.1 %, indicating their excellent suitability for the present work. Other substrates and internal standards gave the same results. Table I summarizes the suitable m/z values of the substances for SIM in the CI mode (isobutane).

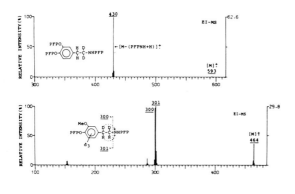

Fig. 1. Mass spectra of PFP derivates of [^{2}H$_2$]-dopamine and [^{2}H$_5$]-3-methoxytyramine (upper: [1,1-^{2}H$_2$]-dopamine; lower: [1,2,2',5',6'-^{2}H$_5$]-3-methoxytyramine).

Biological isotope effect

The ion-cluster technique was used to determine the biological deuterium isotope effect of the enzymatic reaction. An equimolar mixture of unlabelled and deuterated catecholamine was incubated, and the products were isolated and purified as described above. When $[^2H_0]$- and $[^2H_3]$-NE had been used as substrates, the biological isotope effect was calculated from the ratio of the peak areas of the resulting $[^2H_0]$- and $[^2H_3]$-NM. Table 2 indicates the biological isotope effects on the enzymatic turnover of the deuterated substrates of COMT.

TABLE 1
Suitable m/z values of the deuterated products and internal standards for selected ion monitoring in CI mode using isobutane as a reagent gas

substrate	product and internal standard	m/z value
HO-C6H3(OH)-CH2CD2NH2	MeO-C6H3(OH)-CH2CD2NH2	462
	(for IS MeO-C6H3(OH)-CHCHNH2, D3/D D)	465
HO-C6H3(OH)-CDCD2NH2, OH	MeO-C6H3(OH)-CDCD2NH2, OH	461
	(for IS MeO-C6H3(OH)-CDCD2NH2, D3, OH)	464
HO-C6H3(OH)-CDCD2NHCH3, OH	MeO-C6H3(OH)-CDCD2NHCH3, OH	475
	(for IS MeO-C6H3(OH)-CDCD2NHCH3, D3, OH)	478

TABLE 2
Biological isotope effect of the various substrates as relative peak area in MS of the products after incubation with COMT

Substrate	peak area
norepinephrine	1
$[1,1,2-^2H_3]$-norepinephrine	0.94
epinephrine	1
$[1,1,2-^2H_3]$-epinephrine	0.81
dopamine	1
$[1,1-^2H_2]$-dopamine	0.71
$[2',5',6'-^2H_3]$-dopamine	0.61

The isotope effect on the ring labelled $[2',5',6'-{}^2H_3]$-DA was greater than that on the side chain labelled $[1,1-{}^2H_2]$ isotopomer. The fact that even substrates labelled with deuterium in the side chain exhibit a biological isotope effect, suggests an interaction between the side chain of the substrate and COMT. However, as the substrates labelled in the side chain showed the smallest biological isotope effect, they were selected for further investigations.

COMT activity by the multi-substrate system

The method is based on the determination of $[{}^2H_2]$-3-MT, $[{}^2H_3]$-NM and $[{}^2H_3]$-MN by GC-MS after the incubation of a mixture of $[{}^2H_2]$-DA, $[{}^2H_3]$-NE and $[{}^2H_3]$-E with COMT. The supernatant of a rat liver homogenate was used as enzyme source, and in an extract from 20-100 µl incubation medium, the peak area ratio of products to internal standards exhibited a linear relationship. The COMT activities in a rat liver were calculated to be 64.9 nmol/min/g tissue for $[{}^2H_2]$-3-MT, 31.8 nmol/min/g tissue for $[{}^2H_3]$-NM and 16.5 nmol for $[{}^2H_3]$-MN/min/g tissue.

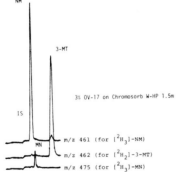

Fig. 2. Selected ion recording of $[{}^2H_3]$-NM(m/z 461, $[{}^2H_2]$-3-MT(m/z 462) and $[{}^2H_3]$-MN(m/z 475) obtained from the enzymatic reaction. (Substrate: mixture of $[{}^2H_2]$-DA, $[{}^2H_3]$-NE and $[{}^2H_3]$-E; enzyme: rat liver COMT).

Fig. 2 shows a representative selected ion recording obtained from an incubation using a mixture of $[{}^2H_2]$-DA, $[{}^2H_3]$-NE and $[{}^2H_3]$-E as a multisubstrate. The resulting products were completely separated under the selected GC-MS conditions.

Substrate characteristics of COMT are listed in table 3. Although epinephrine has been exclusively used in conventional fluorometric or radioisotopic methods, the above results indicate that epinephrine has the lowest affinity to this enzyme.

TABLE 3

Substrate characteristics of COMT in a rat liver homogenate

Substrate	K_m (mM) single	multi	V_{max}[+) single	multi
dopamine	4	2	100	38
norepinephrine	9	3	75	27
epinephrine	9	3	41	14

[+) The maximum reaction velocity rates were estimated from double reciprocal plots and presented relative to dopamine (100 %) in a single substrate system.

The new technique was applied to determine the change of COMT activity in tissues of rats after stress. Fig. 3 shows the COMT activities, obtained by the multi-substrate test, in liver and brain of rats which had been stressed by restraint and water immersion, in comparison to that in the corresponding tissues of control rats. The COMT activity for NE in the brain of the stressed rats was decreased significantly with regard to that in the corresponding tissue of the control rats. In the single substrate system, however there were no statistivally significant difference between the COMT activities of the brain enzyme activity of the two groups.

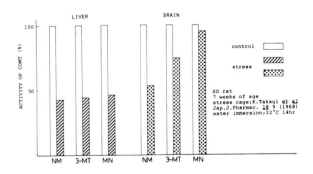

Fig. 3. Effect of stress (restraint and water immersion) on the activity of COMT in rat liver and brain.

REFERENCES

1 Y. Hashimoto and H. Miyazaki, J. Chromatogr., 168(1979)59-68.
2 Y. Hashimoto, K. Kimoto, M. Ishibashi and H. Miyazaki, Abstract, Symposium on Analytical Chemistry for Biological Substances, Tokyo, 1973, p.51.

III. BIOMEDICAL APPLICATIONS
B. CLINICAL DIAGNOSIS

THE APPLICATION OF THE STABLE ISOTOPES OF OXYGEN TO BIOMEDICAL RESEARCH AND NEUROLOGY

D. SAMUEL

Isotope Department, The Weizmann Institute of Science, Rehovot, Israel

ABSTRACT

There is increasing evidence that changes in the level or turnover of the biogenic amine neurotransmitters (dopamine, norepinephrine and serotonin) are involved in a wide range of mental disorders, such as the schizophrenias, mania and depression, and neurological diseases including Parkinsonism. The major difficulties in the study of the biological bases of these diseases are the ethical and practical limitations to the study of the chemistry of the brain and the absence of suitable models in experimental animals. During the past five years a collaborative project between the NIH, Bethesda, the Karolinska Hospital, Stockholm and the Weizmann Institute has successfully developed the use of the stable isotope of oxygen, oxygen-18 in enriched air for the in-vivo labelling of the amines and their metabolites. The rate of incorporation of ^{18}O into homovanillic acid (HVA) and other metabolites in cerebrospinal fluid (CSF) blood and urine can then be followed by mass-fragmentography. The results indicate that in Parkinsonism patients, in which the dopaminergic neurons in the nigro-strital pathway are impaired, there is an increased turnover of dopamine, due probably to a compensatory mechanism. A comparison of the rate of turnover of dopamine in the ventricles and lumber CSF of other neurological patients has also been studied. This approach affords a safe and useful method of studying the metabolism of biogenic amines in patients with various disorders of the brain and central nervous system.

INTRODUCTION

The central nervous system utilises about one fifth of all oxygen inhaled, although the brain of most mammals consists of only 1 or 2 percent of the total body weight. This constant and abundant supply of oxygen is required for three main functions. First, as in all biological systems, oxygen is needed in order to provide the necessary energy required for the synthesis of molecules needed for the development and maintenance of brain cells of all types. A second function, unique to the nervous system is the provision of energy required for the ion pumps located in the membranes of excitable cells - to transport sodium, potassium and other ions for the transmission of nerve impulses. The consequence of any reduction in oxygen supply is a rapid drop in the normal functions of the brain, which can lead within a few minutes to irreparable damage and eventually to death. The third role of molecular oxygen in the brain, is as a source of oxygen for the hydroxylation of the three of the key neurotransmitters (Fig. 1), serotonin (5HT), dopamine (DA) and norepinephrine (NE). Chemical imbalances between the levels of these amines are implicated in many mental disorders and neurological diseases - such as Parkinsonism [1], manic-depression [2] and schizophrenia [3] although the precise relationship between the level and turnover of these amines in the brain and in these disorders is not yet fully understood.

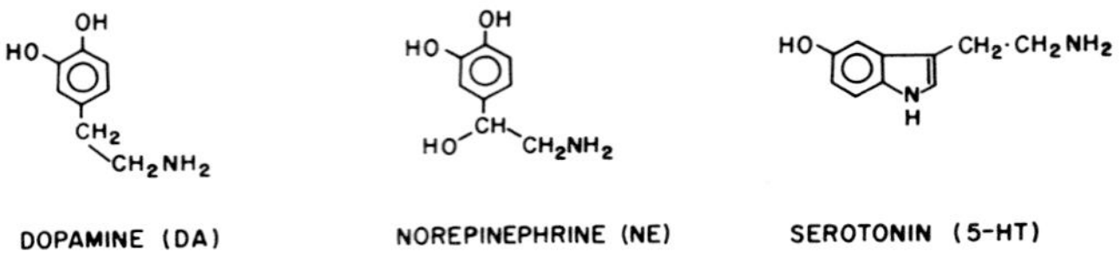

Fig. 1. Some key transmitters

   The indirect evidence for this relationship has been discussed extensively, but experimental difficulties have prevented a proper study of biogenic amine turnover in man. These difficulties stem from the inaccessibility of the human brain to direct chemical and pharmacological analysis. Therefore, the usual approach consists of the isolation, identification and measurement of low concentrations of the major metabolites of the biogenic amines (various aromatic acids and alcohols) in blood, urine and to a more limited extent in the cerebrospinal fluid (CSF) of patients. It should however be mentioned in passing that the recent advances in tomography, and particularly emission tomography may one day provide another useful approach to the study of the brain. Currently the major metabolites of interest are (a) homovanillic acid (HVA) produced from dopamine by a series of enzymatic reactions, (b) methoxy hydroxyphenylene ethylene glycol (MHPG or MOPEG) from norpinephrine and (c) 5-hydroxyindole acetic acid (5HIAA) from serotonin (Fig. 2). The origin and time course of the formation of these metabolites in the brain has not yet been entirely elucidated as will be discussed later [4].

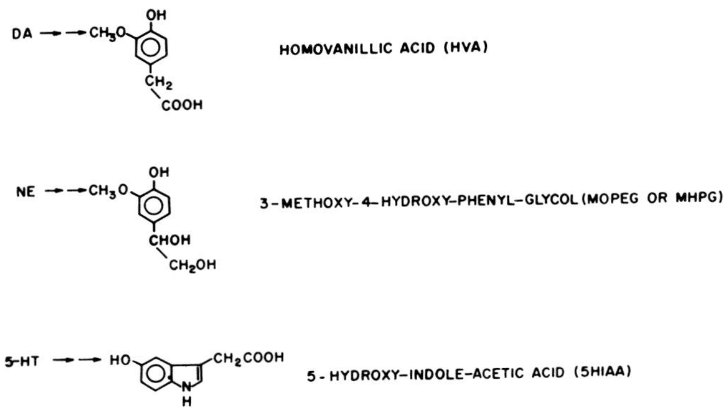

DA → → CH₃O — [benzene ring with OH] — CH₂—COOH   HOMOVANILLIC ACID (HVA)

NE → → CH₃O — [benzene ring with OH] — CHOH—CH₂OH   3-METHOXY-4-HYDROXY-PHENYL-GLYCOL (MOPEG OR MHPG)

5-HT → → HO — [indole ring] — CH₂COOH   5-HYDROXY-INDOLE-ACETIC ACID (5HIAA)

Fig. 2. The major metabolites

A great deal of basic research has been done on the central nervous system of experimental animals - the rat, the cat and the monkey. Various analytical methods have been used, with some success, to study the in vivo rates of the turnover and transport of biogenic amines of these animals, including the use of inhibitors of amine biosynthesis, various precursors labelled with radioactive isotopes and extensive studies on the effect of different psychoactive drugs.

The application of these techniques to study mental and neurological disorders in man is limited by the restrictions on the use of enzyme inhibitors and of precursors labelled with radioactive isotopes in human subjects. However, labelling with stable isotopes (of hydrogen, carbon, nitrogen or oxygen) provides an alternative approach. In fact, the use of gas-chromatographic mass spectrometry (GC/MS) methods for the quantitative analysis of nanogram quantities of biogenic amines and their metabolites extracted from tissues and body fluids, has already been very successful in many areas of biomedical and psycho-pharmacological research [5]. Using GC-MS or mass fragmentography (MF), the detection and determination of complex mixtures of volatile organic compounds separated by gas chromatography, has enabled both the biogenic amines and their metabolites to be measured in a single small sample of CSF or other body fluids using deuterated standards.

An important, and fortunate, fact in the biochemistry of all three amines (DA, NE and 5-HT) is that they are synthesized in the brain from amino acid precursors by enzymes which use molecular oxygen and appropriate cofactors to hydroxylate the substrate (Figs. 3 and 4). Since oxygen is readily transported to all parts of the brain by cerebral blood flow, use of oxygen labelled with one of the two stable isotopes of oxygen (^{17}O or ^{18}O) presents a method for the in vivo labelling of the amines in both animals and man. Oxygen-17 and oxygen-18 can be concentrated by distillation of water from which oxygen gas containing over 90% ^{17}O or ^{18}O can be prepared.

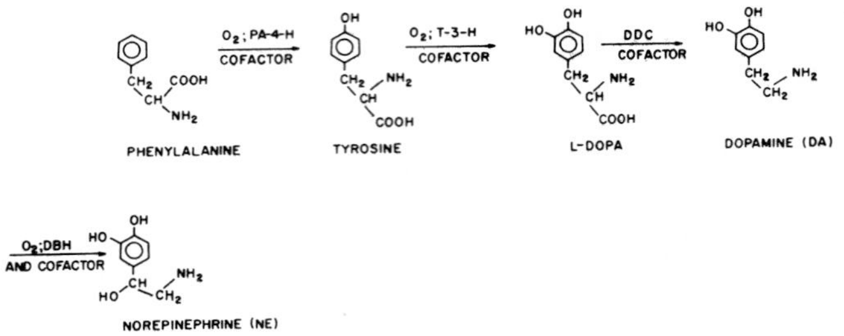

Fig. 3. The biosynthesis of dopamine and norepinephrine.

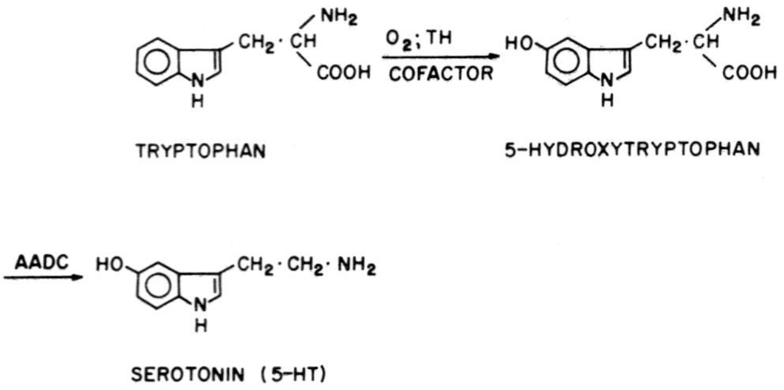

Fig. 4. The biosynthesis of serotonin.

RESULTS AND DISCUSSION

Using an adaptation of the **original apparatus used by Priestley** over 200 years ago for studies of oxygen consumption in mice, we have placed rats in a specially designed apparatus (Fig. 5) in which the circulating oxygen is gradually replaced by highly enriched oxygen-18. This labelled air is inhaled, transported to the brain and incorporated into the biogenic amines. In the course of extensive experiments it has been shown that mice at least can breath an oxygen-18 enriched atmosphere for long periods of time - up to 60 days - without any apparent ill effects to their health [6].

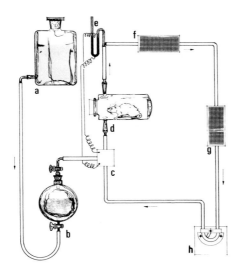

Fig. 5. Oxygen-18 breathing apparatus for rodents.

The ^{18}O label introduced into the hydroxyl groups of dopamine, norepinephrine and serotonin is an extremely stable one [7], and like all phenolic groups, does not undergo isotopic exchange with water, or any other oxygen-containing materials in the brain or body. This label is retained throughout all the catabolic processes which produce the various metabolites. Compounds, such as steroids, which are also labelled *in vivo* by other oxygenases are formed on an entirely different time scale, are readily separated during the chemical processing and present no problem in isotopic analyses. It should be noted however, that following extensive exposure to $^{18}O_2$, all "body water" and many substances that undergo isotopic exchange such as sugars, phosphates, etc. also contain some oxygen-18 but these also do not affect the results in any way.

Using the technique described, ^{18}O-labelled dopamine, HVA and other metabolites have been measured [8,9] in extracts from the brains of rats and the rate of turnover of dopamine determined. As an example of the results, it was found that an i.p. injection of the antipsychotic drug chlorpromazine (CPZ) (10 mg/kg) has been found to cause a threefold increase in the rate of incorporation of ^{18}O into HVA in the rat brain. This supports the suggestion that CPZ increases the rate of conversion of dopamine to HVA, by blocking the appropriate receptors or otherwise increasing the rate of release of the amine neurotransmitter from its stores.

In order to examine the feasibility of using this technique with human patients, an extensive series of ^{18}O-labelling experiments was also conducted in non-human primates [10]. Young male olive baboons (Papio anubis) (weighing 18-20 kg) were first anesthetised by an intramuscular injection, a tracheal catheter was inserted into the throat, connected by a T-junction through a CO_2- absorber and a drying tube, to a respiratory rubber balloon and to a large glass vessel containing highly isotopically enriched oxygen gas (over 90% ^{18}O). As the oxygen was consumed by the monkey, the carbon dioxide and water vapour were absorbed and replaced by enriched $^{18}O_2$ gas introduced into the system by hydrostatic pressure. Samples of lumbar CSF, blood and urine were flown to Stockholm for analysis by MF at the Karolinska Institute. The biogenic amines, DA, NE, 5HT and their main metabolites (HVA, MOPEG, DOPAC and 5-HIAA) were converted to volatile derivatives, appropriate deutero-standards added and the mixture analysed for its isotopic content. As an example of one of many experiments, the time course of the rise and fall of the ^{18}O content of HVA and DOPAC and 5-HIAA in the urine of a baboon is shown in Fig. 6. The ^{18}O concentration in the first two metabolites reaches a maximum in between one to two hours after the start of the experiment, whereas the 5-HIAA maximum occurs somewhat later.

Similar results have also been obtained for HVA and MOPEG in the CSF, obtained by lumbar puncture. Again the effect of CPZ (1.5 mg/kg given in two doses) in the monkey, is as expected from the experiments in the rat. i.e. the rate of incorporation of ^{18}O into HVA is faster than that in controls indicating an enhancement of DA turnover or metabolism.

Repeated experiments on baboons did not show any ill effects on their health or behaviour, demonstrating that ^{18}O can be breathed without risk. In addition to the extended and repeated experiments on $^{18}O_2$ breathing in rats [6] and baboons [10], pairs of breeding mice were kept in a highly ^{18}O enriched atmosphere and produced three successive litters without any unusual effects. The successful outcome of the animal experiments and the lack of toxicity of the ^{18}O isotope enabled us to apply the ^{18}O technique for monoamine turnover estimates in human subjects [11]. The protocol of this study was approved by the Ethical Committee of the NINCDS, Bethesda, Maryland, USA. Six Huntington chorea patients and four parkinsonian patients gave informed consent for participation in these experiments. All the patients had been off drug treatment for at least a week before the study. Each patient was allowed to breathe from a closed respirometer system initially

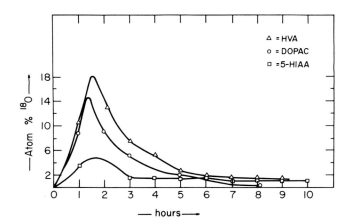

Fig. 6. Oxygen-18 content of metabolites in urine of monkey (baboon).

filled with air. As oxygen was consumed or absorbed within the system, highly ^{18}O-enriched gas (95% ^{18}O) was introduced. Inhalation continued for 1 hour, and after 3, 6 and 9 and 12 hours about 5 ml of CSF was removed by lumbar puncture. The total concentrations of the ^{18}O, and ^{16}O, labelled monoamine metabolites HVA, MOPEG, 5-HIAA were determined by MF in Stockholm. None of the patients reported any discomfort in relation to or after the ^{18}O inhalation and conventional laboratory blood and urine analyses did not show any apparent deviations. As shown in Fig. 7 the isotopic ratio of ^{18}O in HVA in lumbar CSF rose rapidly to reach a maximal level after 3 to 6 hours after the beginning of the inhalation. After the initial peak the ratio declined with a half-life of roughly 3-4 hours. The ^{18}O-content of MOPEG and 5-HIAA rose more slowly and reached peak levels that were less than half of that of HVA.

The two patient groups examined showed a significant difference in regard to the time course and maximal level of isotopic labelling of HVA and total HVA concentrations in CSF. These data indicate that the Parkinsonian patients seem to have a smaller pool of dopamine which is turning over at a more rapid rate than in the Huntington chorea patients. This is compatible with the view [1] that Parkinsonian patients have a reduced number of functioning dopamine neurons in the brain. This deficit is apparently partially compensated for an accelerated transmitter metabolism in the remaining neurons.

Certain patients have been diagnosed as having normotensive (or normal pressure) hydrocephalus (NPH) caused by trauma, hemorrhage and variour other neurological conditions leading to mental deterioration, dementia or persistent depression.

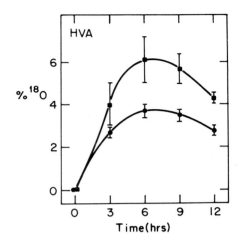

Fig. 7. Oxygen-18 content of HVA in lumbar CSF of patients with Parkinsonism and Hungtington's chorea.

One method of treatment is to install ventriculo-peritoneal or ventriculo-atrial shunts surgically. A cannula is permanently implanted into a ventricle and an internally located plastic tube removes excess CSF pressure to the stomach or heart cavities. This cannula affords an unusual opportunity of examining the biogenic amine metabolites in CSF in the ventricles. At the same time, with the patients consent, CSF from the lumbar region can also be obtained by lumbar puncture (LP). Volunteer patients were given highly ^{18}O-enriched air to breath for one hour. During the following 24 hours samples of ventricular CSF were taken every hour from the cannula and lumbar CSF was taken by L.P. every three hours. The experiments were done at Ichilov Municipal Hospital, Tel-Aviv, in accordance with the rules laid down by the appropriate "Helsinki Committee". The samples of CSF (as well as of blood and urine) were frozen in dry ice, and flown to the Karolinska Hospital in Stockholm for isotopic analysis. The isotopic results for HVA and 5HIAA are shown in Figs. 8 and 9.

It is evident from these Figs. that the shape of the curves for ^{18}O incorporation for both metabolites in the ventricular and lumbar regions is very similar, a maximum being reached in about 8 hours for HVA and slightly earlier for 5HIAA. If the major part of the metabolites were derived from the brain, one would expect the peaks of incorporation in the lumbar region to be reached several hours later than those in ventricular CSF owing to the time required to flow down the spinal cord. The results obtained can be explained by assuming that there is a substantial amount of labelling (i.e. synthesis or turnover) of DA and 5HT in the lumbar

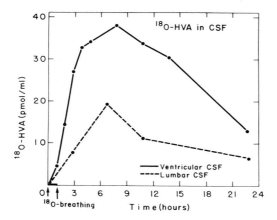

Fig. 8. Oxygen-18 content of HVA in ventricular and lumbar CSF of NPH patients.

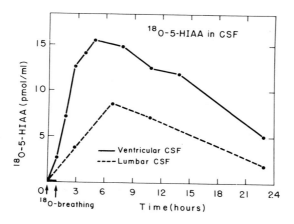

Fig. 9. Oxygen-18 content of 5HIAA in ventricular and lumbar CSF of NPH patients.

region. An alternative explanation is that these patients do not have normal CSF dynamics. However the results strongly suggest that there is considerable synthesis of DA in the spinal cord, as has earlier been suggested [4] for 5HT and NE.

These experiments demonstrate the feasibility of a new method for studying the kinetics of monoamine transmitter metabolism in the human central nervous system which can be applied to other neurological disorders and mental diseases.

This research was part of a collaborative project involving G. Sedvall and coworkers at the Department of Psychiatry of the Karolinska Institute, Stockholm; T.N. Chase and coworkers at the National Institute of Neurological Communicative Diseases and Stroke, Bethesda, A. Bartal and E. Geller, Ichilov Municipal Hospital Tel Aviv and E. Benhar, V.E. Grimm and I. Wasserman of the Weizmann Institute of Science, Rehovot.

REFERENCES

1. O. Hornykiewicz, Pharmacol. Revs., 18 (1966) 925-964.
2. J. Schildkraut, Amer. J. Psychiatr., 22 (1965) 509-522.
3. S.H. Snyder, S.P. Banergee, H.I. Yamamura and D. Greenberg, Science, 184 (1974) 1243-1253.
4. M. Bulat, Brain Res., 122 (1977) 388-391.
5. G.G. Swahn, B. Sandgärde, F.A. Wiesel and G. Sedvall, Psychopharmacol., 48 (1976) 147-152.
6. D. Wolf, H. Cohen, A. Meshorer, I. Wasserman and D. Samuel, in E.R. Klein and P.D. Klein (Eds.), Proceedings of the Third International Conference on Stable Isotopes, Academic Press, 1979, pp. 353-360.
7. D. Samuel, in O. Hayaishi (Ed.), Oxygenases, Academic Press, 1962, pp. 31-87.
8. A. Mayevsky, B. Sjoquist, C.G. Fri, D. Samuel and G. Sedvall, Biochem. Biophys. Res. Commn., 51 (1973) 746-755.
9. G. Sedvall, A. Mayevsky, D. Samuel and C.G. Fri, in E. Usdin and S. Snyder (Eds.), Frontiers in Catecholamine Research, Pergamon Press, 1973, pp. 1071-1075.
10. G. Sedvall, O. Beck, E. Benhar, E. Geller, V. Grimm, D. Samuel and I. Wasserman, in O. Almagren, A. Carlsson and J. Engel (Eds.), Chemical tools in Catecholamine Research, Vol. II, North Holland Publishing Co., 1975, pp. 17-24.
11. T.N. Chase, A. Neophytides, D. Samuel, G. Sedvall and C.G. Swahn, Presented at the International Catecholamine Symposium, Asilomar, Calif., 1978.

^{13}C-LABELLED VALPROIC ACID PULSE DOSING DURING STEADY STATE ANTIEPILEPTIC THERAPY FOR PHARMACOKINETIC STUDIES DURING PREGNANCY

W. WITTFOHT, H. NAU, D. RATING[1] and H. HELGE[1]

Institut für Toxikologie und Embryonalpharmakologie, [1]Kinderklinik der Freien Universität Berlin, Berlin (G.F.R.)

ABSTRACT

Using 1,2-$^{13}C_2$-labelled valproic acid, detailed pharmacokinetic studies were performed in a pregnant epileptic patient on valproic acid (VPA) therapy. Kinetic parameters during the first trimester were similar to those of nonpregnant adults. During the third trimester, the total clearance, the hepatic clearance ($^{13}CO_2$-exhalation), the renal clearance and the volume of distribution increased, resulting in a drop of serum levels. It was shown – for the first time – that pharmacokinetic studies can be carried out during pregnancy and steady state therapy under ethically acceptable conditions with the use of stable isotope methodology.

INTRODUCTION

Blood levels of some drugs drop during pregnancy. Reduced absorption or protein binding, increased distribution volume, hepatic or renal clearance, or poor patient compliance may cause the low blood levels observed.

In the present study we measured the half life, the distribution volume, the total as well as "hepatic" and renal clearance values of VPA in a pregnant epileptic patient using ^{13}C-labelled VPA.

METHODS

The usual morning dose of an epileptic patient (600 mg VPA) was partially substituted by 100 mg 1,2-$^{13}C_2$-VPA (KOR-Isotopes, Cambridge, USA; 90% label) once during the first trimester and 300 mg 1,2-$^{13}C_2$-VPA once during the third trimester. Blood and urine samples were analyzed for VPA and metabolites as well as their ^{13}C-labelled analogues by GC-MS-computer using selected ion monitoring [1-3]. The $^{13}CO_2$-exhalation was measured by an isotope ratio mass spectrometer.

RESULTS AND DISCUSSION

Quantitative methods using GC-MS-selected ion monitoring have been developed for the simultaneous determination of VPA and various metabolites which are shown in Fig. 1:

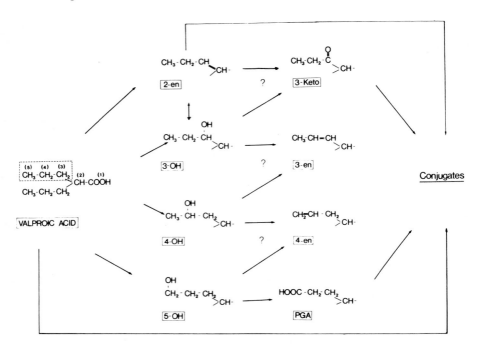

Fig. 1. Structures of VPA and metabolites which have been assayed.
VPA, valproic acid (2-propylpentanoic acid); 2-en, 2-propyl-2-pentenoic acid (trans); 3-en, 2-propyl-3-pentenoic acid (trans); 4-en, 2-propyl-4-pentenoic acid; 3-OH, 3-hydroxy-2-propylpentanoic acid; 4-OH, 4-hydroxy-2-propylpentanoic acid; 5-OH, 5-hydroxy-2-propylpentanoic acid; 3-Keto, 3-oxo-2-propylpentanoic acid; PGA, 2-propylglutaric acid.

The levels of VPA and the two main plasma metabolites 2-en and 3-Keto in a pregnant epileptic patient during the third trimester are shown in Fig. 2 in addition to the levels of the ^{13}C-labelled analogues originating from a 3oo mg pulse dose of 1,2-^{13}C$_2$- labelled VPA.

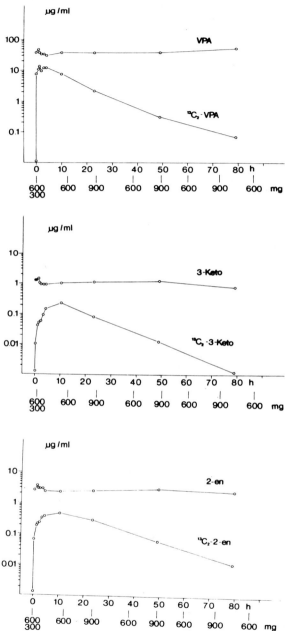

Fig.2 Steady state concentrations in serum of VPA and two metabolites 2-en and 3-Keto in an epileptic patient during the third trimester of pregnancy (1.5g VPA/day). At time 0 a portion of the usual morning dose of 9oo mg was substituted by 3oo mg 1,2-$^{13}C_2$ VPA (9o% label); the concentration-time curves of the ^{13}C-labelled substances in the serum are also shown.

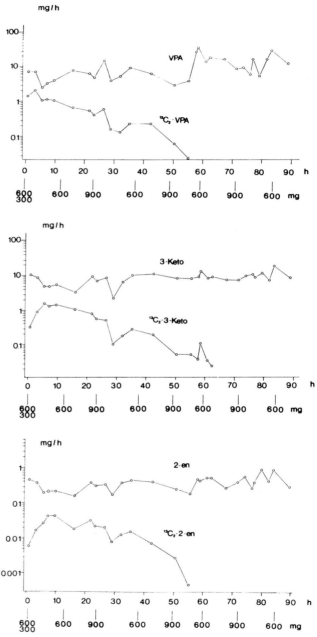

Fig. 3

Urinary elimination of VPA and two metabolites 2-en and 3-Keto in an epileptic patient during the third trimester of pregnancy (1.5 g VPA/day). At time 0 a portion of the usual morning dose of 900 mg was substituted by 300 mg $1,2\text{-}^{13}C_2\text{-}$VPA. The urinary elimination of the substances were measured following enzymatic treatment of the urine samples.

Measurement of the ^{13}C-labelled drugs showed that VPA (as sodium salt) was rapidly absorbed orally (y_{max} = 1.9 h). The two metabolites 2-en and 3-Keto reached maximum serum levels between 8 and 9 h. Half lives of VPA and 3-Keto were between 9 and 1o h, while the half life of 2-en was longer (12.5 h). The total clearance values of VPA calculated from the steady state concentrations of unlabelled VPA (Cl_{tot} = 24.5 ml/h/kg) as well as from the AUC of the ^{13}C-pulse label (Cl_{tot} = 24.3 ml/h/kg) were very close, indicating that the pharmacokinetic properties of VPA and the ^{13}C-labelled VPA used were indistinguishable from each other.

Although the daily dose did not change throughout pregnancy, blood levels in this patient were lower, and consequently total clearance values higher, during the third trimester than during the first trimester (Table 1).

TABLE 1
Comparison of valproic acid pharmacokinetics of a pregnant epileptic patient[1] with nonpregnant adults

	Cl_{tot} (ml.h^{-1}.kg^{-1})	Cl_{ren} (ml.h^{-1}.kg^{-1})	$t_{1/2}$ (h)	V_d 1.kg^{-1}
1st Trimester	9.2	0.019		
3rd Trimester	24.5	0.046	1o	0.36
Nonpregnant[2]	8.2 ± 1.6		14 ± 4	0.18

[1] steady state therapy (1.5 g VPA/day)
[2] Bowdle et al. (198o)

The total clearance values during the first trimester were similar to those of nonpregnant adults (Table 1). Renal clearance values of VPA and metabolites also increased during the third trimester (Table 1). Total clearance of a drug is a function of both the half life ($t_{1/2}$) and volume of distribution (V_d). Since the $t_{1/2}$ values were similar during the first and third trimester, the increased V_d during the third trimester was mainly responsible for increased total clearance values of VPA.

The exhalation of $^{13}CO_2$ has also been measured following the ^{13}C-pulse dosing in the same patient (Fig. 3).

Maximal exhalation rates were reached between 2o-3o minutes, thus earlier than the maximal serum levels of VPA (1.9 h). The cumulative ^{13}C-exhalation values (expressed as % of dose) were 4 % (after 3 h), 5 % (after 5 h) and 5.3 % (after 6 h) following the pulse dosing experiment during the third trimester. The corresponding values during the first trimester (2-3 %) after 5 h were lower than during the third trimester.

These results indicate that the low VPA blood levels observed during the third trimester were caused at least in part by increased hepatic and renal clearance and increased volume of distribution.

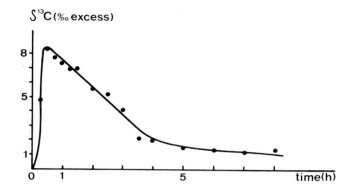

Fig. 4 $^{13}CO_2$-exhalation rates in a pregnant patient during the third trimester.

ACKNOWLEDGEMENTS

This work was supported by grants of the DFG to the SFB 29. The preparation of the manuscript by Mrs. R. Ebbinghaus and the donation of pure VPA metabolites by Dr. H. Schäfer (Hamburg) are gratefully acknowledged.

REFERENCES

1 T.A. Bowdle, I.H. Patel, R.H. Levy and A.J. Wilensky, Valproic acid dosage and plasma protein binding and clearance, Clin. Pharmacol. Therap., 28,486-492 (1980).
2 H.Nau, H.Schäfer, D.Rating, C.Jakobs and H.Helge in D.Janz et al. (Eds.), Epilepsy, Pregnancy and the Child, Placental Transfer and Neonatal Pharmacokinetics of Valproic Acid and Metabolites, Raven Press, in press.
3 H.Nau, W.Wittfoht, D.Rating, C.Jakobs, H.Schäfer and H.Helge, Pharmacokinetics of Valproic Acid and Metabolites in a Pregnant Patient, ibid., in press.
4 H.Nau, W.Wittfoht, H.Schäfer, C.Jakobs, D.Rating and H.Helge, Valproic Acid and Several Metabolites: Quantitative Determination in Serum, Urine, Breast Milk and Tissues by Gas Chromatography-Mass Spectrometry Using Selected Ion Monitoring, J. Chromatogr., in press.

THE APPLICATION OF DEUTERATED VALPROIC ACID (VPA) IN CHRONICALLY TREATED EPILEPTIC PATIENTS UNDER MONOTHERAPY AND POLYPHARMACY.

W.KOCHEN, B.TAUSCHER, M.KLEMENS, and E.DEPENE
University of Heidelberg (G.F.R.), Children's Hospital and Chemical Institute.

ABSTRACT

The synthesis of $[^2H_7]$ VPA and $[^2H_1]$ VPA is described. The urinary metabolites of $[^2H_7]$ VPA found in a patient under monotherapy are presented. Some pharmacokinetic data of VPA under monotherapy and comedication are given.

INTRODUCTION

Valproic acid (VPA) is an effective drug for treatment of epilepsy. The doses are in the range of 10 - 100 mg/kg·day. Many patients must be treated in addition with other antiepileptic drugs. The blood level of VPA is then considerably decreased under comedication. The kinetics of VPA under long-term therapy have not yet been thoroughly investigated. The interactions of other antiepileptic drugs with VPA are unknown to a large extent. To gain pharmacokinetic data without interrupting the therapeutic treatment it is necessary to perform a pulse dose experiment using the deuterated drug.

METHODS

Gaschromatographic micro-assay of VPA in blood

To 50 μl serum are added in an Eppendorf vial 10 μg of the internal standard (2-ethyl-2-methylcapronic acid) and 50 μl 10% $HClO_4$. Homogenisation with a vortex mixer for 30 s, then addition of 70 μl CCl_4, and further homogenisation (60 s) are carried out. The sample is then centrifuged (60 s) and the organic phase is transferred with a Hamilton syringe to a microvial with septum (content 200 μl, i.d. 2 mm). The acid is esterified with 5 μl of the silylating agent MSTFA (Machery & Nagel, Düren).

Micro-assay of VPA (free and conjugated) in urine. To 50 μl urine 50 μl 0.5 M citrate buffer (pH 5.5) and 10 μl glucuronidase/arylsulfatase (10^5 fishman units/ml, Serva, Heidelberg) are added. The mixture is incubated for 3 hrs at 37 °C. The sample is then acidified to pH 1 - 2 and the products are extracted.

Urinary metabolites. 5 ml urine are enzymatically hydrolized and then acidified to pH 1 - 2 with 1 N HCl and immediately extracted three times with acetic acid ethylester. The organic phase is dried with Na - sulfate, concentrated to 50 -

100 μl, and silylated with MSTFA.

Gaschromatographic conditions. Metabolites: Siemens gaschromatograph L 350, Deans valveless column switching. Capillary pre-column 10 m OV-1, capillary main-column 50 m SE-54 (Jaegi,Trogen,Switzerland). Temperature program 75 °C (10 min), 3 °C/min up to 200 °C, P_i 1.2 bar (N_2), split 1 : 10.

Single assay of VPA. 25 m capillary column SE-54, isothermal 100 °C, split 1 : 15.

Mass-spectrometric conditions. Mass-spectrometer Du-Pont 21-492 B, open interface combination, ionisation energy 70 eV, ion source temperature 250 °C, interface 210 °C. Gaschromatography see above.

Synthesis of 2-[$^2H_7$]propylpentanoic acid

$C_4H_9COOH \longrightarrow C_3H_7-\underset{Li}{CH}-COOLi \longrightarrow C_3H_7-\underset{C_3D_7}{CH}-COOLi \longrightarrow C_3H_7-\underset{C_3D_7}{CH}-COOH$

0.0759 mole n-butyllithium in hexane is added to a solution of 7.7 g (0.077 mole) diisopropylamine in 55 ml THF and cooled below 0 °C. After adding dropwise 3.6 g (0.0359 mole) pentanoic acid to the solution and stirring for 15 min, 14.2 ml (0.079 mole) hexamethylphosphoramide are added. The reaction is completed by stirring for 15 min at 5 °C. 5 g (0.038 mole) of [$^2H_7$]-1-bromopropan are poured into the solution and stirring is continued for 2 hrs. The mixture is then worked up using common methods. $b_{p\ 14}$: 116 - 118 °C. The mass-spectrum is presented in Fig.1.

^{1}H-NMR data: CH   2.3 - 2.4 ppm (1H), CH_2-CH_2 1.2 - 1.7 ppm (4H),
                CH_3 0.8 - 1.0 ppm (3H).

^{13}C-NMR data: C_2 45.17 ppm singlet; C_3 34.48 ppm and C_3' 33.34 ppm multiplet;
                   C_4 20.73 ppm and C_4' 19.54 ppm multiplet;
                   C_5 14.02 ppm and C_5' 12.85 ppm multiplet.

Synthesis of 2-propyl-2-[$^2H_1$]pentanoic acid-[$^2H_1$]

$\underset{C_3H_7}{C_3H_7}>C<\underset{COOH}{COOH} \longrightarrow \underset{C_3H_7}{C_3H_7}>C<\underset{COOK}{COOK} \longrightarrow \underset{C_3H_7}{C_3H_7}>C<\underset{COOD}{COOD} \longrightarrow \underset{C_3H_7}{C_3H_7}>CD-COOD$

To 8.8 g (0.0468 mole) dipropylmalonic acid 15 ml 40% KOD/D_2O are added. The mixture (HOD, D_2O, H_2O) is evaporated to dryness and the procedure is repeated. The remaining salt is dissolved in D_2O/DCl (pH 1) and the free acid is heated at 160-170 °C for decarboxylation for 24 hrs. After extraction with diethylether the acid is worked up. $b_{p\ 15}$: 117 - 118 °C. For the mass spectrum see Fig.2. The spectrum of the non-deuterated VPA is given in Fig.3.

^{1}H-NMR data: CH_2-CH_2 1.2 - 1.7 ppm (8H); CH_3 0.8 - 1.0 ppm (6H).

^{13}C-NMR data: C_2 45.13 ppm multiplet; C_3 34.67 ppm singlet; C_4 20.94 ppm singlet; C_5 14.08 ppm singlet.

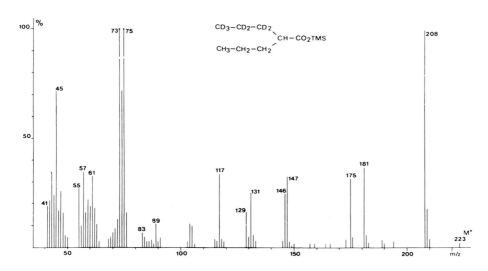

Fig. 1. Mass spectrum of [^{2}H$_7$]VPA · TMS

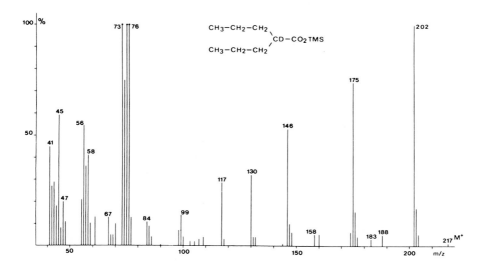

Fig. 2. Mass spectrum of 2-[^{2}H$_1$]VPA · TMS

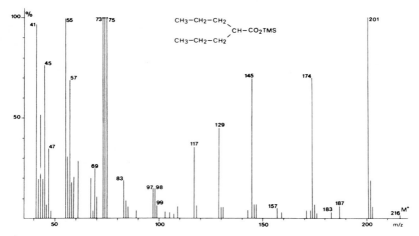

Fig.3. Mass spectrum of VPA·TMS

RESULTS

Isotope effect of $[^2H_7]$VPA. VPA and its deuterated isotopomer were simultaneously administered orally (200 mg each) to a healthy volunteer. The parallel decay in plasma (Fig.4) proves that there is no isotope effect on absorption of $[^2H_7]$VPA, whereas this is the case with the urinary excretion of the deuterated VPA which is reduced by 5 % within 48 hrs.

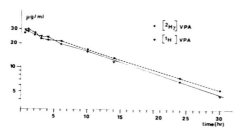

Fig.4. Plasma concentration decay of VPA and deuterated VPA after simultaneous application in a control.

urinary excretion in % of the dose

	VPA	$[^2H_7]$VPA
1st 24-hr	24.7	20.3
2nd 24-hr	18.0	17.5

Patient under VPA monotherapy.
A daily dose of 3 X 300 mg VPA had been given for 1 month to this patient. The morning intake of 300 mg was substituted by 292.3 mg $[^2H_7]$VPA (Na-salt), whereupon the therapy was interrupted for 2 days. The half-life time $t_{1/2}$ of $[^2H_7]$VPA in plasma was determined to be 10.0 hrs, and that of the non-deuterated form as 14.0 hrs (Fig.5).

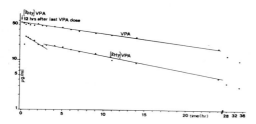

Fig.5. Plasma concentration of VPA and deuterated VPA in the patient.

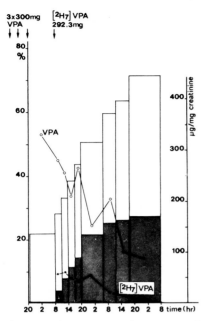

Fig.6. Urinary excretion of total VPA after the last dose of the daily intake (open beams) and of deuterated VPA (dashed beams) in % of the dose. The curves are related to 1 mg creatinine.

The urine was collected at 3-hr-intervals on the first day, then at greater intervals. Fig.6 illustrates the excretion rates of both forms of VPA in 9 urine fractions in % of the dose (daily VPA intake 900 mg), and additionally in μg VPA and $[^2H_7]$ VPA resp. per mg creatinine. The total excretion of free and conjugated $[^2H_7]$ VPA amounted 27.3 % of the dose within 48 hrs, whereas 70.1 % of the daily VPA dose (900 mg) were eliminated in 60 hrs after the last application on the day before.

The following deuterated urinary metabolites were found by their TMS derivatives (1). The ratio of the differently deuterated forms of a metabolite was determined by the intensity of the $(M^+ - 15)$ peak: E-2-en-VPA ($^2H_7:^2H_6:^2H_5=3:1:0.8$), Z-2-en-VPA ($^2H_7:^2H_6=3:1$), 3-en-VPA (only $^2H_7$), 4-en-VPA ($^2H_7:^2H_5=3:1$), 5-OH-VPA ($^2H_7:^2H_6=4:1$), 2-propylglutaric acid ($^2H_7:^2H_4=5:1$), 4-OH-VPA ($^2H_7:^2H_6=4:1$), 3-OH-VPA could not be found, 3-OH-2-en-VPA ($^2H_7:^2H_5=2:1$), OH-en-VPA(no.2) ($^2H_7:^2H_5=3:1$), OH-en-VPA(no.1) (only $^2H_6$).

The last 2 metabolites contain a double bond and a hydroxy group. The position of these functional groups has not been exactly elucidated (2). In this patient under monotherapy the most important metabolites are E-2-en-VPA, 3-OH-2-en-VPA (the enol form from 3-keto-VPA) and the OH-en-VPA no.2. The quantitative results are given in Tab.1. Within 48 hrs after application of the deuterated VPA 27.1 % of the dose were excreted in the unchanged form of VPA and 52.5 % as metabolites. The qualitative and quantitative urinary excretion pattern of the non-deuterated VPA metabolites differed slightly from that of the deuterated metabolites. A distinctly isotope effect was found in the metabolism of the deuterated propylic side chain. The excretion of non-deuterated VPA and of 8 metabolites amounted within 60 hrs to 156.3 % of the last daily dose of 900 mg.

Some pharmacokinetic data of VPA under monotherapy and comedication.

The following example demonstrates that in a patient under chronic comedication (carbamazepine, ethosuximide, and phenytoin) $t_{1/2}$ of $[^2H_7]$ VPA was reduced to 3.8 hrs.(Fig.7). The volume of distribution V_d was determined to 0.13 l/kg and the

Tab.1. Urinary excretion of deuterated VPA and 8 metabolites after a pulse dose $[^2H_7]$ VPA. Values are given in mmole/urine fraction.

Urine portion (hr)	08 - 11	11 - 14	14 - 17	17 - 20	20 - 06	06 - 12	12 - 18	18 - 08	sum	%
$[^2H_7]$ E-2-en + $[^2H_6]$ + $[^2H_5]$	0.019	0.025	0.035	0.021	0.036	0.020	0.030	0.032	0.218	21.4
$[^2H_7]$ Z-2-en + $[^2H_6]$	0.006	0.009	0.012	0.011	0.027	0.007	0.002	0.008	0.008	0.8
$[^2H_7]$ 5-OH + $[^2H_6]$	0.006	0.000	0.001	0.0035	0.0164	0.009	0.0054	0.005	0.046	4.5
$[^2H_7]$ 4-OH	0.000	0.000	0.000	0.0013	0.0043	0.0024	0.0032	0.0038	0.015	1.5
$[^2H_6]$ OH-en (1)	0.0029	0.014	0.01	0.0074	0.0105	0.020	0.008	0.008	0.082	8.0
$[^2H_7]$ 3-OH-2-en + $[^2H_5]$	0.0088	0.0062	0.0079	0.0353	0.0366	0.060	0.0427	0.116	0.313	30.7
$[^2H_7]$ OH-en (2) + $[^2H_5]$	0.0041	0.0085	0.023	0.030	0.0244	0.031	0.023	0.058	0.2089	20.5
$[^2H_7]$ 2-PG + $[^2H_5]$	0.0027	0.0103	0.0046	0.002	0.0389	0.0276	0.021	0.021	0.128	12.6
	0.050	0.073	0.094	0.129	0.194	0.177	0.135	0.252	1.020	52.5
$[^2H_7]$ VPA	0.0755	0.0766	0.0722	0.0511	0.144	0.056	0.033	0.0182	0.5278	27.17

clearance Cl_p to 26.0 ml/min. The mean daily blood level of VPA amounted to 35 µg/ml on a daily dose of 30 ml/kg. According to our experience such a dose would result under monotherapy in a mean value of about 115 µg/ml (3). Long-term comedication induces a considerable decrease in VPA blood level.

A patient under monotherapy of 25 mg/kg VPA (Fig.8) had a mean value of 102 µg/ml and $t_{1/2}$ was determined as 14.0 hrs with V_d 0.14 l/kg and Cl_p 8.9 ml/min. Evidently the decreased $t_{1/2}$ caused by comedication is not based upon a change in V_d but rather upon a change in metabolism followed by an increased Cl_p. In further studies we could demonstrate a distinctly different excretion pattern of VPA metabolites in the urine of patients under monotherapy and polypharmacy.

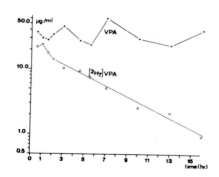

Fig.7. Blood levels after a pulse dose of $[^2H_7]$ VPA in a patient on polypharmacy.

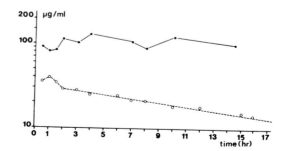

Fig.8. Blood levels after a pulse dose of $[^2H_7]$ VPA in a patient on monotherapy.

References
(1) W.Kochen and E.Scheffner: Antiepileptic Therapy:Advances in Drug Monitoring (eds.S.J.Johannessen,P.L.Morselli et al.),pp.111-117,Raven Press New York 1980;
(2) W.Kochen et al.:Biochem.Pharmacol.1982 (i.p.); (3) W.Kochen et al.: Annals of Neurology 1982 (i.p.).

IN VIVO MEASUREMENT OF ENZYMES WITH DEUTERATED PRECURSORS AND GC/SIM

H.-CH. CURTIUS
Division of Clinical Chemistry, Department of Pediatrics, University of Zurich, Steinwiesstr. 75, CH-8032 Zurich (Switzerland)

ABSTRACT

Deuterated phenylalanine, tyrosine and tryptophan have been applied for the in vivo measurement of enzymes by loading patients and healthy controls. The deuterated precursors and the deuterated products formed have been analyzed by SIM. The amount of incorporated deuterium is an index for the respective enzyme activity. The analyses were made from urine or blood. Using this method, an in vivo determination of the enzymes phenylalanine-4-hydroxylase, tyrosine-3-hydroxylase, tryptophan-5-hydroxylase and tryptophan-pyrrolase is possible.

INTRODUCTION

The administration of radioactive isotopes is not without risk to human subjects and, therefore, only rarely suitable for in vivo metabolic studies. Usually in vitro tests are performed on biopsy tissues or on tissue cultures. These in vitro procedures can, however, often not replace in vivo studies.

Nonradioactive stable isotopes such as D, ^{13}C, ^{18}O and ^{15}N are perfectly suited for the in vivo study of metabolic pathways in man without any risk to the subject. Nevertheless, the use of stable isotopes may show some difficulties. The first problem is the synthesis of the stable isotope labelled compounds. ^{13}C or ^{18}O are expensive and the synthesis of ^{13}C labelled compounds is very time consuming and costly. The prices of deuterated reagents, such as D_2O, DCl, etc. are more reasonable, and the deuterated compounds can more easily be synthesized. On the other hand, the application of deuterium labelled compounds also shows some disadvantages. The isotopic effect of the C-D bond is much greater compared to that of the ^{13}C-^{12}C bond. In addition, there is always the possibility of a loss of the deuterium when it is located in a labile position

The first in vivo metabolic studies with deuterium were done by Schoenheimer and Rittenberg in 1935 (1). For a long time, only a few in vivo experiments with nonradioactive stable isotopes, such as ^{13}C, ^{15}N or deuterium, were performed, since their determination posed certain analytical problems. The most suitable instrument for detection of stable isotopes is the mass spectrometer. Separation and purification for mass spectrometric analysis is, however, elaborate and often gives poor yields. Since the introduction of combined instruments for gas chromatography-mass spectrometry, this problem has been considerably simplified. Complex biological mixtures with components in the nanogram range can be separated by gas chromatography and their isotope contents can be analyzed directly by mass spectrometry. The separation and sensitivity of the method can be improved significantly by the use of glass capillary columns (2).

Other instruments can also be used after the isolation by chromatography, such as NMR, IR or a high frequency emission technique for the measurement of ^{15}N after the transforming of the nitrogen containing compound into gaseous N_2 (3).

Based upon the early pioneering work of Schoenheimer and Rittenberg, we investigated the phenylalanine-tyrosine-dopa, leucine, tryptophan and steroid metabolism in vivo, by loading healthy subjects and patients with various diseases with deuterated L-phenylalanine-d_5 (4), deuterated L-tyrosine-d_2 (5,6), deuterated tryptophan-d_5 (7), deuterated leucine (8) and deuterated cholesterol, pregnenolone and progesterone (9). The analysis of their respective metabolites in plasma and urine was performed using gas chromatography-mass spectrometry (GC-MS) and selective ion monitoring (SIM).

The stable isotope GC-MS technique was also used by our group for the in vivo measurement of enzymes, by loading patients and healthy controls with deuterated precursors. The deuterated product formed was subsequently analyzed by GC-MS (10,11,12, 13). The amount of incorporated deuterium is an index for the respective enzyme activity. The analyses were made from urine or blood. Some examples of our investigations are presented below:

1. Phenylalanine-4-hydroxylase (E.C. 1.14.16.1)(Phenylalanine-4-monoxygenase). The tetrahydrobiopterin (BH_4) dependent liver enzyme phenylalanine-4-hydroxylase is responsible for the conversion of phenylalanine into tyrosine in metabolic pathways. In phenylketonuria (PKU), one of the most common inborn errors of human metabolism, this enzyme shows no, or only partial activity.

Due to this enzyme defect, the phenylalanine blood level in untreated phenylketonuric patients increases to approximately 50 mg/100 ml. One can differentiate between: classical PKU, where the enzyme activity is only about 2-3%, various forms of hyperphenylalaninemia, with residual activities of about 7-17%, and patients with atypical PKU due to a defect in tetrahydrobiopterin (BH_4) biosynthesis. In order to evaluate or to determine an increase in the remaining enzyme activity after therapeutic treatment, we have developed an in vivo measurement of phenylalanine-4-hydroxylase. We determined the in vivo "overall activity" of the phenalalanine-4-hydroxylase system by loading patients with deuterium labelled L-phenylalanine-d_5 and measuring the deuterated tyrosine-d_4 formed in their plasma. This is shown in Fig. 1.

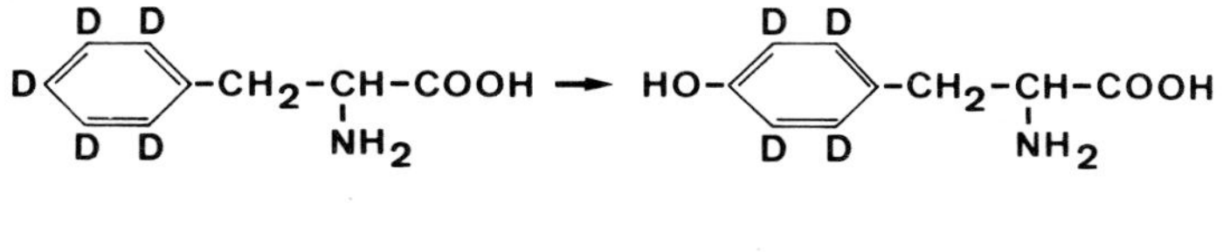

D-⟨D,D⟩-CH₂-CH(NH₂)-COOH (D,D) → HO-⟨D,D⟩-CH₂-CH(NH₂)-COOH (D,D)

PHE-d$_5$ **TYR-d$_4$**

Fig. 1. Conversion scheme of phenylalanine-d$_5$ to tyrosine-d$_4$.

The deuterated phenylalanine as substrate and the deuterated tyrosine formed as product in the serum were analyzed by mass fragmentography. After the addition of DL-phenylalanine-d$_1$ and L-tyrosine-d$_7$ as internal standards in 0,5 ml of the plasma samples, the solution was deproteinized with sulfosalicylic acid and fractionalized by ion exchange chromatography. Out of the several derivatives investigated, the N-, (O-),-trifluoroacetyl methyl esters were found to be the most suitable for our purposes. The ratio of deuterated product to substrate is a measure for the overall activity of phenylalanine-4-hydroxylase in the body. Some of our results are shown in Fig. 2. The logarithm of tyrosine-d$_4$ over phenylalanine-d$_5$ is plotted against the time of blood collection after the phenylalanine-d$_5$ loading.

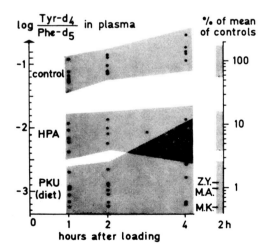

Fig. 2. Logarithm of deuterated phenylalanine over deuterated tyrosine in plasma after loading with deuterated phenylalanine-d$_5$ (200 mg/kg).
Above, control persons: hyperphenylalaninemics (HPA);
below, patients with phenylketonuria under diet.
The % scale is based on the values at 2 hours after loading.

Healthy controls formed a lot of tyrosine, about 1/10 of the concentration of deuterated phenylalanine. To the right of Fig. 2 a percent scale is plotted based on the controls. As we can see on Fig. 2 children with hyperphenylaninemia reach only about 10% of the control values, and patients with PKU only about 1% or less.

The results of the in vivo assay were compatible with those of the in vitro experiments on liver biopsy material. The in vivo assay with deuterated phenylalanine proved to be a powerful tool for the diagnosis and treatment of the PKU.

2. Tyrosine-3-hydroxylase (E.C. 1.14.3.a). Tyrosine-3-hydroxylase is the key enzyme of the tyrosine-dopa-norepinephrine metabolism. Several years ago we found that, in vivo, tyrosine-3-hydroxylase and also tryptophan-5-hydroxylase were competitively inhibited by high concentrations of L-phenylalanine in PKU patients. This leads to very low concentration of dopamine in the brain of these patients and is most probably responsible for the mental retardation and the neurological symptoms in these patients (13a). This is demonstrated in Fig. 3.

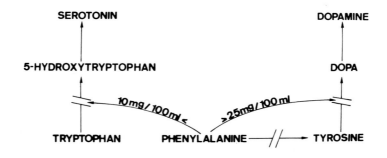

Fig. 3. Competitive inhibition of tryptophan-5-hydroxylase and tyrosine-3-hydroxylase by L-phenylalanine.

In order to prove the hypothesis of the inhibition of tyrosine-3-hydroxylase in PKU, we needed an in vivo estimation of the enzyme. For this purpose, PKU patients and healthy controls were loaded orally with deuterated L-tyrosine-d_2, and the deuterated dopamine-d_1 in their urine was analyzed. This is shown on Fig. 4.

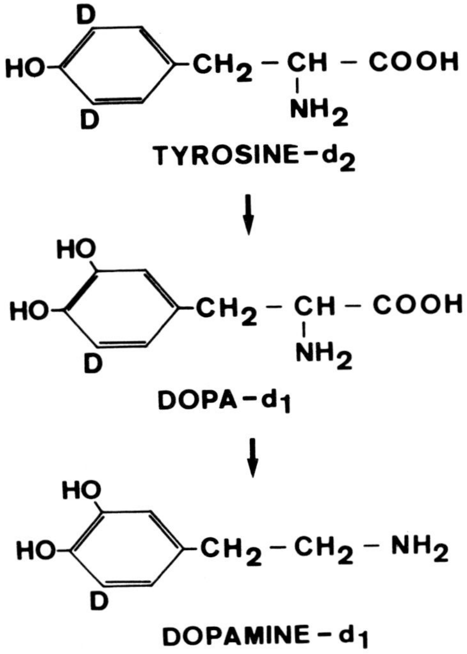

Fig. 4. Conversion scheme of tyrosine-d_2 to dopamine-d_1

The analytical procedure was as follows: After addition of the internal standard dopamine-d_4 to 10 ml of the 24 h urine, the catechols are adsorbed on aluminium oxide at pH 8.4. Dopamine is eluted with 0.25 N acetic acid and analyzed as trifluoroacetyl derivative by SIM. The detailed procedure has been described elsewhere (14).

Tests were performed on patients with high (25-35 mg/100 ml) or low (<4 mg/100 ml) plasma phenylalanine blood levels. Our results are shown in Fig. 5. At high phenylalanine blood levels the excretion of dopamine was low. When plasma phenylalanine was low, a large increase in the amount of urinary dopamine was observed. Since dopamine is an important transmitter substance in the brain, the in vivo determination of the key enzyme tyrosine-3-hydroxylase is of great importance.

3. Tryptophan-5-hydroxylase (E.C. 1.14.16.4). The tetrahydrobiopterin (BH_4)-dependent enzyme tryptophan-5-hydroxylase is known to be the rate limiting enzyme of the biosynthesis of serotonin from tryptophan. Normally about 1% of the tryptophan contained in the food intake is converted to serotonin. Tryptophan is first hydroxylated to 5-hydroxy-tryptophan, followed by decarboxylation to 5-hydroxy-

tryptamine. The second enzymatic reaction is about 100 times faster. The enzyme that catalyses the conversion from 5-hydroxytryptophan to serotonin is the relatively unspecific aromatic L-amino acid decarboxylase which is also involved in the synthesis of the catecholamines. In Fig. 6 the tryptophan metabolism is shown.

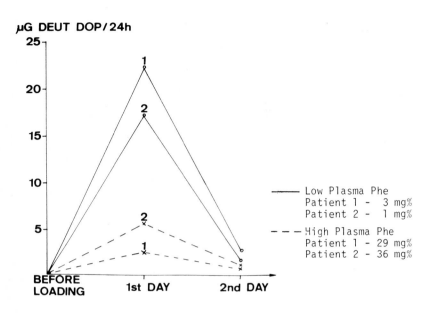

Fig. 5. Loading with L-tyrosine-d_2 (150 mg/kg); SIM-determination of deuterated dopamine in urine.

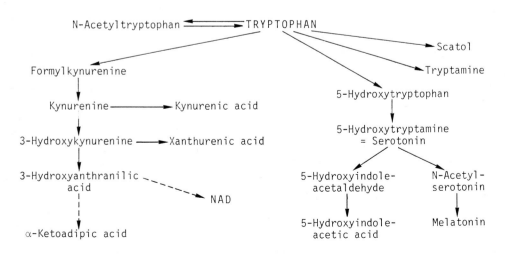

Fig. 6. Tryptophan metabolism.

Tryptophan-5-hydroxylase occurs mainly in the enterochromaffin cells, in the neurons and in the rodent mast cells. It is assumed that the tryptophan-5-hydroxylase in the brain is not saturated by its substrate and that the in vivo synthesis of serotonin depends to some extent on the tryptophan concentration in blood. It has not yet been ascertained whether BH_4 is the natural cofactor in brain, but it is suspected to be a pterin.

Serotonin possesses a variety of pharmacological effects and acts as a very important neurotransmitter in the brain. In mammals about 90% of the serotonin present in the body, which in an adult human probably totals up to 10 mg, can be found in the gastrointestinal tract, mainly in enterochromaffin cells and enterochromaffin-like cells. A few such serotonin-containing cells are also present in other tissues. Most of the remaining serotonin is present in platelets and in the brain. Several workers have reported that disorders of the tryptophan metabolism might play an important role in affective disturbances, e.g. mania, endogenous depression, schizophrenia and neurological diseases. Pare et al. (15), Berendes et al. (16) and Matsuda et al. (17) reported for the first time a decreased serotonin blood level and a decreased 5-hydroxyindoleacetic acid excretion in phenylketonuria (PKU) patients. We have recently shown by in vivo investigations that the conversion of tryptophan to 5-hydroxytryptophan by the enzyme tryptophan-5-hydroxylase is inhibited by elevated phenylalanine concentrations in blood and tissues (18).

In patients with atypical PKU suffering from a BH_4 deficiency we have recently shown that the serotonin excretion was reduced to only 10% compared with the values of normal controls (12). With BH_4 substitution the serotonin excretion increased about four-fold. For these reasons, an in vivo measurement of the tryptophan-5-hydroxylase activity is of great importance. An indirect in vivo measurement is possible by loading patients with deuterated L-tryptophan and subsequently measuring the deuterated serotonin formed in urine by using a gas chromatography-selective ion monitoring (GC-SIM) method. A scheme for the conversion of tryptophan-d_5 to serotonin-d_4 is shown in Fig. 7.

The following method was applied: After addition of our internal standard serotonin-d_4 to 5 ml of 24 h urine, the solution was extracted with n-butanol and the butanol phase was washed with 50 mM ammonia solution containing 30 mg/100 ml of ascorbic acid and 1 mg/100 ml of creatinine sulfate. After addition of cyclohexane the solution was re-extracted with 0.5 M formic acid and evaporated to dryness. After the formation of the pentafluoropropionate derivatives serotonin was analyzed by SIM. The detailed procedure is described elsewhere (19).

The results of one of the loading experiments is shown in Fig. 8.

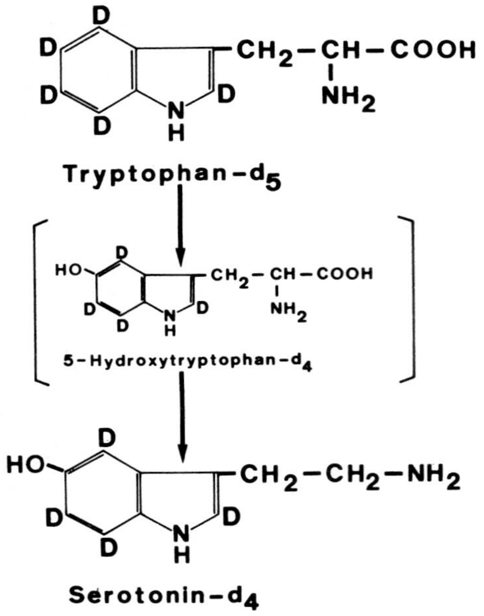

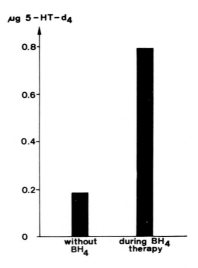

Fig. 7. Conversion scheme of tryptophan-d_5 to serotonin-d_4.

Fig. 8. Urinary excretion of serotonin-d_4 by a patient with BH_4 deficiency in 24 h after oral loading with L-tryptophan-d_5, without and during BH_4 theraphy.

4. Tryptophan pyrrolase (L-tryptophan oxygen 2,3-oxidoreductase (decyclizing), E.C. 1.13.11.11). Several authors have demonstrated that the concentration of free tryptophan in plasma is decreased in patients suffering from endogenous depression (20). Comparative studies of the kynurenine excretion on patients with depression also show an increased excretion of the metabolites of kynurenine (21). This leads to the assumption that the decreased level of free tryptophan in patients with depression is due to the increased activity of the tryptophan pyrrolase. Finally, according to another hypothesis, schizophrenia is the result of a disorder in the tryptophan-niacin-pathway (22). For these reasons an in vivo measurement of the tryptophan pyrrolase activity was of great importance.

An in vivo determination of the tryptophan pyrrolase activity is possible by loading the subjects with 50 mg/kg deuterated L-tryptophan-d_5 and subsequently measuring the deuterated L-kynurenine-d_4 formed and the residual L-tryptophan-d_5 by SIM. For the SIM determination of tryptophan and kynurenine we used N-pentafluor-propionyl methyl ester derivatives. The serum levels of deuterated kynurenine-d_4, after loading with deuterated tryptophan-d_5, are plotted against the time of blood collection. The slope of this curve is an index for the enzyme activity.

Our analytical procedure was the following: After addition of the internal standards tryptophan-d_8 and kynurenine-d_2 to 1 ml of plasma and 1 ml of 0.9% NaCl solution the plasma is deproteinized with sulphosalicylic acid and separated on a Dowex 50 W-X2 (H^+) ion exchange column. After rinsing with 10% acetic acid, water and 25 mM pyridine-formic acid buffer (pH 3.0) tryptophan and kynurenine were eluted with 25 mM pyridine-formic acid buffer (pH 4.55). The total procedure is described elsewhere (13).

These examples clearly demonstrate the usefulness of the in vivo determination of an enzyme activity with the loading of deuterated precursors and the analysis of the deuterated product by SIM.

ACKNOWLEDGEMENTS

This study was supported by the Swiss National Science Foundation, Grant no. 3.919-0.80.

REFERENCES

1 R. Schoenheimer and D. Rittenberg, J. Biol. Chem., 111(1935)163.
2 J.A. Völlmin, Clin. Chim. Acta, 34(1971)207.
3 M. Zachmann, M. Zagalak, J.A. Völlmin et al., Clin. Chim. Acta, 77(1977)147.
4 H.-Ch. Curtius, J.A. Völlmin and K. Baerlocher, Clin. Chim. Acta, 37 (1972)277.
5 H.-Ch. Curtius and M. Mettler, J. Chromatogr., 126(1976)569.
6 M. Fuchs-Mettler, H.-Ch. Curtius, K. Baerlocher and L. Ettlinger, Eur. J. Biochem., 108(1980)527.
7 Helen Wegmann, H.-Ch. Curtius, R. Gitzelmann, A. Otten, Helv. paed. Acta, 34(1979)

497.
8 H.-Ch. Curtius, J.A. Völlmin and K. Baerlocher, Anal. Chem., 45(1973)1107.
9 H.-Ch. Curtius, J.A. Völlmin, M.J. Zagalak and M. Zachmann, J. Steroid Biochem., 6(1975)677.
10 H.-Ch. Curtius, M.J. Zagalak, K. Baerlocher et al., Helv. paed. Acta, 32(1977)461.
11 H.-Ch. Curtius, M.J. Zagalak, M. Mettler et al., Mass Spectrometry and Combined Techniques in Medicine, Clinical Chemistry and Clinical Biochemistry, Symposium in Tübingen, FRG, 14-15th November, 1977.
12 H.-Ch. Curtius, H. Farner-Wegmann, A. Niederwieser et al., Proceedings of the Third International Meeting of the International Study Group for Tryptophan Research, Kyoto, Japan (1980)281-291.
13 H. Wegmann, H.-Ch. Curtius and U. Redweik, J. Chromatogr., 158(1978)305.
13a H.-Ch. Curtius, K. Baerlocher and J.A. Völlmin, Clin. Chim. Acta, 42(1972)235.
14 M. Wolfensberger, Dissertation: Gaschromatographische und massenfragmentographische Methoden zur Bestimmung von Dopamin, Noradrenalin, Adrenalin, 6-Hydroxydopamin, 5-Hydroxytryptophol und 5-Methoxytryptophol in biologischem Material: Anwendung in der klinischen Biochemie, ETH, Zurich, 1976.
15 C.M.B. Pare, M. Sandler and R.S. Stacey, Lancet I(1957)551.
16 H. Berendes, J.A. Anderson, M.R. Ziegler and D. Ruttenberg, Amer. J. Dis. Child, 96(1958)430.
17 I. Matsuda, M. Sugai, S. Arashima and M. Anakura, Clin. Chim. Acta, 34(1971)491.
18 H.-Ch. Curtius, A. Niederwieser, M. Viscontini et al., Current Aspects of Neurochemistry and Function, Plenum, New York, in press.
19 H.-Ch. Curtius and Helen Wegmann, J. Chromatogr., 199(1980)171.
20 A. Coppen, E.G. Eccleston and M. Peet, Lancet 2(1973)60.
21 G. Curzon and P.K. Bridges, J. Neurol. Neurosurg. Psychiatry, 33(1970)698.
22 L. Gilka, Acta Psychiatr. Scand. Suppl. no. 258(1975).

PRENATAL DIAGNOSIS OF PROPIONIC AND METHYLMALONIC ACIDEMIA BY STABLE ISOTOPE DILUTION ANALYSIS OF METHYLCITRIC AND METHYLMALONIC ACIDS IN AMNIOTIC FLUIDS

L. SWEETMAN[a], G. NAYLOR[a], T. LADNER[a], J. HOLM[a], W.L. NYHAN[a], C. HORNBECK[b], J. GRIFFITHS[b], L. MORCH[c], S. BRANDANGE[c], L. GRUENKE[d] and J.C. CRAIG[d]

[a]University of California, San Diego, La Jolla, California 92093, USA
[b]Veterans Administrative Medical Center, San Diego, California 92161, USA
[c]Arrhenius Laboratory, University of Stockholm, Stockholm, Sweden
[d]University of California, San Francisco, San Francisco, California 94143, USA

ABSTRACT

Rapid, sensitive and accurate assays were developed for the measurement of methylcitric and methylmalonic acids in amniotic fluids for the prenatal diagnosis of pregnancies at risk for the inherited metabolic disorders propionic acidemia and methylmalonic acidemia. These assays utilize deuterium labeled internal standards with liquid partition chromatography and chemical ionization gas chromatography-mass spectrometry. The levels of acids have given the correct prenatal diagnosis in 17 pregnancies.

INTRODUCTION

Propionic acidemia and methylmalonic acidemia are inherited disorders of the metabolism of propionic acid which result in severe, often lethal, illness early in life (ref. 1). Propionyl-CoA is carboxylated by propionyl-CoA carboxylase to methylmalonyl-CoA and this enzyme is deficient in propionic acidemia. Methylmalonyl-CoA is isomerized to succinyl-CoA by methylmalonyl-CoA mutase and this enzyme is deficient in methylmalonic acidemia. Because of the life-threatening nature of these disorders it is important to provide prenatal diagnoses for parents at risk of having an affected child. This has been done by the measurement of deficient activity of the enzymes in fetal cells from amniotic fluid which requires many weeks to obtain sufficient cells for analysis (refs. 2-10). More rapid prenatal diagnosis is possible by measuring elevated levels of metabolites in the amniotic fluid immediately after amniocentesis. Propionyl-CoA is metabolized to two diastereoisomers of methylcitric acid (MCA) by patients with either disorder (refs. 11-13). MCA was found to be highly elevated in the amniotic fluid of fetuses with propionic acidemia using liquid partition chromatography (LPC) followed by GC amd GCMS (refs. 3,7). A more sensitive method for MCA in amniotic fluid using D_3-MCA as internal standards for LPC and chemical ionization (CI) GCMS showed that MCA is a normal constituent of amniotic fluid and is elevated with fetuses with methylmalonic acidemia (ref. 14).

Methylmalonic acid (MMA) has been shown to be elevated in amniotic fluids of fetuses with methylmalonic acidemia using GC analysis (refs. 6,10,15-17). This paper describes a new method for measurement of MMA using D_3-MMA as an internal standard for LPC and CI GCMS.

METHODS AND RESULTS

(2S,3S)-MCA and (2S,3R)-MCA (ref. 18) and (R*,R*)-D_3-MCA and (R*,S*)-D_3-MCA (ref. 14) were synthesized as described. D_3-MMA was synthesized by the methylation of the diethyl ester of malonic acid with D_3-methyl iodide. Normal amniotic fluids were obtained between 16-18 weeks of pregnancy by amniocentesis and stored at -20°C. Fluids from pregnancies at risk were sent for analysis as sterile fluids or lyophilized powders at ambient temperature or frozen in dry ice.

100 nmoles each of D_3-MMA, (R*,R*)-D_3-MCA and (R*,S*)-D_3-MCA were added to 1-5 ml of amniotic fluid, lyophilized, acidified with 0.3 ml of 0.5 N H_2SO_4 and absorbed onto silicic acid. A new LPC column, 0.6 cm I.D. x 45 cm, of silicic acid hydrated with H_2SO_4 (50 ml of 0.1 N H_2SO_4 per 92 g of dried 100 mesh silicic acid) was prepared for each sample and rinsed with 9% (v/v) 2-methyl-2-butanol in $CHCl_3$. The sample was placed on the column and eluted with this solvent at 2.5 ml/min and 3 min fractions collected. Fractions between 52.5-75 ml contained MMA and were pooled and dried. For the purpose of showing the elution profile of acids in Fig. 1, the effluent from the column was continuously mixed with 0.5 ml/min of 5 mmol/l of Na o-nitrophenolate in 95% ethanol and the titration of this acid indicator monitored spectrophotometrically at 340 nm. For the usual isolation of MMA, after the elution of MMA without the acid indicator, the indicator was added beginning at 75 ml to verify the elution of lactic acid and MCA. To elute MCA, the eluting solvent was changed at 90 ml to 30% 2-methyl-2-butanol in $CHCl_3$. Fractions containing MCA (usually 157.5-187.5 ml) were pooled and dried.

The samples of MMA were acidified with 1 ml of 0.1 N HCl in methanol, transferred with methanol to 1 ml serum bottles and redried. The samples of MCA were acidified with 1 ml 0.1 N HCl in methanol and redried. For methylation, diazomethane was generated from Diazald (Aldrich) (ref. 19). The MMA samples were dissolved in 0.2 ml of 10% methanol in diethyl ether, saturated with diazomethane, and sealed. Any evaporation of the ether-diazomethane solution resulted in large losses of the volatile dimethyl malonate. The MCA samples were dissolved in 1-2 ml of 10% methanol in diethyl ether, saturated with diazomethane, left 10 min, transferred to 1 ml serum bottles with methanol, dried, redissolved in 0.2 ml of methanol and sealed. Standard mixtures of the acids and deuterated standards were prepared and carried through the entire procedure. The methylated standards were stable when stored at $-20°C$ and could be used for GCMS standard curves for at least a year. Two to four µl were used for GCMS analysis.

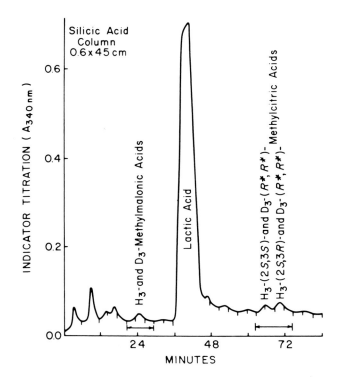

Fig. 1. Liquid partition chromatogram of 2 ml of normal amniotic fluid with 100 nmoles each of D_3-methylmalonic acid $(\underline{R}^*,\underline{R}^*)$-$D_3$-methylcitric acid and $(\underline{R}^*,\underline{S}^*)$-$D_3$-methylcitric acid. The absorbance at 340 nm is proportional to the amount of acid eluted. Downward deflections indicated fraction changes.

The GCMS analyses were done with a Finnigan 4021 quadrapole with the INCOS Data System. The GC column was 3% ECNSS-M on 100/120 mesh Gas-Crom Q, 2 mm x 180 cm with 25 ml/min of helium as carrier gas. The injector was at $255°C$, the direct transfer line at $280°C$, and the ion source at $220°C$. The column temperature was $90°C$ for dimethyl methylmalonate and $155°C$ for trimethyl methylcitrate. Ammonia was used as the reagent gas for CI, giving similar intensities of the $(M+1)^+$ and $(M+18)^+$ ions as the only significant peaks. The ammonia pressure in the source was 9 Pa and the ionizing voltage 70 eV. The protonated molecular ions were chosen to obtain selected ion chromatograms with a dwell time of 100 msec at each $\underline{m/z}$.

Figure 2 shows selected ion chromatograms for dimethyl methylmalonate. Peak areas were calculated by the data system after the operator visually selected the baseline points.

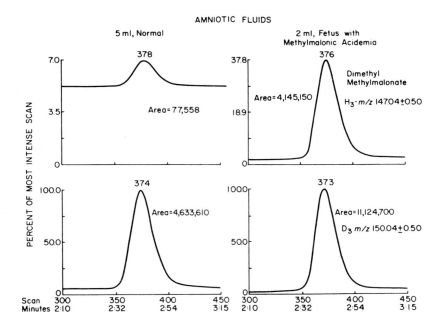

Fig. 2. Selected ion chromatograms for the protonated molecular ions of dimethyl-methylmalonate at $\underline{m/z}$ 147 and dimethyl-D_3-methylmalonate at $\underline{m/z}$ 150 obtained with ammonia chemical ionization GCMS.

The standard curve in Figure 3 shows good linearity of the ratio of areas at $\underline{m/z}$ 147 to $\underline{m/z}$ 150 versus the amount of standard MMA with a constant 100 nmoles of D_3-MMA. The analysis of MCA was done as previously described (ref. 14).

The levels of MMA and total MCA in the amniotic fluids are summarized in Table 1. Duplicate values indicate two different fluids from the same pregnancy. The diagnosis of an affected or not affected (either normal or heterozygous) fetus was confirmed by assay of propionyl-CoA carboxylase, methylmalonyl-CoA mutase (M.J. Mahoney, personal communication) on the birth of a clinically normal or affected child. The normal level of MMA was 0.29±0.08 (SD) μmol/l. The normal levels of MMA previously reported with GC methods were undetectable (refs. 15,17), 0-0.85 μmol/l (ref. 6), less than 8μmol/l (ref. 4) and 6.4±3.2μmol/l (ref. 16). Our results are consistent with all but the 22 times higher levels reported by Nakamura et al (ref. 16). Their method used isolation of MMA by ion exchange chromatography and quantitation with GC. Without GCMS verification that their method was specific for MMA, it is likely that they measured a coeluting compound. The levels of MMA in amniotic fluids of 4 fetuses with methylmalonic acidemia ranged from 17.6-80.7 μmol/l, or 60-280 times the normal mean (Table 1). The levels are comparable to those of 18.6-266μmol/l previously reported for affected fetuses using GC methods (refs. 4,6,15-17). The levels of MCA in amniotic fluids form 4 fetuses with propionic acidemia ranged from 14 to 24 times the normal mean (Table 1). For 5 fetuses with methylmalonic acidemia, the levels of MCA were somewhat less elevated at 5 to 8 times the normal mean. For all of the 17 cases with independent confirmation of the diagnosis, the stable isotope dilution analysis of MCA in amniotic fluid has given the correct diagnosis as has the analysis of MMA for the eight cases at risk for methylmalonic acidemia.

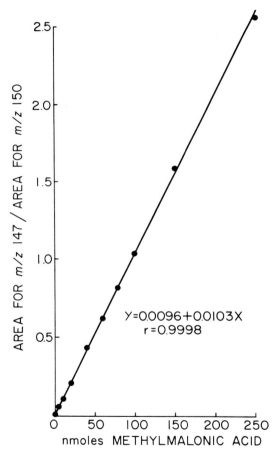

Fig. 3. Standard curve of the ratio of peak areas from CI GCMS for dimethyl-methylmalonate at $\underline{m/z}$ 147 and dimethyl-D_3-methylmalonate at $\underline{m/z}$ 150 vs nanomoles of H_3-methylmalonic acid with a constant 100 nmoles of D_3-methylmalonic acid.

DISCUSSION

The stable isotope dilution assays for MMA and MCA in amniotic fluid are rapid, sensitive and accurate. The deuterium labeled standard serves as a carrier during preparation of the sample for GCMS and measurement of the ratio of nonlabeled to labeled compound corrects for any losses during the procedures. Measurement of the internal standard verifies that the preparation of the sample was suitable and enhances the confidence in the results. Based on the results of 17 correct diagnoses for propionic or methylmalonic acidemia using the stable isotope dilution assays for the acids in amniotic fluids, this appears to be a reliable method for prenatal diagnosis.

TABLE 1

Amniotic fluid levels of acids

	Methylmalonic Acid µMolar	Total Methylcitric Acid µMolar
Normal Mean ± S.D.(n)	0.29±0.08(9)	0.38±0.10(8)
At risk for propionic acidemia		
Not affected	–	0.36
	–	0.20
	–	0.21
	0.32	0.42
	–	0.34
	–	0.33[a]
Affected	–	8.94
	–	7.87
	–	5.20
	–	5.27
At risk for methylmalonic acidemia		
Not affected	–	0.19
	0.59	0.31
	0.46	0.29
Affected	–	1.79
	74.0, 54.2	2.79, 2.55
	17.6	2.73
	53.6, 80.7	2.57, 3.05
	19.41	2.68

[a] Confirming diagnosis pending.

MMA and MCA are stable, nonvolatile compounds which permit amniotic fluid samples to be sent for analysis as sterile fluids at ambient temperature, frozen, or lyophilized while shipment of sterile cultures of living amniotic cells for enzyme analysis is much more difficult. The analysis is rapid, requiring only 2-3 days as compared with many weeks for the cultivation of sufficient amniotic cells for assays of enzyme deficiencies. Additional problems with enzyme assays in cultured amniotic cells are that a significant percentage of cultures fail to grow or become contaminated with microorganisms. A significant problem can arise if the amniotic cells are overgrown by maternal cells (ref. 7). Until more experience is gained we recommend that the rapid prenatal diagnosis provided by the analysis of MMA or MCA in amniotic fluid be confirmed when possible with assays for the enzyme activity in cultured amniotic cells. Provided that the metabolite levels continue to provide the correct diagnosis, it should be practicable to utilize this method alone. It is likely that the analysis of elevated metabolites in amniotic fluid utilizing stable isotope dilution analysis can be employed for the prenatal diagnosis of a large number of severe metabolic diseases.

ACKNOWLEDGEMENTS

We thank the many physicians who provided amniotic fluids from pregnancies at risk for propionic or methylmalonic acidemia, Dr. O.W. Jones for providing normal amniotic fluids and Dr. M.J. Mahoney for his results for methylmalonyl-CoA mutase. This study was supported in part by US Public Health Service Grant No. HD-04608 from the National Institute of Child Health and Development and Grants No. GM-17702 and GM-27583 from the National Institute of General Medical Science, National Institutes of Health, Bethesda, MD, USA.

REFERENCES

1 L.E. Rosenberg, in J.B. Stanbury, J.B. Wyngaarden and D.S. Fredrickson (Eds.), Metabolic Basis of Inherited Disease, McGraw-Hill, New York, 1978, pp. 411-429.
2 D. Gompertz, P.A. Goodey, H. Thom, G. Russell, A.W. Johnston, D.H. Mellor, M.W. MacLean, M.E. Ferguson-Smith and M.A. Ferguson-Smith, Clin. Genet. 8 (1975) 244-250.
3 L. Sweetman, W. Weyler, T. Shafai, P.E. Young and W.L. Nyhan, J. Am. Med. Assoc. 242 (1979) 1048-1052.
4 M.J. Mahoney, L.E. Rosenberg, B. Lindblad, J. Waldenström and R. Zetterström, Acta Paediatr. Scand. 64 (1975) 44-48.
5 M.G. Ampola, M.J. Mahoney, E. Nakamura and K. Tanaka, N. Engl. J. Med. 293 (1975) 313-317.
6 D. Gompertz and P.A. Goodey, Pediatrics 54 (1974) 511-513.
7 P.D. Buchanan, S.G. Kahler, L. Sweetman and W.L. Nyhan, Clin. Genet. 18 (1980) 177-183.
8 G. Morrow, B. Revsin, C. Mathews and H. Giles, Clin. Genet. 10 (1976) 218-221.
9 H.F. Willard, L.M. Ambani, A.C. Hart, M.J. Mahoney and L.E. Rosenberg, Hum. Genet. 32 (1976) 277-283.
10 G. Morrow, B. Revsin, J. Lebowitz, W. Britt and H. Giles, Clin. Chem. 23 (1977) 791-795.
11 S. Brandänge, S. Josephson, A. Mahlén, L. Mörch, L. Sweetman and S. Vallén, Acta Chem. Scand. B31 (1977) 628.
12 T. Ando, K. Rasmussen, J.M. Wright and W.L. Nyhan, J. Biol. Chem. 247 (1972) 2200-4.
13 S. Brandänge, O. Dahlman and L. Mörch, J. Chem. Soc. Chem. Comm. (1980) 555-556.
14 G. Naylor, L. Sweetman, W.L. Nyhan, C. Hornbeck, J. Griffiths, L. Mörch and S. Brandänge, Clin. Chim. Acta 107 (1980) 175-183.
15 G. Morrow, R.H. Schwartz, J.A. Hallock and L.A. Barness, J. Pediatr. 77 (1970) 120-3.
16 E. Nakamura, L.E. Rosenberg and K. Tanaka, Clin. Chim. Acta 68 (1976) 127-140.
17 H.D. Bakker, A.H. Van Gennip, M. Duran and S.K. Wadman, Clin. Chim. Acta 86 (1978) 349-352.
18 S. Brandänge, S. Josephson, L. Mörch and S. Vallén, Acta Chem. Scand. B31 (1977) 307-312.
19 H. Schlenk and J.L. Gellerman, Anal. Chem. 32 (1960) 1412-1414.

MASS FRAGMENTOGRAPHIC DETERMINATION OF METHYLMALONIC ACID IN MATERNAL URINE AND AMNIOTIC FLUID USING $^2H_3$-METHYLMALONIC ACID AS INTERNAL STANDARD

F.K. TREFZ[1], B. TAUSCHER[2], and W. KOCHEN[1]

[1] University Children's Hospital, Im Neuenheimer Feld 150, D-6900 Heidelberg, W.-Germany

[2] Chemical Institute, University of Heidelberg, D-6900 Heidelberg, W.-Germany

ABSTRACT

A sensitive and reliable method for the trace analysis of methylmalonic acid in amniotic fluid and urine is described using deuterated methylmalonic acid as internal standard and capillary gas chromatography/mass fragmentography. The application of the method for the prenatal diagnosis of methylmalonic acidemia is demonstrated in three pregnancies at risk. In two pregnancies the fetuses were affected by methylmalonyl-CoA-mutase deficiency. Correspondingly, the excretion of methylmalonic acid in the maternal urine was elevated as early as at the 12/13th week of gestation, reaching its highest level shortly before abortion at the 19/20th week: 156.7 and 172.8 µmoles/24 h (excretion in normal pregnancies: 38.9±7.6 µmoles/24 h, n=8). In addition, the concentration of methylmalaonic acid in amniotic fluid at the 16th week (13.4 and 33.8 µM/l, normal range 0.31±0.10 µM/l, n=8) strongly suggested that the fetuses were affected. In the third pregnancy no increase of the methylmalonic acid excretion in maternal urine at 11-17 weeks of gestation could be found (42.3±9.3 µmoles/24 h, n=5). The cultured amniotic cells of this fetus showed normal enzyme activity. Nevertheless abortion was initiated without further biochemical investigation because of an elevated α1-fetoprotein value in the amniotic fluid. The fetus was anencephalic. The data suggest that it is possible to make a reliable prenatal diagnosis of methylmalonic acidemia even in those cases where cultured amniotic cells are not available.

In 1967 Oberholzer [1] and Stokke [2] described the clinical and biochemical features of methylmalonic acidemia. Since then a number of cases has been published with different enzyme defects leading to high concentrations of methylmalonic acid in plasma and urine (for review see [3]). The incidence of this entity has been considered to be about 1:100 000 [4]. It is, however, likely that a number of cases has still been missed because the diagnosis is overlooked in the severely ill newborn.

The treatment of the disease - especially of the vitamin B_{12} unresponsive form - is difficult and prognosis in respect to a normal psychomotoric development is dubious. It is generally accepted that interruption is justified in pregnancies with an affected fetus as proven by prenatal diagnosis. The latter can be performed directly by enzyme measurement in cultured amniotic cells [5]. Many efforts have also been made to measure the concentration of methylmalonic acid in amniotic fluid and maternal urine during pregnancies at risk [5,6,7,8,9]. These additional data were expected to support the results of the enzymatic assay to help the clinician in his difficult decision to initiate artifical abortion. Unfortunately, the data given in the literature for the excretion of methylmalonic acid in the urine of mothers carrying a fetus at risk are contradictory during early pregnancy; in addition an elevation of methylmalonic acid of 3-4 times above normal does not allow a definite diagnosis. This may be due mainly to insufficient methods for the trace analysis in biological material (i.e. colorimetric methods, gas chromatography with packed columns). To overcome this difficulty we developed an isotopic dilution method by using deuterated methylmalonic acid as internal standard and capillary gas chromatography/mass fragmentography. We want to communicate the results of this method in three pregnancies at risk in comparison to normal values of methylmalonic acid in amniotic fluid and urine of healthy pregnant women.

MATERIALS AND METHODS

All reagents were of analytical grade and purchased from Merck (Darmstadt, W.-Germany). N-methyl-trimethyl-silyl-trifluoracetamide was used as silylating reagent (Machery&Nagel, Düren, W.-Germany).

Methylmalonic acid was determined by using a simple extraction method: 2ml of urine or 1 ml of amniotic fluid are acidified to pH 1-2 with 1 N HCL and saturated with NaCl. 10 µl of the internal standard solution (42.4 µmoles of $^2H_3$-methylmalonic acid in 5.0 ml of methanol) are added. The sample is extracted twice with 4 ml of ethylacetate in a reagent tube. After centrifugation (3000 rpm for 5 min) the organic phases are combined, dried over Na_2SO_4 for 1 h and evaporated to dryness under reduced pressure. The residue is brought in a reaction vial with ca. 0.5 ml of ethylacetate, dried under a stream of nitrogen and finally dissolved in 50 µl of methyl-trimethylsilyl-trifluoracetamide/chloroform (1:1 v/v) shortly before use. 1 µl is injected.

Synthesis of $^2H_3$-methylmalonic acid (internal standard)
1) synthesis of $^2H_3$-methyldiethylmalonate:
To a solution of 2.3 g Na (0.1 mole) in 50 ml of absolute ethanol, 15.55 g (0.097 mole) of diethylmalonate are added. A well cooled ampoule containing 10 g

(0.097 mole) of $^2H_3$-methylbromide is connected with a bubble counter and with the reaction vessel. The gas inlet tubing ends above the surface of the vigorously stirred reaction mixture, which is heated up to 50°C. The addition of gaseous $^2H_3$-methylbromide (70 bubbles/min) is regulated by warming up the ampoule to room temperature. After completing the reaction by refluxing during a period of 3 hours, the main quantity of ethylalcohol is destilled and the mixture is worked up by common methods.

2) Synthesis of $^2H_3$-methylmalonic acid

15 g (0.12 mole) crude $^2H_3$-methyl-diethylmalonate boiled under reflux with 7.9 g (0.14 mole) KOH and a mixture of 42.1 ml of ethylalcohol and 21 ml of water acidified to pH 1 and worked up. Recrystallization from benzene/ethylacetate gives a product with a melting point of 103-105°C. The yield amounts to 87 %. The identity is confirmed by nuclear magnetic resonance (NMR). ^{1}H-NMR data: COOH 12.66 ppm broad, CH 3.24 (and 3.29 ppm) singlet. ^{13}C-NMR-data: COOH 174.14 (and 170.81 ppm) singlet, HC 45.72 ppm, CD_3 12.15 ppm (and 40.4 ppm) multiplet. Further evaluation is given by mass spectroscopic data of the di-silylated compound. The purity was proven to be 98.5 %.

The mass spectrometer (Du Pont 21-492) was modified for use with capillary gaschromatography [10]. The gaschromatograph (Varian 2700) is combined via an open allglass interface to the mass spectrometer. A 50 m x 0.3 mm i.d. capillary column SE 54 (Jaeggi, Trogen, Switzerland) is used. Operating conditions are as follows: helium flow rate 2 ml/min, initial column temperature 100°C for 5 min, temperature program at 4°/min to 180°C, injector, interface and source 200°C, electron impact energy 70 eV. Split ratio 1:10. Quantification is performed by measuring the peak heights.

The details of the enzymatic assays are described elsewhere [11,12,13].

RESULTS

The calibration curve for the selected ion monitoring of $^1H_3$-methylmalonic acid using $^2H_3$-methylmalonic acid as internal standard gives a linear correlation (r=0.996) between the ratio of the peak heights at m/z 247 and m/z 250 versus the amount of injected sample (range: 0.1-10 nmol). The detection limit of endogenous methylmalonic acid is 0.05 nmol; the coefficient of variation for individual measurements is 2% for a sample containing 0.5 µmol (n=8). A typical mass fragmentogram of a urine containing 35.5 µM/l of methylmalonic acid is given in Fig. 1.

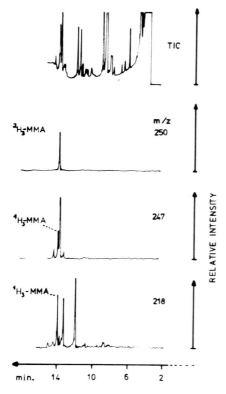

Fig. 1: Mass fragmentograms of a control urine containing 35.5 μM/l of methylmalonic acid. TIC=total ion current, $^2H_3$-MMA= deuterated standard of methylmalonic acid, $^1H_3$-MMA=endogenous methylmalonic acid. Conditions see text.

$^1H_3$-methylmalonic acid is not completely separated from a byproduct which contains the fragment at m/z 247. This problem does not arise when measuring the fragment at m/z 218. Nevertheless quantification of $^1H_3$-methylmalonic acid is possible by measuring the m^+-15 ions at m/z 247 and 250 as well as the fragments at m/z 218 and 221 respectively.

The fetus of mother P.G. (table as well as fetus 1 of mother S.G. are affected with methylmalonyl-CoA-mutase deficiency as indicated by deficient propionate utilization found in P.G.'s cultured amniotic cells and undetectable mutase activity of S.G.'s cultured amniotic cell homogenate [12]. Correspondingly in both pregnancies, the concentration of methylmalonic acid in amniotic fluid is about 40 times higher than in normal controls of the same gestational ages (table). Similar results are obtained for the excretion of methylmalonic acid in the maternal urines. There is an increase detectable as early as at the 12/13th week of gestation [12]. Shortly before the interruption the highest values are reached (156.7 and 172.8 μmoles/24 h) (table). These values are about four times higher than in normal pregnancies as well as in the second pregnancy of mother S.G. (table). Unfortunately in this fetus 2, ultrasonographic investigation in early pregnancy indicated that the fetus might be anencephalic. After confirming this diagnosis by finding an elevated α1-fetoprotein value in amniotic fluid interruption was initiated without further investigation. The cultured rib fibroblasts of the aborted anencephalic fetus showed normal activity of methylmalonyl-CoA-mutase (data not shown).

TABLE

Methylmalonic acid concentration in amniotic fluid and excretion rate in maternal urine in three pregnancies at risk and in normal controls.

PATIENT/ FETUS	AFFECTED	METHYLMALONIC ACID IN AMNIOTIC FLUID AT 16th WEEK (μM/l)	METHYLMALONIC ACID EXCRETION IN MATERNAL URINE (19th/20th WEEK) (μmoles/24 h)
P.G.	+	13.4	156.7
S.G. FETUS 1	+	33.8	172.8
FETUS 2	−	not done [+)]	42.3 ± 9.3 [++)]
NORMALS (n=8)		0.31 ± 0.10	38.9 ± 7.6 [+++)]

[+)] α1-fetoprotein elevated, interruption performed because of an anencephalic fetus, [++)] 11-17 weeks, [+++)] 32-40 weeks of gestation

DISCUSSION

To our knowledge the quantification of methylmalonic acid in biological fluids with the deuterated compound as internal standard has not yet been described. Recently, Naylor et al.[14] published a similar method for the determination of methylcitrate in amniotic fluid for the prenatal diagnosis of propionic acidemia and methylmalonic acidemia. It seems that the determination of methylcitrate in amniotic fluid for the prenatal diagnosis of methylmalonic acidemia offers no advantage over the measurement of methylmalonic acid itself because there is only a 5-6-fold increase above normal values [15]. By this the discrimination power of the method described by Naylor is not better than the determination of methylmalonic acid by conventional gaschromatography [9]. With our method the difference between normal values of methylmalonic acid in amniotic fluid and those of mothers with affected fetuses is at least 40-fold. The sensitivity of the method is high enough to readily detect small amounts such as those found in amniotic fluid of nonaffected pregnancies. The value of metabolite analysis in amniotic fluid as compared to enzyme assay in cultured amniotic cells was demonstrated recently: because of the overgrowth of the fetal cells by contaminating maternal fibroblasts a prenatal diagnosis of propionic acidemia was missed [15]. For the future, analysis of abnormal metabolites in maternal urine during a pregnancy at risk is a promising approach for the prenatal diagnosis of certain metabolic disorders, since it avoids the risks of amniocentesis. Concerning methylmalonic acidemia this approach seems to be possible. Our data suggest that with a most sensitive method as mass fragmentography is one can find an increase of the methylmalonic acid excretion in maternal urine as early as at the 12/13th week of gestation. Unfortunately we did not collect enough urine specimens between the 10-20th week to make a reliable statistical trend analysis. It is our opinion that by analyzing methylmalonic acid excretion in the maternal urine from the 10th week of gestation at weekly intervals one can find enough data to make a reliable prenatal diagnosis of methylmalonic acidemia before the 20th week.

ACKNOWLEDGEMENTS

We thank G. Pecht for skilful technical assistance, Dr. M. Wolf, Department of Clinical Genetics, University of Ulm, and Dr. K. Sprenger, Bad Schönborn, for their kind cooperation.

REFERENCES
1. V.G. Oberholzer, B. Levin, E.A. Burgess, and W.F. Young.
   Arch.Dis.Child. 42, 1967, 492-496.
2. O. Stokke, L. Eldjarn, K.R. Norum, J. Steen-Johnson, and S. Halvorsen.
   Scand.J.Clin.Lab.Invest. 20, 1967, 313.
3. D. Leupold.
   Klin.Wschr. 55, 1977, 57-63.
4. H.L. Levy, J.T. Coulombe and V.E. Shih in: H. Bickel, R. Guthrie, and G. Hammersen (Eds.): Neonatal screening for inborn errors of metabolism. Springer Verlag Berlin, Heidelberg, New York, 1980, p. 89.
5. M.J. Mahoney, L.E. Rosenberg, B. Lindblad, J. Waldenström, and R. Zetterström.
   Acta Paediatr.Scand. 64, 1975, 44-48.
6. H.D. Bakker, A.H. van Gennip, M. Duran and S.K. Wadman.
   Clin.Chim.Acta 86, 1978, 349-352.
7. D. Gompertz, P.A. Goodey, J.M. Saudubray, C. Charpentier, A. Chignolle.
   Pediatrics 54, 1974, 511-513.
8. G. Morrow III, R.H. Schwarzt, J.A. Hallock, and L.A. Barness.
   J. Pediat. 77, 1970, 120-123.
9. E. Nakamura, L.E. Rosenberg, and K. Tanaka.
   Clin.Chim.Acta 68, 1976, 127-140.
10. W. Kochen.
    Chromatographia 1981, in press.
11. E.R. Baumgartner, H. Wick, J.C. Linnel, G.E. Gaull, C. Bachmann, and B. Steinmann.
    Helv.paediat.Acta 34, 1979, 483-496.
12. F.K. Trefz, H. Schmidt, B. Tauscher, E.Depène, R. Baumgartner, G. Hammersen, and W. Kochen.
    Eur.J.Ped., 1981, in press.
13. H.F. Willard, and L.E. Rosenberg.
    Hum. Genet. 34, 1976, 277-283.
14. G. Naylor, L. Sweetman, W.L. Nyhand, C. Hornbeck, J. Griffiths, L. Mörch, and S. Brandänge.
    Clin.Chim.Acta 107, 1980, 175-183.
15. P.D. Buchanan, S.G. Kahler, L. Sweetman and W.L. Nyhan.
    Clin.Genet. 18, 1980, 177-183.

THE FATE OF ORALLY ADMINISTERED TESTOSTERONE-19-d_3 AND ITS INFLUENCE ON THE PLASMA LEVELS OF ENDOGENOUS TESTOSTERONE IN HUMANS

S. BABA, Y. SHINOHARA AND Y. KASUYA
Tokyo College of Pharmacy, 1432-1 Horinouchi, Hachioji, Tokyo 192-03, Japan

ABSTRACT

A mass fragmentographic method employing stable isotopically labeled testosterone was employed in humans to examine the absorption and disposition of orally administered testosterone and to clarify the influence of exogenous testosterone on the plasma levels of endogenous testosterone. It became apparent that the orally administered testosterone (20 mg of testosterone-19-d_3) was rapidly absorbed and appeared in the circulating blood in small amounts. Furthermore only a very small percentage of the testosterone given orally was excreted in urine as such. The urinary excretion of the two main testosterone metabolites, androsterone glucuronide and etiocholanolone glucuronide, was also followed at various urine collection periods after oral administration of testosterone-19-d_3 to two healthy male volunteers. The results obtained in the present study indicate that the extensive metabolism of testosterone in the liver could be the reason for the relative lack of oral testosterone compared with intravenous testosterone; even though testosterone was completely absorbed from the gastrointestinal tract when given orally.

INTRODUCTION

We have developed an MF technique from our need for a simple, reliable and rapid method for the determination of plasma testosterone levels in a biological sample [1]. As the internal standard, we used testosterone-19, 19,19-d_3 (testosterone-19-d_3) synthesized in our laboratory [2]. The method was then uniquely used to examine the absorption and disposition of oral testosterone and to clarify the influence of exogenous testosterone on the plasma levels of endogenous testosterone after the administration of 20 mg of testosterone-19-d_3 to two healthy male volunteers [3]. Furthermore, we examined the urinary excretion of the metabolites of orally administered testosterone in order to clarify why only a small amount of orally administered testosterone appeared in the circulating blood [3]. The results obtained from these studies are discussed in relation to the absorption and rapid liver metabolism which cause the low concentration of the exogenous testosterone after oral administration of testosterone-19-d_3.

MATERIALS AND METHODS
Chemicals

Testosterone-19,19,19-d_3 (testosterone-19-d_3) was synthesized in our laboratory as described previously [2]. The isotopic composition was 99.0% deuterium atoms (d_3;97.8%, d_2;2.2%, d_1;0.0%). Androsterone and etiocholanolone were purchased from Tokyo Kasei Kogyo Co., Ltd, Japan and Sigma Chemical Co., Missouri, respectively.

Mass fragmentography

Mass fragmentographic (MF) measurements were made with a Shimadzu LKB-9000B gas chromatograph-mass spectrometer equipped with a Shimadzu high speed multiple ion detector-peak matcher 9060S. The GC-MS analysis was performed under the conditions described previously [1]. The GC column was a glass column (2 m x 3 mm id) packed with 1.5% OV-1. The column and flash heater temperatures were 230 and 250°C, respectively, and helium was used as a carrier gas at a flow rate of 30 ml/min. Mass spectra were obtained at an accelerator voltage of 20 eV. The separator and ion source temperatures were 250 and 270C, respectively. The multiple ion detector was focused at the molecular ions of the trifluoroacetyl (TFA) derivatives of testosterone (d_0;m/e 480, d_3;m/e 483), and androsterone and etiocholanolone (d_0;m/e386, d_3;m/e 389) to obtain peak height ratios.

Protocol

The subjects were two healthy adult male volunteers, 36 (subject 1) and 25 (subject 2) yr of age, each weighing 64 kg. They urinated completely just before drug intake and received a single oral dose of 20 mg of finely powdered testosterone-19-d_3 in a gelatin capsule with 100 ml water after an overnight fast. No food was permitted for 4 h after the administration of the labeled steroid. Ten-milliliter heparinized blood samples were taken 5 min before and 15, 30, 45, 60, 75, 95, 105, 120, 135, 150, 180, 240, 480, 600, 720, and 840 min after dosing. Plasma was separated by centrifugation and kept in a frozen state at -20C until analysis. Complete collections of urine were made at 0-2, 2-4, 4-8, 8-12, and 12-24 h after dosing. The urine volumes and pH were measured and the samples were kept frozen at -20C until analysis.

Sample preparation for MF

To make simultaneous measurements of testosterone-19-d_3 and endogenous testosterone in plasma, plasma samples were subjected to the MF analysis, as described previously [3]. Samples for determination of testosterone-19-d_3, androsterone-19-d_3, etiocholanolone-19-d_3, and their respective endogenous steroids in urine were prepared as described previously [3,4]. The sample size for GC-MS analysis was 1-3 μl n-hexane solution.

Calculation

Testosterone and testosterone-19-d_3 levels were determined as assayed by the double isotope dilution method [3]. Androsterone and androsterone-19-d_3, and etiocholanolone and etiocholanolone-19-d_3 levels were determined as assayed also by the double isotope dilution method [4] and the calculations were basically the same as those used for testosterone and testosterone-19-d_3.

RESULTS AND DISCUSSION

Foss and Camb [5] were the first to suggest that orally administered testosterone is pharmacologically ineffective. This ineffectiveness is considered to be caused mainly by the rapid metabolism of testosterone during its first pass through the liver. Nieschlag et al. [6] measured total plasma testosterone by RIA in normal men after oral administration of 63 mg testosterone. Plasma samples were obtained at 1-h intervals for 10 h starting 2 h after drug administration. No significant increases in total plasma testosterone levels were observed. With the aid of the double isotope dilution assay employed in the present study, however, orally administered testosterone was rapidly detected in the circulating blood of normal men as shown in Fig. 1.

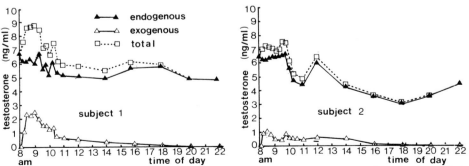

Fig. 1 Endogenous and exogenous plasma testosterone levels after oral administration of 20 mg of testosterone-19-d_3.
Maximal plasma testosterone levels were observed in the samples taken 30 min to 1 h after oral administration of testosterone-19-d_3. Exogenous testosterone

continued to contribute to total plasma levels for up to 12 h. The maximum plasma values of the exogenous testosterone, however, were only 2.45 and 0.95 ng/ml for subjects 1 and 2, respectively. Furthermore, exogenous testosterone disappeared from the plasma at a very rapid rate and was no longer detectable in samples taken 14 h after drug administration. Analysis of the two curves in Fig. 1 suggests that a 20-mg dose of testosterone has little if any influence on the regulation of endogenous testosterone. Higher doses, however, could influence the biosynthesis of endogenous testosterone.

In an effort to evaluate the extent to which urinary excretion of testosterone may regulate plasma testosterone levels after oral administration of testosterone, we examined the amounts of testosterone-19-d_3 and unlabeled testosterone in timed urine samples (Fig. 2).

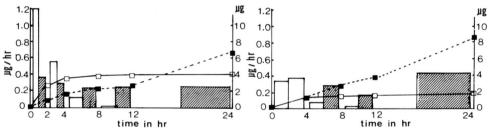

Fig. 2 Average renal excretion and cumulative renal excretion of endogenous and exogenous testosterone after oral administration of testosterone-19-d_3. Excretion rate is expressed as micrograms per h (▨, endogenous; ☐, exogenous). Cumulative excretion is expressed in micrograms (■----■, endogenous; ☐——☐, exogenous).

We found that the fraction of unchanged testosterone-19-d_3 excreted in the 24-h urine was very small (0.2% of the administered dose in subject 1 and 0.084% in subject 2). These results are comparable to those reported by Camacho and Migeon [7] on the urinary excretion of ^{14}C-labeled testosterone administered iv. The low plasma levels of testosterone-19-d_3, therefore, may not be attributable to renal clearance of the unchanged drug. The amount of endogenous testosterone excreted in the 24-h urine determined after treatment with β-glucuronidase was less than 10 μg in the two subjects investigated. Since the administration of testosterone-19-d_3 might suppress endogenous testosterone secretion, testosterone was analyzed by the present method in the urine of subject 2 collected for 24 h before the administration of exogenous testosterone. The amount of testosterone present in this 24-h urine sample was 6.80 μg, not significantly different from that measured in the 24-h urine sample collected after the administration of testosterone-19-d_3. Therefore, it is very unlikely that the urinary excretion of endogenous testosterone is suppressed by exogenous testosterone.

Baulieu and Mauvais-Jarvis [8,9] have shown extensive metabolism of radioactive testosterone administered iv to man and found that about 20% of the administered radioactivity was radioactive androsterone glucuronide and etiocholanolone glucuronide excreted in the urine. Sandberg and Slaunwhite, Jr. [10,11] have reported that after the iv administration of testosterone-^{14}C to human subjects, the radioactivity was excreted in the 48-h urine almost quantitatively. The excretion of the radioactivity was very rapid, and about 50% of the administered radioactivity was excreted in the urine within 4 h. Hellman et al. [12] studied the absorption of testosterone from the intestinal tract and demonstrated that after the oral administration of testosterone-^{14}C to human subjects, testosterone directly entered the human liver via the portal circulation. Moreover, these authors showed that the lack of biological effect after the oral administration of testosterone compared with the parenteral route was unrelated to the rate and extent of absorption, since the excretory patterns were the same in the human subjects after either oral or iv labeled testosterone. These findings tempted us to examine the urinary excretion of the metabolites of orally administered testosterone in order to clarify why only a small amount of orally administered testosterone appeared in the circulating blood.

The average urinary excretion and the cumulative urinary excretion were calculated from the results obtained by the MF analysis of the urinary concentrations of androsterone-19-d_3 glucuronide and etiocholanolone-19-d_3 glucuronide after oral administration of testosterone-19-d_3. The data are shown in Figs. 3 and 4.

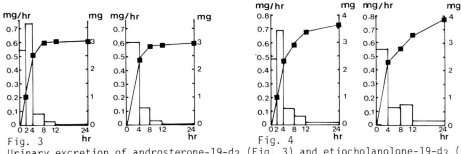

Fig. 3   Fig. 4
Urinary excretion of androsterone-19-d_3 (Fig. 3) and etiocholanolone-19-d_3 (Fig.4) after the oral administration of testosterone-19-d_3. ☐, Average; ■—■ cumulative.

We found that more than 30% of the administered testosterone-19-d_3 dose was excreted in the 24-h urine as androsterone-19-d_3 glucuronide and etiocholanolone-19-d_3 glucuronide. These percentage values were a little higher than those obtained by Baulieu and Mauvais-Jarvis [8,9]. Moreover, most of the cumulative amounts of these metabolites in the 24-h urine were excreted during the first 4 h. It is then not unreasonable to assume that orally administered testosterone was quantitatively absorbed and that metabolic formation of androsterone and etiocholanolone from the oral testosterone occurred more rapidly and more extensively than from iv testosterone. As discussed above, the orally administered testosterone appeared in the circulation only in small amounts. Furthermore, only a very small percentage of the testosterone given orally was excreted in the urine as such. A possible explanation of these findings could be that the extensive metabolism of testosterone in the liver causes the relative lack of activity of oral testosterone compared with iv testosterone, even though testosterone was completely absorbed from the gastrointestinal tract when given orally.

REFERENCES

1. S. Baba, Y. Shinohara and Y. Kasuya, J. Chromatogr., 162(1979)529.
2. S. Baba, Y. Shinohara and Y. Kasuya, J. Label. Compd. Radiopharm., 14(1978) 783.
3. S. Baba, Y. Shinohara and Y. Kasuya, J. Clin. Endocrinol. Metab., 50(1980) 889.
4. Y. Shinohara, S. Baba and Y. Kasuya, J. Clin. Endocrinol. Metab., 51(1980) 1459.
5. G.L. Foss and M.B. Camb, Lancet, 1(1939)502.
6. E. Nieschlag, J. Mauss, A. Coert and P. Kićovič, Acta Endocrinol. (Kbh), 79 (1975)366.
7. A.M. Camacho and C.L. Migeon, J. Clin. Invest., 43(1964)1083.
8. E. Baulieu and P. Mauvais-Jarvis, J. Biol. Chem., 239(1964)1569.
9. E. Baulieu and P. Mauvais-Jarvis, J. Biol. Chem., 239(1964)1578.
10. A.A. Sandberg and W.R. Slaunwhite, Jr., J. Clin. Invest., 35(1956)1331.
11. W.R. Slaunwhite, Jr. and A.A. Sandberg, J. Clin. Endocrinol. Metab., 18(1958) 1056.
12. L. Hellman, H.L. Bradlow, E.L. Frazell and T.F. Gallagher, J. Clin. Invest., 35(1956)1033.

IN VIVO STUDIES OF STEROID-METABOLISM USING DEUTERATED PREGNENOLONE

TH. KUSTER, M. ZACHMANN AND B. ZAGALAK
Department of Pediatrics, Division of Clinical Chemistry, University of Zurich, Steinwiesstr. 75, CH-8032 Zurich, Switzerland

ABSTRACT

Two healthy subjects were loaded intravenously with deuterated pregnenolone to study its metabolism and to compare the pattern with that of patients suffering from defective steroid biosynthesis (3β-hydroxysteroid dehydrogenase, 21-hydroxylase, 11β-hydroxylase deficiencies).

Determination of the metabolites in urine was achieved by gas chromatography-mass spectrometry and gas chromatography-mass fragmentography. Deuterated metabolites found were: pregnenediol, pregnanediol and allo-pregnanediol. For all three compounds, the maximum excretion of the labelled species has been found between two and four hours after loading with deuterated precursor. The results allow to determine the secretion rates of these metabolites.

INTRODUCTION

Stable isotopes have only rarely been used to study the steroid metabolism in humans. Pinkus et al. [1,2] have used deuterated estrogens in pregnant and postmenopausal women to determine estrogen secretion rates and we have previously used deuterated cholesterol [3].

The obvious advantages of stable isotopes versus radioactive material are that they can be used without danger of irradiation in children and pregnant women. The disadvantages are that deuterated compounds are relatively difficult to obtain, that relatively large quantities have to be administered and that the techniques of analysis are difficult and time consuming.

Theoretically, deuterated steroids might be useful in children 1) to determine secretion and/or metabolic clearance rates of a specific steroid and 2) to evaluate enzyme defects in steroid biosynthesis by administering precursors and by evaluating the labelled products of the respective enzymes.

The ultimate goal would be to study the steroid metabolism in more detail and to analyze the degree of the enzyme defect in cases of congenital adrenal hyperplasia due to 21- and 11β-hydroxylase and 3β-hydroxysteroid dehydrogenase deficiency, as well as in patients with male pseudohermaphrodism (17,20-desmolase and 17-ketosteroid-reductase deficiency).

EXPERIMENTAL

Reference steroids were purchased from Makor Chemicals Ltd., Jerusalem. Deuterated pregnenolone was prepared by acidic exchange of the enolic hydrogens in position 17 and 21 with D_2O/D_2SO_4/dioxane according to [4]. All solvents were of analytical grade and were distilled prior to use.

Loading

The subjects were loaded intramuscularly with 2 mg Tetracosactrin (ACTH, 1 mg/m^2 body surface area) and one hour later with 17,21,21,21-d_4-pregnenolone (10 mg intravenously). Four urine portions were collected (6,12,18 and 24 hours for control 1 and 2,4,6 and 24 h for control 2).

Cleaning

20 ml of urine were extracted with three 20 ml portions of ethyl acetate. The aqueous phase was adjusted to pH 4.6 and enzymatically hydrolized for 24 h at 37°C with 40 mg Helicase (20'000 units β-glucoronidase and 300'000 units sulfatase) in 10 ml 0.1 M acetate buffer (pH 4.62). The free steroids were extracted with three 20 ml portions of ethyl acetate and the organic layers were washed with 30 ml 0.1 N NaOH and 30 ml water until they reached pH 7. Further purification has been performed on a cation exchange column (sulfoethyl Sephadex LH-20 [5], eluant methanol), an anion exchange column (diethylaminohydroxypropyl Sephadex LH-20 [6], eluant methanol/chloroform/water 9/2/1), a Lipidex 5000 column (Packard, eluant methanol/chloroform/water 9/2/1) and a silica gel column (eluant methanol).

Derivatization

The eluate of the silica gel column was concentrated under nitrogen to 1 ml, 100 μl evaporated to dryness and with 100 μl 10% methoxylamine-hydrochloride in pyridine methoximated at room temperature for 15 h.

The pyridine was evaporated with a stream of nitrogen and the sample was silylated with 100 μl N,O-Bis-(trimethylsilyl)-trifluoroacetamide at 60°C for 1 h.

Gas-chromatography

A Carlo Erba Fractovap 2350 with a glass capillary column (25m x 0.3 mm i.d.) coated with OV-1 was used. The carrier gas was hydrogen at an inlet pressure of 0.6 bar, the injector temperature was 300°C. Temperature program: 1 min. at 165°C, then programmed to 265°C with a rate of 3°C/min.

Mass Spectrometer

The gas chromatograph (same column and temperatures, carrier gas: Helium at 1.1 bar) was coupled to a VG-16F mass spectrometer by a fused silica capillary interface at a temperature of 250 °C. Samples were analyzed at 25eV ionizing voltage, 220 °C ion source temperature, 50 µA emission current and 4 kV accelerating voltage either by scanning the mass range from m/z 100 to 650 (cycle time 1.8 sec.) magnetically or by multiple ion detection between m/z 117 and 120 (cycle time 0.437 sec.) by switching the accelerating voltage with the magnet set at m/z 110. Data acquisition and processing was performed by a Finnigan INCOS 2000 data system.

RESULTS

Fig. 1 shows the total ion current chromatogram of a urine portion collected 2-4 h after loading with deuterated pregnenolone.

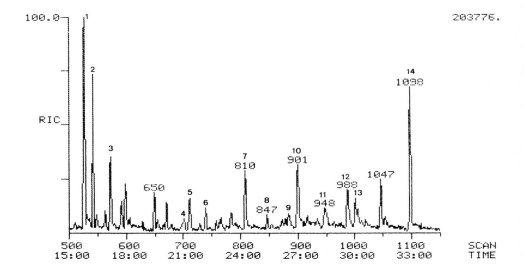

Fig. 1. Total ion current chromatogram of urine portion 2 from control 2

Abbreviations: 1) androsterone = 3α-hydroxy-5α-androstane-17-one, 2) etiocholanolone = 3α-hydroxy-5β-androstane-17-one, 3) dehydroepiandrosterone = 3β-hydroxy-androst-5-ene-17-one, 4) allo-pregnanediol = 3α, 20α-dihydroxy-5α-pregnane, 5) pregnanediol = 3α, 20α-dihydroxy-5β-pregnane, 6) pregnenediol = 3β, 20α-dihydroxy-pregn-5-ene, 7) pregnanetriol = $3\alpha,17\alpha,20\alpha$-trihydroxy-$5\beta$-pregnane, 8) pregnenetriol = $3\beta,17\alpha,20\alpha$-trihydroxy-pregn-5-ene, 9) pregnanetriolone = $3\alpha,17\alpha,20\alpha$-trihydroxy-$5\beta$-pregnane-11-one, 10) tetrahydro-S-$3\alpha,17\alpha$, 21-trihydroxy-5β-pregnane-20-one, 11) tetrahydrocortisone = 3α, $17\alpha,21$-trihydroxy-5β-pregnane-11,20-dione, 12) tetrahydrocortisol = $3\alpha,11\beta,17\alpha,21$-tetrahydroxy-$5\beta$-pregnane-20-one, 13) cortolone = $3\alpha,17\alpha,20\alpha,21$-tetrahydroxy-$5\beta$-pregnane-11-one, internal standard = cholesterylbutyrate.

Investigation of the spectra reveals 5 compounds which are partially labelled, namely pregnenediol, pregnanediol, allo-pregnanediol and 2 not yet clearly identified steroids. Although the control persons were loaded with ACTH which stimulates the excretion of corticosteroids, these biochemically highly interesting compounds could not be found as labelled species. This is probably due to the longer metabolic pathway needed to generate these steroids (see Scheme 1).

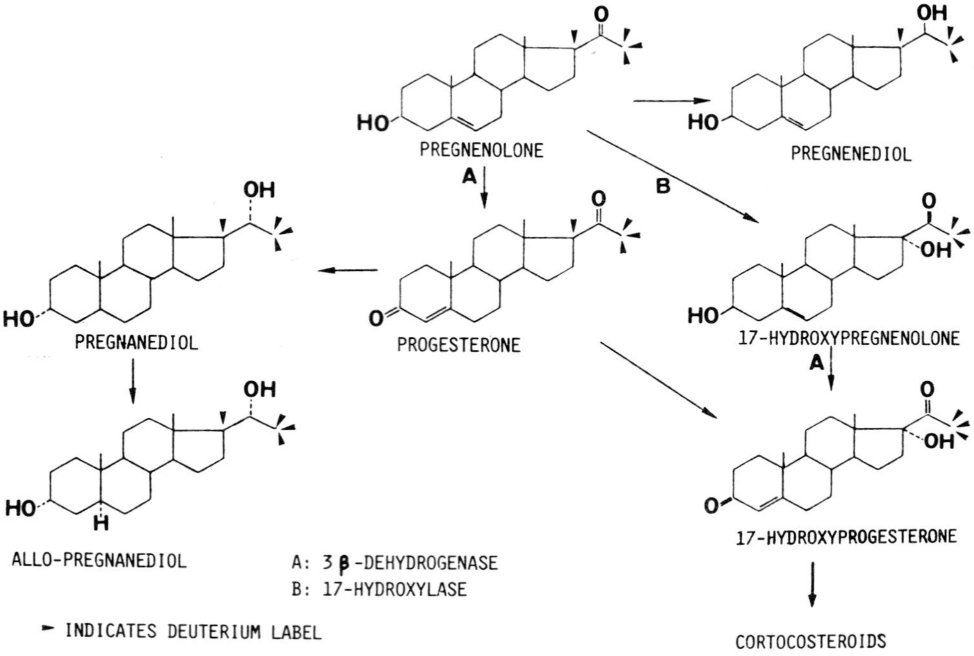

Scheme 1. Part of metabolic pathways in steroid biosynthesis (simplified).

For measuring the corticosteroids, the subjects should be loaded with, for instance, 17α-hydroxy-progesterone.

All 5 labelled compounds found could be identified by the ion m/z 120 which is formed by cleavage of the side chain in position 17 (m/z 117 in the unlabelled compound, see Fig. 2).

The measurement of the amount of the labelled steroids was performed by selected ion measuring of the masses 117 and 120. Fig. 3 shows the mass chromatograms of portion 2 from control 2. The figure shows that there is a further confirmation of the deuterium content of these compounds: since deuterated compounds have shorter retention times on capillary columns, mass 120 reaches its maximum before mass 117 (see scan 3057 against 3065 for pregnanediol). The broader peak shapes for the mass chromatogram arise from the content of unlabelled m/z 120 which has been determined to 4 - 6 % using unlabelled references.

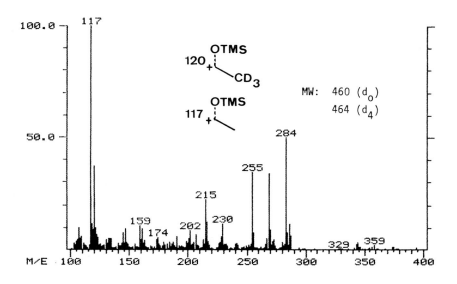

Fig. 2. Electron impact mass spectrum of pregnanediol (scan 708-714 in Fig. 1).

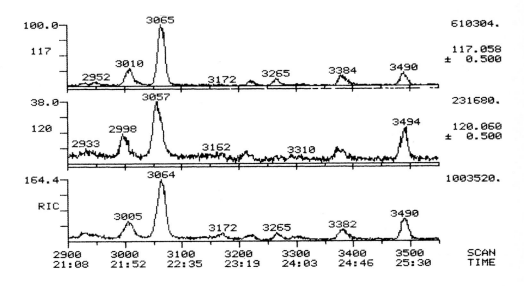

Fig. 3. Mass chromatogram (masses 117 and 120) of urine portion 2 from control 2 obtained by selected ion monitoring.

Fig. 4 and 5 show the ratio m/z 120/117 over the collection periods for the two control persons which represent the percentage of excreted labelled metabolites. The maximum amount is excreted between 2 and 4 h after loading and then falls slowly down to reach the normal values after 24 h.

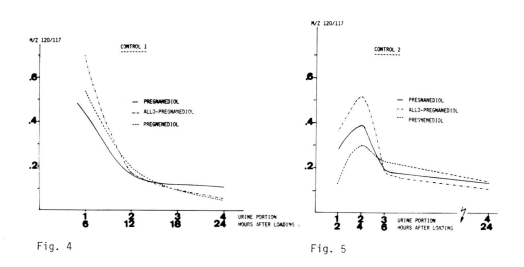

Fig. 4   Fig. 5

Ratio m/z 120/117 (measured by selected ion monitoring) as a function of time after loading with deuterated pregnenolone. Fig. 4: from control 1, Fig. 5: from control 2.

The results suggest that it should be possible to analyze 3β-dehydrogenation and to determine the secretion rates of these specific steroids.

ACKNOWLEDGEMENT

This work was supported by the Swiss National Foundation (grant No. 3.883.0-77) and the skilful technical assistance of Miss Bianca Kempken and Mr. Erich Wetzel is gratefully acknowledged.

REFERENCES

1. J.L. Pinkus, D. Charles and S.C. Chattoraj, J. biol. Chem., 246 (1971)633.
2. J.L. Pinkus, D. Charles and S.C. Chattoraj, Horm. Res., 10(1979)44.
3. H.-Ch. Curtius, J. Völlmin, M.-J. Zagalak and M. Zachmann, J. Steroid. Biochem., 6(1975)677.
4. B. Zagalak, H.-Ch. Curtius, R. Foschi, G. Wipf and U. Redweik, Experientia 34(1978)1537.
5. K. Setchell, B. Alme, M. Axelson and J. Sjövall, J. Steroid Biochem., 7(1976)615.
6. B. Alme and E. Nyström, J. Chromatogr., 59(1971)45.

STUDIES OF RENAL AND EXTRARENAL URIC ACID EXCRETION DURING A DIETARY PURINE LOAD USING ^{15}N-URIC ACID

W. LÖFFLER, W. GRÖBNER, R. MEDINA and N. ZÖLLNER
Medizinische Poliklinik der Universität München, Pettenkoferstraße 8a, 8000 München 2 (G.F.R)
Lehrstuhl für Allgemeine Chemie und Biochemie der Technischen Universität München, Freising-Weihenstephan (G.F.R.)

ABSTRACT

Uric acid metabolism was studied in three normal subjects ingesting a purine-free diet (Experiment I) as well as an additional oral supplement of purines (Experiment II) using ^{15}N-labelled uric acid.

It was found that 1. the turnover rate of the uric acid pool is enhanced by dietary purines, 2. there is no constant change in the percentage of renal and extrarenal excretion of uric acid when purines are added to the diet, and 3. there is no inhibition of endogenous uric acid synthesis by dietary purine supplementation.

INTRODUCTION

In 1949, Benedict et al. [1] described an experiment, which allows the estimation of uric acid pool size and turnover rate as well as the size of renal and extrarenal excretion after intravenous injection of isotopic uric acid. Sorensen [2] was able to show that that part of the uric acid turnover, which was not excreted renally, was excreted into the gut, and subsequently metabolized by the intestinal flora.

In man uric acid is derived from two sources: from endogenous production on one hand, and from dietary purines on the other hand. When purines are added to a low purine or a purine-free diet, the renal uric acid clearance is enhanced in normal subjects. However, no conclusions can be drawn from the changes in renal clearance concerning the changes in the total body uric acid clearance, as long as the size of the intestinal excretion of uric acid is unknown.

METHODS

We used the method of Benedict et al. [1] to study the influence of dietary purines on pool size and turnover rate of uric acid. In this method, after intravenous injection of isotopic uric acid, the turnover of the uric acid pool is calculated from the decline of the isotope content of the urinary uric acid, assuming that at any time this was the same as that of the plasma uric acid. A semilogarithmic plot of the isotope concentration versus time gives the turnover rate k (pools/day). The isotope concentration of the body uric acid immediately after injection can be derived from the slope by extrapolation to t=0. From the amount (a) and isotope abundance of the uric acid injected (I_i) as well as from the isotope abundance within the uric acid pool immediately after injection (I_o) the size of the uric acid pool can be calculated:

$$A = a (I_i/I_o - 1)$$

The limitations of this method have been clearly discussed previously [3, 4].

Three normal subjects were studied on two occasions. During the first period an isoenergetic, purine-free liquid formula diet was given alone, during the second part of the experiment purines were added to this diet in the form of mononucleotides, which are completely absorbed from the gut. The dosage administered was 1 g of adenosine-5'-monophosphate and 1 g of guanosine-5'-monophosphate per 70 kg of body weight. The first ten days of each period were used to allow the subjects to reach steady state conditions. 1,3-[$^{15}N_2$]-uric acid (Merck, Sharpe and Dohme, Montreal; 95 percent enrichment in both positions; 120 to 156 mg per experiment) was injected intravenously after these ten days, and urine was collected in 12 h periods for isolation of the uric acid by ion exchange chromatography and subsequent purification according to Johnson and Emmerson [5].

The purified uric acid was degraded by a modified Kjeldahl procedure. Molecular nitrogen was produced from the ammonium chloride by oxidation with lithium hypobromite. The ^{15}N-enrichment of the nitrogen was measured in a VG Micromass 903 mass spectrometer.

RESULTS AND DISCUSSION

Figure 1 shows the semilogarithmic plots of the isotope enrichment of the urine uric acid in the three subjects studied. k_I gives the results during control conditions (purine-free diet), k_{II} during dietary purine supplementation. It is obvious from this figure, that the uric acid pool, which is increased during the administration of dietary purines, is turned over faster than the small pool, which exists during control conditions.

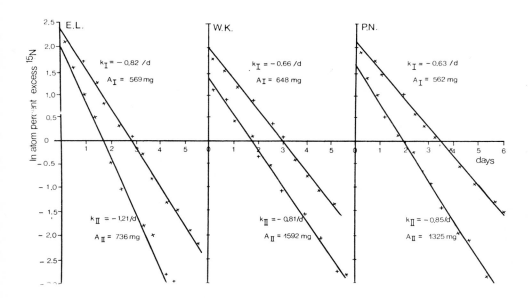

FIGURE 1: Semilogarithmic plot of the urinary uric acid isotope enrichment versus time in three normal subjects. A = pool size of uric acid (mg), k = turnover rate (pools/day).

A similar experiment was performed by Bowering et al. [6] in two normal subjects, who were of the same age, weight, and sex. In these subjects the addition of 4 g of ribonucleic acid to their diet resulted in an increase of the uric acid pool from 767 to 1480 mg, and from 1060 to 2127 mg, respectively. The turnover rate of the uric acid pool, however, rose from 0.65 to 0.92/day in the first subject, and was only slightly enhanced in the second (0.61 and 0.65

per day, respectively).

Table 1 shows the percentage of intestinal uric acid excretion in our subjects as well as in those studied by Bowering et al. [6]. As can be seen from these data, there is a wide range of individual response to dietary purines, but there is no constant change in the percentage of renal and extrarenal uric acid excretion.

TABLE 1: Intestinal uric acid excretion in 5 normal subjects (percent of turnover).

Reference (Purine supplement)	control	oral purine loading
Bowering et al., 1969 (RNA)	22	25
	54	25
Present experiment (AMP + GMP)	49	36
	39	41
	21	38

To study intestinal uric acid excretion directly, we gave high doses of antibiotics to normal subjects and measured uric acid in the feces during a purine-free diet as well as during an additional administration of ribonucleic acid [7]. In this experiment the maximum change in renal and extrarenal uric acid excretion induced by dietary purines was two percent. From the studies cited it can be concluded that in normal subjects the intestinal uric acid clearance is changed by dietary purines to the same direction and to the same degree as is the renal clearance.

From investigations using different doses of oral purines in the same subject it was concluded that there is no feedback inhibition of endogenous uric acid production by dietary purines [8]. In the present investigation, endogenous uric acid production can be calculated from the first part of the experiments, while the second part gives the total production derived from endogenous as well as dietary sources. Taking into account the amount of uric acid administered in the form of dietary purines, endogenous uric acid production can be calculated during oral administration of purines from isotope dilution experiments. Results show that there is no suppression of endogenous uric acid production by dietary purines.

REFERENCES

1 J.D. Benedict, P.H. Forsham, D. Stetten, J. Biol. Chem., 181 (1949) 183-191.
2 L.B. Sorensen, Scand. J. Clin. Lab. Invest., 12 (1969) Suppl. 54.
3 Ch. Bishop, W. Garner, J.H. Talbott, J. Clin. Invest., 30 (1951) 879-888.
4 N. Zöllner, Ergebn. inn. Med. Kinderheilk., 14 (1960) 321-389.
5 L.A. Johnson, B.T. Emmerson, Clin. Chim. Acta, 41 (1972) 389-393.
6 J. Bowering, D.H. Calloway, S. Margen, N.A. Kaufmann, J. Nutr., 100 (1969) 249-261.
7 W. Löffler, W. Gröbner, N. Zöllner, in: Fortschr. Urol. Nephrol., 16 (1981) in press.
8 N. Zöllner, W. Gröbner, CIBA Found. Symp., 48 (1977) 165-178.

ESTIMATION OF PROTEIN TURNOVER IN PATIENTS WITH LIVER DISEASES USING ^{15}N-LABELLED GLYCINE

H. FAUST, P. JUNGHANS, R. MATKOWITZ, W. HARTIG and K. JUNG
Academy of Sciences of GDR, Central Institute for Isotope and Radiation Research and Hospital St. Georg, Surgical Clinic, Leipzig (GDR)

ABSTRACT

The ^{15}N excretion after a single oral administration of ^{15}N-labelled glycine to adult patients with different liver diseases (acute hepatitis, cirrhosis etc.) was studied. For the quantitative description of the nitrogen metabolism we used a three-pool model. The rate of protein synthesis was estimated by means of a graphical method, which is based on the evaluation of area under the ^{15}N abundance-time curve.

It was found that the rate of protein synthesis is decreased in patients with acute hepatitis.

INTRODUCTION

Much work is concerned with the determination of the protein turnover in man using isotopic methods. In the last years increasing interest has been devoted to investigating special influences such as nutrition, age, diseases etc. on the protein metabolism. The main results are comprehensively reviewed by Waterlow et al. (ref. 1).

In the present work the protein metabolism is studied in patients with different liver diseases. This work aims to find differences between the normal and pathological liver function in regard to its nitrogen metabolism with the help of the ^{15}N excretion kinetics of the total nitrogen in urine.

More detailed presentations are given by Junghans et al. (ref. 2) and Matkowitz et al. (ref. 3).

THEORETICAL CONSIDERATIONS

The ^{15}N tracer data were evaluated by means of a 3-pool model created by Sprinson and Rittenberg (ref. 4) (Fig. 1).

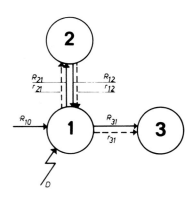

Fig. 1 3-pool model of human protein metabolism: 1) Non-protein nitrogen (NPN); 2) Protein nitrogen; 3) Total nitrogen in urine (Symbols: R_{10} rate of nitrogen intake, R_{31} rate of nitrogen excretion, R_{21} rate of protein synthesis, R_{12} rate of protein degradation; r_{12}, r_{21} and r_{31} are the corresponding ^{15}N flows (unit: mol/h or mol/(h kg)); D ^{15}N dose (unit: mol or mol/kg)

The mathematical formalism for a single pulse dosage of ^{15}N into pool 1 using the concept of the compartment theory (ref. 5) was given in a previous paper (ref. 6).

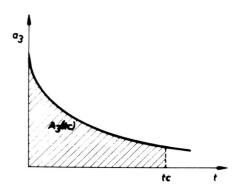

Fig. 2 Graphic estimation of the rate of protein synthesis from the area A_3 under the N-15 excess abundance-time curve of total nitrogen in urine (a_3: N-15 excess abundance in pool 3).

For the graphic estimate of the rate of protein synthesis R_{21} the area A_3 under the ^{15}N excess abundance-time curve was taken (Fig. 2).
The rate of protein synthesis R_{21} can be calculated from the following equation:

$$R_{21} = \frac{D}{A_3} - R_{31} \qquad D: {}^{15}N \text{ dose}$$

The area A_3 was determined with the help of a planimeter.

Furthermore a direct graphic procedure for the estimate of R_{21} is suggested, as shown in Fig. 4. The area under the ^{15}N excess abundance-time curve normalized to the ^{15}N dose A_3/D as a function of the N excretion rate R_{31}, gives hyperbolic curves representing equal values of R_{21}. That means, if A_3/D and R_{31} are known R_{21}, can be easily read to a good approximation.

MATERIAL AND METHODS

Fifteen patients with different liver diseases and nine healthy persons were studied (Table 1). The test subjects received 0.167 mmol ^{15}N/kg body weight as $[^{15}N]$glycine (95 at.%) dissolved in water. The patients were given a special diet which provided 0.8 g protein/(d·kg) and about 100 kJ/(d·kg). Urine samples were collected for six days after the ^{15}N administration. Total nitrogen was determined after Kjeldahl and the ^{15}N abundance was measured by emission spectrometry.

RESULTS

The estimates using the area under the ^{15}N abundance-time curve up to 48 h (a) and 144 h (b) after ^{15}N administration yield the following results.

The rate of protein synthesis R_{21} is decreased in patients with liver diseases (except one patient with alcohol cirrhosis) in case (a) by 13 % ($\approx$0...48 %) with $p < 0.05$ (n = 14) and in case (b) 40 % (18.....58 %) with $p < 0.001$ (n = 10) in comparison to the mean of R_{21} in healthy adult man (n = 9).

In patients with acute hepatitis the corresponding values are 29 % (11...48 %) with $p < 0.01$ (n = 5) in case (a) or 48 % (36...63 %) with $p < 0.001$ (n = 3) in case (b) (Fig. 3).

The decrease of the rate of protein synthesis is connected with the reduction of the rate of protein breakdown (Fig. 3).

The individual values of R_{21} can be easily read from the special graph shown in Fig. 4.

TABLE 1

Characteristics of healthy man and patients with various liver diseases

Subject	Sex	Age (yr.)	Body mass (kg)	Clinical diagnosis
02...09 [a] (without 05) C109, C110	3m. [b] 6f.	41-64	55-94	Healthy
B 10	f.	55	68	Acute hepatitis (Hepatitis infectiosa icterica)
B 13	m.	28	83	
B 14	m.	35	69	
B 16	f.	52	88	
B 17	f.	18	57	
B 02	m.	49	71	Chronic aggressive hepatitis
B 05	m.	63	73	
B 06	f.	57	73	
B 04	f.	52	82	Obstructive jaundice with cholelithiasis
B 07	f.	49	84	
B 08	f.	43	72	Obstructive jaundice
B 01	m.	42	70	Decompensated liver cirrhosis with ascites
B 03	m.	46	68	Suspected liver cirrhosis
B 11	m.	49	94	Compensated liver cirrhosis
B 15	f.	25	57	Alcohol cirrhosis

[a] Hartig et al. (ref. 7), [b] m. = male, f. = female

DISCUSSION

The results show that there are significant differences between healthy man and patients with liver diseases as to the N or ^{15}N excretion kinetics of the total N in urine, reflecting the whole body nitrogen or protein metabolism.

The decrease of the synthesis rate R_{21} of the total body protein in patients with liver disorders could be assigned to the decrease of the protein synthesis in the liver including the liver-produced plasma proteins. This explanation is emphasized by the fact that the concentration of various plasma proteins is simultaneously reduced (ref. 8) and the albumin synthesis is diminished (ref. 9) in several hepatic disorders.

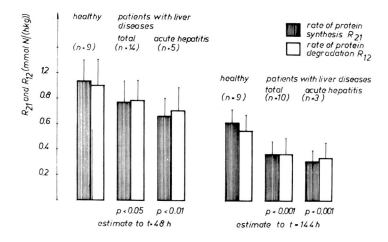

Fig. 3 Rates of protein synthesis and degradation R_{21} and R_{12} resp. calculated from the area under the ^{15}N excess abundance-time curve up to (a) t = 48 h and (b) t = 144 h.

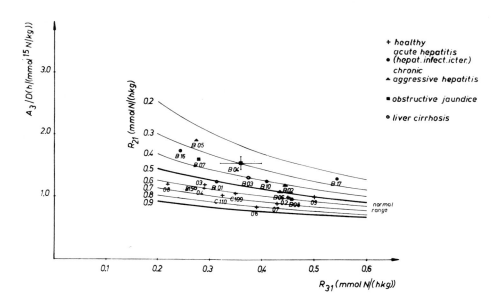

Fig. 4 Relationship between the normalized area under the ^{15}N excess abundance-time curve A_3/D up to t = 144 h and the rate of the total nitrogen excretion in urine R_{31}.

CONCLUSIONS

Significant differences in the rate of protein synthesis and breakdown between healthy man and various patients with liver diseases can be shown by means of the ^{15}N excretion kinetics in urine on the basis of the given 3-pool model.

The estimate of R_{21} corresponding to the method shown in Fig. 4 is very appropriate for a routine technique.

An exact assessment of the liver damage or the distinction between liver diseases is only possible in connection with appropriate clinical parameters as well as by means of histological investigations.

REFERENCES

1 J.C. Waterlow, P.J. Garlick and D.J. Millward, Protein Turnover in Mammalian Tissues and in the Whole Body, Elsevier, Amsterdam, 1978.
2 P. Junghans, H. Faust, R. Matkowitz and W. Hartig, ZfI-Mitt. (in preparation).
3 R. Matkowitz, P. Junghans, H. Faust and W. Hartig, Dtsch. Gesundheitswesen (in preparation).
4 D.B. Sprinson and D. Rittenberg, J. Biol. Chem. 180(1949)715-726.
5 M. Berman, M.F. Weiss and E. Shahn, Biophys. J. 2(1962)289-316.
6 K. Wetzel, P. Junghans, H. Faust and B. Dudek, in E.R. Klein and P.D. Klein (Eds.), Proc. Third Int. Conf. Stable Isotopes, Academic Press, New York, 1979, pp. 591-598.
7 W. Hartig, H. Faust, H.-D. Czarnetzki and E. Winkler, in Proc. Tech. Comm. Meet., IAEA, Vienna, 1977, pp. 335-341.
8 T. Kawai, Clinical Aspects of the Plasma Proteins, Igaku Shoin Ltd. Tokyo, 1973, pp. 236-256.
9 P. Wilkinson and C.L. Mendenhall, Clin. Sci., 25(1963)281-293.

ISOLATION OF ^{15}N LABELLED HUMAN SERUM PROTEINS AND NON-PROTEIN NITROGEN COMPOUNDS OF SERUM AND URINE

H. FAUST, H. BORNHAK and K. HIRSCHBERG
Academy of Sciences of GDR, Central Institute for Isotope and Radiation Research, Leipzig (GDR)

ABSTRACT

Simple chromatographic methods were devised for the separation of non-protein nitrogen compounds from serum and urine, and of serum proteins after ^{15}N tracer experiments. The procedures described are adapted to the special requirements of ^{15}N analysis and are suitable for routine analysis in clinical chemistry.

Results of an application of the separation methods are presented from a tracer experiment with a healthy human.

INTRODUCTION

Recently, the use of the stable isotope ^{15}N as a tracer in biomedical research has increased markedly. The ^{15}N technique is well known as a powerful tool for metabolic studies and investigations of nitrogen and protein metabolism. Tracer-kinetic studies for the determination of the turnover of the nitrogen and protein pool in the body require as diverse information as possible on the distribution of the applied ^{15}N in metabolically relevant nitrogen fractions of the urine and serum.

With the automatic ^{15}N analyzer "Isonitromat" it becomes possible to carry out tracer experiments with essentially higher frequencies of sample taking and to extend the number of fractions to be analyzed (ref. 1). Besides it becomes possible to describe the nitrogen turnover in a more differentiated manner by multi-compartment models so that the experimental worker in medicine has the chance to recognize pathologically caused disturbances of the metabolism by means of ^{15}N tracer experiments. For the biochemical processing of such experiments, the application of standardized separation methods for NPN compounds and proteins which are easy to put into analytical practice and which are adapted to the special requirements of the ^{15}N analysis, is necessary.

Following is a description of separation methods for both NPN compounds of the serum and urine and serum proteins which are usable also in routine laboratories.

RESULTS

The separation of non-protein nitrogen (NPN) compounds of urine and serum was accomplished by chromatography on ion exchanger resins. Applying strongly acidic ion exchangers of cross linked polystyrene sulfonic acid in the H-form, it is possible to use, besides ion exchange effects, also effects of a different adsorption of uncharged NPN compounds on the resin matrix. The use of both separation effects permits the fractionation of the most important NPN compounds of urine and serum by means of a continuous separation procedure. The urine fractionation was carried out by passing the urine through a short column of Dowex 50WX8 (H-form, 200-400 mesh) at room temperature. Such a short column is suitable to fractionate the NPN compounds of a urine sample containing about 60 mg nitrogen. Fig. 1 demonstrates a typical elution pattern obtained by elution of uric acid, hippuric acid, and urea with water, and of the cationic components, with a citrate buffer-step gradient. The stronger adsorption of the hippuric acid in comparison to the uric acid is obviously due to the hydrophilic interaction of its benzoyl group with the resin, and the considerably delayed elution of urea affects the excellent separation of this main component of urine. Because the ionic sorption and hydrolytic desorption of urea is connected with the appearance of an isotopic effect (ref. 2) it is necessary to carry out the determination of the ^{15}N abundance with an aliquot amount of the total urea fraction.

This short separation programme does not allow a fractionation of amino acids (AA), but is suitable to separate the "glycine fraction" (AA-I fraction) which besides glycine, glutamine, and asparagine contains the acidic AA and the major part of the neutral AA, from the more broadly distributed AA-II fraction consisting mainly of basic AA and ammonia. The separation of ammonia from the AA was carried out by means of steam distillation after alkalisation of the corresponding effluent volume with magnesium oxide. The higher ionic strength and pH of the buffer for the elution of creatinine avoids a tailing of this fraction. Urea, ammonia, the amino acids, and creatinine were detected in the eluate by means of well known colorimetric methods. The detection of uric acid and hippuric acid was performed by their characteristic UV - absorptions.

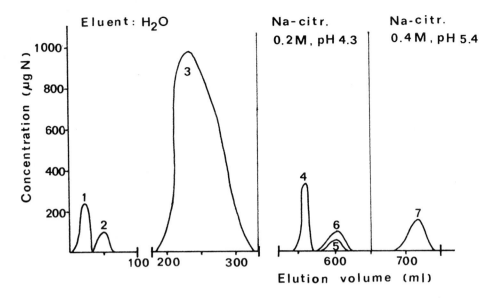

Fig. 1 Separation of NPN compounds from urine (5 ml) on a column of Dowex 50WX8, 10x170 mm, H-form; eluent: water, sodium citrate buffer 0.2 M, pH 4.3 and 0.4 M, pH 5.4. Peaks: (1) uric acid; (2) hippuric acid; (3) urea; (4) glycine-fraction (AA-I fraction); (5) AA-II fraction; (6) ammonia; (7) creatinine.

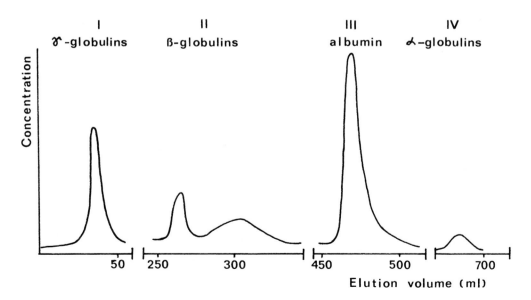

Fig. 2 Stepwise gradient elution of human serum on DEAE-Sephadex A-50.

The chromatographic separation of the NPN compounds of the serum was performed after deproteinisation of the sample with 0.6 M TCA (volume ratio 1:1). The use of 5-7 ml serum per sample is recommended in order to obtain sufficient nitrogen from each component for the isotopic analysis. The fractionation procedure corresponds to the urine fractionation. As a consequence of the very low concentration of hippuric acid in the serum it is impossible to detect this compound (apart from this the elution pattern of serum NPN compounds corresponds to Fig. 1). Because of the importance of glutamine in metabolic reactions of nitrogen, the amide nitrogen was separated from the AA-I fraction by steam distillation after alkalisation of the AA-I fraction.

For the separation of the proteins from the NPN compounds, the procedure of gel chromatography on Sephadex G-25 (Pharmacia, Sweden) was used because the speed of separation is high and the recovery of proteins and low molecular N compounds usually approaches 100 %. The total protein (1st peak) of 1-2 ml serum was transferred to the appropriate starting buffer (0.015 M phosphate buffer, pH 6.6) for subsequent ion exchange chromatography and the NPN compounds were pre-fractionated (amino acids, urea, uric acid). The protein separation after gel chromatography by means of ion exchange chromatography on DEAE-Sephadex A-50 (1 g dry gel per column; bed dimensions after equilibration with starting buffer: 1.5 x 26 cm) into 4 main fractions was achieved by using a varying NaCl concentration superimposed on a 0.015 M phosphate buffer (pH 6.6). Each protein group was eluted by 200 ml of the phosphate buffer of different ion strength (Fig. 2). The γ-globulins are not adsorbed on the DEAE-Sephadex and merely pass straight through the column (fraction I). The other ionically bound protein groups (β-globulins, albumin, α-globulins) are removed by stepwise increasing the ion strength of the buffer with NaCl (β-globulins - 0.07 M NaCl, albumin - 0.21 M NaCl, α-globulins - 0.38 M NaCl).

This separation procedure can be performed at a relatively low expense. By means of an automatic procedure it was possible to separate 12 samples at the same time.

The separated protein fractions were characterized by polyacrylamide gel electrophoresis. Fraction I contains the γ-globulins without any contamination. The main constituent of fraction II is transferrin. This fraction is free of albumin. The glycoprotein contamination (mainly α_2-globulins) of the albumin fraction (III) is removed by subsequent adsorption on Concanavalin A-Sepharose (Pharmacia, Sweden, ref. 3). For this purpose 1 ml of the albumin fraction, concentrated about 50 fold, was applied to a small column (0.9 x 10 cm) filled with

Con A-Sepharose (bed dimensions 0.9 x 5 cm). The pure albumin was eluted by about 20 ml of 0.015 M phosphate buffer (pH 6.6) containing 1 M NaCl. After accomplishing the albumin elution the glycoproteins were desorbed with 30-40 ml of the same buffer containing 0.2 M glucose. In fraction IV α-globulins and prealbumin were detected. All chromatographic procedures for protein separations were carried out at 4 °C.

The ^{15}N analysis demands a conversion of the sample nitrogen into ammonia, and therefore the quantitative determination of the NPN compounds and the proteins was carried out by steam distillation of ammonia after the Kjeldahl digestion.

APPLICATION AND DISCUSSION

To obtain kinetic data on the distribution of the tracer element after multiple oral application of $[^{15}N]$ glycine a healthy patient was submitted to a long term tracer experiment.

Using the described chromatographic procedures it was possible to investigate the time course of the tracer distribution for the different N fractions (ref. 4). The renal ^{15}N excretion allows the evaluation of parameters of the protein metabolism (ref. 5). The simple separation of hippuric acid is valuable for a calculation of the glycine pool. Taking a 6-pool model as a basis the turnover rate of glycine was estimated to be 1.3 h^{-1}. The separated protein fractions were labelled significantly a few hours after the first tracer application. The relatively highest amounts of tracer nitrogen in the serum proteins were found in the fractions of the α-globulins (including the glycoproteins) and β-globulins (transferrin). We detected the absolutely highest amounts of the nitrogen tracer in the albumin fraction representing more than 60 % of the total serum protein. The course of labelling of the γ-globulins is different from that of other protein fractions synthesized in the liver. Contrary to the latter the γ-globulins still showed a ^{15}N enrichment of 0.02 at.% after 6500 h which is significantly higher than that which we found in a previous study (ref. 6) of natural variations of the ^{15}N abundance of serum protein fractions measured by precision mass spectrometry.

The methods presented here allow their use as routine procedures for the separation of serum and urine samples after ^{15}N tracer experiments. A fractionation of the metabolic pool, the excretion pool, and of the serum proteins gives the possibility to extend the concept of the 3-pool model of Sprinson and Rittenberg (ref. 7) to a multi-compartment model, and thus provides the basis for an evaluation of the protein turnover which is better adapted to the existing biochemical facts. Moreover abnormal courses of labelling of individual fractions should give an opportunity to indicate a pathological change of the nitrogen turnover.

REFERENCES

1 H. Faust and R. Reinhardt, in Stable Isotopes in the Life Sciences: Proc. Tech. Comm. Meet., IAEA, Vienna, 1977, pp. 179-187.
2 K. Hirschberg, P. Krumbiegel and H. Faust, Isotopenpraxis, 17(1981) 178-182.
3 Affinity Chromatography, Principles and Methods, Handbook from Pharmacia Fine Chemicals, Sweden, June 1979.
4 H. Faust, H. Bornhak, K. Hirschberg, K. Jung, P. Junghans, P. Krumbiegel and R. Reinhardt, ZfI-Mitt., 36(1981)3-205.
5 K. Wetzel, P. Junghans, H. Faust and B. Dudek, in E.R. Klein and P.D. Klein (Eds.), Proc. Third Int. Conf. Stable Isotopes, Academic Press, New York, 1979, pp. 591-598.
6 H. Faust, H. Bornhak, K. Hirschberg and H. Birkenfeld, 2. Arbeitstagung "Isotope in der Natur", Leipzig, Nov. 1979, ZfI-Mitt., 29 (1980)217-225.
7 D.B. Sprinson and D. Rittenberg, J. Biol. Chem., 180(1949)715-725.

DYNAMIC ASPECT OF AMINO ACIDS IN NEONATES, PROBED BY NITROGEN-15 AND GCMS

A. LAPIDOT[1], J. AMIR[2] and S.H. REISNER[2]
[1] Isotope Department, The Weizmann Institute of Science, Rehovot, Israel, and
[2] Department of Neonatology, Beilinson Medical Center, Petah Tikva, Israel

ABSTRACT

This report makes use of a recent developed method employing stable isotope and gas-chromatography mass spectrometry (GC-MS) to determine directly the nitrogen-15 enrichment of plasma glycine and L-alanine in term and preterm infants during the first month of life.

Significant differences in the dynamic parameters of intravenously administered labelled amino acids were noted after few days of life in more mature infants as compared to the first day of life. At the age of 3 to 4 weeks all infants showed kinetic data similar to adults.

INTRODUCTION

In view of the increasing use of parenteral nutrition in premature and sick newborn infants, knowledge of the metabolism of infused amino acids is most important.

There are some known time-dependent changes in plasma amino acid levels during the first few hours and days of life and changes according to dietary intake (1,2), but there are relatively few studies on the dynamic aspects of amino acid metabolism in the newborn (3). This is because it has not been technically possible to measure stable isotope enrichments of amino acids in small samples of plasma and measurements have instead been made on the excretion of the isotope in the form of urinary urea.

The rate of disappearance of a labelled amino acid from the circulation in the first few minutes following the intravenous administration reflects the transport of circulating amino acids from the extracellular space to the intracellular pool (4). Although radiolabelled amino acids were used to study glycine metabolism in children (5), this method was not further applied in studies in children due to health hazards associated with the use of radioactive tracers.

A rapid, convenient, non invasive stable isotope (and GC-MS) method capable of measuring the rate of uptake of amino acid and the pools of extracellular amino acid in whole body was recently developed (5,6,7,8,9) and applied to studies in humans (10,11,12,13).

EXPERIMENTAL

Subjects and diet

Our first study subjects were only infants that were appropriate for gestational age and were well at the time of the studies (12) whereas current studies include other premature infants. The preterm infants received only I.V. glucose prior to the first study.

Experimental design

$[^{15}N]$ amino acid kinetic measurements were studied three times in each preterm infant and twice in the full-term infants. The first study was carried out in all infants during the first day of life. The second study was performed on the 3rd day of life and the third study, included only the preterm infants, was at the age of 3-4 weeks.

An indwelling catheter was inserted into a peripheral vein for blood sampling. A single dose of $[^{15}N]$ amino acid (95% ^{15}N, E0-100 μmoles/kg body weight) was administered via another peripheral vein over ∼15 sec. Blood samples (∼0.8 ml) were taken before and at intervals of 5 minutes after intravenous administration.

Sample preparation, GC-MS and data analysis have been detailed in previous publications (9,10,11).

RESULTS AND DISCUSSION

Plasma glycine or alanine levels were relatively constant during the kinetic studies. No indication that these amino acids could leak out as a result of the administered ^{15}N-labeled amino acid was noticed.

No significant differences were observed in glycine or L-alanine kinetic parameters in parallel experiments when the administered dose was reduced by half or when a combined dose of ^{15}N glycine and ^{15}N-L-alanine (100 μmole/kg body wt of each amino acid) was administered to the infant subject (see Figures).

The isotope disappearance curves were linear in all studies carried out. Hence, it was assumed (9,10,11,12) that within the first hour following administration of the labelled amino acid, the disappearance of the $[^{15}N]$ glycine or L-alanine from the circulation can be described by unidirectional irreversible flux of glycine or L-alanine through a single compartment with a pool size (A), unidirectional influx (F) and efflux (E). The isotope enrichment (I) at time (t) following a single dose (w, μ equivalents ^{15}N) of labelled compound is: $I_t = W/(A+W) \exp(kt) \times 100$, where k is the first order rate constant for efflux (E = k.W). The turnover rate constant of glycine or L-alanine in the pool is

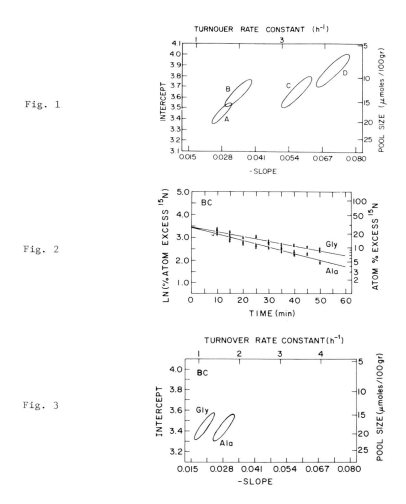

Fig. 1. Summary of the joint confidence region map of glycine dynamics in premature and full-term infants as a function of age of life and gestational age. Region map (A) corresponds to all infants on day of birth; (B) third day of life of young premature, (C) for 'old' premature and (D) 3-4 weeks old, all preterm infants.

Fig. 2. Plasma ^{15}N-glycine and ^{15}N-L-alanine isotope decay curves (represented by non weighted least squares lines) in an infant on day of birth after I.V. administration of a mixture of ^{15}N glycine and ^{15}N-L-alanine (100 μmoles/kg body wt each).

Fig. 3. Joint confidence region maps of ^{15}N-glycine and ^{15}N-L-alanine, turnover rate constants and pool sizes of the above decay curves. Each decay curve is shown as an elipse consisting of points representing the slopes and intercepts of each least squares line that describes the decay curves at the limit 95% confidence (above). The dynamic parameters of ^{15}N-glycine and ^{15}N-L-alanine, administered in one dose in one day old infants (BC) as compared to the summary of glycine dynamics in all preterm and full-term studied infants. (A, all infants on day of birth). (Fig. 1).

represented by k (hr^{-1}), which is obtained directly from the slope of the isotope enrichment time decay curve. The pool size, A (mole/100g body wt), I_d is the isotope enrichment of the administered amino acid and I_o is the isotope enrichment of plasma glycine or alanine extrapolated to time zero. The errors in the slope and intercept were evaluated by standard statistical methods. It was found that the joint confidence region map (slope vs. intercept) is a powerful and accurate method for visually evaluating the significance of differences in pool sizes, rate constants and flux in a population of humans or individual human in different metabolic states. Furthermore certain areas on the joint confidence region maps can be associated with different metabolic and pathological states.

Statistically significant high glycine (or alanine) pool sizes were noted in all infants studied on day of birth when compared with one month of life. A decrease in pool sizes was observed on the third day of life in the preterm and term infants.

The volume of glycine pool obtained in this study varied from day of birth to 3 days and 3-4 weeks of life; 25 to 40% decline in pool volumes were observed in preterm infants at 3-4 weeks of life. Mean values of 530 ml per kg body wt at day of birth and 360 ml per kg body wt at the age of 3-4 weeks are compatible with the extracellular volumes at the corresponding period of life.

Significantly, increases in turnover rate constants were noted on the third day of life in most infants. In the youngest preterm infants, turnover rate constants on the third day of life did not change significantly from day of birth, whereas on 3-4 weeks of life the turnover rate constant increase significantly ($p<0.01$). The turnover rate constants at 3-4 weeks of life increased to 2-3 fold as compared to day of birth. Glycine or alanine fluxes remained unchanged in most cases studied during the first month of life.

Insignificant differences were found for all glycine dynamic parameters, pool size, and turnover rate constants between the different groups on the first day of life. Significant differences were found between 'young' and 'old' preterm on the third day of life although pool sizes of the two groups were almost identical. Turnover rate constants were significantly elevated in the 'old' preterm infants and as a result of these the fluxes in this group were significantly higher. At 3 to 4 weeks of life statistically differences in pool size and turnover rate constants were obtained, when compared to day of birth and to 3-4 days of life. The pool size declined, whereas the turnover rate constant increased and the resulting flux remained unchanged. Factors such as enzyme maturation, hormonal changes and transport mechanisms probably influence the dynamic parameter of the amino acids metabolism It is well known that metabolic processes are impaired in the first days of life as a result of low activity of liver enzymes. The hormonal activities of insulin, glucagon and growth hormone change rapidly after birth (14).

Our ^{15}N amino acids kinetic parameters of day of birth coincide with the enzymes' and hormones' activities on day of birth. But this is the first time that a dynamic assay has been used to monitor the utilization of amino acid in preterm or term infants.

TABLE 1

Variations of Glycine Dynamic Parameters as a Function of Age

Groups and numbers of subjects	Pool (μmoles/100g body wt) (mean)	Turnover rate constant (hr^{-1})	Flux (μmoles hr^{-1}/100g body wt) (mean)
A (n = 8)	18.0±2.1	1.68±0.18	30.2±2.0
B (n = 2)	13.3±2.2	2.07±0.24	27.4±3.1
C (n = 4)	13.1±2.5	3.43±0.28	44.3±7.2
D (n = 6)	9.0±2.2	4.27±0.31	37.8±8.9

A. All infants on day of birth
B. 3-day-old, preterm infants after 29-31 weeks of gestation
C. 3-day-old, preterm infants after 33-35 weeks of gestation
D. 3-4 weeks old, all preterm infants.

CONCLUSION

The kinetic study of individual amino acids during the first few days of life in newborn infants might help in determining the correct usage of intravenous amino acids during this period in infants requiring parenteral nutrition.

We have shown in this study that there are significant differences in the dynamic parameters of intravenously administered labelled glycine (or L-alanine) during the first few days of life in both preterm and term infants. By the age of 3-4 weeks, all these infants showed similar kinetic data as in adults. Current studies extended with L-alanine demonstrate the special role of alanine. Significant increase of turnover rate constants were observed in comparison to glycine turnover whereas the overall dynamic parameters, as a function of age of life or gestational age were not significant from those obtained for glycine (15).

Further studies are now in progress in neonates which are not appropriate for gestational age. These studies demonstrate significant differences between amino acids dynamic parameters during the first weeks of life in comparison to the first group studied.

ACKNOWLEDGEMENTS

This research was supported by grant from the Israel Ministry of Health. We acknowledge the generous supply of [^{15}N]HNO$_3$ by the Isotope Separation Plant of the Weizmann Institute of Science.

REFERENCES

1 J.C. Dickinson, H. Rosenblum and P.B. Hamilton, Pediatrics, 45 (1970) 606.
2 S.E. Snyderman, L.E. Holt Jr., P.M. Morton, E. Roitman and S.V. Phansalker, Pediatr. Res. 2 (1968) 131.
3 P.B. Pencharz, W.P. Stefee, W. Cochran, N.S. Scrimshaw, W.M. Rand and V.R. Young, Clin. Sci. Mol. Med. 52 (1977) 485.
4 H.N. Munro, in H.N. Munro (Ed.), Mammalian Protein Metabolism, Vol. 4, Academic Press, New York (1970) pp. 299.
5 W.I. Nyhan, and B. Childs, J. Clin. Invest. 43 (1964) 2404.
6 C.S. Irving, I. Nissim, and A Lapidot, Biomed. Mass Spectrom. 5 (1978) 117.
7 C.S. Irving, I. Nissim and A.Lapidot in O.Sperling and A. de Vries (Eds.), Monographs in Human Genetics, Vol. 9 (1978) pp. 50-55.
8 I. Nissim and A. Lapidot, A. J. Phys. 237 (1979) (5) E418.
9 A. Lapidot, I. Nissim, C.S. Irving, Israel J. Chem. 77 (1978) 209.
10 A. Lapidot, I. Nissim, C.S. Irving, U.A. Liberman and R. Samuel, in E.R. Klien and P. Klien (Eds.), Stable Isotopes, Proceeding of the Third International Conference, Academic Press, New York (1979) pp. 599-609.
11 A. Lapidot and I. Nissim, Metabolism, 29 (1980) 230.
12. J. Amir, S.H. Reisner, and A. Lapidot, Pediatr. Res., 14 (1980) 1238.
13 A. Lapidot and I. Nissim, in A. Quapli (Ed.), Adv. Mass Spectrom., Hyden, London, Vol. 8B (1980) 1142.
14 B.S.Lindblad and A. Baldesten, Acta Pediat. Scand. 54 (1969) 252.
15 A. Lapidot, I. Zelikovitz and S.H. Reisner (manuscript in preparation).

USE OF WATER LABELLED WITH DEUTERIUM FOR MEDICAL APPLICATIONS

E. Roth[+], G. Basset[++], J. Sutton[+++], M. Apfelbaum[++++], and J. Marsac[+++++]

[+]   Division d'Etudes de Séparation Isotopique et de Chimie Physique - CEN.Saclay, 91191 Gif sur Yvette Cedex, France
[++] Faculté de Médecine BICHAT - Départment de Physiologie, 16, rue Huchard, 75018 Paris, France
[+++] Département de Physico Chimie - CEN.Saclay, 91191 Gif sur Yvette Cedex, France
[++++] Laboratoire de Nutrition - UERX Bichat, 16, rue Huchard, 75018 Paris, France
[+++++] Clinique de Pneumo-Phtisiologie - Hôpital LAENNEC, 42, rue de Sèvres, 75007 Paris, France

ABSTRACT

Whenever a corporal function experiences a disturbance, reflected either by changes in metabolic activities or modifications of the importance of pools of certain molecules, the possibility exists of making use of isotopes in diagnosis. This paper discusses the use of deuterium to measure total body water and extravascular water in the lungs, and gives examples of clinical applications [1,2].

INTRODUCTION

The use of stable isotopes in medicine opens possibilities in three main fields of diagnosis:
- Identification of an element, a molecule, or a fragment of a molecule along its biological pathway.
- Quantification of biological pools by isotopic dilution.
- Measurement of metabolisation rates, and more generally of clearances.

Apart from the restrictions connected to the use of radioactivity, neither hydrogen and carbon, nor oxygen and nitrogen have radioactive isotopes of half life times convenient for medical purposes, and some of them are only available for hospitals that can use accelerator irriadiation facilities. Stable isotopes are thus of special interest in medicine, and their use avoids irradiation of patients. Why then are they not more widely used?

The answer lies mainly in the problems connected with their analysis, as compared with radioisotopes, which are easily analyzed.

a) Thus a lower sensitivity may necessitate the use of larger quantities of drugs and lead to higher costs, and even harmful doses.

b) In the most cases it is necessary to isolate the species to be isotopically analysed. This increases the response time, which together with the cost of the analytical equipment, contributes also to make the competition financially difficult.

Although the absence of radioactivity is the major advantage of stable isotopes, they possess other valuable attributes.

a) They are always ready for immediate use, as the labelled molecules undergo no radiolysis.

b) They may be more easily used to tag a specific chemical species, and for such a use their detection may become much more sensitive.

We consider here the use of deuterium labelled water for the measurement of body water pools and associated quantities. The low cost of D_2O and the sensitive and rapid analytical methods developed to this end, make this tracer attractive both technically and ecomically. We will illustrate its use in the determination of total body water, extravascular water in the lungs and other specific pools, with reference to selected cases.

Two stable isotopic forms of water are available, DHO and $H_2{}^{18}O$. Normally deuterium appears to be the most convenient tracer for body water pools, except in cases where the fast diffusion rate of the isotope may make it unsuitable, e.g. for the investigation of a specific organ. In general, administered as D_2O, it has found a wide application for the rapid measurement of total body water and intracellular water under diverse conditions (obesity, oedema, kidney deficiency), and this demand has led to the development of new, simpler analytical techniques.

When the exchange of deuterium between water and hydrogen containing groups (e.g. NH_2) may prove disadvantageous, $H_2{}^{18}O$ may be used as tracer. The exchange of oxygen isotopes is slow, and thus $H_2{}^{18}O$ may be used in these cases for determining total body water and specific pools. However, although the analytical sensitivity for D and ^{18}O is about the same, namely 1 ‰ of the natural abundance, the use of D_2O as the tracer of choice is favoured by the fact that its natural abundance is only 0.15 ‰ compared with about 2 ‰ for ^{18}O, and that the cost of D_2O is about 1000 times less than that of $H_2{}^{18}O$.

EXPERIMENTAL

A) Analytical techniques

Two methods for the ^{18}O analysis have been developed in our laboratories. The first is the direct determination of ^{18}O by mass spectrometry in microgram samples of water [3], with a sensitivity $\delta < 1$ ‰. A conversion of organic samples to water is required and also the elimination of interfering volatile organic material. Biological samples can also be analysed without chemical treatment using the gamma activation processes $^{16}O(\gamma,n)^{15}O$ and $^{18}O(\gamma,p)^{17}N$ and measuring the ratio of the ^{15}O and ^{17}N signals. This technique is rapid but less sensitive [4] and so entails the use of large quantities of $H_2^{18}O$ in addition to having access to a suitable particle accelerator.

Direct highly sensitive mass spectrometric determination of D in water samples has also been developed using the same instrument as for $H_2^{18}O$. Blood samples must be deproteinized, using zinc sulfate dissolved in water of known deuterium content, followed by double distillation in a flow of helium gas for the isolation of the water [5].

The increasing interest in the use of D_2O as a tracer led to the need for a rapid method for measuring the DHO content of body fluids without purification of the water. Infra red spectrometry is the only alternative technique capable of furnishing the necessary precision, and we have therefore explored this field. The infrared absorption spectrum of blood presents a "window" around 2500 cm^{-1}, which allows the measurement of the band due to the OD vibration when D_2O is present in the blood. Two methods have been devised for measuring DHO at this frequency.

The first involves a modified infrared spectrograph, and is well suited to individual static samples. This apparatus allows 50 ppm excess deuterium concentration to be detected in water samples resulting from lyophilisation. Custom built apparatus can provide better performances. Measurements on blood and other body fluids, principally urine, necessitate a pre-treatment of the samples. When blood is taken at intervals, each sample must be analysed immediately or suitably treated for later observation, as the absorption at 2500 cm^{-1} varies with time. Usually the required result involves a comparison of the deuterium content over a few hours. The techniques used at present for infrared analysis of discrete blood samples involve separating the serum by centrifugation, and storing it in a refrigerator prior to lyophilisation just before analysis [2]. Urine cannot be analysed directly, but must be lyophilised immediately to obtain the required water sample. Other conditioning techniques are not satisfactory [7].

The second method depends on a double colorimeter of original design, which can operate on line with blood, the fluid investigated in the main medical applications. This apparatus was designed for the continuous simultaneous measurement of D_2O and cardiogreen (indocyanine-green) concentrations in flowing blood. Filters are used to select the wavelengths, 4 µ for DHO measurements and 0.8 µ for cardiogreen, in two light beams derived from a single source. Details of the construction are given in references [2,6,8]. The actual sensitivity of the apparatus corresponds to a detection limit of 3 mg of D_2O per liter of blood (3 ppm), i.e. $\frac{1}{50}$ of the natural abundance, and of 0.08 mg of cardiogreen per liter of blood. The volume of the absorption cell is a few tens of microliters, and the blood flow-through is a few ml per minute. The total measurement takes three or four minutes, but could even be reduced, as the transit time between the injection point and the analytical point is less than half a minute, with a difference between cardiogreen and D_2O transit times of about 3 seconds.

Using this apparatus, analyses can be performed at the patient's bedside, and some results are discussed below. The apparatus can also be used for determining total body water.

B) Sampling and Calculations

Total body water: When a patient ingests a known quantity of heavy water (> 99.8 % D_2O, usually 0.15 g/kg body weight) his total body water may be determined from the D_2O abundance in his blood or urine after three hours, a period after which the isotopic equilibrium is complete, after correction of the heavy water excreted during this period.

$$\text{Total body water} = \frac{D_2O \text{ dose} - D_2O \text{ excreted}}{\text{concentration of } D_2O \text{ in blood or urine}}$$

This equation simply expresses the deuterium balance. The precision of the results is about 2 %.

Water in the lung tissue: The measurement of the total or partial water pool of a specific organ necessitates the injection of the tracer into the blood upstream of the organ and the determination of its concentration downstream as a function of time. The labelled water is immediately diluted in the blood, mixing first with the organ water, and later with the total body water.

In order to determine the extravascular lung water, one injects simultaneously, into a peripheral vein for example, D_2O plus a second tracer, which gets diluted only in the blood volume, usually cardiogreen. Both tracers are measured in the arterial blood downstream of the lungs before irrigation of any other tissue can occur [9,10,11]. The transit times of both tracers are very small, and therefore continuous analysis is to prefer in contrast to the fact, that measurements on discrete samples are still performed in most research units. The extracellular volume of lung water is evaluated from the following equations:

$$v = \dot{Q} \times \Delta \bar{t} \qquad (1)$$

where v is the volume irrigated by the diffusible tracer D_2O,
   $\dot{Q}$ is the flowrate of the blood,
   $\Delta \bar{t}$ is the difference in the transit times between the injection point and the analytical point for D_2O and cardiogreen.

For each of the tracers:

$$\dot{Q} = q / \int_0^\infty c(t) \, dt \qquad (2)$$

$$\bar{t} = \int_0^\infty t\, c(t) \, dt / \int_0^\infty c(t) \, dt \qquad (3)$$

where $\bar{t}$ is the transit time,
   q is the quantity of tracer injected,
   c(t) is the excess concentration of the tracer over its natural level, as a function of time in the blood at the sampling point.

Units of concentration are chosen to match those in which q is expressed.

The physical assumption implied in these equations are:

a) The measured volume v is uniformly swept by the flow of labelled water during the time $\bar{t}$ (preferential paths will lead to erroneous results).

b) At the downstream sampling point the flow of blood in the artery is uniform, and the instantaneous tracer concentration c (t) in the sample is identical to that in the artery.

c) The tracer is quantitatively recovered. In case of losses (e.g. in oedema by aerial exit), the equation for $\dot{Q}$ is no longer valid. Other mathematical treatments of the data have been proposed [12].

In practice for each measurement on man, a sterile solution containing 2 g of D_2O and 5 mg of cardiogreen is injected. Three measurements are made at 15 minute intervals, the flow in the analytical circuit being about 15 ml per minute. On animals, dogs or rats, tracer amounts are decreased depending on total body weight and specific lung tissue proportion.

RESULTS

Total body water determinations

Total body water measurements have been made in a variety of clinical cases and have in some of them furnished unequivocal answers to the questions posed. We quote here a few case studies.

Total body water measurements on children and young adults help to establish diets. In normal young subjects total body water has been estimated to represent 55 $\pm$ 4 % of total body weight.

In the case of a pair of twin sisters, probably monozygotic, one was found to have overweight (Service de Nutrition, Hôpital BICHAT, [13]). Was the weight difference genetic or due to diet? The two were subjected to the same restrictive diet (87 g of protein/day, total energy 1400 calories/day) during which, after an initial loss, their weights and their arterial pressures stabilized (Table 1). In attempting to determine the origin of the residual weight difference, the total body water was measured in both cases. From the identic results it was concluded that neo-natal diet was the origin of this difference in weight.

TABLE 1
Data concerning the case of twin sisters P and C with overweight and normal weight before and after restrictive diet [13].

	P	C
Pre-diet		
Weight at birth	2.4 kg	1.7 kg
Weight at examination	60 kg	52 kg
Height at examination	1.53 m	1.53 m
Bi-iliac diameter	26 cm	25.5 cm
Bi-trochanderian diameter	33 cm	33.5 cm
Fat mass (skin crease)	21.31 %	20.6 %
Post diet		
Weight	50 kg	45 kg
Arterial pressure	110/60 mm Hg	110/60 mm Hg
Total body water	26.6 l	26.6 l

At the Hopital NECKER D_2O as a tracer for total body water is simultaneously used with inulin, an indicator for the extracellular water. The amount of intracellular water is then calculated by difference, and the lean weight, directly proportional to the latter, is then evaluated. Children undergoing hemodialysis in context with different therapies are frequently examined in this way in order to fix their diets [14].

The literature [15] cites a case where total body water variations in a premature baby with gastro schisis, ileal atresia and secondary short gut syndrome, was measured with D_2O, to monitor the use of peripheral hyperalimentation. It was observed that over a period of 4 months simultaneously with a 14 g per day increase in weight the percentage of body water decreased from 77.1 % to 60.5 %.

This showed conclusively that the mechanism of weight gain was tissue accretion rather than fluid retention.

Extravascular water in the lung

It has been shown (Hôpital TENON - Hôpital BICHAT [2,5]) on several dozen patients and a hundred animals that the D_2O technique provides a sensitive method of measuring extra-vascular water. In some cases it revealed developing oedema well before other techniques of investigation. Indeed correlations between radiological signs and pulmonary water accumulation show that at the stage of distension of the upper pulmonary vessels the extra-vascular water excess may be as much as 25 %, and that it has increased to 40 to 45 % when radiological signs of interstitial oedema become flagrant.

The comparison between such radiological data and quantitative measurements of water by D_2O dilution carried out on a certain number of patients has proved in many cases the superiority of the second method. This is specially true for patients suffering from chronic kidney deficiency together with pulmonary and total body water accumulation, who showed an increase of 50 % or more in pulmonary water without radiological signs. The same observation was made on two patients suffering from encephalitis with coma.

The excellent coincidence, necessary to apply the tracer dilution method, between $\dot{Q}$ mearured using cardiogreen and D_2O tends to suggest, that D_2O could adequately replace the former for cardiac flowrate measurements. It involves the same injection and sampling techniques, is less costly, less toxic and does not induce peripheral colour. Up to now the use of D_2O for measuring the cardiac flowrate has been mainly considered as a possible reference method, able of improving the cardiogreen test. Besides thermodilution is currently prefered, as it does not involve an arterial puncture. However, one could use D_2O dilution by sampling and analysing blood in the pulmonary artery, installing a catheter in place of a thermal probe. The total cost of a determination would be reduced since thermal probes are expensive and, above all, the accuracy of the method would be far greater.

CONCLUSIONS AND PROSPECTS

Total body water measurements provided by D_2O are already part of routine checks in several hospitals in Paris, especially in the cases of child patients, and of hemodialysis subjects. Such measurements are performed on discrete samples. Continuous D_2O measurements are easy to perform with the special instrument constructed by R. CAPITINI [5,11]. In that device

    a) used alone, D_2O can compete with cardiogreen in particular for cardiac flowrate measurements.

    b) used in combination with cardiogreen, in our research D_2O has provided a tool for the measurement of extravascular water in the lungs, that has proved more sensitive than conventional X ray measurements; however, extravascular lung water measurements are still not generally introduced in clinical diagnosis. Further development of these methods is to expext for the coming months.

ACKNOWLEDGEMENTS

We thank INSERM, DGRST and CEA for providing partial support for this work, Dr. R. CAPITINI for his very helpful contribution to the improvement and maintenance of the instruments. The case of the two sisters was studied by Dr. F. BAIGTS in Prof. APFELBAUM's unit at Hôpital BICHAT. At Hôpital NECKER - Enfants malades - the use of D_2O to follow intracellular water in children under treatment is developed by Prof. BROYER. Mme BOTTER, Prof. POCIDALO, Dr. GAUDEBOUT, Dr. LEICKNAM, M. CECCALDI contributed many suggestions in the course of the work, and Mme BOTTER carefully reread the manuscript.

REFERENCES
1 E. Roth, Possibilités et conditions de l'emploi des isotopes non radioactifs dans le domaine médical et en pharmacologie, Note CEA 2186 (1980).
2 Compte rendu de la table ronde tenue le 9 mai 1979 à l'UER Bichat sur l'utilisation des isotopes stables en médecine, Note CEA 2179 (1980).
3 M. Majzoub and G. Nief, "Advances in Mass Spectrometry", Institute of Petroleum, 4(1968)511-590.
4 Ch. Engelmann, G. Filippi, J. Gosset, and F. Moreau, Radioanalyt. Chem. 37(1977)559-570.
5 H. Nivet, Thesis, Faculté de Médicine, Tours (1976).
6 G. Basset, G. Martet, F. Buchannet, J. Marsac, J. Sutton, F. Botter and R. Capitini, J. Appl. Physiol. 50(1980)1367-1371.
7 M. Apfelbaum, M. Brigant, M. Ceccaldi, and M. Riedinger, J. Biophys. Méd. Nucl. 4(1980)91-94.
8 R. Capitini, Thesis, University Paris Sud, Orsay (1980).
9 G. Basset, F. Moreau. J. Marsac, R. Capitini, and F. Botter, Méthodes actuelles de mesure de l'eau pulmonaire chez l'homme, in F. Vachon and G.G. Pocidalo (Eds.), Journées de réanimation, Hôpital Cl. Bernard, Arnette, Paris (1979)225-231.
10 G. Basset, F. Moreau, J. Marsac. M. Scaringella, M. Ceccaldi, and F. Botter, Bull. Europ. Physiopath. Resp. 14(1978)431-445.
11 F. Botter, M. Scaringella, R. Capitini, J. Marsac, F. Moreau, and G. Basset, Continuous Analysis of Deuterated Water in the Blood Circulation: Application to the Determination of the Pulmonary Extravascular Water, in E.R. Klein and P.D. Klein (Eds.), Stable Isotopes, Proceedings of the Third International Conference, Academic Press, New York 1979, p. 107-119.
12 P. Martin and D. Yudilevich, Amer. J. Physiol. 207(1964)162-168.
13 F. Baigts, private communication.
14 C. Drifford, private communication.
15 A.G.J. Rhodin, A.G. Coran, W.H. Weintraub, and J.R. Wesley, Surg. Gynecol. Obstet. (USA), 148(1979)196-200.

III. BIOMEDICAL APPLICATIONS
C. BREATH TESTS AND LUNG FUNCTION TESTS

THE COMMERCIAL FEASIBILITY OF ^{13}C BREATH TESTS

PETER D. KLEIN and E. ROSELAND KLEIN

Stable Isotope Laboratory, Children's Nutrition Research Center, Department of Pediatrics, Baylor College of Medicine, Houston, Texas 77030

ABSTRACT

The factors limiting the expansion of the use of ^{13}C in medical research and diagnosis are of fundamental concern to the producers of enriched ^{13}C, to manufacturers of labeled compounds, and to the developers of instrumentation for isotope ratio measurements. Each group stands to benefit from commercially successful applications that are accepted in the medical community and that broaden the base of users. Examined herein is the hypothesis that such expansion will depend upon the recovery of the cost of stable isotope applications from health care payments of either private or public origin.

INTRODUCTION

When the decision was made by the United States Atomic Energy Commission to expand production of ^{13}C in 1968, it was reasoned that greater production would stimulate greater use, especially in medical applications and that this widespread use would make possible a significant reduction in the price of ^{13}C. Today, the medical uses of ^{13}C in research and diagnosis have increased, but isotope production continues to exceed consumption. Annual consumption is an estimated 2.5 kg/yr versus world production of approximately 15 kg/yr. The chief reason for the limited medical use is that the purchase of labeled materials and equipment or the analysis of isotopic abundance is supported almost entirely by research funds from grants awarded on a competitive basis. The essential factor in the growth potential of medical applications is that the cost of stable isotope usage must be recovered, and the recovery will be determined by the willingness of third party insurance carriers in the United States to reimburse the patient for the costs of the tests. This paper explores the processes and requirements involved in establishing procedures such as breath tests as accepted components of health care delivery.

BREATH TEST COMPONENTS

There are a number of essential components in the development and implementation of a ^{13}C breath test. There must be a basic production facility for the enrichment of ^{13}C in a form or forms that can be converted into labeled compounds through organic or biosynthetic procedures. A means for determining the isotopic composition of exhaled CO_2 is required, with a minimum precision of 1 $^o/_{oo}$, although 0.3 or 0.1 $^o/_{oo}$ may be necessary under certain circumstances. There must be a clinical condition that causes release of ^{13}C from the labeled compound, such that measurement of the isotopic abundance of excess $^{13}CO_2$ in respiratory CO_2 will provide significant diagnostic information about the patient. Finally, such information must have utility in the clinician's assessment and management of the patient.

DEVELOPMENTAL STAGES OF BREATH TESTS
Conceptual Design

The development of diagnostic breath tests, originally with ^{14}C, but more recently with ^{13}C, includes three stages: conceptual design, clinical validation, and application to patient care. The design and synthesis of the substrate take place in the first stage, and the basal testing conditions are established. In the case of ^{13}C breath tests an awareness is necessary of the requirements for an invariant (or nearly invariant) baseline of $^{13}CO_2$ production during the period of the test. The baseline is established by fasting the patient or by "clamping" the ^{13}C abundance of endogenous metabolic fuels by using substrates that have a natural abundance of ^{13}C close to those utilized in the fasting condition (ref. 1). The sensitivity of the breath test and the amount of substrate required to separate normal function from abnormal function must be determined. In the instance where the normal response is the release of label from the substrate, sufficient substrate must be given to insure that the lower limits of $^{13}CO_2$ production in the severely diseased patient are distinguishable from the baseline abundance (ref.2). In those cases where the normal subject does not produce labeled CO_2 from the substrate [e.g., cholylglycine (ref. 3)], it is important to establish the minimal levels of disease that are to be detected by the test and to provide sufficient substrate to make such detection feasible, given respiratory isotopic baseline stability and instrumental precision.
Clinical Validation

During the second stage of development, clinical validation, it becomes possible to dissociate the clinician from the measurement process. That is, accessory components now permit the test to be conducted in a clinical setting or in an outpatient environment, and the subsequent measurement processes on gas samples need not concern the physician, nor do they have to be carried out in the immediate vicinity of the test performance. This is important because often it is neither convenient nor desirable to bring the patient to the instrument or vice versa. During this stage, the earliest breath tests with ^{13}C also were validated against the prior ^{14}C tests to demonstrate the equivalence of the results obtained. Since none of the results of the tests showed any significant differences between the two isotopes of carbon, this practice now appears to be redundant and has been discontinued.

An important component of the clinical validation process is the validation of the breath test against other accepted means of diagnosis of the disease state under consideration. The ability of the ^{13}C breath test to detect the presence or absence of disease must be compared with techniques known to discriminate with good sensitivity and selectivity between healthy subjects and patients with disease. This includes determination of the number of false positive and false negative results that are obtained when the test is applied to two preselected groups that represent clearcut alternatives. A further refinement of the validation process includes the prospective study of patients whose subsequent clinical course is followed and correlated with the outcome of the test. Two examples may be cited as illustrations of the validation process. In a test of the ability of triolein-^{14}C to detect fat malabsorption in patients with steatorrhea resulting from pancreatic insufficiency as well as a variety of intestinal syndromes, Newcomer et al. (ref. 4) compared the peak excretion of ^{14}C with the amount of fat recovered in 72-h stool collections and found (by fat excretion measurements) that the normal individuals could be separated completely from abnormal subjects on the basis of the peak level of $^{14}CO_2$ in the breath after ingestion of the labeled triglyceride. There were 4% false positives and no false negatives. A similar study, using ^{13}C, in children with pancreatic insufficiency, bile salt deficiency, and mucosal disorders who were compared with normal children has been carried out by Watkins and has shown identical discriminative ability (ref. 5). In studies of patients with alcoholic liver disease, Schneider et al. (ref. 6) investigated the value of the aminopyrine breath test in identifying patients who were at high risk of dying and those whose probability of dying was low. They found that when compared with a normal cumulative 2-h excretion of 13.2% of dose, patients who had values above 1% had a 97% probability of survival; patients, however, whose values were below 1% had greatly increased probabilities of death within 3 weeks. Moreover, when compared to independent clinical assessment of patient status, improvement in the aminopyrine breath test by an increase of 100% or greater was correlated strongly with the patient's clinical improvement. Finally, in a comparison with the ability of serum albumin values, bilirubin levels, or prothrombin times to predict the severity of liver injury as measured independently by liver biopsy and microscopic examination, only the aminopyrine breath test was able to demonstrate a significant change in response from mild to moderate liver injury.

Such clinical validations are essential for the third stage of breath test development, namely the application to patient care. The importance of the validation is that it underlies the cost effectiveness of the test; that is, the information gained by the physician must be commensurate with the costs of conducting the test, and the test must indeed provide useful, consistent, and predictable information about the patient's condition. This is the current status of ^{13}C breath tests in the United States.

Application to Patient Care

Unfortunately, further requirements must be met before these tests may be used as a part of routine patient care. The United States Food and Drug Administration requires that certain legal procedures be carried out and documentation be provided before such tests may be used in health care delivery. In regard to the manufacturing processes of substrates, for example, there must be documentation of the synthetic route of substrate preparation, of the sources of materials used in the synthesis and purification, of the process controls (including assay of chemical purity), and of the manner in which lot numbers are assigned and the inventory is maintained. This information is covered in the Investigation of New Drug form (IND) that must be filed with the Food and Drug Administration before human studies may be undertaken. The second form, called New Drug Application (NDA), requires further documentation of the clinical efficacy of the test, as well as safety and toxicological studies if relevant. Such documentation, if approved by the FDA, permits the substrate to be sold for use in a breath test. To date, no substrates have progressed to this stage in the United States, but it is our hope that certification will be achieved within the next 2 to 5 years for several of the most important substrate candidates.

There are other important steps in the application process. The laboratory that performs the tests must be certified by the American College of Pathologists. Certification is based on the accuracy and precision of the measurements, the ability to maintain adequate patient records, and the establishment of quality controls on the procedures. These requirements assure that legal and medical criteria for accountability in performance of the tests is established, and marks a departure (for the mass spectrometrist) from the somewhat less rigorous documentation permissible in the developmental phase of breath tests. Finally, the economical aspects of the test cost structure must be assembled and used to demonstrate to insurance companies the cost savings (or improved information) that justify the reimbursement of the procedural costs.

ECONOMIC ASPECTS OF BREATH TESTS

Having satisfied the legal requirements that permit the distribution of a ^{13}C substrate for breath tests is but one determinant of the ultimate commercial viability of such procedures. Three economic aspects will be reviewed briefly: material and operating costs, documentation costs, and finally the revenue potential of breath tests.

Under the category of material and operating costs are subsumed all items from the price of raw material (i.e., the ^{13}C precursor used in the synthesis of substrate); the costs involved in the initial synthesis and cost reduction by scale-up in the amount of material synthesized; the patient age category for whom the substrate is intended (substrates are administered in mg/kg body weight and an expensive substrate used in newborn infant studies would be economical, whereas use in adult studies would not); the ancillary equipment required for the test which includes face masks, inflatable bags,"Vacutainers®", etc; and finally the isotope ratio measurement instrumentation. Under this last category, we must address questions with regard to sample preparation requirements, the degree to which the process can be automated, and the costs associated with the purchase, operation and maintenance of a given instrument. Many of these costs are difficult to determine with accuracy, and therein lies the speculative basis behind the commercialization of the breath tests.

Documentation costs as an economic factor in the development of these tests include initialization procedures described above under the IND and NDA procedures, (which may involve two or three scientist/years for completion) as well as the documentation requirements for routine service. Many of the costs in the latter category depend upon the preexisting data management system in operation in the clinical center and whether new tracking procedures for patient care can be added to existing systems without substantial costs being incurred.

Finally, the revenue potential of the breath tests must be assessed. The size of the potential population to whom the tests would be applied must be determined, not only from the standpoint of the disease frequency and the ability of the test to screen the general population, but also to determine the feasibility of the test as a means of monitoring patient progress in serial measurements. Any planning for distribution of tests, such as the ^{13}C breath tests, must take into account the test market area. This includes the effective radius of operation for the sample analysis center, the number of hospital beds included, estimates of the speed and realiability of sample delivery, turnaround time of test results required for effective use of the information by the physician, and finally an estimation of the ability of the facility to meet peak sample loads as they occur in the expansion of the test utilization.

The absence of radioactivity is usually an inadequate justification for the use of ^{13}C breath tests since the cost effectiveness of alternative tests includes the possible hazards associated with the use of ^{14}C. One must know the cost and information source alternatives presently in use and whether or not they require hospitalization. The most significant areas in which stable isotope tests may be successful today are those in which the hospitalization costs can be eliminated by ^{13}C breath tests administered in an outpatient setting.

All of the factors necessary for commercial development of ^{13}C breath tests now can be visualized and identified. Fulfillment of the remaining legal requirements will make the commercial propagation of such tests and the diffusion of this testing capability to hospitals and physicians at large a challenging opportunity for future economic development.

ACKNOWLEDGMENTS

This work is supported by USDA/SEA, Children's Nutrition Research Center, Department of Pediatrics, Baylor College of Medicine and Texas Children's Hospital and by a grant from NIH #AM 28129.

REFERENCES

1   D.A. Schoeller, P.D. Klein, J.B. Watkins, T. Heim, and W.C. MacLean, Jr. Am. J. Clin. Nutr. 33(1980)2375-2385.
2   J.B. Watkins, D.A. Schoeller, P.D. Klein, D.G. Ott, A.D. Newcomer and A.F. Hofmann, J. Lab. Clin. Med. 90(1977)422-430.
3   N.W. Solomons, D.A. Schoeller, J.B. Wagonfeld, D. Ott, I.H. Rosenberg and P.D. Klein. J. Lab. Clin. Med. 90(1977)431-440.
4   A.D. Newcomer, A.F. Hofmann, E.P. Dimagno, P.J. Thomas and G.L. Carlson, Gastroenterology 76(1979)6-13.
5   J.B. Watkins, P.D. Klein, D.A. Schoeller, B. Kirschner, R. Park and J.A. Perman, Gastroenterology (1981) Submitted for publication.
6   J.F. Schneider, A.L. Baker, N.W. Haines, G. Hatfield and J.L. Boyer, Gastroenterology 79(1980)1145-1150.

DECONVOLUTION OF ^{13}C-BREATH TEST DATA BY COMPARTMENTAL MODELING

C.S. IRVING, D.A. SCHOELLER[*], K. NAKAMURA[+], AND P.D. KLEIN
Stable Isotope Laboratory, Children's Nutrition Research Center, Department of Pediatrics, Baylor College of Medicine, Houston, Texas 77030, [*]Department of Medicine, University of Chicago, Chicago, Illinois 60637, and [+]Sankyo Co. Ltd., Tokyo, Japan

ABSTRACT

A compartmental modeling method is described for the generation of a computer cloned patient population (CCPP) that provides a facsimile patient group with selective, calibrated alterations in otherwise normal physiological and metabolic processes. When a CCPP was constructed for the ^{13}C-aminopyrine breath test, it was possible to calibrate both clinical and computer-generated breath tests with respect to alteration in the rates of gastrointestinal output and hepatic metabolism of aminopyrine. This method can be used to assess the ability of various scoring methods to distinguish between changes in absorption and hepatic function.

INTRODUCTION

The type of breath tests to be examined employs a substrate that is administered orally in clinical studies. The substrate undergoes absorption, tissue uptake, cleavage of a methyl group, release of a C_1 fragment that subsequently undergoes further oxidation and excretion in breath. The use of this type of breath test as a diagnostic tool is based on the principle that a change in the rate of one of these processes, namely demethylation, affects the rate of appearance of labeled carbon in breath. The output of labeled carbon in breath has been quantitated by various investigations using the empirical parameters shown in Fig. 1. These parameters will be called breath test (BT) scores. Correlations between breath test scores and particular metabolic and physiological processes have been demonstrated by: 1) the effects of drugs, hormones, or surgery on breath test scores in animal models, 2) good separation of control subjects from patients in scattergrams of breath test scores, and 3) statistically significant correlations of breath test scores with conventional liver function tests. The validation of the sensitivity and specificity of breath tests by the above methods has led to two problems. By basing the predictive power of breath tests on empirical correlations with conventional function tests, the breath tests do not offer the clinician new types of information that supplement information provided by conventional function tests. A second and more serious consideration is that the disease state seldom is restricted to a single process and the manner in which changes in ancillary processes affect breath test scores is difficult to separate from the major effect.

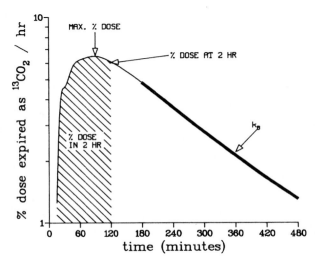

Fig. 1. Breath test scores that have been used to quantitate the output of labeled CO_2.

In order to determine the manner in which the sensitivity and selectivity of a breath test respond to alterations in a particular process, one needs an ideal patient population. Such a population would exhibit selective, calibrated alterations in otherwise identical normal metabolic and physiological processes. Unfortunately, such an ideal patient population is never seen in the real world and if it were, the accumulation of the type of data base needed would be an enormous task. We recently have been exploring methods to obtain a facsimile of such an ideal patient population and the result has been designated as a computer-cloned patient population. It was obtained by designing a pharmacokinetic protocol that could be executed in normal subjects to obtain a set of kinetic curves. The kinetic curves then were used to build a compartmental model representing substrate metabolism in the normal subject. Once

the normal subject can be described, a particular pathological state can be
simulated by increases or decreases in a specific set of rate constants.
Following is a description of the procedure used to analyze the aminopyrine
breath test in such a progression.

The aminopyrine breath test is based on the principle that the amount of
labeled CO_2 that appears in breath following oral administration of $[^{13}CH_3]$-
dimethylaminoantipyrine (DMAAP) reflects the rate of sequential demethylation of
DMAAP by mixed-function oxidases in hepatic microsomes (ref. 1). The test has
been used successfully to monitor the amount of functional parenchymal tissue in
the liver that has sustained an injury (ref. 2,3).

The first step in the generation of an electronic patient population is the
design of a pharmacokinetic protocol for normal subjects. To do this we first
drew a scheme that represented our current understanding of aminopyrine
metabolism and transport. The scheme can be used in the design of tracer studies
to locate suitable routes for drug administration and sites for sampling.
Following oral administration (Fig. 2), DMAAP is believed to enter an hepatic
compartment where it is demethylated to yield monomethylaminoantipyrine (MMAAP)

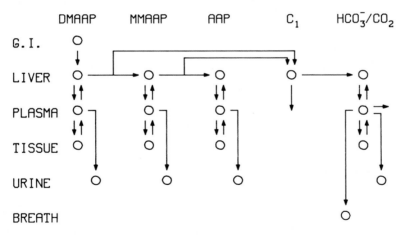

Fig. 2. Compartmental model representing current understanding of aminopyrine
metabolism and transport for use in design of pharmacokinetic protocol. Abbreviations defined in text. G.I. = gastrointestinal.

and a C_1 fragment. The DMAAP that is not demethylated on the first pass enters a
plasma pool that is in equilibrium with other tissue pools. Urinary excretion
can occur from the plasma compartment. Similarly, hepatic MMAAP can undergo a
second demethylation reaction to yield aminoantipyrine (AAP) and a C_1 fragment,
or it can leave the liver in an unchanged form and be distributed between plasma
and tissue. The C_1 fragments can enter into biosynthetic pathways and be lost
irreversibly or undergo oxidation to HCO_3^-. Bicarbonate will mix rapidly with a
central pool which is in equilibrium with a deeper pool. In addition to

excretion in breath, $\overline{HCO_3}$ can be lost irreversibly to bone and biosynthetic pathways or it can be excreted into urine.

The pharmacokinetic protocol that we selected to elucidate these processes consisted of the oral administration of ^{13}C-DMAAP (and simultaneous i.v. injection of ^{14}C-DMAAP) with sampling of plasma DMAAP and MMAAP and breath $^{13}CO_2$ and $^{14}CO_2$. Bicarbonate kinetics were studied on an alternative day by the intravenous administration of ^{13}C-bicarbonate and collection of breath samples. The protocol was carried out on five subjects. The five kinetic curves of the first subject were used to construct a compartmental model for the aminopyrine breath test using SAAM 27 (model code 10) (ref. 4). Linear, first order, non-saturated, steady state kinetics were assumed. A model was built based on the amounts of tracer administered and the points on the kinetic curves, together with their analytical standard deviations. No a priori constraints were imposed and the step-wise addition of compartments and additional rate constants was carried out until the addition of new parameters did not improve the fit and began to increase the uncertainty in the estimated parameters. The simplest compartmental model that reproduced quantitatively all five kinetic curves in a single subject is shown in Fig. 3. The quantitative aspects of the model will

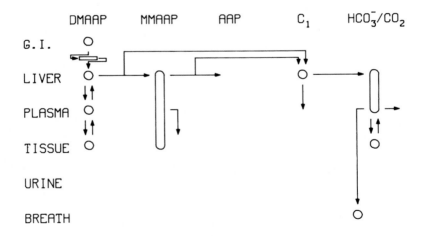

Fig. 3. The simplest compartmental model of the aminopyrine breath test.

not be discussed in this paper, however, all the rate constants and volumes of distribution could be obtained with less than 30% standard deviations and the model accommodated additional subjects in the study without additional modification. Some of the qualitative aspects of the model that evolved from the protocol will be discussed. First, the model is remarkably similar to the ideal model used to design the protocol. However, the computer model includes a delay between the G.I. and liver compartments and requires no urinary outputs for either DMAAP or MMAAP. The model has lumped all MMAAP compartments together and has an additional elimination pathway for MMAAP, which does not generate C_1 fragments. It also is interesting that the data indicated that only half of the C_1 fragments underwent oxidation to HCO_3^-.

Our model of a normal subject now could be used to generate a computer-cloned patient population by the simple expedient of either increasing or decreasing certain rate constants and letting the computer generate a breath test curve for that set of conditions. The rate constants that we chose to alter initially were the G.I. emptying rate, the first demethylation rate, and the first and second demethylation rates simultaneously. In each case, the individual rate constants were increased 2-, 5-, and 10-fold and decreased by 50, 75, 90, 95, and 99% to account for both enzyme induction and liver damage. To simplify interpretation, only one rate constant was studied in a given population.

The simulated breath test curves then were generated and scored by conventional methods. When peak % dose/h was plotted against cumulative % dose expired in 3 h (as has been done for actual aminopyrine breath tests on normal subjects and patients with liver disease by Schneider et al. [ref. 3]) we found that the regression line that passed through the points for the tests obtained from the electronic patient population was virtually identical to the one reported for actual aminopyrine breath test scores by Schneider. This result supported our original belief that electronic patient populations could be constructed that would be reasonable facsimiles of real patient populations.

We have used the electronic patient population to calibrate the breath test scores with respect to changes in rate constants for various processes. We began with normal subjects and found that a 50% reduction in any of the three types of processes considered placed the score in the region of primary biliary cirrhosis, whereas further attenuations to between 75% and 90% moved the score into the cirrhotic region, and attenuations to 90-95% placed the score in the region of patients with short-term survival (ref. 3). When the selectivity of this particular scoring method was considered, it was determined that the attenuation of G.I. output and demethylation rates had qualitatively and quantitatively the same effects on the particular scoring method used by Schneider, et al. (ref. 3) which provided only a gross picture of damage.

Henry et al. (ref. 5) have suggested that a more refined picture of liver function can be obtained by using the terminal decay constant, k_b, and have plotted % dose expired in 2 h as a function of k_b for normals, patients with hepatic disease, and epileptics on drug therapy with induced liver function. Examination of the breath test scores of our electronic patient population showed that only the electronic patients with both first and second demethylation rate accelerations resembled epileptics. Although some divergence of the curves representing the three types of damage was seen, it was not sufficient to distinguish between changes resulting from G.I. output and hepatic function. We currently are using our computer-cloned patient population to try other less conventional and more mathematically involved methods of scoring.

In conclusion, the breath test scores of the electronic populations that we have generated reproduce actual clinical study scores. The use of electronic patients allows us to calibrate conventional scoring methods for the first time, and points out the need to find more advanced scoring methods that can unlock further information already contained in an aminopyrine breath test curve.

ACKNOWLEDGMENTS

This work is a publication of the USDA/SEA, Children's Nutrition Research Center, Department of Pediatrics, Baylor College of Medicine and Texas Children's Hospital and also is supported by NIH Grant #AM 28129.

REFERENCES

1 G.W. Hepner and E.S. Vessel, N. Engl. J. Med. 26 (1974) 1384-1388.
2 R. Carlisle, J.T. Galambos, and W.D. Warren, Dig. Dis. Sci. 24 (1979) 358-361.
3 J.F. Schneider, A.L. Blake, N.W. Haines, G. Hatfield and J.L. Boyer, Gastroenterology, 79 (1980) 1145-1150.
4 M. Berman and M.F. Weiss, SAAM Manual, US DHEW Publ. No. (NIH), 78-180 (1978).
5 D.A. Henry, G. Sharpe, S. Chaplain, S. Cartwright, G. Kitchingman, G.D. Bell and M.J.S. Langman, Br. J. Clin. Pharmacol. 8 (1979) 539-545.

THE EFFECT OF ACARBOSE ON SUCROSE ABSORPTION MEASURED BY THE ^{13}C-SUCROSE BREATH TEST [1]

D. RATING, N. GRYZEWSKI[2], W. BURGER, C. JAKOBS, B. WEBER and H. HELGE
Universitäts-Kinderklinik - Kaiserin Auguste Victoria Haus - Freie Universität Berlin (G.F.R.)

ABSTRACT

The digestion of carbohydrates is brought about mainly by α-glucoside-hydrolysis which can be inhibited by acarbose. Intestinal sucrose absorption with and without a concomitant administration of acarbose was studied in healthy children by means of a ^{13}C-sucrose breath test and concomitant breath hydrogen determination. In healthy children maximal $^{13}CO_2$-exhalation appears $2^1/4$ h after an oral load of ^{13}C-enriched sucrose. After 5 h, 30,5% of the ^{13}C-dose administered are cumulatively eliminated by breath. The failure of breath hydrogen to increase above fasting levels ($\bar{x}$ = 5,3 ppm) confirms that absorption of sucrose is complete in healthy children. Thus velocity, rate and time course of sucrose absorption as measured by breath tests seem to be in the same range as those of ^{13}C-glucose. Administration of 50 mg of acarbose flattens the $^{13}CO_2$-exhalation curve and delays its peak ($\bar{x}$ = $2^3/4$ h). The cumulative ^{13}C-elimination is lowered, especially during the first two hours, when 50% inhibition is found. After 5 hours the ^{13}C-elimination is reduced by 25% only. These delayed higher ^{13}C-elimination rates during the 3rd to 5th h, following acarbose administration may represent colonic absorption of sucrose metabolites produced by bacteria. Breath hydrogen concentrations after sucrose plus acarbose are markedly increased ($\bar{x}$ = 74,1 ppm instead of 5,4 ppm in controls; $p < 0,005$) proving sucrose malabsorption after acarbose. In conclusion, sucrose absorption can be easily studied non-invasively by combining a ^{13}C-sucrose breath test and breath hydrogen determinations. Acarbose apparently prevents sucrose absorption during the first two hours. However, $^{13}CO_2$-exhalation from the 3rd to 5th h after a sucrose load in presence of acarbose may include $^{13}CO_2$ from hydrolysis products of unabsorbed sucrose, thus underrating the inhibitory effect of acarbose on the absorption of sucrose.

[1] By support of the DFG to the Sonderforschungsbereich Embryonalpharmakologie (SFB 29).
[2] This study includes work done by N.G. for his doctoral thesis.

INTRODUCTION

In juvenile diabetics insulin secretion by pancreatic β-cells is deficient and can not cope with the postprandial blood glucose increase. One way to handle this discrepancy between insulin demand and its availibility is to limit the intestinal carbohydrate absorption. For this purpose dietary fibers and enzyme inhibitors, reducing intestinal absorption or decomposition of carbohydrates, were recently introduced into the treatment of diabetics (1-4). Acarbose, a pseudotetrasaccharide (Fig. 1) of microbial origin (5), is an α-glucoside hydrolase inhibitor preventing the breakdown of starch, dissacharides, and sucrose by the enzyme saccharase (6,7).

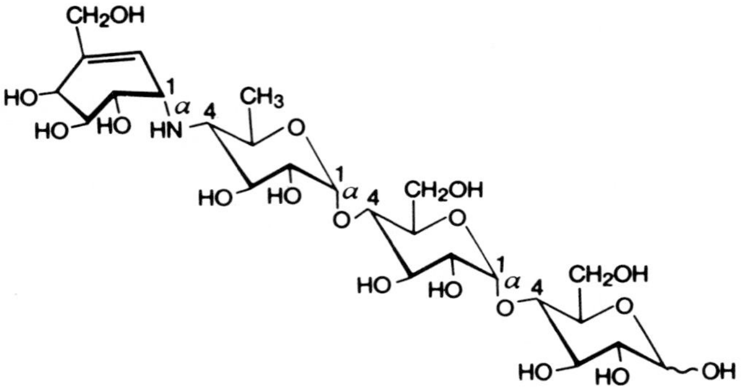

Fig. 1 Acarbose, a pseudotetrasaccharide

Different groups reported a blunting of postprandial blood glucose and insulin increase after acarbose (8-16). However, Maruhama et al. (17) failed to demonstrate any effect of acarbose on oral glucose tolerance in maturity onset diabetics. Moreover, Cook et al. (18) were unable to demonstrate weight loss in obese patients on a maintenance diet following acarbose administration.

The aim of this study was to investigate rate and time course of sucrose absorption in healthy children and to study the effect of a single dose of acarbose on sucrose absorption. Similar studies in diabetic children are in progress.

METHODS

Sucrose absorption was studied by using the ^{13}C-sucrose breath test and additionally by determining breath hydrogen after an oral sucrose load.

Orally administered uniformly labelled [^{13}C]-sucrose ([^{13}C]-UL-sucrose) will be converted to [^{13}C]-UL-glucose and [^{13}C]-UL-fructose at the site of the jejunal brush border. [^{13}C]-UL-glucose and [^{13}C]-UL-fructose will then be absorbed and mainly metabolized to $^{13}CO_2$ and H_2O. $^{13}CO_2$-exhalation was measured and regarded as an index of sucrose absorption.

Breath hydrogen is produced in the human intestine by bacterial fermentation from unabsorbed carbohydrates. Hydrogen diffuses into the blood stream and is eliminated via the lungs. Under normal conditions, this takes place in the large intestine only, and always reflects carbohydrate malabsorption (19).

The tests were performed in 7 healthy children aged $4\frac{3}{4}$ to $8\frac{1}{2}$ years. All parents gave informed consent. All children were tested twice within a week with and without a single dose of 5o mg of acarbose. Thus each child served as its own control. Tests started between 8 and 9 a.m. after an overnight fast; during the test the children stayed at rest and no food intake was permitted. Within 2 min. they drank a solution of 42.75 g per m² of sucrose (2o% solution) plus 2.85 mg per kg [^{13}C]-UL-sucrose (Stohler Isotopes Chemicals). Immediately before the intake of sucrose the children received a single dose of 5o mg of acarbose or a dummy tablet, respectively. Breath samples were collected twice before the administration of [^{13}C]-sucrose and at intervals of 3o to 6o min for 5 hours thereafter. The children exhaled directly into plastic bags, and 5o ml aliquots of these breath samples were then processed for $^{13}CO_2$ analysis as described previously (2o).

The $^{13}CO_2/^{12}CO_2$ ratio was determined using a mass spectrometer (Varian MAT 23o) equipped with a double inlet system and a double ion collector. Data are expressed as $^o/_{oo}$-increase of the $^{13}CO_2/^{12}CO_2$ ratio above basal levels, and calculated as cumulative ^{13}C-elimination in % of the ^{13}C-dose administered, corrected for body weight and endogenous CO_2 production (2o).

For hydrogen analysis, expired breath was immediately transferred from plastic into aluminium stratified bags (Linde AG) to prevent leakage of hydrogen. Breath hydrogen was analyzed by gas chromatography (Varian M 37oo) (column molecular sieve 5 A, 36-6o mesh; carrier gas argon), and its concentrations are expressed as parts per million (ppm) and as the area under the curve.

RESULTS

Following the ingestion of the ^{13}C-enriched sucrose load, $^{13}CO_2$ concentrations in the breath rise gradually to peak at about $2\frac{1}{4}$ h (range 2 to $2\frac{1}{2}$ h) (Fig.2). After 3 hours, 4 hours, and 5 hours 18,7%, 25,5% and 3o,5% of the ^{13}C-dose, respectively, are cumulatively eliminated by breath (Tab. 1).

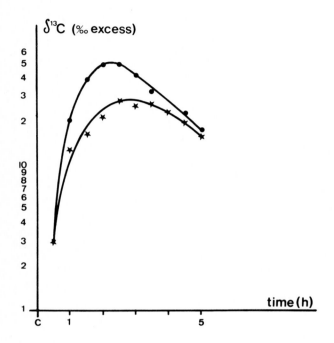

Fig. 2 Influence of 5o mg of acarbose on $^{13}CO_2$-exhalation after an oral ^{13}C-enriched sucrose load (42,75 g sucrose/m² plus 2,85 mg/kg ^{13}C-sucrose) in a healthy boy (F.R. 4 3/4 y; ●——● placebo, ✶——✶ acarbose).

TABLE 1

^{13}C-sucrose breath test: cumulative ^{13}C-elimination[1] in healthy children

	cumulative ^{13}C-elimination[2] after			
	2 h	3 h	4 h	5 h
placebo	9,5	18,7	25,5	30,5
acarbose	5,8[3]	11,9[3]	18,3[3]	23,7[3]

[1] mean of 7 children
[2] in % of ^{13}C-dose administered
[3] p < 0,005 compared to placebo

The intake of a single dose of 5o mg of acarbose delays the $^{13}CO_2$ peak for about half an hour (peaking time 2 3/4 h, range 2 1/2 to 3 1/2 h) (Fig. 2); the curve is flattened and the peak lower. One hour after administering sucrose plus acarbose, the cumulative ^{13}C-elimination reaches only 5o% of the concentrations in the placebo tests; after 3 hours, the values amount to 6o%, and after 5 hours to 75%. Thus the absorption of labelled material is impeded mostly during the first three hours.

Breath hydrogen levels do not increase after sucrose ingestion in any of the children during the placebo tests. The fasting level of breath hydrogen is 5,3 ppm (range 2 to 12 ppm) (Tab. 2) and breath hydrogen concentrations stay at low levels (Fig. 3). The mean maximal level after sucrose is 5,4 ppm (range 3 to 17 ppm).

TABLE 2
Breath hydrogen after an oral sucrose administration in healthy children (n = 7)

	breath hydrogen concentrations (ppm[1])	
	fasting	maximal[2]
placebo	5,3 (2 - 12)	5,4 (3 - 17)
acarbose	8,3 (3 - 22)	74,1[3](38 -1oo)

[1] in parts per million (ppm); mean (range)
[2] mean of maximal level after sucrose administration
[3] $p < 0,005$ compared to placebo

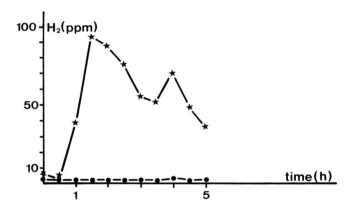

Fig. 3 Influence of 5o mg of acarbose on breath hydrogen after sucrose administration (42,75 g/m²) in a healthy boy (F.R. 4 3/4 y; ●——●placebo, ★——★acarbose).

After adding acarbose to the sucrose the hydrogen concentrations are significantly increased (Tab. 2, Fig. 3), up to a mean maximal level of 74,1 ppm (range 38 to 1oo ppm) ($p < 0,005$). The area under the curve is markedly enlarged (mean: 191,7 ppm x h compared to 18,2 ppm x h in the placebo test; $p < 0,005$). Breath hydrogen peaks at about 2½ h after the ingestion with a range of 1½ to 3 hours (Fig. 3). However, in most of the children a biphasic pattern of hydrogen increase is to be seen. In all children taking acarbose, hydrogen concentrations are significantly above control levels at 9o min. Thus, the bulk of unabsorbed carbohydrates has already reached the colon 6o to 9o min after the ingestion.

DISCUSSION

Using ^{13}C-sucrose the absorption of this carbohydrate can be studied in a convenient and non-invasive manner. The feasibility of this method to study sugar absorption was previously demonstrated using ^{13}C-glucose (2o). Following glucose ingestion, $^{13}CO_2$ in breath peaked at about one hour when small amounts of ^{13}C-glucose were used, but considerably later, i.e. at about $2\frac{1}{2}$ hours, when a load of ^{13}C-enriched glucose (4o g glucose per m²) was applied. It is evident from this study, that velocity, degree and time course of sucrose absorption are comparable to glucose absorption for the loading as well as for the tracer tests (data not shown). The similar timing of the $^{13}CO_2$ peaks in ^{13}C-glucose and ^{13}C-sucrose breath tests favors the assumption of a rapid breakdown of sucrose by the enzyme saccharase at the site of the jejunal brush border and an immediate absorption of the monosaccharides thereafter. According to this test fructose absorption has to be equally effective as glucose absorption. No increase of breath hydrogen was seen following the ingestion of 42,75 g sucrose per m², proving that in healthy children an almost complete absorption of this amount of carbohydrate had taken place.

Acarbose was shown to reduce significantly the postprandial increase of blood glucose (8-15) and insulin (11-15) in healthy adults. In insulin-dependent diabetics the postprandial rise of blood glucose was lowered (9,14,21,22), and the daily insulin requirements were substantially reduced in studies using an artificial B-cell system (14,21,22). However, under routine conditions a reduction of the daily insulin dose by acarbose could not be demonstrated by others (16-18). The suppression or even prevention of the postprandial rise of blood glucose by acarbose can well be explained by our finding of a markedly reduced ^{13}C-elimination after acarbose during the first 3 hours, when $^{13}CO_2$-exhalation reaches only 5o-6o% of the individual control values.

Bond et al. (23) reported that carbohydrates not absorbed from the small bowel may be salvaged to an important extent in the large intestine. It was demonstrated (24,25) that unabsorbed carbohydrates may be split by fecal bacteria to form short chain organic acids such as acetic, propionic and butyric acid, which at least in animals, are rapidly absorbed across the colonic mucosa (26-28).

Breath hydrogen studies are regarded as a good instrument to prove carbohydrate malabsorption (19). Increased hydrogen exhalation following acarbose administration was first reported by Caspary (8) and confirmed by others (11,17). In adults 2oo mg of acarbose can provoke the maldigestion of approximately 4o% of a 1oo g sucrose load (8), and Jenkins (11) calculated an almost complete inhibition of sugar absorption in a 5o g carbohydrate load after 2oo mg of acarbose. These estimations were performed comparing the areas under the curves after applying other carbohydrates and a lactulose load in the same subject. Lactulose is regarded to be not absorbed at all, and the rates of hydrogen liberated from lactulose, sucrose, maltose and lactose by fecal bacterial fermentation have been shown to be comparable (19). The rather high hydrogen production rate observed in this study after acarbose, therefore, suggests that sucrose malabsorption in fact constitutes an important factor of the acarbose effect.

Therefore, we have to consider that after acarbose a breakdown to ^{13}C-enriched short chain organic acids may have happened in the large intestine of our test subjects. These ^{13}C-compounds may have been absorbed and metabolized to $^{13}CO_2$. The calculation of cumulative $^{13}CO_2$-exhalation in these healthy children after acarbose application may include $^{13}CO_2$ production from sucrose metabolites produced by bacterial fermentation, especially during the latter part of the test. An extrapolation of our data to determine the inhibition of sucrose absorption might then be permissible only for the first 2 hours of the test period.

REFERENCES

1 H.C. Trowell, in D.P. Burkitt and H.L. Trowell (Eds.) Refined Carbohydrate Foods and Fibre. Academic Press, London, 1975, pp 227-25o.
2 D.J.A. Jenkins, T.M.S. Wolever, R.H. Taylor et al., Brit.Med.J. 1(198o) 1553-4.
3 R.J. Walton, I.T. Sherif, G.A. Noy and K.-A. Alberti, Brit.Med.J. 1 (1979) 22o-21.
4 D.J.A. Jenkins, D. Reynolds, B. Slavin, A.R. Leeds, A.L. Jenkins and E.M. Jepson, Americ.J.Clin.Nutr. 33 (198o) 575-81.
5 D.D. Schmidt, W. Frommer, B. Junge, L. Müller, W. Wingender, E. Trusch, Naturwissenschaften 64 (1977), 535-36.
6 L. Müller, B. Junge, W. Frommer, D. Schmidt and E. Truscheit, in U.Brodbeck (Ed), Enzyme Inhibitors, Verlag Chemie (198o), pp 1o9-122.
7 B. Junge, H. Böshagen, J. Stoltefuß, L. Müller, in U. Brodbeck (Ed), Enzyme Inhibitors, Verlag Chemie (198o), pp 123-37.
8 W. F. Caspary, Lancet 1 (1978), 1231-33.
9 R.J. Walton, I.T. Sherif, G.A. Noy, K.G.M.M. Alberti, Brit.Med.J. 1 (1979), 22o-21.
1o R.H. Taylor, D.J.A. Jenkins, H.M. Barker, H. Fielden, D.V. Goff, J.J. Misiewicz et al., "Effect of acarbose on the 24-hour blood glucose profile and pattern of carbohydrate absorption", "submitted for publication".
11 D.J.A. Jenkins, R.H. Taylor, D.V. Goff, H. Fielden, J.J. Misiewicz, D.L. Sarson, St.R. Bloom, K.G.M.M. Alberti, "Scope and specificity of acarbose in slowing carbohydrate absorption in man", "submitted for publication".
12 W. Puls, U. Keup, H.P. Krause, G. Thomas and F. Hoffmeister, Naturwissenschaften 64 (1977), 536.
13 W. Puls, U. Keup, H.P. Krause and G. Thomas, Diabetologia 13 (1977) 426.
14 P. Berchtold, I. Hillebrand, K. Boehme, R. Aubell, L. Nagel, H. Schulz, La Clinica Dietologica 7 (198o) 426.
15 I. Hillebrand, K. Boehme, G. Frank, H. Fink and P. Berchtold, Res.Exp.Med.(Berl) 175 (1979), 87-94.
16 D. Sailer and G. Röder, Arzneim.-Forsch./Drug Res.3o (II) (198o) 2182-85.
17 Y. Maruhama and Y. Goto, Tohoku J. exp. Med., 13o (198o), 243-52.
18 R.F. Cook, A.N. Howard and I.H. Mills, "The Failure of an Alpha-Glucosidehydrolase Inhibitor to Produce Weight Loss in Obese Patients on a Maintenance Diet", "submitted for publication".
19 J.H. Bond and M.D. Levitt, J. Clin. Invest. 51 (1972) 1219-25.
2o H. Helge, B. Gregg, C. Gregg, S. Knies, I. Nötges-Borgwardt, B. Weber and D. Neubert,* MacMillan Press Ltd., London (1978), 227-34.
*in T.A. Baillie (Ed) Stable Isotopes. Applications in pharmacology,toxicology and

clinical research.
21 S. Raptis, G. Dimitriadis, K. Karaiskos, J. Rosenthal, Chr. Zoupas and S. Moulopoulos, Diabetes 29, suppl. 2, 41A, (1980).
22 H. Jäger, U. Krause, E. Wolf, U. Cordes and J. Beyer, 84. Tagung Dtsch. Ges.Inn. Med. 1978 (Abstract 394).
23 J.H. Bond, B.E. Currier, H. Buchwald and M.D. Levitt, Gastroenterology 78 (1980), 444-47.
24 J.H. Bond and M.D. Levitt, J. Clin. Invest. 57, (1976), 1158.
25 R.A. Argenzio, M. Southworth, C.E. Stevens, Am. J. Physiol. 226,(1974), 1o43.
26 C.E. Stevens, Am. J. Clin. Nutr. 31, (1978), 161-68.
27 R.A. Argenzio, M. Southworth, J.E. Lowe, C.E. Stevens, Am. J. Physiol. 233 (1977) 469-78.
28 R.A. Argenzio, N. Miller, W. von Engelhardt, Am.J. Physiol. 229 (1975) 997-1oo2.

APPLICATION OF ^{13}C-FATTY ACIDS BREATH TESTS IN MYOCARDIAL METABOLIC STUDIES

M. SUEHIRO, K. UEDA, M. IIO
Tokyo Metropolitan Geriatric Hospital, Tokyo
J. MORIKAWA, M. NAKAJIMA, R. OHSAWA
Eiken Immunochemical Laboratory, Tokyo, Japan

ABSTRACT

^{13}C-fatty acids breath tests were evaluated as a diagnostic tool to detect or predict heart diseases or myocardial dysfunctions. Breath tests were performed by oral administration of 3.5 mg/kg body weight of ^{13}C-octanoate, followed by breath collection at 15 minute intervals for 1 hour and 30 minute intervals for the following 2 hours. $^{13}CO_2$ in expired air was analyzed by a mass spectrometer. Two cases of myocardial infarction, and four of heart failure caused by valvular diseases were tested and compared with 10 healthy adults and 3 in-patients who had neither cardiac nor hepatic diseases. When comparing cumulative $^{13}CO_2$ recoveries for 3 hours between patients with heart diseases and the normal controls, no significant differences were observed except for the cases of myocardial infarction. In myocardial infarction, 48-63 % higher $^{13}CO_2$ recoveries were observed. Exercise studies were also performed by healthy volunteers. Thirty minutes after oral administration of ^{13}C-octanoate, they started jogging or skipping a rotating rope. During exercise, breath was collected. $^{13}CO_2$ analysis in expired air revealed that during exercise $^{13}CO_2$ production was depressed drastically. The cumulative $^{13}CO_2$ recoveries during 3 hours of exercise tests were 46-65 % of those obtained in resting experiments.

INTRODUCTION

It is well known that fatty acids are metabolized by the myocardium and are utilized as major energy sources. On the basis of this fact, in Nuclear Medicine, the myocardium can be visualized by labeled fatty acids such as ^{11}C-fatty acids [1, 2] or fatty acid analogues such as ^{18}F-, ^{131}I-labeled fatty acids [3, 4] etc.: After injection, the tagged fatty acids are observed to accumulate into the normally functioning myocardium and be metabolized quickly into water-soluble metabolites. Then, the metabolites are transported away by blood. On the other hand, the myocardium with infarction fails to take up the fatty acids.

Since CO_2 is produced as one of the final metabolic products, it is presumable that the analysis of the tagged CO_2 in the breath after administration of tagged fatty acids such as ^{13}C-octanoate, palmitate, oleate etc. might help to predict or detect heart diseases or myocardial dysfunctions. In this study a way to use ^{13}C-fatty acids breath tests in myocardial metabolic studies was tested.

METHOD
Substrate

^{13}C-octanoate was chosen as the substrate for the breath test, because this chort-chain fatty acid is absorbed faster from the gastrointestine when orally administered, and $^{13}CO_2$ recovery in the breath is higher than with long-chain fatty acids such as palmitate or oleate. In addition, ^{13}C-labeled octanoate is much cheaper than other ^{13}C-fatty acids.

^{13}C-octanoate breath test

Before administration of ^{13}C-octanoate, expired air was collected for basal $^{13}CO_2$ determination. Then 3.5 mg/kg body weight of ^{13}C-octanoate (90 % atomic excess, Kor Incorp.) were given to a patient or a volunteer who had been fasting for 15-16 hours. After administration, breath collection was performed at 15 minute intervals for one hour, then at 30 minute intervals for the following 2 hour period.

$^{13}CO_2$ in the expired air was analyzed by a mass spectrometer (Nuclide RMS) equipped with a dual inlet and with a dual collector. The ^{13}C abundance in the sample CO_2 was expressed as a permil ($^o/oo$)-increase from the basal $^{13}CO_2/^{12}CO_2$, or as a %-dose/mM CO_2, %-dose/3 hrs, %-dose/hr. %-dose/3 hrs or %-dose hr was calculated assuming that CO_2 excretion rate is 9 mM CO_2/kg hr.

Patients

Two cases of myocardial infarction (50 y.o. female and 72 y.o. female) and four of heart failure caused by valvular diseases (54-86 y.o., ♀2, ♂2) were tested and compared with 10 healthy adults (average 29.7±6.2 y.o., ♀3, ♂7) and 3 in-patients (50-78 y.o., ♀1,

62) who had neither cardiac nor hepatic diseases.

Exercise Studies

In order to assess the relationship between octanoate metabolism and the cardiac load, exercise studies were also performed. Thirty minutes after oral administration of ^{13}C-octanoate, healthy volunteers jogged or skipped a rotating rope continuously for 20-30 minutes. During exercise, breath was collected at 4-5 minute intervals.

Before $^{13}CO_2$ analysis, CO_2 partial pressure was measured to evaluate the endogenous CO_2 increase caused by exercise. The endogenous CO_2-production was taken into account when %-dose was calculated.

RESULTS AND DISCUSSION

Breath tests

A result of the breath tests is shown in Fig. 1. The closed circles show the typical healthy adult without malabsorption: $^{13}CO_2$-production reached a maximum 30 minutes after administration and declined exponentially with two, fast and slow, components. The open circles represent a 86 y.o. male patient with heart failure: $^{13}CO_2$ production stayed at maximum for 1 hour and then showed a slow decline. The full squares represent a 50 y.o. female with myocardial infarction: High $^{13}CO_2$ recovery in expired air was observed.

Between the patients with cardiac disease and the normal control, there appeared to be a difference in the octanoate metabolism. However, comparing these apparent abnormal cases with a few cases which were supposed to be normal, for example, some in-patients without cardiac diseases, the difference was not obvious.

Figure 2 shows the summary of the cumulative $^{13}CO_2$ recovery for 3 hours after administration. From the left, the first column shows 10 normal controls, the second 4 cases of heart failure, the third 2 cases of myocardinal infarction and the last, 3 in-patients without neither cardiac nor hepatic diseases. The open circle on the second column shows a 63 y.o. female patient who has found to have a liver tumor as well as cardiac disease.

Average cumulative $^{13}CO_2$ recoveries of the patients with heart failure and with myocardial infarction were 42.7±4.9 and 50.7±3.3 %-dose/3 hrs, respectively, and those of the normal controls and 3 in-patients are 32.6±7.4 and 42.2±6.5, respectively. The ^{13}C-cumulative recoveries of the patients with cardiac diseases tend to be higher than the controls, however, except for the cases of myocardial infarction, the differences are not significant.

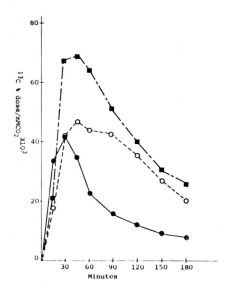 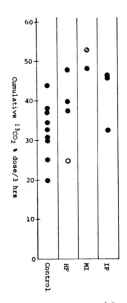

Fig. 1. Time course of $^{13}CO_2$-production after oral administration of ^{13}C-octonate.
●—● : 36 y.o. healthy male,
○--○ : 86 y.o. male with heart failure, ■--■ : 50 y.o. female with myocardial infarction.

Fig. 2. Cumulative $^{13}CO_2$ recovery through 3 hours after oral administration of ^{13}C-octanoate.
(HF: Heart Failure, MI: Myocardial Infarction, IP: In-patients without heart diseases).

The result of the cases with myocardial infarction, which showed 48-63 % higher cumulative $^{13}CO_2$ recoveries than the average of normal controls, suggests that the ^{13}C-octanoate is metabolized in the liver through the citric cycle to give $^{13}CO_2$.

On the other hand, the result of the cases of heart failure which showed little difference from normal controls or in-patients without cardiac diseases indicates that fatty acids consumption on the myocardium with heart failure is neither significantly high nor low in spite of their enlarged heart volume compared with normal hearts.

Between these results of breath tests and the blood concentration of triglyceride, free fatty acids, or cholesterol, no correlations were observed.

Exercise studies

Figure 3 shows a result of the exercise studies. As represented by the solid line, as soon as exercise started, $^{13}CO_2$ production was depressed drastically compared with the rest state which is expressed by a dotted line. This result was reproducible with different volunteers, and cumulative $^{13}CO_2$ recoveries during 3 hours of exercise studies were 46-65 % of those obtained at rest (Fig. 4).

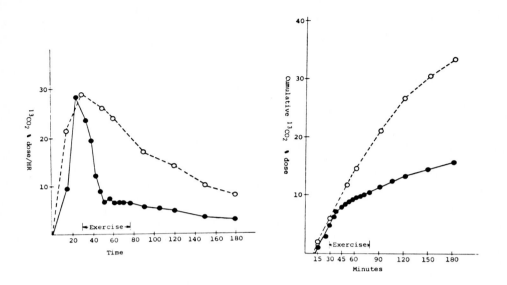

Fig. 3. $^{13}CO_2$ production during exercise and at rest.

Fig. 4. Cumulative $^{13}CO_2$ production during exercise and at rest.

Exercise started 30 minutes after administration of ^{13}C-octanoate.
●—● : Exercise test
○--○ : Resting experiment.

When assuming that $^{13}CO_2$ which is recovered in the breath following administration of ^{13}C-labeled compounds reflects $^{13}CO_2$ produced in the tissue, these observations suggest that during exercise, a) ^{13}C-octanoate fails to be taken up into the myocardium, or b) ^{13}C-ketone body formation blocked the citric acid cycle from producing $^{13}CO_2$, or c) competitive consumption between glucose and fatty acids resulted in a decrease in $^{13}CO_2$ production from the ^{13}C-fatty acid, or d) gastrointestinal absorption of the fatty acid was blocked.

Akioka et al. [5] who measured arterial and venous blood concentration of FFA in patients with and without heart diseases observed a 37-51 % decrease in the myocardial uptake of fatty acids during exercise in the cases of heart diseases and a 6.9-16 % decrease in the cases without heart diseases. On the other hand Costill et al.

[6] reported that after administration of ^{14}C-labeled glucose $^{14}CO_2$ was recovered in expired air 6.5 times more rapidly during exercise than at rest. Taking their observations into account, we presume that during exercise a) and c) took place, and as a result ^{13}C-ketone bodies were formed via a metabolic by-path such as suggested by Stern et al. [7] or by Green et al. [8], then the citric acid cycle which produces $^{13}CO_2$ was blocked.

Regarding c), if d) were the case, the $^{13}CO_2$ production should be recovered at the end of ecercise. In fact no recovery was observed through 2 hours after the end of exercise. However to clarify this the breath tests performed by intravenous injection instead of by oral administration would be a better choice: With intravenous injection we can avoid the problem of absorption. Generally, when breath tests are used in metabolic studies, intravenous injection would give us clearer figures about the metabolism, in which we are interested.

CONCLUSION

Although, in this preliminary study, breath tests performed following oral administration of ^{13}C-octanoate could not reveal significant differences between the patients with heart diseases and the normal controls, it is still worth while trying to evaluate the ^{13}C-fatty acid breath tests performed by intravenous injection, and/or with exercise, and/or with other ^{13}C-fatty acids such as palmitate or oleate, which might be able to give us clearer figures on fatty acids metabolism in the myocardium. In particular we are interested in comparing ^{11}C-fatty acids studies to ^{13}C-breath tests studies.

REFERENCES
1. E.J. Hoffman, M.E. Phelps and E.S. Weiss, J. Nucl. Med., 18(1977) 57-61.
2. R.A. Coldstein, M.S. Klein and M.J. Welch, J. Nucl. Med., 21(1980) 342-348.
3. E.J. Knust, C.H. Kupfernagel and G. Stöcklin, J. Nucl. Med., 20 (1979)1170-1175.
4. N.D. Poe, G.D. Robinson and N.S. MacDonald, Proc. Doc. Exp. Biol. Med., 148(1975)215-218.
5. H. Akioka, Y. Higashihara, N. Kijima and K. Kondo, Jap. Heart J., 7(1975)267-274.
6. D.L. Costill, A. Bennett, G. Branam and D. Eddy, J. Appl. Physiol., 34(1973)764-769.
7. J.R. Stern, M.J. Coon and A.D. Campillo, Nature, 171(1953)28-30.
8. D.E. Green, D.S. Goldman, S. Mii and H. Beinert, J. Biol. Chem. 202(1953)137-150.

and the Evolution of Life: An Overview *(M. Schidlowski)*. Isotope Geochemistry of Carbon *(J. Hoefs)*. Carbon Isotope Fractionation Factors of the Carbon Dioxide-Carbonate System and their Geochemical Implications *(L.E. Maxwell and Z. Sofer)*. The Isotopic Fractionation During the Oxidation of Carbon Monoxide by Hydroxylradicals and its Implication for the Atmospheric Co-Cycle *(H.G.J. Smit et al.)*.

III. Biomedical Applications. A. Pharmacology and Drug Metabolism. Applications of Stable Isotopes in Pharmacological Research *(T.A. Baillie et al.)*. Application of ^{13}C-Labelling in the Bioavailability Assessment of Experimental Clovoxamine Formulations *(H. de Bree et al.)*. Measurement of the Pharmacokinetics of DI-([15,15,16,16-D4]-Linoleoyl)-3-sn-Glycerophosphocholine After Oral Administration to Rats *(A. Brekle et al.)*. **B. Clinical Diagnosis.** The Application of the Stable Isotopes of Oxygen to Biomedical Research and Neurology *(D. Samuel)*. In Vivo Measurement of Enzymes with Deuterated Precursors and GC/SIM *(H.-Ch. Curtius)*. Prenatal Diagnosis of Propionic and Methylmalonic Acidemia by Stable Isotope Dilution Analysis of Methylcitric and Methylmalonic Acids in Amniotic Fluids *(L. Sweetman et al.)*. Use of Water Labelled with Deuterium for Medical Applications *(E. Roth et al.)*. **C. Breath Tests and Lung Function Tests.** The Commercial Feasibility of ^{13}C Breath Tests *(P.D. Klein and E.R. Klein)*. Application of ^{13}C-Fatty Acids Breath Tests in Myocardial Metabolic Studies *(M. Suehiro et al.)*. Breath Test Using ^{13}C Phenylalanine as Substrate *(D. Glaubitt and K. Siafarikas)*. Use of the $^{13}C/^{12}C$ Breath Test to Study Sugar Metabolism in Animals and Man *(J. Duchesne et al.)*. **IV. Life Sciences, Agriculture, and Environmental Research.** Stable Isotopes in Agriculture *(H. Faust)*. Balance of ^{15}N Fertilizer in Soil/Plant System *(N. Sotiriou and F. Korte)*. Nitrogen Isotope Ratio Variations in Biological Material - Indicator for Metabolic Correlations *(R. Medina and H.-L. Schmidt)*. Possibilities of Stable Isotope Analysis in the Control of Food Products *(J. Bricout)*. **V. Methods. A. Analytical Developments.** Recent Applications of ^{13}C NMR Spectroscopy to Biological Systems *(N.A. Matwiyoff)*. Quantitative Mass Spectrometry with Stable Isotope Labelled Internal Standard as a Reference Technique *(I. Björkem)*. Application and Measurement of Metal Isotopes *(K. Habfast)*. Stable Isotopes in Biomedical and Environmental Analysis by Field Desorption Mass Spectrometry *(W.D. Lehmann and H.-R. Schulten)*. **B. Isotope Separation and Synthesis of Labelled Compounds.** The Production of Stable Isotopes of Oxygen *(I. Dostrovsky and M. Epstein)*. Separation of the Stable Isotopes of Chlorine, Sulfur and Calcium *(W.M. Rutherford)*. The Synthesis of Mono-and Oligosaccharides Enriched with Isotopes of Carbon, Hydrogen and Oxygen *(R. Barker et al.)*. New Approaches in the Preparation of ^{15}N Labeled Amino Acids *(Z.E. Kahana and A. Lapidot)*.

Send your order to **your bookseller** or
ELSEVIER SCIENCE PUBLISHERS
P.O. Box 211, 1000 AE Amsterdam, The Netherlands

Distributor in the U.S.A. and Canada:
ELSEVIER SCIENCE PUBLISHING CO., INC.
52 Vanderbilt Ave., New York, N.Y. 10017

Continuation orders for series are accepted.

Orders from individuals must be accompanied by a remittance, following which books will be supplied postfree.

The Dutch guilder price is definitive. US$ prices are subject to exchange rate fluctuations.
Prijzen zijn excl. B.T.W.

Chemistry, Life Sciences, Biochemistry, Molecular Biology, Pharmacology, Pharmaceutical Sciences, Earth Sciences, Environmental Research

Stable Isotopes

Proceedings of the Fourth International Conference held in Julich, March 23-26, 1981

edited by H.-L. SCHMIDT, H. FÖRSTEL *and* K. HEINZINGER

ANALYTICAL CHEMISTRY SYMPOSIA SERIES, 11

**1982 xvii + 758 pages
Price: US $127.75 / Dfl. 275.00
ISBN 0-444-42076-2**

The 85 contributions and 15 reviews in this volume make it a comprehensive overview of the state of investigations on stable isotopes of the main bioelements as they occur in nature. The conference emphasized stable tracer applications in medicine, pharmacology, agriculture and biochemistry, and added the dimension of the theory and consequences of isotope effects, and the importance of stable isotopes in geochemistry, cosmochemistry and environmental research.

The papers compare recent results and methodology in the different disciplines obtained on the basis of isotope measurements. Interaction between the disciplines was facilitated by a common methodology which revealed that a particular problem could be approached from various angles, and stimulus given for further investigations. Examples of new possibilities are the use of NMR in biological research and the introduction of "naturally labelled" compounds in nutrition physiology. The book is essential to all those working on stable isotopes but even scientists not familiar with isotope research may benefit from the demonstration of the wide possibilities isotope application can have as a common tool in most biosciences from agriculture to zoology.

A SELECTION OF THE CONTENTS: **I. Isotope Effects, Theory and Consequences.** The Theoretical Analysis of Isotope Effects *(M. Wolfsberg).* Vapour Pressure Isotope Effects of Acetonitrile *(Gy. Jakli et al.).* Heavy-Atom Isotope Effects on Enzyme-Catalyzed Reactions *(M.H. O'Leary).* Isotope Effects on Each, C- and N-Atoms, as a Tool for the Elucidation of Enzyme Catalyzed Amide Hydrolyses *(R. Medina et al.).* **II. Geochemistry and Cosmochemistry.** Stable Isotopes

ELSEVIER SCIENTIFIC PUBLISHING COMPANY

Amsterdam *and* New York

^{13}C-GLYCOCHOLATE BREATH TESTS IN CHILDREN WITH CYSTIC FIBROSIS

K. SIAFARIKAS[1], D. GLAUBITT[2] and H. STEINHAUER[1]
Departments of Pediatrics[1] and Nuclear Medicine[2], Academic Teaching Hospital, D-4150 Krefeld (Federal Republic of Germany)

ABSTRACT

Breath tests using ^{13}C or ^{14}C labeled glycocholate provide information about the intestinal deconjugation of conjugated bile acids. ^{13}C glycocholate should replace ^{14}C glycocholate as substrate of such breath tests. The practicability of the ^{13}C-glycocholate breath test was confirmed by us in 11 patients with cystic fibrosis. The largest deviation from the normal range was observed in a female juvenile suffering from this disease complicated by chronic bronchitis whereas 2 girls with cystic fibrosis presented slightly pathologic results. The other patients showed normal ^{13}C-glycocholate breath tests.

INTRODUCTION

Disorders of the enterohepatic circulation of bile acids may play an important role in various diseases. Great diagnostic progress was made by the introduction of the ^{14}C-glycocholate breath test [1,2] which may detect both bacterial overgrowth and malabsorption of bile acids. Since ^{13}C glycocholate has become available as substrate for a corresponding breath test [3 - 5] and as ^{14}C glycocholate causes a radiation burden, the use of ^{14}C glycocholate for breath tests appears to be no more justified at least in children, adolescents, and women who are pregnant or at risk of pregnancy.

^{13}C-glycocholate breath tests were carried out by us in children and adolescents with cystic fibrosis in order to assess the practicability of the test as well as the frequency of disorders of the intestinal deconjugation of conjugated bile acids in these patients. Only a small dose of highly purified ^{13}C-glycocholate was applied so that there was little probability of significant interference with metabolic regulations of the organism.

PATIENTS AND METHODS

Two boys and 5 girls aged 4 to 14 years as well as 2 male and 2 female juveniles aged 16 to 18 years were examined. These 11 patients were treated symptomatically with drugs; in 3 of them (1 boy and 2 girls) the breath test was repeated following suspension of the treatment on the preceding day. One boy and 4 girls aged 4 to 14 years who had no apparent disease served as controls. Sodium glycocholate($1-^{13}C$) with 99,3 atom per cent ^{13}C (obtained from Prochem BOC Ltd., London SW19 3UF, Great Britain, through Amersham Buchler GmbH & Co KG, D-3300 Braunschweig) was administered orally as substrate of the breath test to all persons after they had fasted for 12 - 14 hours. Two samples of expired air each were collected in evacuated tubes twice prior to the administration of the substrate as well as at 10 to 12 different times during the following 7 hours. The isotope ratio in the exhaled air was measured by mass spectrometry using a ^{13}C standard. Further details of the methods used have been published elsewhere [6].

We are highly indebted to Professor H.-L. Schmidt (Chair for General Chemistry and Biochemistry, Technical University of Munich, D-8050 Freising-Weihenstephan) for mass spectrometry in his institute.

RESULTS

The reproducibility of the ^{13}C-glycocholate breath test was moderate as was found in an 8-year-old boy in whom the test had been repeated within 6 months and who had normal test results (Fig. 1). The first breath test showed a larger variation of single values than the second did.

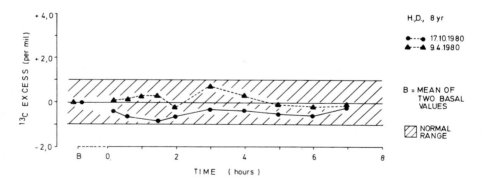

Fig. 1. Moderate reproducibility of normal ^{13}C-glycocholate breath test in a school-boy without metabolic disease.

A similar variation of the ^{13}C excess during a normal ^{13}C-glycocholate breath test was noted in some patients, among them a 4-year-old girl (Fig. 2).

In an 11-year-old boy with cystic fibrosis, the ^{13}C excess fell below the normal range at the end of the ^{13}C-glycocholate breath test (Fig. 3). The reason is still open to question. Dietary factors cannot be excluded.

In 3 female patients with cystic fibrosis, a pathologic result of the ^{13}C-glycocholate breath test was observed.

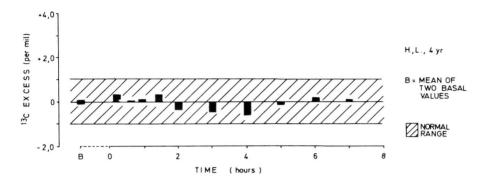

Fig. 2. Normal ^{13}C-glycocholate breath test in a preschool-girl without metabolic disease.

We attempted to elucidate whether there would be different results of ^{13}C-glycocholate breath tests in patients with cystic fibrosis who received symptomatic treatment or in whom this treatment had been suspended.

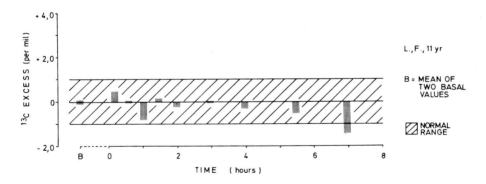

Fig. 3. ^{13}C-glycocholate breath test with final reduction of ^{13}C excess in a school-boy with cystic fibrosis.

In fact, a uniform tendency towards a rise or a decline of the ^{13}C excess during ^{13}C-glycocholate breath tests was not recognized in our patients after their symptomatic treatment had been suspended for one day. (With regard to the ^{13}C-glycocholate breath test, probably this period of time is too short as to create corresponding effects.)

In a 14-year-old girl who had cystic fibrosis and was treated with pancreatic enzymes, the ^{13}C excess during the ^{13}C-glycocholate breath test was normal with the exception of 1 value which was very slightly raised (Fig. 4). Following the suspension

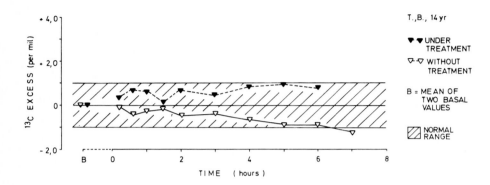

Fig. 4. Distinctly diverging results of ^{13}C-glycocholate breath tests before and after suspension of symptomatic treatment in a school-girl with cystic fibrosis.

of symptomatic treatment, the ^{13}C excess continuously declined until it was below the normal range at the end of the test.

Contrary to this finding, a 12-year-old boy with cystic fibrosis who was also treated with pancreatic enzymes had a ^{13}C excess within the lower part of the normal range whereas during the ^{13}C-glycocholate breath test performed after symptomatic treatment

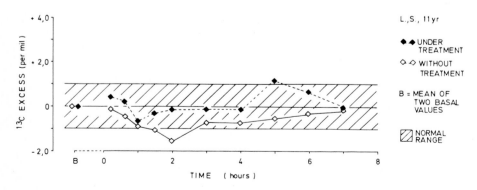

Fig. 5. Different results of ^{13}C-glycocholate breath tests before and after suspension of symptomatic treatment in a school-girl with cystic fibrosis.

had been suspended, the ^{13}C excess rose into the upper part of the normal range.

In a 11-year-old girl who suffered from cystic fibrosis and was treated with a mucolytic drug, the ^{13}C excess during the ^{13}C-glycocholate breath test varied markedly but, with the exception of 1 elevated value, kept within the normal range (Fig. 5). Subsequent to the suspension of symptomatic treatment, the ^{13}C excess in the corresponding ^{13}C-glycocholate breath test fell below the normal range within 2 hours but rapidly climbed reaching the normal range after another hour.

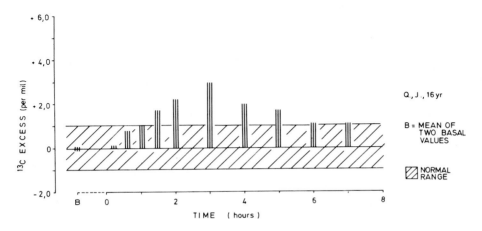

Fig. 6. Highly pathologic result of ^{13}C-glycocholate breath test in a female juvenile with cystic fibrosis.

A highly pathologic result of the ^{13}C-glycocholate breath test was noted in a 16-year-old female patient with cystic fibrosis complicated by chronic bronchitis (Fig. 6). The test result matched with the clinically recognizable severity of the disease and the increase of the integrated turnover of ^{13}C glycine in the total body of this patient. Her 19-year-old sister, however, who had a mild cystic fibrosis presented a normal ^{13}C-glycocholate breath test.

DISCUSSION

According to our studies the ^{13}C-glycocholate breath test may show an increased ^{13}C excess as a pathologic result in children and juveniles suffering from cystic fibrosis. Previous investigations in 8 male patients aged 11 to 24 years (with a mean age of 18 years) having cystic fibrosis with pancreatic and pulmonary involvement favoured the assumption that malabsorption of bile acids was only minimal; the cumulative $^{14}CO_2$ exhalation within 6 hours of a ^{14}C-glycocholate breath test performed at least 1 week after suspension of antibiotic treatment was even lower in these patients than in normal individuals [7]. The reason of the discordance with the results obtained in our patients who had a lower mean age and probably less severe pulmonary complication remains uncertain. The small number of patients and normal individuals in both series does not permit a statistic evaluation for comparison.

It cannot be discerned by ^{13}C-glycocholate breath tests alone whether a raised ^{13}C excess in the expired air is due to intestinal bacterial overgrowth or to intestinal malabsorption of bile acids [6]. Therefore it is open to question in the female juvenile patient with cystic fibrosis to which extent the elevated ^{13}C excess during the ^{13}C-glycocholate breath test reflected intestinal bacterial overgrowth. In this context it may be mentioned that in fasting healthy adults with normal gastric acid, the upper small intestine was virtually free from bacteria whereas the numbers of bacteria isolated in the lower small intestine rose as the terminal ileum was approached, the bacterial flora of which resembled that of the feces [8].

The evaluation of ^{13}C-glycocholate breath tests has to consider that the intermediary metabolism may influence $^{13}CO_2$ exhalation [9]. Distinct variations of the ^{13}C excess as were found in some persons examined by us might have biochemical reasons recognizable only by additional studies. The ^{13}C-glycocholate breath test is not specific for certain diseases. Its clinical value appears to lie chiefly in intraindividual follow-up studies provided that the tests are carried out under constant metabolic conditions.

REFERENCES

1 H. Fromm and A.F. Hofmann, Lancet, 2 (1971) 621-625.
2 H.P. Sherr, Y. Sasaki, A. Newman, J.G. Banwell, H.N. Wagner, Jr., and T.R. Hendrix, New Engl. J. Med., 285 (1971) 656-661.
3 N. Solomons, D. Schoeller, J. Wagonfeld, D. Ott, I. Rosenberg and P. Klein, Clin. Res., 23 (1975) 520 A.
4 N. Solomons, D.A. Schoeller, P.D. Klein, J. Wagonfeld, D. Ott, F. Viteri and I. Rosenberg, Clin. Res., 24 (1976) 291 A.
5 N.W. Solomons, D.A. Schoeller, J.B. Wagonfeld, D. Ott, I.H. Rosenberg and P.D. Klein, J. Lab. Clin. Med., 90 (1977) 431-439.
6 D. Glaubitt, K. Siafarikas and H.-L. Schmidt, in H.A.E. Schmidt, F. Wolf and J. Mahlstedt (Eds.), Nuklearmedizin, Nuklearmedizin im interdisziplinären Bezug, 18. Int. Annual Meeting of the Society of Nuclear Medicine - Europe, Nürnberg, September 9-12, 1980, F.K. Schattauer Verlag, Stuttgart, New York 1981, p. 796-800.
7 R.J. Roller and F. Kern, Jr., Gastroenterology, 72 (1977) 661-665.
8 B.S. Drasar, M. Shiner and G.M. McLeod, Gastroenterology, 56 (1969) 71-79.
9 D. Glaubitt, in E.R. Klein and P.D. Klein (Eds.), Proc. Second Int. Conf. on Stable Isotopes (CONF - 751027), Oak Brook, Ill./USA, October 20-23, 1975, National Technical Information Service, U.S. Department of Commerce, Springfield, VA, USA, 1976, p. 219-245.

BREATH TEST USING ^{13}C—PHENYLALANINE AS SUBSTRATE

D. GLAUBITT[1] and K. SIAFARIKAS[2]

Departments of Nuclear Medicine[1] and Pediatrics[2], Academic Teaching Hospital, D-4150 Krefeld (Federal Republic of Germany)

ABSTRACT

Breath tests after oral administration of DL-(1-^{13}C)phenylalanine were performed in 8 individuals without metabolic disease and in 2 children with cystic fibrosis under symptomatic treatment. In both children with cystic fibrosis the ^{13}C excess in the expired air was reduced during the first 2 hours of the test. The intraindividual reproducibility of the breath test with DL-(1-^{13}C)phenylalanine was unsatisfactory in 1 child who had no metabolic disease. The role of breath tests using ^{13}C phenylalanine, preferably the physiologically occurring ^{13}C L-phenylalanine, remains to be clarified in clinical medicine.

INTRODUCTION

Breath tests following intravenous injection of essential amino acids in 12 patients with cirrhosis of the liver disclosed that the cumulated exhalation of $^{14}CO_2$ was

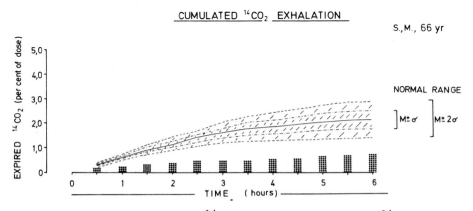

Fig. 1. Breath test using L-(U-^{14}C)phenylalanine: cumulated $^{14}CO_2$ exhalation in a 66-year-old man with cirrhosis of the liver.

reduced after L-(U-^{14}C)phenylalanine (Fig. 1) but normal after L-(methyl-^{14}C)methionine whereas the results were less uniform after L-(U-^{14}C)lysine monohydrochloride or L-(1-^{14}C)methionine [1]. Phenylalanine is an essential glucoplastic and ketoplastic amino acid and a precursor of tyrosine which is involved in the formation of thyroid hormones, adrenalin, melanin, fumarate, and acetoacetate.

We investigated to which extent adults or children without metabolic disease as well as children with cystic fibrosis had the capability of intestinal absorption and utilization of orally administered ^{13}C phenylalanine, with subsequent exhalation of ^{13}CO$_2$, and whether diagnostic information could be gained from this test. When these examinations were performed it was not possible to obtain sterile and pyrogen-free L-(1-^{13}C)phenylalanine so that DL-(1-^{13}C)phenylalanine was used.

PATIENTS AND METHODS

The study comprised one 52-year-old male volunteer without metabolic disease, 7 children (1 boy, 6 girls) aged 4 to 14 years who had no metabolic disease and, moreover, 1 boy aged 12 years and 1 girl aged 11 years who had cystic fibrosis with pancreatic and pulmonary involvement and were symptomatically treated with drugs. Informed consent had been obtained from the parents of the children. Breath tests were carried out with DL-(1-^{13}C)phenylalanine with 98 atom per cent ^{13}C (supplied by Prochem BOC Ltd., London SW19 3UF, Great Britain, through Amersham Buchler GmbH & Co KG, D-3300 Braunschweig). The substrate (3,0 mg/kg of body weight) was administered orally to the individuals who had fasted for 12 to 14 hours. Two samples of expired air each were taken twice prior to as well as 5, 10, 15, 30, 45, 60, 90, 120, 180, 240, 300, 360, and 420 minutes after the test had begun. Using mass spectrometry with a ^{13}C standard, the isotope ratio was determined in the exhaled air, the volatile ^{13}C compounds of which may be assumed to consist of carbon dioxide almost solely. General details of the methodology were reported elsewhere [2].

We are highly indebted to Professor H.-L. Schmidt (Chair for General Chemistry and Biochemistry, Technical University of Munich, D-8050 Freising-Weihenstephan) that mass spectrometry of the samples was performed in his institute.

RESULTS

All normal persons showed the peak of the ^{13}C excess in the expired air 30, 45, or 60 minutes after oral administration of DL-(1-^{13}C)phenylalanine.

In the 52-year-old male volunteer without metabolic disease, the ^{13}C excess during the breath test was found to fit into the normal range established for children (Fig. 2). Thus it might be assumed that the results of breath tests using DL-(1-^{13}C)phenylalanine share the same normal range for adults and children.

In 1 girl each aged 4 and 8 years who had no metabolic disease, the ^{13}C excess

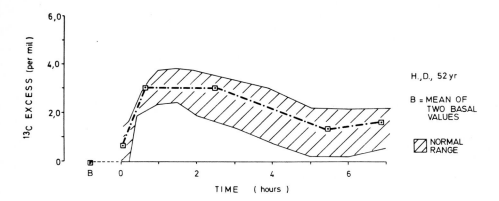

Fig. 2. Normal DL-(1-^{13}C)phenylalanine breath test in a man without metabolic disease.

in the exhaled air during the DL-(1-^{13}C)phenylalanine breath test demonstrated a marked variation within the normal range (Fig. 3).

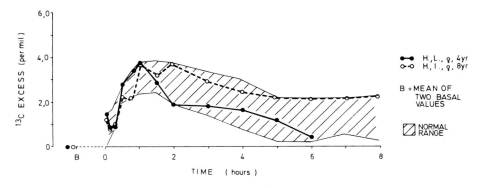

Fig. 3. Normal result of the DL-(1-^{13}C)phenylalanine breath test in a preschool-girl and in a school-girl without metabolic disease.

When in a 12-year-old girl the DL-(1-^{13}C)phenylalanine breath test was repeated within 3 months, the ^{13}C excess remained within the normal range in both tests but presented values with small deviation from each other only between 30 and 90 minutes after oral administration of the substrate (Fig. 4). Despite the relatively long time between the first and second test, the reproducibility of the DL-(1-^{13}C)phenylalanine test in this child had to be considered unsatisfactory. The reason of this finding is not known. Metabolic changes which might have been due at least to nutritional

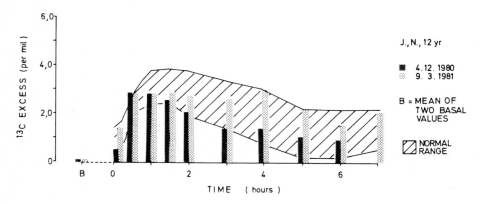

Fig. 4. Unsatisfactory reproducibility of the DL-(1-^{13}C)phenylalanine breath test in a school-girl without metabolic disease.

influences cannot be excluded.

Both children with cystic fibrosis showed a very small ^{13}C excess and a delayed occurrence of its peak during the DL-(1-^{13}C)phenylalanine breath test. It is open to question whether this finding reflected a slow turnover of the test substrate. A 12-year-old boy had a reduced ^{13}C excess in the expired air; the peak was observed not until 420 minutes (Fig. 5). In an 11-year-old girl suffering from cystic fibrosis

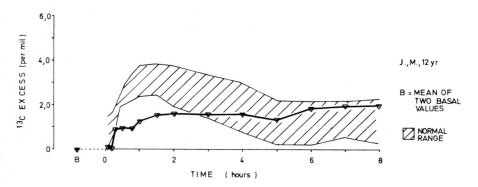

Fig. 5. Marked retardation of the increase of ^{13}C excess during the DL-(1-^{13}C)phenylalanine breath test in a school-boy with cystic fibrosis.

the ^{13}C excess reached only low values like those in the boy with the same disease but the peak occurred after 120 minutes (Fig. 6). Both children were under treatment

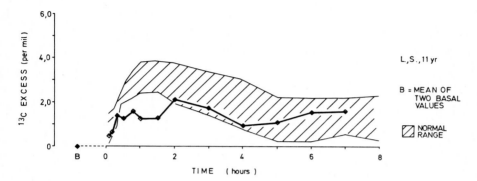

Fig. 6. Distinct diminution and delay of the rise of ^{13}C excess during the DL-(1-^{13}C)phenylalanine breath test in a school-girl with cystic fibrosis.

with pancreatic enzymes which had been interrupted in the morning when the DL-(1-^{13}C)phenylalanine breath test was performed.

DISCUSSION

Phenylalanine plays an important role among the essential amino acids involved in metabolism and nutrition. The daily nutritional requirement of single essential amino acids is highest for L-phenylalanine and - at the same level - for L-leucine [3,4]. L-phenylalanine has considerable meaning also in parenteral nutrition [5-7]. Patients with renal or hepatic diseases may present disorders of the metabolism of L-phenylalanine; if the plasma level of this amino acid is reduced in the patients, adequate substitution may be necessary [7].

Among the breath tests that are performed after oral administration of the substrate, the DL-(1-^{13}C)phenylalanine breath test has not been used to our knowledge, contrary to the well-known ^{13}C-glycocholate breath test which was applied also in patients with cystic fibrosis [8]. A breath test with ^{14}C labeled phenylalanine was reported; the effect of the load and route of administration of (1-^{14}C)phenylalanine was studied in normal persons and in patients with phenylketonuria in whom, according to the results after oral or intravenous injection of the substrate, a decreased oxidation of this amino acid was assumed [9]. It is not clear to which extent breath tests after oral administration of L-(1-^{13}C)phenylalanine would provide results discordant with those obtained in corresponding DL-(1-^{13}C)phenylalanine breath tests.

The evaluation of our results has to consider that breath tests carried out after oral administration of the substrate may be influenced by gastrointestinal absorption and subsequent utilization of the substrate or metabolites yielding carbon dioxide [10]. Thus many metabolic variables appear to be reflected in the result of the DL-(1-^{13}C)phenylalanine breath test. The biochemical pathways leading to pathologic findings in single patients cannot be disclosed without additional examinations being performed.

Our observations in 2 children with cystic fibrosis do not allow generalizing conclusions. Further breath tests, preferably with L-(1-^{13}C)phenylalanine, are mandatory to elucidate whether there are significant differences of the ^{13}C excess between healthy children and children suffering from cystic fibrosis, especially those with marked pulmonary or pancreatic involvement. It is not known whether in patients with cystic fibrosis the ^{13}C/^{12}C ratio of exhaled carbon dioxide may be affected by isotopic effects [11] subsequent to variations of the intermediary metabolism.

Breath tests using L-phenylalanine as a physiologic amino acid and ^{13}C as label of this substrate might have some value for follow-up studies in patients with cystic fibrosis, phenylketonuria and with renal or hepatic diseases, if the metabolic conditions do not change essentially during the period of observation.

REFERENCES

1 D. Glaubitt and E. Moebes, Congressus Quartus Societatis Radiologicae Europaeae, Hamburg, September 4-8, 1979, Abstract N 19.
2 D. Glaubitt, K. Siafarikas and H.-L. Schmidt, in H.A.E. Schmidt, F. Wolf and J. Mahlstedt (Eds.), Nuklearmedizin, Nuklearmedizin im interdisziplinären Bezug, 18. Int. Annual Meeting of the Society of Nuclear Medicine - Europe, Nürnberg, September 9-12, 1980, F.K. Schattauer Verlag, Stuttgart, New York, 1981, p. 796-800.
3 D.M. Hegstedt, Med. Ernährung, 11 (1970) 225-229.
4 W.C. Rose, Nutr. Abstr. Rev., 27 (1957) 631.
5 J.P. Striebel, Krankenhausarzt, 52 (1979) 678-692.
6 G. Kleinberger, W. Druml, A. Gaßner, H. Lochs and M. Pichler, Krankenhausarzt, 54 (1981) 20-32.
7 K. Pistor, Dtsch. Med. Wochenschr., 106 (1981) 1759-1760.
8 D. Glaubitt, H.-L. Schmidt and H. Krüger, in H. Hornbostel, G. Strohmeyer and E. Schmidt (Eds.), XVth International Congress of Internal Medicine, Hamburg, August 18-22, 1980, Excerpta Medica International Congress Series No 536, Excerpta Medica, Amsterdam, Oxford, Princeton, 1980, p. 87, Abstract C 22.2.
9 M.B. Fish, M. Pollycove, J. DeGrazia, P. Cohen and K. Fleury, J. Nucl. Med., 9 (1968) 317.
10 D. Glaubitt, in E.R. Klein and P.D. Klein (Eds.), Proc. Second Int. Conf. on Stable Isotopes (CONF - 751027), Oak Brook, Ill./USA, October 20-23, 1975, National Technical Information Service, U.S. Department of Commerce, Springfield, VA, USA, 1976, p. 219-245.
11 J. Duchesne, M. Lacroix and F. Mosora, in E.R. Klein and P.D. Klein (Eds.), Proc. Second Int. Conf. on Stable Isotopes (CONF - 751027), Oak Brook, Ill./USA, October 20-23, 1975, National Technical Information Service, U.S. Department of Commerce, Springfield, VA, USA, 1976, p. 55-60.

METABOLIC RATE OF ^{13}C LABELLED AMINOPYRINE APPLIED TO THE
EXPLORATION OF HEPATIC FUNCTIONAL ACTIVITY.

F. BOTTER, M. DRIFFORD, J. SUTTON
DPC/SCM, CEN.Saclay, 91191 GIF SUR YVETTE CEDEX
F. DEGOS, J.P. BENHAMOU
Hôpital Beaujon, 100, Bd. du Gal. Leclerc 92110 CLICHY

ABSTRACT

The ^{13}C labelled aminopyrine breath test has been carried out on 14 controls and 19 patients suffering from liver disorders. A critical analysis of the data indicates that a new parameter, the slope at the inflexion point of the cumulative aminopyrine metabolism curve, is a much better diagnostic criterion than those previously employed. A theoretical treatment based on a two pool model justifies the choice of this scoring parameter.

INTRODUCTION

The use of isotopically labelled 4,4'-dimethyl aminoantipyrine (aminopyrine or pyramidon) as a test for hepatic insufficiency is now well established even though the overall mechanism of its metabolism is not completely known [1,2,3]. The important step is the oxidative demethylation at position 4 giving labelled CO_2, and previous studies [5,6,7] using ^{14}C and ^{13}C labelled aminopyrine have already demonstrated that the progressive impairment of this process is accompanied by comparable decreases in both the peak concentration of labelled CO_2 expired and its cumulative production over a short time (2 or 3 hours).

The present work aims at validating a simple diagnostic test for hepatic microsomal malfunction using ^{13}C labelled aminopyrine. In view of the variations in the existing data the choice of score parameter was re-examined with a view to optimising the conditions for applying the test.

EXPERIMENTAL

^{13}C aminopyrine, 85-90 % enriched in the labile methyl groups was prepared by the "Service des Molécules Marquées, CEN.Saclay". It was administered orally or by injection, after a 12 hour fast in a dose of 2 mg/kg body weight, to a series of healthy subjects and others with known liver disorders. Breath samples were collected at various times after administration. The CO_2 was separated by a standard freezing technique and isotopically analysed using a V.G. Micromass model 602 C double collection mass spectrometer. The ^{13}C isotopic concentrations were expressed as the difference δ in parts per mil. from the $^{13}C/^{12}C$ ratio in the conventional PDB (Pee Dee Belemite limestone) standard.

Two groups of experiments were carried out on 33 subjects. The first group consisted of 34 tests in which the drug was administered orally, and 7 or 8 breath samples were collected over a period of 2 hours (31 experiments) or 8 hours (3 experiments). The second group consisted of 6 tests in which labelled aminopyrine was administered rapidly by intravenous injection : in these cases 10 samples were collected from each subject over a period of 8 hours.

Four types of subject were included : 14 healthy subjects supposedly free from liver disease. 19 cases with histologically proven hepatic disease of whom 7 were cirrhotic ; 4 were treated before the test with phenobarbital to increase the hepatic function.

RESULTS

The δ values of $^{13}CO_2$ for typical subjects are plotted against time in Fig. 1. The typical curve for a healthy subject shows a period of induction followed by a more or less sharp rise to a peak, then an approximate exponential decay. Both the height and appearance-time of the peak are poorly reproducible even in healthy subjects. In the case of severe liver diseases the curves are considerably flattened and often without trace of a maximum. Thus, although the height of the peak is related to the hepatic state, the dispersion of the data is too great for the simple use of this parameter as a valid diagnostic criterion, and the appearance time of the peak is even less representative.

Continuing the search for a valid criterion, the cumulative quantity of aminopyrine metabolized expressed as a percentage of the initial dose was calculated as a function of time for the different subjects.

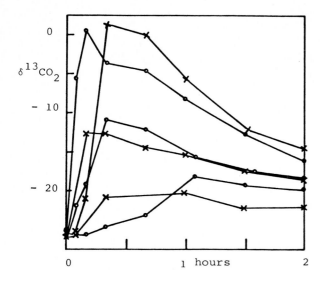

Fig. 1. Typical $^{13}CO_2$ breath test data. $\delta^{13}CO_2$ versus time after injection. O : Controls, X : Subjects with liver disease.

Typical curves, of sigmoid form which flatten out well below the 100 % metabolisation level, are shown in Fig. 2. The asymptotic limiting values are in general much lower in the case of a liver disease. These plots enable, the cumulative fraction metabolized after a given time (2 hours), to be determined and compared. Whereas both the total fraction of the initial dose, and the fraction metabolized after a given time gave some degree of diagnostic correlation it was found that the slope of the integrated curves measured at the inflexion point, (i.e. the maximum slope), gave a much better correlation. The scatter diagrams, Fig. 3, show the distribution of the controls and the patients as a function of this slope m expressed as a percentage of the initial aminopyrine dose metabolized per hour and of $q(t)$ the cumulative fraction metabolized after 2 hours. The statistical distribution of m gives for the controls 9.0 ± 2.5 %/h and, for the patients, 2.2 ± 2.0 %/h.

The sample of subjects considered here may be grouped into those for whom $m > 9.0 - 2.5 = 6.5$ which included most of the controls but also 2 with liver malfunction, and those for whom $m < 6.5$ which includes all the other patients plus some controls. Thus the value of $m \simeq 6.5$ %/h appears to define a critical level which differentiates reasonably well the sick from the healthy and so provides a diagnostic aid. However it seems clear that when the results fall inside the zone

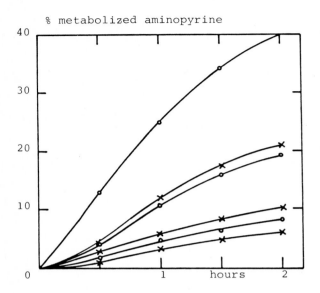

Fig. 2. Cumulative curves of percentage aminopyrine metabolized as a function of time ⊙ : controls, X : subjects with liver disease.

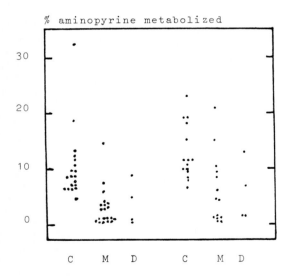

Fig. 3. Scatter diagrams using as score parameters a) the slope of the cumulative curves at the inflexion points - left hand side, b) the 2 hour cumulative percentage of metabolized aminopyrine - right hand side: C = controls, M = patients, D = patients treated with phenobarbital.

m = 6.5 ± 0.5 %/h other independent tests will be necessary in order to establish the true hepatic state of the patient.

Mathematical model. Since the only correlation studied here is that of m with the clinical state of the patient it was decided to examine the use of this parameter in the light of a mathematical two pool model shown in Fig. 4.

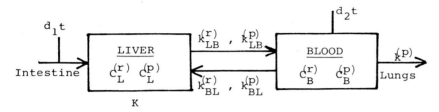

Fig. 4. The two pool model.

$c_L^{(r)}$ and $c_L^{(P)}$ are the concentrations of reagent (aminopyrine) and product (CO_2) in the liver cells and $c_B^{(r)}$ and $c_B^{(P)}$ are the corresponding values in the blood pool. The rates of transfer from one pool to the other are represented by k values with appropriate sub- and superscripts. $k^{(P)}$ represents the rate of transfer of $^{13}CO_2$ to the pulmonary air in which the partial pressure p(t) is related to $c_B^{(P)}$ by Henry's law. $d_1 t$, and $d_2 t$ represent the rates of transfer of aminopyrine to the 2 pool system following oral administration and intravenous injection respectively. The rate of enzymatic formation of $^{13}CO_2$ from aminopyrine is represented by K. It is considered to be a first order process as are all the transfer processes implicated in the model.

On the basis of this model five equations may be formulated of which four represent the mass balance of the two important constituents, reagent and product, in the two pools whilst the fifth represents the equilibrium between the blood pool CO_2 concentration and its partial pressure, p(t), in the pulmonary gas.

$$\frac{dc_L^{(r)}}{dt} = - K\, c_L^{(r)} - k_{LB}^{(r)} \cdot c_L^{(r)} + \frac{d_1(t)}{V_L} + k_{BL}^{(r)} \cdot c_B^{(r)}$$

$$\frac{dc_L^{(P)}}{dt} = 2 K\, c_L^{(r)} + k_{BL}^{(P)}\, c_B^{(P)} - k_{LB}^{(P)}\, c_L^{(P)}$$

$$\frac{dC_B^{(r)}}{dt} = \frac{d_2(t)}{V_B} + k_{LB}^{(r)} C_L^{(r)} - k_{BL}^{(r)} C_B^{(r)}$$

$$\frac{dC_B^{(P)}}{dt} = k_{LB}^{(P)} C_L^{(P)} - k_{BL}^{(P)} C_B^{(P)} - k^{(P)} C_B^{(P)}$$

$$p(t) = \alpha C_B^{(P)}$$

The resolution of this system of equations in order to obtain the variation of $p(t)$ as a function of time may be obtained using convolu-convolution algebraic methods [8] but in the present context an analytical approach has been adopted introducing the following symplifying assumptions : 1) the rates of transfer of aminopyrine and $^{13}CO_2$ between the two pools are much greater than the rate K of the metabolism of aminopyrine. 2) the rate of transfer of CO_2 from the blood pool to the pulmonary air is much greater than the rate of transfer of CO_2 between the liquid pools.

This treatment leads to a general expression for $p(t)$ given by

$$p(t) = \alpha K k_{LB}^{(P)} k_{BL}^{(r)} \cdot C_B^r t^3 \cdot \frac{e^{-k_{LB}^{(P)} \cdot t}}{3}$$

where α is the Henry's law constant.

The plot of $p(t)$ versus time passes through a maximum value given by

$$p_{max} = \alpha K \frac{k_{BL}^{(r)}}{(k_{LB}^{(P)})^2} \cdot 9 \, e^{-3}$$

at a time $t_{max} = 3/k_{LB}^{(P)}$, the values of these two parameters varying from one subject to another even though $\alpha k_{BL}^{(r)}$ and $k_{LB}^{(P)}$ should remain relatively constant. Thus it is evident that the lower the rate of metabolism, K, the flatter the curve will be.

The curves giving the fractional conversion of aminopyrine $q(t)$ as a function of time are related to the $p(t)$ - time curves by

$$q(t) = A \, d \int_0^t p(t) dt$$

where A combines the numerical multiplying constants, d is the pulmonary CO_2 output rate for the individual.

These curves, of sigmoid form, correspond to those found from the experimental data and which are plotted in **Fig. 3**. The slope at the inflexion point is given by

$$m = 2\, A\alpha dK \cdot \frac{k_{BL}^{(r)}}{(k_{LB}^{(P)})^2} \cdot 9\, e^{-3}$$

Throughout these considerations, administration of a constant unit dose of antipyrine has been assumed : if a different value is used then this factor will appear in the denominator of the expression for $q(t)$.

Thus the model though simple, justifies the empirical finding that the significant parameter for breath test diagnosis is the maximum rate of metabolization of aminopyrine as measured by the slope at the inflexion point of the integrated $^{13}CO_2$ curve. It shows moreover that this parameter depends not only on the metabolic rate of aminopyrine but also on the transfer rate of the drug from the blood to the liver pool, the transfer rate of $^{13}CO_2$ from the liver to the bloodpool and on the pulmonary output of CO_2. We have assumed in this model that this last quantity depends on a direct exchange of CO_2 between the bloodpool and the pulmonary air. In fact it involves exchange via the blood bicarbonate pool which a more sophisticated model would take into account.

We express our gratitude to the "Direction Générale à la Recherche Scientifique et Technique" for a grant to aid this work.

REFERENCES

1 A.F. Hofmann and B.H. Lauterburg, J. Lab. Clin. Med. 90, 1977, 405-411.
2 B.B. Brodie and J. Axelrod, J. Pharmacol. Exp. Ther 99, 1970, 171-184.
3 R. Platzer, R.L. Galeazzi, G. Karlaganis and J. Bircher, Europ. J. Clin. Pharmacol. 14, 1978, 293-299.
4 G.W. Hepner and E.S. Vesell, N. Engl. J. Med. 291, 1974, 1384-1388.
5 J.F. Schneider, D.A. Schoeller, B. Nemchausky, J.L. Boyer and P. Klein, Clin. Chim. Acta. 84, 1978, 153-162.
6 G.W. Hepner and E.S. Vesell, Clin. Pharm. Exp. Ther. 20, 1976, 654-660.
7 J. Birch r, A. Kupfer, I. Gikalov and R. Preisig, Clin. Pharm. Ther. 20, 1976, 484-492.
8 M. Drifford, P. Loison and T. Robin, to be published.

COMPARISON OF NATURALLY AND ARTIFICIALLY LABELLED GLUCOSE UTILIZATION TO STUDY GLUCOSE OXIDATION BY MEANS OF $^{13}C/^{12}C$ BREATH TEST.

M. LACROIX, N. PALLIKARAKIS and F. MOSORA
Institute of Physics, University of Liège, Sart-Tilman, B-4000 Liège, (Belgium).

ABSTRACT.

An isotope effect which occurs in photosynthesis, enriches most plants in ^{12}C but this effect is less pronounced for maize, sugar cane and other C4 plants. As a result maize glucose is relatively enriched in ^{13}C in comparison with common foodstuffs, so can be used as a natural tracer for glucose metabolism studies. This enrichment is nevertheless low ($\sim 10^{-4}$) and near the limit of the sensitivity of the method (double collector mass spectrometry). As a possible way of increasing the limits of sensitivity, mixtures of natural and artificially labelled ^{13}C-glucose were studied. It was thus hoped to calculate the amount of labelled substance necessary to achieve a predetermined increase in sensitivity. As an example of this, glucose consumption during muscular exercise, using both natural and artificial ^{13}C-glucose, was studied.

INTRODUCTION.

The measurement of the $^{13}C/^{12}C$ ratio of exhaled CO_2, resulting from the oxidation of naturally labelled substances, remains very appropriate as a method for breath tests (1-7) when using substances such as maize glucose or alcohol which are rapidly metabolized. However insufficient sensitivity results when large dilutions of these substances occur within the body. Furthermore, even in some studies performed with maize glucose, as in the cases of obeses and of obeses with chemical diabetes etc., the small differences obtained for the $^{13}C/^{12}C$ ratio of exhaled CO_2, in comparison with a normal man, require an increase of the sensitivity of the method in view of obtaining conclusive calculations for exogenous glucose oxidation. It is also evident that this method of investigation, which is very useful in the important field of glucose metabolism, is only applicable in the regions where corn is not utilized as common food, such as in Europe, because the $^{13}C/^{12}C$ ratio of naturally labelled maize glucose must remain higher that that of common food to obtain sufficient sensitivity.

However, the utilization of artificially labelled substances, in spite of giving a better sensitivity, is an expensive method which allows only a limited number of experiments.

In order to study this problem, we present here a possible alternative methodology by using a mixture of naturally and artificially labelled substances. In our examples we combined the maize glucose naturally enriched in ^{13}C with about 1.10% of ^{13}C atoms, mixed with a glucose uniformly labelled with 79% of ^{13}C atoms.

MATERIALS AND METHODS.

To estimate the very small differences in the isotope ratio of the samples (the difference between the ^{13}C content of common food and maize glucose is only 0.02%), we used the method of double collector mass spectrometry which compares the ratio of peak heights of mass 45 ($^{13}C\ ^{16}O\ ^{16}O$) and mass 44 ($^{12}C\ ^{16}O\ ^{16}O$) of the sample studied with that of a CO_2 standard. The results are expressed as $\delta^{13}C$ (7,8). One unit of $\delta^{13}C$ corresponds to a change of about 1.10^{-5} in the ratio $^{13}C/^{12}C$ and the accuracy of a measurement is ± 0.2 units of $\delta^{13}C$. The method of collection of CO_2 expired or of CO_2 resulting from glucose combustion, is described with details in previous papers (1,2,3).

The quantities of the various glucoses to be mixed in order to obtain a certain $\delta^{13}C$ of CO_2 can be easily calculated from their values and the self-evident formulas *. In the range of $\delta^{13}C$ of common food, we use potato glucose (1.0679 atom % ^{13}C, $\delta^{13}C=3.7$); as naturally enriched glucose, we take maize glucose (1.1008 atom % ^{13}C, $\delta^{13}C=18.0$) and as labelled glucose, ^{13}C-U-glucose from

---
*We utilised the equality of 13C atom weight conservation law:
$$\frac{6.13}{MW} \cdot (X_1 A_1 + X_2 A_2) = \frac{6.13}{MW} \cdot X_3 A_3$$
where A_1, A_2, A_3 are respectively, the ^{13}C isotope abundance of artificially labelled glucose, of naturally labelled glucose and of their mixture; X_1, X_2, X_3 are respectively, their weights; MW is the molecular weight of glucose. A_3 is obtained from the $\delta^{13}C$ formulation by means of the following evident relation.
$(1 + 10^{-3} \cdot \delta^{13}C) \cdot (^{13}C/^{12}C)$ standard = $(^{13}C/^{12}C)$ desired = $A_3/1-A_3$.

Stohler Isotopes (79 atom % ^{13}C, $\delta^{13}C$ =343000). The standard used here has a $\delta^{13}C$ of 0.8 relatively to NBS 21 (8). The uniformity of labelling of the Stohler ^{13}C-glucose was verified by ^{13}C-Magnetic Resonance and its ^{13}C concentration measured by simple collector mass spectrometry.

The human subjects were 21-25 years old and were of normal weight (2), none having a family history of diabetes or post-prandial glycosuria. These volunteers fasted overnight, received different quantities of glucose and were tested during a 7 hour period at rest, or during a 2 or 3 hour period of physical exercise. The physical exercise was performed on a bicycle ergometer (Jaeger), with a work load of between 80 and 110 watts, to obtain a approximate oxygen consumption of about half of the individual maximum aerobic power.

In the case of physical exercise, simultaneous measurements of the $^{13}C/^{12}C$ ratio and of the volume of exhaled CO_2 (VCO_2), permitted a quantitative evaluation of exogenous glucose consumption to be obtained (3,4). In addition, the respiratory quotient (RQ) was obtained by calculating the ratio between the expired CO_2 and the simultaneous O_2 utilization.

RESULTS.

We can observe in Fig. 1 that the $\delta^{13}C$ of the expired CO_2 has the same evolution profile if a normal subject at rest received firstly, 100 g of maize glucose alone ($\delta^{13}C$ = 18.4) and secondly, 100 g of maize glucose mixed with 46 mg of artificially labelled glucose. The $\delta^{13}C$ of this mixture, obtained after its dilution in water, lyophilisation and combustion of an aliquot, is 52.00 $\pm$ 0.2 (calculated value : 51.63). The maximum of curve A, obtained after the ingestion of maize glucose alone, peaks at 4 ‰, relative to the basal value of the subject. Taking into account that biological isotope fluctuations can reach 1 ‰ (5,6), the sensitivity of the procedure is low. The curve B, obtained after glucose mixture ingestion, presents a rise of 16 ‰ and the relative sensitivity is thus greatly increased. It is evident that in this latter case, the utilization of exogenous glucose is not complete after 7 hours, a hypothesis which was not conclusive after the ingestion of maize glucose alone.

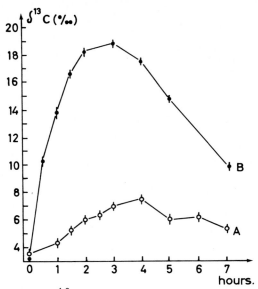

Fig. 1. Evolution of $\delta^{13}C$ of expired CO_2 in the case of one subject at rest who received two separate oral glucose loads : (A) 100 g of maize glucose alone of $\delta^{13}C = 18.4$; (B) a mixture of $\delta^{13}C = 52.0$ obtained from 100 g of maize glucose and from 46 mg of artificially labelled glucose (UL - 79 at %)

Significant results are obtained in Fig. 2, in the case of two subjects who performed physical exercise for 2 hours. After 15 minutes of adaptation to exercise, at time zero, they received firstly, 100 g of maize glucose ($\delta^{13}C = 18.0$) and secondly, 100 g of maize glucose mixed with 47 mg of artificially labelled glucose. The $\delta^{13}C$ of this mixture obtained after its combustion is 52.03 ± 0.08 (calculated value : 51.95).

During the physical exercise, we obtained about the same gain in sensitivity for ^{13}C as in the case of the subject at rest. Moreover, the consumptions of exogenous maize glucose (calculated from $\delta^{13}C$ and VCO_2) and of total glucose (calculated from RQ) remain similar in the two experiments. The exogenous glucose oxidations during the physical exercise are 23 g and about 32 g for the two subjects respectively, for the glucose mixture as well as for the maize glucose experiments. Both the subjects have 118 g as a value for their total glucose oxidation after the glucose mixture ingestion and have 105 and 99 g when maize glucose was taken alone. These results account for the independence of this method relative to the isotope content of the glucose used (no measurable effect of isotope dilution in the natural glucose pool).

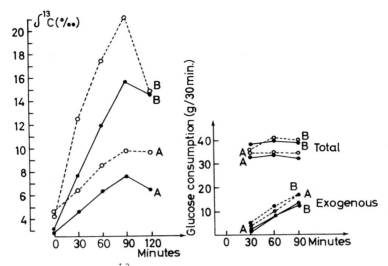

Fig. 2. Evolution of $\delta^{13}C$ of expired CO_2 (left) and consumption of total and exogenous glucose (right) during the physical exercise of two subjetcs (———, - - - - - -) who received each two separate oral glucose loads : (A) 100 g of maize glucose alone of $\delta^{13}C = 18.0$; (B) a mixture of $\delta^{13}C = 52.0$ prepared with 100 g of maize glucose and with 47 mg of artificially labelled glucose (UL-79 at %).

In an other experiment, we followed the carbohydrate consumption (exogenous and total) in the case of 4 subjects who performed physical exercise and who received, after 15 minutes of adaptation to exercise, at time zero, 19 mg of ^{13}C uniformly labelled glucose of $\delta^{13}C = 343.10^3$ (79 at %), with no other load. The amount of artificial label was chosen to give a $\delta^{13}C$ of expired CO_2 situated in the same range as that obtained in our previous studies (1,2,3,4). The results are represented in Fig. 3.

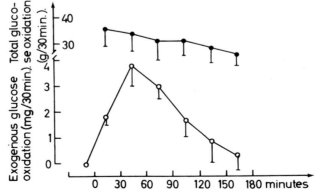

Fig. 3. Carbohydrate metabolism during exercise after ingestion of 19 mg of ^{13}C uniformly labelled glucose of $\delta^{13}C = 340.10^3$ (79 at %). Means and standard deviations for four subjects are shown.

We can observe that the shape of the curve is the same whether the labelled glucose is associated with a load (4) or not. Total carbohydrate oxidation decreased gradually during the exercise in relation to the absence of an external glucose supply and in contradiction to the experiments of Fig. 2., where this parameter remains constant. The total glucose oxidation in the 3 hours of this muscular exercise is 186.35 ± 40.35 g and the oxidation of exogenous glucose is 12.3 ± 0.7 mg from the 19 mg administrated. Thus we can state the recovery of 2 thirds of the initial amount of exogenous glucose. This last result concerned with the trace amount of glucose ingested, can give some information about the glucose pool and in particular about the importance of the quantity of the glucose (one third) that does not contribute to the energy expenditure, for instance, by retro-conversion into storage, lactate formation, etc.

ACKNOWLEDGEMENTS.

We wish to express thanks to F. Pirnay M.D. for his useful discussion and his assistance concerning the VCO_2 and RQ measurements.

REFERENCES.

1. M. Lacroix, F. Mosora, M. Pontus, P. Lefebvre, A. Luyckx and G. Lopez-Habib, Science, 181 (1973) 445-446.
2. F. Mosora, P. Lefebvre, M. Lacroix, J. Duchesne, A. Luyckx and F. Pirnay, Metabolism, 25 (1976) 1575-1582.
3. F. Mosora, P. Lefebvre, F. Pirnay, M. Lacroix, A. Luyckx and J. Duchesne, in A.P. De Leenheer and R.R. Roncucci (Eds.), Proc. 1st Int. Symp. Quantitative Mass Spectrometry in Life Science, Ghent, June 16-18, 1976, Elsevier, Amsterdam,1977, pp 229-236.
4. F. Pirnay, M. Lacroix, F. Mosora, A. Luyckx, P. Lefebvre, and J. Duchesne, Appl. Physiol. (U.S.A.) 43 (1977) 258-261.
5. M. Lacroix and F. Mosora (Mémoire Commandé, IAEA-SM 191/29), in Proceed. of a Symp. on Isotope Ratios as Pollutant Source and Behaviour Indicators, Vienna, 18-22 Nov. 1974, IAEA FAO, Vienna pp 343-358.
6. F. Mosora, IRE, 2 (1977) 2-15.
7. J. Duchesne, M. Lacroix, F. Mosora in H.-L. Schmidt, M. Förstel and K. Heinzinger (Eds.),Proc. 4th Int. Symp. on Stable Isotopes, Jülich, March 23-27, 1981, Elsevier, Amsterdam.
8. H. Craig, Geochimica Cosmochimica Acta 12 (1957) 133-149.

USE OF THE $^{13}C/^{12}C$ BREATH TEST TO STUDY SUGAR METABOLISM IN ANIMALS AND MEN

J. DUCHESNE, M. LACROIX and F. MOSORA
Institute of Physics, University of Liège, Sart-Tilman-B-4000 Liège (Belgium).

ABSTRACT.

As a consequence of their particular photosynthetic pathway some plants, such as maize and sugar cane, are slightly richer in ^{13}C than most common foodstuffs. This small but significant enrichment (0.02%) allows the use of maize products and especially of maize glucose, as natural non radioactive tracers in metabolism research by measuring the $^{13}C/^{12}C$ ratio of the CO_2 exhaled. This provides a convenient method to quantitatively presume the metabolism of an oral glucose load, in rats and in men.

By the method the human glucose metabolism was evaluated in normal subjects : at rest in relation to the size of the loading dose, during a prolonged physical exercise and during a recovery period after this effort, as a function of different intensities of muscular exercise, and after a dose fractionated in 4 hourly loads during the muscular exercise. The pathological glucose metabolism in men was evaluated in insulino-dependent diabetics, insulin-injected diabetic subjects during a prolonged muscular exercise, obeses without and with diabetes and on the drug influence on diabetic subjects.

The potential of this method is very great : for instance, it is appropriate in the study of glucose metabolism in pregnant women and children and it is also valid for other naturally or artificially labelled substances used in CO_2 breath tests.

INTRODUCTION.

Measuring the $^{13}C/^{12}C$ ratio in expired CO_2, following administration of carbon labelled sugars, can provide useful information about the metabolism of these substances in animals and men. However, artificial enrichment is not always necessary. Because of their particular photosynthetic pathway, some plants, such as maize and sugar cane, are slightly richer in ^{13}C than are most common foodstuffs. This small but significant enrichment (0.02%) allows the use of maize derivatives and especially of maize glucose as natural and nonradioactive tracers by simply measuring, by double collector mass spectrometry, variations of the $^{13}C/^{12}C$ ratio of the CO_2 exhaled.

This method, convenient for following the complete conversion of an oral glucose load into CO_2, was developed by our group in 1971 (1,2). First applied to the rat, this method giving qualitative information (1-5) was then extended to man (6) under various experimental or clinical conditions (7,8) and in a quantitative form (9-21).

METHODS.

1. <u>Subjects, collection of CO_2</u>

Volunteers aged 21-25 years and of normal body weight without glycosuria or diabetes precedents, after having fasted overnight, received 100 g of maize glucose in about 400 ml of water. They were tested during a seven hour period at rest or during a four hour period of physical exercise. This consisted of walking on a treadmill with a ten percent upward slope at a speed of about four and a half km/h, which corresponds to a rate of metabolism of about half the maximum individual oxygen consumption.

During both tests, the air breathed was sampled to determine the $^{12}C/^{13}C$ ratio and to measure the amount of exhaled CO_2. Samples of expired air were collected into one-liter rubber balloons, two samples at time zero (ingestion of glucose), one every hour during the next seven hours at rest and one every half-hour during the next four hours of physical exercise. CO_2 and water were immediately separated from the air by being trapped in liquid nitrogen using a vacuum pump. It is also possible to use trapping by barium hydroxide, but special care must be taken to trap the total CO_2 in order to avoid isotope fractionation prevailing in this chemical reaction. We obtain good results by strictly controlling the flow rate of the gas and the saturation of the hydroxide solution. The carbonate is easily reconverted into CO_2 by means of phosphoric acid.

For rats, a CO_2-free mixture of oxygen (20%) and nitrogen (80%) passes through a 2 litre air-tight metabolism cage, and the CO_2 is immediately trapped by liquid nitrogen (flow rate of about 0.5 litre/min).

2. Determination of the isotopic ratio.

The CO_2 is expanded into the mass spectrometer while the water remains trapped in a mixture of methanol and dry ice (-70°C). The measurements were made on a double collector mass spectrometer (MAT-CH5) and the results expressed in per mil as $\delta^{13}C$:

$$\delta^{13}C = 10^3 \cdot \left[(^{13}C/^{12}C) \text{ sample} / (^{13}C/^{12}C) \text{ standard} - 1 \right]$$

We referred our results to the standard NBS 21 so as to have positive values. They can be related to the PDB scale by subtracting 28‰. The precision of the determination is about 0.2‰.

3. Determination of the amount of exhaled CO_2.

To determine the amount of exhaled CO_2 in men, each half-hour, after two minutes of adaptation, the subjects breathed through a double low resistance valve for a period of 5 minutes and the total amount of exhaled air was collected in a Douglas bag and measured by a Tissot gasometer. Once the water vapor had been trapped, an aliquot of this collected air was analyzed for its CO_2 content in a respiration mass spectrometer (quadrupole Centronic or multiple collectors MAT-M3) which has been calibrated before each sample determination with two air standards of accurately known composition (different CO_2 contents). Determinations were also made by using infrared absorption of CO_2, combined to a gas flowmeter.

4. Calculation of glucose metabolized.

The exogenous glucose consumed during each period (1.0 or 0.5 hour) following the ingestion of the load was calculated from the total amount of CO_2 expired and its $\delta^{13}C$ value, the latter lying between the value of the expired CO_2 before the glucose load (about 4‰) and the $\delta^{13}C$ value of the glucose itself (about 18‰). Thus, if the $\delta^{13}C$ of the CO_2 in the expired air is x‰, the contribution (%) by the exogenous glucose is given by $(x-4)/(18-4).100$. From this value and the total amount of CO_2 produced, the metabolized glucose can be easily calculated.

RESULTS.

1. Natural variations of carbon isotopes in expired CO_2.

Before giving results with maize glucose, we will first recall some previous observations, obtained in rats, concerning the natural isotope variations in CO_2 from metabolism. They are mainly due to endogenous carbon sources, however a 5 day period of starvation eliminates these effects of stored food (5-14).

A small decrease in the $\delta^{13}C$ value of the expired CO_2 observed in the starvation period comes from the oxidation of lipids. By the use of hormones (3,14) we can enhance this effect, especially in pancreatectomized rats where the hormonal regulation is not possible. A secondary increase after the 3rd day expresses a protein contribution after the fat elimination. An interesting result was observed when we compared old and young rats during the starvation period. The old rats (3 years) showed no lipid contribution to their CO_2 but a major protein contribution (4,14).

2. Maize glucose as a tracer to study glucose oxidation.

Maize glucose is advantageous for the control of glucose metabolism in normal and sick men. An application is shown in Fig. 1 : on the right side we have the blood glucose level of normal, obese

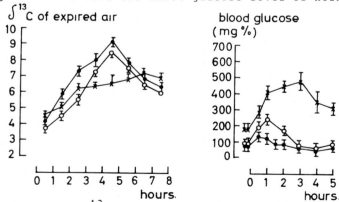

Fig. 1. Variation of $\delta^{13}C$ of the expired air (left) and of blood glucose level (right) after oral administration of 100 g maize glucose (8) : ● in 11 normal weight controls and in 14 obese subjects with normal glucose tolerance, O in 8 obese patients with chemical diabetes, X in 5 insulin dependent diabetics.

and diabetic subjects after a glucose load, at the left the corresponding $\delta^{13}C$-values, which are less distinct, indicating that there may interfere storage processes in the utilization. We must keep in mind in the case of a diabetic patient, a dilution of the exogenous glucose by the circulating glucose pool will occur which can, in part, explain the lowering of the $\delta^{13}C$ curve. As the $\delta^{13}C$ CO_2 curve reflects the final fate of the applied glucose it will differ from the common clinical measurements.

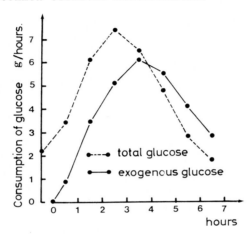

Fig. 2. Consumption of total glucose (calculated from the respiratory quotient -RQ- values) and of exogenous maize glucose (calculated from $\delta^{13}C$ and V_{CO_2} values of expired CO_2) after oral administration of 100 g glucose in eleven subjects (normal weight controls and obeses with chemical diabetes)(8). The points represent the per hour utilization of glucose. The total consumption during a 7 hour period gives 33.1 ± 15.2 g for RQ and 28.4 ± 4.8 g for $\delta^{13}C$. The curves represent the mean for the eleven subjects; the individual variations around the mean curve are three times more important for the RQ than for the $\delta^{13}C$ curve.

In Fig. 2. a comparison between the total glucose consumption (calculated from the RQ values) and the exogenous glucose (calculated from the $\delta^{13}C$ and V_{CO_2} values) is made. The crossing of the curve can indicate the contribution of other processes, such as interconversions, between carbohydrates and lipids, which affect the RQ but are not distinguished by the $\delta^{13}C$ which follows only the fate of the carbon independently of the pathway. It can also originate from larger imprecision of RQ measurements at rest.

If the blood glucose level is not affected by the amount of the load, glucose concentration and utilization are proportional. The same observation is obtained in rats. Therefore it appears that the $\delta^{13}C$ method provides original results which are complementary to the classical ones : respiratory quotient or glycemia measurements.

3. The consumption of exogenous glucose during muscular exercise.

It is known that the effect of a glucose load on glycemia is observed after one hour and that exercise lowers the glucose blood level. Quantitative effects of an ingestion of glucose during exercise have often been discussed but only recently some information has been obtained, by means of ^{14}C-labelled compounds, about the use of this glucose by the organism. Our own results, obtained with maize glucose, give important information about this problem.

Determinations of $\delta^{13}C$, V_{O_2} and V_{CO_2} made possible the evaluation of the carbohydrate metabolism during periods of exercise (11), while for subjects at rest after a load of 100 g glucose, the amount oxidized to CO_2 averaged 29.6 $\pm$ 1.0 g (mean $\pm$ SEM) in 7 h, during physical exercise, almost the total load given was metabolized after 4 h, i.e. 94.8 $\pm$ 1.6 g (mean $\pm$ SEM) (10). The utilization of the exogenous glucose in physical exercise began as early as within the first ten minutes and reached a maximum rate of 40 g/h between the first and second hours, whereas at rest this maximum rate was only 6 g/h and occurred between the third and fourth hours (10).

The quantity of exogenous glucose consumed during muscular exercise was a function of the intensity of effort, and it increased from 0.3 to 0.7 g/min, when the consumption of O_2 varied from 951 to 2 204 ml/min. In more intensive muscular exercises no supplementary utilization of exogenous glucose was observed. However, during prolonged muscular exercise, orally administered glucose was readily oxidized. Fractionnation of the 100 g load into 4 loads of 25 g resulted in a more stable contribution of the exogenous glucose until about 12-15 % of the total energy requirement. However, the exogenous glucose oxidation was reduced by 29 % during recovery after a prolonged muscular exercise, and this findings are in agreement with the concept that the reconstitution of carbohydrate reserves represents a notable share of the glucose metabolism in this condition.

4. Glucose utilization during exercise in normal and diabetic subjects.

As we are living in a period where sports play an important part in everyday life, we decided to investigate with our method the effects of muscular exercise on insulin-dependent diabetics.

TABLE 1.

Total and exogenous glucose oxidation during prolonged muscular exercise in normal and diabetic subjects.

	Total glucose oxidation	Exogenous glucose oxidation
Normal subjects (at rest)	32 ± 2	16.8 ± 0.6
Diabetic subjects (at rest with intravenous insulin)	32 ± 4	----
Normal subjects (in muscular exercise)	242 ± 18	95 ± 2
Diabetic subjects (in muscular exercise with intravenous insulin)	238 ± 19 p <0.005	84 ± 8 p< 0.025
Diabetic subjects (in muscular exercise without intravenous insulin)	177 ± 14	42 ± 11

The total carbohydrate oxidation was determined from the non protein RQ and the exogenous glucose oxidation after a load of 100 g "naturally labelled ^{13}C-glucose" by $\delta^{13}C$-value measurements in the expired CO_2. The experiments were performed at rest and during prolonged muscular exercise (45% V_{O_2} max for 4 hours) in normal volunteers and in juvenile-type diabetics with or without intravenous insulin. The results are presented in Table 1. Thus, total carbohydrate oxidation is similar in normal and insulin-infused diabetics at rest, and it is equally increased during exercise in both, however it is markedly less increased, when insulin is not infused. The oral load of 100 g glucose is similarly oxidized during exercise in normal and insulin-infused diabetics, but not in non-insulin-infused diabetics. Therefore glucose can be given orally to well-insulinized diabetics during prolonged exercise since it is readily oxidized (20).

At present, work is in progress to study the effect of training on normal young men by comparing the glucose metabolism, at rest and in muscular exercise, before and after a six-week period of daily training.

5. Perturbation of the exogenous glucose consumption by drugs.

The maize glucose can be also used to test the effects of some drugs affecting glucose metabolism. We studied two oral anti-diabetic drugs, butylbiguanide and benfluorex. After administration of biguanide, the oxidation curves of the exogenous glucose were similar in shape and magnitude to those obtained with untreated persons. Slightly higher rates of oxidation were recorded between the 2nd and the 4th hours of the test (12). Treatment with benfluorex had for consequence a reduced oxidation of exogenous glucose (16). We also studied with this method an indirect adipokinetic effect of amphetamine which can, by the intermediary of cathecholamines, affect the carbohydrate metabolism (15).

Another direction of application of this method in the study of pathological glucose metabolism in men is research on hereditary diseases. Our first investigation concerned the evaluation of glucose oxidation during muscular exercise in a patient with Mc.Ardle's disease. The myopathie described by Mc.Ardle is characterized by the lack of muscle phosphorylase. This lack has for consequence the impossibility of the utilization of glycogene for the supply of energy, which results in a great difficulty to perform muscular exercise. Our investigations demonstrated that oral glucose administration improves muscular performances (the duration of exercise being doubled) in that the exogenous glucose oxidation begins early and contributes to about 1/3 to the total energy expenditure (21).

These results show only a few examples for the potentialities of the use of the $^{13}C/^{12}C$ breath test to study glucose metabolism in animals and men. In fact, the possibilities of this method are very large : for instance, it is very appropriate to study glucose metabolism in pregnant women and children and it is also valid for metabolic studies with other naturally labelled substances (alcohol, etc).

The method can also be employed by utilizing a mixture of naturally and artificially labelled products with a view to increasing the sensitivity. A discussion of this last procedure, with illustrative examples is reported by us in another paper presented at this meeting (19).

REFERENCES.

1. F. Mosora, M. Lacroix and J. Duchesne. C.R. Acad. Sci. Paris, Série D, 273 (1971) 1423-1425.
2. F. Mosora, M. Lacroix and J. Duchesne. C.R. Acad. Sci. Paris, Série D, 273 (1971) 1752-1753.
3. F. Mosora, M. Lacroix, M. Pontus, and J. Duchesne. Bull. Acad. Roy. Belg., 5ème série, 58 (1972) 565-576.
4. F. Mosora, J. Duchesne and M. Lacroix. C.R. Acad. Sci. Paris, Série D, 278 (1974) 1119-1122.
5. M. Lacroix and F. Mosora. (Mémoire Commandé, I.A.E.A.-SM-191/29), Proceed.of a Symposium on "Isotope Ratios as Pollutant Source and Behaviour Indicators", Vienna, Nov 18-22, 1974, IAEA-FAO, Vienna, pp. 343-358.
6. M. Lacroix, F. Mosora, M. Pontus, P. Lefebvre, A. Luyckx and G. Lopez-Habib. Science, 181 (1973) 445-446.
7. P. Lefebvre, F. Mosora, M. Lacroix, A. Luyckx, G. Lopez-Habib and J. Duchesne. Diabetes, 24 (1975) 185-189.
8. J. Duchesne, F. Mosora, M. Lacroix, P. Lefebvre, A. Luyckx and F. Pirnay. in R. Klein, P. Klein (Eds), Proceed. of the 2nd Inter. Conf. on Stable Isotop. Oak Brook, Ill., Oct. 20-24, 1975, USERDA Conf.-751027, Illinois, 1977, pp. 282-286.
9. F. Mosora, P.Lefebvre, M. Lacroix, J. Duchesne, A. Luyckx and F. Pirnay. Metabolism, 25 (1976) 1575-1582.
10. F. Mosora, P. Lefebvre, F. Pirnay, M. Lacroix, A. Luyckx and J. Duchesne, in A.P. De Leenheer and R.R. Roncucci (Eds), Proc. 1st Int. Symp. on Quant. Mass Spectr. in Life Science, Ghent, June 16-18, 1976, Elsevier, Amsterdam, 1977, pp. 229-236.
11. F. Pirnay, M. Lacroix, F. Mosora, A. Luyckx and P. Lefebvre. Europ. J. of Appl. Physiol., 36 (1977) 247-254.
12. P. Lefebvre, A. Luyckx, F. Mosora, M. Lacroix and F. Pirnay. Diabetologia, 14 (1978) 39-45.
13. F. Pirnay, M. Lacroix, F. Mosora, A. Luyckx and P. Lefebvre. J. Appl. Physiol. (U.S.A.), 43 (1977) 258-261.
14. F. Mosora, I.R.E., 2 (1977) 2-15.
15. J. Duchesne, M. Lacroix, F. Mosora, N. Pallikarakis and F. Pirnay, in R. Klein, P. Klein (Eds), Proceed. of the Third Inter. Conf. on Stable Isotop., Oak Brook, Ill., May 23-26, 1978, Academic Press, 1979, pp. 527-532.
16. G. Krzentowski, N. Pallikarakis, M. Lacroix, F. Pirnay, F. Mosora, A. Luyckx and P. Lefebvre. Thérapie, 34 (1979) 445-455.
17. P. Lefebvre, A. Luyckx, G. Krzentowski, F. Pirnay, M. Lacroix, F. Mosora and N. Pallikarakis. Revue Médicale de Liège, 35 (1980) 249-254.
18. N. Pallikarakis, J. Duchesne, M. Lacroix, F. Mosora, F. Pirnay, G. Krzentowski, A. Luyckx and P. Lefebvre, in A. Frigerio, M. McCamish (Eds), Proceed. 6th Int. Symp. Mass Spectrometry in Biochem. and Medicine, Venice, June 21-22, 1979, Elsevier, Amsterdam, 1980, pp. 163-172.
19. M. Lacroix, N. Pallikarakis and F. Mosora, in H.L. Schmidt, H. Förstel and K. Heinzinger (Eds), Proceed. 4th Intern. Symp. on Stable Isotopes, Jülich, March 23-27, 1981, Elsevier, Amsterdam.
20. P.J. Lefebvre, G. Krzentowski, A. Luyckx, N. Pallikarakis, F. Pirnay, F. Mosora, M. Lacroix. Diabetes, 28 (1979), 423.
21. G. Krzentowski, N. Pallikarakis, F. Pirnay. Acta Clinica Belgica, 34 (1979) 151-157.

ASSESSMENT OF PULMONARY FUNCTION: APPLICATION OF STABLE ISOTOPES TO MEASUREMENT OF LUNG DIFFUSING CAPACITY

M. MEYER, P. SCHEID and J. PIIPER

Abteilung Physiologie, Max-Planck-Institut für experimentelle Medizin, Göttingen (F.R.G.)

ABSTRACT

The pulmonary diffusing capacity D, defined as the amount of gas transferred from the alveoli to the capillary blood per unit time as a function of the mean gas/blood partial pressure difference, is a characteristic index to quantitate the ability of the lung to conduct gas from the alveoli to the capillary blood. This communication summarizes recent developments of techniques to measure lung diffusing capacity in man with particular reference to the application of the stable isotopes $^{18}O_2$, $^{13}CO_2$, $C^{18}O$ and $^{13}C^{18}O$. The physiological implications derived from studies of the uptake kinetics of these isotopes in the lung are the following: (1) Alveolar-to-capillary O_2 transfer is slightly diffusion-limited in resting conditions, but may be appreciably diffusion-limited in hypoxia or exercise. (2) For CO_2, alveolar-capillary equilibration is mainly limited by $CO_2/HCO_3^-/H^+$ equilibration between red cells and plasma. (3) Transfer of CO (and O_2) in the lung is by passive diffusion alone, the recently invoked concept of a carrier-mediated transfer mechanism being not supported by experimental evidence.

INTRODUCTION

The basic mechanisms involved in O_2 uptake and CO_2 elimination by the lung are gas transfer by diffusion across the alveolar-capillary membrane and chemical reactions in blood. Membrane resistances and finite reaction kinetics may therefore constitute limiting factors for equilibration of O_2 and CO_2 between alveolar gas and blood by the time blood leaves the pulmonary capillaries. A characteristic quantitative index for pulmonary gas transfer is the gas-blood transfer conductance or diffusing capacity of the lung. Generally, it is held that diffusional equilibrium of CO_2 is much faster than that of O_2 on the basis of the ratio of Krogh's diffusion constants. However, equilibration of alveolar gas with capillary blood is not determined by the diffusive conductance D (= diffusing capacity) alone, but by the ratio of diffusive (D) to perfusive conductance $\dot{Q}\beta_b$ ($\dot{Q}$ = flow rate, β_b = effective solubility of gas in blood). Moreover, recent experimental evidence suggests that CO (and O_2) transfer in the lung is enhanced, above that by passive diffusion, by a particular facilitated transfer mechanism [1]. This paper summarizes the conclusions derived from recent measurements of the pulmonary diffusing capacity for O_2, CO_2 and CO in normal men assessed from rebreathing equilibration kinetics of the stable isotopes $^{18}O_2$, $^{13}CO_2$, $C^{18}O$ and $^{13}C^{18}O$. A detailed account of this work is presented elsewhere [2,3,4].

THEORY

General principle of blood-gas equilibration in lungs

Gas transfer in the lung can be analyzed on the basis of a simple lung model comprising a homogeneous alveolar gas space and a blood perfused pulmonary capillary, both separated by a "membrane", the diffusive conductance of which is generally termed diffusing capacity D. During capillary transit, partial pressure in each volume element of blood varies between mixed venous ($P_{\bar{v}}$) and end-capillary ($P_{c'}$) approaching equilibrium with alveolar gas (P_A) exponentially:

$$\frac{P_A - P_{c'}}{P_A - P_{\bar{v}}} = e^{-\frac{D}{\dot{Q}\beta_b}} \tag{1}$$

According to eq. (1) the degree of gas/blood equilibration as indicated by the ($P_A - P_{c'}$) difference depends on the diffusive to perfusive conductance ratio $D/(\dot{Q}\beta_b)$. The capacitance coefficient β_b equals physical solubility for all inert gases and is equivalent to the slope of the blood dissociation curve for the respiratory gases O_2 and CO_2.

Fick's law of diffusion may be used to quantitate the rate of transfer ($\dot{M}$) across a plane membrane per driving partial pressure difference (ΔP).

$$\dot{M} = \underbrace{d \cdot \beta_m \cdot \frac{A}{l}}_{D} \cdot \Delta P \tag{2}$$

(d = diffusion coefficient, β_m = solubility of gas in membrane, A = surface area, l = thickness of diffusion barrier)

With this relationship the following expression for the $D/(\dot{Q}\beta_b)$ ratio is obtained.

$$\frac{D}{\dot{Q}\beta_b} = d \cdot \frac{\beta_m}{\beta_b} \cdot \frac{A}{l} \cdot \frac{1}{\dot{Q}} \tag{3}$$

Hence the $D/(\dot{Q}\beta_b)$ ratio for different gases is dependent on a) the diffusion coefficient d and b) the solubility ratio between barrier and blood (β_m/β_b).

For all inert low and high solubility gases the β_m/β_b ratio is close to unity and equilibration between alveolar gas and blood is complete whithin capillary transit time of blood. Transfer of these gases is therefore not limited by diffusion, the rate of uptake being determined by perfusion.

Conversely, transfer of CO is exclusively diffusion limited because $ß_b$ much exceeds $ß_m$ on account of the high affinity of hemoglobin for CO.

Transfer of O_2 and CO_2 is limited by both diffusion and perfusion and may be considered intermediate between the extreme cases outlined above.

METHODS

Rebreathing technique

In the rebreathing technique the subject breathes in a closed system, which is made up of a rubber bag and the lungs of the subject. Test gases initially contained in the bag are homogeneously distributed throughout the system and taken up by pulmonary capillary blood according to their physical characteristics (see above). In practice, the diffusing capacity D for O_2, CO_2 and CO is calculated from the rate of disappearance of these gases from the system assessed from continuous monitoring of partial pressure changes by a fast-responding respiratory mass spectrometer sampling close to the mouth of the subject.

Advantage of stable isotopes

In principle, D can be determined from rebreathing equilibration kinetics of the naturally abundant isotopes of O_2, CO_2 and CO. However, the use of stable isotopes as test gases provides distinct advantages that are practically important.

1. For rare stable isotopes, mixed venous partial pressure ($P_{\bar{v}}$), which is the asymptotic value for equilibration of alveolar partial pressure (P_A) during rebreathing, is practically zero. By contrast, for the abundant isotopes of O_2 and CO_2 (and for CO particularly in smokers) $P_{\bar{v}} > 0$, and precise determination is rendered difficult by non-invasive techniques.

2. The use of isotopic CO, e.g. $C^{18}O$ or $^{13}C^{18}O$, is indispensable because of the mass interference encountered in conventional respiratory mass spectrometers when the abundant component $^{12}C^{16}O$ is measured in the presence of atmospheric N_2.

3. The most important adventage for the use of $^{18}O_2$ and $^{13}CO_2$ as test gases instead of their naturally abundant components results from the fact, that under certain experimental conditions the effective solubility or effective slope of the blood dissociation curve $ß_b$ (= dC/dP) is constant, while it normally varies for the abundant isotopes between $P_{\bar{v}}$ and P_a according to the shape of the dissociation curve. If there is partial replacement of the naturally abundant isotope by the rare isotope, e.g. replacement of $^{16}O_2$ by $^{18}O_2$, such that the total number of O_2 molecules both physically dissolved in plasma and bound to hemoglobin remains constant, the slope of the dissociation curve effective for uptake of the rare component is constant for all values of $P^{18}O_2$. This condition is experimentally achieved, and equally applies for CO_2, e.g. replacing $^{12}C^{16}O_2$ by $^{13}C^{16}O_2$, provided the abundant component is in rebreathing equilibrium (i.e. zero net-transfer) and P_A for the rare isotope is small compared to the naturally abundant isotope.

Experimental procedure

Three series of experiments were performed in normal young subjects. The difference between the series was mainly due to the difference in test gas mixtures and isotopic species.

1. Simultaneous determination of D_{O_2} and D_{CO}

The subject rebreathes for about 15 sec a gas mixture containing 0.07% $^{18}O_2$, 0.07% $C^{18}O$, 1% He, 1% C_2H_2, 1-3% $^{16}O_2$ and 8% CO_2. The addition of an insoluble inert gas (He) is required for determination of lung volume and rebreathing ventilation. C_2H_2 is used to estimate pulmonary capillary blood flow. The gas concentrations of $^{16}O_2$ and CO_2 are selected to establish blood/gas equilibrium for these gases during the rebreathing period. Before onset of rebreathing the subject breathes a mixture of 21% O_2 in N_2 which serves to wash out disturbing ^{36}Ar from lung gas. All gases are simultaneously measured by a respiratory mass spectrometer (redesigned Varian M3), [3,5].

2. Determination of CO_2 diffusing capacity

Procedures and gas mixtures were essentially the same as for determination of D_{O_2} and D_{CO}, except that 0.07% $^{13}CO_2$ was used instead of $^{18}O_2$ and $C^{18}O$. The relative natural abundance of ^{13}C (1.11 atom % of total carbon isotopes) that contributes to the measured signal at mass 45 was taken into account by recording $^{12}CO_2$ at mass 22 to derive an equivalent signal to compensate for its contribution at mass 45. With this correction $^{13}CO_2$ can be used as a test gas at a low concentration [4].

3. Simultaneous determination of D_{CO} with $^{12}C^{18}O$ and $^{13}C^{18}O$

The experimental design of this series was established to test the hypothesis of a particular facilitated transfer mechanism for CO (and O_2) in the lung by measuring the CO diffusing capacity of the lung at various alveolar CO levels. While the concentration of $^{13}C^{18}O$ in the rebreathing gas mixture was kept constant at low levels (160 ppm) for reference purposes the other component $^{12}C^{18}O$ was varied at random in the range of 0 to 2240 ppm. A special double collector was constructed for separation of mass 30 and mass 31 [2].

RESULTS

The results that have been obtained in a group of healthy young men both at rest and exercise are summarized in Table 1. The results of measurements of D_{CO} at various alveolar CO levels showed no dependence of D_{CO} with the alveolar CO concentration which is at variance with the results and conclusions by Mendoza et al. [1].

DISCUSSION

Relationship of D_{O_2} and D_{CO}

The ratio for the diffusive conductance D_{O_2}/D_{CO} can be compared to predictions derived on theoretical grounds considering two mechanisms as limiting factors:

a) diffusion resistance of the alveolar-capillary membrane, b) rate of uptake of O_2 and CO by red cells.

a. If the membrane resistance is considered a limiting factor for O_2 and CO uptake in the lung, the diffusing capacity ratio is expected to equal the ratio of Krogh's diffusion constants which yields a value of 1.23 on the basis of the different physical properties of these gases with regard to diffusivity and solubility.

b. If the kinetics of O_2 and CO uptake constitute the major limitation, the D_{O_2}/D_{CO} ratio is calculated to be in the range of 1.1 - 3.0 on the basis of in vitro reaction kinetic measurements. Yet there is no agreement as to the specific gas uptake conductances by red cells invalidating the use of a precise figure for this ratio.

Our experimental values for the D_{O_2}/D_{CO} ratio are thus compatible with both limiting cases. At present, it is therefore not possible to assess the relative importance of membrane and blood processes as limiting factors for O_2 uptake in the lung.

TABLE 1
Results from healthy young subjects. Diffusing capacity D in $ml \cdot min^{-1} \cdot Torr^{-1}$

	Rest	Exercise	$\frac{Exercise}{Rest}$
D_{O_2}	54	63	1.2
D_{CO}	47	52	1.1
D_{CO_2}	180	305	1.7
D_{O_2}/D_{CO}	1.14	1.20	
D_{CO_2}/D_{O_2}	3.3	4.8	

Relationship of D_{CO_2} and D_{O_2}

A similar analysis can be applied to elucidate the limiting mechanisms for CO_2 elimination by the lung. Considering the CO_2/O_2 ratio of Krogh's diffusion constants in water and tissue, which is about 20, one may infer from our much lower experimental values that other mechanisms than diffusion of molecular CO_2 predominate. It is well known that the release of CO_2 from blood involves a number of processes of which dehydration of carbonic acid, bicarbonate/chloride exchange between red cells and plasma and re-equilibration of the $H^+/HCO_3^-/CO_2$ system in plasma are known to be potentially slow. This is in line with computations which show that by partitioning the total CO_2 transfer resistance into a membrane and a blood component the blood component clearly prevails.

Blood-gas equilibration efficiency of the lung

The left hand term of eq. (1) designates the relative equilibration deficit during capillary transit. Using standard data on cardiac output (Q) and the slope of the dissociation curve together with the experimental D values for O_2 and CO_2, one can calculate the blood-gas partial pressure differences, i.e. $(P_a - P_A)_{CO_2}$ and $(P_A - P_a)_{O_2}$. According to these calculations there is almost complete equilibration for both O_2 and CO_2 at rest. In heavy exercise however, a sizable partial pressure difference between alveolar gas and arterial blood may occur.

Evidence against a facilitated transfer mechanism for CO

Mendoza et al. [1] have recently observed that the pulmonary diffusing capacity for CO, as measured by the steady state technique increased to a maximum when end-tidal CO concentration was about 200 ppm, and upon transition to higher CO levels decreased to a constant value. These authors have attributed this finding to saturation kinetics of a specific carrier, possibly cytochrome P_{450}, that should likewise be functionally operative for O_2. We have reinvestigated this problem by measuring D_{CO} at various alveolar CO levels using a technique of simultaneous measurement of two stable carbon monoxide isotopes. We were unable to reproduce this finding since no apparent dependence of D_{CO} with the alveolar CO level could be detected. Our results therefore do not support the concept of a particular facilitated transfer mechanism in alveolar-capillary gas transfer above that by passive diffusion.

CONCLUSIONS

The main conclusions derived from several experimental series are summarized as follows:

1. Alveolar-capillary equilibration of O_2 and CO_2 is almost complete during pulmonary capillary transit time. The equilibration deficit is slight at rest but may be appreciable during heavy exercise.
2. For O_2, diffusion may be the limiting process rather than chemical reactions in blood.
3. For CO_2, the kinetics of red cell/plasma ion exchange appear to be the limiting factor.
4. The hypothesis of a particular facilitated transfer mechanism for CO (and O_2) above that by passive diffusion is not supported.

REFERENCES

1. C. Mendoza, H. Peavy, B. Burns and G. Gurtner, J. Appl. Physiol., 43(1977)880-884.
2. M. Meyer, W. Lessner, P. Scheid and J. Piiper, J. Appl. Physiol., (in press).
3. M. Meyer, P. Scheid, G. Riepl, H.-J. Wagner and J. Piiper, J. Appl. Physiol., (submitted).
4. J. Piiper, M. Meyer, C. Marconi and P. Scheid, Respir. Physiol., 42(1980)29-41.
5. P. Scheid, M. Meyer and H. Slama, Bull. Eur. Physiopath. Resp., 15(1979)11P-14P.

A CONSTANT-FLOW SINGLE-EXHALATION MEASUREMENT OF REGIONAL LUNG FUNCTION*

C. HOOK and M. MEYER

Abteilung Physiologie, Max-Planck-Institut für experimentelle Medizin, Göttingen (F.R.G.)

ABSTRACT

Pulmonary blood flow ($\dot{Q}$) and carbon monoxide diffusing capacity (DL_{CO}) per unit accessible lung volume (V_A) were measured in normal subjects during a single exhalation at constant flow rate after inspiration of a test gas mixture containing 0.07% $^{13}C^{18}O$, 1% He and 1% C_2H_2 in air. Test gas concentrations were continuously measured by two respiratory mass spectrometers, one sampling from close to the lips (Whole lung test), the other, by inserting a flexible fiberoptic bronchoscope, from the main or lobar bronchi of the lung (Regional test). Absolute values for the whole lung for both DL_{CO} and $\dot{Q}$ were thus obtained in the Whole lung test, whereas for different lung regions the DL_{CO}/V_A and $\dot{Q}/V_A$ ratios were compared with the corresponding Whole lung data. The results in normal subjects indicate a significant vertical gradient for DL_{CO} and $\dot{Q}$ from the upper to the lower lung lobes which is slightly attenuated at moderate exercise. The technique appears to be suitable for the study of topographical variations of perfusion and diffusing capacity in patients undergoing diagnostic bronchoscopy.

INTRODUCTION

The measurement of the diffusing capacity of the lung for carbon monoxide (DL_{CO}) and pulmonary blood flow ($\dot{Q}$) constitutes a widely used clinical test for assessment of pulmonary function. Common to the various techniques available is that their results reflect average values as a mean of individual lung units (lobes and segments). Consequently, the lung is formally treated as an ideal one-compartment model with homogeneous gas mixing and diffusion across the alveolar membrane. With the technique recently developed in our laboratory the topographical distribution of DL_{CO} and $\dot{Q}$ within the lung can be measured and compared to the overall value for the whole lung.

---

* Supported by Europäische Gemeinschaft für Kohle und Stahl.

TECHNIQUES

In the single-exhalation maneuver the subject inspires a standard test gas mixture holds his breath at full inspiration for about 4 sec and expires slowly at a constant rate of about 0.2 l·sec^{-1}. The inspired test gas mixture comprises 0.07% $^{13}C^{18}O$ (or $^{12}C^{18}O$), 1% He, and 1% C_2H_2 in air. The application of the stable isotopes $^{13}C^{18}O$ (mass 31) or $^{12}C^{18}O$ (mass 30) eliminates the problems of mass interference encountered in conventional respiratory mass spectrometers with standard resolution, when the abundant component $^{12}C^{16}O$ is measured in the presence of atmospheric N_2, and renders possible a negligible back pressure in mixed venous blood. Test gas partial pressures are continuously monitored by two respiratory mass spectrometers (modified Varian M3, cf. Scheid et al. [1], the first sampling close to the mouth of the subject, the second sampling through a thin polyethylene tubing that is guided to the different lung lobes through a flexible fiberoptic bronchoscope. With this technique, test gas partial pressure disappearance curves are simultaneously obtained from the whole lung (Whole lung test) and from different parts of the lung (Regional test). Instead of using two respiratory mass spectrometers the delay-loop-technique of Davies and Denison [2] may be used with one mass spectrometer only.

The functional lung parameters that can be determined from the rate of partial pressure changes during slow exhalation depend on the test gas species used (cf.[3]).

1. <u>Low-Solubility Inert Gas (e.g. He)</u>. A plateau will be observed in the expired trace the magnitude of which is to the inspired concentration depends on the inspired ventilation per unit accessible lung volume.

2. <u>High-Solubility Inert Gas (e.g. C_2H_2)</u>. Soluble inert gases are taken up by pulmonary capillary blood. The rate of disappearance is thus related to pulmonary capillary blood flow ($\dot{Q}$).

3. <u>Carbon Monoxide (CO)</u>. Due to the high affinity of CO for hemoglobin, the rate of CO uptake is limited by diffusion across the gas/blood interface and hence allows estimation of lung diffusing capacity for CO (DL_{CO}).

RESULTS

Preliminary results obtained in normal subjects at rest and during 50 W exercise yield the following:

1. Overall values for $\dot{Q}$ and DL_{CO} (Whole lung test) were in agreement with those obtained with conventional techniques (e.g. rebreathing).

2. Both DL_{CO} and $\dot{Q}$ slightly increase by 20% upon transition to moderate exercise.

3. In the lateral decubitus positions both perfusion and diffusing capacity are higher in the dependent than in the upper lung.

4. In the upright posture regional distribution of DL_{CO} and $\dot{Q}$ shows a vertical

gradient from the upper to the lower lung lobes which may be attributed to the effects of gravitational forces.

5. During exercise, the topographical variations of perfusion and diffusing capacity are slightly reduced.

CONCLUSION

The constant-flow single-exhalation technique appears to be suitable for the measurement of pulmonary blood flow and diffusing capacity. In particular, lobar and segmental lung function can be studied by intrapulmonary gas sampling during fiberoptic bronchoscopy. The test is simple and requires little cooperation by the subject and is thus suitable for clinical use in patients subjected to diagnostic bronchoscopy.

REFERENCES

1   P. Scheid, M. Meyer and H. Slama, Bull. Eur. Physiopath. Resp., 15(1979)11P-14P.
2   N.J.H. Davies and D.M. Denison, Respir. Physiol., 37(1979)335-346.
3   J. Piiper and P. Scheid, in J.B. West (Ed.), Pulmonary Gas Exchange, Vol. I, Academic Press, London, New York, San Francisco, 1980, pp. 131-171.
4   D.M. Denison, N.J.H. Davies, M. Meyer, R.J. Pierce and P. Scheid, Respir. Physiol., 42(1980)87-99.

A detailed account on part of this work was published by Denison et al. [4].

IV. LIFE SCIENCES, AGRICULTURE AND ENVIRONMENTAL RESEARCH

STABLE ISOTOPES IN AGRICULTURE

H. FAUST

Academy of Sciences of the GDR, Central Institute for Isotope and Radiation Research, Leipzig (GDR)

ABSTRACT

A review of the literature on the use of stable isotopes in soil-plant relations research, in animal nutrition and physiology, environmental research, and further disciplines related to agricultural problems is presented. The literature pertaining to this review covers the period from 1977 to 1981 and is researched for trends of application. Among the stable isotopes of biological interest ^{15}N is characterized by an exclusive use in agricultural studies. Recent progress and lines of development in sample treatment and analysis of ^{15}N are described. A great number of experiments have shown that from the technical as well as the biological point of view ^{15}N is the ideal isotope for these investigations, nevertheless expense and analytical problems are considerable. The combined efforts of theorists and experimenters alike will be required for continued advances in the theory and use of stable isotopes as tracers.

INTRODUCTION

In 1977 and 1978 two important international meetings on the application of stable isotopes took place (refs. 1, 2). At the IAEA Technical Committee Meeting on Modern Trends in the Biological Applications of Stable Isotopes we discussed the use of stable isotopes in agriculture (ref. 3). Since this time a lot of publications have appeared documenting the broad interest in the use of tracer techniques with stable isotopes.

Obviously, only a short and general presentation of main research lines can be given here and the particular examples chosen are merely intended as illustrations. Accordingly, it is the objective of this paper to give an introduction into the latest results of ^{15}N tracer research in agriculture.

For a detailed presentation of the wide range of research which has been and is being undertaken in this field reference is made to an extended review with a comprehensive bibliography (ref. 9).

DEVELOPMENT OF RECENT LITERATURE ON ^{15}N

A graph representing the annual number of scientific publications clearly shows that since 1962 the number of papers has steadily increased and today is several times as large as in that year (Fig. 1).

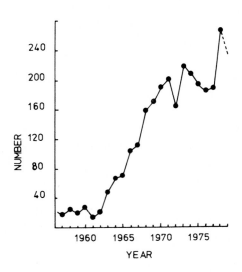

Fig. 1 Publications with ^{15}N: Isotope analysis and bioscientific application

The approximate proportions of the total number of publications for the individual fields in the period from 1978 to 1980 are calculated and summarized in Figure 2. It is shown that tracer experiments with the stable nitrogen isotope ^{15}N, which are designed to solve problems of the nitrogen metabolism in the soil-plant-animal system, play an important part within the framework of isotope applications in agriculture. The growing interest in environmental and animal nutrition research is of special importance.

In contrast to clinical research where ^{13}C and deuterium are tantamount to ^{15}N, isotope work in agronomic sciences is characterized by the exclusive use of ^{15}N as demonstrated in Figure 3. Stable isotopes of nitrogen and sulfur in environmental research are of less importance than the other isotopes indicated.

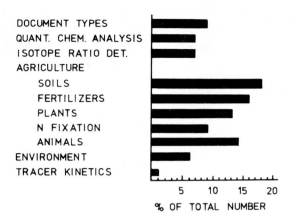

Fig. 2 Publications with ^{15}N in special research fields: 1978-1980

	^{15}N	^{13}C	^{2}H	^{18}O	^{34}S
AGRICULTURE	●				
MEDICINE	●	●	●		
ENVIRONMENT	(●)	●	●	●	(●)

Fig. 3 Predominant use of different stable isotopes in bioscientific research

Among the recent publications those papers reviewing special fields of agricultural research will be mentioned and summarized in Table 1. It is remarkable that the increasing concern about nitrogen isotope techniques has led to a number of national and international research programmes (refs. 10-12). A major part of the present know-

TABLE 1
Recent literature on stable isotopes: Document types

Type	Subject	Refs.
Reviews	Agriculture, life sciences	2 - 7
Bibliographies	Agriculture, biomedicine	8, 9
Proceedings	N fixation, soils research	10 - 14
Manuals	N-15: Agriculture, medicine	15, 16
Lectures	N-15: Soil-plant research	17
Recommendations	Biosciences, terminology	2, 11, in 15

ledge about the nitrogen metabolism in the soil-plant system in developing countries originates from work undertaken within such programmes including the organization of international training courses (refs. 15, 17) where manuals, lectures and recommendations are of special importance.

PROGRESS IN QUANTITATIVE CHEMICAL ANALYSIS AND ISOTOPE RATIO DETERMINATION

Considerable progress has been made in ^{15}N tracer methodology. The main results are summarized in Table 2.

TABLE 2
Recent progress in ^{15}N methodology

Step	Subject	Refs.
Sampling	Fertilization studies	18
Chemical treatment	Native amino acids, proteins	19-21
	Inorganic N compounds	18, 22-24
Sample chemistry	Conversion methods, gas preparation and purification	18, 22-26, 30
Isotope analysis	Emission spectrometry	
	Methodology	18, 27-29, 32
	New instruments	24, 30-32
	Mass spectrometry	
	Methodology	26, 33, 34
	Instrument coupling	25, 33-35

The improvements achieved are found along four lines
   a) special isolation procedures for ^{15}N labelled nitrogen compounds (mainly by chromatographic techniques)
   b) new methods for the preparation of molecular nitrogen
   c) ^{15}N analysis with very small amounts of nitrogen
   d) extended use of emission spectrometry and increasing use of GC-MS techniques in the analysis of biological fluids (blood etc.)

NMR and infrared spectroscopy for the isotope analysis of nitrogen has not yet been introduced in agricultural research. On the basis of the general trend towards automation and data processing in analytical chemistry one can conclude that it is necessary to improve and simplify further the instrumentation for ^{15}N analysis at a level of precision, accuracy, and reliability needed in most tracer studies.

RECENT ^{15}N APPLICATIONS IN AGRICULTURE
Soils, fertilizer efficiency, plants

Isotopically labelled fertilizers provide a unique and extremely valuable tool for evaluating such yield determining factors as the time of fertilization, the relative merits of different sources of nitrogen and the effect of irrigation and fertilizer utilization under the different soil and climatic conditions in various countries.

Apart from methodological studies for designing field experiments with ^{15}N (refs. 36, 48) much interest is devoted to questions of nitrogen immobilization-mineralization processes in the soil (refs. 37-41) and the main sources of losses after fertilization of mineral nitrogen (ref. 38, 42-47).

Together with the fertilizer saving effect of a high nitrogen fixation capacity of legumes, studies of nitrogen fertilizer efficiency are of high importance especially in the light of the dramatic increase of N fertilizer use in some developing countries and the problems of soil N conservation.

New possibilities in research are opened by the use of variations in the natural ^{15}N abundance as an indicator for nitrogen fixation and by new field measurement techniques (refs. 9, 10) or as an aid in tracing fertilizers nitrogen transformation (refs. 76, 77).

TABLE 3
Nitrogen fertilizer efficiency studies

N Balance	Effect	Importance	Refs.
Plant	Plant yield / ha Protein yield / ha N yield / ha	Economic evaluation Crop quality Animal nutrition Fertilizer utilization	11, 49-61 51, 89 78, 81, 89 49-69, 76, 77
Soil	N Remaining in soil	Fertilizer aftereffect Soil fertility Soil organic matter Pollution	50, 60 6, 7, 12 7, 50, 52 11
Loss	Release to environment (atmosphere, ground water)	Environmental release Pollution	11, 97, 98

As illustrated in Table 3 the best way to evaluate quantitatively the efficiency of fertilizer N is to draw up a balance sheet. Due to different points of view (economic, physiological, nutritional, agronomic, environmental) the definition of the term "fertilizer efficiency" is always important.

Nitrogen fixation studies

From the quantitative input factors in the soil-plant system and the present literature it can be deduced that dinitrogen fixation research is one of the most important, rapidly expanding and competitive areas of agricultural and biological sciences.

Great activities have been developed by international and national organisations to initiate interdisciplinary research programmes as shown in Table 4.

TABLE 4
^{15}N in nitrogen fixation studies

	Subject	Refs.
Research programmes	National and international studies (IAEA, FAO, UNEP)	10, 49
Methodology	N yield difference method, natural abundance method, low N-15 enriched soil organic matter method	10, 17 70, 75, 76
	Isotope effects	75, 76
Agricultural studies	Field measurement techniques	72, 74
	Fertilizer efficiency	73
	Inoculation studies	71
	Comparative evaluation	72

Animal nutrition and physiology

Tracer studies with ^{15}N of non-protein nitrogen and protein metabolism in animals are of growing interest and have demonstrated the potential of isotope techniques as a means of studying quantitative nitrogen transformations. Research with monogastrides and ruminants illustrates the prevailing interest in the optimum utilization of NPN compounds (refs. 81, 86, 87, 92-94), amino acids and proteins (refs. 79-85) and physiological problems (ref. 88-91, 93).

The construction of models for nitrogen metabolism and their use in the study of the quantitative and dynamic behaviour of nitrogen utilization has helped to gain deeper insight into the metabolic processes of animals.

Figure 4 shows an example modelling the nitrogen metabolism of growing pigs. It was found in this extensive study that the quantification of the total amino acid pool would not provide real conditions and must be rejected because of the varying shares and functions of the individual amino acid in the pool (ref. 78).

In planning tracer experiments with ^{15}N the prediction of the isotope distribution by computer simulation avoids the use of insuf-

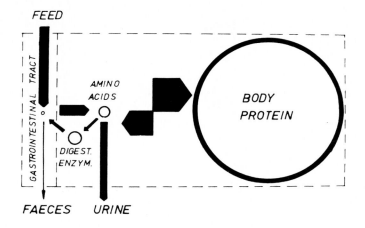

Fig. 4 Nitrogen metabolism in growing pigs: Important quantitative relations

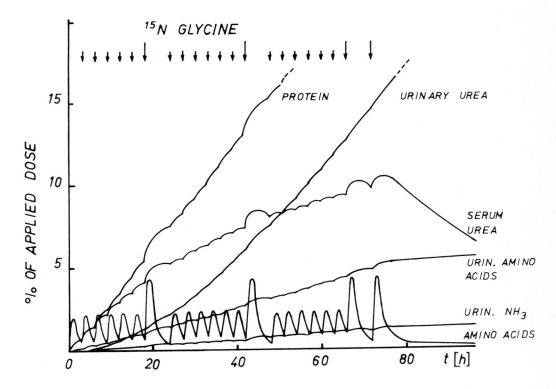

Fig. 5 Simulation of the ^{15}N distribution in different compartments after multiple pulse tracer application according to a 10-pool model

ficiently enriched ^{15}N labelled material or difficulties in the interpretation of the results. For this purpose we calculated the appearance and disappearance of the isotope in different body compartments according to a 10-pool metabolic model as depicted in Figure 5 and compared the calculated values with the experimental ones (ref. 16). First results indicated the general validity of this model.

ENVIRONMENTAL RESEARCH

It is necessary to consider the intensification of agriculture involving the rapidly increasing use of and dependance upon nitrogen fertilizer (refs. 95-99) also against the background of environmental pollution. The use of ^{15}N is of special importance in identifying the fate of unrecovered agricultural nitrogen residues and in finding out ways for more economic fertilizer use with a minimum of pollution. However, the problem of soil N conservation and useful maintenance in the crop root zone seems to be more important than that of ground water pollution.

OUTSTANDING PROBLEMS

Agricultural production for feed, food and industrial purposes during recent decades is characterized by an increasing intensification (e.g. increase in the use of highly energy dependent mineral fertilizers of artificial irrigation, specialized animal husbandry, and intensified crop farming).

Against this background the knowledge of quantitative data on the fate of applied fertilizer nitrogen (reliable balance data), an efficient soil nitrogen management and high utilization of nitrogen compounds for protein synthesis in farm animals seem to be some of the principal problems in the near future. The wide application of the isotope technique is a promising way to meet these goals and to achieve results of higher efficiency for practice.

REFERENCES

1 E.R. Klein and P.D. Klein, Stable Isotopes. Proc. 3rd Int. Conf. 1978, Academic Press, New York, 1979.
2 Stable Isotopes in the Life Sciences: Proc. Tech. Comm. Meet., IAEA, Vienna, Austria, 1977.
3 D.A. Korenkov and H. Faust, Stable Isot. Life Sci., Proc. Tech. Comm. Meet. Mod. Trends Biol. Appl. Stable Isot., 1977:345-361, IAEA, Vienna, Austria.
4 M. Kralova and I. Havassy, Acta Univ. Agric., Fac. Agron. (Brno), 27(1979)7-23.
5 H.-L. Schmidt and E. Schmelz, Chemie in unserer Zeit, 14(1980)26-34.
6 A.M.L. Neptune and T. Muraoka, Rev. Bras. Cienc. Solo, 2(1978)151-163

7   E.A. Paul and J.A. Van Veen, Comm. Eur. Communities, EUR 1979, EUR 6361, Modell. Nitrogen Farm Wastes, 75-132.
8   E.R. Klein and P.D. Klein, Biomed. Mass Spectrometry, 12(1979) 515-545.
9   H. Faust, ZfI-Mitt. 1981, in press.
10 Isotopes in Biological Dinitrogen Fixation, Proc. Adv. Group Meet. Nov. 1977, IAEA, Vienna, Austria, 1978.
11 Soil Nitrogen as Fertilizer or Pollutant, Proc. Res. Coord. Meet. July 1978, IAEA, Vienna, Austria, 1980.
12 5. Wiss. Konf. "N-Transformation im Boden und Effektivität der N-Düngung", Proc., Sofia, 1978.
13 R. Letolle and A. Mariotti, ZfI-Mitt. 5(1976)260-270.
14 R.A. Olson, Isotope studies on soil and fertilizer nitrogen, in Proc. Symp. Isot. Rad. Res. Soil Plant Rel. IAEA,1978(1979) pp.3-32.
15 H. Faust (Ed.), FAO/IAEA Interregional Training Course on the Use of N-15 in Soil Science and Plant Nutrition: Training Manual, ZfI-Mitt. 1981, No. 38, pp.1-181.
16 H. Faust, H. Bornhak, K. Hirschberg, K. Jung, P. Junghans, P. Krumbiegel and R. Reinhardt, Klinisch-chemische und isotopenanalytische Methoden zur Untersuchung des Stickstoff-Stoffwechsels mit N-15 beim Menschen: Methodenkatalog, ZfI-Mitt. 1981, No.36, pp.1-210.
17 H. Faust (Ed.), FAO/IAEA Interregional Training Course on the Use of N-15 in Soil Science and Plant Nutrition: Lectures, ZfI-Mitt. 1980, No. 32, pp. 1-134.
18 H. Lippold, W. Ackermann and W. Teske, Isotopenpraxis,15(1979)62-65.
19 K. Hirschberg and H. Faust, VII. Symp. "Chromatographie in der klinischen Biochemie", 28.-30.5.1980, Leipzig.
20 H. Bornhak and K. Jung, VII. Symp. "Chromatographie in der klinischen Biochemie", 28.-30.5.1980, Leipzig.
21 K. Krawielitzki and T. Völker, Isotopenpraxis, 14(1978)152-156.
22 W.A. O'Deen and L.K. Porter, Anal. Chem., 52(1980)1164-1166.
23 R.J. Volk and W.A. Jackson, Anal. Chem., 51(1979)463-464.
24 J.D.S. Goulden and D.N. Salter, Analyst (London), 104(1979)756-765.
25 M. Potts, W.E. Krumbein and J. Metzger, Environ. Biogeochem. Biomicrobiol., Proc. Int. Symp., 3rd, 1977, 3(1978)753-769.
26 H. Faust, H. Bornhak, K. Hirschberg and H. Birkenfeld, ZfI-Mitt. 26(1979)40-41, 29(1980)217-225.
27 K.R. Reddy and W.H. Patrick, Soil Sci. Soc. Am. J., 42(1978)316-318.
28 J.C. Burridge and I. Hewitt, Anal. Chem. Acta, 118(1980)11-28.
29 J.C. Burridge and I.J. Hewitt, Commun. Soil Sci. Plant Anal., 9(1978)865-872.
30 G.K. Dudich, P.N. Zanadvorov, V.V. Kidin, G.S. Lazeeva, Yu.M. Loginov, A.A. Petrov, Yu.V. Frolov, R.V. Khomyakov and V.Ya. Shevchenko, Izv. Timiryazevsk. S-kh. Akad., 1980, pp. 112-119.
31 H. Gerstenberger and G. Meier, Isotopenpraxis,17(1981)150-155.
32 G. Guiraud, J.C. Fardeau, F. Gueye, C. Marol and M.C. Vincent, Analysis, 8(1980)148-152.
33 D.D. Focht, N. Valoras and J. Letey, J. Environ. Qual., 9(1980) 218-223.
34 C.S. Irving, I. Nissim and A. Lapidot, Biomed. Mass Spectrom., 5(1978)117-122.
35 J. Metzger, Fresenius' Z. Anal. Chem., 292(1978)44-45.
36 F.E. Broadbent and A.B. Carlton, J. Environ. Qual., 9(1980)236-242.
37 C.G. Kowalenko, Soil Sci., 129(1980)218-221.
38 W. Thies, K.W. Becker and B. Meyer, Landwirtsch. Forsch., Sonderh., 34(1978)55-62.
39 J.N. Ladd and M. Amato, Soil Biol. Biochem., 12(1980)185-190.
40 J.M. Jones and B.N. Richards, Soil Biol. Biochem., 10(1978)161-168.
41 R.L. Westerman and T.C. Tucker, Soil Sci. Soc. Am. J., 43(1979) 95-100.
42 C.J. Smith and P.M. Chalk, Soil Sci. Soc. Am. J., 44(1980)288-291.

43 D.E. Rolston, F.E. Broadbent and D.A. Goldhamer, Soil Sci. Soc. Am. J., 43(1979)703-708.
44 H. Lippold and I. Foerster, Arch. Acker- Pflanzenbau Bodenkd., 24(1980)85-90.
45 W. Matzel, H. Mouchova, J. Apltauer and H. Lippold, Arch. Acker-Pflanzenbau Bodenkd., 23(1979)421-427.
46 J. Letey, N. Valoras, A. Hadas and D.D. Focht, J. Environ. Qual., 9(1980)227-231.
47 D.D. Focht, Nitrogen Environ., 2(1978)433-490.
48 R.V. Olson, Soil Sci. Soc. Am. J., 44(1980)428-429.
49 M. Saxena, IAEA-R-1653, Oct. 1978.
50 V.V. Kidin and L.A. Ivannikova, Izv. Timiryazevsk. S-kh. Akad., 0/5(1979)190-195.
51 H. Peschke, G. Markgraf, O. Schmidt and U. Selle, Tagungsber., Akad Landwirtschaftswiss. DDR, 155(1978)63-73.
52 P.M. Smirnov, V.V. Kidin and O.N. Ionova, Izv. Timiryazevsk. S-kh. Akad., 5(1980)65-70.
53 G.E. Ham and A.C. Caldwell, Agron. J., 70(1978)779-783.
54 S.Y. Daftardar, D.L. Deb and N.P. Datta, J. Nucl. Agric. Biol., 8(1979)94-97.
55 H. Bemkenstein, H. Nehl, H. Peschke and P. Richter, Arch. Gartenbau (Berlin), 26(1978)323-332.
56 N. Atanasiu, A. Westphal and M. Silva, Z. Acker-Pflanzenbau, 146(1978)165-177.
57 G. Dev and D.A. Rennie, Aust. J. Soil Res., 17(1979)155-162.
58 I. Watanabe, K. Ikegaya and S. Hiramine, Study tea, 0/57(1979)32-37
59 R.V. Olson, Soil Sci. Soc. Am. J., 44(1980)514-517.
60 K.R. Reddy and W.H. Patrick, in Proc. Symp. Isot. Rad. Res. Soil Plant Rel. IAEA 1978(1979), pp. 607-617.
61 B.L. Vasilas, J.O. Legg and D.C. Wolf, Agron. J., 72(1980)271-275.
62 A. Oaks, I. Stulen and I.L. Boesel, Can. J. Bot., 57(1979)1824-1829
63 G.A. Peterson, F.N. Anderson, G.E. Varvel and R.A. Olson, Agron. J. 71(1979)371-372.
64 J. Matula, Sci. Agric. Bohemoslov., 11(1979)25-32.
65 S. Mori and N. Nishizawa, Soil Sci. Plant Nutr., 25(1979)51-58.
66 K. Mengel and M. Viro, Soil Sci. Plant Nutr., 24(1978)407-416.
67 W. Merbach and G. Schilling, Arch. Acker- u. Pflanzenbau u. Bodenkd 24(1980)39-46.
68 H.M. Helal and K. Mengel, Plant Soil, 51(1979)457-462.
69 H. Beringer and K. Koch, Z. Pflanzenernaehr. Bodenkd., 143(1980) 449-456.
70 N. Amarger, A. Mariotti, F. Mariotti, J.C. Durr, C. Bourguignon and B. Lagacherie, Plant Soil, 52(1979)269-280.
71 W. Merbach and G. Schilling, Zentralbl., Bakteriol., Parasitenkd., Infektionskr., Hyg., Abt.2, Naturwiss.: Mikrobiol. Landwirtsch., Technol. Umweltschutzes,135(1980)99-118.
72 K.M. Goh, D.C. Edmeades and B.W. Robinson, Soil Biol. Biochem., 10(1978)13-20.
73 E. Ikonomova, Pochvozn. Agrokhim., 13(1978)88-97.
74 D.C. Edmeades and K.M. Goh, Commun. Soil Sci. Plant Anal., 10(1979) 513-520.
75 D.H. Kohl and G. Shearer, Plant Physiol., 66(1980)51-56.
76 G. Shearer, D.H. Kohl and S.-H. Chien, Soil Sci. Soc. Am. J., 42 (1978)899-902.
77 R.E. Karamanos and D.A. Rennie, Soil Sci. Soc. Am. J., 44(1980) 57-62.
78 Rolle und Umfang des Aminosäure-Pools bei Schwein und Geflügel im Wachstum. Abschlußbericht. Herausgeber: Akademie der Landwirtschaftswissenschaften der DDR, Berlin, 1980.
79 G. Gebhardt, R. Köhler and T. Zebrowska, Arch. Tierern., 28(1978) 603-608.

80 K. Gruhn and D. Glotz, Arch. Tierern., 29(1979)781-786.
81 A.N. Kosharov, N. Shmanenkov and M.D. Aitova, S-kh. Biol., 14(1979)449-451.
82 K. Krawielitzki, I. Voelker, J. Wünsche, U. Hennig, H.D. Bock, L. Buraczewska and J. Zebrowska, Arch. Tierern., 29(1979)771-780.
83 M. Shioya, S. Katsumata, M. Noma and K. Sasaki, Bull. Coll. Agric. Vet. Med. Nihon. Univ., 35(1978)290-299.
84 O. Simon, K. Adam and H. Bergner, Arch. Tierern., 28(1978)609-617.
85 H. Bergner, O. Simon and K. Adam, Arch. Tierern., 28(1978)21-29.
86 E. Deguchi, M. Niiyama, K. Kagota and S. Namioka, J. Nutr., 108(1978)1572-1579.
87 J. Varady, K. Boda and K. Kosta, Zivocisna Vyroba, 23(1978)537-544.
88 H. Bergner, U. Bergner and K. Adam, Die Nahrung, 24(1980)227-238.
89 U. Bergner, K. Adam and H. Bergner, Arch. Tierern., 28(1978)585-602.
90 I.V. Nolan and S. Stachiw, Brit. J. Nutr., 42(1979)63-80.
91 R. Roets, R. Verbeke and G. Peeters, J. Dairy Sci., 62(1979)259-269.
92 D.N. Salter, K. Daneshvar and R.H. Smith, Brit. J. Nutr., 1(1979) 197-209.
93 R.C. Siddons, D.E. Beever, J.V. Nolan, A.R. McAllan and J.C. Macrae, Ann. Rech. Vet., 10(1979)286-287.
94 E.L. Syvaoja and M. Kreula, J. Sci. Agric. Soc. Finl., 51(1979) 497-505.
95 C.W. Kreitler, J. Hydrol., 42(1979)147-170.
96 T. Yoneyama, A. Hashimoto and T. Totsuka, Soil Sci. Plant Nutr. (Tokyo), 26(1980)1-7.
97 F. Botter and Y. Belot, Bull. Inf. Sci. Tech. (Paris), 230/231 (1978)33-38.
98 T.J. Wolterink, H.J. Williamson, D.C. Jones, T.W. Grimshaw and W.F. Holland, U.S. Environ. Prot. Agency, Off. Res. Dev. EPA 1979.
99 H.D. Freyer, Tellus, 30(1978)83-92.

BALANCE OF ^{15}N FERTILIZER IN SOIL/PLANT SYSTEM

N. SOTIRIOU and F. KORTE

Gesellschaft für Strahlen- und Umweltforschung mbH München
Institut für Ökologische Chemie, D-8042 Neuherberg (G.F.R.)
Technische Universität München
Institut für Chemie, D-8050 Freising-Weihenstephan (G.F.R.)

ABSTRACT

A number of long-term experiments were conducted to study the fate of urea N under field conditions in varying soil types. A fertilizer N balance under the above conditions was calculated. The results show that the loss of ^{15}N enriched N fertilizer applied to soil is not significantly due to downward migration but is mainly the result of volatilization and denitrification. The utilization of applied N by plants increased during the experimental period. In these amounts only a very small percentage of N from previous applications was present.

INTRODUCTION

Extensive use of isotope techniques in agriculture and biology during the last three decades has greatly contributed to a better understanding of the problems of fertilizing and N-nutrition in plants. The fate of fertilizer N can be traced and a N balance consequently be drawn using ^{15}N enriched N fertilizer materials.

All experiments were carried out in wooden lysimeters (60 x 60 x 60 cm^3) as described in an earlier publication (ref. 1). Common West-European plants were used in experiments, where the crop rotation practice: root vegetables - cereals was followed. N fertilizer application at the beginning of the experiments and every second year was in the form of ^{15}N urea (47% ^{15}N). ^{15}N abundance assay was performed by mass spectrometry (ref. 2).

RESULTS

^{15}N utilization by plants

The significance of fertilizer N for plants becomes evident when N uptake by plants is compared with the amount of N added with the fertilizer and the soil N pool. The crop yield from a fertilized soil is substantially increased in comparison to the yield from the same soil with a high N pool but without fertilizer application. Whereas in the first year of application, when only ^{15}N was applied, the uptake of ^{15}N by plants was about 24 %, after reapplication of ^{15}N urea four years later about 50 %, of total N-uptake (N-pool+N-fertilizer). This suggests that the relative uptake of N from the soil N pool (N pool : N fertilizer) was significantly reduced. The increase in the uptake of N from fertilizer was still higher in the loess soil (Fig. 1).

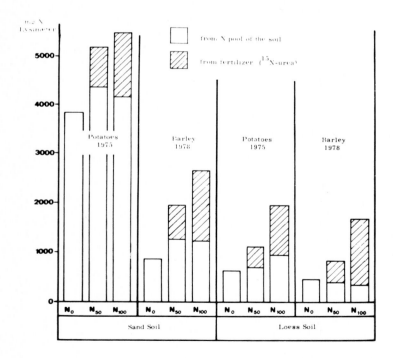

Fig. 1. Nitrogen uptake by plants from soil N-pool and ^{15}N-fertilizer (N_{50} = 50 kg N/ha; N_{100} = 100 kg N/ha; 1/2 at seeding time and 1/2 at the blossom stage)

The utilization rates of fertilizer N by plants during the whole experimental period (1975 - 1979) are summarized in Table 1. A very significant amount of the applied N was incorporated in the fruits as compared to that accumulated in the vegetative parts of the plants. An increase in the utilization of applied N by plants was observed throughout the 5-year experiment. After five years, for example, in the experiment with the sandy soil, N utilization was 57 %. Only a very small percentage of N from previous applications is included in the above amount. The same trends were obtained in the loess soil as well, thus - allowing for the 2 - 5% residues from previous applications - amounting to 45 % in the fourth year.

The N uptake by sugar-beet plants was abnormally low, probably reflecting the abnormally low yields of the crop.

TABLE 1

Percent utilization of ^{15}N urea by plants (100 kg N/ha, 1/2 at seeding time and 1/2 at the blossom stage)

Experiment/Year			Sand	Loess
Potatoes	1975	Leaves	8	2
		Potatoes	26	26
		Total	34 +	28 +
Winter Barley	1976	Straw	10	<1
		Grain	34	1
		Total	44 +	1
Sugar-Beets	1977	Leaves	1	2
		Sugar-beets	1	2
		Total	2	4 +
Summer Barley	1978	Straw	7	5
		Grain	35	40
		Total	42 +	45 +
Potatoes	1979	Leaves	19	1
		Potatoes	38	3
		Total	57 +	4

+) Fertilization with ^{15}N urea

In treatments where the first ^{15}N application was not followed by a new ^{15}N application the utilization rates of the residual fertilizer ^{15}N over the years in the sandy and the loess soil were 1 - 2% and 2 - 5% respectively.

Residual ^{15}N in soil

Most of the applied fertilizer ^{15}N remained in the top 20 cm of the soil throughout the duration of the experiment. Soil samples were obtained in the 0-20 cm, 20-30 cm, 30-40 cm and 40-50 cm depth every year at harvest time. The levels of the exchangeable fraction NH_4^+-N were 3 - 5% of those of total N. However NO_3^--N was often in non-detectable levels in both the sandy and the loess soil.

Urea, applied to soil is transformed to NH_3 and CO_2 by urease, which in cultivated soils is a product of the soil microbial activity (ref. 3 - 4). Nitrification can start operating immediately following urea application. As the soil organic matter content drastically decreased below 30 cm depth (ref. 5) it can be assumed that most of the fertilizer N which moved to deeper layers, was fixed by clay minerals (ref. 6).

More than 80% of residual fertilizer N was found in the 0-30 cm depth. Considerable amounts of labelled N in greater depths especially between 40 and 50 cm were detected only in the sandy soil and only after the first year's experiment. The residual fertilizer N data expressed as total applied ^{15}N urea are presented in Table 2.

TABLE 2
Soil residual fertilizer ^{15}N at harvest time of the 1975 potato experiment. Results are expressed as % of applied ^{15}N urea (100 kg N/ha at seeding time and 1/2 at the blossom stage)

Soil Depth (cm)	Sand	Loess
0 - 20	15.9	21.8
20 - 30	5.0	3.4
30 - 40	2.2	0.9
40 - 50	2.4	<0.1
Total	25.5	26.2

The largest amounts of residual fertilizer N were found in the top 20 cm of the loess soil. However there was no significant difference between the fertilizer ^{15}N total residues in sandy and loess soil (0 - 50 cm).

Balance of applied ^{15}N urea in soil/plant system

The major loss of fertilizer N applied to soil is a result of volatilization and denitrification rather than downward migration (leaching). The measured ^{15}N values of NO_3^--N in leached water are only less than 1% of the applied fertilizer N. The ^{15}N balance after the first application of ^{15}N-labelled urea is shown in Table 3.

TABLE 3
Balance of ^{15}N after single application of ^{15}N-urea in % of applied amount (100 kg N/ha, 1/2 at seeding time and 1/2 at the blossom stage)

Assay	Sand	Loess
Plant Utilization	35	29
Soil Residues	26	26
Leaching Water	<1	<1
Losses	39	45

The utilization of applied N by plants was better in sandy soil than in loess in which the loss of N in the first year after application was higher than in sandy soil. A considerable part of the fertilizer N (26%) was still in the soil.

During the vegetation of the second year (barley) the plants took up only small amounts of the residues N from the year before (1 - 2% in sandy soil, 2 - 5% in loess soil, table 1). The percentage of fertilizer N recovered was 10 - 20%, and 20 - 30% in the sandy and loess soil, respectively. These amounts were most probably bound to organic matter and clay minerals of the soil (ref. 7).

REFERENCES

1 F. Korte and N. Sotiriou, in Soil Nitrogen as Fertilizer and Pollutant, Proc. and Report of a Research Coordination Meeting, Piracicaba, July 3 - 7, 1978, IAEA, Vienna, 1980, pp. 105 - 125.

2 R. Medina, W. Hoppe and H.-L. Schmidt, Fresenius Z. Anal. Chem., 292 (1978) 403-407.
3 J.P. Conrad, Soil Sci. Soc. Amer. Proc. 5 (1940) 238-241.
4 T. Pfanneberg, W.R. Fischer and E.A. Niederbudde, Pflanzenernähr. und Bodenkd., 141 (1978) 469-477.
5 H. Fleige and K. Baeumer, Agro-Aecosystems, 1 (1974) 19-29.
6 H. Nishita and R.M. Haug, Soil Sci. 116 (1973) pp. 51-58.
7 A.L. Allen, F.S. Stevenson and L.T. Kurtz, J. Environmental Quality, 2 (1973) pp. 120-124.

NITROGEN-15 FOR THE CONTROL OF NITROGEN UTILIZATION IN THE
ACTIVATED SLUDGE PROCESS

H. NOGUCHI, K. MURAOKA, S. NODA, T. MATSUZAKI and T. MORISHITA
Hikari Kogyo Co Ltd, Tokyo, Japan

ABSTRACT

When industrial waste water is treated by the Active Sludge Process, nitrate and phosphate are added as nutrient salts for the decomposing organisms in the ratio $BOD^{+)}$: N:P = 200:5:1. The main ingredient of the treated sludge consists of the biomass of the decomposing microorganisms. If the concentration of toxic substances in the sludge is under the permissible limit, the product can be utilized as an useful organic fertilizer. The nitrogen of the bacterial protein is coming from the added nutrient and from suspended organic solids. ^{15}N was used to label the added nitrogen. The ^{15}N-content was measured by emission spectroscopic technique. The mechanism and the efficiency of nitrogen absorption were studied under different conditions. The uptake of nitrogen by the decomposers was influenced remarkably by the kind and the amount of the added carbon source. In order to trace the distribution of P and N in a plant, a doubly labelled sludge (^{32}P, ^{15}N) was prepared and used for the cultivation of rice plants. In the seedling ^{32}P is preferably taken up by the growing parts of the plant. At the time of harvest the grain contains a large amount of ^{15}N.

INTRODUCTION

The Activated Sludge Process is generally used by various industrial factories for the elimination of organic pollutions from sewage water. When water with organic pollutions from industrial effluents is treated by the Activated Sludge Process, nitrate and phosphate are added as nutrient salts for the decomposing organisms (ratio in respect to Biological Oxygen Demand: BOD/N/P = 200/5/1).

$^{+)}$ Abbreviations used: BOD = Biological Oxygen Demand

After the treatment the biomass of the sludge is mainly consisting of the decomposing microorganisms, and the nitrogen in the bacterial protein is originating in part from organic compounds in the sewage water and in part from the added nitrate. In spite of the general use of the Activated Sludge Process the quality of the sludge as a nitrogen and phosphorous fertilizer has not been deeply investigated, yet in the case when the content of the sludge in toxic substances would be within or under the permissible limit, it could be used as a valuable organic fertilizer, contributing simultaneously to the recycling of natural compounds and to resolve pollution problems. Therefore the present investigation with ^{15}N and ^{32}P deals with the investigation on the uptage of inorganic N and P by sludge under various conditions and with the use of the organic fertilizer by growing rice plants.

METHODS
Incubation and production of labelled sludge

After the acclimatisation time of 7 days 50 ml/10 l culture medium was added to a sludge originating from human treatment waste. The culture medium contained nitrogen and phosphorus in the ratio 3,5 or 8 N and 2 P/200 BOD. The nitrogen compounds ammonium sulfate ammonium nitrate and urea contained 15 atom-% ^{15}N. Phosphorus was supplied as $H_3^{32}PO_4$ with an activity of 27.9 µC/g. Various industrial sludges were simulated by different carbon sources: acetaldehyd, butadien rubber, cellulose, glucose and palmitic acid. Polyacrylamid was used as coagulant. The experimental apparatus is schematically demonstrated in fig. 1. The mixture of sludge and nutrient solution is incubated at 25-35 °C, 8 ppm O_2, pH 7-8 and an effluent volume of 1 m^3/h. The dried sludge was mixed with soil. Rice plants were grown on this product under greenhouse conditions.

Isotope analyses

Nitrogen bound in the material was converted into gaseous N_2 (usually 1-5 µg nitrogen). The evacuated sealed tubes were supplied with CuO. The ^{15}N content was estimated in discharge tubes (4-8 mm diameter Pyrex vessels) by an emission spectroscopic analyzer.

The ^{32}P activity was measured by a G.M. counter and located by autoradiography.

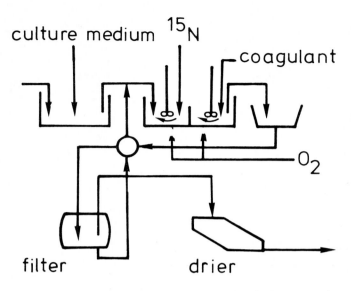

Fig. 1. Scheme of the experimental design to produce labelled sludge for the rice cultivation test.

RESULTS

Factors affecting the nitrogen uptake by the sludge

The total nitrogen content of sludges from various industrial sources is different (Tab. 1).

TABLE 1
N-content of various industrial sludges

origin	N-content (%)
starch (sugar factory)	6.6 - 7.4
paper (pulp factory)	5.5 - 6.4
processed animal (fish) factory	4.2 - 4.9
synthetic acetic acid factory	6.2 - 7.1
synthetic rubber factory	4.0 - 4.6

It is generally accepted, that the N-uptake into sludges is determined by the N-content of the effluent liquids and to the relation to their BOD alone, but our results with ^{15}N-labelled urea show, that the ^{15}N-uptake is only partially proportional to the offered concentration of nitrogen, and also largely affected by the kind of the carbon source present (Fig. 2). Glucose and acetaldehyde were found to be most efficient. Aeration of the sample decreases the amount of carbon source needed to bind a given amount of N (Fig. 3).

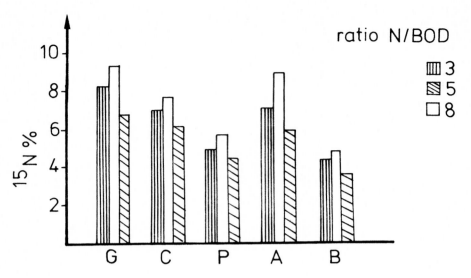

Fig. 2. N-uptake into sludge dependent on the ratio of N-concentration/BOD using different carbon sources (A: acetaldehyde, B: butadien rubber, C: cellulose, G: glucose, P: palmitic acid).

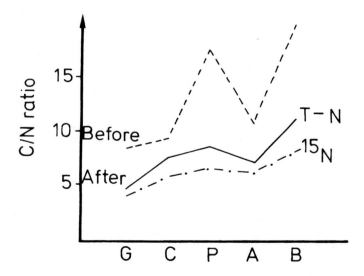

Fig. 3. Variation of the C/N-ratio before and after aeration treatment. Abbreviations see Fig. 2.

The variation of ^{15}N-uptake may be due to a difference between the organisms present after the supply of various substrates (Tab. 2).

TABLE 2
Dominating Protozoa at various kinds of carbon source

C-source	Main Prior Geneses
control	Opecularia, Vorticella
acetaldeyde	Vorticella, Opecularia
butadien rubber	Arcella, Difflugia, Vorticella, Opecularia
cellulose	Vorticella, Opecularia
glucose	Monas, Opecularia
palmitic acid	Vorticella, Epistylis, Aspidsca

Availability of N and P bound by sludge as fertilizer

The uptake of ^{15}N applied in different forms from the soil into plants is shown in Table 3. The ^{15}N-label is mainly concentrated in the grain. Fig. 4 supports the results about uptake for the radioactive phosphorus.

TABLE 3
Uptake of nitrogen (labelled with ^{15}N) from a sludge from the soil into rice plants. Yield from artificial fertilizer given in %.

^{15}N source	grain	stalk	upper leaf blade	lover leaf blade	leaf sheath
^{15}N sludge	61.79	4.92	16.21	4.82	12.22
(^{15}NH$_4$)$_2$SO$_4$	55.42	8.42	21.07	3.95	11.08
^{15}N urea	58.96	5.89	19.32	5.05	10.78
(^{15}NH$_4$)$_2$SO$_4$ and manure	63.56	5.09	14.42	5.52	11.41

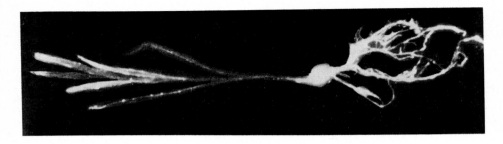

Fig. 4. Radioautography of ^{32}P uptake into a young rice plant. The phosphorus is transported preferably into the developing parts of the plant.

CONCLUSION

The sludge produced by the process described above can be serve as a useful fertilizer. The nitrogen is taken up rapidly by the plant. The phosphorus seems to be transported mainly to the growing tissues of the young plants.

AKNOWLEDGEMENTS

We thank to K. Kumazawa and S. Yamamuro for the analyses of the ^{15}N-samples, and S. Aso and A. Takenaza for the cultivation of rice plants.

REFERENCES

1 K. Kumazawa and S. Yamamuro, in Study for Heavy Nitrogen Utilisation, Academic Press Center, 1980, p. 17-63.
2 R. Sudo, Chem. Engin., 37 (1973), 886.

^{15}N IN THE STUDY OF NITROGEN TRANSFORMATIONS AND AMMONIUM EXCHANGE REACTIONS IN SOIL

MARIE KRÁLOVÁ, K. DRAŽĎÁK, J. KUBÁT[1] and M. EBEID[2]
Institute of Experimental Botany Czechoslovak Academy of Sciences, Praha (CSSR)
Institute of Crop Production, Praha - Ruzyně (CSSR)[1]
Department of Botany, Qatar University, Doha (Qatar)[2]

ABSTRACT

^{15}N- and ^{14}C-techniques were employed in the study of nitrogen mineralization-immobilization processes connected with denitrification and NO_3^--assimilation and affected by the presence or absence of glucose and by constant or fluctuating temperatures. To study ammonium exchange reactions of nitrogen in soil solution and within interlayers of the mineral lattice of trioctahedral soil clay minerals, the isotopic dilution procedure has been used. The ability of soil clay minerals to bind nitrogen compounds and their contribution to the transformation process of nitrogen has been studied.

INTRODUCTION

In this paper, ^{15}N- and ^{14}C-isotopic techniques were employed to investigate processes of nitrogen mineralization-immobilization, denitrification and nitrate assimilation. The effect of glucose was also studied. Moreover, studies of ammonium exchange reactions and ammonium fixation by soil components were carried out.

MATERIAL AND METHODS

Soil material

The soil used was clay loam slightly leached chernozem (clay content 65.4%) with a pH of 6.9, contents of carbon 1.42%, nitrogen 0.147% and a moisture content of 20 % (w/w).

Incubation procedure
a) 50 g soil samples (2-4mm) were treated with 10 ml of a solution containing 500 mg glucose and 89.9 mg ($^{15}NH_4$)$_2$SO$_4$ labelled with 90 atom % ^{15}N. The effect of temperature 20 and 30°C and of a fluctuating 20-30-20°C temperature cycle in a 24 hour period was investigated under laboratory conditions for 4 days.
b) 50 g soil samples (2-4 mm) were treated with a 10 ml solution containing 40 mg K^{15}NO$_3$ labelled with 96.2 at % ^{15}N and 600 mg glucose. The effect of temperature (28°C) in a 90 day period on nitrogen assimilation was investigated.
c) 50 g portions of soil aggregates (2-4 mm) were remoistened with 10 ml of a solution containing 600 mg [1-^{14}C]- or [6-^{14}C] glucose and 113.2 mg (NH$_4$)$_2$SO$_4$. The incubation flasks were attached to an open circuit radiorespirometer [1], and the incubation was conducted at constant (20°, 30°C) or sinusoidally fluctuating temperatures between 20° and 30°C in 24 h cycle for 3 days.
d) Samples of 50 g soil were presaturated to a maximum cation exchange capacity with ^{15}NH$_4$Cl labelled with 4.74 ^{15}N. After 30 minutes the samples were centrifugated and leached until negative Cl$^-$-reaction. The presaturated soil sample was dried to a moisture content of 20 % (w/w), then incubated at 28°C for 90 days.

The moisture content of all samples was maintained at the same level by weight control. The experiments were performed in triplicate and the results were expressed on the basis of oven-dry soil (105°C).

Methods

For the chemical determinations, methods described elsewhere [2-5] were used. The abundance of ^{15}N was determined by a mass spectrometer after separation of N$_2$ from the sample [5]. For the extraction of exchangeable NH$_4^+$-N two techniques were used [6].

RESULTS AND DISCUSSION

Mineralization-immobilization

A biological mineralization-immobilization turnover in the soil was confirmed by our experiments. The mineralization-immobilization process was enhanced by increasing the temperature and fluctuating the temperature were more effective [7-8]. The total nitrogen showed a slight decrease in its content after 24 h of incubation at 20° or 30°C (Fig.1). This behaviour is attributed to ammonium exchange reactions and N-chemical fixation by soil clay minerals as has been

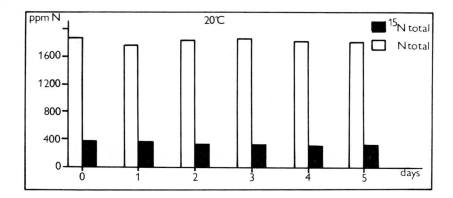

Fig.1. Relationships of total N and ^{15}N after incubation with $(^{15}NH_4)_2SO_4$ during a 5-day incubation at $20°C$. The analytical results were calculated on the oven-dried soil, the remoistened soil was amended with glucose.

confirmed by X-ray analysis. On the other hand, the total nitrogen content increased after 48 h of $20°C$ incubation. This increase was associated with the increase of Azotobacter growth and nitrogenase activity [7].

During incubation at $20°C$ for 10 h the evolution of $^{14}CO_2$ from [1-^{14}C]-labelled glucose covered the total production of CO_2. This result suggests the involvement of the pentose phosphate (PP) pathway in the primary glucose decomposition. The maximum evolution rate of $^{14}CO_2$ from [6-^{14}C]-glucose was recorded after a 45 h incubation at the same temperature probably due to a glucose degradation by the Embden-Meyerhof-Parnas (EMP) pathway. The relative consumption of [1-^{14}C]- and [6-^{14}C]-glucose was found to be 62.2 % and 43.2 %, respectively [9].

The processes of nitrate dynamics in the soil, particularly denitrification and nitrate assimilation into organic compounds were also studied (Fig. 2). Our results showed that under the same experimental conditions (incubation for 24 h at $28°C$), the reduction of ^{15}N-labelled nitrate either in presence or absence of glucose was 65 and 4 %, respectively [10]. The reduction of nitrates was a result of a biological immobilization.

Ammonium exchange reactions

The amount of exchangeable NH_4^+ (replacement by treating the soil with 1N KCl solution) in dependence on the incubation time is shown in Fig. 3. After 30 minute contact 87 % of 0.77 mg NH_4^+-^{15}N

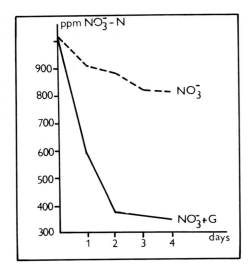

Fig.2. Reduction of nitrate at 28°C in incubations with ^{15}N-potassium nitrate in presence or absence of glucose in soil sample.

in 1 g soil was an exchangeable. After 90 days decreased up to 64 %. The extract contained about 6 % more ammonium when the soil was pretreated with ethylene glycole. This result is attributed

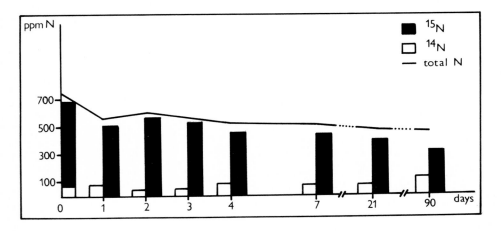

Fig.3. Exchangeable ammonium-N extracted with 1N KCl from soil sample incubated at 28°C during 90 day period.

to a release of ammonium ions physically entrapped within interlayers of the lattice of clay minerals[6].

Thus one can assume that the exchangeable ammonium consists of two differently bound portions, one ionic bound which can be replaced by treatment with 1N KCl solution, the second probably dissolved in water layers which can be removed by the pretreatment of the soil with ethylene glycole.

Ammonium fixation by soil clay minerals

The non-exchangeable ammonium ions are fixed by soil clay minerals and this strong binding was defined as chemical fixation [11]. This phenomenon is most common in clays with trioctahedral structure.

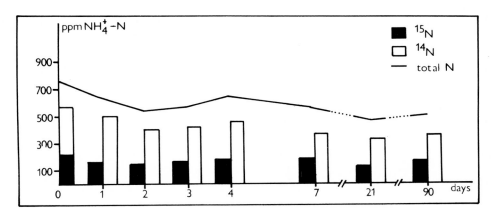

Fig.4. Fixed nitrogen in the lattice of trioctahedral soil clay minerals. (50% H_2SO_4 and 1N HF procedure).

In determinations of the fixed ammonium by the soil complex used we found 6 % of the ammonium added (mg NH_4^+-^{15}N/ 1 g soil) was fixed after 30 minutes contact. During short-term incubations (i.e. 0-4 days) at 28°C, we observed a slight change in the amount fixed, which was in agreement with our previous results concerning N-transformation processes [7]. Long-term incubation (i.e. till 90 days) at the same temperature showed a decrease in fixed N probably due to the mobilization of the fixed ammonium. The biological activity was found to start after more than 21 days [6].

The quantitative descriptions of some nitrogen cycle processes permit a first approximation of nitrogen balance in productive agricultural systems. The use of ^{15}N techniques have made possible to control these processes.

REFERENCES
1 J.Kubát, M. Králová, F. Kysela, K. Draždák and B. Novák,

Scientia agric. bohemoslovaca, 11 (1979), 21-24.

2. M. Králová, M. Ebeid and K. Draždák, Scientia agric. bohemoslovaca, 12 (1980) 157-161.
3. M. Králová, K. Draždák, P. Stránský and J. Kubát, in Proc.Techn. Committee Meeting Stable Isotope in the Life Science, Leipzig, February 14-18, 1977, IAEA, Vienna, 1977, pp. 393-397.
4. M. Králová, Scientia agric. bohemoslovaca, 9 (1977) 1-5.
5. M. Králová, Rostlinná výroba, 18 (1972) 107-111.
6. M. Králová, K. Draždák and J. Kubát, Scientia agric. bohemoslovaca, (in press).
7. M. Králová, J. Kubát, K. Draždák and B. Novák, Scientia agric. bohemoslovaca, 12 (1980) 1-7.
8. J. Kubát, M. Králová and B. Novák, Zbl. Bakt. II., Abt.134 (1979) 229-236
9. J. Kubát, M. Králová, B. Novák, K. Draždák and F. Kysela, Scientia agric. bohemoslovaca, 11 (1979) 91-96.
10. M. Králová, K. Draždák, J. Kubát and B. Novák, Scientia agric. bohemoslovaca, 11 (1979) 83-89.
11. S.L. Jansson, Kugl. Lantbr. Ann., 24 (1958) 101-361.

ISOTOPIC FRACTIONATION IN SOYBEAN NODULES[1]

DANIEL H. KOHL, BARBARA A. BRYAN, LORI FELDMAN, PETER H. BROWN and GEORGIA SHEARER
Department of Biology, Washington University, St. Louis, Missouri 63130 (USA)

ABSTRACT

The nodules of soybean (Glycine max [L] Merrill) plants are consistently found to be considerably more enriched in ^{15}N than the rest of the plant; other plant parts are much more homogeneous in ^{15}N abundance. The ^{15}N enrichment of soybean nodules increased with time during the growing season and with N_2-fixing efficiency of associations formed with different strains of Rhizobium japonicum. These results demonstrate that a process, other than N_2-fixation itself, which is associated with N_2-fixing activity, also causes elevation of ^{15}N abundance in soybean nodules. Other legumes we measured also showed ^{15}N enriched nodules, while non-leguminous N_2 fixers did not.

Denitrification experiments with the symbiont strains we used did not support the hypothesis that denitrification (or dissimilatory NO_3^- reductase activity) is the cause of ^{15}N enrichment of nodules. Another mechanism of ^{15}N enrichment of nodules is consistent with all of our results. This mechanism involves isotopic fractionation associated with synthesis of compounds rich in N which are then exported from the nodule, leaving behind ^{15}N enriched metabolites for the synthesis of nodule tissue. The degree of enrichment of ^{15}N of nodules may indicate the degree to which N recently fixed by the same nodule is used for nodule synthesis.

INTRODUCTION

The N of whole soybean plants which are grown with atmospheric N_2 as the sole source of N_2 has an isotopic composition within 1‰ of atmospheric N [1,2]. In contrast, we have consistently observed that soybean nodules are significantly enriched in ^{15}N [3]. This observation has been recently confirmed by Turner and Bergersen [4]. The difference in $\delta^{15}N$ between nodules and whole plants ranged from +2.8 to +12.8, with an average (for 59 observations) of +8.3. The ^{15}N abundance of other plant parts (roots, stems, foliage, pods, seeds) was much more uniform [3]. The largest difference between any of the other plant parts was only about two

---

[1] This work was supported by the National Science Foundation, Grant #PCM 7823270.

$\delta^{15}N$ units, a modest difference in comparison with the usually quite large difference between nodules and other plant parts. The homogeneity of isotopic composition of non-nodular tissue persisted throughout the growing season, even during times of massive mobilization and transport of N from vegetative to reproductive tissues.

In this paper, we expand on the earlier observations. Our objective is to investigate the cause of the ^{15}N elevation in nodules, motivated by the conviction that details of the metabolic pathways of fixed N will be revealed by such an exploration.

RESULTS AND DISCUSSION

Figure 1 shows the elevation of ^{15}N in nodules as a function of time during the growing season for three experiments done under different conditions. In one experiment (Panel A, Fig. 1), plants were grown hydroponically in N-free medium [3]; in the second (Panel B, Fig. 1), the plants were grown in the greenhouse in a nutrient poor soil (700 µg N/g soil) [3]; and in the third (Panel C, Fig. 1), the plants were grown in the field in a nutrient rich soil (2000 µg N/g soil) [5,6]. In all three experiments, there was a significant correlation between $\Delta\delta^{15}N$ ($\delta^{15}N_{nodules}$ minus $\delta^{15}N_{plants}$) and time after planting (r=0.924, 0.990 and 0.997 for experiments A, B and C respectively; and p<0.025, <0.005 and <0.005 respectively).

In a fourth experiment, soybeans infected with six strains of Rhizobium japonicum, were grown hydroponically in N-free medium, and harvested at a single time (mid-bloom). The associations formed with these bacterial strains were variably effective in N_2-fixing capability. One of the associations was totally ineffective. The ^{15}N abundance of these ineffective nodules was virtually identical to the rest of the plant. The ^{15}N enrichment of the nodules ($\Delta\delta^{15}N$) was highly correlated (r=0.985, p<0.001) with N_2-fixing efficiency, calculated as the quantity of N_2 fixed per quantity of N in the nodules (see Fig. 2). This experiment suggests a direct connection between N_2-fixation in soybeans and some process which leads to partitioning of N isotopes between nodules and the rest of the plant.

We also have a few data on the ^{15}N abundance of foliar and nodule tissue in other N_2-fixing plants. These data are summarized in Table 1. The results show that ^{15}N enrichment of nodules is not restricted to soybeans. The ^{15}N content of the nodules of two other legumes examined was also elevated. The results also suggest a difference between legumes and non-legumes with non-legumes showing little enrichment and sometimes depletion of ^{15}N in their nodules.

The unique ^{15}N enrichment of nodules of soybeans and certain other legumes is dependent on N_2-fixing activity. This strongly suggests that some metabolic event in the overall process of N_2-fixation by legumes leads to an elevation of ^{15}N abundance in the pool of intermediates used for synthesis of active nodules, but not in the pool of intermediates used for the synthesis of any other plant part (including ineffective nodules).

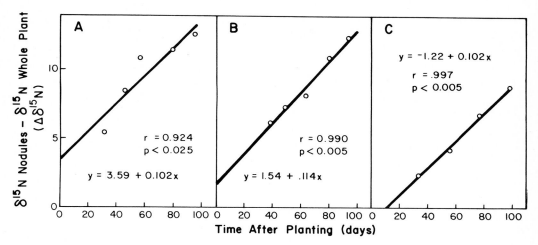

Fig. 1. ^{15}N enrichment in nodules versus time after planting. Soybeans were grown hydroponically with N-free medium (panel A), in nutrient poor soil in the greenhouse (panel B), and in the field in a nutrient rich soil (panel C). Each data point is a mean result from at least 4 replicate plants.

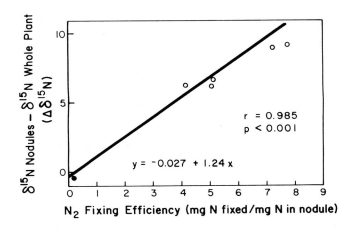

Fig. 2. ^{15}N enrichment in nodules versus N_2 fixing efficiency. Soybeans were inoculated with 6 strains of *Rhizobium japonicum*, including one that was totally ineffective (filled circle), and grown hydroponically in N free medium. Plants were harvested at a single time (mid-bloom). Each data point is a mean of results from 5 replicate plants.

TABLE 1

The ^{15}N Abundance of Nodules Minus the ^{15}N Abundance of Leaves ($\Delta\delta^{15}N$) in Several N_2-Fixing Plants

Plant	Number of Specimens	$\Delta\delta^{15}N$ mean ± s.e.
LEGUMES		
Glycine max (L) Merrill[a]	58	+9.1 ± 0.7
Vigna sinesis (L) Endl.[b]	4	+6.7 ± 0.5
Prosopis glandulosa[c]	6	+7.7 ± 1.9
NON-LEGUMES		
Eleagnus angustifolia[d]	1	+1.1 ---
Myrica pensylvanica (loisel)[d]	1	-5.8 ---
Alnus maritima[d]	1	+2.8 ---
Alnus rubra[e]	6	+1.0 ± 0.2
Ceanothus velutinus[e]	4	-0.9 ± 0.3

a) From Shearer et al [3].
b) Samples were supplied by J.C. Burton (Nitragin Co., Milwaukee, WI 53209).
c) Samples were supplied by Dr. Ross Virginia who grew the plants in the greenhouse from seed collected in the Sonoran (California) desert. Plants were grown in native soil.
d) From Shearer et al [7].
e) Samples were supplied by Drs. Kermit Cromack and Sue Conard (Dept. of Forestry, Oregon State University, Corvallis, Oregon) who collected them from the field.

One possibility is that active nodules are synthesized from a pool which includes residual substrate of the denitrification process. A significant isotope effect (∿2%[8]) is known to accompany denitrification. Anaerobic nitrate dependent N_2-fixation and denitrification have been reported in free living bacteria and soybean bacteroids metabolizing under strictly anaerobic conditions [9,10,11,12,13]. Denitrification (or nitrate respiration) could conceivably contribute energy for N_2-fixation. To test the possibility that denitrification is the cause of ^{15}N elevation in nodules, we studied NO_3^- reduction with pure culture of the Rhizobium japonicum strains which we used to inoculate soybeans (see Fig. 2). Cultures were grown aerobically to mid-log phase, at which time flasks were sealed and NO_3^- and acetylene were aseptically injected (in order to block reduction of N_2O to N_2 [14,15]). The cultures were allowed to gradually become anaerobic as they used the atmospheric oxygen in the head space of the flask. The incubation was continued for three days: gas samples were analyzed for N_2O and the medium was analyzed for NO_3^- and NO_2^-. The results are shown in Table 2. Strain 61A 24 fixed almost no nitrogen, produced nodules with $\Delta\delta^{15}N$ values close to zero, but was competent to denitrify. On the other hand, strain USDA 33 was a quite capable nitrogen fixer, produced nodules with an elevated $\delta^{15}N$ value, but was incapable of reducing NO_3^-. Clearly then these results are not consistent with the hypothesis that denitrification or nitrate respiration is responsible for ^{15}N enrichment in soybean nodules. (However, the possibility that the nodule has denitrifying capabilities that are different from

TABLE 2

Denitrification and NO_3^- Respiration by Six Strains of Rhizobium japonicum Which Exhibit Variably Effective Symbioses with Soybeans

Strain	Rhizobium - Soybean Symbioses		Denitrification Experiments-Pure Cultures		
	N_2 - fixing[a] Efficiency	$\Delta\delta^{15}N$[b]	Residual NO_3^--N	NO_2^--N	N_2O-N
			% of NO_3^--N added		
61A 24	.07	−0.4	4	n.d.[c]	96
USDA 33	5.24	+6.6	100	n.d.	n.d.
61A 92	4.14	+6.2	15	84	0.8
61A 93	5.05	+6.0	24	76	n.d.
USDA 136	7.26	+8.9	8	n.d.	92
USDA 138	7.71	+9.1	0	n.d.	101

a) mg N fixed per mg nodule-N.
b) $\delta^{15}N$ of nodule minus $\delta^{15}N$ of plant.
c) n.d. indicates non-detectable levels.

those of the free living bacteria cannot be ruled out by this experiment. Against this possibility we plan an investigation of denitrification with intact nodules that are infected with different symbiont strains.)

Another mechanism of ^{15}N enrichment of nodules is consistent with all of our results. This mechanism involves isotopic fractionation associated with synthesis of compounds rich in N (to be called transport compounds) which are exported from the nodule. This fractionation would leave behind ^{15}N enriched metabolites for the synthesis of additional nodule tissue. This mechanism is illustrated by Fig. 3, which is a simplified representation of certain N transformations within a N_2-fixing plant, and more especially within an individual nodule. This diagram shows two separate pools of N-containing metabolites: one, external to the nodule (pool I_e); and one, internal to the nodule (pool I_n). The sources of N to pool I_e are transport compounds from all nodules, N added exogenously, and N from the seed. This pool is used for synthesis of plant tissue (including nodules) via paths 6´ and 6, with no resultant alteration in isotopic composition (since plant tissue, including ineffective nodules are uniform in isotopic composition). The source of N to pool I_n is N that is recently fixed within the same nodule. The bulk of this N is used to synthesize transport compounds (via path 3), but a small fraction of it can be used to synthesize new nodular tissue within the same nodule (via path 4). If the isotopic fractionation for synthesis of transport compounds is greater than that for synthesis of nodule tissue, a mechanism is provided for elevating the ^{15}N abundance of nodules compared to other plant tissues. (In addition, nodule decomposition, via path 5, could cause additional ^{15}N enrichment of nodules.)

This mechanism of ^{15}N enrichment would require the following: (a) since nodule tissue synthesis via path 4 is possible only after nodules are competent to fix significant amounts of N, newly developing nodules must receive N entirely from the external pool (pool I_e) via path 5 and should therefore have a value of $\delta^{15}N$ close

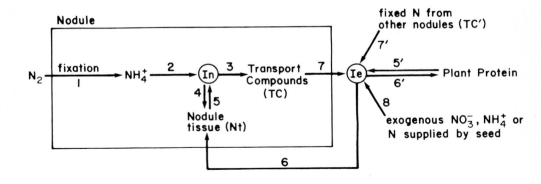

Fig. 3: Simplified diagram of N transformations which could lead to ^{15}N enrichment of N_2-fixing nodules.

to that of the rest of the plant. Since there is a higher proportion of young nodules early in the growing season, one would expect the enrichment of the total pool of nodules to increase as the plant matures. This expectation is consistent with the time course data shown in Fig. 1. (b) If exogenous N is supplied, the size of the external pool, I_e, should increase relative to that of the internal pool, I_n. This should increase the contribution of N from pool I_e (compared to I_n) and reduce the ^{15}N enrichment of nodules relative to other plant parts. This expectation is also supported by the results seen in Fig. 1, which shows that the difference in ^{15}N abundance between nodules and whole plants was greatest in the hydroponic experiment (in which all of the N, except for N supplied in the seed, was fixed), and smallest in the field experiment, in which a large fraction of the plant's N was derived from the soil (50 to 95% [5,6]). (c) Symbiont strains which have a decreased ability to fix N, without a proportional decrease in nodule tissue synthesis, should receive an increased relative contribution of N from pool I_e, and should thus show a smaller ^{15}N enrichment in nodules. This expectation is also consistent with our data which show a strong correlation between the elevation of ^{15}N in nodules and N_2-fixing efficiency.

Implicit in the mechanism of ^{15}N enrichment of soybean nodules illustrated by Fig. 3 is the assumption that the N isotope fractionation associated with transport compounds is greater than that for nodule synthesis. Thus the ^{15}N enrichment may well depend on the particular compounds which are exported from the nodule. In soybeans for example, a large fraction of the N compounds exported are ureides [14] whereas in Alnus rubra, citrulline is the major transport compound [15]. Such differences may account for the difference in $\Delta\delta^{15}N$ values between legumes and non-legumes seen in Table 1. When we measure a larger selection of legumes, we may also find differences in ^{15}N abundance between them, resulting from the synthesis of different kinds of compounds for export from the nodules. Alternatively in non-legumes there may be no further synthesis of nodule tissue after nodules are competent to fix N_2 with the consequence that path 4 from the internal pool, I_n, to nodule tissue would not be operational.

Also implicit in this mechanism is the assumption that a major source of N for nodule tissue synthesis is N that is recently fixed within the same nodule. Thus, the magnitude of ^{15}N enrichment should serve as an index of the degree to which nodules synthesize their own tissue from N fixed within the same nodule, if this mechanism proves to be correct. We plan to carry out a series of experiments designed to rigorously test the hypothesis that this is the mechanism responsible for nodule ^{15}N elevation.

REFERENCES

1. D.H. Kohl and G. Shearer, Plant Physiol. 66 (1980) 51-56.
2. N. Amarger, A. Mariotti, F. Mariotti, J.C. Durr, C. Bourguignon and B. Lagacherie, Plant Soil 52 (1977) 269-280.
3. G. Shearer, D.H. Kohl and J.E. Harper, Plant Physiol. 66 (1980) 57-60.
4. G.L. Turner and F.J. Bergersen, Poster, Fourth International Symposium on Nitrogen Fixation, Canberra, Australia, December, 1980.
5. D.H. Kohl, G. Shearer and J.E. Harper, in E.R. Klein and P. Klein (eds) Stable Isotopes: Proceedings of the Third International Conference, pp. 317-325, Academic Press, New York.
6. D.H. Kohl, G. Shearer and J.E. Harper, Plant Physiol. 66 (1980) 61-65.
7. G. Shearer and D.H. Kohl, in A. Frigerio (ed) Recent Developments in Mass Spectrometry in Biochemistry and Medicine, pp. 623-640, Plenum Press, New York.
8. R.P. Wellman, F.D. Cook and H.R. Krouse, Science 161 (1968) 269-270.
9. J. Rigaud, F.J. Bergersen, G.L. Turner and R.M. Daniel, J. Gen. Microbiol. 77 (1973) 137-144.
10. C.A. Neyra and P. Van Berkum, Can. J. Microbiol. 23 (1977) 306-310.
11. R.M. Zablotowicz and D.D. Focht, J. Gen. Microbiol. 111 (1979) 445-448.
12. C.A. Neyra, J. Döbereiner, R. Laland and R. Knowles, Can. J. Microbiol. 23 (1976) 300-305.
13. R.M. Zablotowicz, D.L. Eskew and D.D. Focht, Can. J. Microbiol. 24 (1978) 757-760.
14. W.L. Balderston, B. Sherr and W.J. Payne, Appl. Environ. Microbiol. 31 (1976) 504-508.
15. T. Yoshinari and R. Knowles, Biochem. Biophys. Res. Comm. 69 (1976) 705-710.
16. D.F. Herridge, C.A. Atkins, J.S. Pate and R.M. Rainbird, Plant Physiol. 62 (1980) 495-498.
17. A.D.L. Akkermans, W. Roelofsen and J. Blom, in J.C. Gordon, C.T. Wheeler, D.A. Perry (eds) Symbiotic Nitrogen Fixation in the Management of Temperate Forests, School of Forestry, Oregon State University, Corvallis, April 2-5, 1979, pp. 160-174.

EXPERIMENTAL DETERMINATION OF KINETIC ISOTOPE FRACTIONATION OF NITROGEN ISOTOPES DURING DENITRIFICATION

A. MARIOTTI[1], J.C. GERMON[2], A. LECLERC[2], G. CATROUX[2], R. LETOLLE[1]

[1]Université P.M.Curie, Laboratoire de Géologie dynamique, 4 pl.Jussieu, 75005 PARIS.
[2]Laboratoire de Microbiologie des Sols, INRA, 7 rue Sully, 21034 DIJON cedex.

ABSTRACT

The fractionation of the nitrogen isotopes ^{14}N and ^{15}N in the denitrification process has been studied in laboratory experiments. We measured the isotopic fractionation for the step $NO_2^- \rightarrow N_2O$. This study has been carried out on natural soils under anaerobic conditions. ^{15}N is enriched in the substrate during denitrification. The isotopic enrichment factor changes with experimental conditions from about -30 to -10 ‰. Greatest enrichment is obtained for the lowest reduction rate. For high rates of denitrification, the isotopic fractionation decreases. An exponential relation is found between isotopic enrichment and the reaction rate.

INTRODUCTION

As previously established [1],[2],[3],[4],[5], an isotope effect is associated with the denitrification reaction. This effect results in an ^{15}N enrichment of the substrate (NO_3^- or NO_2^-).

ISOTOPIC TERMINOLOGY

i- isotopic composition is expressed in delta units :

$$\delta^{15}N \text{ ‰} = \left[\frac{^{15}N / ^{14}N_{sample}}{^{15}N / ^{14}N_{standard}} - 1 \right] \cdot 1000$$

ii- the isotope fractionation factor in a reaction such as substrate S → product P will be through convention : $\alpha_{p/s} = R_{pi} / R_s$ (1)

where R_{pi} is the "instantaneous" isotopic ratio of the dP quantity of product yielded during a dt period and R_s is the isotopic ratio of the residual substrate at the same time. Then follows :

$$(\alpha_{p/s} - 1) \cdot \ln f = \ln \frac{10^{-3}\delta_s + 1}{10^{-3}\delta_{s,o} + 1} \quad (2)$$

where f is the unreacted fraction of substrate, δ_s the isotopic composition of substrate at time t, $\delta_{s,o}$ this isotopic composition at t = 0.
An isotopic enrichment factor can be introduced : $\varepsilon_{p/s} = (\alpha_{p/s} - 1) \cdot 1000$

PURPOSE OF THE WORK

This work mainly aims at measuring the magnitude of the isotope fractionation during denitrification. We chose the reduction of NO_2^- into N_2O. It is usually agreed that biological denitrification occurs as the following reaction sequence [6]:

$$NO_3^- \rightarrow NO_2^- \rightarrow NO \rightarrow N_2O \rightarrow N_2$$

No accumulation of NO was observed in the experiments reported here: we therefore consider the transformation $NO_2^- \rightarrow N_2O$ as a single step unidirectional reaction. This study dealt with waterlogged soil samples (soil weight/water weight = 1) incubated under a He atmosphere. We changed the experimental conditions to vary the reaction rate: -i- temperature variation (10°C, 20°C, 28°C) - ii- variation of soil preparation before incubation: we used either an air-dried soil or a soil with the same water content as at sampling (14% humidity) -iii- some soil samples have been enriched in glucose which increases the denitrification rates.

MATERIALS AND METHODS

Most experiments have been carried out on soils sampled in a surface horizon of a calcareous brown soil. These soils were - either kept as they were in the field (14% moisture content) and directly put under incubation, with water added in order to obtain the soil-water ratio of 1, - or air dried. 100 g of dry soil was put in 600 ml flasks. N-serve (nitrapyrine = 2-chloro-6-(trichloromethyl)-pyridine) was added (concentration of 20 ppm of soil). This product inhibits the activity of nitrifying bacteria and prevents the production of nitrites and nitrates due to the mineralisation. 10 ml of a solution containing about 20 mg of NO_2^--N as $NaNO_2$ was then added. For the experiments with added glucose, 0.5 of this carbohydrate was also added. The flasks were evacuated and filled with helium at one atmosphere. 50 ml of helium were then removed from the flasks and replaced by 50 ml of acetylene which will stop denitrification at N_2O step [7].

During incubation, the isotopic composition and concentration of the residual NO_2^- were measured in flasks removed at regular time intervals. The nitrogen isotopic composition was determined by a previously reported methodology [8]. The precision of the measurement of $\delta^{15}N$-NO_2^- is about 0.2‰.

The isotopic fractionation factor was estimated experimentally from the isotopic composition of the residual substrate which was measured at least 7 times during the reaction progress. Calculation of the isotopic fractionation factor was based on measurements of samples collected between time zero and a time at which no less than 15% of the substrate remained. In practice for all of the results reported here, f will never be less than 0.17 and the isotopic fractionation factors will be therefore calculated between f = 1 and f = 0.17 (in some experiments we allowed the reaction to proceed to a lower value of f. The results will be discussed below).

RESULTS

The results are presented in figure 1.

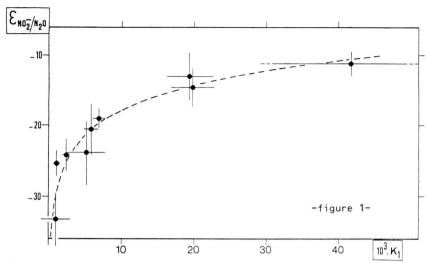

Fig.1 Relation between denitrification rate and isotopic enrichment factor.

It can be first stated that the incubation conditions (temperature variation, drying or not of soil before incubation, glucose addition) had a considerable influence on the denitrification rate. The influence of temperature [9] or glucose [10] is well known : the effect on the denitrification rate of soil drying before incubation has only recently been proved [11],[12]. The increase in the denitrification rate following drying is probably due to the release of carbonaceous compounds from the microbial biomass killed during dessication.

Isotope enrichment factor $\varepsilon_{p/s}$ was determined as the slope of the straight line of $10^3 \ln \{10^{-3} \delta_s + 1 / 10^{-3} \delta_{s,o} + 1\}$ against $\ln f$ (equation 2). The correlation coefficient is always significant at probability of 0.01). The estimation interval of $\varepsilon_{p/s}$ (p = 0.05) was also calculated.

It is noted that the values of the isotopic enrichment factor varied greatly, ranging from -33 to -11‰. An increase in the incubation temperature induced a decrease in the isotope fractionation. At the same temperature, the soil kept at the field sampling humidity before incubation, presented a greater isotope fractionation than the soil dried before incubation. On the other hand, the soil supplied with organic carbon as glucose presented the lowest isotope fractionation. There was a reciprocal relation between the denitrification rate and the isotope fractionation. We calculated the fit of experimentally determined reaction rate to expectation arising from assumptions of first order ($Q = Q_0 e^{-k_1 t}$, where Q = substrate concentration). If we plot $\varepsilon_{p/s}$ as a function of the rate constant k_1 for the 9 experiments, we find (figure 1) that there is an excellent correlation between these two parameters. Of the simple functions tested, the best fit of the data were to an exponential curve :

$$\varepsilon_{p/s} = 5.78 + 5.14 \ln k_1 \quad , \quad n = 9 \quad , \quad r = 0.949 \text{ (significant at p = 0.01)}$$

INTERPRETATION

Kohl and Shearer [5] have proposed, for denitrification, a model with two consecutive steps :
- uptake of substrate into the cell, or formation of an enzyme substrate complex, steps which both present a low isotopic fractionation.
- reduction of the substrate with breaking of a N-O bond which produces a large isotope effect (theoretical value -65.9% [13]). This step is saturable, according to the well-known equation of Michaelis-Menten.

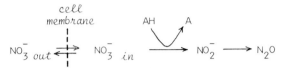

or : NO_3^- + ENZYME $\underset{desorption}{\overset{adsorption}{\rightleftarrows}}$ ENZYME-SUBSTRATE COMPLEX $\longrightarrow NO_2^- \longrightarrow N_2O$
with AH, A

From this hypothesis, Kohl and Shearer deduced that in theory,

1) the overall reduction rate and the isotopic fractionation should increase with the initial concentration of substrate : indeed they foresaw that at low concentration of substrate there would be a sufficient amount of the reductant, AH, to promote reduction as compared to the egress (or desorption) rate of substrate in the medium : in this case, isotope fractionation should be low. When the concentration of substrate in the medium is increased, it is logical to expect that the concentration of substrate within the cell will increase and reach the saturation concentration : the concentration of reductant is then no longer sufficient to sustain maximal reduction rates. The internal nitrate (NO_{3in}^-) accumulates, egress becomes more successful compared to the reduction rate and the isotopic fractionation increases.

2) By increasing the concentration in organic carbon (and consequently of AH) and with a substrate concentration proved to be sufficient for reaching saturation, it can be predicted that besides an increase in the overall rate of NO_3^- disappearance there will be a decrease in isotopic fractionation : at a low concentration of carbon this fractionation should be high because the egress rate is fast compared to the reduction rate. When the concentration of electron donor increases, the reduction is relatively promoted in comparison with the egress rate and the isotopic fractionation consequently decreases.

Such a model allows us to fully understand the decrease in the isotopic fractionation with the increase in the concentration of reduced organic carbon. In our experiments this increase was either linked with glucose supply, or with soil drying before

incubation causing the release of carbon of the bacterial biomass destroyed. On the contrary, Kohl's expectation concerning the variation of the isotopic fractionation factor with the variation of the concentration of substrate in the medium (increase of the isotopic effect with concentration) raises a question : if it were so, then as the reaction progressed, the isotopic fractionation should decrease. In our experiments, this phenomenon is not observed up to f = 17%. It must be noticed however :
- that the initial substrate concentrations employed were always very high (200 mg N/l),
- that the results obtained in some experiments where the extent of reaction had passed beyond f = 15% (data that we deliberately did not take into account for the determination of the isotopic fractionation) reveal that the measured isotopic composition of the substrate at these extents of reaction presents sometimes a lower value than that expected according to the isotopic enrichment factor calculated up to f = 15%. These results are consistent with the conclusion that below a certain concentration in substrate, fractionation decreased. This phenomenon would take place at f values below 17% (that is, at high extents of reaction). Since the initial substrate concentration was about 200 mg N/kg soil, the concentration below which isotopic fractionation would decrease is about 30 mg N/kg soil under the experimental conditions chosen. This concentration would be that at which the second step of the model (reduction) would cease to be saturated. Above this concentration the rate of egress of substrate from the cell would be low compared to that of the reductive step. However the results obtained for very low values of f are too few and may be insufficiently precise to draw final conclusions : they only suggest tendencies.

CONCLUSIONS

It is clearly established that the isotopic fractionation during the $NO_2^- \rightarrow N_2O$ denitrification step is fundamentally dependent on two parameters : temperature and abundance of electron donor in the reduction : the variations of the isotopic fractionation caused by soil drying before incubation appear only as an effect of the latter parameter. The influence of the concentration of substrate is not clearly settled : only a tendency to a decrease in the isotopic fractionation with a decrease in the substrate concentration was observed. The rather high nitrite concentrations chosen at the beginning of these incubation experiments do not easily allow the study of the influence of this parameter.

In the studied range of variations of these parameters, the isotopic enrichment factor varies between -33 and -11 ‰ . It must however be noticed that the natural phenomenon of denitrification in the field should present a large isotope fractionation under the following conditions : (a) average temperature ; (b) no exceptional concentration of organic carbon ; (c) humid or saturated conditions (conditions under which denitrification is most likely to take place). These conditions represent those which generally prevail in natural environments. All of these conditions would con-

tribute to a large isotope effect ($\varepsilon_{p/s} < -20‰$).
The model we proposed after Kohl and Shearer [5] has two steps, one being reversible with little fractionation, and the other unidirectional, saturable and highly fractionating (since it involves the breaking of a N-O bond). The study of figure 1 seems to indicate that the minimum fractionation is about -10 or -11 ‰. The model described here predicts that the lowest fractionation is that of the first step which is then limiting and could be the NO_2^- uptake through the cell membrane.
Lastly, we wish to note the great similarity of the phenomena described here with those published concerning sulfur isotopic geochemistry [14],[15],[16].

REFERENCES

1 Wellman R.P., Cook F.D. and Krouse H.R., *Science*, *161*, *1968*, *269-270*.
2 Delwiche C.C. and Steyn P.L., *Environm. Sci., Techn.*, *4*, *1970*, *929-935*.
3 Blackmer A.M. and Bremner J.M., *Soil Biol. Biochem.*, *9*, *1977*, *73-77*.
4 Chien S.H., Shearer G. and Kohl D.H., *Soil Sci. Soc. Am. Proc.*, *41*, *1977*, *63-69*.
5 Kohl D.H. and Shearer G., *"Recent developments in mass spectrometry in biochemistry and medicine"*, *A. Frigerio Ed., Plenum Press, N.Y., 1978*.
6 Payne W.J., *Denitrification. Trends in biochemical Sciences*, *1*, *10*, *1976*, *220-222*.
7 Federova R.I., Milekhina E.I. and Il'Yukhina N.I., *Izv. Akad. Nauk. SSSR Ser. Biol.*, *6*, *1973*, *797-806*.
8 Mariotti A. and Létolle R., *Analusis*, *6*, *1978*, *421-425*.
9 Garcia J.L., *Bull. Inst. Pasteur*, *73*, *1975*, *167-193*.
10 Bowman R.A. and Focht D.D., *Soil Biol. Bioch.*, *6*, *1974*, *297-301*.
11 Letey J., Hadas A., Valoras N. and Focht D.D., *J. Environm. Qual.*, *9*, *1980*, *232-235*.
12 Patten D.K., Bremner J.M. and Blackmer A.M., *Soil Sci. Soc. Am. J.*, *44*, *1980*, *67-70*.
13 Urey H.C., *J. Chem. Soc.*, *1947*, *562*.
14 Harrison A.G. and Thode H.G., *Trans. Faraday Soc.*, *54*, *1958*, *84-92*.
15 Kemp A.L.W. and Thode H.G., *Geoch. et Cosmochim. Acta*, *32*, *1968*, *71-91*.
16 Kaplan I.R. and Rittenberg S.C., *J. Gen. Microbiol.*, *34*, *1964*, *195-212*.

NITROGEN ISOTOPE RATIO VARIATIONS IN BIOLOGICAL MATERIAL, INDICATOR FOR METABOLIC CORRELATIONS?

R. MEDINA and H.-L. SCHMIDT

Institut für Chemie Weihenstephan, Techn. Universität München, D-8050 Freising (FRG)

ABSTRACT

$\delta^{15}N$-values of nitrogen containing biological material indicate, although within large limits, an enrichment of ^{15}N in food chains. Kinetic isotope effects and fractionation factors of processes in the nitrogen metabolism are in line with these findings, however, they do not explain the extend and the differences of δ-values found in some cases. Thus amino acids from different pools of the same organism and even the different N-atoms of basic amino acids may differ in isotope abundance. Comparisons show that the main enrichments are due to ramifications of the metabolism and to recycling of N-atoms. Thus excretion products are highly enriched in ^{15}N. In the soil the urease reaction, NH_3-evaporation and nitrification have for consequence an enrichment of ^{15}N in NO_3^-. The uptake and the reduction of NO_3^- by plants is accompanied by isotope discriminations, and NO_3^- accumulated in plants is highly enriched in ^{15}N. Because of the very complicated metabolic intercorrelations between the different nitrogen pools an assignment of nitrogen containing material to its origin is only possible with large restrictions. However, δ-value determinations may be useful for the elucidation of metabolic recycling in vivo.

INTRODUCTION

The fundamentals and the consequences of the different carbon isotope discrimination by C_3- and C_4-plants are well understood, and organic material, even after having passed a food chain, can be unequivocally assigned to one of these plant groups by means of its $\delta^{13}C$-value* [1].

On the other hand, due to the complicated intercorrelations between the different pools in the nitrogen cycle, an assignment of organic material to its origin on the basis of $\delta^{15}N$-values* shall only be possible with difficulty. However, it should be

---

*The natural variations of isotope abundances are designed as δ-values. These are expressed as relative isotope ratio differences in regard to a standard, in the case of the nitrogen isotopes to atmospheric N_2:

$$\delta^{15}N\ [‰] = \frac{([^{15}N]/[^{14}N]_{sample} - [^{15}N]/[^{14}N]_{standard})}{[^{15}N]/[^{14}N]_{standard}} \times 1000$$

possible to establish some general aspects by means of a systematic study of isotope discriminations in the metabolism of nitrogen containing compounds, and by comparisons of δ^{15}N-values reported for organic material from well defined origins. Vice versa, variations of δ^{15}N-values should demonstrate correlations in the nitrogen cycle and indicate the origin of nitrogen containing compounds. Therefore this report summarizes data from the literature and from own experimental work, with the aim of finding general interdependences of practical value.

RESULTS AND DISCUSSION

Variations of δ^{15}N-values are caused by isotope discriminations connected to chemical reactions or to differences in thermodynamic equilibria. In a chemical or an enzyme-catalyzed reaction the ratio of the rate constants for the turnover of the molecules with the different N-isotopes is called the kinetic isotope effect k^{14}/k^{15}. Isotope discriminations by complicated in vivo-systems are expressed by fractionation factors α or β, [2,3], which express the isotope ratio of substrates or/and products as a function of turnover. For qualitative comparisons, they may be regarded as numerically equivalent to kinetic isotope effects.

TABLE 1
Mean δ^{15}N-values of biological material. After [4-6].

	δ^{15}N-value [‰]		δ^{15}N-value [‰]
landplants	- 6.5 ... +10	phytoplankton	+ 0.4**... + 5
land animals	+ 5 ... +15	zooplankton	+ 2** ... + 7.5
urea	+12 ... +20	seaweed	+ 4 ... + 8
soils	0 ... +17	seafish	+10 ... +17

** symbiotic nitrogen fixation

A global review of the isotope abundances of biological material (Tab. 1) shows, that δ^{15}N-values of land plants are slightly positive. Land animals show an enrichment of ^{15}N up to 15 ‰, and their excretion product urea is even more enriched in heavy nitrogen. Similary, seafish (δ^{15}N up to 20 ‰) is enriched in ^{15}N in regard to phytoplankton, zooplankton and seaweed. This demonstrates an enrichment of heavy nitrogen through the links of food chains, while the variations within the groups indicate the very complicated intercorrelations between the different pools in the nitrogen cycle, due to overlapping of catabolism and metabolism.

Exact isotope effects or fractionation factors have only been determined for a few reactions or processes contributing to nitrogen isotope discriminations in the biosphere (Tab. 2). Neither the biological nor the industrial nitrogen fixation seem to proceed with an appreciable isotope effect [3,7-12]. In plants NH_3 is incorporated into organic material through the glutamine synthetase-glutamate synthase pathway or through the glutamate dehydrogenase reaction; the latter was found to have a very large isotope effect in vitro [13]. Moreover one can expect, that transaminations, aminations and deaminations are accompanied by isotope discriminations. Isotope discriminations are also implied to protein metabolism, as can be seen from the chrymotrypsin and the papain reactions [17-20], and to nitrogen excretion processes, demonstrated by the isotope effects of the arginase and urease reactions [21,22]. Bacterial processes in the soil, e.g. nitrification [3,14] and

TABLE 2

Fractionation factors (α or β) and isotope effects (k^{14}/k^{15}) of reactions involved in nitrogen isotope discriminations in the biosphere.

Process (reaction)	k^{14}/k^{15}	α or β	References
biological fixation	-	0.998-1.004	[3,7-12]
glutamate dehydrogenase	1.047	-	[13]
chrymotrypsin	1.010	-	[17,18]
papain	1.024	-	[19,20]
arginase	1.010	-	[21]
urease	1.007	1.007	[21,22]
nitrification	-	1.017-1.026	[3,14-16]
denitrification	-	1.017-1.020	[3,15,16]
nitrate reductase	1.015	1.005-1.011	[10,23]

denitrification [3,15,16] cause large isotope discriminations as investigated so far, and also the uptake of nitrate by plants and its reduction is accompanied by isotope effects, as has been shown recently by Shearer et al. [10] and in our laboratory [23].

Thus, any incomplete reaction in the metabolism of a given compound can have for consequence an ^{15}N-enrichment in the non-metabolized molecules. Especially in the case of amino acids, the molecules used for the biosynthesis of body proteins should be enriched in ^{15}N as compared to the isotope content of the nitrogen source from which they are originating. Caebler et al. [24] compared $\delta^{15}N$-values of amino acids from proteins of different sources, and as shown in Tab. 3, amino acids in leguminoses are only slightly enriched in ^{15}N, because their nitrogen is provided from the nitrogen fixation reaction, while the same amino acids in plants receiving their nitrogen from fertilizers in the soil show higher $\delta^{15}N$-values. A further enrichment seems to occur by the metabolism of amino acids in animals.

Among these amino acids from plants and from liver, two groups can be distinguished representing two different pools of turnover and N-recycling. In plants glycine, serine, phenylalanine and tyrosine are amino acids directly

TABLE 3

δ^{15}N-values of amino acids in proteins of different sources. From [24].

	δ^{15}N-value [‰] in amino acids from proteins of		
	plants		rat liver
	with N-fixation	without N-fixation	
glycine	2.5	12.8	10.6
serine	-0.6	8.0	9.6
tyrosine	3.2	11.9	9.5
phenylalanine	-	-	9.8
mean value	1.7	10.9	9.9
alanine	4.3	10.3	14.7
arginine	1.2	7.4	15.1
aspartic acid	5.5	11.2	13.7
glutamic acid	4.5	10.4	15.4
amid-N	6.7	12.9	14.2
mean value	4.4	10.4	14.6

connected to the carbohydrate metabolism, and they are not largely involved in transamination reactions, as this is the case with aspartic acid and glutamic acid. In animals the latter amino acids belong to pools implied in transamination and nitrogen excretion (ornithin cycle). Differences in the δ^{15}N-value of a given amino acid from different organs of the same organism reflect the specific recycling rate within these organs and the nitrogen flux between organs [25,26]. Diffusion processes in the kidney and isotope effects of the urea cycle reactions may be responsable for the particular ^{15}N-enrichment found for the amino acids of liver protein. Thus δ^{15}N-values reflect the protein turnover in and between organs.

In a study on the time of nitrogen equilibration between diet and body protein, we made tracer experiments with "naturally labelled" proteins on chicken (Fig. 1). When the animals received soya and fish meal as a protein diet, an equilibration of the egg nitrogen with the diet nitrogen was obtained within a few days. The half life time of the proteins in the white of the egg appeared to be much shorter than that of the nitrogen compounds in the egg yolk, which corresponds to the development of eggs. A certain time after the feed of fish meal the δ^{15}N-values of both egg fractions even exceeded that of the diet. This is probably the expression of the isotope discrimination implied in the nitrogen metabolism. The result corresponds well to our finding of differences between the carbon isotope ratio in human diet protein and human hair protein [27].

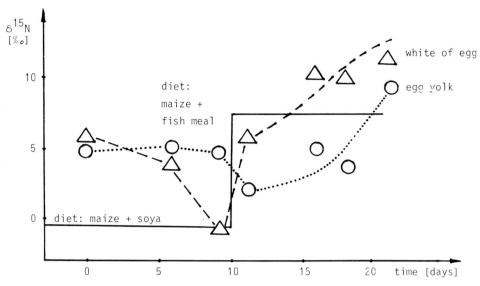

Fig. 1. δ^{15}N-values of egg yolk and white of egg in dependence on diet.

As already pointed out, the influence of metabolic recycling on ^{15}N-enrichment is most apparently seen in the case of amino acids involved in the urea cycle. The large enrichment of arginine from rat liver (Tab. 3) is in line with the isotope effect we have found for the arginase reaction [21].

TABLE 4
δ^{15}N-values of different N-atoms in arginine from bacteria.

N on C-atom	2	5	side chain	mean value
δ^{15}N ‰	0	33	40	27 (28)*

*calculated

In context with investigations on the kinetic isotope effect of this reaction we have developed a method for the determination of the isotope abundance of the different N-atoms of this amino acid. As free arginine from animal liver is commercially not available, we have made preliminary investigations on arginine from bacteria and determined the isotope distribution within the molecule. According to Tab. 4 the highest δ^{15}N-values were found for the guanido group (side chain), and no enrichment was observed for the amino group on C-atom 2. This is again an expression for the different recycling of nitrogen in distinct positions.

Urea, the nitrogen excretion product of higher animals, is highly enriched in ^{15}N [4]. In addition, Shearer et al. [28] found high $\delta^{15}N$-values in other fractions of natural fertilizers, e.g. in ammonia-N (14-17 ‰), nitrate-N (11-38 ‰) and reduced-N (8-11 ‰). Nevertheless, urea is the main nitrogen containing compound in natural fertilizers and thus one of the main primary sources for nitrate in soils. A further enrichment of ^{15}N in the soil must be due to isotope effects of the urease reaction [21,22], NH_3-evaporation [30] and nitrification [3,14]. Nitrogen from mineral fertilizers is submitted to the same reactions, however, as the absolute values of the initial material are not far from 0 ‰ [28,29], the δ-values of the fractions from soil are less positive.

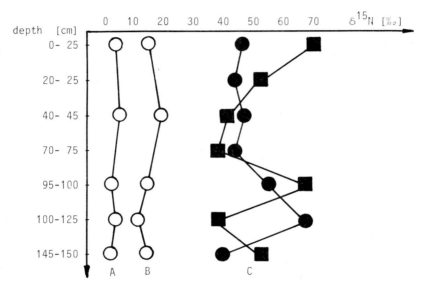

Fig. 2. $\delta^{15}N$-values of NO_3^- (circles) and NH_4^+ (squares) versus depth after mineral or animal dung fertilization. Open marks from [30], full marks own results. A: mineral fertilizer, Texas; B: pasture, Texas; C: repeated fertilization with liquid manure, Germany.

On the base of these differences Kreitler tested the possibility for the evaluation of sources of nitrate in ground water [31]. As shown in Fig. 2, he found distinct and constant $\delta^{15}N$-values for NO_3^- in different depths of the soils with mineral fertilization and under cattle pastures. In supplement to these investigations we determined $\delta^{15}N$-values in soils with intensive natural fertilization and found values for NO_3^- up to 50 ‰, while the isotope abundance of NH_4^+ was even higher. Variations within the profile and differences between the $\delta^{15}N$-values of NH_4^+ and NO_3^- may be attributed to isotope effects of adsorption and desorption of NH_4^+ to clay minerals, which have been demonstrated to occur by Karamanos and Rennie [32] and in own investigations [33]. The $\delta^{15}N$-value of

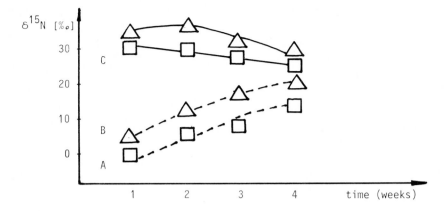

Fig. 3. δ^{15}N-values of nitrate (———) and organic N (- - -) in lettuces after fertilization with NO_3^- ( ) or NH_4^+ ( ). Indication of the δ-values of NH_4^+-fertilizer (A), NO_3^--fertilizer (B) and NO_3^- in soil (C). Fertilization with 0.15 g N/kg soil.

nitrate from ground water was very close to the mean δ-value of the nitrate in the soil. Thus one can expext, that under certain conditions δ^{15}N-values could be useful to detect sources of nitrate in ground and drinking waters, however, because of overlapping of the ranges of δ-values and because of secondary mixing of ground water an absolute assignment will not be possible.

From the facts mentioned before one can assume, that plants in general assimilate NO_3^- enriched in ^{15}N, and that therefore the δ^{15}N-values of the organic material produced should reflect that of the NO_3^- present as fertilizer. However, in experiments with lettuces and spinach the uptake and the reduction of NO_3^- was accompanied by isotope discriminations (Fig. 3). The δ-value of NO_3^- accumulated in plants exceeded that of the soil-NO_3^- before fertilization, while the organic nitrogen was relatively depleted in ^{15}N. During the growth of the plants the δ^{15}N-values of the two pools converged, probably because of enhanced accumulation of NO_3^- from the fertilizer and slower turnover in the reduction to NH_3. Similar results were obtained with ammonium as fertilizer, which indicates, that the nitrification in the soil was very rapid and quantitative. The isotope discrimination of the NO_3^--reduction in vivo calculated from these data was about 1 %, which is in agreement with preliminary results on the vitro kinetic isotope effect of the nitrate reductase reaction (k^{14}/k^{15} = 1.015, enzyme from spinach, unpublished results). Correspondingly an isotope discrimination of 0.5 % was found by Shearer et al. for the nitrate reduction in soybeans [10].

CONCLUSIONS

The determination of the variation of natural abundances is indispensable in investigations on geochemical and ecological questions in the nitrogen cycle, and in studies of kinetic isotope effects. However in many tracer experiments on metabolic reactions and their mechanisms these measurements cannot replace the application of labelled material. On the other hand experiments with labelled material in biological systems have to take into consideration the isotope discriminations by such systems.

The examples given prove the possibility of using "naturally labelled" nitrogen containing material for general studies of protein metabolism in human and animal nutrition physiology, however they also indicate, that the δ^{15}N-value of a biological sample does not necessarily indicate its origin, because it is not only depending on its precursor but also on its recycling within a certain pool. With this restriction, assignments of nitrogen containing material to a distinct pool may be possible. This is mainly true for the assignment of some amino acids to a given organ or of NO_3^- in ground water or lakes to mineral or biological origin. Further determinations of δ-values of biological material and of isotope discriminations by biological processes will help to develop a more sophisticated understanding of in vivo systems of nitrogen metabolism. We suppose, that especially the determination of the nitrogen isotope distribution in complex molecules like purines and pyrimidines will be a tool for the study of their dynamic state in vivo.

REFERENCES
1. F.J. Winkler and H.-L. Schmidt, Z. Lebensm. Unters. Forsch., 171(1980)85-94.
2. J.Y. Tong and P.E. Yankwich, J. Phys. Chem., 61(1957)540.
3. C.C. Delwiche and P.L. Steyn, Environ. Sci. Technol., 4(1970)929-935.
4. W. Geren, A. Bendich, O. Bodansky and G.B. Brown, J. Biol. Chem., 183(1950)21.
5. Y. Myaka and E. Wada, Recs. of Oceanog. Works, 9(1967)37-53.
6. E. Wada, T. Kadonaga and S. Matsuo, Geochem. J., 9(1975)139-148.
7. T.C. Hoering and H.T. Ford, J. Amer. Chem. Soc., 82(1960)376-378.
8. N. Amarger, A. Mariotti and F. Mariotti, C.R. Acad. Sci. Paris, 284(Ser D)(1977) 2179-2182.
9. A. Mariotti, F. Mariotti, N. Amarger, G. Pizelle, J.-M. Ngambi, M.L. Champigny and A. Moyse, Physiol. Veg., 18(1980)163-181.
10. D.H. Kohl and G. Shearer, Plant. Physiol., 66(1980)51-56.
11. G. Shearer, D.H. Kohl and J.E. Harper, Plant Physiol., 66(1980)57-60.
12. R.J. Rennie, D.A. Rennie and M. Fried, Proc. Symp. on Isotopes in Biological Dinitrogen Fixation. I.A.E.A.-AG 92/7 Vienna 1978 pp 107-133.
13. M.I. Schimerlik, J.E. Rife and W.W. Cleland, Biochemistry, 14(1975)5347-5354.
14. H.D. Freyer and A.I.M. Aly, Proc. Symp. on Isotope Ratios as Pollutant Source and Behaviour Indicators. I.A.E.A. S.M. 191/9 Vienna 1978, pp 21-33.
15. R.P. Wellman, F.D. Cook and H.R. Krouse, Science, 161(1968)269-270.
16. A.M. Blackmer and J.M. Bremner, Soil Biol. Biochem., 9(1977)73-77.
17. M.H. O'Leary and M.D. Kluetz, J.Am. Chem. Soc., 92(1970)6089-6090.
18. M.H. O'Leary and M.D. Kluetz, J.Am. Chem. Soc., 94(1972)3585-3589.
19. M.H. O'Leary, M. Urberg and A.P. Young, Biochemistry, 13(1974)2077-2081.
20. M.H. O'Leary and M.D. Kluetz, J. Am. Chem. Soc., 94(1972)665.
21. R. Medina, R. Olleros-Izard and H.-L. Schmidt, in H.-L. Schmidt, H. Förstel, K. Heinzinger (Eds.), Proc. 4th Int. Conference on Stable Isotopes, Jülich, March 23-27, 1981.
22. H.D. Freyer, Pageoph., 116(1978)393-404.
23. R. Medina, T. Olleros-Izard and H.-L. Schmidt, unpublished results

24 O.H. Gaebler, H.C. Choitz, T.C. Vitti and R. Vukmirovich, Cand. J. Biochem. Physiol., 41(1963)1089-1097.
25 O.H. Gaebler, T.G. Vitti and R. Vukmirovich, Cand. J. Biochem., 44(1966)1249-1257.
26 P. Felig, Ann. Rev. Biochem., 44(1975)933-955.
27 N. Nakamura, D.A. Schoeller, F.J. Winkler and H.-L. Schmidt, Biomed. Mass Spectrom. (to be published).
28 G. Schearer, D.H. Kohl and B. Commoner, Soil Science, 118(1974)308-316.
29 H.D. Freyer and A.I.M. Aly, J. Environ. Quality, 3(1974)4.
30 I. Kirschenbaum, J.S. Smith, T. Crowell, J. Graff and R. McKee, J. Chem. Physics, 15(1947)440-446.
31 Ch.W. Kreitler, Report of Investigation N° 83, Bureau of Economic Geology, The University of Texas at Austin, Texas (1975).
32 R.E. Karamanos and D.A. Rennie, Cand. J. Soil, Sci., 58(1978)53-60.
33 T. Pfanneberg, W.R. Fischer, E.A. Niederbudde and R. Medina, J. Soil. Sci., 32(1981)409-418.

NATURAL VARIATIONS OF ^{15}N-CONTENT OF NITRATE IN GROUND AND SURFACE WATERS AND TOTAL NITROGEN OF SOIL IN THE WADI EL-NATRUN AREA IN EGYPT

A.I.M. ALY, M.A. MOHAMED and E. HALLABA
Nuclear Chemistry Departm., Atomic Energy Establishm., Cairo, Egypt

ABSTRACT

An inverse correlation was found between the nitrate content and its nitrogen isotopic composition in groundwater of the southern sector of Wadi El-Natrun. The average nitrate concentration has greatly increased in the northern sector, compared to a previous study 1976, whereas much lower nitrate concentrations were measured in the southern sector with no significant difference between the two studies. The homogeneity of the nitrogen isotopic composition of groundwater nitrate in the northern sector supports a possible inflow of groundwater from the Nubarya project. The drainage stream has a low δ^{15}N-value indicating a major contribution of nitrogen fertilizers.

INTRODUCTION

The industrial fixation of nitrogen in the form of fertilizers is almost doubling every six years [1] leading to a disturbance of the natural equilibrium of the nitrogen cycle and - through leaching from soil - to accumulation of nitrate in ground and surface waters.

Kohl et al. [2] were the first to investigate the possibility of using the natural variations of the ^{15}N-content of nitrate in surface water as an indicator of the source of pollution. Their method is based on the assumption that the measured ^{15}N-content of nitrate in surface water represents a result of mixing between soil-derived and fertilizer-derived nitrates with different isotopic compositions. This method was, however, discussed by several authors [2-6]. So the δ^{15}N of soil derived nitrate should depend on the type of soil and the time of incubation [4]. Feigin et al. [7] also indicated that long term incubation of soil yields nitrate of a similar isotopic composition as that of soil total nitrogen.

Freyer and Aly [8] found an inverse correlation between the concentration and ^{15}N-content of nitrate in groundwater under cultivated areas, whereas no such correlation was observed in groundwaters under forest areas.

The purpose of this study is to investigate the possibility of utilizing the natural variations of ^{15}N to give some useful information concerning the source of nitrate pollution of ground and surface waters in Egypt.

AREA OF INVESTIGATION

The investigated area is called Wadi El-Natrun, which is an elongate depression located 75 km northwest of Cairo and 40 km west of the River Nile. Intensive land reclamation has been conducted in this area. The area is irrigated by groundwater from different shallow wells. The soils are sandy and contain silt clay. The groundwater aquifer is being continuously recharged from the S, SE, N and NE directions [9]. However, mixing with old water from deeper aquifers cannot be excluded on the basis of ^{18}O-measurements [10].

Near the northern sector of Wadi El-Natrun the Nubarya project is located. This project represents 40 % of the new land whose reclamation has been made possible by the construction of the Aswan high dam. Due to intensive land reclamation, the groundwater level is steadily increasing resulting in reversal of the previous groundwater flow direction . A flow of water from Nubarya project to the Wadi El-Natrun area cannot be ruled out [11].

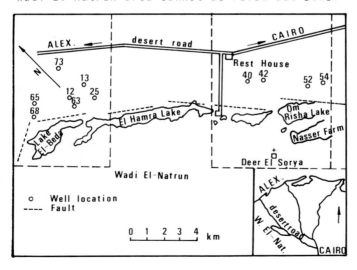

Fig. 1. Location of Wadi El-Natrun and the positions of the water samples.

EXPERIMENTAL

Water samples of wells (location see Fig. 1) and of the drainage stream were collected in polyethylene bottles and preserved by 1 ml/l concentrated sulphuric acid. Soil samples close to the wells were also collected.

The O_2-content of water samples was determined by a modified Winkler method [12]. To determine NH_3, the samples were brought to pH 3 and evaporated under vacuum to a volume less than 20 ml. Samples for the determination of nitrate were adjusted to pH 7. A steam distillation was applied for each seperate fraction [13]. The total nitrogen of soil was determined using the Kjeldahl method [14]. Extracts of the soil by 2 N KCL were also determined by the vacuum distillation technique [15].

The nitrogen gas has been purified and measured in a V.G. Micromass 602 D mass spectrometer. The $\delta^{15}N$-values are reported as permil deviation and are compared to a sample of purified atmospheric (air) nitrogen:

$$\delta^{15}N = \frac{(^{15}N/^{14}N)_{sample} - (^{15}N/^{14}N)_{air}}{(^{15}N/^{14}N)_{air}} \times 1000 \quad [^o/oo].$$

RESULTS AND DISCUSSION

Variations of ^{15}N-content of nitrogen fertilizers produced in Egypt

The main artifical nitrogen fertilizers produced in Egypt are ammonium nitrate, calcium nitrate and urea. The ^{15}N-content of these fertilizers has been determined by the authors [16]. The average $\delta^{15}N$ for fertilizer-ammonium is $-2.1 \pm 1.1\ ^o/oo$, whereas fertilizer-nitrate has an average $\delta^{15}N$ of $-0.7 \pm 2.3\ ^o/oo$. Urea has a $\delta^{15}N$ value of $-2.6\ ^o/oo$. The mean $\delta^{15}N$ value for the different Egyptian fertilizers is $-1.8\ ^o/oo$.

Isotopic composition of nitrate in ground and surface waters as well as total nitrogen of soils in the Wadi El-Natrun area

Table 1 shows a comparison between the nitrate concentrations of ground water samples in January 1976 [17] and those in September 1979, as well as $\delta^{15}N$ values for nitrate at the later date. The concentration of nitrate in the groundwater has greatly increased in the Wadi El-Natrun area. In the northern sector, the average nitrate content in groundwater samples in September 1979 is 3.75 mg NO_3-N/l compared to only 0.76 mg NO_3-N/l at January 1976. In the southern sector the average nitrate content has only slightly increased from 0.4 to 0.45 mg NO_3^--N/l. Hazzaa et al. [18] reported in 1967 an average nitrate content of groundwater in Wadi El-Natrun of not more than 0.2 ppm. The greater increase of nitrate concentration in the northern sector could be due to the following facts: Firstly, the northern sector is more intensively cultivated than the southern sector. Secondly, the proximity of the northern sector is more close to the Nubarya project.

TABLE 1

Nitrate concentration of Jan. 1976 and Sept. 1979 and characterisation of the water samples during the second observations period.

Well	mgNO_3-N/litre		Ratio	Water samples 1979		
	1976	1979	$\frac{1979}{1976}$	δ^{15}N-NO_3	pH	mg O_2/l
Northern sector						
12	1.51	1.40	0.93	3.0	8.11	6.99
13	0.36	8.00	22.20	4.1	8.17	6.93
25	0.32	1.60	5.00	---	9.05	7.17
63	----	1.70	----	3.6	8.35	7.22
65	0.32	7.00	21.88	3.3	8.33	7.19
68	1.22	6.00	4.92	2.9	8.05	7.17
73	0.81	0.56	0.69	2.9	8.59	7.20
Average	0.76	3.75	----	3.36	----	----
Southern sector						
40	0.34	0.70	2.06	3.5	8.53	6.50
42	0.50	0.80	1.60	6.0	8.22	7.06
52	0.36	0.00	0.00	---	8.05	6.94
54	0.38	0.30	0.79	14.5	8.33	7.19
Average	0.40	0.45	----	8.0	----	----
Drainage stream		2.1	----	0.7	----	8.22

^{15}N-data indicate an inverse correlation between the nitrate content and its isotopic composition in groundwater in the southern sector in accordance with Kohl et al. [2] and Freyer and Aly [8]. No such relation is found in the northern sector. The homogeneity of the ^{15}N-content of groundwater nitrate in the northern sector suggests, that the major contribution to nitrate pollution comes from outside of the area, possibly from the groundwater flow of the Nubarya project. The $\delta^{15}N$ of nitrate in the drainage stream is +0.7 °/oo, which is much lower than values measured in groundwater, indicating a greater contribution from fertilizer origin. Preliminary investigations of soil samples in the northern sector of Wadi El-Natrun gave a $\delta^{15}N$ value of +17 °/oo for ammonium and +6.6 °/oo for nitrate. The total soil nitrogen in the northern sector is more enriched in ^{15}N as compared to soils of the southern sector (see Tab. 2). An increase of the total nitrogen content is accompanied by an increase of $\delta^{15}N$-values. This effect was experimentally observed [8], and has been predicted theoretically [19]: Immobilisation in soils of high organic matter content should lead to an enrichment of ^{15}N.

TABLE 2
Total content of nitrogen and $\delta^{15}N$-value of water samples from locations covered with different cultivation plants.

Well No.	Kjeldahl-N mg/100 g soil	$\delta^{15}N$	pH	Cultivation
Northern sector				
12	15	6.8	9.35	Olives
13	175	1.4	8.11	Citrus
25	80	0.5	8.63	Citrus
63	0	---	9.14	Citrus
65	90	23.0	8.32	Maize
68	35	5.9	9.07	Olives, Citrus
73	84	4.9	8.98	Beans, Oats
Average	68.4	7.1	8.80	
Southern sector				
40	35	2.7	9.04	Olives
42	0	---	9.08	Olives
52	30	0.9	8.78	Water melons
54	31	3.4	8.75	Sesame
Average	24	2.3	8.91	
Drainage stream	60	6.3	9.21	Sesame

The more positive $\delta^{15}N$ values of nitrate in some groundwater samples can be also attributed to the following effects:

a. Partial volatilization of soil- and fertilizer-ammonia leaving the residual ammonia and nitrate isotopically enriched in ^{15}N. The basic nature of the soil in Wadi El Natrun (pH range, 8.1 to 9.4) favours this process. Furthermore the isotopic exchange reaction:

$$^{14}NH_4^+{(aq)} + {}^{15}NH_3{(g)} \rightleftharpoons {}^{15}NH_4^+{(aq)} + {}^{14}NH_3{(g)}$$

which has an isotopic equilibrium constant of 1.034 at 298.1 ^{o}K [20] leads to an enrichment of ^{15}N in the ammonium ion. High $\delta^{15}N$ values for nitrate in basic soils have been also measured by Kreitler [21].

 b. Denitrification could also offer another alternative explanation. Denitrification gives rise to significant isotopic fractionation resulting in enrichment of ^{15}N in the substrate [22, 23]. In the Wadi El-Natrun area, aneorobic conditions necessary for denitrification are not met.

CONCLUSIONS

1) The great increase of nitrate content in groundwater in the northern sector of the Wadi El-Natrun area may be the result of intensive cultivation and groundwater flow from the Nubarya project.

2) The nitrogen-15 variations in soil water systems could provide a basis to distinguish between the sources of nitrate contamination. However, quantitative estimates for fertilizer contribution to groundwater nitrate - based on nitrogen-15 natural variations are not easy due to the complexity of the nitrogen cycle.

REFERENCES

1 C.C. Delwiche, Scientific American,223(1970)137-146.
2 D.H. Kohl, G.B. Shearer and B. Commoner, Science, 174(1971)1331-1334.
3 R.D. Hauck, W.V.Bartholomew, J.M. Bremner, F.E. Broadbent, H.H. Cheng, A.O. Edwards, D.R. Keeney, J.O. Legg, S.R. Olson and L.K. Porter, Science, 177(1972)453-454.
4 J.M. Bremner and M.A. Tabatabi, J. Environ. Quality, 2(1973)363-365.
5 A.P. Edwards, J. Environ. Quality, 2(1973)382-387.
6 R.D. Hauck, J. Environ. Quality, 2(1973)317-327.
7 A. Feigin, D.H. Kohl, G. Shearer and B. Commoner, Soil Sci. Soc. Am. Proc., 38(1974)90-95.
8 H.D. Freyer an A.I.M. Aly, Nitrogen-15 studies on identifying fertilizer excess in environmental systems, in: Isotope ratios as pollutant source and behaviour indicators FAO/IAEA (1975) pp. 21-33.
9 F.M. Swailem, M.S. Hamza, A.A. Abdel-Monem, M.S. El-Manharawy and A. Nada, Isotope & Rad. Res. 12, 1(1980)17-23.
10 M.S. Hamza, F.M. Swailem, A.A. Abdel-Monem, M.S. El-Manharawy and A. Nada, Isotope & Rad. Res., 11, 2(1979)111-118.
11 F.E. Schulze and N.A. de Ridder, The rising water table and related problems in the west Nubarya area, United Arab Republik, Egypt, UNESCO report international institute for land reclamation and improvement, Wageningen, The Netherlands (1973).
12 H.L. Golterman and R.S. Clymo, Methods for chemical analysis of fresh waters, IBP Handbook No. 8, Blackwell scientific publications, Oxford and Edinburgh (1969), pp. 124-131.
13 J.M. Bremner and D.R. Keeney, Anal. Chim. Acta, 32(1965)485-495.
14 J.M. Bremner, "Total nitrogen", Ch. 83, Methods of soil analysis, part 2, in C.A. Black (Ed.), American Society of Agronomy, Wisc. (1965).

15 A.I.M. Aly, N-15-Untersuchungen zur anthropogenen Störung des natürlichen Stickstoffzyklus, Ph.D. Thesis, Aachen University, West Germany (1975).
16 A.I.M. Aly, M.A. Mohamed and E. Hallaba, Mass spectrometric determination of the nitrogen-15 content of different Egyptian fertilizers, submitted for publication.
17 A.A. Nada, Determination of oxygen and hydrogen isotopes in some Egyptian water samples, M.Sc. Thesis, Al-Azhar University (1978).
18 I.B. Hazzaa, K.F. Saad, R.K. Girgis and F. Wahby, J. Egyptian Medical Association, 50, 11/12(1967).
19 G. Shearer, J. Duffy, D.H. Kohl and B. Commoner, Soil Sci. Soc. Am. Proc., 38(1974)315-322.
20 I. Kirchenbaum, J.S. Smith, T.Crowell, J. Graff and R. McKee, J. Chem. Phys., 15(1947)440-446.
21 C.W. Kreitler, J. Hydrology, 42(1979)147-170.
22 C.L. Delwiche and P.L. Steyn, Environ. Sci. Technol., 4(1970)929-935.
23 S.H. Chien, G. Shearer and D.H. Kohl, Soil Sci. Soc. Am. Proc., 41(1977)63-69.

POSSIBILITIES OF STABLE ISOTOPE ANALYSIS IN THE CONTROL OF FOOD PRODUCTS

J. BRICOUT

Pernod Ricard Research Laboratories - 120, ave. Foch - 94015 Créteil Cedex (F)

ABSTRACT

For many years research has been conducted on the mechanism of isotope fractionation in biological systems and it has been shown that generally kinetic isotopic effects lower the heavy isotope content of biosynthesized products. On the other hand, evapotranspiration in plants enriches the water of leaves and fruits in the heavy isotopes of hydrogen and oxygen. Food products are synthesized by living system, therefore investigations were conducted in order to establish if the stable isotope ratios could confirm the occurence of particular biochemical pathways and consequently the origin and the labelling of a given food product.

An application deals with water from fruit juices, which shows an enriched level of deuterium and oxygen-18 as compared to tap water and this fact is used to distinguish natural fruit juices from rediluted concentrates. Different isotope analyses of organic substances were also performed and demonstrated the usefulness of the $^{13}C/^{12}C$ and $^{2}H/^{1}H$ ratios for the characterization of the origin of some food constituents.

INTRODUCTION

The role of the food chemist is to ensure the quality of the different foods and beverages which are offered to the consumer. In many countries one can observe that the consumer would prefer natural products, and one of the main objectives of the food chemist is to control the origin and the label of the product. This task is generally solved by chemical analysis, and by determination of the relative concentration of different constituents. But in some cases this type of analysis is no longer reliable enough and it would be very useful to find a relationship between the isotopic composition of a substance and its origin and history.

^{14}C analysis was first used to differentiate a biological substance from its synthetic counterpart. But this powerful analysis is restricted to a limited number of cases, i.e when the synthetic processes start from coal or oil.

So the possibility of using stable isotopic ratios as a proof of the origin of some food constituents was investigated.

STABLE ISOTOPIC COMPOSITION OF WATER

Water being one of the main constituents of beverages, an investigation was conducted to determine whether its isotopic composition could be used to ensure the authenticity of some beverages like fruit juices. There are four main different isotope species of water : $H_2{}^{16}O$; $H_2{}^{18}O$; $HD{}^{16}O$; $H_2{}^{17}O$ ($H_2{}^{17}O$ will not be considered as there is no easy way to measure its relative abundance).

The relative abundance of the stable isotopes of hydrogen and oxygen are expressed as difference in ‰ relative to a standard (Vienna S M O W, Standard Mean Ocean Water) which is representative of the main reservoir of water (1)

$$\delta^2H = \left[\frac{{}^2H/{}^1H \text{ (sample)}}{{}^2H/{}^1H \text{ (SMOW)}} - 1 \right] \times 1000 \text{ (‰)}$$

$$\delta^{18}O = \left[\frac{{}^{18}O/{}^{16}O \text{ (sample)}}{{}^{18}O/{}^{16}O \text{ (SMOW)}} - 1 \right] \times 1000 \text{ (‰)}$$

The different isotope species of water have different vapour pressures and the water vapour in equilibrium with the liquid phase is, at ambient temperature, depleted in ^2H by 75 ‰ and in ^{18}O by 9 ‰ (2). Due to these fractionation factors, rain water is lighter than ocean water and the isotope composition of the precipitation varies according to climatic conditions. However there is a correlation between ^2H and ^{18}O content : $\delta^2H = 8 \delta^{18}O + 10$ (3). For mediterranean rain water the correlation is slightly modified (4) : $\delta^2H = 8\delta^{18}O + 22$.

Figure 1 shows the results of the isotope analysis of tap water and orange juice water, which is highly enriched in heavier isotopes, as compared to rain water (5). If an orange juice is concentrated and then rediluted with tap water, the isotopic composition of the water of this reconstituted juice is very close to that of the water used ; and thus deuterium or oxygen 18 analysis is an easy way to differentiate natural orange juice from reconstituted juices. This assumption was proved to be valid for other fruit juices : grapefruit, pineapple, apple (6), tomato (7).

In table I we report the isotopic composition of water from different grape juices (8). In any case we observe an enrichment in deuterium and oxygen 18 of the grape juice water as compared to rain water ($\delta^2H = -38,5$ ‰ ; $\delta^{18}O = -6,7$ ‰ for Bordeaux area), but we can also notice that grape juices from the South of the Mediterranean Basin (Marocco, Algeria, Turkey) have the

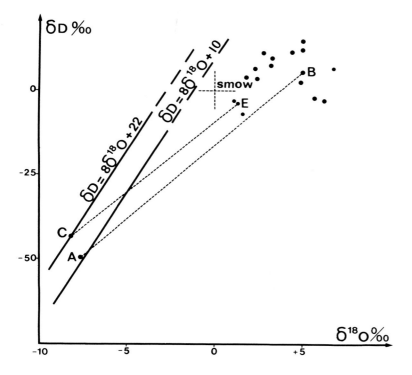

Fig. 1. $^2H/^1H$ and $^{18}O/^{16}O$ ratios of rain water and orange juice water
- tap water : (C) Corsica,   (A) Brazil
- orange juice water : (E) Corsica, (B) Brazil, other origins (•)

highest deuterium and oxygen 18 concentrations. As oxygen 18 concentration in water is not modified during fermentation, there is a possibility of using this type of analysis to control the origin of wine (9).

TABLE 1
Isotope composition of the water of grape juices

Origin of the grape juices		δ^2H (‰) SMOW	$\delta^{18}O$ (‰) SMOW
Bordeaux area			
Sauternes	1972	− 13.2	+ 1.0
Premières Côtes de Bordeaux	1972	− 16.1	+ 0.2
Morvan			
Pouilly-sur-Loire	1972	− 25.4	− 1.8
Pouilly-sur-Loire	1973	− 15.5	+ 0.4
Bourgogne area			
Beaune	1972	− 20.9	+ 1.4
Pommard	1973	− 11.6	+ 3.0
Volnay	1973	− 12.6	+ 2.4
Languedoc area			
Aigues-Mortes	1971	− 2.8	+ 1.6
Aigues-Mortes	1972	− 2.6	+ 2.6
Marocco (Grenache)	1972	+ 1.2	+ 3.3
Algeria N°1	1971	+ 6.5	+ 4.6
N°2	1973	+ 5.4	+ 3.3
Turkey N°1	1972	+ 3.8	+ 5.3
N°2	1972	+ 8.9	+ 7.0

STABLE ISOTOPE COMPOSITION OF CARBON

Vegetable products are the main source of carbohydrates in our food and many different plants are cultivated for this purpose. The source of carbon for terrestrial plants is atmospheric CO_2 which contains the different stable isotope species $^{12}C\ ^{16}O_2$ (98.426 %) ; $^{13}C\ ^{16}O_2$ (1.095 %) ; $^{12}C\ ^{16}O\ ^{18}O$ (0.195 %) ; $^{12}C\ ^{16}O\ ^{17}O$ (0.079 %).

The $^{13}C/^{12}C$ ratio of atmospheric CO_2 is constant particularly in rural areas (10). Results of $^{13}C/^{12}C$ analysis are expressed as the difference in ‰ relative to an international standard PDB (carbonate from the fossil skeleton of Belemnitella americana from the Pee Dee formation of South Carolina) (11) :

$$\delta^{13}C = \left[\frac{^{13}C/^{12}C\ (sample)}{^{13}C/^{12}C\ (PDB)} - 1 \right] \times 1000\ (‰)$$

The $^{13}C/^{12}C$ analysis of terrestrial plant tissue has revealed that carbon isotope fractionation is associated with photosynthesis. The fractionation factor falls into two groups, and it was shown that these two different fractionation factors are associated with two different pathways of CO_2 photoassimilation.

The first known CO_2 assimilation pathway was described by Calvin and Bassham (12) (C_3-pathway). Atmospheric CO_2, which shows a constant $\delta^{13}C$ of -7 ‰ PDB, penetrates the leaf through the stomata and is dissolved into the cytoplasm. It is then bound to ribulose-1,5-bisphosphate to give two molecules of phosphoglyceric acid which are reduced by NADPH to phosphoglyceraldehyde. Further enzymatic steps lead to different carbohydrates. There is an initial small fractionation factor of some ‰ in the dissolution of CO_2 but ribulose-1,5-bisphosphate carboxylase has a discrimination factor against $^{13}CO_2$ of about 17 ‰. Thus C_3-plant organic matter shows a range of variation of $\delta^{13}C$ between -24 and -30 ‰ PDB (13).

Some plants assimilate CO_2 according to a different pathway (C_4-pathway) which has been described by Hatch and Slack (14). CO_2 is fixed in the mesophyll by phosphoenolpyruvate to give oxaloacetic acid which is then converted to malic or aspartic acid. These acids are translocated in the bundle sheath cells, where they are decarboxylated. The liberated CO_2 is completely refixed by ribulose-1,5-bisphosphate to give 2 molecules of phosphoglyceric acid. The other biochemical steps, which are then the same as in the C_3-pathway, lead to various carbohydrates. Yet as the isotope discrimination by phosphoenol pyruvate carboxylase is 2-5 ‰ and as there is no further fractionation associated with carboxylation of ribulose-1,5-bisphosphate in the bundle sheath, the organic matter of these C_4-plants shows a range of variation of $\delta^{13}C$ between -9 and -14 ‰ PDB (13). C_4-plants are not predominant in the plant world but are present in different families. Most of the plants cultivated for food supply are C_3-plants, however with two major exceptions : corn and cane. With the development of enzymatic technological processes, corn and cane are now the major sources of glucose, fructose and sucrose. It appears very easy to detect an addition of any sugar from corn or cane in a food or beverage by measuring the $^{13}C/^{12}C$ ratio of the sugars.

TABLE 2
Range of $\delta^{13}C$ of carbohydrates according to their vegetal origin

	C_4-plant ($-9 < \delta^{13}C < -12$ ‰PDB)	C_3-plant ($-24 < \delta^{13}C < -30$ ‰)
Sucrose	cane	beet maple
Glucose + Fructose	corn	potato apple honey orange

This method of analysis was applied to control the authenticity of different foods and beverages :
- fruit juices (or concentrates) from apple (15), orange (16), grape (9), grapefruit
- maple syrup (Acer saccharum) (17)
- honey (18) from nectar of flowering plants, mostly C_3-plants

The sugar of all these foods or beverages shows low $\delta^{13}C$ values : less than -24 ‰ PDB and illicit addition of corn or cane glucides with high $\delta^{13}C$ values can be easily detected by $^{13}C/^{12}C$ measurements. However, this method does not allow the detection of the addition of sugars obtained from beet or potato.

Plants also constitute the food source for animals, and food chain investigations have shown that $^{13}C/^{12}C$ ratios of animal tissues are closely related to the $^{13}C/^{12}C$ of the cattle feed sources. Based on this observation, Gaffney et al. (19) have proposed the use of $^{13}C/^{12}C$ analyses for the characterization of mixtures of soybean protein (C_3-plant) and pork and beef protein (table 3), which in the United States of America show high $\delta^{13}C$ values consistent with the use of corn as feed for these animals. But this favorable situation does not exist in Europe, where C_3-grasses are the major feed source and this fact illustrates that $^{13}C/^{12}C$ analyses are of value only in particular situations, which can vary from one country to the other.

TABLE 3

$^{13}C/^{12}C$ ratio of protein

Origin of protein	$\delta^{13}C$ ‰ PDB
Beef	- 13.1
Pork	- 12.5
Soyprotein	- 23.6

Plant carbohydrates are also used for the manufacture of alcohol and it was shown that there is no ^{13}C fractionation furing fermentation by Saccharomyces. The $^{13}C/^{12}C$ ratios of alcohols from different origins were determined (20, 21) (Table 4).

TABLE 4
Range of $\delta^{13}C$ values of alcohol of different origins

C_3-plant	C_4-plant
$-25 < \delta^{13}C < -30$ ‰ PDB	$-9 < \delta^{13}C < -12$ ‰ PDB
Grape Beet Malt Potato	Corn Cane

Thus the $^{13}C/^{12}C$ analysis allows the detection of the addition of corn or cane sugar in fermented beverages like wine (9) or beer (22, 23), as Vitis species and barley are C_3-plants. Similarly $^{13}C/^{12}C$ ratio can be used to determine the percentage of malt whisky in blended whisky as grain whisky is obtained exclusively from corn (24).

Ethanol is also used for the manufacture of acetic acid in vinegar by fermentation, however acetic acid is easily prepared by total synthesis, so it is necessary to detect the addition of synthetic acetic acid in vinegar. ^{14}C measurement can be used for this purpose but the recent finding of Rinaldi et al. (25) seems more promising : carbon isotopic fractionation during fermentation by Saccharomyces and Acetobacter strains leads to acetic acid (Table 5), which shows a higher $^{13}C/^{12}C$ ratio in the carboxyl group than in the methyl group. This molecular isotopic order is reversed in synthetic acetic acid.

TABLE 5
Intramolecular isotopic order of acetic acid

Acetic acid	$\delta^{13}C$ ‰ PDB	
	CH_3	COOH
Synthetic	- 15.1	- 44.4
Biological	- 34.9	- 16.5

This intrinsic property of biological acetic acid was used by Schmid et al. (26), who devised a rapid method of $^{13}C/^{12}C$ measurement in the carboxyl group of acetic acid. Acetic acid with a $\delta^{13}C < -24$ ‰ PDB in the carboxyl group is thus suspected to contain synthetic acetic acid.

Some plants are cultivated not directly as food supply but as a source of characteristic flavor. Among food flavors, vanilla beans are particularly prized. Vanilla planifolia is a very peculiar plant. It is an Orchidaceae, originating from Mexico, which grows in very limited parts of the world : Madagascar,

La Réunion and Tahiti. Many orchids show a third pathway of carbon dioxide assimilation called Crassulacean Acid Metabolism (27) : in the light CO_2 is fixed by ribulose-1,5-bisphosphate, however CO_2 can also be fixed in the dark by phosphoenolpyruvate to give malic acid, which accumulates during the night. The next day malic acid is decarboxylated and the liberated CO_2 is refixed by ribulose-1,5-bisphosphate. Thus CAM plants show a range of variation of δ^{13} C-value from -10 ‰ PDB to -30 ‰ PDB, according to the relative importance of each pathway of carbon flow (28). In vanilla beans the mean flavouring constituent is vanillin which can be synthetized from wood lignin (C_3-plants) or from guaiacol (fossil oil). Thus, $^{13}C/^{12}C$ analysis was tried as a means of detecting the addition of synthetic vanillin to vanilla extract and was proved to be very effective (29, 30). As vanilla beans are very expensive, it was suggested by Calabretta and Keppel (31) that the $^{13}C/^{12}C$ analysis would not be sufficient as one can easily prepare ^{13}C enriched synthetic vanillin by methylating dihydroxybenzaldehyde with I $^{13}CH_3$. We tried this synthesis in our laboratory and we obtained synthetic ^{13}C-enriched vanillin which was mixed at a concentration of 1 % with lignin vanillin. The resulting vanillin with $\delta^{13}C = + 92.3$ ‰ PDB was demethylated by the procedure of Williard and Fryhle (32) and the recovered dihydroxybenzaldehyde after purification by TLC showed $\delta^{13}C = -25.4$ ‰ PDB. Thus, the procedure of $^{13}C/^{12}C$ analysis of dihydroxybenzaldehyde obtained from vanillin allows the detection of synthetic vanillin even when this product is artificially enriched in ^{13}C (33). In table 6 we summarize the $^{13}C/^{12}C$ measurements of vanillin from different origins and of its demethylated product, 3,4-dihydroxybenzaldehyde.

TABLE 6

$^{13}C/^{12}C$ ratios of vanillin and its demethylated product

Origin of the sample	$\delta^{13}C$ ‰ PDB vanillin	$\delta^{13}C$ ‰ PDB dihydroxybenzaldehyde obtained from vanillin
Natural vanillin from vanilla beans		
Madagascar	- 20.5	- 20.4
Comores	- 20.0	- 19.6
Synthetic vanillin		
from lignin (Europe)	- 27.0	- 26.7
from lignin U.S.A	- 27.8	- 28.2
from guaiacol	- 29.5	- 29.2
Synthetic vanillin + 1 % ^{13}C artificially enriched vanillin	+ 92.3	- 25.4

We have tried to use $^{13}C/^{12}C$ measurement for the characterization of other synthetic flavouring substances but this analysis is not specific enough as natural products from C_3-plants do not differ significantly from synthetic products. The same situation occurs in the case of ethanol. ^{14}C analysis is of limited use as some synthetic processes start with natural raw material (turpentine for example) and the addition of an inexpensive synthetic labelled compound is simple and was proved to occur in one alcoholic product (34). Therefore we have investigated the possibility of using deuterium analysis as a means of characterization of some synthetic substances.

$^2H/^1H$ RATIO OF ORGANIC FOOD CONSTITUENTS

First we investigated the $^2H/^1H$ ratio of different volatile compounds and found that this ratio is quite constant for different classes of volatile substances isolated from plants growing in different areas. We have compared these results with the $^2H/^1H$ ratio of identical compounds of synthetic origin (35).

For terpenic compounds we found lower $^2H/^1H$ ratios in natural products than in their synthetic counterparts (table 7).

TABLE 7

$^2H/^1H$ ratio in flavouring substances extracted from plants or synthetized

Origin	δD ‰ SMOW
Linalol	
Thymus vulgaris (France)	- 257
Coriandrum sativum (France)	- 269
Petit Grain (Paraguay)	- 244
Synthesis	- 170
Citral	
Citrus aurantifilia	- 258
Cymbopogon citratus	- 276
Litsea cubeba	- 251
Synthetic	- 174
Menthol	
Mentha piperita (France)	- 394
L-menthol natural	- 358
synthetic Europe	- 196
synthetic U.S.A	- 242
Trans-anethole	
Foeniculum vulgare var. vulgare (France)	- 91
Illicium verum (China)	- 96
Synthetic U.S.A	- 45

For a phenylpropanoïd compound like anethole, the $^2H/^1H$ ratio is also lower in the natural product than in its synthetic equivalent, although the manufacture of synthetic anethole only involves the isomerization of natural methylchavicol.

It was also shown that ethanol, which is produced by fermentation of sugar, is highly depleted in ^2H as compared to sugar or water. Consequently, the water of wine becomes enriched. By contrast, this isotope discrimination does not occur in the synthesis of ethanol from ethylene, and $^2H/^1H$ of synthetic ethanol is 100 ‰ higher than in fermentation alcohol (20, 21).

CONCLUSION

These examples have shown that stable isotope analysis represents a very interesting possibility to determine the origin of different food constituents, and now this method of analysis must be included with other chemical analysis in order to ensure better control of our food.

The limitations of this method are due to the minimum quantity of substance necessary to perform an analysis : about 5 to 10 mg. We can expect that this limitation will be reduced in the near future with technical improvement in the methods of isotope analysis. Particularly, a great deal of effort must be made to develop a technique to monitor isotope ratios of gas chromatographic effluents. This new development of mass spectrometry would be very useful for the food chemist.

REFERENCES

1 H. Craig, Science, 133 (1961) 1833-1834
2 M. Majoube, J.Chim.Phys., 68 (1971) 1423-1436
3 H. Craig, Science, 133 (1961) 1702-1703
4 A. Nir, in : Development of isotope methods applied to ground-water hydrology. Proceeding Symposium on isotopes techniques in the hydrologic cycle, A.G.U Monographs, series N°11, (1967) 109
5 J. Bricout, J.Ch. Fontes and L. Merlivat, C.R.Acad.Sci.Paris, 274 (1972), 1803-1806
6 J. Bricout, L. Merlivat and J.Ch. Fontes, Ind.Alim.Agric., 90 (1973) 19-22
7 A. Marell, A. Carisiano, M. Riva, R. Thillio, Adv.Mass.Spectrom., 7A (1978) 523-527
8 J. Bricout, J.Ch. Fontes and L. Merlivat, Connaissance de la Vigne et du Vin, 2 (1974) 161-170
9 J. Bricout, J. Koziet, D.G.R.S.T. Compte-rendu de fin de contrat N°75-70-370 (action concertée Technologie Agricole et Alimentaire) (1979)
10 C. Keeling, Geochim.Cosmochim.Acta, 24 (1961) 277-298
11 H. Craig, Geochim.Cosmochim.Acta, 12 (1957) 133-149
12 M. Calvin and J.A Bassham, in : the Photosynthesis of Carbon Compounds, Benjamin, New York, 1962
13 B.N Smith and S. Epstein, Plant Physiol., 47 (1971) 380-384
14 M.D Hatch and C.R Slack, Biochem.J., 101 (1966) 103-111

15 L.W Doner, H.W Krueger, R.H Reesman, J.Agric.Food Chem., 28 (1980) 362-364
16 A. Nissenbaum, A. Lifshitz, Y. Stepek, Lebensm.Wiss.Technol. 7 (1974) 152-154
17 O. Carro, C. Hillaire-Marcel, M. Gagnon, J.Ass.Off.Anal.Chem., 63 (1980) 840-844
18 J.W White, L.W Doner, J.Ass.Off.Anal.Chem. 61 (1978) 746-750
19 J. Gaffney, A. Irsa, L. Friedman and E. Emken, J.Agric.Food Chem. 27 (1979) 475-478
20 J. Bricout, J.Ch. Fontes, L. Merlivat, Ind.Alim.Agric., 92 (1975) 375-378
21 P. Rauschenbach, H. Simon, W. Stichler and H. Moser, Z.Naturforsch, 34c (1979) 1-4
22 H.L Schmidt, U. Kunder, F.J Winkler and H. Binder, Brauwissenschaft, 33 (1980) 124-126
23 M. Benard, J. Bricout, R. Scriban, to be published
24 J. Koziet and J. Bricout, Ann.Nutr.Alim. , 32 (1978) 941-946
25 G. Rinaldi, W.G Munschein and J.M Hayes, Biomed.Mass.Spect., 1 (1974) 412-414
26 E.R Schmid, I. Fogy and P. Schwarz, Z. Lebensm.Unters.Forsch., 166 (1978) 89-92
27 A. Moyse in "Travaux dédiés à L. Plantefol", Masson, Paris (1965) p. 21
28 M. Bender, I. Rouhani, H. Vines, Black Jr.C, Plant Physiol., 52 (1973) 427
29 J. Bricout, J.Ch. Fontes and L. Merlivat, J.Ass.Off.Anal.Chem., 57 (1974) 713-715
30 P.G Hoffman, M. Salb, J.Agr.Food Chem., 27 (1979) 352-355
31 J.P Calabretta and F.J Keppel, Communication to the VIIIth International Congress on Essential Oils(1980)
32 P.G Williard and G.B Fryhle, Tetrahedron Letters, 21 (1980) 3731-3734
33 J. Bricout, J. Koziet, M. Derbesy and B. Beccat, Ann.Fals.Exp.Chim.,to be published
34 P. Resmini, G. Volonterio, L. Cecchi, Personal communication
35 J. Bricout, J. Koziet, In "Flavor of Foods and Beverages", Academic Press, New York, 1978, 199-208

DETECTION OF ADDED WATER AND SUGAR IN NEW ZEALAND COMMERCIAL WINES

J. DUNBAR

University of Waikato, Hamilton, New Zealand.

ABSTRACT

Water $^{18}O/^{16}O$ ratios and ethanol $^{13}C/^{12}C$ ratios in both standard and commercial wines from New Zealand have been measured. A plot of the δ-values against each other showed that in some cases there had been large additions of cane sugar and of water made at some stage during the wine making process. This was particularly so in the case of wines in which the grape type and vintage were not shown on the label, sparkling wines and sherries.

INTRODUCTION

Most plants including the grape vine, employ the C-3 or Calvin cycle of photosynthesis which discriminates strongly against the carbon isotope ratio of CO_2, so that the carbon incorporated into the sugars present in grapes has approximately 20‰ (parts per thousand) less ^{13}C than atmospheric CO_2.

Alternatively, a small group of plants of which sugar cane is a member, employ the C-4 or Hatch-Slack cycle of photosynthesis. These plants show only a small discrimination against the carbon isotope ratio of CO_2, thus sugar cane produces sugars which are depleted by only 5‰ in ^{13}C (relative to atmospheric CO_2). The primarily fixed isotope ratios are modified only very slightly by the process of wine making [1] hence can be used to determine the fraction of C-4 (e.g. cane sugar, the only sugar available in New Zealand) to C-3 (e.g. grape sugar) derived ethanol or other components in a wine.

This difference in $^{13}C/^{12}C$ ratio between C-3 and C-4 plants has often been used to determine the proportion of products from the two plants present in particular foods. For example White and Doner [2] developed a technique to detect the addition of corn syrup to commercial honey and Koziet and Bricout [3] used it to study blended malt and bourbon whiskies.

It is also known that the water in fruits can become enriched in ^2H and ^{18}O due to evapotranspiration within the plant which is the basis for analytical differentiation between the water from the fruit and the ground water in which the plant grew [4-6].

This principle has thus been used to determine the source of water in commercial fruit drinks. Bricout and Mouaze [7] using D/H ratios, developed a method to distinguish between pure orange juice and orange juice which had been prepared by redilution of a concentrate. The same technique was also used for the analysis of commercial apple juices [8].

$^{18}O/^{16}O$ ratios of the water in New Zealand wines have been measured [9] with an enrichment of approximately 7‰ above the ground water of the vineyard being found. Also a yearly variation in the mean ^{18}O enrichment as well as a location effect were observed. No difference in $^{18}O/^{16}O$ ratio between red and white wines was found and the variation between bottles of the same batch of wine was small. The fermentation process as well as the use of concentrated grape juices were found to have little effect on the $^{18}O/^{16}O$ ratio of the resultant wine.

By combining the results of $^{18}O/^{16}O$ determinations of the water with the $^{13}C/^{12}C$ determinations of the ethanol of a wine, it was hoped to develop an analytical technique which would indicate the source of water and sugar in that wine.

METHODS

$^{18}O/^{16}O$ ratios were determined by equilibrating CO_2 with the grape juice or wine using the method as described by Epstein and Mayeda [10]. A 10ml sample of liquid was equilibrated with 20ml of CO_2 at 25°C for 48 hours after which the CO_2 was separated from the liquid by fractional distillation through glass sinters and then admitted into an isotope mass spectrometer.

In the case of carbon isotope measurements, ethanol was distilled off a 25µl sample of wine (10µl in the case of sherries) and was combusted with oxygen over copper oxide in a vacuum line. The carbon dioxide was frozen out at liquid nitrogen temperature, and after removal of trace contaminants and water by fractional distillation, the $^{13}C/^{12}C$ ratio was determined.

The results of the isotope determinations are reported in the δ-notation where:

$$\delta^{18}O = \left[\frac{(^{18}O/^{16}O)\ sample}{(^{18}O/^{16}O)_{SMOW}} - 1 \right] \times 1000‰ \qquad \delta^{13}C = \left[\frac{(^{13}C/^{12}C)\ sample}{(^{13}C/^{12}C)_{PDB}} - 1 \right] \times 1000‰$$

$(^{18}O/^{16}O)_{SMOW}$ is the $^{18}O/^{16}O$ ratio of the international standard SMOW (Standard Mean Ocean Water) and $(^{13}C/^{12}C)_{PDB}$ is the $^{13}C/^{12}C$ ratio of the international standard PDB (Pee Dee Belemnite).

$\delta^{13}C$ values can be replicated to within ±0.02‰ (1σ) and $\delta^{18}O$ values to within ±0.05‰ (1σ), however the natural variation in wines are very much larger than these so that it is not necessary to measure them to better than ±0.2‰. All samples were measured in duplicate.

CALCULATIONS

For calculating the amount of added water present in a commercial wine, the following equation was used:

$$\left[(100-Y) \cdot \delta^{18}O_{std.wine}\right] + \left[Y \cdot \delta^{18}O_{water}\right] = 100 \cdot \delta^{18}O_{comm.wine}$$

where Y = % water in the commercial wine, std.wine = standard wine, comm.wine = commercial wine.

Therefore after measuring the $\delta^{18}O$ values of the standard and commercial wines and also that of the water, the % water addition (Y) can be calculated.

RESULTS

A comparison of the $^{18}O/^{16}O$ ratio of standard and commercial wines produced in the same area

The standard wines used in the experiments were produced at the Ministry of Agriculture and Fisheries Government Wine Research Unit at Te Kauwhata, a wine producing area which is also the centre of several commercial vineyards. Because the vineyards are all subjected to the same climatic conditions and water supplies and the wines produced are often from the same grape varieties, it was decided to compare the $^{18}O/^{16}O$ ratios of the wines produced by a commercial company against the control wines produced by the Government research centre. The varieties chosen were Riesling Sylvaner and Golden Chasselas.

The results in Table 1 show that in 6 out of 7 cases the $^{18}O/^{16}O$ ratio of the commercial wine is lighter than that of the corresponding standard wine. The history of each control wine is known and in no cases has water been added during the wine making process, however this does not seem to be so for the commercial wines. By measuring the $^{18}O/^{16}O$ ratio of the groundwater of the region (i.e. 0% grape juice) and using the $^{18}O/^{16}O$ ratio of the control wine (100% grape juice), the relative position of the commercial wine between these two values can be calculated. This then indicates the probable degree of water addition to the commercial wine.

TABLE 1: $\delta^{18}O$ values of commercial and standard wines produced in the same area.

	Year	TK.H$_2$O	Std.wine		Comm.wine		% water addition to commercial wine
			$\delta^{18}O_{SMOW}$ ‰				
Riesling Sylvaner	1979	-7.6	-2.8	(+1.8)	-4.3	(+3.3)	31±8
	1978	-7.0	+0.1	(+6.9)	-1.1	(+5.9)	17±5
	1977	-6.7	-1.2	(+5.5)	-1.7	(+5.0)	8±7
	1976	-7.1	+1.3	(+8.4)	-0.9	(+6.2)	26±5
Golden Chasselas	1979	-7.6	-3.2	(+4.4)	-3.9	(+3.7)	16±9
	1978	-7.0	+0.1	(+7.1)	-0.1	(+6.9)	3±5
	1977	-7.6	-0.2	(+6.5)	-0.1	(+6.6)	-1 (iv)

Note: (i) Values shown in brackets are the enrichments of the wines compared with the corresponding tap/ground waters. (ii) TK.H$_2$O = Te Kauwhata ground and tap water.

TABLE 1 (cont.)

(iii) For explanation of calculation method see eqn (1). (iv) Commercial wine heavier in ^{18}O than standard wine. (v) Errors calculated assuming a 0.2‰ uncertainty in each $^{18}O/^{16}O$ determination.

In 1977 the commercial Golden Chasselas wine had a 0.1‰ more positive $^{18}O/^{16}O$ ratio than the control wine, indicating the limits of natural biological variation. As a result of this it was decided to allow ±0.2‰ on each wine measurement to take this error into account. It therefore seems likely that the only commercial wines which have had no water addition are the 1977 and 1978 Golden Chasselas and possibly the 1977 Riesling Sylvaner.

An error could also be introduced into these values if the winemaker blended the wines with varieties other than those specified on the label or if he used juices from other vintages (years). However for commercial wines tested there was no mention of this on the label.

A comparison of the $^{18}O/^{16}O$ and $^{13}C/^{12}C$ ratios of control and commercial wines

For the development of a general control method both the $^{18}O/^{16}O$ ratio of the water and the $^{13}C/^{12}C$ ratio of the ethanol of commercial and standard wines were measured with the results being shown in Figures 1, 2 and 3.

In each of these figures certain fixed lines have been introduced. Firstly the maximum and minimum δ-values of the tap water (which is also the ground water), surveyed from the major wine producing regions over several years are shown. This information is necessary when comparing commercial wines against the standards so that an estimate of the possible water additions can be made.

For estimating cane sugar additions the $^{13}C/^{12}C$ ratio for ethanol derived from pure cane sugar is shown. (Line at $\delta^{13}C$ = -12.4‰). Also shown is the mean ethanol $\delta^{13}C$ value for the control wines studied which is based on the values shown plus those of an extra 30 wines additionally measured. The sugar addition to a commercial wine can therefore be calculated by determining the position of the wine relative to the value for pure cane sugar and the mean of the control wines.

A line corresponding to a 40% cane alcohol addition to the standard wine (mean at -21.7‰) has been introduced because of the following possibility. If a wine with a final alcohol concentration of 10% is to be made from a juice with a theoretical minimum sugar content of 12%, then a further 6% sugar must be added (18% sugar in total). Thus in this case no more than 33% of the total sugar, and hence 33% of the alcohol should have come from cane sugar. If the same juice were to be made into a sweet wine then perhaps an additional 1 to 6% sugar (for residual sugar) would need to be added. Therefore in an <u>extreme</u> case a wine made from a natural juice with only the addition of sugar could have a composition such that up to 40% of the ethanol originated from cane sugar. Higher additions of sugar would therefore have to be accompanied by simultaneous water additions.

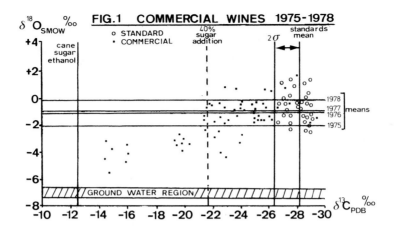

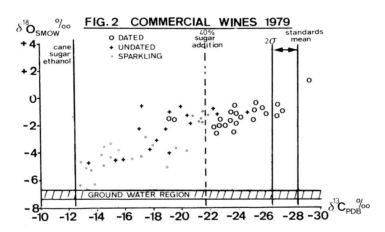

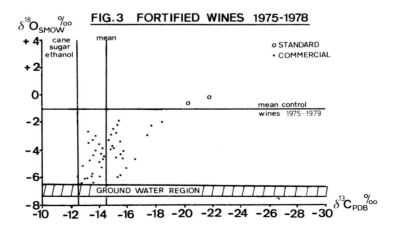

In Figure 1, four additional lines corresponding to the mean $\delta^{18}O$ values of 50 control wines from 1975, 76, 77 and 78 have been shown. This was done so that water addition to the commercial wines could be estimated.

Figure 1: As can be seen from the graph there is a wide spread of values for the commercial wines studied ranging from those with no added sugar or water to those which contain large additions of both. The commercial wines studied can thus be divided into three groups. i) 10% with no added sugar; ii) 61% with acceptable sugar addition; iii) 29% with sugar addition such that tap water must also have been added.

Figure 2: This figure serves to illustrate the difference in quality between wines in which the grape variety and vintage are shown on the label and those in which little information is given. The three catagories of wine shown are: i) 1979 wines; ii) Wines purchased in 1979, hence presumably produced then, but with no date printed on the label; iii) Sparkling wines also undated but presumably produced in 1979.

There is an area of overlap between the groups (i) and (ii) but generally the "dated" wines show less sugar and water additions than do those of the cheaper "undated" group. The sparkling wines have obviously been made with large additions of sugar, with some showing the sugar source as being close to 100% cane sugar. As all except one of the samples are to the left of the 40% cane alcohol line this clearly indicates large amounts of water are added to sparkling wines.

Figure 3: This figure shows data from commercial fortified wines and two control sherries (no added water).

Fortification involves the addition of potable spirit to a wine. Traditionally this alcohol should be produced by fermentation of the lees (skins and pips etc) to which sugar has been added. In New Zealand commercial wineries however, this alcohol is almost exclusively made by the fermentation of a sugar and water solution. Therefore fortification is usually carried out with pure cane alcohol. The New Zealand Food and Drug Regulations 1973 228(1) state that fortified wines should not contain more than 22.9% alcohol, so the production of a fortified wine would normally involve the addition of between 9 and 14% ethanol. (Not all fortified wines would have an alcohol content as high as 22.9%). The $^{13}C/^{12}C$ value of a sherry that has been made from 100% pure grape juice with no added sugar, and has then been fortified with cane sugar ethanol, would therefore have a $\delta^{13}C_{PDB}$ value of approximately -20‰.

Apart from the two control sherries, the fortified wines show very large shifts in the $^{13}C/^{12}C$ and $^{18}O/^{16}O$ ratios. It appears that the unfortified wine base of most of the samples has been made from a heavily sugared and diluted grape juice ferment and five in fact are barely distinguishable from fermented cane sugar and tap water solutions.

CONCLUSIONS

Combined $^{18}O/^{16}O$ plots can be very useful for the categorising of commercial wines. The results presented show that a considerable diversity between samples exists and that in some cases large quantities of sugar and tap water have been used in the wine making process. This is especially so of: 1) Wines without a date or wine type printed on the label; 2) Sparkling wines and 3) Fortified wines e.g. sherries and ports.

A technique such as this is especially suitable for a country such as New Zealand because the only source of commercial sugar is C-4 cane sugar and the isotopic variation of the water supplies in the grape growing areas is small.

ACKNOWLEDGEMENTS

The author wishes to thank the University Grants Committee of New Zealand for financial support.

REFERENCES

1. J. Bricout, J.C. Fontes and L. Merlivat, Ind. Agr. Alim., 92(1975)37, 375.
2. J.W. White and L.W. Doner, J.A.O.A.C., 61(1978)746.
3. J. Koziet and J. Bricout, Ann. Nut. Alim., 32(1978)195.
4. J. Bricout, Ann. Fals. Exp. Chim., 66(1973)195.
5. J. Bricout, J.C. Fontes and L. Merlivat, Acad. Sci. Paris Compt. rend. Ser.D., 274(1974)161.
6. J. Bricout and L. Merlivat, Acad. Sci. Paris Compt. rend. Ser.D., 273(1971)1021.
7. J. Bricout and Y. Mouaze, Fruits 26(1971)77.
8. J. Bricout, L. Merlivat and J.C. Fontes, Ind. Agr. Alim., 90(1973)19.
9. J. Dunbar and A.T. Wilson (submitted for publication).
10. S. Epstein and T. Mayeda, Geochim. Cosmochim. Acta, 4(1953)213.

$^{18}O/^{16}O$-RATIO OF WATER IN PLANTS AND IN THEIR ENVIRONMENT (RESULTS FROM FED. REP. GERMANY)

H. FÖRSTEL

Institute of Radioagronomy, Nuclear Research Center (KFA), P.O. Box 1913, 5170 Jülich (F.R.G.)

ABSTRACT

The water turnover of plants, an important part of the energy and mass balance of ecosystems, can be characterized by the oxygen isotope ratios of the fluids. In Jülich (F.R. Germany) the isotope ratios of precipitation, soil moisture, groundwater and xylem sap agree well. Only in the leaf water one observes a remarkable enrichment of ^{18}O and a rapid exchange with the water vapour of the air.

INTRODUCTION

The report summarizes results of the $^{18}O/^{16}O$-ratio in water samples from the local ecosystem, observed for ten years. The KFA is situated at the plain of river Rur within a forest.

In an ecosystem one should distinguish between two water cycles (fig. 1):
1. non-living components (precipitation, infiltration into the soil, evaporation back to the air humidity) (step 1, 2, 3),
2. transport through the living organisms (water transport via the roots and the xylem, transpiration of leaves) (step 4, 5).

The plants are the most dominant part of the living biomass in an ecosystem. Animals and decomposers can be neglected. Because of the light requirement of photosynthesis, the plants offer a large surface area to the surrounding air, and consequently, a large water turnover seems to be necessary.

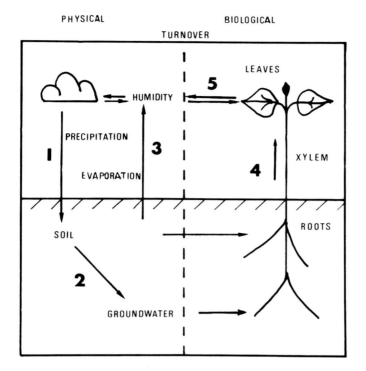

Fig. 1. Scheme of water fluxes in an ecosystem.

METHODS

This report discusses only results of $^{18}O/^{16}O$-ratio measurements of water (precipitation, groundwater). Air humidity was condensed out of an air stream into traps at $-77°$ (dry ice/acetone). Usually water of plant material was obtained by complete vacuum distillation. For the final measurement 1 ml of water is equilibrated with a certain amount of gaseous CO_2 overnight. Afterwards the CO_2 is separated from the fluid by a cool trap ($-77°$) between exchange vessel and mass spectrometer inlet system (micromass 602, VG Isogas, Winsford, U.K.). All $^{18}O/^{16}O$-ratios are reported as δ-values related to Vienna-SMOW:

$$\delta\ [°/oo] = (\frac{R_{Sample}}{R_{Standard}} - 1) \cdot 10^3. \qquad (1)$$

RESULTS

Local precipitation (step 1)

The δ-value of local precipitation depends mainly on its geographical position. Using the data of the IAEA network [1], Dansgaard emphasized the correlation between the δ-value and the mean annual temperature. We tested the correlation between the oxygen isotope ratios of precipitation and five other parameters by a multicorrelation analysis (IAEA-data of 1966 and of stations $20°$ and $65 °N$). The result of the analysis is expressed in percent:

        D/H-ratio of precipitation   47.5 %,
        mean annual temperature   24.1 %,
        T-content of precipitation   15.0 %,
        amount of precipitation   13.1 %,
        relative humidity of air   0.3 %.

Besides the well-known correlation between the isotopes of oxygen and hydrogen, the other most important factors are the mean annual temperature and the precipitation. Two other influences should be mentioned also: the distance from the sea and the altitude above sea level. At a first glance single precipitations have large variations [2], but the monthly, or better, yearly mean values are more constant.

Infiltration into soil and groundwater (step 2)

The mean values of precipitation (monthly average 1974-1980: $- 7.6 \pm 2.0 °/oo$), tapwater (- 7.7) and samples of the local water supply (-7.6) agree well. This is in accordance with the ideas of Eichler [3]: The successive precipitations infiltrate into subsequent layers of the soil contineuously and are mixed during their way down through the soil column. Only directly at the soil surface one could expect an enhanced ^{18}O-content (fig. 2 and 3), but in the field this effect is observed only under very dry conditions (meadow, fig. 3). Generally, a negligible amount of enriched water at the surface does not change the isotopic ratio of soil moisture. All observations can be explained by the statement: The precipitation is deposited into the soil and groundwater without any change of its δ-values.

Relation between precipitation and air humidity (step 3)

Short term observations demonstrate a very rapid change of δ-values of water vapour in the air humidity corresponding to the local weather conditions, but in Jülich (fig. 4) and Chicago [4] the annual mean δ-values of air humidity correspond to an equilibrium between the liquid and gaseous phase. These results may only be valid at these locations within the temperate climate, but have been repeatedly found [5].

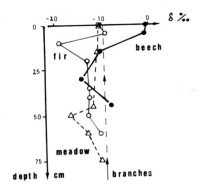

Fig. 2. δ-values of soil water (beech and fir forest, meadow; IBP-station Solling), dependent on depth. The arrow and broken line indicate the δ-value of branch water, which usually is in accordance with that of the soil humidity and precipitation.

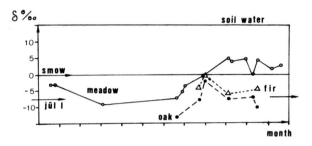

Fig. 3. δ-values of soil water (fir and oak wood, meadow; KFA Jülich) at the surface. The local standard JÜL I represents the mean of precipitation and groundwater (arrow).

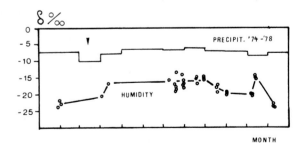

Fig. 4. Annual course of δ-value in precipitation and in the water of the air (humidity) in Jülich. The arrow indicates the minimum during spring.

Transport of water within the plants (step 4)

During the physically driven transport of water in plants no fractionation of oxygen isotopes of water was observed (ref. 6). To test the δ-values of water in the leaves an annual course of twig segments was measured (fig. 5).

During the growth period, the δ-values of water in soil and twigs are comparable. Only after the fall of the leaves the δ-value increases, possibly due to the evaporation of water via the bark. But directly after the onset of the individual growth period the water in the xylem is renewed from the soil reservoir. After this experience one could assume, that the δ-values of water within the transport system of trees really reflect the mean δ-value of mean local precipitation.

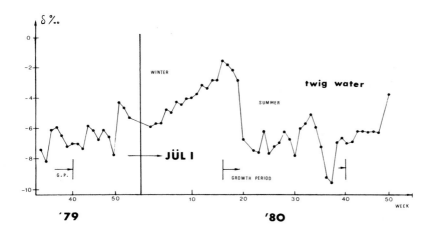

Fig. 5. δ-values of twig water, observed in eight tree species in Jülich during the year (mean values). The mean of local precipitation and groundwater is given also (JÜL I).

Water vapour exchange leaf water/air humidity (step 5)

Measurements of δ-values in leaf water under field conditions do not only report a fractionation, but also a distinct variation during the day [7]. The δ-values of water in the leaves can be explained by a fractionation between a thin water surface and surrounding air. The resulting δ-value of leaf water (δ_L) under stationary conditions:

$$\delta_L = \delta_s(1-h) + \varepsilon_q + \varepsilon_k + (\delta_a - \varepsilon_k) \cdot h \quad [°/oo]. \tag{2}$$

h: relative humidity/ε: fractionation constants (α-1), index (s, q, k, a): (soil, equilibrium, kinetic, air).

The equation (2) can be tested by its boundary conditions: Under total dryness (h=0) the enrichment δ_L is determined by the isotopic composition of the soil water and the fractionation processes alone; otherwise, if the surrounding air is saturated by water vapour (h=1), only the equilibrium effect and the δ-value of air water vapour are responsible for the final enrichment in leaf water. One can see, that the relative humidity of air is the most important parameter, which is responsible for the fractionation of oxygen isotopes.

Field observations have confirmed the theory: The δ-value of leaf water is parallel to the diurnal course of air temperature and conversely to the relative humidity. The δ-value between different types and species of plants does not vary markedly [5]. The observation of rapid changes of δ-values under natural conditions corresponds to the results in growth chambers: A sudden change of environmental conditions was responded to by the leaves even within one hour (fig. 6).

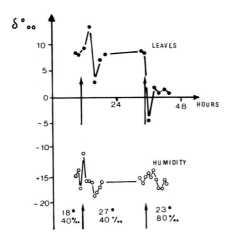

Fig. 6. Variation of δ-values in leaf water (*Prunus spec.*) and air humidity after sudden changes of the environmental conditions (experiment done in a growth chamber, climatic conditions are given at the bottom of the figure).

CONCLUSION

Figure 7 summarizes the δ-values of water in the local ecosystem in Jülich. The mean δ-value of precipitation seems to be very constant, and consequently the oxygen isotope ratio of wet soil, groundwater and fluid of the transport system of the plant, also. The most distinct variations of the δ-value are observed in the leaf water. The oxygen isotope content of this water pool seems to be the result of very rapid exchanges between fluid and gaseous phase. Recent tracer experiments support this idea [8].

Generally, one is able to predict the δ-values of water in an ecosystem, and may use the results for application in studies of plant physiology and of the climate during the past.

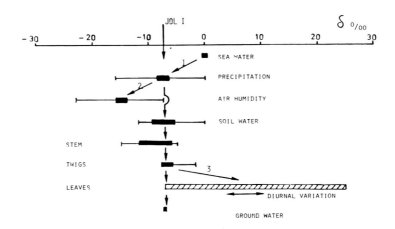

Fig. 7. Summary of the variations of δ-values in our local ecosystem at Jülich. The number 1 represents evaporation and condensation, number 2 evaporation and number 3 transpiration of plants.

REFERENCES

1. IAEA (Ed.), Technical Report Series 96 (1969), 117 (1970), 129 (1971), 147 (1973), 165 (1975), 192 (1979).
2. H. Förstel, H. Hützen, Jül-Bericht 1524 (1978), p. 16.
3. R. Eichler, Geol. Rdsch. 55 (1965), 144-159.
4. B.D. Holt, R. Kumar, O.T. Cunningham, Atmosph. Environm., in Press.
5. H. Förstel, Rad. Environm. Biophys. 15 (1978), 323-344.
6. H. Förstel, H. Hützen, Jül-Bericht 1595 (1979), p. 41.
7. G. Dongmann, H.W. Nürnberg, H. Förstel, K. Wagener, Rad. Environm. Biophys. 11 (1974), 41-52.
8. H. Förstel, H. Hützen, Adv. Mass Spectrometry, Proc. 7th Int. Symp. Mass Spectrometry, Milano 1980, in press.

USE OF WATER WITH DIFFERENT ^{18}O-CONTENT TO STUDY TRANSPORT PROCESSES IN PLANTS

H. FÖRSTEL and H. HÜTZEN
Nuclear Research Centre (KFA), Radioagronomy, P.O.B. 1913,
D 5170 Jülich, Fed. Rep. Germany

ABSTRACT

The movement of liquids in plants can be studied by the use of labelled water. The progress of mass spectrometry enables one to observe even small variations of oxygen isotope ratios in water. Apple seedlings were grown in gravel and supplied by a nutrient solution prepared from the local tap water. The label was introduced by a quick change of the common to the new tracer solution. In the first experiment the common nutrient solution was replaced by a fluid, prepared from distilled sea water. The apple seedlings were observed under field conditions. The second test was made in a climate box under controlled conditions, using arteficially ^{18}O-enriched water as a label. The results of both tests demonstrate, that the upward stream of labelled solution is accompanied by a rapid lateral exchange of water between the different tissues of the stem. The water of the leaves exchanges rapidly with the water vapour of the surrounding air. This can be seen distinctly only, if arteficially ^{18}O-enriched water is used.

INTRODUCTION

Formerly transport processes in plants were observed by the movement of dyes, which are transported rapidly from the lower to the upper parts of the herbs. This experience supported the idea, that the upward transport of fluid is mainly directed through the conducting vessels of the xylem. Contrary to these results were the conclusions of the application of labelled water: The tracer can be detected only after a manifold turnover of water. There seems to be a confusion of the terms "transport velocity" and "velocity of water movement". The dye and the thermal method are able to show only the general tendency of the fluid flow, but do not represent the complete transport of water. One should expect, that only labelled water can be a true tracer of the water transport processes.

Formerly the application of stable isotopes was limited by technical and economical reasons, but today the measurement of even small natural variations of the oxygen isotope ratios is a very convenient procedure. Consequently one can test the suitability of water of different geographic origin or of an artifical ^{18}O-enrichment as a tracer for the fluid movement in plants.

MATERIAL AND METHODS

Apple seedlings of about 1.5-2.0 m were grown in gravel and supplied with a Long-Ashton nutrient solution, which was prepared from local tap water (^{18}O/^{16}O-ratio: -7.65 ^{o}/oo [2][+]).

In the first experiment water of different geographic origin was used as a tracer (common solution, tap water: -7.65 ^{o}/oo/tracer solution, distilled sea water: -2.06 ^{o}/oo). At the start of the experiment the common solution around the roots was replaced by the tracer solution. One seedling was not treated with the new nutrient solution and used as a control. The plants were left under field conditions. Fig. 1 reports these conditions, including the ^{18}O/^{16}O-ratio of the air water vapour and the transpiration of the control plant. The air humidity was collected into cool traps at -77 ^{o}C (dry ice/acetone). The transpiration (water turnover of the plant) was continously recorded by the analogue signal of a balance. The plants after the labelling were harvested successively during the light period of the day. The controll plant was harvested as the last one in the evening.

In the second test the plants were kept under constant conditions in a climate box (dark period: 10°, 95 %/light period: 22°, 65 %, 33.000 Lux). At 9.00 a.m., one hour after the start of the light period, the usual nutrient solution was changed quickly. The labelled solution was prepared from a water of about double enriched ^{18}O-content (about +500 ^{o}/oo). The linearity of the δ-value determination was confirmed by test measurements. Each plant was tested during separate days and harvested at half-hour intervals after labelling. The ^{18}O/^{16}O-ratio of air water vapour in the climate box remained at its natural level.

[+] The ^{18}O/^{16}O-ratios are reported as δ-values related to Vienna-SMOW [3].

The $^{18}O/^{16}O$-ratio of water was determined by our routine method of mass spectrometry [2].

RESULTS AND DISCUSSION
Experiment with natural varying ^{18}O-content under field conditions

The experimental conditions are reported in fig. 1a-c, the oxygen isotope ratios of air water vapour in fig 1d. Fig. 2a demonstrates three different states of label uptake after 1.5, 5.75 and 8.5 hours. Fig. 2b compares the last labelled tree, harvested 10 hours after the labelling. The unlabelled control was harvested parallely 10.5 hours after the start of the experiment.

As a result one can see, that the xylem tissue of the stem is labelled rapidly at the lower parts, but more slowly in the upper region (fig. 2a). The enrichment of the $^{18}O/^{16}O$-ratio in the leaf water is in agreement with former observations. In fig. 2b one can see the distinct difference between labelled and non labelled material. The change of common and labelled fluid was incomplete, probably.

Experiment with labelled water under constant environmental conditions

The results after the labelling vary distinctly, if one observes the uptake of arteficially ^{18}O-enriched water. Fig. 3a shows the time-dependent increase of the ^{18}O-content in the xylem within five segments of the stem. The increase of the label concentration behaves like a typical increase of a tracer concentration in subsequent compartments. Fig. 3b demonstrates the delayed uptake of water into the water compartment of the leaves. The most important fact is the low ^{18}O-concentration in the leaf water, even after 2.5 hours. After arrival of the labelled solution in the leaves the new equilibrium seems to be established very quickly. One should assume a rapid exchange of water between the pool of fluid within the leaf and the unlabelled water vapour in the surrounding air.

CONCLUSION

The results of the root uptake from the nutrient solution, which was prepared from a natural water (sea water), could lead to wrong conclusions. In this case the enrichment of ^{18}O during the fractionation processes of transpiration governs the ^{18}O-content of leaf water. During the transport of water through the whole plant no important fractionation can be observed. Only at the boundary leaf

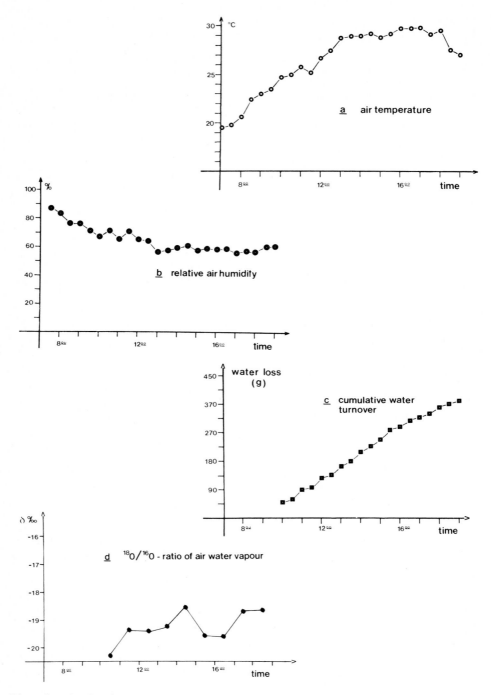

Fig. 1a-d. Environmental data of May 20, 1979 at the experimental site.

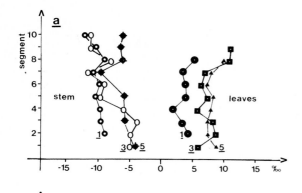

Fig. 2a-b. Uptake of labelled nutrient solution made from distilled sea water into apple seedlings (tree 1: 1.5 h/tree 2: 5.75 h/tree 5: 8.5 h/tree 6: 10 h/tree 7: unlabelled control, harvested 10.5 h after start).

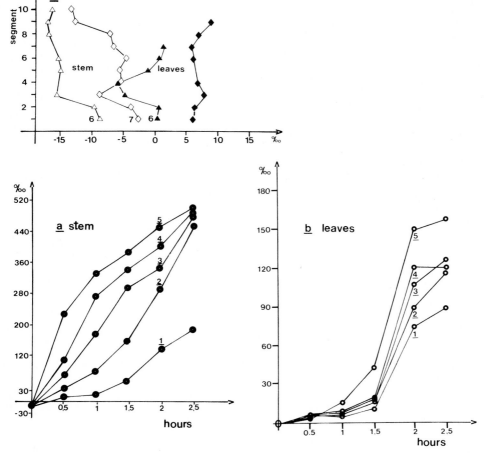

Fig. 3a-b. Time dependence of tracer concentration in the segments of apple seedlings (from bottom to top). (1: 0-10 cm above roots/ 2: 20-30 cm/3: 40-50 cm/4: 60-70 cm/5: 100-120 cm).

water-air a distinct fractionation can be observed, corresponding
to the oxygen isotope fractionation between liquid and gaseous phase.
Consequently the $^{18}O/^{16}O$-ratio of water in the leaves is enhanced.
Its enrichment is determined mainly by the relative humidity of the
air [4]. In the first test the natural fractionation exceeds the differences of labelled and nonlabelled solution. The use of water of
only small oxygen isotope variations is limited and should be replaced by enriched material, especially under field conditions.

The observation of the tracer concentration of enriched water
gives more distinct results. One can demonstrate, that the upward
flow of water is accompanied by a rapid lateral exchange within the
stem. If the natural fractionation can be neglected, the rapid exchange of water between leaves and air can be seen distinctly. This
low ^{18}O-concentration of leaf water must be the result of rapid exchange processes. Reversely, this fact should be useful to label the
the water of leaves or fruits.

The present status of techniques enables one to combine the use
of low enriched, that means cheap material, and a very sensitive
technique. If one uses water of different origin as a tracer of
water movements in plants, fractionation processes during the transpiration must be taken into account carefully.

REFERENCES

1 G. Hübner, Flora, 148(1966)549-594.
2 H. Förstel and H. Hützen, Jül-1595(1979), p. 18.
3 H. Craig, Science, 133(1961)1833-1834.
4 H. Förstel, Rad. and Environm. Biophys., 15(1978)323-344.

TRACER STUDIES ON THE OXYGEN EVOLUTION IN PHOTOSYNTHESIS

H. METZNER, K. FISCHER and O. BAZLEN
Institut für Chemische Pflanzenphysiologie der Universität,
Tübingen (G.F.R.)

ABSTRACT

Irradiated chlorophyll-containing plant cells reduce carbon dioxide to carbohydrates. They gain the sufficient reducing power by the light-induced production of NADPH and ATP. The final source of the necessary electrons is apparently the water.

Energetic considerations make an oxidative H_2O decomposition unlikely: the removal of an electron from a water molecule requires more energy than the 1.8 eV, which are contained in a (red light) photon. Furthermore a water splitting must - like electrolysis and chemical decomposition reactions - give a significant oxygen isotope effect. This should lead to a measurable ^{16}O enrichment in the released O_2. Mass spectrometric determinations of the photosynthetic oxygen set loose from H_2O/CO_2 systems of natural isotope abundances reveal just the reverse: the collected O_2 is always enriched in the heavier isotope.

There are some photosynthesis models which - like an aqueous AgCl crystal suspension - allow to simulate the light-induced oxygen evolution. An analysis of this system demonstrates that it requires the presence of catalytic amounts of bicarbonate anions. Mass spectrometric experiments with $NaHC^{18}O_3$ indicate, that in this case the source of the released oxygen is CO_2, not water.

Similar tests with living plant cells suffer from the fast isotope equilibration between $C^{18}O_2$ molecules (respectively $HC^{18}O_3^-$ anions) and $H_2^{16}O$ molecules. Much more reliable data can be obtained with thylakoids which during their isolation lost the responsible enzyme (carbonic anhydrase). Like AgCl suspensions they require the addition of small amounts of carbonic acid. If this is given in the form of ^{18}O-labelled sodium bicarbonate, the released O_2 shows a time-dependent $^{18}O/^{16}O$ ratio:

the earlier the oxygen is collected, the higher is its ^{18}O content. This observation speaks against a direct oxidation of - primarily unlabelled - water molecules.

It has to be discussed, whether the meanwhile existing data justify to assume a participation of CO_2 in the formation of the real precursor of the photosynthetic oxygen.

INTRODUCTION

To the physicist "photosynthesis" means the sequence of events, by which the green plant converts energy of electromagnetic fields into chemical energy. The chemist will describe these reactions as light-dependent reduction of CO_2 to carbohydrates. The balance equation can be given by

$$\begin{bmatrix} 3\ CO_2 + 6\ H_2O \\ \updownarrow \\ 3\ HCO_3^- + 3\ H_3O^+ \end{bmatrix} \xrightarrow{n \cdot h\nu} C_3H_6O_3 + 3\ O_2 + 3\ H_2O \qquad (1)$$

The sequence of chemical steps between CO_2 and the first analytically identified carbohydrate (triose) molecule could be mainly elucidated by the use of radioactive carbon. It starts with the (dark) incorporation of carbon dioxide into an acceptor molecule of the plant cell:

$$R-H + CO_2 \longrightarrow R-COOH \qquad (2)$$

It is the primarily formed carboxyl group, which is subjected to reduction by a strong reducing agent:

$$R-COOH + X-H_2 \longrightarrow R-CHO + X + H_2O \qquad (3)$$

The annual yield of biomass production corresponds to $\sim 2 \cdot 10^{11}$ tons of dry matter. Nearly half of the incorporated atoms are hydrogen atoms. The only candidate for the <u>ultimate</u> source of this element seems to be the water. This statement does not say, however, that H_2O molecules are the <u>immediate</u> source of the oxygen. At any case the term "water decomposition" should be used with the utmost reservation.

So far we have no conclusive informations on the path of oxygen in photosynthesis. We cannot even say, whether the released O_2 has to be derived from water or carbon dioxide. At first sight one might perhaps expect that mass spectrometric experiments with ^{18}O-labelled carbon dioxide or $H_2^{18}O$ should help to decide, which one of the two possible substrates is the real "oxygen precursor".

RESULTS AND DISCUSSION

Since the pioneer studies of Ruben et al. (ref. 1) isotope determinations have repeatedly been performed. The obtained data indicate that - irrespective of the ^{18}O content of the added CO_2 - the isotope composition of the photosynthetic oxygen always reflects the $^{18}O/^{16}O$ ratio of the suspension water. This observation, however, cannot be taken as argument for the hypothesis that green plant photosynthesis includes a decomposition of water molecules: In all physiological systems the H_2O molecules outnumber the CO_2 molecules by about five orders of magnitude. By the equilibration of the two substrates (see eq. (1)) the oxygen isotopes become distributed among all oxygen-containing reaction partners. It is self-evident that the labelling degree of the quantitatively predominating substrate - water - determines the ^{18}O content of all molecules and ions (ref. 2). In the absence of catalysts this equilibration is a rather slow process: every hydration-dehydration step takes the order of minutes. Living cells contain, however, a very efficient enzyme, the carbonic anhydrase, which enhances the CO_2-H_2O interaction with an extremely high turnover number. This fact invalidates all mass spectrometric determinations with intact cells after gas sampling times of several minutes.

Energetic considerations make a water splitting rather unlikely: The primary photochemical reaction between an excited chlorophyll a (Chl) molecule and the - not yet unequivocally identified, but presumably structurally related to chlorophyll (ref. 3) - primary electron acceptor

$$Chl^* + Acc \longrightarrow Chl^+ + Acc^- \qquad (4)$$

leaves chlorophyll as oxidized radical ion, which has to regain its missing electron. There is convincing evidence that the replacement by the <u>final</u> donor requires the interaction of a complex redox system.

To calculate the energy requirement for a water decomposition most authors use the potential of the oxygen electrode. Its E_o value of + 1.23 V - corresponding to $E_o' = + 0.81$ V - characterizes a <u>reversible</u> <u>two-electron</u> transfer, whereas for photosynthesis an <u>irreversible</u> <u>one-electron</u> exchange has to be postulated. The step $H_2O \longrightarrow H_2O^+$, however, asks for a definitely higher (> 2 eV) energy supply (ref. 4). Even disregarding this apparent erratum, it is not appropriate to discuss the secondary reactions of the chlorophyll radical by a comparison of redox potentials: At the decisive

reaction site only one chlorophyll radical and one secondary donor interfere. The value (Red)/(Ox) can therefore only switch between zero and ∞. We better avoid any data of statistical value and emphasize instead that the ionization energy of the relevant electron donor has to be smaller than the electron affinity of the oxidized chlorophyll. Since the ionization energy of water - in its liquid state - is ~ 3 eV (ref. 5), the electron affinity of Chl^+ would have to exceed this value. However, in the frequency region of red light the energy content of absorbed photons is ~ 1.8 eV only. To explain this discrepancy we may regard three possibilities:
1. There could be a type of "bound water" (ref. 6, 7) with quite different properties (especially a rather low ionization energy);
2. the plant cell may perform a two-step oxidation of an intermediate redox catalyst, which enables the system to exert an - energetically less "expensive" - concerted two-electron transfer;
3. an oxygen-containing compound with a lower ionization energy than water (but in equilibrium with H_2O) may be the presursor of the photosynthetic oxygen.

In view of these difficulties it looked reasonable to replace the complicated in vivo system by an artificial photosynthesis model. Aqueous suspensions of AgCl crystals release O_2, if irradiated in their absorption band (ref. 8). A careful analysis of this simple system demonstrated that this reaction requires the presence of bicarbonate anions (ref. 9). Since in this case the isotope exchange between CO_2 and H_2O may be regarded as rather slow, AgCl suspensions recommended themselves for mass spectrometric studies. Their light-induced O_2 production was started by the injection of ^{18}O-labelled sodium bicarbonate solutions into previously degassed crystal suspensions in $H_2^{16}O$. Mass spectrometric determinations of the $^{18}O/^{16}O$ ratio demonstrated that the primarily released oxygen has a high ^{18}O content (ref. 10). This ratio was significantly higher than that of the suspension water - even higher than the end value which the medium could ever attain after complete isotope exchange. Furthermore the ^{18}O percentage decreased with extended irradiation time. This means that - under the experimental conditions chosen - water cannot be the precursor of the released oxygen; the O_2 must have come - at least partly - from the HCO_3^- anions.

This result asked for new mass spectrometric experiments with plant cells. To avoid the problems connected to the enzymatically enhanced exchange between differently labelled substrates, it seemed easier to supply substrates with their natural isotope abundances. Due to small differences in the reaction rates of isotopically labelled molecules and ions, the constituents of an equilibrium system slightly differ in their $^{18}O/^{16}O$ ratio. Regarding the H_2O-CO_2 system we obtain the following isotope abundances (ref. 11):

$$CO_2 + 2\ H_2O \rightleftharpoons HCO_3^- + H_3O^+$$
$$1.041 \quad 1.000 \quad\quad 1.026 \quad 1.000$$

This equilibrium isotope effect thus produces slight but significant differences in the labelling degree of water and bicarbonate. Provided a sufficiently accurate instrument is available, the $^{18}O/^{16}O$ ratio of the released oxygen can be compared to that of the possible precursors.

To determine the ^{18}O content of photosynthetic oxygen a simple round-bottom glass vessel has been used, in which suspensions of unicellular algae or isolated chloroplasts (respectively thylakoids, see below) can be connected to a vacuum line (ref. 12). To remove most of the dissolved air the suspensions were degased by partial evacuation. After this procedure the vigorously agitated preparations remained (for 5 min) in darkness. Following this equilibration period the suspensions were irradiated up to 45 min; control samples remained in darkness. After different time intervals a tap between the reaction vessel and an attached preevacuated (high vacuum) sampling glass was opened for 10 s. By this means ~ 60% of the evolved oxygen could be recovered.

To get rid of both water vapour and CO_2, the obtained gas mixture was cooled to liquid nitrogen temperature. The remaining O_2-N_2 mixture (together with small amounts of argon) was then directly introduced into the inlet system of an isotope mass spectrometer (Varian MAT 250) with a triple collector. The instrument was connected to a desk computer which printed the δ values of the collected oxygen. For the evaluation only the $^{18}O/^{16}O$ ratios were considered.

The recorded data were corrected both against dark samples and values obtained for DCMU-poisoned suspensions (which do not release any oxygen). After this correction the δ values for unicellular algae demonstrated a strong time dependence. Within short light periods (up to ~ 1 min) the δ values more or less reflect the ^{18}O content of still present (by evacuation not removed) air oxygen ($\delta = +23.7$ ‰). Besides this "contamination" there seem to be at least two other oxygen "fractions": the released photosynthetic oxygen and a form of "bound oxygen" which might be either dissolved in cell lipids or adsorbed (or loosely bound) to an as yet unidentified molecule. In the very first stages of photosynthesis the newly developed O_2 will be partly mixed with this "bound" oxygen, which possesses a rather low ^{18}O content (ref. 12). It is only after prolonged illumination periods that we observe a more or less stable δ value near -2‰ (related to SMOW). The same result was obtained by the extrapolation of the δ values as function of the released O_2 volume. At the same time the δ value of the suspension medium (water) was measured to be -14.2‰. This means that the photosynthetic oxygen is enriched in ^{18}O by a $\Delta\delta$ of > 12‰.

If water - i.e. a mixture of $H_2^{16}O$ and $H_2^{18}O$ molecules - becomes decomposed, the $^{16}O-H$ bond - due to its smaller activation energy - should be preferably split. This should give a measurable ^{16}O enrichment in the released oxygen. This expected isotope discrimination has in fact repeatedly described both for chemical (ref. 13) and electrolytic (ref. 14) H_2O decomposition. If the photosynthetic experiments show just the reverse we have to look for another oxygen precursor. If, instead of water, bicarbonate would be decomposed, the resulting oxygen should have a δ value > -14‰.

We have, however, to consider that green plants respire both in darkness and light. It looks conceivable to assume that isotope effects during the respiratory oxygen consumption overcompensate the "real" isotope effect of the photosynthetic process. This has repeatedly been claimed (ref. 15, 16). The measured isotope discrimination of the mitochondrial respiration is, however, too small to explain a $\Delta\delta$ of > 10‰ (ref. 2). But we still miss any information on possible isotope effects of photorespiration. To exclude this uncertainty, all oxygen-consuming reactions should be strictly excluded. Fortunately enough photosynthesis and respiration are bound to different cell organells. Photosynthesis is only performed within the chloroplasts, organells - in higher plant cells ~ 5-7 μm in diameter - which can easily be isolated by fractional centrifugation of cell homogenates. Due to their semipermeable membrane they suffer an osmotic shock if resuspended in distilled water. This disrupts them to small chlorophyll-containing vesicles, so-called thylakoids. If these particles are supplied with a suitable electron acceptor - e.g. hexacyanoferrate(III) anions or a quinoid dye - they use photons to deliver electrons to this artificial sink. These so-called Hill reactions enable us to study the decisive steps of oxygen liberation in a comparatively simple system.

It was originally assumed that Hill reactions do not require carbon dioxide. This was once taken as evidence against a necessary participation of CO_2 in the oxygen-evolving process. Careful examinations with CO_2-depleted chloroplasts (see below) - primarily performed by Warburg (ref. 17) but meanwhile well established in several laboratories - have, however, shown that chloroplasts and thylakoid preparations ask for a small amount of carbon dioxide. That this dependence has been overlooked so long, reflects the strong CO_2 capacity of these organells. Carbon dioxide is apparently bound in (at least) two different "fractions" (ref. 18). To remove the loosely bound form - at least partially - the preparations have to be washed with formate solutions (ref. 19). After this pretreatment their photochemical activity is nearly lost. To regain effective oxygen evolution these CO_2-depleted thylakoids must be supplied with bicarbonate which can be given in the form of either $NaHC^{16}O_3$ or $NaHC^{18}O_3$.

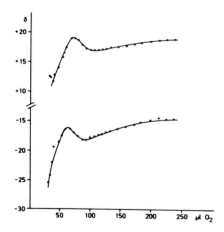

Fig. 1: Relation between the recorded δ values and the amount of oxygen samples released from illuminated thylakoid suspensions. a (lower trace): without, b (upper trace): with supply of labelled bicarbonate (Data corrected against dark controls, see text).

If suspensions of unwashed thylakoids are illuminated, the primarily released oxygen possesses a rather low δ value (< -25 ‰, see Fig. 1, lower trace). It increases pretty fast and - after a depression - approaches a more or less stable value. If these suspensions are supplied with $NaHC^{18}O_3$, the whole trace is shifted to more positive δ values without changing its characteristics (Fig. 1, upper trace).

A different result is obtained if the thylakoids are first CO_2-depleted. In this case (at least part of) the loosely bound form of carbon dioxide is removed. This prevents a strong dilution of the externally given bicarbonate. After addition of unlabelled bicarbonate the first oxygen samples show again a rather low δ value (~ -25 ‰, see Fig. 2, lower trace). If the depleted preparations receive, however, ^{18}O-labelled bicarbonate, the primarily liberated

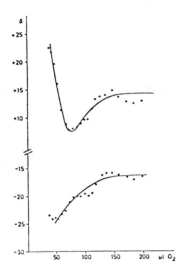

Fig. 2: Relation between recorded δ values and the amount of oxygen samples collected from illuminated thylakoid suspensions after CO_2 depletion by formate treatment.   a (lower trace): after readdition of unlabelled,   b (upper trace): after readdition of labelled bicarbonate (Data corrected against dark controls, see text).

oxygen has a definitely higher ^{18}O content ($\delta > +20$ ‰) than samples collected after extended light periods (Fig. 2, upper trace)

The experimental technique used is in the danger of serious artifacts. Due e.g. to small differences in gas solubility an oxygen isotope fractionation between the suspension medium and the gas phase cannot be excluded. Furthermore there might be a difference in the solubility of the two oxygen isotopes in lipids (respectively adsorption to cell constituents). These purely physical discriminations would invalidate the determined δ values. This can, however, be prevented if instead of the directly recorded data

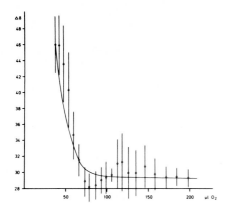

Fig. 3: Difference curve between the traces 2b and 2a. The vertical bars represent the standard error of the mean.

($NaHC^{18}O_3$ - $NaHC^{16}O_3$) difference curves (ref. 20) are calculated (Fig. 3). They clearly demonstrate that the first shares of the released photosynthetic oxygen are enriched in ^{18}O, whereas samples taken after extended illumination times have significantly lower $^{18}O/^{16}O$ ratios. The sampling technique used collects the total oxygen which is produced between the start of the illumination period (t_o) and the actual sampling time (t). The measured δ value therefore represents an averaged value for all oxygen "fractions" released between t_o and t. To obtain the "real" δ value for any selected moment we have to draw the first derivative of the difference

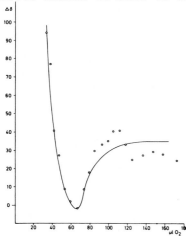

Fig. 4: Corrected difference curve (δ values for differential samples collected at the given volumes, details see text).

curve (Fig. 4). The trace confirms that the earliest oxygen samples have a strongly positive δ value ($> + 90‰$). If extrapolated to t_o the resulting δ value would correspond to an ^{18}O content of $\sim 5\%$. Regarding the dilution by a high amount of "bound" $C^{16}O_2$, a higher labelling rate could hardly be expected.

In the system $HC^{18}O_3^-/H_2^{16}O$ the water has at t_o a low ^{18}O content ($\delta \approx - 14 ‰$). If the primarily released O_2 shows nevertheless a strong ^{18}O enrichment, water cannot be the immediate oxygen precursor. This does, on the other hand, not mean that inorganic bicarbonate has to be regarded as the real "photolyte". It seems conceivable to assume a kind of "bound" bicarbonate $X-HCO_3^-$. If this should function as electron donor to the oxidized chlorophyll (via an intermediate redox system Z) we would expect the formation of "bound bicarbonate" radicals:

$$Chl^+ + X-HCO_3^- \xrightarrow{(Z)} Chl + X-HCO_3\cdot \qquad (5)$$

In electrolytic cells free bicarbonate anions (at sufficiently high local concentrations) dimerize to the peroxidicarbonic acid (ref. 21):

$$2\ HCO_3\cdot \longrightarrow HOOC-O-O-COOH \qquad (6)$$

Whereas the potassium salt of this compound is only sparely water soluble, the free acid decomposes to CO_2, H_2O and oxygen. The liberated oxygen atoms can be delivered to unsaturated compounds like carotenoids (ref. 22). At the present state of our knowledge it remains, however, an open question whether intermediates of peroxidic nature occur in vivo.

The decomposition of the peroxi acid would give CO_2 which together with surrounding water could restore the necessary bicarbonate anions. So we would obtain a cyclic process by which the released CO_2 is permanently reintroduced into the reaction sequence. This makes CO_2 a kind of "catalyst" which allows H_2O a light-induced electron transfer to oxidized chlorophyll molecules (ref. 23). Immediately after supply of $NaHC^{18}O_3$ the released carbon dioxide is highly labelled. By every reaction with unlabelled water its ^{18}O content is diminished by a factor of 2/3. This explains the fast "desaturation" of the heavy oxygen isotope in the precursor.

REFERENCES

1. S. Ruben, M. Randall, M. Kamen and J. Hyde, J. Am. Chem. Soc. 63(1941)877-879.
2. H. Metzner, J. Theor. Biol. 51(1975)201-231.
3. V.V. Klimov, A.V. Klevanik, V.A. Shuvalov and A.A. Krasnovsky, FEBS Lett. 82(1977)183-186.
4. P. George, in T.E. King and M. Morrison (Eds.), Oxidases and Related Redox Systems, Vol. I, Wiley, New York, 1965, pp. 3-36.
5. A.J. Frank, M. Grätzel and A. Henglein, Ber. Bunsenges. physik. Chem. 80(1976)593-602.
6. V.M. Kutyurin, Izv. Akad. Nauk SSSR., Ser. Biol. (1970)570-580.
7. G. Renger, in H. Metzner (Ed.), Photosynthetic Oxygen Evolution, Academic Press, London, New York, San Francisco, 1978, pp.229-248.
8. H. Metzner and K. Fischer, Photosynthetica 8(1974)257-262.
9. H. Metzner, K. Fischer and G. Lupp, Photosynthetica 9(1975)327-330.
10. H. Metzner and R. Gerster, Photosynthetica 10(1976)302-306.
11. H.C. Urey, J. Chem. Soc. (1947)562-581.
12. H. Metzner, K. Fischer and O. Bazlen, Biochim. Biophys. Acta 548(1979)287-295.
13. A.E. Cahill and H. Taube, J. Am. Chem. Soc. 74(1952)2312-2318.
14. M. Anbar and H. Taube, J. Am. Chem. Soc. 78(1956)3252-3255.
15. A.P. Vinogradov, V.M. Kutyurin, M.V. Ulubekova and I.K. Zadorozhny, Dokl. Akad. Nauk SSSR. 134(1960)1486-1489.
16. G.A. Lane and M. Dole, Science, N.Y. 123(1956)574-576.
17. O. Warburg and G. Krippahl, Z. Naturforsch. 15b(1960)367-369.
18. A. Stemler, in H. Metzner (Ed.), Photosynthetic Oxygen Evolution, Academic Press, London, New York, San Francisco, 1978, pp.283-293.
19. K. Fischer and H. Metzner, Photobiochem. Photobiophys. 2(1981) 133-140.
20. H. Metzner, K. Fischer and O. Bazlen, in G. Akoyunoglou (Ed.), Proc. V. Internat. Photosynthesis Congr., Halkidiki 1980 (in press).
21. E.J. Constam and A. v. Hausen, Z. Elektrochem. 3(1896)137-144.
22. C. Bodea and M. Florescu, Rev. de Chim. Roum. 1(1956)105-114.
23. H. Metzner and K. Fischer, Photobiochem. Photobiophys. 3(1981) (in press).

QUANTIFICATION OF INDOLE-3-ACETIC ACID BY GC/MS USING DEUTERIUM LABELLED INTERNAL STANDARDS

J.R.F. ALLEN, J.-E. REBEAUD, L. RIVIER and P.-E. PILET
Institut de biologie et de physiologie végétales, Université de Lausanne, Place de la Riponne, I005 Lausanne, Suisse

ABSTRACT

The variety of deuterium-labelled IAA species available, substituted in the side chain and indole ring are discussed. A procedure for the quantification of IAA in plant tissues, using ($^{2}H_{5}$)-IAA as an internal standard, by GC/MS analysis of the N-heptafluorobutyryl-ethyl ester is described.

INTRODUCTION

Indole-3-acetic acid (IAA) has been the subject of wide spread investigation since it occupies an important position in the metabolism of both plants and animals. In higher plants it occurs per se as the principal representative of the auxin hormones which regulate aspects of growth and differentiation, whilst interest in its role in animals has centred around it being a principal metabolite of tryptophan and tryptamine. Quantification of IAA is vital, to examine its role, in plants, in tropic responses such as root gravi-reaction (I) and in animals, as an indicator of tryptophan and biogenic indoleamine metabolism in the central nervous system (CNS) and peripheral organs (2).

A wide variety of methods have been utilised for the identification and quantification of IAA, including bioassays, spectrometric and colorimetric assays, GC with FID and ECD, HPLC with UV and fluorimetric detection, spectrophotofluorimetry, radioimmunoassay and GC/MS. The endogenous IAA levels in biological samples are usually in the pmole - nmole per g fresh weight range within a general matrix of organic metabolites. Of the methods currently available, quantification by GC/MS using selected-

---

Abbreviations used: GC/MS = gas chromatography-mass spectrometry; IAA = indole-3-acetic acid; SIM = selected-ion-monitoring; HFB = heptafluorobutyryl

ion-monitoring (SIM) offers the advantages of both high sensitivity with the specificity required to discriminate from impurities and is therefore of particular value for the accurate analysis of IAA.

The high levels of impurities, particularly in plant extracts makes a preliminary purification of samples prior to analysis obligatory. Losses of IAA during sample purification can be substantial (3) and thus for accurate quantification, the use of an internal standard is mandatory.

Isotope dilution methods using (^{3}H)- and (^{14}C)- IAA have been used, but can give rise to errors from low specific activity, loss of label during purification and the use of separate methods for the determination of labelled and combined labelled and unlabelled compounds. The most widely used internal standards for quantification by GC/MS are stable isotope labelled analogues (4), which possess the principal advantage of permitting a specific determination of labelled and unlabelled compound by the same detector, simultaneously. For reasons of convenience and cost, ^{2}H labelled IAA offers the greatest potential as an internal standard for GC/MS analyses

DEUTERIUM LABELLED IAA

a) (^2H_I)-IAA. A method for the synthesis of (^2H_I)-IAA by desulfuration of 2-(2,4-dinitrophenylthio)-IAA has been described which provides >95% ^{2}H incorporation, at the C_2-position. The usefulness as internal standards for GC/MS, of mono-deuterated compounds, is limited by the interference between endogenous compound, and internal standard arising from natural isotopes (principally ^{13}C) and the presence of residual unlabelled species as a contaminant of the (^2H_I)-IAA internal standard.

b) (β,β-$^2H_2$)-IAA. Base catalysed ^{1}H/^{2}H exchange in (^{2}H)-NaOH or KOH in (^{2}H)-H_2O can be carried out with relative ease (6-9). Under alkaline conditions IAA is sufficiently stable and soluble to permit high yields to be obtained. The resultant product is labelled predominantly in the methylene group of the side chain, with small amounts of additional incorporation at the C_2 position on the indole ring. The isotopic purity of the labelled product is typically: $^2H_0$ 0.1-4%, ^2H_I 4-7%, $^2H_2$ 70-82% and $^2H_3$ 8-15% (10). Using ($^2H_2$)-IAA as an internal standard for GC/MS a small amount of interference between the $^2H_0$ and $^2H_2$ channels is therefore observed. No ^{2}H/^{1}H back exchange takes place during storage or sample purification providing that extremes of pH are avoided. The principal advantages of using ($^2H_2$)-IAA as an internal standard for GC/MS are the ease of preparation and low cost.

c) $(4,5,6,7-{}^2H_4)$-IAA and $(2,4,5,6,7-{}^2H_5)$-IAA. The use of acid catalysed deuterium exchange in $({}^2H)$-HCl in $({}^2H)$-H_2O, leading to the formation of a product of heterogeneous isotopic purity has been reported (11). This method is unsatisfactory, however, because under acidic conditions IAA is virtually insoluble and also losses due to degradation and polymerisation take place. Substitution of HCl by $({}^2H)$-CF_3COOH and the use of $({}^2H_4)$-CH_3OH as a solvent also failed to provide acceptable yields (10).

Alternative procedures, for the preparation of $(4,5,6,7-{}^2H_4)$- and $(2,4,5,6,7-{}^2H_5)$-IAA by organic synthesis from $({}^2H_5)$-aniline have been described by Magnus et al.(9). The isotopic purity of the products of the syntheses is typically 70-82% of the principal ${}^2H$ species sought, 4-7% with an additional substitution on the side chain and 11-15% with one deuterium less than that sought. The non labelled species is normally less than 0.1% and thus there is no interference during GC/MS analyses between labelled and unlabelled species.

The principal advantages with the use of ring labelled IAA are the stability of deuterium label in the molecule during alkaline hydrolysis of plant extracts and the capability of using lowered MS resolution during analysis to enhance sensitivity (9). Substantial ${}^2H/{}^1H$ back-exchange at the C_2 position may occur at extremes of pH either during hydrolysis in 1N NaOH (9) or during derivatisation in acid anhydrides (12), and thus particular caution against exposure to extremes of pH should be taken when using $({}^2H_5)$-IAA.

d) $(4,5,6,7,\beta,\beta-{}^2H_6)$-IAA and $(2,4,5,6,7,\beta,\beta-{}^2H_7)$-IAA. Sequential synthesis, using the methods described above can be carried out to raise the deuterium incorporation to the maximal value obtainable.

USE OF $({}^2H)$-IAA AS INTERNAL STANDARDS FOR QUANTIFICATION BY GC/MS

A number of suitable derivatives for GC/MS analysis can be prepared, involving esterification, with or without additional N-acylation. The introduction of additional fluorinated alkyl groups is advantageous since the derivatives formed are stable and the resultant increase in molecular weight permits monitoring of higher m/z values, for which the probability of interference from contaminant impurities is reduced (4). Fluorinated esters and N-acyl derivatives can be prepared using fluorinated anhydrides (12,13) and imidazoles (14,15).

The electron impact mass spectra of the N-heptafluorobutyryl-ethyl ester of IAA formed by esterification with diazoalkane followed by an N-acylation with heptafluorobutyryl imidazole (HFBI) is shown in Fig.1.

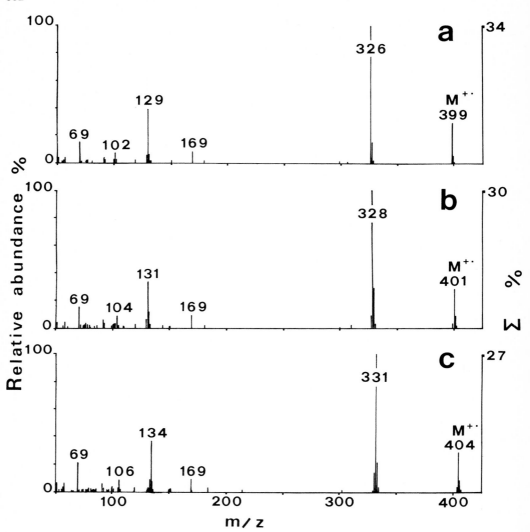

Fig I. Electron impact mass spectra of the N-heptafluorobutyryl-ethyl-ester of a) IAA b) ($^2H_2$)-IAA and c) ($^2H_5$)-IAA.

The ^2H labelled compounds may be used as internal standards for the quantification of IAA by GC/MS by adding a known quantity of the labelled compound to the sample, which is purified and analysed using selected-ion-monitoring to determine the isotope ratio between the label led and endogenous compound (I5, The quantity of endogenous compound in the sample can be derived from calibration plots prepared from the M^+ or fragment ion ratios determined from standards (Fig.2).By monitoring two or more ions derived from the same species an estimate of the probability of errors arising from contaminants can be made.

Within this laboratory a procedure has been developed for the routine quantification of IAA in plant tissues using ($^2H_5$)-IAA as an internal standard (15). In this procedure, 1-3 g fresh weight plant tissue is homogenised in an organic solvent with 125-500 pmole ($^2H_5$)-IAA and a 10 fold excess of 5-MeIAA to act as a carrier through the purification procedure. Following reduction to the aqueous phase, the extract is subjected to preliminary purification by solvent partitioning. The crude acid-ether soluble phase is alkylated with ethereal diazoethane and subjected to preparative TLC using 5-MeIAA-Et as marker spots at the side which are observed under UV-light. The appropriate fractions are eluted from the TLC support and derivatised with HFBI in pyridine (1:1 v/v) which provides a quantitative yield, and analysed by GC/MS (Fig. 3). Recovery of IAA in the initial extract is typically 40-60% through the entire procedure.

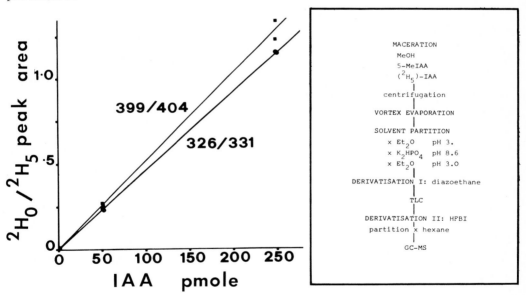

Fig 2. Calibration plot for the quantification of IAA derived from the M^+ and β-cleavage fragment ions peak area ratios from GC/MS analysis of the HFB-Et derivatives. 250 pmole ($^2H_5$)-IAA is used as the internal standard.

Fig 3. Procedure for the extraction, purification and analysis of endogenous IAA in plant tissues.

Using capillary columns and a low resolution quadrupole MS, the limit of detection was ∼ 3 pmole IAA per sample. This value is set by the confidence interval derived from the calibration plots and is thus determined by the precision of the method (coefficient of variation of 2-4% at the 250 and 500 pmole level when n=4-6), and the quantity of internal standard added. This procedure has been used to determine the levels of endogenous IAA in dark- and light-grown maize and pea seedlings (15). It has been found that the principal sources of error in determination of IAA in the tissues using this method is due to natural variation within the endogenous content of IAA in the plant tissues.

ACKNOWLEDGEMENTS

J.R.F. Allen wishes to thank the Royal Society for support from a European Fellowship.

REFERENCES

1 P.E. Pilet, in F. Skoog (Ed.), Plant Growth Substances, 1979, Springer, Berlin, 1980, pp. 450-461.
2 L. Bertilsson and L. Palmér, Science, 177 (1972) 74-76.
3 C.H.A. Little, J.K. Heald and G. Browning, Planta 139 (1978) 133-138.
4 A.P. de Leenheer and A.A. Cruyl, in G.R. Waller and O.C. Dermer (Eds.) Biochemical Applications of Mass Spectrometry, Suppl. Vol.I, Wiley, New York, 1980, pp. 1169-1207.
5 R.K. Raj and O. Hutzinger, J. Labell. Cmpds., 6 (1970) 399-400.
6 J.A. Hoskins and R.J. Pollitt, J. Chromatog., 109 (1975) 436-438.
7 J.L. Caruso, R.G. Smith, L.M. Smith, T.-Y. Cheng and G.D. Daves, Plant Physiol., 62 (1978) 841-845.
8 J.R.F. Allen, A.M. Greenway and D.A. Baker, Planta, 144 (1979) 299-303.
9 V. Magnus, R.S. Bandurski and A. Schulze, Plant Physiol., 66 (1980) 775-781.
10 J.R.F. Allen, D Phil. Thesis, University of Sussex (1980).
11 F.A.J. Muskiett, H.J. Jeuring, C.G. Thomasson, J. van der Meulen and B.G. Wolthers, J. Labell. Cmpds., 14 (1978) 497-505.
12 J.R.F. Allen and D.A. Baker, Planta, 148 (1980) 69-74.
13 E. Watson, S. Wilk and R. Roboz, Anal. Biochem., 59 (1974) 441-451.
14 L. Rivier and P. E. Pilet, Planta, 120 (1974) 107.
15 J.R.F. Allen, L. Rivier and P.E. Pilet, submitted for publication (1981).

H.-L. Schmidt, H. Förstel and K. Heinzinger (Editors), *Stable Isotopes*

QUANTIFICATION OF ABSCISIC ACID AND ITS 2-TRANS ISOMER IN PLANT TISSUES USING A STABLE ISOTOPE DILUTION TECHNIQUE

L. RIVIER and P.E. PILET
Institute of Plant Biology and Physiology of the University,
6 Pl. Riponne, 1005 Lausanne, Switzerland.

ABSTRACT

The use of hexadeuterated abscisic acid is described for the quantitative determination of endogenous abscisic acid (ABA) in *Zea mays* var. LG 11 primary root segments. At the same time, its 2-trans isomer level is accurately calculated by the control of the extent of the photoisomerisation which may occur easily during the extraction and purification steps. No endogenous methyl or ethyl ester of ABA could be detected.

INTRODUCTION

Abscisic acid (ABA) [1] is a plant growth inhibitor which occurs naturally in a number of plants (ref. 1). It plays an essential role in the control of root elongation (ref. 2). Precise and accurate measurements of this hormone on small pieces of plant tissues are indispensable for a better understanding of the mode of action of ABA. As it occurs in pmole per g fr wt level in higher plants, the quantification of trace amounts of ABA in tissue extracts relative to impurities is difficult. Measurement of the ABA present in plants always depends on preliminary purification steps usually done by partitioning between solvents followed by silica gel chromatography.

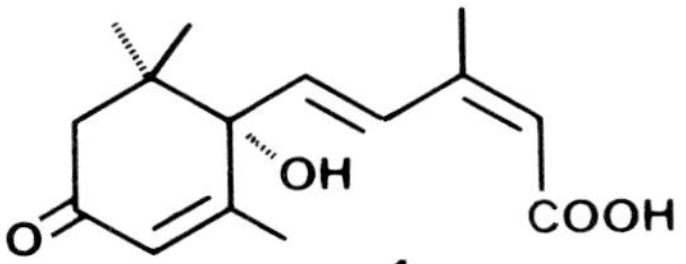

1

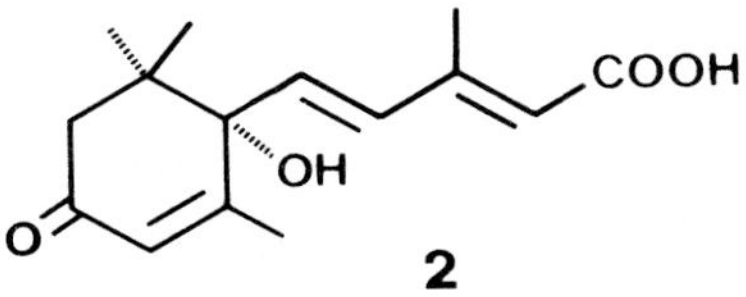

2

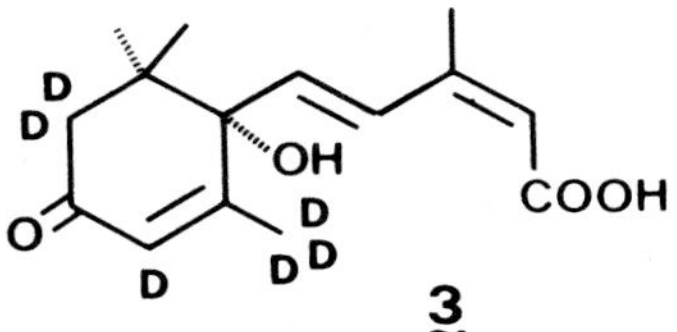

3

A further derivatisation is necessary to obtain the ester for GC analyses. All these processes cause losses which are very important when working with trace amounts of ABA (ref. 3). The proportion of isolated ABA can be augmented by the use of a suitable carrier, provided that extraction efficiency and rate of destruction of both compounds are identical. Most important, all the manipulations have to be carried out in very dim light, in order to prevent the rapid interconversion of ABA into its 2-trans isomer [2] (ref. 4).

The biological activity of trans-ABA is known to be much weaker than that of ABA (ref. 5). Consequently trans-ABA seems to play no important physiological role. However, it is recognized that trans-ABA occurs naturally (ref. 6). For all these reasons, it is very important to monitor the rate of isomerisation during the analytical procedure in order to obtain the correct picture of the levels of ABA and 2-trans ABA in the plant tissue.

Such control can be obtained by the addition of ^{14}C-ABA and unlabelled trans-ABA as internal standards. Any ABA isomerized to the trans isomer would thus carry this label (ref. 4). The main draw-back of this method is the need to use two separate techniques for the determination of the amounts of ABA and trans-ABA. Furthermore, for safety reasons, it is never advisable to use radio-labelled tracers for routine analytical procedure.

The aim of this paper is to describe a suitable routine procedure for the quantification of ABA and its trans isomer in plant tissues by using a single technique of measurement only.

EXPERIMENTAL

ABA (3-methyl-5-[1-hydroxy-4-oxo-2,6,6-trimethyl-2-cyclohexenyl]-2-cis, 4-trans-pentadienoic acid) was obtained from Fluka and trans-ABA was a gift from Prof. J. MacMillan, Bristol, England. Synthesis of deuterated ABA was performed with 1N NaOD in 98% D_2O (ref. 7). The reaction was repeated 3 times in order to obtain the highest rate of substitution of the labile hydrogen sites. The octadeuterio-ABA was dissolved in H_2O at pH = 6. The resulting hexadeuterated ABA (ABA-D_6) [3] was extracted with $CHCl_3$ and its purity was tested by GC/MS as methyl ester.

Diazomethane and diazoethane in Et_2O were prepared by synthesis from N-nitroso-N-methyl-p-toluolsulfonamide and N-ethyl-N-nitroso-urea, respectively (ref. 8).

All other solvents were freshly distilled before use.

The complete extraction and purification procedure has been previously described (ref. 9). Quantitative determinations were performed by using the technique first used by Rivier, Milon and Pilet in 1977 (ref. 7). Only the free ABA was measured to the exclusion of the bound ABA (conjugates). The isolation of ABA from the plant tissue was obtained by rapid partitioning of $CHCl_3$ and acetate buffer at pH = 4. Further purification steps were obtained by TLC after esterification with diazoalkane. When using ethyl acetate and hexane (3:2 v/v) as a developing solvent mixture, ABA-Me, ABA-D_6-Me and trans-ABA-Me moved to the same Rf. The eluate from the corresponding spot allowed the measurement of all three compounds at the same time.

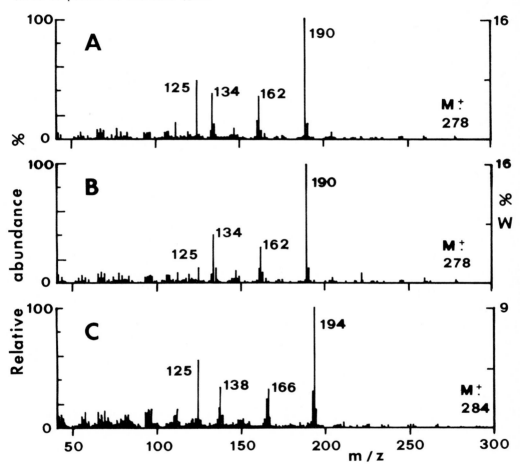

Fig. 1. Electron Impact (70 eV) mass spectra of the methyl esters of :
a) cis, trans-abscisic acid (ABA), b) trans, trans-ABA, c) hexadeuterated cis, trans-ABA (ABA-D_6)

Analysis was carried out on a low resolution GC/MS instrument under computer control. Splitless injections were made from 3 µl onto a 25 m x 0.3 mm OV 101 fused silica WCOT capillary column connected by an open interface to the ion source. Typical GC operating conditions were : injector, 250° C; oven 100°

for 1 min, heated at 25°/min up to 240°; He flow, 2 ml/min; GC/MS interface, 260°. For quantitive analysis, the MS was operated in the Selected Ion Monitoring (SIM) mode at m/z 190 for ABA and trans-ABA esters, and m/z 194 for ABA-D_6 esters, with measuring dwell time of 100 msec each.

Quantification of the endogenous ABA and trans-ABA in the extracts was made by reference to calibration plots derived from a series of standards routinely subjected to the same purification procedure. Standards containing a fixed amount of ABA-D_6 together with known amounts of 0 - 50 ng ABA showed a linear relationship between the amount of ABA and the ratio derived from the base peaks of the two compounds. The same relation was obtained with trans-ABA.

RESULTS AND DISCUSSION

The mass spectrum of ABA-Me is shown in the Figure 1 with the mass spectra of trans-ABA-Me and ABA-D_6-Me.

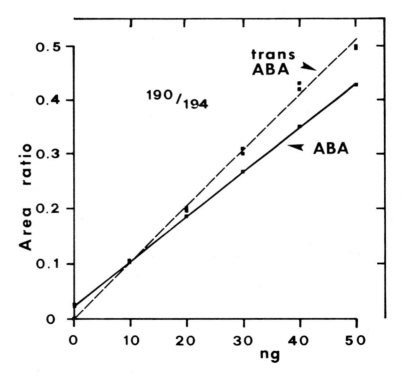

Fig. 2. Calibration plot for the quantification of ABA and trans-ABA derived from the base peak area ratios from GC/MS analysis of methyl esters derivatives. 100 ng of ABA-D_6 was used as internal standard.

ABA cannot be differentiated from its 2-trans isomer by its mass spectrum only, as they are almost identical except for the intensity of m/z 125. However, under the GC conditions used, they separate completely. On the contrary, ABA-D_6-Me emerges at the same Rt as ABA-Me, but it can be discriminated from the latter from its different mass spectrum, and was therefore used as internal standard. The base peaks were chosen for SIM analyses.

The isotope composition of the ion at m/z 194 region of ABA-D_6-Me indicated a small contribution to the ion at m/z 190. For this reason, the calibration line for ABA does not pass through the origin (Fig. 2).

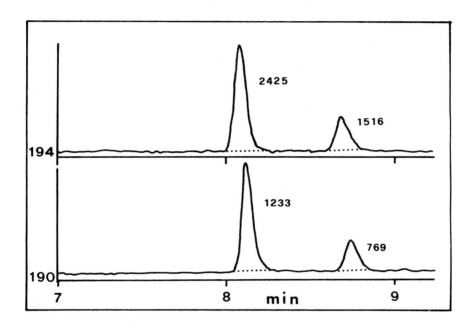

Fig. 3. SIM chromatogram of standards after inducing photoisomerisation. 100 ng of ABA-D_6 and 50 ng ABA have been used and extracted as the plant material. The peaks at Rt 8.2 correspond to the methyl esters of ABA (m/z 190) and ABA-D_6 (m/z 194) and those emerging at Rt 8.7 correspond to their respective 2-trans isomers.

Normally, when a mixture of ABA-D_6 and ABA is extracted and purified under the conditions used in this study, no isomerisation occurs. In one experiment, the starting solution was illuminated for 20 min under sun light and extracted. A mixture of cis and trans isomers was found (Fig. 3). The amount of trans-ABA is exactly the same as that of trans-ABA-D_6, indicating that no isotope effect exists in the photoisomerisation process. The amount of endogenous trans-ABA can thus be calculated simply, even if some photoisomerisation has occurred during work-up.

As an illustration the SIM chromatogram of the methylated extract of 50 10 mm long primary roots of Zea mays grown in darkness for 3 days at 20° C is given in Figure 4A. No trans-ABA-D_6-Me has been detected. The peak at m/z 190 at the Rt of trans-ABA-Me indicates the original level of endogenous trans-ABA in the plant tissue. When photoisomerisation was induced on a similar extract (Fig. 4B), the equivalent amount of trans-ABA-D_6-Me had to be subtracted from that of trans-ABA-Me to obtain the correct endogenous level of this isomer. This is possible as the fragmentation intensities of both standards are almost equivalent (see Fig. 1 A and B).

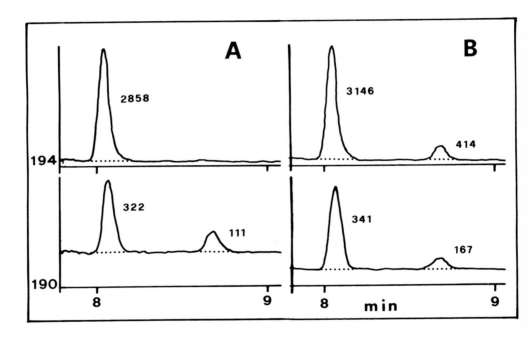

Fig. 4A. SIM chromatogram from maize primary root extract to which 100 ng of ABA-D_6 has been added as internal standard at the beginning of the analytical procedure. No photoisomerisation occurred as indicated by the absence of peak at the Rt corresponding to trans-ABA-D_6-Me.

Fig. 4B. SIM chromatogram from a similar extract as in Fig. 4A after working under strong white light during extraction. The extent of photoisomerisation could be monitored by the measurement of the peak corresponding to trans-ABA-D_6-Me.

The presence of ABA in the extract is based on the detection of the base peak of its mass spectrum at the correct retention time relative to the internal standard. Furthermore, the use of diazoethane instead of diazomethane allows the differentiation between the free acid and any endogenous ABA-Me, as the ethyl ester of endogenous ABA is then obtained. The fragmentation pattern of this compound showed also a base peak at m/z 190 (proceeding from a retro Diels-Alder reaction and the loss of ethanol) (ref. 10). However, the ABA-Et had a longer Rt than ABA-Me but shorter than that of trans-ABA-Me, and all three compounds can be measured separatly. When a plant extract was treated with diazoethane, the corresponding ethyl-ester of ABA was detected by SIM with no indication of the presence of ABA-Me. Furthermore, when the extract was not treated with diazoalkane at all, no evidence of traces of ABA-Me or ABA-Et could be found. This indicates that only ABA is present in the material used in this study, together with about 10% of 2-trans ABA, thus confirming previously obtained data (ref. 11).

In conclusion, the use of GC/MS in the SIM mode together with the addition of $ABA-D_6$ as internal standard allows precise measurements at the picogram level of ABA and its 2-trans isomers. Corrections may be performed if necessary to prevent errors due to interconversion by photoisomerisation of the two isomers. The SIM traces obtained from maize root tissues are very clean and appeared to be the true levels of the endogenous free acids as no naturally occurring methyl or ethyl esters have been found during control experiments.

REFERENCES

1. J.R. Bearder, in J. MacMillan (Ed.), Hormonal Regulation of Development I, Encyclopedia of Plant Physiology, Vol. 9, Springer, Berlin, 1980, p. 9.
2. P.E. Pilet, in F. Skoog (Ed.), Plant Growth Substances 1979, Springer, Berlin, 1980, pp.450-461.
3. C.H.A. Little, J.K. Heald and G. Browning, Planta, 139(1978)133-138.
4. B.V. Milborrow and R. Mallaby, J. exp. Bot., 26(1975)741-748.
5. B.V. Milborrow, Ann. Rev. Plant Physiol., 25(1974)259-307.
6. P. Gaskin and J. MacMillan, Phytochemistry, 7(1968)1699-1701.
7. L. Rivier, H. Milon and P.E. Pilet, Planta 134(1977)23-27.
8. K. Blau and G.S. King, Handbook of derivatives for chromatography, Heyden, London 1978.
9. P.E. Pilet and L. Rivier, Plant Sci. Lett., 18(1980)201-206.
10. R.T. Gray, R. Mallaby, G. Ryback and V.P. Williams, J. Chem. Soc. Perkin II, (1974)919-924.
11. L. Rivier and P.E. Pilet, Phytochemistry, 20(1981)17-19.

THE BIOSYNTHETIC ORIGIN OF THE SULFUR ATOMS IN LIPOIC ACID

R. H. WHITE

Department of Biochemistry and Nutrition

Virginia Polytechnica Institute and State University

Blacksburg, Virginia 24061 (U.S.A.)

ABSTRACT

The mechanism for the incorporation of the sulfur atoms into octanoic acid during the biosynthesis of lipoic acid has not been determined as yet. As a first step in solving this problem, the chemical nature of the sulfur donor used in the biosynthesis had to be established. This was done using sulfur-34. The approach used in solving this problem consisted of two phases. In the first phase, methods were developed to determine the sulfur-34 abundance in the various sulfur containing compounds present in E. coli, i.e., cysteine, methionine, hydrogen sulfide, and sulfane sulfur. In the second phase, either through metabolic manipulation or the use of mutants, the sulfur-34 incorporated in vivo into these compounds by cells grown with either $^{34}SO_4^=$ or ^{34}S-sulfane labeled thiocystine were independently altered. Then, by determining the sulfur-34 incorporated into the lipoic acid under these different growth conditions and comparing it to that found in the above compounds, it was possible to establish which of these compounds most likely supplied the lipoic acid sulfur. The results of this work clearly show that cysteine supplies the sulfur for lipoic acid biosynthesis in E. coli.

INTRODUCTION

Lipoic acid (1,2-dithiolane-3-valeric acid) is a well-established accessory growth factor for a wide variety of microorganisms [1], an established coenzyme for α-ketoacid dehydrogenases [2], and a coenzyme in the glycine cleavage system isolated from vertebrate livers [3] and from bacteria [4]. Its structure has been known for almost 30 years yet its biosynthesis has been investigated only recently. These recent investigations have demonstrated, primarily through the use of stable isotopes, that lipoic acid is biosynthesised in E. coli from octanoic acid [5,6]. It has also been demonstrated, using d_{15} octanoic acid, that each sulfur atom introduced into the octanoic acid is inserted at each of the saturated carbons with the loss of only that hydrogen which each sulfur replaces [6]. In addition, sulfur introduction at C-6 has been shown to most likely involve an overall inversion at this carbon [6,7]. These observations are consistent with a pathway for lipoic acid biosynthesis which would involve hydroxylated octanoic acids. Recent research [8], however, has shown that these compounds are not involved in the biosynthesis which suggests that living systems have evolved a method for the introduction of sulfur directly at saturated carbons. This type of reaction, at least in biological systems, is not unprecedented; the incorporation of the sulfur atom during the biosyntheses of biotin and penicillin is an apparent further example of this type of reaction.

A first step in understanding the mechanism of this reaction is to firmly establish the chemical nature of the sulfur used in this type of biosynthetic transformation. In this paper methods are described which use sulfur stable isotopes to attack this problem.

MATERIALS AND METHODS

Materials

Sodium sulfate 90 atom % ^{34}S was obtained form Prochem Isotopes, U.S. Services, Inc., Summit, New Jersey. ^{34}S-sulfane labeled thiocystine was prepared by Mr. Eddie Demoll from cysteine and elemental sulfur (90 atom % ^{34}S) as described by Fletcher and Robson [9]. Cysteine mutants were supplied by Dr. B. J. Bachmann from the E. coli Genetic Stock Center, Department of Human Genetics, Yale University.

Growth of Organism

E. coli B was grown on 100-200 ml of a defined liquid medium containing glucose and all of the amino acids except for cysteine and methionine as previously described [10]. The $MgSO_4$ in the original medium was replaced by 100 mg/liter of $MgCl_2 \cdot 6H_2O$ in order to generate a medium that was free of sulfur. Stock cultures of the E. coli were maintained on agar slants of the same medium but with the addition of 100 mg/liter Na_2SO_4. Cysteine, methionine and ^{34}S-enriched sodium sulfate were added to this medium to give the concentrations indicated in Table 1. The sulfate concentration used was selected because it was the lowest concentration to permit an acceptable growth rate. Under these conditions, the growth rate and final yield of cells was 95% of that obtained with the complete medium.

Analysis of Isotope Incorporation

The analysis of the isotopic distribution of ^{34}S in the bound cellular cysteine and methionine was performed by gas chromatography-mass spectometry of the N-trifluoroacetyl n-butyl ester derivative of methionine and the S-methyl N-trifluoroacetyl n-butyl ester derivative of cysteine. The preparation of these derivatives and the use of the m/e 61 ion ($CH_3SCH_2^+$) in each of their respective mass spectra to determine the atom % ^{34}S in the cysteine and methionine has been previously described [11]

Lipoic acid was extracted from the cells, converted into its dibenzyl methyl ester derivative and assayed by gas chromatography-mass spectrometry as previously described [6]. Ion intensities were measured from the intense M^+-91 ion at m/e 311 in the mass spectra of this derivative. Mathematically the intensities of the isotope peaks in this fragment, which contains both of the sulfur atoms of the lipoic acid, will fit the coefficients generated by expanding the expression

$$(a + b)^2 = a^2 + 2ab + b^2$$

a^2 = relative intensity of m/z 311 ion
$2ab$ = relative intensity of m/z 313 ion
b^2 = relative intensity of m/z 315 ion

where the ratio $b/(a + b)$ is the atom % ^{34}S in the sample. By solving for the value of b which generates the observed isotope peak intensities, the atom % ^{34}S in the lipoic acid was found.

RESULTS

Table 1 shows the results of competition experiments for the incorporation of $^{34}SO_4^=$ into cysteine, methionine, and lipoic acid by ^{32}S containing compounds.

TABLE 1

Competition of $^{34}SO_4^=$ incorporation into cysteine, methionine and lipoic acid in E. coli by ^{34}S containing compounds

Experiment number and growth conditions[a]	Atom % ^{34}S in		
	Cysteine m/z 61/62	Methionine m/z 61/63	Lipoic acid m/z 311/313
Experiment 1	87.0 ± 0.8	86.7 ± 0.3	87.6 ± 0.8
Experiment 2 0.4 mM met	85.4 ± 0.9	5.8 ± 0.4	87.3 ± 0.7
Experiment 3 0.2 mM cys	62.7 ± 0.8	62.5 ± 0.5	68.4 ± 0.9
Experiment 4 1.9 mM S_8	58.7 ± 0.8	61.0 ± 0.8	65.6 ± 0.9

[a]The growth medium contained glucose and all of the common protein amino acids with the exception of cys and met which were added as indicated. The medium also contained 0.39 mM $^{34}SO_4^=$ with 90 atom % ^{34}S.

Experiment 1 served as a control to show that all the listed metabolites can be labeled to the same extent when cells are grown on a medium containing $^{34}SO_4^=$ as the sole sulfur source. The small reduction from the 90 atom % ^{34}S in the sulfate added to the medium most likely indicates a slight contamination by non^{34}S-enriched sulfate already present in the medium.

When methionine was added to the growth medium there was a sharp reduction of ^{34}S in the methionine but no reduction of ^{34}S in the cysteine or lipoic acid (Experiment 2, Table 1). This clearly indicates that methionine, or a metabolite derived from methionine such as homocysteine, was not the metabolic source for the lipoic acid sulfur. The lack of a significant dilution of the cysteine ^{34}S levels by the ^{32}S containing methionine is consistent with the inability of bacteria to convert methionine into cysteine [12].

In Experiment 3, the addition of 0.2 mM cysteine to the growth medium produced a significant reduction in the ^{34}S incorporated into each of the three metabolites assayed. The reduction of the ^{34}S in both the cysteine and the lipoic acid can be viewed as supporting evidence that the cysteine sulfur supplies the sulfur for lipoic acid.

The cysteine sulfur could have been transformed into the lipoic acid sulfur by several routes. One route would be the direct introduction of the cysteine sulfur into the octanoic acid with no exchange of cysteine sulfur with any inorganic reduced sulfur pools in the cells. These pools could consist of either H_2S or sulfane sulfur. Alternatively, the cysteine sulfur could be released into these pools to be used subsequently for the lipoic acid biosynthesis.

To test the possible involvement of an inorganic sulfane sulfur pool, a hydrophilic suspension of sulfur was fed and, as can be seen from the data of Experiment 4, no selective lowering of the ^{34}S abundance was observed in lipoic acid. This would indicate that sulfane sulfur is not involved in the biosynthesis. This experiment, however, is complicated by the possible nonbiological oxidation of the sulfur prior to its utilization by the cells. This complication was overcome by growing cells with ^{34}S-sulfane labeled thiocystine. In addition to labeling the sulfane sulfur pool, this material permits the selective enrichment of the ^{34}S in the H_2S in the cells to about twice that of the cysteine sulfur, i.e., approximately 50% ^{34}S [13], and thus allows for the clear distinction as to whether the lipoic acid sulfur originates from these inorganic pools or from cysteine.

TABLE 2

Incorporation of ^{34}S from ^{34}S-sulfane labeled thiocystine into cysteine, methionine and lipoic acid by E. coli

Exp.	Strain used	Concentration of thiocystine fed (μm)	Atom % ^{34}S			
			Cysteine	Methionine	Lipoic acid	H_2S^a
1.	E. coli B	21	25.5	25.3	26.7	50-80
2.	E. coli B	104	27.9	27.5	27.8	
3.	E. coli K-12 JM 15, cysE	83	7.6	10.0	9.4	50-80
4.	E. coli K-12 JM 39, cysE	83	12.1	12.8	12.3	50-80
5.	E. coli K-12 JM 246, cysI	83	26.4	25.7	28.3	50-80

^{a}Calculated from E. coli K-12 and B serine mutants grown under the same conditions [13].

From the results of Experiments 1 and 2 (Table 2), it is clear that only the cysteine has a ^{34}S abundance which is consistent with its sulfur being the origin of lipoic acid sulfur. These results are somewhat complicated by the fact that some of the sulfane sulfur is clearly being incorporated back into the cysteine. This incorporation, however, should be blocked or at least greatly reduced by using mutants which are unable to incorporate H_2S into cysteine. This idea is born out by the results of Experiments 3 and 4 (Table 2) which show a sharp drop in the ^{34}S present in both the cysteine and lipoic acid in the two cysE mutants tested. That the observed reduction is not an artifact of a cysteine requirement is confirmed by Experiment 5 in which a cysI mutant, which can incorporate H_2S into cysteine but cannot reduce sulfate to H_2S, is still able to incorporate the labeled sulfane sulfur of thiocystine into cysteine with the same distribution of ^{34}S in its cysteine and lipoic acid as in the wild type strain.

REFERENCES

1 E. L. R. Stokstad, G. R. Seaman, R. J. Davis and S. H. Hutner, Methods Biochem. Anal., 3(1956)23-47.
2 M. Koike and K. Koike, in D. M Greenberg (Ed.), Metabolic Pathways, 3rd edn., Vol. VII, Metabolism of Sulfur Compounds, Academic Press, New York, 1975, Ch. 4, pp. 87-97.
3 K. Fujiwara, K. Okamura and Y. Motokawa, Arch. Biochem. Biophys., 197(1979) 454-462.
4 J. R. Robinson, S. M. Klein and R. D. Sagers, J. Biol. Chem., 248(1973) 5319-5323.
5 R. J. Parry, J. Am. Chem. Soc., 99(1977) 6464-6466.
6 R. H. White, Biochemistry, 19(1980) 15-19.
7 R. J. Parry and D. A. Trainor, J. Am. Chem. Soc., 100(1978) 5243-5244.
8 R. H. White, J. Am. Chem. Soc., 102(1980) 6605-6607.
9 J. C. Fletcher and A. Robson, Biochem. J., 87(1963) 553-559.
10 R. H. White and F. B. Rudolph, Biochim. Biophys. Acta, 542(1978) 340-347.
11 R. H. White, Anal. Biochem., (1981) in press.
12 M. Flavin, in D. M. Greenberg (Ed.), Metabolic Pathways, 3rd edn., Vol. VII, Metabolism of Sulfur Compounds, Academic Press, New York, 1975, Ch. 11, pp. 457-503
13 R. H. White, submitted for publication.

PLANT UPTAKE OF ISOTOPICALLY ENRICHED SO_2 and NO_2 PRESENT AT SUBNECROTIC CONCENTRATIONS IN THE ATMOSPHERE.

M. DUBOIS, F. BOTTER, M. DRIFFORD, J. SUTTON
DPC/SCM, CEN.Saclay, 91191 GIF SUR YVETTE CEDEX (France)
J. GUENOT
DPR/SERE, CEN.Far., B.P. n° 6, FONTENAY AUX ROSES 92260.

ABSTRACT

The rate of absorption by bean plants (Phaseolus vulgaris) of subnecrotic concentrations of the atmospheric pollutants SO_2 and NO_2 used singly, successively or together has been measured using the labelled molecules $^{34}SO_2$ and $^{15}NO_2$. It is observed that the uptake of the pollutants is generally more important in young leaves than in old, and that atmospheric humidity appears to favor SO_2 fixation when this gas alone is present. The results may be explained, in so far as the single pollutant is concerned, in terms of the mechanisms of the opening and closing of the stomata, whereas they furnish some evidence for a synergistic effect, when both pollutants are applied.

INTRODUCTION

Stable isotopes enable very precise measurements of the transfer rates to plants of airborne pollutants present at very low concentrations and during very short contact times. In a previous contribution, namely the measurement of the rate of fixation of sulfur dioxide by pine needles using $^{34}SO_2$ [1], we exposed the plant in its natural habitat to air containing 1 ppm of SO_2. The present study deals with the transfer to bean plants of SO_2 and NO_2 present at subnecrotic concentrations ($\leqslant 0.1$ ppm) in the atmosphere.

In a wind tunnel, we have studied the influence of various parameters such as the hygrometric state of atmosphere, the concentration of pollutants, their action, when used alone, successively or simultaneously etc... on the rate of their absorption.

EXPERIMENTAL

The plants are exposed in a specially designed wind tunnel [2] to a filtered air current (about 100 m^3 NTP/h) into which a mixture of 5000 ppm of the pollutant in nitrogen is introduced at a known rate. The flow rates are selected to give concentrations of 0.05 and 0.1 ppm of each pollutant. The relative humidity of the air may be increased by injecting water vapour. The CO_2 content was kept constant at its natural level. The experimental section is illuminated by a light equivalent to 100 watt per square meter of foliage.

The sulfur dioxide enriched in ^{34}S was prepared at the CENS. Nitric oxide ^{15}NO was purchased from APC, Toulouse. Mixtures containing 5000 ppm of $^{34}SO_2$, ^{15}NO or $^{34}SO_2$ + ^{15}NO in nitrogen were stored under pressure in stainless steel containers. The oxidation of ^{15}NO to $^{15}NO_2$ with oxygen was carried out at the moment of introduction into the exposure chamber.

The exact concentration present in the air around the plant was determined and adjusted by means of two gas analysers. - For SO_2 a COSMA analyser was used - For NO_2 determinations, an ENVIRONMENT A C 3 chimiluminescence analyser was used, which determines both NO and NO_2.

The initial unlabelled NO concentration in the air (there was no NO_2) was measured before admitting the isotopically enriched gas, which after oxidation consisted of 90 % $^{15}NO_2$ and 10 % ^{15}NO.

Bean plants (Phaseolus vulgaris) grown in vermiculite watered with a nutritive solution (Y. COIT (INRA) nutritive solution) were used in all experiments. The plants were between 3 and 6 weeks old, so that the nitrogen and sulfur contents varied, decreasing with the age of the plant.

After measurement of the degree of stomatic opening(only a few measurement were possible), by means of a porometer, leaves were removed from the plants at different exposure times, frozen at - 196°C to stop metabolic and exchange processes and then ground under liquid nitrogen prior to drying by lyophilisation.

On account of the low concentration of sulfur relative to nitrogen present in the plants each element was extracted separately in order to determine its quantity and isotopic composition.

For the nitrogen measurements about 100 mg of the dried, powdered leaves were burned under oxygen in a loop the combustion section of which was situated in an induction furnace. The gaseous products were transferred by a current of helium into a vacuum line for separation [2]. Finally, the volume and the isotopic composition of pure nitrogen were measured. This whole process took 30 minutes.

For sulfur determination 400 mg of dried leaf powder were digested with nitric acid in the presence of magnesium nitrate^x. The sulfate formed was then precipitated as $BaSO_4$ which was separated, and converted to SO_2 [1]. After removal of CO_2 the volume of SO_2 and its isotopic composition were determined.

The initial isotopic contents of the $^{34}SO_2$ (31 %) and ^{15}NO (99,6 %) tracers were measured with a mass spectrometer with a single collector.

The isotope analyses of the gases extracted from the leaves were performed with a double collection mass spectrometer operating in the natural abundance range (Micromass model 602) ; we took into account the correction of variation of ^{33}S.

The natural standards used for comparison with the gases extracted from the plants were, for nitrogen, the air (3663 ppm ^{15}N) and, for sulfur, the Dyablo Canyon standard (95,02 % ^{32}S ; 0.75 % ^{33}S ; 4.21 % ^{34}S ; 0.02 % ^{36}S). Before exposure to the pollutants the bean leaf composition was + 7δ‰ ^{xx} for N and + 8δ‰ ^{xx} for S.

RESULTS AND DISCUSSION

The figures show the amounts transferred Δq, plotted against exposure time t, and tables I and II present results corresponding to short duration experiments expressed in units which facilitate their comparison with literature values.

Pollutants used singly.

At 50 % humidity the NO_2 transfer rate is of the same order of magnitude as that of SO_2 and is situated between 0.03 and 0.006 $cm.s^{-1}$, for, respectively, young and old leaves. The absorption rate obtained for SO_2 is reasonably close to certain values found in the literature [3]. Furthermore, other authors give a transfer rate for NO_2 which is slightly slower than that of SO_2 [4] Published data refer to experiments in which the conditions nevertheless were often quite different (longer exposure times, nature of plants etc...). In water saturated air, the rate of transfer of SO_2 is much greater than in dry air. For "young leaves" no such effect of water vapour is observed on the absorption of NO_2.

---

x Treatment kindly indicated by Mrs. A. BONVALET, Centre de Recherches de Montpellier, GERDAT.

xx $\delta‰ = \dfrac{\text{sample titer} - \text{standard titer}}{\text{standard titer}} \cdot 10^3$

Pollutants used successively (NO_2 followed by SO_2).

During the first 3 hours the results correspond to those found for NO_2 in the previous section ; in the following 3 hours when the NO_2 has been replaced by SO_2, the former gas attains a stationary, or slightly decreasing concentration in the leaves of the plant.

During the last 3 hours one observes a transfer of SO_2 which is similar to that found when SO_2 is used alone and this is observed in air at 50 % humidity and in water-saturated air.

Pollutants used simultaneously.

At 50 % humidity the observed rate of absorption of each pollutant is slightly less than when the pollutants are used singly. In water-saturated air the rate of transfer of SO_2 is maintained and becomes insensitive to the presence of water.

This ensemble of results calls for the following comments :

Our data on the pollutants used singly can be explained, in relating the overall results to the mechanism of opening and closing of the stomata. The bean leaves exposed to light would have their stomata relatively open but in the presence of NO_2 or SO_2 in air at 50 % humidity these latter would rapidly partially close. On the other hand, in the presence of water-saturated air containing SO_2 the closing of the stomata would be slower and less complete [5] , resulting in a greater transfer rate of SO_2.

The results obtained with the pollutants employed successively can be compared with those quoted by De Cormis [6]. We do not confirm a rate of absorption of SO_2 which is really more important after the plant has been exposed to NO_2, thus we have no evidence of a synergitic effect in contradiction to the "protective" effect with respect to high concentrations of SO_2 which is produced by pre-exposure to a subnecrotic concentration of SO_2 [7].

When the pollutants are employed together the results obtained may be compared with those of certain studies by Tingey [8] and De Cormis [9] especially in that we have found for the NO_2 and SO_2 mixture, rates of transfer which are slightly lower then those obtained by adding the two values of the pollutants used singly.

Taking into account the insensitivity of the SO_2 transfer rate to humidity this result could be related to the dominant role of NO_2 in imposing the closing of stomata but it might also be more complex and depend on the oxidising nature of the pollutant.

De Cormis [6] has shown, using plants with stomata on both leaf surfaces, that the necrosis produced (at a lower limit) for NO_2 and SO_2 mixtures was due to a different mechanism from that produced by SO_2 or NO_2 used alone.

We express our gratitude to the "Ministère de l'Environnement et Cadre de la Vie" for a grant to aid this work.

REFERENCES

1 Y. Belot, J.C. Bourreau, M. Dubois, C. Pauly, Symposium on Isotope Ratios as pollutant Source and behaviour indicators. I.A.E.A., Vienne, 18-22 Nov., 1974. Proceedings, 1975, I.A.E.A.-SM-191/18, 403.
2 F. Botter and Y. Belot, B.I.S.T., 230/231 (1978), 33.
3 D.J. Spedding, Nature, 224, (1969) 1229.
4 A.C. Hill, Journal of the Air Pollution Control Association, 2 (1971) 6, 341.
5 J. Bonte and P. Louguet, Physiol. Vég. 13 (1975) 527.
6 L. De Cormis and M. Lutringer, Pollution atmosphérique, 70 (1976) 119.
7 M. Pierre, Physiol. Vég. 15 (1) (1977), 195.
8 D.T. Tingey and al. Phytopathology, 61, 12, (1971) 1506.
9 L. De Cormis and M. Lutringer, Pollution atmosphérique, 75 (1977) 245.

TABLES I AND II.

Pollution of bean plants by NO_2 and pollution of bean plants by SO_2.

Sample	C	Transfer	NO_2 alone humidity		NO_2 followed by SO_2 humidity		$NO_2 + SO_2$ humidity	
			50%	Satd.	50%	Satd.	50%	Satd.
YOUNG LEAVES	0.1	Va			0.019 0.039	0.030 0.480	0.020 0.011	0.019
		ρ			52.1 25.8	33.4 20.9	50.5 92.6	55.2
	0.05	Va	0.031	0.036			0.015	0.012
		ρ	31.8	27.6			66.2	84.0
OLD LEAVES	0.1	Va			0.005 0.006	0.008 0.005	0.009 0.005	0.006
		ρ			180.9 159.4	123.8 204.4	104.5 199	0.006 169.5
	0.05	Va	0.006	0.012			0.004	0.004
		ρ	169.4	83.6			241.2	250.8

Sample	C	Transfer	SO_2 alone humidity		NO_2 followed by SO_2 humidity		$SO_2 + NO_2$ humidity	
			50%	Satd.	50%	Satd.	50%	Satd.
YOUNG LEAVES	0.1	Va	0.03		0.03 0.05	0.22	0.03 0.018	0.023
		ρ	34.8		33.6 19.9	4.6	33.3 56.8	42.6
	0.05	Va	0.043 0.018	0.23		0.21	0.04	0.026
		ρ	23.2 54.6	4.2		4.8	25	38.6
OLD LEAVES	0.1	Va	0.02		0.016 0.02	0.17	0.021 0.012	0.01
		ρ	45.2		60.5 46.8	5.9	47.2 80.6	102
	0.05	Va	0.02 0.020	0.12		0.11	0.032	0.018
		ρ	41.7 50.5	8.1		8.9	31.4	57.8

$Va = \dfrac{\Delta q}{t.S.C.}$ (cm.s^{-1}) ; $\rho = \dfrac{1}{Va}$ (s.cm^{-1})

Δq : pollutant absorbed per gram of dried leaf.

t : exposure time (s).

C : concentration of pollutant in air (vpm). Satd. = Saturated

S : foliage upper surface area (300 cm^2.g^{-1} dried young leaf – 350 cm^2.g^{-1} dried old leaf).

Figures : Transfer, Δq, of pollutants as a function of exposure time. Δq in mmole of N or S transferred, measured per gram of dried leaf ; time in hours. – – – 0.1 vpm pollutant concentration ; ——— 0.05 vpm polluant concentration. d : 50 % humidity ; s : water-saturated air. left side : "young leaves"(level 3 and 4); right side : "old leaves" (2 leaves from the base).

NITROGEN TRANSFER FROM NO_2 ALONE

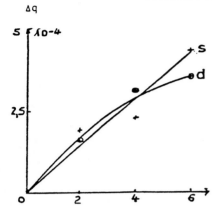
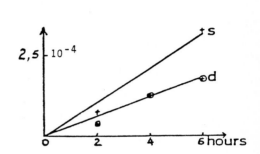

SULFUR TRANSFER FROM SO_2 ALONE

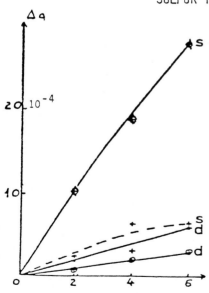
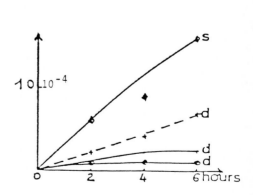

S AND N TRANSFER: NO_2 AND SO_2 USED SUCCESSIVELY

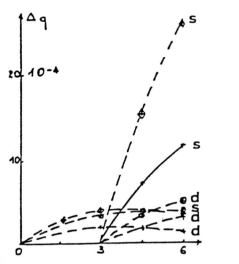
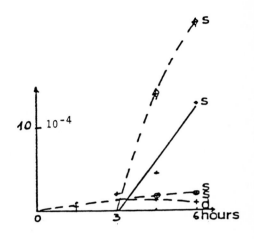

NITROGEN TRANSFER: NO_2 AND SO_2 USED SIMULTANEOUSLY

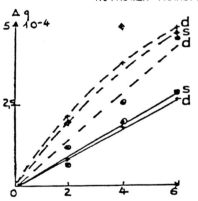
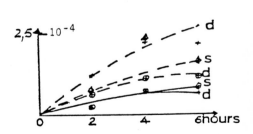

SULFUR TRANSFER: NO_2 AND SO_2 USED SIMULTANEOUSLY

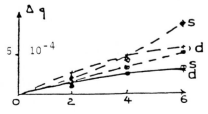

DYNAMIC STUDY ON ANIMAL EXPERIMENTS USING ^{15}N-LABELED NITROGEN DIOXIDE

Y. OHTA, M. YAMADA, Y. YONEYAMA, A. SUZUKI, I. WAKISAKA
National Institute for Environmental Studies, Yatabemachi, Tsukuba, Ibaraki 305, Japan

ABSTRACT

The metabolic pathway of inhaled nitrogen dioxide, labeled with ^{15}N, is reported. The experiments were organized at the level of low NO_2 doses. After exposure of rats to labeled NO_2 the different organs and the blood of the rats were lyophylized and digested thereafter by a semimicro Kjeldahl-method. The ^{15}N-content was determined by mass spectrometry, substracting the natural ^{15}N-content. After the application of $^{15}NO_2$ an enrichment was observed in urine, blood plasma and kidney. The ^{15}N-content of brain, trachea and heart has not been enriched significantly. Results of the ^{15}N-content after perfusion with physiological saline solution are reported, too.

The nitrogen dioxide is distributed within the blood pool immediately, and consequently it is excreted rapidly also. The ^{15}N-content of plasma depends on the concentration and the exposure time of NO_2, which should be converted to nitrite. The concentration of nitrite depends on the dose of nitrogen dioxide applied.

INTRODUCTION

The major source of man-made atmospheric emissions of nitrogen oxides originates from the combustion of fossil fuels. Other sources include special industrial processes, such as the manufacture of nitric acid and explosives, and appliences as smoking, gas firing and oil stoves.

The toxicity of an air pollutant depends on its distribution within the respiratory system. Some pollutants, e.g. sulfur dioxide, react chemically within the respiratory tract. The study of the interaction between nitrogen dioxide and the respiratory tract will enable one to predict damages. Some studies on the fate of nitrogen dioxide indicate, that the major part of the gas inspired is adsorbed probably within the distal regions of the pulmonary system. The nitrogen dioxide is adsorbed by the respiratory system and transported into the body. Studies of the absorption and excretion are important to estimate the effects at the level of the molecular reaction, the cell or the whole organism.

The experiments were done by a tracer technique using ^{15}N enriched nitrogen dioxide. About 3.0 liter per minute $^{15}NO_2$ flowed contineously on the rats. Its concentration was monitored by a chemiluminescence method. After the exposure the different organs and tissues, the blood and the urine were isolated and lyophylized. The samples were afterwards analysed in Kjeldahl-flasks by a chemical procedure. The solution was then divided into two aliquots: one part to determine the ^{15}N-content, the other part to measure the total nitrogen content. The natural ^{15}N-content of organs, tissue, blood and urine of untreated rats was determined, too. Moreover, one group of animal was perfused by physiological salt solution alone. During this experiment the nitrite concentration of plasma was measured colorimetrically [1-6].

RESULTS
Total nitrogen and ^{15}N-content

The content of total nitrogen, the weight of the lyophilized samples, the ^{15}N-content without and after exposure are summarized in table 1. The natural ^{15}N-content is necessary to estimate the excess after exposure. The natural ^{15}N-content of the animal tissue is slightly enhanced, as can be seen in table 2.

TABLE 1
Total nitrogen and ^{15}N-content after 18.6 ppm $^{15}NO_2$ exposure for 52.5 hours

Samples	Weight after lyophilization (g)	Total nitrogen (%)	^{15}N content (atom-%)	^{15}N Excess (%)	^{15}N-content excess (µg)
Lung	0.211	11.6	0.382	0.016	3.92
Trachea	0.0195	11.1	0.378	0.010	0.216
Liver	1.63	11.9	0.388	0.020	38.8
Heart	0.102	12.4	0.375	0.007	0.885
Spleen	0.104	13.0	0.380	0.013	1.76
Kidney	0.272	12.4	0.386	0.019	6.41
Red cell	0.581	14.9	0.373	0.006	
Plasma	0.0880	11.7	0.431	0.063	

TABLE 2
^{15}N natural value of rats

Samples	Wistar[+)] (^{15}N atom-%)	Donryu[+)] (^{15}N atom-%)
Lung	0.366	0.366
Trachea	0.368	0.368
Liver	0.368	0.367
Heart	0.368	0.367
Spleen	0.367	0.366
Kidney	0.367	0.367
Brain	0.368	0.368
Red cell	0.367	0.366
Plasma	0.368	0.366

[+)] n = 5

Relation between the total amount of exposed ^{15}NO$_2$ and ^{15}N-content of samples

Fig. 1 demonstrates one example of the ^{15}N content in organs and tissues after a 5 h exposure of 26.2 ppm ^{15}NO$_2$. It was found, that the ^{15}N-content in urine, plasma and kidney was considerably high. In liver, heart, brain and trachea the ^{15}N-content remained low. A correlation between the exposure to ^{15}NO$_2$ and the ^{15}N-content in plasma, lung and kidney can be seen in fig. 2, 3 and 4.

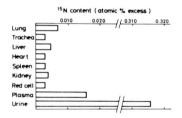

Fig. 1. ^{15}N-content after 26.2 ppm ^{15}NO$_2$ exposure for 5 hours.

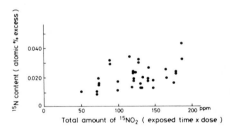

Fig. 2. ^{15}N-content in plasma after exposure.

Effects on ^{15}N-content after perfusion

After the exposure with ^{15}NO$_2$ some animals with high ^{15}N level of plasma were perfused with physiological salt solution. It was the aim of this treatment to eliminate the blood within the vessels of organs to get the true ^{15}N uptake into the tissues. Rats were exposed to 15.9 ppm ^{15}NO$_2$ for 24 h. After that time, one group was used as a control, the other treated by perfusion. Both groups are compared in Fig. 5. The control group is taken as reference. In lung and kidney the ^{15}N-content was lower compared to the control. In liver and heart the perfusion does not change the ^{15}N level after exposure.

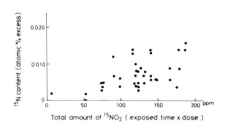

Fig. 3. ^{15}N-content in lung after exposure.

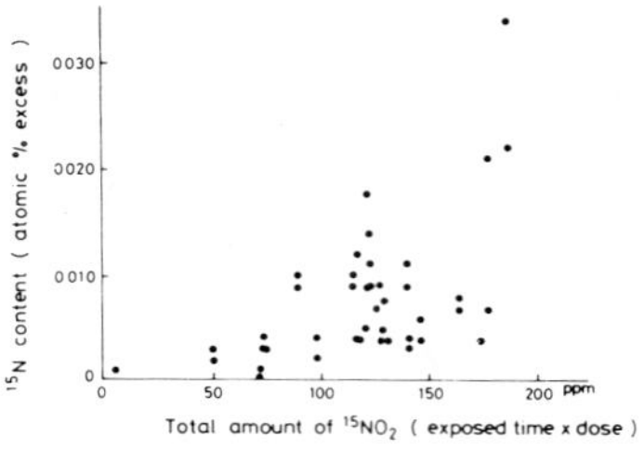

Fig. 4. ^{15}N-content in kidney after exposure.

Fig. 6 reports an experiment with a short exposure (1.5 h). The ^{15}N uptake was low, and the perfusion with physiological solution showed a marked effect especially for the lung.

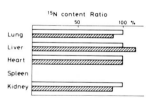
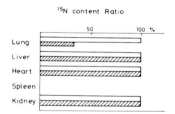

Fig. 5. ^{15}N-content after perfusion with physiological saline solution. This case was exposed 15.9 ppm $^{15}NO_2$ for 24 hrs.
☐ : control group
▨ : perfusion group

Fig. 6. $^{15}NO_2$ content after perfusion with physiological saline solution. This case was exposed 14.6 ppm $^{15}NO_2$ for 1.5 hr.
☐ : control group
▨ : perfusion group

Nitrite concentration in plasma

In the blood inhaled nitrogen dioxide is converted to nitrite [7]. The nitrite content in plasma is measured colorimetrically at 540 nm. A correlation between the nitrite concentration of plasma and the applied nitrogen dioxide dose was found. Fig. 7 shows the correlation of the two groups of test animals.

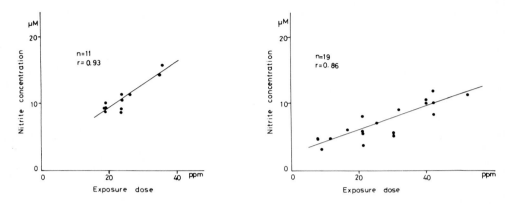

Fig. 7. Correlation between nitrite concentration in plasma and $^{15}NO_2$ exposed dose (left: group 1, right: group 2).

REFERENCES

1 I. Tyuma, Japan Soc. of Air Pollution, 12(1977)407-413.
2 K. Yoshida, K. Kasama, M. Kitabatake and M. Imai, Japan Soc. of Air Pollution, 14(1979)375-381.
3 E. Goldstein, N.F. Peek, N.J. Parks, H.H. Hines, E.P. Steffey and B. Tarkington, Amer. Rev. of Respir. Disease, 115(1977)403-412.
4 Y. Ohta and I. Wakisaka, Jap. J. Ind. Health, 21(1979)707.
5 Y. Ohta, M. Yamada and I. Wakisaka, Jap. J. Ind. Health, 22(1980) 623.
6 A. Steyermark, B.E. McGee, E.A. Bass and R.R. Kaup, Anal. Chem., 30(1958)1561-1563.
7 N.R. Schneider and R.A. Yeary, Am. J. Vet. Res., 34(1973)113-135.

GC-MS ANALYSIS OF ^{2}H- AND ^{15}N-LABELED ANALOGUES OF N-NITROSO-2(3',7'-DIMETHYL-2',6'-OCTADIENYL)AMINOETHANOLS IN ORGANIC TISSUES

S.L. ABIDI[*] and A. IDELSON[**]
National Fishery Research Laboratory, P.O.Box 65, La Crosse, Wisconsin (U.S.A.)[*]
Isotope Labeling Corp., Whippany, New Jersey (U.S.A.)[**]

ABSTRACT

A sensitive method has been developed for the determination of the N-nitrosamines given in the title in fish tissues utilizing a stable isotope internal standard. Enriched samples were quantitated by GC-CIMS with selective ion monitoring at m/e (M+1) and/or m/e (M-103). Results of comparative experiments using a series of ^{2}H- and ^{15}N-labeled analogues of the nitroso-compounds of interest indicated that the highest sensitivity attained in the assays was 5 to 10 pg at a S/N ratio of 4, when the trideuterio-O-methyl ether derivative of N-nitroso-2(3',7'-dimethyl-2',6'-octadienyl)aminoethanol-1',1'-^{2}H$_2$ was used as internal standard. The GC-MS method allows greater selectivity in detection and requires a less tedious sample purification process than the GC-electron capture detection method. Some aspects of the application of ^{13}C and ^{15}N NMR spectroscopy to the identification of various isomers of these N-nitrosamines are briefly discussed.

INTRODUCTION

Because of their carcinogenic and mutagenic properties and ubiquity in the environment (ref. 1), interest continues unabated in the biological and chemical evaluation of hazardous N-nitrosamines detected in   environmental samples as a consequence of the usage of pesticidal chemicals in the agricultural and aquatic fields. Recently we have demonstrated the presence of new mutagenic cis- and trans-N-nitroso-2(3',7'-dimethyl-2',6'-octadienyl)aminoethanols (I)(Fig. 1) in fish treated with nitrite and commercial formulations of 2-bis(3',7'-dimethyl-2',6'-octadienyl)-aminoethanol—a fish control agent (ref. 2,3). The accurate quantification of these isomeric substances at residue levels in aquatically derived food materials is of particular importance.

EXPERIMENTAL

Materials

The unlabeled pure and mixed isomers of I, the corresponding 1,2-^{14}C$_2$-compounds

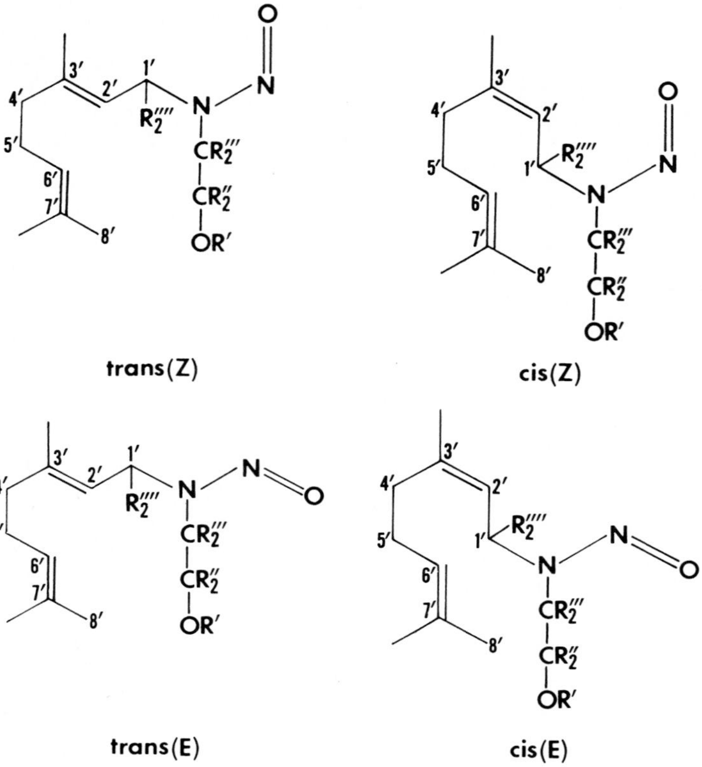

trans(Z)   cis(Z)

trans(E)   cis(E)

	R'	R"	R'''	R''''
I	H	H	H	H
II	H	^2H	H	H
III	H	H	^2H	H
IV	H	H	H	^2H
V	C^2H_3	^2H	H	H
VI	C^2H_3	H	^2H	H
VII	C^2H_3	H	H	^2H

Fig. 1. Structures of cis- and trans-N-nitroso-2(3',7'-dimethyl-2',6'-octadienyl)-aminoethanols and their deuterium labeled analogues.

(37 mci/mmol), the $^2H_2$- and $^2H_5$-labeled N-nitrosamines (II-VII)(Fig. 1), and their ^{15}N-labeled analogues were prepared according to our recently published methods (ref. 4).

Sample preparation

Fortified tissue samples containing a stable isotope internal standard and a ^{14}C-radiolabeled tracer were subjected to cleanup procedures similar to those described in detail in our earlier report (ref. 2) except with some simplification. The method involved exhaustive extraction of tissue homogenates with ether - methanol (1:1), adsorption chromatography on a silica gel column and concentration of radioactive eluates (hexane - ether, 2:3). Subsequent derivatization with CH_3I and NaH in dimethyl sulfoxide yielded analytical samples suitable for GC-MS-SIM quantitation. The same procedures were employed for chemically treated tissues.

Instrumentation

A Finnigan 4021T GC-MS-data processing system was used to obtain up to four simultaneous selected ion traces. For GC, a glass column (200 cm x 2 mm id) was packed with 3% OV-101 on 100/120 mesh Gas Chrom Q. Helium was the carrier gas (40 mL/min.). The injection port and oven temperatures were held at 210°C and 180°C respectively. The quadrupole mass spectrometer was equipped with a dual EI/CI source operated in the CI mode. Isobutane was used as the reagent gas.

RESULTS AND DISCUSSION

While the EI mass spectrum of the O-methyl ether of I shows few usable ion signals for analytical purposes due to lack of detectable molecular ions and other intense mass ions, its CI mass spectrum (Fig. 2, A) displays an abundant pseudo molecular ion (M+1) at m/e 241, a base peak ion at m/e 137, and two other minor peak signals of structural significance at m/e 223 and m/e 210. The mass spectrum of the pentadeuterated analogue VII is shown in Fig. 2, B. In comparison with the spectrum above (Fig. 2, A), the shifts to higher mass by 5 amu for the (M+1), (M-OH), and (M-NO) ions, and by 2 amu for the ($C_{10}H_{15}{}^2H_2$) ion are explicable in terms of the fragmentation mechanism proposed previously (ref. 2).

Although the cis and trans isomers of I are separable by GC, the mass spectra of the individual pure isomers exhibit identical spectral features. In regard to the probable difficulties implied by the existence of isomers of I, we have examined the ^{13}C and ^{15}N NMR spectra of the pure isomers. A small portion of the ^{13}C NMR spectral data (ref. 5) are recorded here to illustrate the distinct differences in chemical shift values between the isomeric pairs (Table 1). The difference in nitrogen chemical shifts between the E and Z isomers is 2.1 ppm for the ^{15}N-enriched nitrosyl nitrogen. These results present some informative examples of the potential applicability of ^{13}C and ^{15}N NMR spectroscopy to the study of the structures of this type of unsaturated N-nitrosamines and other related compounds (ref. 5).

Using the technique of stable isotope internal standardization, the purified and derivatized samples of tissue extracts were analyzed for the cis- and trans-I by GC-MS-SIM with multiple ion detection. The ions selected for the methylated I were m/e 241 and/or m/e 137. The stable isotope labeled internal standards were monitored at m/e (M+1) and/or m/e 139, depending upon the specific internal standard used. An example of the selective ion recording of tissue samples containing

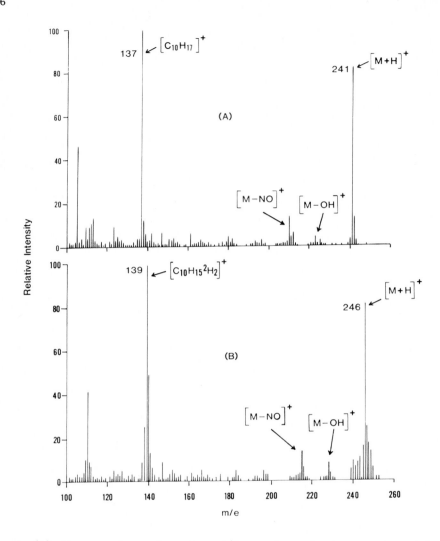

Fig. 2. (A). Mass spectrum of a selected isomer (trans) of O-methyl-I. (B). Mass spectrum of a selected isomer (trans) of VII (internal standard).

O-methyl-I and VII (internal standard) is given in Fig. 3. An excellent linear relationship was obtained for each set of the analytes and an internal standard in the region of 1–50 ng. The minimum detection limit for a tissue sample is near 60 pg with a S/N ratio of greater than 4. For a sample of the pure nitroso-compounds, the detection limit is about 5–10 pg at a S/N ratio of 4. Table 2 summarizes statistical evaluation data for five replicate determinations of 25 ng of I added to fish tissues by selective ion monitoring at m/e 241 using one of the internal standards ((a)–(l)) listed for each set of replicate analyses. It is noteworthy that attainment of good reproducibility occurred in those analyses where the pentadeuterio-N-nitrosamines ((d)-(f) and (j)-(l)) were used. Equally reproducible results were obtained, when the dideuterated ^{15}N-nitroso-compounds ((g)-(i)) were

TABLE 1

Comparison of ^{13}C-carbon resonances for cis-trans and E-Z isomeric pairs of I in chloroform with tetramethylsilane as the reference

Carbon	Chemical shift (ppm)				Δδ (Trans-Cis)	Δδ (E-Z)
	Trans		Cis			
	E*	Z*	E*	Z*		
C-1	58.9	60.2	58.9	60.2	0.0	-1.3
C-2	46.3	53.9	46.3	53.9	0.0	-7.6
C-1'	51.7	42.4	51.7	42.4	0.0	+9.3
C-2'	117.5	115.8	118.1	116.4	-0.6	+1.7
C-3'	131.8	131.7	131.8	131.7	0.0	+0.1
C-4'	39.7	39.7	31.6	31.6	+8.1	0.0
C-7'	143.1	142.0	142.3	141.2	+0.8	+1.1
C-3'-$\underline{CH_3}$	16.5	16.5	22.9	22.9	-6.4	0.0

*E=the nitrosyl group is cis to the hydroxyethyl group; *Z=the nitrosyl group is trans to the hydroxyethyl group.

TABLE 2

Reproducibility data for GC-MS-SIM analyses (m/e 241) of I (25 ng) using various stable isotope labeled internal standards

Internal standard	Selected ion m/e	Base peak ion* m/e	Coefficient of variation, %
(a) O-methyl-II	243	(137)	13.71
(b) O-methyl-III	243	(137)	11.23
(c) O-methyl-IV	243	139	14.98
(d) V	246	(137)	2.45
(e) VI	246	(137)	4.03
(f) VII	246	139	3.16
(g) O-methyl-II-^{15}N	244	(137)	3.84
(h) O-methyl-III-^{15}N	244	(137)	6.39
(i) O-methyl-IV-^{15}N	244	139	4.14
(j) ^{15}N-V	247	(137)	4.70
(k) ^{15}N-VI	247	(137)	7.17
(l) ^{15}N-VII	247	139	2.93

*The base peak ions in parentheses are not suitable to be utilized for SIM quantitation of I due to peak overlaps at m/e 137.

employed as internal standards. Mass fragmentation of the dideuterated analogues ((a)-(c)) affords in all instances the quantifiable mass ion (M+1) at m/e 243. The use of the latter three compounds as internal standards produced somewhat less reproducible results as noted by the values of coefficient of variation (11-15%). The problems could probably arise from the interference of co-eluting mass components derived from unknown sources. Among the better suited stable isotope internal

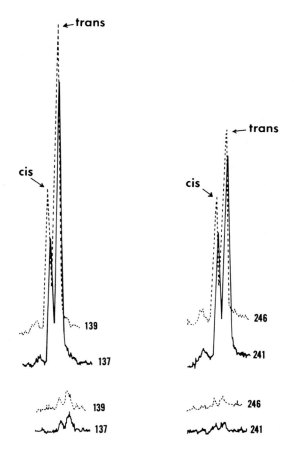

Fig. 3. Selective ion chromatograms of O-methyl-I (solid curves, cis:2.75 min., trans: 3.22 min.) with VII as internal standard (dotted curves, cis:2.74 min., trans:3.21 min.) obtained from purified tissue samples. Right: cis/trans, 201:329 pg (top); 39:62 pg (bottom). Left: cis/trans, 145:315 pg (top); 22:50 pg (bottom).

standards discussed, VII (f) and its ^{15}N-labeled analogue (1) proved to be the most suitable internal standards. Their versatile applicability permits simultaneous GC-MS-SIM quantitation of O-methyl-I by multiple ion detection at m/e 241, m/e 137, m/e 246 (or m/e 247 for (1)), and m/e 139. By doing so, qualitative structural confirmation of the analyte nitrosamines can be achieved. Unless otherwise stated, VII was used as the internal standard for all the assays performed throughout this work. For analysis of trace amounts of I, the number of ions selected was two.

A comparison of the analytical results obtained by GC-MS-SIM and GC-electron capture detection (ECD) methods for the determination of I in fish tissues treated with the fish toxicant and nitrite is shown in Table 3. The numerical data show a fairly good agreement between the two methods, although the GC-MS-SIM technique tends to be more selective and specific in detecting and quantifying I at trace

TABLE 3

Results of analyses of I in chemically treated fish tissues by GC-MS-SIM and GC-ECD methods

Sample	Concentration used in treatment, µg/g		Quantitation method (ng/g)			
	Toxicant*	Nitrite	GC-MS-SIM		GC-ECD	
			Cis	Trans	Cis	Trans
#1	10	20	1.6	3.2	1.7	3.4
#2	20	40	1.8	3.5	2.1	4.2
#3	30	60	2.1	3.8	2.4	4.6
#4	40	80	2.3	4.7	2.4	4.9
#5	50	100	3.0	5.6	3.3	6.3

*A formulated mixture of geometrical isomers of the toxicant was used.

levels in the complex sample matrix of organic tissues.

In conclusion, the method developed in the present study is much less tedious and less time-consuming than the GC-ECD method. It is suitable for analyzing the isomeric N-nitrosamines of concern in organic tissues in general and may be applicable to the quantification of possible metabolites related to these nitroso-compounds.

REFERENCES

1. E.J. Olajos, Ecotoxicology and Environmental Safety, 1(1977)175-196, and references therein.
2. S.L. Abidi, 180th National Meeting of the American Chemical Society, Las Vegas, Nevada, U.S.A., August 24-29, 1980, AGFD 11.
3. S.L. Abidi and R.A. Finch, Environmental Mutagenesis, in press.
4. S.L. Abidi and A. Idelson, J. Labeled Compounds and Radiopharmaceuticals, 18(8)(1981), in press.
5. S.L. Abidi, 180th National Meeting of the American Chemical Society, Las Vegas, Nevada, U.S.A., August 24-29, 1980, ANAL 141.

V. METHODS
A. ANALYTICAL DEVELOPMENTS

RECENT APPLICATIONS OF ^{13}C NMR SPECTROSCOPY TO BIOLOGICAL SYSTEMS

N. A. MATWIYOFF

Los Alamos National Laboratory, University of California, Los Alamos, New Mexico 87545, U.S.A.

ABSTRACT

Carbon-13 nuclear magnetic resonance (NMR) spectroscopy, in conjunction with carbon-13 labeling, is a powerful new analytical technique for the study of metabolic pathways and structural components in intact organelles, cells, and tissues. The technique can provide, rapidly and nondestructively, unique information about: the architecture and dynamics of structural components; the nature of the intracellular environment; and metabolic pathways and relative fluxes of individual carbon atoms. With the aid of results recently obtained by us and those reported by a number of other laboratories, the problems and potentialities of the technique will be reviewed with emphasis on: the viscosities of intracellular fluids; the structure and dynamics of the components of membranes; and the primary and secondary metabolic pathways of carbon in microorganisms, plants, and mammalian cells in culture.

INTRODUCTION

Carbon-13 nuclear magnetic resonance (NMR) spectroscopy, in conjunction with carbon-13 labeling, has become an important analytical technique for the study of biological systems and biologically important molecules. The growing list of its well established applications to isolated molecules in solution includes the investigation of: metabolic pathways; the microenvironments of ligands bound to proteins; the architecture and dynamics of macromolecules; the structures of coenzymes and other natural products; and the mechanisms of reactions. Recently interest has been reawakened in the use of the technique for the study of metabolic pathways and structural components in intact organelles, cells, and tissues. The promise and problems in the use of ^{13}C NMR spectroscopy and ^{13}C labeling in such investigations can be illustrated by the results of some early work on suspensions of the yeast, Candida utilis (refs. 1,2).

Reproduced in Fig. 1 is a set of ^{13}C NMR spectra of a thick suspension of C. utilis cells to which 1-^{13}C glucose had been added (ref. 1). These spectra

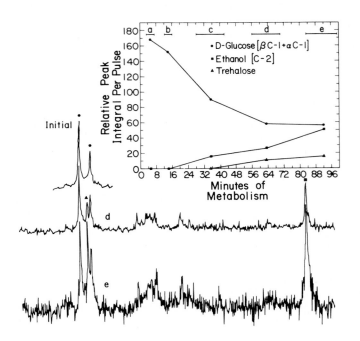

Fig. 1. Proton noise decoupled, Fourier transform ^{13}C NMR (25.2 MHz) spectra obtained during metabolism of the β- and α-anomers of $[1^{13}C]$ glucose (●) to $[2-^{13}C]$ ethanol (■) with the intermediate formation of $[1,1'-^{13}C]$ trehalose (▲) by a suspension of C. utilis cells; and plots of the relative intensities of the labeled glucose, ethanol, and trehalose resonances as a function of time: a) 500 pulses, 3-7 min. after initiation of metabolism; b) 500 pulses, 12-16 min.; c) 1500 pulses, 28-42 min.; d) 1500 pulses, 56-70 min.; e) 1500 min., 83-99 min. Spectra obtained during the initial time period exhibited only the resonances (●) corresponding to the C-1 of the β- and α-anomers of the substrate $[1-^{13}C]$ glucose. Spectra obtained in the time periods (d) and (e) exhibited prominent resonances corresponding to C-1 of trehalose (▲) and C-2 of ethanol (■) together with natural abundance signals of glucose and C. utilis cells, and ^{13}C enriched resonances of transient intermediates. Due to the low sensitivity, these could not be identified unequivocally.

show that, despite the high viscosity and heterogeniety of the suspension, the relaxation times of ^{13}C are favorable enough to allow the time course to be traced for the anaerobic metabolism of the β and α anomers of glucose (●) to the end product ethanol (■). In addition, the spectra revealed that an intermediate metabolite, later identified as trehalose (ref. 3), slowly accumulates and eventually is consumed. The ^{13}C spectrum of the osmotically shocked C utilis cells themselves shown in Fig. 2b is also of remarkably high resolution, allowing the assignment of the major resonances to the fatty acid and choline residues of the lipids in the cell membranes and the glycoside residues of the cell wall (ref. 1). Later ^{13}C relaxation time studies (ref. 2) of the membranes revealed a gradient of increased mobility from the glycerol backbone of the lipids toward the terminal methyl groups of the fatty acids and the choline head groups, and further that protein-lipid interactions make a negligible contribution to the mobility gradient.

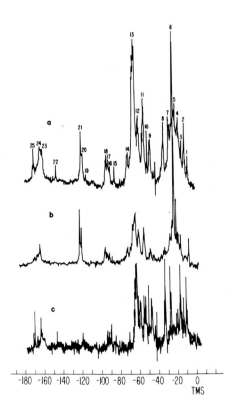

Fig. 2. Proton noise decoupled Fourier transform ^{13}C NMR (25.2 MHz) spectra of <u>Candida Utilis</u> cells enriched to the 20 atom % level with [1,2-^{13}C$_2$ (20 atom %)] acetate: (a) Cells suspended in D$_2$O, 995 pulses; (b) Cellular components remaining after exhaustive extraction with D$_2$O, 10,000 pulses; and (c) Water soluble components released by the osmotic shock of (b), 1082 pulses.

Despite the promise of the technique as revealed by these and other early studies, relatively little work on intact systems was done in the period following. This was due no doubt to a combination of circumstances including: (a) lack of suitably sensitive instrumentation in many laboratories; (b) the scarcity and expense of appropriate ^{13}C labeled precursors; (c) the limited access of NMR spectroscopists to biochemists and microbiologists who could provide guidance on the selective labeling of the extremely complex cells and tissues (vide infra); and (d) NMR spectroscopists with the appropriate instrumentation were directing their attention to developing a data base for the study of biologically interesting macromolecules. As noted earlier, this situation has begun to change and, in the following discussion, we review briefly recent progress in this rapidly growing field.

METABOLIC PATHWAYS

Spurred mainly by the increased availability of sensitive high field ^{13}C NMR spectrometers, the study of metabolic pathways in intact cells has increased dramatically in recent years. Illustrative examples of this work are summarized in Table 1. For the purposes of examining some of the detailed information obtainable in such studies, we will explore some recent work (ref. 13) we have been doing with <u>Microbacterium ammoniaphilum</u> which had its origins in the continuing program at the Los Alamos National Laboratory on the use of microorganisms for the large scale synthesis of natural products, uniformly or specifically labeled with carbon-13. The focus of this research work at present is on those L-amino acids for which efficient organic synthesis methods are not now available and for which there is a need in human metabolic and nutritional studies and in the investigation of the structure and dynamics of enzymes enriched with labeled amino acids. <u>M ammoniaphilum</u> was one of the microorganisms selected for early study because it produces glutamic acid in a 35-40% yield from glucose and will also use acetate as a substrate for glutamate production (ref. 14). In addition to optimizing the incorporation of the ^{13}C label from glucose or acetate into glutamate, a matter of practical concern to us for some mass spectrometric and NMR studies is the degree to which the label becomes randomized in glutamate due to the flux of glucose and acetate metabolites through various pools.

The major metabolic pathways for glucose and acetate leading to glutamate production in this microorganism are summarized in Fig. 3. As noted in the figure caption, the flow of [1-^{13}C] glucose metabolites to α-ketoglutarate and glutamate through part of the TCA or GS cycles will result in specifically labeled products but that those formed from intermediates experiencing one or more turns of these cycles will have a more random and complex distribution of the ^{13}C label. Summarized in Fig. 4 is the time dependence of the ^{13}C spectra of a suspension of <u>M ammoniaphilum</u> containing [1-^{13}C] glucose and near the stationary phase of growth. Since the production of glutamate in large quantities by this microorganism does not begin until that stage is attained (ref. 14), the initial growth was accomplished with natural abundance glucose to minimize loss of the ^{13}C label. In addition to

TABLE 1

Illustrative ^{13}C nuclear magnetic resonance studies of metabolism in intact cells and tissues

System	^{13}C Labeled Substrate	Observations
Soybean	$^{13}CO_2$	Early label appears in sugars and lipids (ref. 4a)
Anacystis nidulans, blue green alga	$^{13}CO_2$	Significant dark respiration contributes to the CO_2 pool available for synthesis (ref. 5a)
Escherichia coli, bacterium; saccharomyces cerevisiae, yeast	[1-^{13}C] glucose and [6-^{13}C] glucose	Different rates for α- and β-glucose translocation; flux of carbon through most glycolysis intermediates to end products evaluated
Rhodopseudomonas spheroides; propionibacterium shermanii	[5-^{13}C] aminolevulinic acid [11-^{13}C] porphobilinogen	Direct demonstration of porphyrinogen intermediates in porphyrin biosynthesis (ref. 8)
Rat hepatocytes, normal and hyperthyroid	[2-^{13}C] glycerol, [1,3-$^{13}C_2$] glycerol, and [3-^{13}C] alanine	Gluconeogenesis from alanine and glycerol; hyperthyroid rat cells show an enhanced gluconeogenesis (ref. 9a)
Escherichia coli	[1-^{13}C] glucose	Uptake and efflux of pyruvate affected by pH of medium and will occur against osmotic gradient (ref. 10)
Erythrocytes	[1-^{13}C] glucose	Label from glucose incorporated into diphosphoglycerate and lactate (ref. 11)
Neuroblastoma X glioma cells	[3[2-^{13}C]-glycine] methionine-enkephalinamide	Cells degrade enkephalinamide to a mixture of peptides and free glycine (ref. 12)

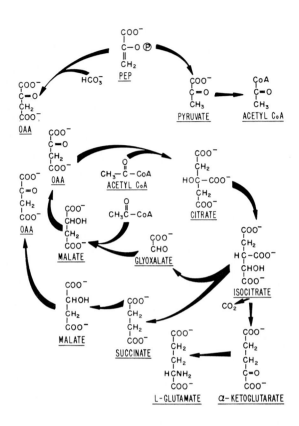

Fig. 3. Major metabolic pathways in M ammoniaphilum involving glucose, acetate, and glutamate. Glucose labeled at C-1 produces [3-^{13}C] pyruvate via the Embden-Meyerhof pathway (EMP) and unlabeled pyruvate via the hexose monophosphate shunt (HMS). [3-^{13}C] pyruvate enters the tricarboxylic acid (TCA) and glyoxylate shunt (GS) cycles as [3-^{13}C] oxaloacetate and/or [2-^{13}C] acetate and can result in the formation of [2-^{13}C] glutamate, [4-^{13}C] glutamate, and [2,4-^{13}C] glutamate via α-ketoglutarate formed in a third of a turn of the TCA cycle. Formation of glutamate after one or more turns of the TCA cycle will tend to randomize the label because of the formation of the symmetrical intermediates succinate and fumarate. In particular, C-4 and C-3 will interchange in one turn, as will C-2 and C-1. Similar scrambling will occur through the flow of citrate through the glyoxylate cycle.

the expected resonances from ^{13}C enriched bicarbonate, glutamate, succinate, and lactate, the spectra in Fig. 4 exhibit prominent peaks from two unusual ^{13}C labeled products which are eventually consumed, trehalose and glucosylamine. The accumulation of trehalose, which presumably functions as a storage carbohydrate, has been observed previously by ^{13}C NMR in growing yeast (refs. 1,3) and differentiating

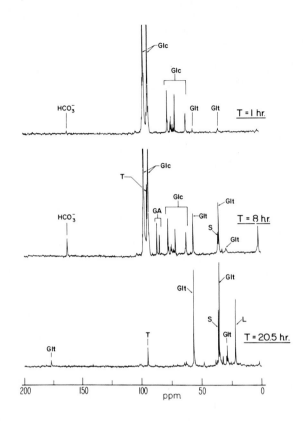

Fig. 4. Time dependence of the proton decoupled Fourier transform ^{13}C NMR spectrum (25.2 MHz) of a suspension of M ammoniaphilum initially containing [1-^{13}C (90 atom %)] glucose: T=1 hr., spectrum obtained during the 0-2 hr. metabolism period; T=8 hr., during the 7-9 hr. period; and T=20.5 hr., during the 19.5-21.5 period. Enriched ^{13}C resonances are: HCO_3-ion; Glc, β (97.0 ppm) - and α(93.4 ppm)-C-1 of [1-^{13}C] glucose; T (94.2 ppm), C-1 of α,α -[1,1'-$^{13}C_2$] trehalose; Glt, C-2 (56.0), C-4 (34.6) and C-3 (28.3) of glutamate; S, C-2,3 (35.3) of succinate; L, C-3 (21.3) of lactate; and GA, B (86.2) and (84.1) C-1 of glucosylamine. In the T=1 hr. and 8 hr. spectrum the glucose C-1 resonances are truncated and the natural abundance glucose resonances are apparent.

amoeba cultures (ref. 15). To our knowledge, glucosylamine formation in cell culture has not been observed and, at present, we do not known whether its synthesis is under enzyme control and what role it plays in the control of glucose or nitrogen metabolism. The formation of lactate and succinate by M ammoniaphilum depends on the oxygen tension which, for the cultures appropriate to Fig. 4, fell sharply in the later stages of glucose consumption. In fully aerated cultures, the lactate and succinate levels are sharply reduced.

The C-2, C-3, and C-4 glutamate resonances in Fig. 4 are predominately singlets suggesting that either: (a) There is a low incorporation of ^{13}C resulting in a small probability of neighboring ^{13}C-^{13}C and ^{13}C-^{13}C-^{13}C occupations whose scalar interactions would result in ^{13}C multiplets; or (b) The ^{13}C incorporation is high and results from those biosynthetic pathways discussed earlier, which introduce the label into specific sites with a minimum of randomization. The proton spectrum reproduced in Fig. 5 demonstrates clearly that the latter choice is the correct one. Indeed, the proton NMR data suggest that the ^{13}C population at C-4 of glutamate (38%) approaches that theoretically possible (45%) if all the ([3-^{13}C (45 atom %)]) pyruvate resulting from [1-^{13}C(90 atom %)] glucose metabolism were

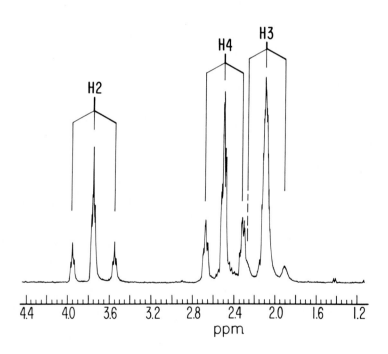

Fig. 5. Proton NMR spectrum (360 MHz) of [1,2,3,4-^{13}C$_4$] glutamate derived from M ammoniaphilum grown on [1-^{13}C (90 atom %)] glucose. For each proton, the center of the multiplet arises from ^{12}C-H moieties with fine structure caused by H-H scalar interactions. The doublets with similar fine structure are due to the ^{13}C-H splitting from moieties containing the ^{13}C label. The ^{13}C populations calculated from the ratio of the doublet to the singlet intensities are: C-2, 34%; C-4, 38%; and C-4, 14%.

incorporated into C-4 via the first third of the Krebs cycle (see captions, Fig. 3). By combining the proton and carbon-13 NMR data, one can calculate that the relative contribution of the metabolic routes to the labeling of glutamate is that summarized in Fig. 6. A noteworthy aspect revealed in this analysis is

POSSIBLE LABELING PATTERNS FOR C-2 C-3 AND C-4
OF L-GLUTAMATE DERIVED FROM D-[1-^{13}C] GLUCOSE

Pathway Designation	Labeling Pattern	Source of Labeling	Pathway %
X	$\overset{*}{C_2} - C_3 - \overset{*}{C_4}$	PEP → OAA → Glutamate via 1st third of Krebs Cycle	60%
Y	$\overset{*}{C_2} - C_3 - \overset{*}{C_4}$ $C_2 - \overset{*}{C_3} - \overset{*}{C_4}$	Single turn of the Krebs or Glyoxylate Cycle	30%
Z	$\overset{*}{C_2} - \overset{*}{C_3} - \overset{*}{C_4}$	Multiple turns of the Krebs or Glyoxylate Cycle	10%

C-2 Intensity = α (X + Y/2 + Z)

C-3 Intensity = α (Y/2 + Z)

C-4 Intensity = α (X + Y + Z)

Normalize   X + Y + Z = 1

Fig. 6. Labeling pathways for the incorporation of ^{13}C into C-2, C-3, and C-4 of L-glutamate derived from D-[1-^{13}C] glucose. Abbreviations are: PEP, phosphoenolpyruvate; OAA, oxaloactate; and the α, fractional population of ^{13}C at the site designated. The data are consistent also with studies of [6-^{13}C] glucose. The results are averages, there being evidence of a time dependence of the relative contributions.

that the simultaneous occupation of ^{13}C at C-2 and C-3 and C-3 and C-4 of glutamate is only possible via multiple turns of the Krebs or glyocylate cycles. The lack of pronounced multiplets in the ^{13}C spectra of labeled glutamate rule out the single or multiple turns as predominant events in the processing of glucose metabolites by the cells. This point, which is emphasized further in Fig. 7,

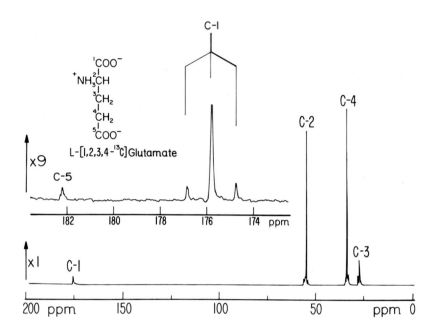

Fig. 7. Proton decoupled ^{13}C NMR spectrum (25.2 MHz) of [1,2,3,4-^{13}C$_4$] glutamate derived from M ammonialphilum grown on [1-^{13}C (90 atom %)] glucose. The spectrum illustrates the nonrandom distribution of the ^{13}C label among the C-2, C-3, and C-4 sites. If the label were distributed randomly with the ^{13}C abundances derived from the spectra in Fig. 5 then, for example, the C-1 and C-2 signals should consist of approximate 1:2:1 multiplets.

should serve as a caution to those who seek to extract labeling probabilities solely from a comparison of multiplet to singlet intensities in ^{13}C NMR spectra - as we have observed previously (ref. 5), the technique generally will work only if the labeling of the carbon atoms giving rise to the multiplet (s) is random.

The studies summarized in Table 1 address a number of interesting and important questions in metabolism but space will allow only a brief discussion of a few illustrative examples. In the study of Deslaurier et al. (ref. 12) on the degradation of ^{13}C labeled enkephalin and enkephalinamide by neuroblastoma X glioma hybrids, the intent was to analyze the conformation of those neuroactive peptides when bound to the opiate receptors of the cells. Instead, and despite reports of successful studies of the binding of eukephalins to neuroblastoma X glioma cells using other techniques, the ^{13}C NMR studies of [3[2-^{13}C] glycine] methionine--enkephalinamide showed unequivocally that the peptide is extensively degraded and the authors concluded that ^{13}C NMR studies of opioid peptide-receptor interactions will require the use of metabolically stable analogs. The study of Orgino et al. (ref. 10) on the metabolic regulations and pyruvate transport in anaerobic E. coli cells is of special interest because only the signals of metabolites which had diffused through the cell membrane and accumulated in the medium were observed, thus allowing the evaluation of the effect of perturbations to the cell on the influx and egress of pyruvate. In one of the extensive series of studies by Shulman and co-workers (refs. 6,7,9), an intact mouse liver filling the radio frequency coil space of a NMR spectrometer, was perfused with nutrients containing [3-^{13}C] alanine and [2-^{13}C] ethanol and the time course of the effects of ethanol on the metabolism of the glucogenic amino acid were studied in detail, including the effect on the flow of the label into various metabolites of: mitochondrial and cytosolic fumarase activity, operation of the pentose cycle; and the activities of glutamine synthetase and pyruvate carboxylase. Considering the utility of the studies discussed above and summarized in Table 1, we expect that studies of the metabolism of ^{13}C labeled substrates in cells, tissues, and organs by ^{13}C NMR spectroscopy will continue to expand rapidly. Looking to additional areas of application of the technique in the future, we felt that a particularly promising one is in the area of biochemical genetic analysis of mutant mammalian cells by ^{13}C tracing experiments to establish the precise site(s) of the biochemical block(s) where ^{13}C intermediate(s) accumulate in mutants produced by radiation and environmental chemicals.

STRUCTURAL COMPONENTS OF INTACT CELLS AND TISSUES

The ^{13}C spectrum of the c. utilis membranes reproduced in Fig. 2b allows one to extract only the average chemical shift and relaxation data for a system which is markedly heterogeneous. Studies of more general utility must make use of selected biosynthetic pathways to label specific sites or structural elements for ^{13}C NMR investigations on intact cells and tissues or on component reconstituted macromolecular assemblages. Recently there has been an increasing amount of work directed to that end (refs. 16-26). Again, space limitations allow only a brief discussion of significant highlights. In an early study (ref. 18), mice were kept on a histidine deficient diet supplemented with [2-^{13}C] histidine for a red blood cell life time with a consequent specific ^{13}C enrichment of the histidine residues of the red blood cell hemoglobin. Subsequent ^{13}C relaxation studies of the hemoglobin within the cell showed that the viscosity of the intracellular fluid is similar to that of water. A similar result was obtained in a related study of frog muscle labeled with ^{13}C enriched glycine (ref. 21).

In an extensive series of studies, Van Deenen and co-workers (ref. 20) were able to highly label the phosphatidylcholine of red blood cell, liver microsome and saccoplasmic reticulum membranes of rats fed a choline deficient diet supplemented with [Me$_3$-^{13}C] choline for a period of eight days. ^{13}C NMR studies of the muscle microsomes revealed an asymmetric distribution of phosphatidyl choline between the inner (60%) and outer (40%) surfaces of the membranes of the sarcoplasmic reticulum. This asymmetry is opposite to that found for erythrocytes and may be related to the fact that the former are endoplasmic membranes whose outer surface corresponds to the inner surface of plasma membranes. That same group has used reconstituted macromolecular assemblages to study the permeability barriers in large unilamellar glycophorin containing vesicles of [Me$_3$-^{13}C] phosphatidyl choline and have found that the barrier properties of glycophorin containing bilayers of phosphatidyl choline can be restored by 10 mol % phosphatidylethanolamine or lysophosphatidylcholine.

A final illustration of the broad range and great promise of the experiments being conducted in this area is the study of de Wit and co-workers (ref. 22) who enriched tobacco mosaic virus and its proteins to the 12 atom % ^{13}C level using ^{13}CO$_2$ as the carbon source. In addition to developing some interesting conclusions about the interaction between RNA and protein, these workers observed that the double disks formed by the proteins in the absence of RNA contain mobile ^{13}C atoms at all sites within the disks which were formed in solutions of 0.1-0.2 ionic strengths. In contrast, X-ray data for crystals of stacks of double disks obtained from 0.8 ionic strength solutions indicate that the motion of the carbon atoms is highly restricted. Thus there are important structural differences between the crystals studied by X-ray and the disks formed in solution near the physiological ionic strength, a conclusion strengthened by the NMR observation that there is an increasing degree of immobilization of the carbon atoms in the double disklike oligomers as the ionic strength increases.

ACKNOWLEDGMENT

The author is grateful for helpful discussions of the glutamate biosynthesis with Drs. R. E. London and T. E. Walker and for their participation in the design Figs. 3-7. This work was done under the auspices of the U.S. Department of Energy.

REFERENCES

1. R. T. Eakin, L. O. Morgan, C. T. Gregg and N. A. Matwiyoff, FEBS Letts., 28 (1973) 259-263.
2. R. E. London, V. H. Kollman and N. A. Matwiyoff, Biochem., 14 (1975) 5492-5000.
3. M. Kainosho, K. Ajisaka, and H. Nakazawa, FEBS Letts., 80 (1977) 385-389.
4. (a) J. Schaefer, E. O. Stejskal and C. F. Beard, Plant Physiol., 55 (1975) 1048; see also (b) J. Schaefer, E. O. Stejskal and R. A. McKay, Biochem. Biophys. Res. Commun., 88 (1979) 274-280.
5. (a) R. E. London, V. H. Kollman, N. A. Matwiyoff and D. D. Mueller, in E. R. Klein and P. D. Klein (Eds.), Proc. Sec. Intl. Conf. Stable Isotopes, U. S.

Energy Research and Development Report CONF-751027, 1976, pp. 470-484; also (b) V. H. Kollman, J. L. Hanners, R. E. London, E. G. Adame and T. E. Walker, Carbohydr. Res., 73 (1979) 193-202.
6. K. Ugurbil, T. R. Brown, J. A. Den Hollander, P. Glynn and R. G. Schulman, Proc. Natl. Acad. Sci. (USA), 75 (1978) 3742-3746.
7. J. A. Den Hollander, T. R. Brown, K. Ugurbil and R. G. Shulman, Proc. Natl. Acad. Sci. (USA), 76 (1979) 6096-6100.
8. A. I. Scott, G. Burton and P. E. Fagerness, J. Chem. Soc. (Chem. Commun.), (1979) 199-202.
9. (a) M. Cohen, S. Ogawa and R. G. Schulman, Proc. Natl. Acad. Sci. (USA), 76 (1979) 1603-1607; see also B. M. Cohen, R. G. Shulman and A. C. McLaughlin, ibid, 76 (1979) 4808-4812.
10. T. Ogino, Y. Arata and S. Fujiwara, Biochem., 19 (1980) 3684-3691.
11. P. Styles, C. Grathwohl and F. F. Brown, J. Magn. Res., 35 (1979) 329-336.
12. R. Deslaurier, H. C. Jarrell, D. W. Griffith, W. H. McGregor and I. C. P. Smith, Int. J. Peptide Protein Res., 16 (1980) 487-493.
13. T. E. Walker, C. H. Han, V. H. Kollman, R. E. London and N. A. Matwiyoff, work in progress.
14. S. Kinoshita, in K. Yamada, S. Kinoshita, T. Tsunoda, and K. Aida (Eds.), The Microbial Production of Amino Acids, Glutamic Acid, John Wiley and Sons, Inc., NY, 1972, Ch. 10, pp. 263-324.
15. R. Deslaurier, H. C. Jarrell, R. A. Byrd and I.C.P. Smith, FEBS Letts., 118 (1980) 185-190.
16. J. C. Metcalfe, N. J. M. Birdsall and A. G. Lee, FEBS Letts., 21 (1972) 335.
17. W. Stoffel and K. Bister, Biochem., 14 (1975) 2841.
18. R. E. London, C. T. Gregg and N. A. Matwiyoff, Science, 188 (1975) 266.
19. I. C. P. Smith, Can. J. Biochem., 57 (1979) 1-14.
20. A. M. H. P. Van Den Besselaar, B. DeKruijff, H. Van Den Bosch and L. L. M. Van Deenen, Biochim. Biophys. Acta, 555 (1979) 193-199; B. De Kruijff, A. M. H. P. Van Den Besselaar, A. Van Den Bosch and L. L. M. Van Deenen, ibid (1979) 181-192; W. J. Gerritsen, E. J. J. Van Zoelen, A. J. Verkley, B. De Kruijff and L. L. M. Van Deenen, ibid, 551 (1979) 248-259.
21. M. C. Neville and H. R. Wyssbrod, Biophys. J., 17 (1977) 255-267.
22. J. L. de Wit, N. C. M. Alma-Zeestraten, M. A. Hemmings and T. J. Schaafsma, Biochem., 18 (1979) 3973-3976.
23. S. Weinstein, B. A. Wallace, E. R. Blout, J. S. Morrow and W. Veatch, Proc. Natl. Acad. Sci. (USA), 76 (1979) 4230-4234; B. A. Wallace and E. R. Blout, ibid, 76 (1979) 1775-1779.
24. B. De Kruijff, A. Rietveld and C. J. A. Van Echteld, Biochim. Biophys. Acta, 600 (1980) 597-606.
25. S. Geller, S. C. Wei, G. K. Shkuda, D. M. Marcus and C. F. Brewer, Biochem., 19 (1980) 3614-3623.
26. P. L. Yeagle, Biochim. Biophys. Acta, 640 (1981) 263-273.

INTER- AND INTRAMOLECULAR ISOTOPIC HETEROGENEITY IN BIOSYNTHETIC ^{13}C-ENRICHED AMINO ACIDS

E. BENGSCH
Centre de Biophysique Moléculaire, C.N.R.S., F-45045 Orléans Cedex, France
J.-Ph. GRIVET
Centre de Biophysique Moléculaire, C.N.R.S., et Département de Physique, Université d'Orléans, F-45045 Orléans Cedex, France
H.-R. SCHULTEN
Institut für Physikalische Chemie der Universität Bonn, Wegelerstr. 12, D-5300 Bonn, F.R.G.

ABSTRACT

The labeling of some biosynthetic amino acids (serine, threonine, aspartic acid and glutamic acid) is neither random nor statistical. Isotope effects, known from the experiences with the natural abundance, can be observed at the level of high enrichment, too.

INTRODUCTION

The rare isotopes of the common elements H, C, N, O and S are often used as labels in biochemical and medical research. Biosynthesis, starting from very simple enriched precursors, is the preferred method for the preparation of complex labeled molecules. Since the starting material is never isotopically pure, the resulting compounds (such as amino acids, proteins, nucleic acids or even viruses) can only be partially enriched. In the case of nuclear magnetic resonance (NMR) studies, the low sensitivity of the technique has led to the use of highly enriched material. It has been assumed in the past [1-3] that labeled compounds of biosynthetic origin are statistically enriched and show an uniform enrichment. These assumptions do not agree with the experiences of isotopic effects in enzymatic reactions [4-6] and with the results on isotopic separation in plants [7, 8]. We have tested the above discussed hypotheses using a combination of field desorption mass spectrometry (FDMS) and ^{13}C NMR. The sample was a set of four amino acids (serine, threonine, aspartic acid and glutamic acid) extracted from the photosynthetic organisms <u>Spirulina maxima</u> and <u>Synechococcus cedrorum</u>, which had been grown on ^{13}CO$_2$ as their sole carbon source [1].

METHODS

We give here a bare outline of the analytical procedure, which is described in detail elsewhere [9]. The n carbon atoms of a particular amino acid are numbered consecutively (starting with the carboxylic group, through to the α carbon atom). Let p be the number of ^{13}C nuclei in this molecule. The output of the mass spectral analysis will be a set of N+1 numbers

$$x_p^{MS} \quad (0<p<n).$$

x_p is the relative abundance of that subset of molecules for which p carbon nuclei out of n are ^{13}C. We call such a subset a p-manifold. Various estimates of the enrichment can be derived from the mass spectrometry (MS) data. We define the mean enrichment factor, α', as:

$$\alpha' = \frac{1}{n} \sum_0^n p \, x_p^{MS}.$$

If the enrichment is uniform and statistical, we expect the abundances to follow the binomial law, with probability α':

$$x_p' = \binom{n}{p} \alpha'^P (1-\alpha')^{n-p}.$$

One may also introduce the "best" (in the least squares sense) enrichment, which is defined by the equations:

$$\begin{cases} x_p'' = \binom{n}{p} \alpha''^P (1-\alpha'')^{n-p} \\ \sigma^2 = \sum_0^n (x_p'' - x_p^{MS})^2, \text{ minimum.} \end{cases}$$

Let subscript i (1≤i≤N) label a particular carbon atom. Because vicinal C-C coupling constants are large, fragments such as $^{13}C_i - ^{13}C_{i+1}$ and $^{13}C_i - ^{12}C_{i+1}$ give non-overlapping patterns in the NMR spectrum. We therefore obtain very simply an approximate value of the relative abundance of ^{13}C at position i+1 from the integral of the NMR signal of atom i. We call y_i the set of numbers so derived. Finally, the individual relative concentrations of isotopomers, denoted as

$$z_{p,k} \quad \left(1 < k < \binom{n}{p}\right),$$

may be determined from a detailed computer simulation of the NMR spectra. To this end, we systematically vary the $z_{p,k}$ until a best fit is obtained. However, since the FDMS results are very accurate, individual concentrations are only adjusted within a p-manifold. That is, the $z_{p,k}$ are varied subject to the constraints

$$\sum_k z_{p,k} = x_p^{MS}$$

RESULTS

Non-statistical enrichment

Theoretical abundances x_p' or x_p'' are only fair approximations to the experimental values x_p^{MS}.

Other α-values did not give better fits. The table 1 below shows the x_p's for threonine. It can be seen that a large part of the discrepancy is due to the "zero-spin" (all ^{12}C) species. To see whether an artifact of the preparation procedure could be involved, we have tested the consequences of setting $x_p^{MS} = 0.0$; the fit is not qualitatively better.

TABLE 1
Comparison of experimental and theoretical abundance of ^{13}C for threonine.

		α'=0.7650	α"=0.7965
	x_p^{MS}	x_p'	x_p''
p		(percentage values)	
0	3.91	0.30	0.17
1	3.24	3.97	2.68
2	15.17	19.39	15.76
3	38.29	42.09	41.14
4	39.39	34.25	40.25

Furthermore, it appears that a negative correlation exists between degree of enrichment and quality of fit to the binomial law; the least enriched species (glutamic acid) also showed the highest discrepancies.

Enrichment dependend on the molecular species

As shown by table 2 below, different amino acids have different approximate enrichment levels.

TABLE 2
Comparison of the approximate enrichment levels of the four amino acids tested

	α'	α''
serine	0.869	0.876
threonine	0.765	0.796
aspartic acid	0.852	0.872
glutamic acid	0.726	0.780

Enrichment dependend on the position

The y_i-values are all different. For the limited set of compounds investigated here, a rough rule is the following. Hydroxylic and carboxylic atoms have a higher ^{13}C content than the mean. In contradiction, CH_n groups are depleted in ^{13}C with respect to predictions based on statistical abundance. This is illustrated, in the case of threonine, in table 3 below.

TABLE 3
Demonstration of an isotope enrichment in threonine dependend on the position of the C-atom. (The raw NMR data have been renormalized to make them directly comparable to α'.)

i	1	2	3	4
y_i	76.6	75.3	76.9	75.7
$y_i - \bar{y}$	+0.5	-0.8	+0.8	-0.5

The last line shows that the C_1 (carbonyl) position is enriched with respect to the mean of all positions, $\bar{y}$. Such an effect has been demonstrated a long time ago, in the natural abundance, using the more difficult methods of chemical degradation [10]. In fact the δ-values found by Abelson [10] correlate well, for our sample, with the $y_i - \bar{y}$.

Different isotopomers

The different isotopomers have different, characteristic abundances. The relative concentrations of isotopomers seem to follow rather simple rules, which are just generalizations of the previous observations. An isotopomer will be favoured (i.e., more concentrated than the binomial law would predict), if carboxylic and hydroxylic atoms are ^{13}C, and if CH_n fragments hold ^{12}C. Two isotopomers which are characterized by the same number of positive factors may be further ordered using the following supplementary rules. Aliphatic and carboxylic positions have more influence than hydroxylic

TABLE 4
Demonstration of the individual relative concentrations of isotopomers. A circle denotes a ^{12}C nucleus; a cross indicates the presence of a ^{13}C nucleus.

p	Structure code by position i				$z_{p,k}$	k	x_p
	1	2	3	4			
4	X	X	X	X	39.40%	1	39.40
3	X	X	X	O	10.80	1	
	X	X	O	O	8.70	2	
	X	O	X	X	10.35	3	
	O	X	X	X	8.45	4	38.30
2	X	X	O	O	2.85	1	
	X	O	X	O	3.15	2	
	X	O	O	X	2.10	3	
	O	X	X	O	2.65	4	
	O	X	O	X	1.95	5	
	O	O	X	X	2.45	6	15.15
1	X	O	O	O	1.25	1	
	O	X	O	O	0.90	2	
	O	O	X	O	0.55	3	
	O	O	O	X	0.55	4	3.25
0	O	O	O	O	3.90	1	3.90
y_i	78.6	75.7	77.8	73.9₅			

atoms, which in turn exert a greater effect than α-carbons. These statements may be verified from a look at table 4, which displays the $z_{p,k}$ for threonine. A comparison of y_i values computed from the $z_{p,k}$ with those of table 3 provides a check on the internal consistency of the method.

CONCLUSION

We have shown that ^{13}C-labeling of some biosynthetic amino acids is neither random nor statistical. Effects which have been known for some time in the case of natural abundances are also operative at very high enrichment levels. In the present case, the use of NMR makes available a large body of detailed, local information. It is then possible to propose a set of tentative rules for the prediction of the sites of preferential labeling. It is reasonable to assume that these will apply to other aliphatic amino acids as well. One would then be able to predict isotopomer concentrations from the sole knowledge of the overall enrichment, a quantity easily measured using FDMS [11, 12]. The same type of effects should be qualitatively similar in the case of ^{14}C labeling. Finally, we should like to point out that our data, in some way, reflect the carbon metabolism of photosynthetic bacteria.

ACKNOWLEDGEMENTS

This work was financially supported by the Deutsche Forschungsgemeinschaft. We thank the Service de Biochimie, Commissariat à l'Energie Atomique, Saclay, for providing the labelled samples.

REFERENCES

1 S. Tran-Dinh, S. Fermendjian, E. Sala, R. Mermet-Bouvier, M. Cohen and P. Fromageot, J. Am. Chem. Soc., 96(1974)1484.
2 R.E. London, T.E. Walker, V.H. Kollmann and N.A. Matwiyoff, J. Am. Chem. Soc., 100(1978)3723,
3 R.E. London, J.M. Stewart, R. Williams, J.R. Cann and N.A. Matwiyoff, J. Am. Chem. Soc. 101(1979)2455.
4 W.W. Cleland, M.H. O'Leary and D.B. Northrop, Isotope Effects in Enzyme Catalyzed Reactions, 6th Harry Steenbock Symposium, 1977.
5 R.D. Gandour and R.L. Schowen (Eds.), Transition States of Biochemical Processes, Plenum Press, New York, 1978.
6 K.D. Monson and J.M. Hayes, J. Biol. Chem., 255(1980)11435.
7 H. Ziegler, Ber. Dtsch. Bot. Ges., 92(1979)169.
8 H.-L. Schmidt, Ber. Dtsch. Bot. Ges., 92(1979)185.
9 E. Bengsch, J.-Ph. Grivet and H.-R. Schulten, Z. Naturforsch., 366(1981)109.
10 P.H. Abelson and T.C. Hoering, Proc. Nat. Acad. Sci. U.S., 47 (1961)623.
11 H.R. Schulten and W.D. Lehmann, Biomed. Mass Spectr., 7(1980)468.
12 U. Bahr and H.R. Schulten, J. Label. Comp. Radiopharm., 18(1981)571.

QUANTITATIVE MASS SPECTROMETRY WITH STABLE ISOTOPE LABELED INTERNAL STANDARD AS A REFERENCE TECHNIQUE.

I. BJÖRKHEM

Department of Clinical Chemistry, Karolinska Institutet, Huddinge Hospital, Huddinge (Sweden)

ABSTRACT

Isotope dilution - mass spectrometry (ID-MS) has become a most powerful tool as a reference technique.

In order to obtain maximal accuracy, certain requirements should be fullfilled. Factors of importance for optimal sensitivity, specificity and precision are discussed here. Under carefully controlled conditions, the most important errors seem to be bound to the GC-MS instrument. The sources and magnitudes of these errors have however not been sufficiently defined.

Results of comparative studies are presented, showing that with usual commercial instruments and with usual pipetting procedures, ID-MS methods in general give reproducible results within a few %.

Some recent applications of ID-MS in quality control are reviewed. In particular the technique has proved to be of great value in connection with evaluation of different commercial kits used for radioimmunoassay. In addition, ID-MS is of value in attempts to define the "accuracy pattern" of routine analyses of different compounds of clinical chemical interest.

GENERAL BACKGROUND

Accuracy should be the most important concept in connection with all types of measurements. In order to obtain maximal accuracy, different types of reference methodologies are necessary. During the last few years, isotope dilution – mass spectrometry (ID-MS) has become a most powerful tool in this connection.

It has been pointed out by several authors that from a theoretical point of view, ID-MS has a higher potential for accuracy than most other techniques (1-4). The most important errors in ID-MS methods are due to contaminations in the reference material used, errors in pipettings or weighings, and variations in the final mass spectrometric determination of the ratio between unlabeled and labeled molecules. All these errors can be kept on a low level, and it seems theoretically possible to determine the concentration of the analyte with a deviation of only a few % or less from the "true" value.

Several groups of researchers have utilized ID-MS as a reference technique (2-8). Recently, an ID-MS method for determination of total serum cholesterol was even proposed as a definitive method (9). In our laboratory, different ID-MS methods (9-20) have been used in different quality control activities more or less routinely since 1974.

The usual mode of performing an ID-MS assay is the following :

1. Addition of an exact amount of stable isotope labeled internal standard to an exact amount of sample
2. Equilibration between internal standard and analyte
3. Purification procedures
4. Derivatization
5. Selected ion monitoring of two ions. One ion should be derived from the unlabeled analyte and one from the stable isotope labeled internal standard.
6. Calculation of the ratio between the two tracings. Calculation of the concentration of the analyte with use of a standard curve.

In order to obtain an optimal ID-MS assay, there are certain requirements with respect to the internal standard:
1. Chemical identity between internal standard and compound to be detected.
2. The isotope label should be introduced in a stable position of the internal standard.
3. The amount of internal standard should be similar to the amount of compound assayed.
4. There should be an optimal difference in mass between the internal standard and the compound to be assayed.

As a general rule, the internal standard should be as similar to the compound to be analyzed as possible. The ideal internal standard for a specific compound is the compound itself labeled with a stable isotope such as ^2H, ^{18}O, ^{15}N or ^{13}C. If the difference in mass between the compound to be determined and the internal standard is only a few mass units, the two compounds have essentially identical properties when extracted and chromatographed. The ratio between the two molecular species will thus not be influenced during these steps.

From an instrumental point of view, it is preferable to use a similar amount of internal standard as the compound to be determined. This point has recently been discussed more in detail by Reiffsteck, Dehennin and Scholler (21). When assaying very small amounts, however, it may be preferable to use a great excess of internal standard. A large amount of internal standard may function as a carrier, and may minimize losses due to adsorption or thermal decomposition on the gaschromatographic column or in the ion source. A prerequisite for a carrier effect is that the two compounds have the same retention time in gas chromatography. If a capillary column is used, there may be a considerable separation between the isotope labeled internal standard and the compound to be determined, even if there is only a small difference in mass. Under such conditions, a packed column may sometimes be preferred.

Synthesis of a specific compound labeled with stable isotope may sometimes be very difficult. Useful ^{14}C- and ^{3}H-labeled compounds with high isotopic excess are sometimes commercially available. At least when working with ^{14}C-labeled material, relatively small amounts of radioactivity can be used in each analysis (13). When available, however, stable isotope labeled internal standard should always be preferred.

In order to obtain maximal accuracy in ID-MS, the method must be optimized with respect to sensitivity, specificity and precision.

SENSITIVITY OF ID-MS PROCEDURES

In the use of ID-MS as a reference methodology, sensitivity may be of less importance, since the amount of the analyte is seldom a limiting factor.

In order to obtain maximal sensitivity, there are certain requirements:

1. The number of ions detected simultaneously should be minimized.
2. An optimal derivative should be prepared, which gives ions of high intensity.
3. The optimal mode of ionization should be choosen. Chemical ionization may sometimes be better than electron impact.
4. Optimal focusing of the desired masses at the collector
5. Optimal opening of the slits on the mass spectrometer
6. Absence of interference
7. Absence of adsorption on the column or elsewhere.

SPECIFICITY OF ID-MS PROCEDURES

In order to obtain maximal specificity and to detect interference, the following points should be considered:

1. The purity of the material analyzed should be high.
2. The mass of the specific ions detected should preferably be high.
3. Several ions should be used.

4. The material should also be analyzed in the absence of internal standard.

The specificity of the selected ion monitoring technique is based on the fact that in general there is little risk for presence of a compound which has the same retention time in gas liquid chromatography and contains the same specific ion in its mass spectrum. The risk for interference decreases with increasing purity of the material. In the development of reference methodologies we try to use several different purification steps, and we are satisfied first after a demonstration that further purification does not change the final results of the assay. In general the risk for interferences decreases with increasing mass number of the fragment choosen for assay. With compounds as cortisol and cholic acid, which have relatively high molecular weights and from which it is relatively easy to obtain intense ions of a high molecular weight, it is possible to get accurate assays on simple unpurified extracts of serum (10,22).

The presence of interference in the analysis can often be detected by using several ions in the mass spectrometric analysis. It is advisable to run a sample without the internal standard in order to show that there is no interference in the channel of the instrument used for detection of the internal standard. Very often, contaminating compounds might interfere when using one specific ion, but not when using another. The ion which is most specific for the compound to be analyzed, may sometimes be of low intensity, and the use of such an ion might thus decrease sensitivity as well as precision. The relative need for specificity and sensitivity determines which ion should be used in each case.

Use of a high-resolution instrument may often increase the specificity considerably.

PRECISION OF ID-MS PROCEDURES

In order to obtain a high precision, the following points should be considered:

1. Relatively large amounts of analyte should be used.
2. A specific ion of high intensity should be used, which is different from ions due to bleeding from the column or from other sources.
3. The instrument should be as stable as possible. Long warm-up periods may be required.
4. The slits of the instrument should be optimally opened.

Most often, a major source of imprecision is the instability of the mass spectrometer. There might be great differences in stability between different instruments, and even between individual instruments of the same type. In our own studies, we have a coefficient of variation due to the instrument which varies between 0.5% and 1% in the case of an LKB 2091 instrument and between 1% and 1.5% in the case of an LKB 9000 instrument. Significantly higher degree of precision has been reported in some studies (6,8).

It may be pointed out that long warm-up periods with the mass spectrometer focused on the desired mass often decreases the drift of the magnetic current. In the early instruments, refocusing was necessary at repeated intervals. Now this problem can be overcome by different means and the focusing can be controlled by a computer.

In general the coefficient of variation in our ID-MS procedure varies between 0.5% and 2.5% (9-20). This degree of imprecision covers everything, also pipettings and weighings. The pipetting procedures in particular are of importance and if gravimetric procedures are used, the imprecision may decrease. An imprecision of 2.5% may appear to be rather high for a reference technique. In principle, however, if the figures represent a true imprecision, and no systematic errors are involved, a low degree of precision can always be compensated for by increasing the number of replicates. In connection with e.g. hormone analysis, when routine procedures may give values which deviate 20-50% or more from the true value (11,23), an imprecision of 2% is of little importance.

Accuracy of ID-MS procedures.

In order to obtain maximal accuracy, there are certain requirements:
1. Optimal specificity.
2. Relatively high degree of precision.
3. Pure reference material and reagents.
4. True equilibration between the internal standard and the compound to be assayed.
5. Optimal procedures for quantitation of internal standard and standard material. Gravimetric methods are better than pipettings.
6. Use of suitable standard curve or a "bracketing" procedure.

A most important point in connection with accuracy, is the equilibration between the internal standard and the analyte. In connection with e.g. assay of steroid hormones in serum, it must be assumed that the isotope labelled internal standard equilibrates with the hormone binding proteins, and if the yield in the extraction from serum is less than 100%, there will be an error.

In view of the fact that the analyte should behave identical with the ideal internal standard, the error due to incomplete equilibration is probably of a very small magnitude. The problem is that it is difficult to exactly quantitate this error. If the ambition is very high to get as close to the true value as possible, there might appear difficult theoretical problems.

A suitable standard curve is of outmost importance in connection with isotope dilution – mass spectrometry. Accurate use of a standard curve requires exact knowledge of its shape. Preferably it should be a straight line which can be calculated by linear regression analysis.

In the proposed definitive procedure for assay of serum cholesterol developed at National Bureau of Standards (8), a type of "bracketing" procedure was used. Thus each individual serum sample was compared with two standard samples, one containing slightly more unlabeled cholesterol than the analyte, and one slightly less. This should give optimal accuracy, but it is very laborious and time-consuming.

REPRODUCIBILITY OF ID-MS METHODS

It is accepted that methods based on isotope dilution – mass spectrometry in general are highly accurate, the methods should also be highly reproducible. Very few studies on this point have been performed as yet, however.

In one recent joint study with National Bureau of Standards (24), we compared our isotope dilution method for assay of serum cholesterol (9) with a similar method developed at National Bureau of Standards (8). The latter method has been proposed as a definitive method. The most important difference between the two modifications is that NBS are weighing instead of pipetting all their solutions. In our procedure, we had reasons to believe that the errors due to pipetting are of the magnitude of about 1%. Consequently, in a comparative study, the difference between our method and the NBS method should not exceed 1% if both methods lack other systematic errors.

In our first preliminary comparisons, we obtained consistently about 1.5% lower values than those obtained at NBS. A careful analysis of our standard cholesterol showed that it contained about 1.5% lathosterol. Under the specific chromatographic conditions used in our laboratory, lathosterol did not separate completely from cholesterol. Furthermore, lathosterol gave the same ion in mass spectrometry as cholesterol, only that the intensity was higher.

After repeating the analyses, using NBS-cholesterol as unlabeled standard and chromatographic conditions which separated lathosterol from cholesterol, the mean difference between the results obtained in our laboratory, and the results obtained at NBS was only 0.2%.

This degree of reproducibility is sufficient for most purposes. In addition to the above joint study, we have done some similar joint studies with other laboratories (23,25). In general, these joint studies have been successful and have shown that it is possible to obtain a reproducibility within a few %.

In one specific joint study, however, specificially aimed to study the importance of the instrumental factor, we obtained a difference between the two laboratories of about 4% (25). This difference was highly significant from a statistical point of view ($p < 0.001$, Student's t-test) and has remained unexplained. The difference might be due to the instrument, the chromatographic conditions or something else. Before an explanation is obtained, it is evident that it is difficult to claim that we have a definitive method.

USE OF ID-MS AS A REFERENCE METHODOLOGY IN CLINICAL CHEMISTRY

Table 1 summarizes different methods developed at our department. Several of these methods have been used more or less routinely in different quality control programs during the last years.

TABLE 1.
Reference methods based on ID-MS developed at the Department of Clinical Chemistry at Huddinge Hospital.

Compound	Internal standard	Reference
Bile acids	$^2H_5$-Bile acids	22
Cholesterol	$^2H_4$-Cholesterol	9
Cortisol	^{14}C-Cortisol	10
Creatinine	$^{15}N_2$-Creatinine	17
1,25-Dihydroxy-vitamin D_3	$^2H_3$-Dihydroxy-vitamin D_3	28
Estriol	$^2H_2$-Estriol	12
Glucose	$^2H_7$-Glucose	15
25-Hydroxy-vitamin D_3	$^2H_3$-25-Hydroxy-vitamin D_3	27
Progesterone	^{14}C-Progesterone	11
Testosterone	^{14}C-Testosterone	13
Triglycerides	$^2H_4$-Triglycerides	14
Urea	$^{15}N_2$-Urea	16
Uric acid	$^{15}N_2$-Uric acid	18

These methods have proved to be most useful tools to select the most accurate routine method available for each given compound of clinical chemical interest.

In particular, this is of importance in connection with radioimmunoassays. There is a great number of different kits on the market for determination of each specific compound. Each kit has its own characteristics, and the antibodies used all have their own pattern with respect to cross-reactions. Accuracy is the most difficult and therefore also the most neglected aspect of quantity control in this connection.

In several cases, we have shown that commercial radioimmunoassay kits may give values which differ by 50% or more from the values obtained by ID-MS (11,23). We have recently shown that a major part of this deviation may be due to defects in the calibration procedure. Thus calibration standards are often used in the kits which have a changed biological matrix. By using human sera in the calibration, which have been analyzed by isotope dilution – mass spectrometry, it is possible to get much more accurate results (23).

Each year we prepare one or two great batches of human serum. These batches are very carefully analyzed by ID-MS with respect to such compounds as cholesterol, creatinine, uric acid, urea, cortisol and glucose (19,20). The analyzed serum pool is then used in our internal quality control. In order to define the "accuracy pattern" with respect to clinical chemical analyses, such a serum was once distributed to all clinical chemical laboratories in Sweden (n = 62), and was analyzed with the different routine methods (cf. ref. 20). In Fig. 1, the results are summarized. It is seen that the "state-of-art" values for cholesterol, glucose, urea and uric acid are very similar to the values obtained by ID-MS. There was a most significant difference, however, between the state of art values for creatinine and cortisol and the values obtained by ID-MS. As discussed above, it is possible to increase the accuracy of radioimmunoassay of cortisol by changing the calibration procedure. Recently, we developed a new method for determination of creatinine in serum,

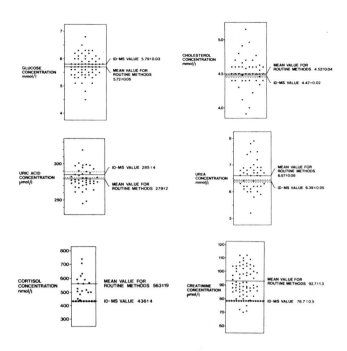

Fig. 1. Comparison between values obtained by different routine laboratories and values obtained by isotope dilution - mass spectrometry in the analysis of a human serum pool (cf. ref. 20). Each point represents the mean of 16 replicate measurements in one specific laboratory. The values given correspond to means ± S.E.M.

which gives values very close to those obtained by ID-MS (26). Thus it should be possible to increase the accuracy of routine determinations of cortisol and creatinine in the future.

To summarize, in spite of some limitations, which must be further investigated, ID-MS is a most promising methodology in connection with quality control in clinical chemistry.

REFERENCES

1 J.P. Cali, Pure & Appl. Chemistry, 45(1976)63-68
2 I. Björkhem, R. Blomstrand, O. Lantto, L. Svensson and G. Öhman, Clin. Chem., 22(1976)1789-1801.
3 L. Dehennin, A. Reiffsteck and R.A. Scholler, J. Steroid. Biochem., 5(1974)81-86.
4 H. Breuer and Siekmann, J. Steroid Biochem., 6(1975)685-688.
5 P. Vestergaard, J.F. Sayegh and J.H. Mowat, Clin. Chim. Acta, 62(1975)163-168.
6 L. Eldjarn, P. Helland, S. Skrede, B. Christophersen and E. Jellum, Proceedings of the 10th Int. Congr. in Clinical Chemistry, Mexico 1978 p. 125 (abstract).
7 P. Gambert, C. Lallemant, A. Archambault, B.F. Maume and P. Padieu, J. Chromatogr., 162 (1979) 1-6.
8 A. Cohen, H.S. Hertz, J. Mandel, R.C. Paule, R. Schaffer, L.T. Sniegoski, T. Sun, M.J. Welch and E.V. White, Clin. Chem., 26(1980)854-860.
9 I. Björkhem, R. Blomstrand and L. Svensson, Clin. Chim. Acta, 54(1974)185-193.
10 I. Björkhem, R. Blomstrand, O. Lantto, A. Löf and L. Svensson, Clin. Chim. Acta, 56(1974)241-248.
11 I. Björkhem, R. Blomstrand and O. Lantto, Clin. Chim. Acta, 65(1975)343-350.
12 I. Björkhem, R. Blomstrand, L. Svensson, F. Tietz and K. Carlström, Clin. Chim. Acta, 62(1975)385-392.
13 I. Björkhem, O. Lantto and L. Svensson, Clin. Chim. Acta, 60(1975)59-66.
14 I. Björkhem, R. Blomstrand and L. Svensson, Clin. Chim. Acta, 71(1976)191-198.
15 I. Björkhem, R. Blomstrand, O. Falk and G. Öhman, Clin. Chim. Acta, 72(1976)353-362.
16 I. Björkhem, R. Blomstrand and G. Öhman, Clin. Chim. Acta, 71(1976)199-205.
17 I. Björkhem, R. Blomstrand and G. Öhman, Clin. Chem., 23(1977)2114-2121.
18 G. Öhman, Clin. Chim. Acta, 95(1979)219-226.
19 I. Björkhem, R. Blomstrand, S. Eriksson, O. Falk, A. Kallner, L. Svensson and G. Öhman, Scand. J. Clin. Lab. Invest., 40(1980)529-534.
20 I. Björkhem, A. Bergman, O. Falk, A. Kallner, O. Lantto, L. Svensson, E. Åkerlöf and R. Blomstrand, Clin. Chem. - in press 1981.
21 A. Reiffsteck, L. Dehennin and R. Scholler, Proceedings of the Fourth International Conference on Stable Isotopes March 23-27, 1981 (abstract).
22 B. Angelin, I. Björkhem and K. Einarsson, J. Lipid Res., 19(1978)527-537
23 O. Lantto, I. Björkhem, R. Blomstrand and A. Kallner, Clin. Chem., 26(1980)1899-1901.
24 R. Schaffer, L.T. Sniegoski, M.J. Welch, E. White, V.A. Cohen, H.S. Hertz, J. Mandel, R.C. Paule, L. Svensson, I. Björkhem and R. Blomstrand, Submitted to Clin. Chem.
25 I. Björkhem, L. Svensson, H. Adlercreutz and S. Skrede, Scand. J. Clin. Lab. Invest., 39, Suppl. 152, (1979) p 22 (abstract).
26 P. Masson, P.Ohlsson and I. Björkhem, Clin. Chem., 27(1981)18-21.
27 I. Björkhem and I. Holmberg, Clin. Chim. Acta, 68(1976)215-221.
28 I. Björkhem, I. Holmberg, T. Kristiansen and J.I. Pedersen, Clin. Chem., 25(1979)584-588.

GC/MS ASSAY OF MYO-INOSITOL (AS HEXAACETATE) IN HUMANS UTILIZING A DEUTERATED INTERNAL STANDARD

E. LARSEN*, J.R. ANDERSEN*, H. HARBO**, B. BERTELSEN**, J.E.J. CHRISTENSEN**, and G. GREGERSEN**

* Chemistry Department, Risø National Laboratory, DK-4000 Roskilde, Denmark
**Central Hospital, DK-6700 Esbjerg, Denmark

ABSTRACT

The isotopic dilution technique was used for determining the content of myo-inositol in human urine, plasma, and hemolysed erythrocyte samples. A hexadeuterated myo-inositol was added as internal standard to the samples at an early stage in the analytical procedure. After separation and derivatisation to the hexaacetate, a minicomputer-controlled gas chromatograph-quadrupole mass spectrometer was employed for the analysis using a capillary column and selected ion monitoring at masses m/z 210 and m/z 214. Calibration curves on water, urine, plasma, and hemolysed erythrocytes show parallel, linear responses in the ratio between analyte and internal standard in the area of interest (0.2-2.0).

INTRODUCTION

Myo-inositol, a member of the vitamin B complex and a natural ingredient in human food, has been associated with diabetes mellitus for some time. Patients with uncontrolled diabetes excrete large amounts of myo-inositol in the urine [1-3], and even with insulin treatment the diabetic has higher levels in the urine than the non-diabetic [4]. Furthermore, studies on diabetic rats suggested a relationship between dietary myo-inositol intake and improved motor-nerve conduction velocities [5]. Myo-inositol is an important component of cell membrane lipids and also seems to take part in the normal hepatic metabolism of triglycerides. It is a necessary constituent of all cell culture media.

So far, myo-inositol analysis has been carried out mostly by means of gas chromatography (GC) on derivatives of the parent molecule [2,6]. To our knowledge, only one method utilizing gas chromatography/mass spectrometry (GS/MS) with an isotopically labelled internal standard has been disclosed in the literature [7]. In the present work, a detailed procedure for the quantitative analysis of myo-inositol in human urine, plasma, and hemolysed erythrocytes is given.

EXPERIMENTAL

The hexadeuterated internal standard was synthesized from inosose-2 by a base-catalysed exchange reaction followed by reduction with D_2. The analyses were performed on a Hewlett-Packard GC/MS System HP 5992B. The chromatographic column was a flexible HP fused silica capillary column, 25 m × 0.20 mm, coated with methyl silicone. Calibration of the method was obtained by adding 100 µl of an aqueous solution of labelled myo-inositol (conc. 0.33 mg/ml) together with known amounts of pristine myo-inositol to 500 µl samples of water, urine, plasma, and hemolysed erythrocytes. After separation and derivatisation to the hexaacetate, the myo-inositol content was assayed by measuring the ratio between the ions m/z 210 and m/z 214. Further experimental details will be published elsewhere [8].

RESULTS AND DISCUSSION

The abundant and characteristic fragment ions at m/z 210 and m/z 214 [9] are suited to quantitative measurements by SIM, as no other polyolacetates have fragments at these m/z values [10]. Furthermore, the preceding GC separation should warrant that only very small amounts of unwanted material elute together with the myo-inositol hexaacetates - an expectation found to be valid by experiment. A representative GC/MS analysis is shown in Fig. 1. The sample size is 10-20 ng. On reducing the sample size to below 1 ng a remarkably high memory effect, and low reproducibility, were seen. The small peaks with retention times different from that of myo-inositolhexaacetate are caused by isomeric material as confirmed by their complete mass spectra. Their parent hexoles are probably formed by isomeriza-

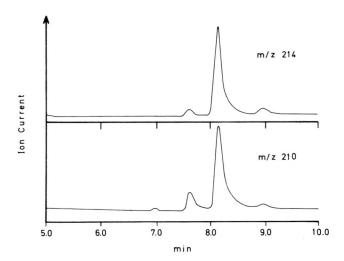

Fig. 1. GC/MS analysis of a plasma sample. Myo-inositol appears after 8.2 min.

tion during the base-catalysed exchange reaction used in the preparation of the deuterated standard.

The validity and applicability of the method was evaluated dy determining calibration curves. Double independent experiments of five different ratios of added myo-inositol and deuterated standard were carried out on samples of water, urine, plasma, and hemolysed erythrocytes. The results are presented in Fig. 2. The correlation coefficients are high, especially for water and urine samples. The slopes of the calibration curves for water, urine, and plasma are in relatively good agreement; a content of naturally occurring myo-inositol besides the added amount is responsible for the different intercepts observed, except in the case of water where a value of 0.023, - found by injecting pure standard, - instead of 0.068 was expected. The reason for this discrepancy is not clear. Memory effects were diminished in all cases by using 10 ng amounts of inositol (myo-inositol + deuterated standard) for the GC/MS analysis. High recoveries were found for water and urine, whereas only a few per cent of the original amounts were detected in plasma samples. Erythrocytes sometimes gave even lower yields.

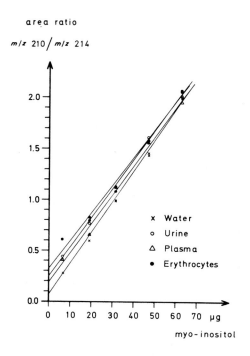

Fig. 2. Calibration curves obtained on water, urine, plasma, and hemolysed erythrocytes. The area ratios between m/z 210 and m/z 214 are plotted versus added myo-inositol to 500 µl samples. The same amount of deuterated standard (33 µg) was added in all cases. The correlation coefficients are r = 0.9987 (water), r = 0.9996 (urine), r = 0.9963 (plasma), and r = 0.9942 (erythrocyte).

There are some drawbacks to the method, two of which are obvious. The first is the low yield of myo-inositolhexaacetate from samples with a high content of dry matter, i.e. plasma and erythrocyte samples. A recovery of a few per cent as with plasma samples is annoying, but acceptable. Erythrocyte samples, on the other hand, represent a more severe problem. In some cases a reasonable recovery of myo-inositol (and internal standard) was obtained, and a GC/MS spectrum as the one in Fig. 1 was the result; in others, spectra of insufficient quality were recorded. As a consequence of this, the analyses from 192 erythrocyte samples gave merely an indication of the level of myo-inositol in the medium, and only in a few cases could the variation of the levels following dietary intake be estimated [11]. The reason for the apparently unsystematic low recoveries in some samples is unknown at present, but it is presumably due to the evaporation procedures performed on a block thermostat at 100°C [8]. Charring of the samples coinciding with loss of inositol were sometimes observed.

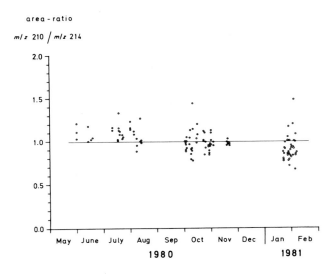

Fig. 3. Variation in GC/MS analysis of a standard sample with a nominal area ratio between m/z 210 and m/z 214 of 1.0 versus time.

The second disadvantage lies in the mass spectrometer, which is a computerized quadrupole instrument in which the ion source, filter, and electron multiplier are located in the center core of the oil diffusion pump. The expectations regarding reproducibility have not been met to date. From time to time during a nine month period myo-inositol analyses were made, and every day a standard sample with an area ratio of 1.0 between m/z 210 and m/z 214 was run at least twice, in the morning and evening. The 125 measurements shown in Fig. 3 have a mean value of 0.994 and a standard deviation of 15%. The point deviating more than 50% cannot be explained. With freshly cleansed ion source, filter, and multiplier, on the other hand, it was possible - in the same period as the four calibration curves were recorded (Fig. 3, Nov.) - to run the same standard sample as above 11 times, one after the other, with a standard deviation of 0.6%, and within a week with a standard deviation of 2.2%.

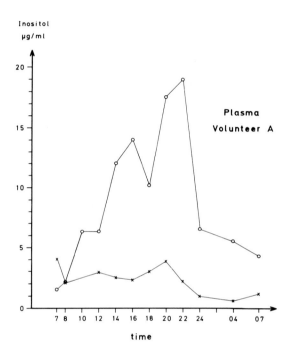

Fig. 4. Content of myo-inositol in plasma from a nondiabetic person. The curves show the variation as a function of time of day. The difference between each is that in the case of the upper curve 2 g of myo-inositol was taken orally in connection with each meal at 7, 12, and 18 hrs.

In conclusion, on the basis of 32 human urine and 192 plasma samples it is safe to claim that the isotopic dilution technique using hexadeuterated myo-inositol as internal standard is adequate for routine analyses on these substrates. The method was also used successfully to assay the content of myo-inositol in the lipophilic as well as in the hydrophilic phase of muscle biopsies [11].

A fast and rather simple GC/MS instrument with a relatively low reproducibility is sufficient for these purposes, as demonstrated in Fig. 4, where the variation of myo-inositol levels in plasma from a healthy volunteer with and without intake of myo-inositol is displayed as a function of time.

REFERENCES

1 W.H. Daughaday and J. Larner, J. Clin. Invest., 33(1954)326-332.
2 E. Pitkänen, Clin. Chim. Acta, 38(1972)221-230.
3 J.F. Aloia, J. Lab. Clin. Med., 82(1973)809-817.
4 R.S. Clements, Jr. and R. Reynertson, Diabetes, 26(1977)215-221.
5 D.A. Greene, P.V. De Jesus and A.I. Winegrad, J. Clin. Invest., 55(1975)1326-1336.
6 R.S. Clements, Jr. and W.R. Starnes, Biochem. Med., 12(1975)200-204.
7 W.R. Sherman, P.M. Packman, M.H. Laird and R.L. Boshans, Anal. Biochem., 78(1977)119-131.
8 J.R. Andersen et al., to be published.
9 B.R. Pettit, G.S. King and K. Blau, Biomed. Mass Spectrom., 7(1980)309-313.
10 S.R. Heller and G.W.A. Milne, EPA/NIH Mass spectral data base, NSRDS-NBS 63, Vol. 2 (1978).
11 G. Gregersen et al., to be published.

ANALYSIS OF I-METHYL-I,2,3,4-TETRAHYDRO-β-CARBOLINE BY GC/MS USING DEUTERIUM LABELLED INTERNAL STANDARDS

JAMES R.F. ALLEN[*] and OLOF BECK
Department of Toxicology, Karolinska Institute, S-I04 OI, Stockholm, Sweden.

ABSTRACT

Analysis of I-methyl-I,2,3,4-tetrahydro-β-carboline (Me-THBC) by GC/MS using ($^2H_8$)-Me-THBC and ($^2H_4$)-tryptamine as internal standards enables simultaneous quantification of Me-THBC and a check against its formation as an artefact during the preparation of samples for analysis. Levels of 0.5-2.I pmoles per ml in the urine of alcoholics and healthy controls and 5-450 pmole per ml in alcoholic beverages were determined by this method. The possible sources and significance of the urinary Me-THBC are discussed.

INTRODUCTION

The β-carbolines are a diverse group of indole derivatives possessing a tricyclic pyrido (3-4,b)indole ring structure, which may be sub-divided into 3 groups on the basis of the hydrogenation state of the pyridyl ring. These compounds occur naturally as alkaloids in a range of plant species but have also been isolated from other sources such as alcoholic beverages, foodstuffs and tobacco smoke (I).

Organic synthesis of the β-carbolines has been extensively studied and it has been shown that tetrahydro-β-carbolines (THBC's) can form readily by a spontaneous condensation of aldehydes with indole amines (2). This observation has stimulated interest into investigating the possibility that they arise by a similar mechanism in animal tissues. Interest in the occurrence of THBC's in animals centres around examination of their effects and involvement in serotonergic neurotransmitter function (see 3).

Evidence for the existence of THBC's in rat brain and human urine

---

[*]Present address: Institut de biologie et de physiologie végétales, Université de Lausanne, Place de la Riponne, I005 Lausanne, Suisse.

Abbreviations: THBC=tetrahydro-β-carboline; TNH_2=tryptamine; Trp=tryptophan; CSF=cerebrospinal fluid; PFPA=pentafluoropropionic anhydride.

has been reported (3) though the biosynthetic pathway by which these compounds arise is not yet fully clear. Particular interest has been expressed in the possibility that they are formed in the body by the condensation of acetaldehyde derived from ethanol metabolism, with endogenous indole amines to produce I-methyl-tetrahydro-β-carbolines (Me-THBC) (4).

Me-THBC

Increased levels of Me-THBC following acute ethanol consumption by healthy human subjects have recently been reported to occur in urine (5) and blood (6). In contrast to these reports however, no significant increase in urinary Me-THBC levels in healthy subjects, following acute ethanol consumption was observed in a study in this laboratory (7).

The multiplicity of naturally-occurring β-carbolines and the low levels of THBC's in animal tissues demand that specific and sensitive methods be used for their analysis. The application of GC/MS enables stable isotope labelled internal standards to be used for the quantification of the endogenous compound by determination of their isotope ratios. A method in which the ($^2H_8$)-labelled internal standard, for quantification of the endogenous compound, is used in conjunction with a ($^2H_4$)-labelled precursor to check for formation during sample preparation, is described and results obtained from analyses of urine and alcoholic beverages are presented. The following account partly summarises data which are reported in more detail elsewhere (7;8).

METHODS

^2H-labelled internal standards

($\alpha,\alpha,\beta,\beta-^2H_4$)-TNH$_2$ was synthesised from indole by the method of Shaw et al.(9) ; (3,3,4,4-$^2H_4$)-Me-THBC was prepared from ($^2H_4$)-TNH$_2$ by the method of Akabori and Saito (10), and (3,3,4,4,5,6,7,8-$^2H_8$)-Me-THBC was prepared by acid catalysed exchange in (^2H)-HCl (7).

Quantification by GC/MS

To samples of urine, CSF and a range of alcoholic beverages were added 49.I pmole ($^2H_8$)-Me-THBC, 508 pmole ($^2H_4$)-TNH$_2$ and 0.4-4.0 g semicarbazide. Following the addition of 0.4 g NaCl and 0.I5 mmole HCOOH, the samples were partitioned against toluene/isoamyl alcohol and the organic phase was discarded. The aqueous phase was made alkaline by the addition of I mmole NaOH and re-partitioned against toluene/isoamyl alcohol. The organic phase was blown to dryness under N_2 and derivatised with PFPA.

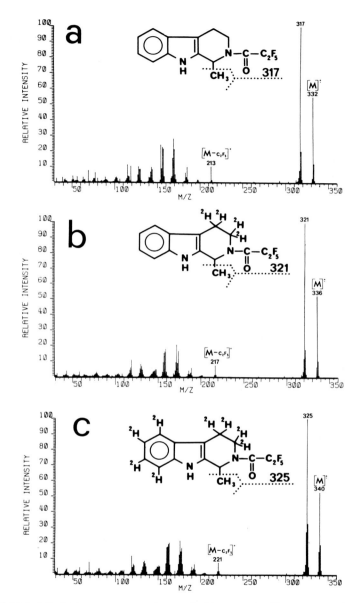

Fig I. Electron impact mass spectra of the pentafluoropropionyl derivative of a) Me-THBC, b) ($^2H_4$)-Me-THBC and c) ($^2H_8$)-Me-THBC.

The GC/MS analyses using selected-ion-monitoring were made on an LKB 209I with 20m x 0.25 mm SE52 WCOT capillary columns. Quantification of the endogenous compound was made by reference to calibration plots derived from standards prepared by the same procedure as extracts. A linear relationship between the $^2H_0/^2H_8$ ratios obtained from the $M^{+\bullet}$ or $(M-15)^+$

ions (Fig I.) was observed in standards containing 0-140 pmole Me-THBC per sample (7;8).

RESULTS AND DISCUSSION

The formation of ($^2H_4$)-Me-THBC as an artefact during sample preparation was routinely checked by monitoring the corresponding $^2H_4$ ion channels during the analyses. No evidence for its formation during

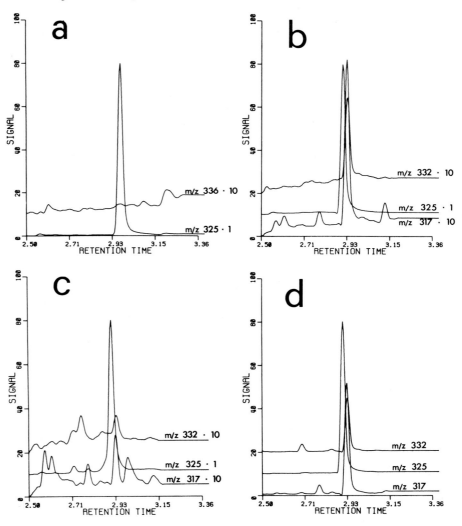

Fig 2. Chromatograms obtained from SIM analysis of a) Urine from intoxicated alcoholic showing absence of ($^2H_4$)-Me-THBC formation b) Presence of endogenous Me-THBC in the same urine sample as (a) c) Urine from a healthy volunteer showing endogenous Me-THBC d) Sample of Swedish beer.

sample storage or work-up using the method described was found. When semicarbazide, a carbonyl trapping agent, was omitted, formation of $(^2H_4)$-Me-THBC from $(^2H_4)$-TNH$_2$ was noted in some beverage samples, presumably due to condensation with aldehydes present in the samples (11).

The identification of endogenous Me-THBC in the urine and beverage samples was based upon the elution of a compound at a similar retention time and with a similar relative intensity of $M^{+\bullet}$ and $(M-15)^+$ ions as the $(^2H_8)$-Me-THBC internal standard (Fig.2:b,c,d). Endogenous Me-THBC was present in the urine of healthy volunteers and chronic alcoholics. The levels in the urine were low, corresponding to an excretion rate of circa 2 nmoles /24 hr, but were significantly higher in the urine of the alcoholics during intoxication than after recovery or in comparison with healthy controls (7). No Me-THBC could be detected in lumbar CSF samples from alcoholics in a state of intoxication (<0.3 pmole per ml CSF; n=10).

TABLE I. Endogenous MeTHBC content in the urine of healthy male subjects and male alcoholics during intoxication and following abstention from alcohol in hospital for one week.

	N	Me-THBC conc.[a] pmole /ml	Me-THBC level/creatinine[a] μmole /mole creatinine
Healthy controls	17	1.6 ± 0.2	0.125 ± 0.025
Alcoholics (day 1)	9	2.1 ± 0.7	0.40 ± 0.15
(day 8)	9	0.5 ± 0.4	0.076 ± 0.013

[a] Values represent mean ± SE.

Relatively high levels of Me-THBC were also found to be present in a range of alcoholic beverages, with values ranging from 9-69 pmole per ml in the beers tested and 5-450 pmole per ml in the wines (8), though was not detectable in distillates. The most likely origin of the Me-THBC in the beverages is from spontaneous condensation, during fermentation and storage, of Trp and TNH$_2$ derived from the ingredients with aldehydes formed during fermentation. Thus the variation in Me-THBC content may reflect the levels of potential precursors and the conditions of treatment and storage. The absence of Me-THBC in the spirits examined is presumably a consequence of their purification by distillation.

A preliminary experiment showed that a high proportion of Me-THBC can be excreted in the urine, unmetabolised, after the consumption of a rich source such as wine (Allen, unpublished results). However, the main source of alcohol for the alcoholic patients in the present study was distilled spirits in which the Me-THBC levels are low and no evidence of a correlation of urinary Me-THBC levels in the alcoholics with the previous intake of wine and beer was apparent. In addition to the alcoholic beverages shown to contain high Me-THBC levels, the possibility also exists of a urinary derivation from other dietary origins, particularly well cooked foodstuffs and tobacco and marijuana smoke in which β-carbolines occur as a result of their formation from Trp by pyrolysis.

REFERENCES

1. J.R.F. Allen and B.R. Holmstedt, Phytochemistry, 19 (1980) 1573-1582.
2. R.A. Abramovitch and I.D. Spenser, in A.R. Katritzky, A.J. Boulton and J.M. Lagowski (Eds.), Advances in Heterocyclic Chemistry, Vol.3, Academic Press, New York, 1964, pp. 69-207.
3. N.S. Buckholtz, Life Sci., 27 (1980) 893-903.
4. W.M. McIsaac, Biochim. Biophys. Acta, 52 (1961) 607-609.
5. H. Rommelspacher, S. Strauss and J. Lindemann, FEBS Lett., 109 (2) (1980) 209-212.
6. P. Peura, I. Kari and M.M. Airaksinen, Biomed. Mass Spectrom., 7 (1980) 553-555.
7. J.R.F. Allen, O. Beck, S. Borg and R. Skröder, European Journ. Mass Spectrom., 1 (1980) 171-177.
8. O. Beck and B. Holmstedt, Fd. Cosmet. Toxicol. (1981) in press.
9. G.J. Shaw, G.J. Wright and G.W.A. Milne, Biomed. Mass Spectrom., 4 (1977) 348-353.
10. S. Akabori and J. Saito, Ber. 63 (1930) 2245.
11. R.B. Holman, G.R. Elliott, K. Faull and J.D. Barchas, in M. Sandler (Ed.), Psychopharmacology of Alcohol, Raven Press, New York, 1980, pp. 155-169.

THE CARRIER EFFECT IN CAPILLARY GAS CHROMATOGRAPHY-MASS SPECTROMETRY AND ITS IMPLICATION ON THE ACCURATE MEASUREMENT OF ION ABUNDANCE RATIOS

L. DEHENNIN, A. REIFFSTECK and R. SCHOLLER

Fondation de Recherche en Hormonologie, 67/77 bd Pasteur, 94260 Fresnes and 26 bd Brune, 75014 Paris (France).

ABSTRACT

Generally attributed to adsorption-reducing agents, the carrier effect is effective even at mole ratios of analyte to carrier with values between 1 and 10. The reversibility of the carrier effect is demonstrated, thus indicating that a labelled compound can act as a carrier for the unlabelled analogue and vice versa. The sample size does not influence significantly the ion abundance ratio of an analyte versus its labelled analogue. The determination by GCMS of isotopic composition of deuterium labelled compounds is biased by a systematic error introduced by the carrier effect.

INTRODUCTION

In gas chromatography-mass spectrometry the so-called "carrier effect" is generally attributed to adsorption-reducing agents. Its existence or effectiveness is still controversial [1]. Some authors [2-5] have advocated the use of deuterium labelled analogues as internal standards, according to the isotope dilution principe, and as carriers to improve the detection limit. This was however challenged by Millard et al. [6]. Other authors have drawn attention to active sites in the mass spectrometer source which can create catalytic decomposition with variation in the mass spectral fragmentation [7] and strong adsorption leading to memory effects [8].

We have presented data [9] indicating that the carrier effect in GCMS is a combined effect whose origin is located in the GC column as well as in the post-column entity (interface, ion source, mass filter and detector).

Abbrevations used : GCMS = gas chromatography-mass spectrometry, SID = selected ion detector, SIM = selected ion monitoring, TMS = trimethylsilylether, MO = O-methyloxime, E_1 = estrone, E_2-17β = estradiol-17β, E_2-17α = estradiol-17α, $E_2 - {}^{14}C$ = [4 - ${}^{14}C$] estradiol-17β , E_1-d_4 = [2,4,16,16-${}^2H_4$] estrone, E_2-d_4 = [2,4,16,16-${}^2H_4$] estradiol-17β.

We intend to report here on further characteristics of the carrier effect and on the accuracy of ratios of ion abundance measurements.

EXPERIMENTAL

GCMS Data System

GCMS with SIM was performed on a quadrupole mass spectrometer (R 10.10, Nermag, France) equipped with a PDP 8-based data system (Digital Equipment, USA). A glass capillary column (25m x 0.25mm), deactivated by persilylation and coated with OV-73 stationary phase had 82,000 theorectical plates (measured on the E_2-17β-TMS peak with $k' = 5.2$). The column was directly coupled to the mass spectrometer source which was operated in the electron impact mode. Injection was performed by a moving glass needle system. The carrier gas was helium (0.8 bar inlet pressure), the column temperature was settled at 230°C, the interface was maintained at 240°C and the ion source at 170°C. The source conditions were: ionizing electron voltage, 70 eV; filament current, 6 V; ion energy voltage, 4-6 V. The isotopic compositions were measured by SIM with adjustment of the integration time on each selected ion channel. This allowed similar numbers of ions to impinge on the detector with comparable ion statistical errors. Computer integration of (Σ ion current voltage x $\frac{\text{number of samplings}}{\text{dwell time}}$) gave peak areas in analogue-to-digital converter units. Scans corresponding to the beginning and the end of the GC peak were determined visually on the display unit.

The absolute detector responses for labelled and unlabelled compounds increased linearly over the range of 100 pg to 20ng.

Deuterium labelled steroids

$[2,4,16,16-^2H_4]$ Estrone and $[2,4,16,16-^2H_4]$ estradiol-17β were synthesized according to labelling techniques published previously together with their corresponding deuterium contents [10].

RESULTS AND DISCUSSION

The carrier effect exerted in the GCMS system and assigned to adsorption-reducing carriers, prevailently stable isotope labelled analogues, has been demonstrated [9] at medium and high mole ratios of carrier versus analyte. Figure 1 illustrates the same effect in the range of lower mole ratios, where E_2-17β-TMS and E_2-d_4-TMS have been used successively as analyte and as carrier.

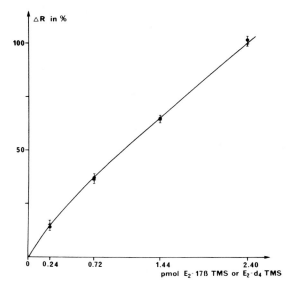

Fig. 1. Reversibility of the carrier effect exerted by E_2-d_4 on E_2-17β (●————●) and by E_2-17β on E_2-d_4 (x ———— x) as their TMS derivatives. ΔR represents the increase in percent of the SID response for 0.24 pmol (= 100 pg) of either the d_0 or the d_4 compound relative to an isomeric internal standard (E_2-17α), at m/z 416 and 420, respectively.

Thus, it should be emphasized that E_2-17β-TMS acts as a carrier for E_2-d_4-TMS in the same way as E_2-d_4-TMS acts as a carrier for E_2-17β-TMS. This suggests that the adsorption sites for E_2-17β-TMS and its deuterium labelled analogue are very similar, if not identical. The contrary would have been unexpected when is taken into account the very close similarity between the chemical structures of E_2-17β and E_2-d_4, the latter being slightly more polar by the electropositive character of deuterium [11]. The statement made by Millard [12] on the selectiveness of adsorption sites for a compound and its deuterated analogue does not seem to be general. The higher SID responses due to the carrier effect are not limited to the molecular ion, fragment ions are similarly increased.

Sample size is without any significant influence on the ratio of ion abundance measurements of equal amounts of analyte (E_2-17β-TMS) and labelled internal standards, E_2-d_4-TMS or E_2-^{14}C-TMS (table 1).

Conversely, the ion abundance ratio of equal amounts of analyte (E_2-17β-TMS) and isomeric internal standard (E_2-17α-TMS) is highly dependent on the sample size. The increase of the ion ratio 416 (17β)/416(17α) with sample size can be explained as a more efficient deactivation of the GCMS system by the higher amount (10 ng) of the compound with the lowest retention time (E_2-17α-TMS).The increase of the ion ratio 420/416(17α) with sample size may be similarly explained.

The results point out that the isotopic internal standards should be preferred to isomeric internal standards for accurate quantitative determinations by capillary GCMS, even though more precise measurements have been claimed by Lee and Millard with SIM [13].

TABLE 1

Influence of sample size on ion abundance ratios.

Sample composition	Sample size (each component)	Ion abundance ratio		
		$\dfrac{416\ (17\beta)}{420}$	$\dfrac{416\ (17\beta)}{416\ (17\alpha)}$	$\dfrac{420}{416\ (17\alpha)}$
E_2-17β-TMS E_2-d_4-TMS E_2-17α-TMS	1 ng 10 ng	1.17 ± 0.02 1.18 ± 0.01	1.16 ± 0.02 1.51 ± 0.01	0.99 ± 0.02* 1.28 ± 0.01
		$\dfrac{416\ (17\beta)}{418}$	$\dfrac{416\ (17\beta)}{416\ (17\alpha)}$	$\dfrac{418}{416\ (17\alpha)}$
E_2-17β-TMS E_2-^{14}C-TMS E_2-17α-TMS	1 ng 10 ng	1.05 ± 0.02 1.05 ± 0.01	1.39 ± 0.02 1.70 ± 0.02	1.33 ± 0.02 1.62 ± 0.01

*Mean ± standard deviation ; n = 3

Another consequence of the carrier effect is inaccuracy in the case where the isotopic composition of highly enriched deuterium labelled compounds has to be determined by GCMS. This source of systematic error was evaluated by the following experimental data.

First was determined the ion ratio 420/416 of E_2-17β-TMS (5ng) and subsequently that of E_2-17β-TMS (5ng) together with E_1-MOTMS (10ng), the latter compound (M^+ = 371) being highly purified and thus exerting no contribution at all at m/z 416 and 420. All retention times involved here were identical (314±1sec). Secondly was measured the ion ratio 416/420 of E_2-d_4-TMS (5ng) and that of E_2-d_4-TMS (5ng) in presence of E_1-MOTMS (10ng). It must be emphasized that the retention time (310±1sec) of E_2-d_4-TMS (M^+ = 420) was significantly shorter than that (314±1sec) of the corresponding d_0 analogue (M^+ = 416).

The results of these ion ratio measurements are summarized in table 2.

TABLE 2

Inaccuracy associated with the measurement of isotopic compositions by capillary GCMS.

	Ion ratio 420/416 in %
E_2-17β-TMS (5ng)	0.727 ± 0.031*
E_2-17β-TMS (5ng) + E_1-MOTMS (10ng)	0.708 ± 0.029
	Ion ratio 416/420 in %
E_2-d_4-TMS (5ng)	0.582 ± 0.019
E_2-d_4-TMS (5ng) + E_1-MOTMS (10ng)	0.722 ± 0.029

* Mean ± standard deviation ; triplicate determinations.

The data from the upper part of table 2 demonstrate that there is no significant difference between the ion ratio 420/416 determined on either E_2-17β-TMS or on the mixture of E_2-17β-TMS and a co-eluting compound (E_1-MOTMS). This can be explained by the fact that the ions at m/z 420 are due predominantly to naturally labelled (^{13}C, ^{29}Si, ^{30}Si) molecules which have a retention time (314±1sec) identical to the parent unlabelled molecules at m/z 416. In this case, the carrier effect exerted by the bulk (∼5ng) of the unlabelled molecules on the small amount (∼36pg) of naturally labelled (^{13}C, ^{29}Si, ^{30}Si) molecules has already reached a maximum value, which cannot be further increased by the addition of supplementary carrier molecules (E_1-MOTMS).

The data from the lower part of table 2 indicate a significant increase (+24%) of the ion ratio 416/420 when the value obtained with E_2-d_4-TMS and with the mixture of E_2-d_4-TMS plus E_1-MOTMS are compared. The explanation for this finding can be given by the fact that the retention time (310±1 sec) of the d_4 molecules is significantly shorter than that (314±1 sec) of the corresponding d_0 molecules. Thus the carrier effect, exerted by the large excess of d_4 molecules (∼5ng) on the small amount (∼30pg) of d_0 molecules, has not yet reached the maximum value. Addition of supplementary carrier molecules (E_1-MOTMS) further increases the ion ratio 416/420.

CONCLUSION

Carriers can increase considerably the SID response when subnanogram amounts of analyte are injected into the GCMS system. The importance of the carrier effect depends on the number of active sites in the GCMS system and on the excess of carrier versus analyte. Accurate quantitative determinations can be performed by isotope dilution-mass spectrometry when the mole ratios of analyte to labelled internal standard are close to 1.

The isotopic composition of compounds naturally labelled with isotopes (^{13}C, ^{29}Si, ^{30}Si), which do not alter the retention time of the parent unlabelled compound, can be determined accurately by GCMS with SIM.

A systematic error is involved in the determination of the isotopic composition of highly enriched deuterium labelled compounds, at least when they are measured by capillary GCMS. This error leads to an underestimation of the less abundant molecule species, particularly of the d_0 molecules.

REFERENCES

1 R. Self, Biomed. Mass Spectrom., 6 (1979) 315-316.
2 B. Samuelson, M. Hamberg and C.C. Sweeley, Anal. Biochem., 38 (1970) 301-304.
3 T.E. Gaffney, C.G. Hammar, B. Holmstedt and R.E. McMahon, Anal. Chem., 43 (1971) 307-310.
4 B.A. Petersen and P. Vouros, Anal. Chem., 49 (1977) 1304-1311.
5 N.J. Haskins, G.C. Ford, S.J.W. Grigson and K.A. Waddell, Biomed. Mass Spectrom., 5 (1978) 423-424.
6 B.J. Millard, P.A. Tippett, M.W. Couch and C.M. Williams, Biomed. Mass Spectrom., 4 (1977) 381-384.
7 A. Wegmann, Anal. Chem., 50 (1978) 830-832.
8 J.H. McReynolds, N.W. Flynn, R.R. Sperry, D. Fraisse and M. Anhar, Anal. Chem., 49 (1977) 2121-2122.
9 A. Reiffsteck, L. Dehennin and R. Scholler, Adv. Mass Spectrom., 8 (1980) 295-304.
10 L. Dehennin, A. Reiffsteck and R. Scholler, Biomed. Mass Spectrom. 7 (1980) 493-499.
11 C.K. Ingold, Structure and Mechanism in Organic Chemistry, Cornell University Press, Ithaca, 1969, P. 1115.
12 B.J. Millard, Quantitative Mass Spectrometry, Heyden, London, 1978, P. 156.
13 M.G. Lee and B.J. Millard, Biomed. Mass Spectrom., 2 (1975) 78-81.

APPLICATION AND MEASUREMENT OF METAL ISOTOPES

K. HABFAST

Finnigan MAT GmbH, Barkhausenstr. 2, 2800 Bremen (F.R.G.)

ABSTRACT

This paper presents a short introduction to isotope dilution mass spectrometry for metals, using thermal ionization.

Isotope dilution mass spectrometry is an analytical method especially suited for the determination of trace concentrations of nearly every element in the periodic table that has at least two isotopes.

Gases, liquids and solids can be used as samples. For metals, the thermal ionization technique is normally used together with a single focusing mass spectrometer.

As compared to other trace methods, the isotope dilution method is an absolute method because it is basically based on a weighing procedure. Therefore, it can be used as a secondary standard method for routine trace determinations.

The problem of the blank value can be solved precisely, error propagation analysis is transparent and straightforward.

The thermal ionization technique requires a clean sample of the element to be analyzed. Therefore, a careful separation and cleaning process has to be applied to the sample before loading it into the ion source. This process is necessary in addition to the well-known incineration or extraction procedures in trace determinations. The paper gives examples of this cleaning process, with special emphasis on ion exchange column techniques. As the isotope dilution method does not require a quantitative extraction of the element in question, the analyst can concentrate on getting a blank value as low as possible, rather than being forced to consider both, quantitative extraction and low blank.

A major part of the paper is describing the mass spectrometric measuring procedures and data evaluation methods to be used for isotope dilution mass spectrometry. Special attention is given to the only systematic error of the method: the fractionation of the sample during evaporation and ionization. A special algorithm to correct for fractionations is given. In practice, fractionation correction in trace determinations is only to be applied to elements which have an atomic weight of less than approx. 50.

The paper concludes with a summary of analytical results which were obtained by several research groups. The given examples include the demonstration of the practical limits of the method, the accurate solution of important analytical problems (like the determination of chlorine in minerals), and the experience with the determination of the analytical blank.

I. INTRODUCTION

It seems to be obvious to everyone who is involved in measuring trace concentrations of metals that there is a demand for higher accuracy and reproducibility of the analytical results. If the analytical methods are producing widely scattering results, it is impossible to take appropriate measures against the seemingly general trend of poisoning our environment by toxic substances like heavy metals.

It is the purpose of this contribution to give a short introduction into the isotope dilution analysis technique, as applied to metals, and to show its analytical power with respect to accuracy, reproducibility and attainable detection limits.

Trace analysis of metals, using isotope dilution techniques, is a very old method in mass spectrometry (refs. 1-4). Nevertheless, it has not found widespread use up to now. Instead, optical methods, like atomic absorption, flame emission or x-ray fluorescence spectroscopy are commonly used (ref. 9). These methods all have their merits in the low permille range, but as soon as the low ppm range or even the ppb range have to be covered, they all suffer from a considerable lack of accuracy, even if the attainable precision might be sufficient (refs 5-7).

In fact, there has been proposed not only a general standardization of the individual methods, but also a management scheme for analytical quality control (ref. 8), in order to get a better comparison of the analytical results between laboratories and between methods.

In an isotope dilution analysis, each analytical result is originally based on a gravimetric determination of the components of the analyte. Moreover, all sources of possible errors are easily identified and very transparent. This method, therefore, can be called a definite analytical method and can be used at least as a secondary standard method to all other methods, if it is not directly applied to each individual sample.

II. THE METHOD

The basic principle of IDMS is shown in figure 1. If the isotopes of an element, having the abundance A_{Bp} and A_{Dp} are mixed with a spike of the same element, but with different isotopic abundances, the calculated isotope ratio of the mixture is given by the equation (1) in fig. 1. The C's and W's are the respective concentrations and weights of the sample and the spike. Solving this equation for C_p results in equation (2) in fig. 1. Thus one can calculate the unknown concentration C_p in the sample, if the concentration of the spike solution and the isotope ratios of the mixture, of the sample and of the spike are known.

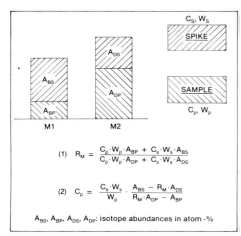

Fig.1 Principle of the isotope dilution method

$A_{BS}, A_{BP}, A_{DS}, A_{DP}$: isotope abundances in atom -%

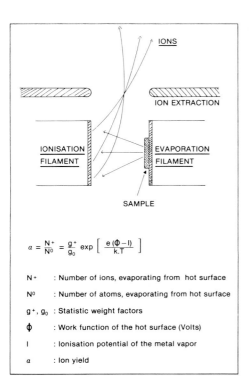

Fig.2 Thermal ionisation source, Saha-Langmuir equation

The isotope ratio of the sample in most cases is known from certified tables of the natural abundances of isotopes (Ref.10). The spike ratio is certified by the manufacturer. All these ratios originally are based on a comparison to standards which are prepared by mixing accurately weighted aliquots of pure isotopes. However, all the required ratios also can be measured accurately enough in the mass spectrometer as it is done with the isotope ratios of the sample/spike mixture.

Isotope dilution can be applied to all elements which have more than one stable or quasi stable isotope, if an ionisation method with sufficient yield and stability is available.

For metal isotopes, sparc source mass spectrometry (ref. 11), field desorption mass spectrometry (FDMS) (ref.12) and thermal ionisation mass spectrometry have been used.

With sparc source techniques one can determine many elements in one analysis by application of a mixed spike of all interesting elements. On the other side, this technique is quite time consuming and also expensive. Thus it is not used very often.

FDMS seems to be very powerful, but there is only very little analytical experience and, therefore, nothing can be told on the reliability of this method up to now.

With thermal ionisation mass spectrometry normally only one element at a time can be measured. But for this method, generally spoken, the best results in trace analysis, concerning detection limits as well as accuracy and precision, have been reported.

We therefore will concentrate on isotope dilution analysis with the thermal ionisation mass spectrometer.

Fig. 2 shows the basic principle of this method.

In the mass spectrometric ion source, the metal of interest first is loaded onto and then evaporated from a hot filament. The metal vapor hits a second filament and is ionized on its surface, following in some way Langmuir's law for surface ionisation.

The ions are accelerated into a mass analyzer which is used to measure the ratios of the ion currents of the different isotopes.

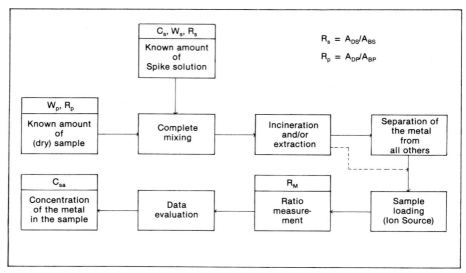

Fig. 3   General scheme of a complete analysis

III. THE ANALYSIS

The general scheme of a complete analysis is shown in fig. 3. A known amount of sample is mixed with a known amount of spike. It is very important that the mixing is done until a homogeneous mixture is reached. Then this mixture is ashed, if necessary, by one of the conventional methods which are used in a similar way for atomic adsorption or flame emission spectroscopy. After complete incineration, the metals normally are available as a mixture of nitrates, oxydes or chlorides. In many cases, instead of complete incineration, one of the well-documented chemical precipitation methods for the metal of interest is applied to the sample.

It is important to mention here that an incomplete recovery of the element during the chemical treatment does not influence the result, because it can be safely assumed that the recovery is the same for the sample as well as for the spike.

Much more important for the final result is the problem of the blank value, as in all trace analysis methods. But, in contrast to other methods, in the isotope dilution technique one can concentrate all efforts to a low blank value without being forced to get 100% recovery at the same time.

This, obviously, is a big advantage for the selection of the appropriate extraction resp. incineration procedure.

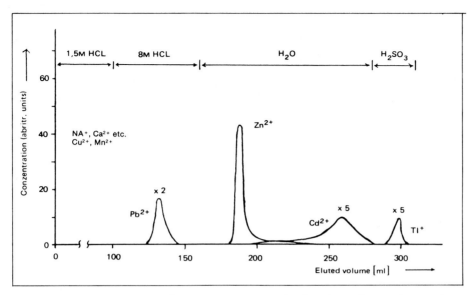

Fig.4　Separation of trace components by ion chromatography

In some cases, after proper precipitation and washing, the samples are ready to be loaded onto the filaments of the thermionic ion source.
In most cases, however, where also other metals than the one of interest are present in the solution, a further separation step is necessary.
Usually the metals are separated by anion exchange (ref.13). A suitable anion exchange column is loaded with the solution, and the different metals are eluted by their appropriate elutant, as shown in fig.4 for the separation process of a plant sample.
The different fractions are sampled and evaporated to dryness. Then they are again soluted in nitric acid to a concentration of 0.5 to 10 µg/µl, and are now ready to be loaded into the ion source.
Other methods for separation include electrolytic deposition (ref.14) of the metal or selective adsorption on small resin beads (ref.15).

Fig.5　1 µg of Gd-sample on the evaporation filament

Fig. 5 shows a gadolinium sample loaded on a rhenium filament. In this case, 1 µg of the sample has been loaded onto the filament in a 1 µg/µl nitrate solution. After drying the sample at approx. 80°C for 15 min., the filament has been heated in open air to a dark red glow. This, probably, produced (yellow) Gd-oxide on the filament, which has a low enough volatility to produce stable evaporation in the vacuum.
The loading procedure itself is a very important step in the whole analysis procedure. The stability and the size of the ion current and to a

certain extent also the measured isotope ratio primarily depend on the proper loading technique of the sample. Therefore, quite a variety of loading techniques have been developed for different elements and it must be mentioned that, concerning different loading techniques, not everything, which happens during the treatment of the loaded filament, has been understood completely and for this reason, loading of the sample sometimes has the odeur of "black magic": Good results are obtained, but nobody completely knows, why.

This, obviously, is a very dangerous situation (at least for very high accuracy runs), and it is the main reason for the commonly applied rule, that everything in the whole procedure has to be performed always in exactly the same way in order to get the best results.

Data collection is done in the well-known single ion detection mode by repetitive peak jumping to the isotopic masses of interest.

Fig. 6 shows schematically the recording of a complete run (ref.16).

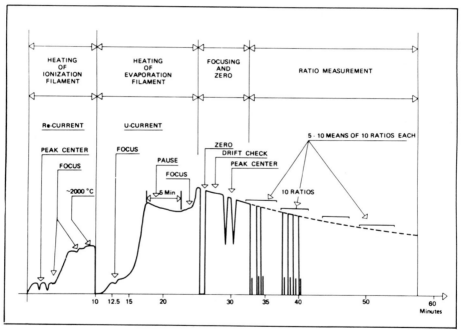

Fig.6 Schematic representation of the automatic procedure for one sample, as recorded on a strip chart recorder

In the first step, the sample is heated until a stable and high enough ion current is obtained.

Then the ratio measurement is started and approximately 10 blocks of data with a total of 100 ratio measurement are taken. All this lasts about one hour, and produces data of very high accuracy and precision (in the range of 1/100 of a percent for the relative accuracy).

This whole procedure can be abbreviated considerably to less than 10 minutes. Under this condition, however, precision and accuracy will only be 1% to 2%, which, anyway, might be sufficient for a lot of practical analysis.

The result of a complete precision measurement is shown in fig. 7 with uranium as an example (ref.16). The measured isotope ratios of 235U/238U have been normalized to the certified value and plotted against time.

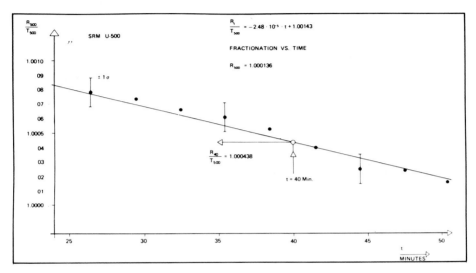

Fig.7   Regression line through the data points of a U-500 sample

As can be seen, the measured ratio decreases with time by some permille and reaches the true value of the isotope ratio after about one hour. This effect is called fractionation and is due to the fact that the lighter isotopic species is evaporating faster than the heavier one. The true isotope ratio cannot be measured directly. Besides sample loading, fractionation is one of the most important sources of error of this method. The degree of fractionation depends on mass, ion yield, sample size and temperature. A quantitative assessment of the degree of fractionation can be derived from Ralleigh's distellation law and is given in fig. 8. This is a simplified notation of a more complex fractionation theory (refs 17-19) and assumes, that only one chemical species is evaporating from the filament. Using this equation one can correct for fractionation.

$R_{obs} = R_0 (a + (a - 1) \ln q)$

R_{obs} : observed isotope ratio (light to heavy mass)

R_0 : true isotope ratio (light to heavy mass)

a : $\sqrt{M_H/M_L} \approx 1 + \frac{\Delta M}{2 M_L}$

q : relative amount of sample on the filament

q = f (temperature, ion yield, loaded sample size)

Fig.8 Fractionation law

A very elegant method (ref.20) can be applied, if the metal to be analysed has more than two natural isotopes. This is the case for 23 metals.

The algorithm used for a three isotope system is shown in fig. 9. It assumes the following relation between the two measured and the two true ratios, which can be deduced from the fractionation law by eliminating q:

$$\frac{r_{obs1}}{R_{o1}} = (\varphi - 1) + \varphi \frac{r_{obs2}}{R_{o2}} \quad ; \quad \varphi = \frac{\alpha_{1,4} - 1}{\alpha_2 - 1}$$

This is a general equation for normalization to an internal standard which can be used for fractionation correction if more than two isotopes are available.

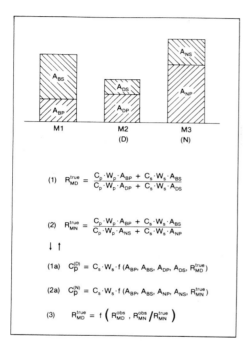

Fig.9 Fractionation correction utilizing a three isotope system

To correct for fractionation in a spiked sample, the isotope dilution equation is applied twice for two isotopically diluted peaks (D) and (N) having different masses.

If no fractionation would exist ($R^{obs} = R^{true}$), the two values $C_P^{(N)}$ and $C_P^{(D)}$ (equ. (1a) resp. (2a) obviously should be the same.

In fact, the calculation results in different values, if R^{obs} is used, due to the mass dependant relation between R^{obs} and R^{true} (equ. (3)).

However, one can use equ. (1a) to compute a first approximation for $C_P^{(D)}$ by assuming that the observed ratio R_{MD}^{obs} is equal to the true ratio R_{MD}^{true}.

This preliminary concentration value is then used to calculate a first approximation for the true ratio R_{MN}^{true} by using equ. (2). With the fractionation law (equ. (3)) now a second approximation of R_{MD}^{true} can be calculated, which is inserted in equ. (1a) to obtain a second approximation of $C_P^{(D)}$ and so on. This iteration process is performed until $C_P^{(D)}$ does not change significantly.

This is, in most practical applications, the case after about 5 iteration steps. The iteration converges very fast.

For metals, which only have two stable or quasi stable isotopes, fractionation correction is not so accurate as with a three isotope system, because it is necessary to run an external standard in a separate measurement exactly under the same conditions. This, indeed, can be very difficult, if not impossible. Thus it is safe to assume that the isotope ratio error contribution to the analytical accuracy is raised by a factor of approx. 5 in such cases as compared to internally normalized results. Experimental experience, however, shows that other sources of error (like blank values) considerably exceed the error resulting from fractionation in almost all practical cases.

IV. EXPERIMENTAL RESULTS

In this section some recent experimental results, which seem to be typical for the isotope dilution method, are reported. More examples, some of which go back more than 30 years, are given in (refs 2-4). Fig. 10 summarizes the results of trace determinations of molybdenum and nickel in a glass standard (SRM 610) resp. in some fossil fuels (ref. 20). These examples originate from the mass spectrometry group of the National Bureau of Standards in Washington/D.C., which not only has specialized in very high accuracy isotope ratio measurements and related topics but which also has pioneered isotope dilution methodology, using thermal ionisation.

MOLYBDENUM IN SRM 610 (GLASS STANDARD)

CONCENTRATION $\mu g/g$ (ppm)	STD-DEV (ppm)	95% CONFID.-LIMIT (ppm)
422.6	± 0.6 (0.14 %)	± 1.4 (0.33 %)

CONCENTRATION OF NICKEL IN FOSSIL FUELS

Sample	CONCENTRATION $\mu g/g$	STD-DEV $\mu g/g$
FUEL OIL	38.1	± 2.6
COAL	14.7	± 0.6
FLY ASH	96.4	± 1.2

Fig.10 Isotope dilution analysis with internally normalized isotope ratios in multi-isotope systems

These experiments show the power of the technique as a standard method; the 95% confidence limit of the analytical result of molybdenum, including chemical pretreatment and sample inhomogeneity is 0.3%. As has been shown, the largest contribution to this error comes from the inhomogeneity of the molybdenum concentration in the glass.

One of the advantages of the isotope dilution method is the very high precision of a single mass spectrometric determination. Therefore, in most cases, the relevant sources of the overall analytical error can be safely identified by applying statistical methods.

In fig. 11, some results of trace determinations (ref.21) of Pb and Tl in some plants and in a milk powder sample are shown.

They originate from a group of analytical chemists at the University of Regensburg (Heumann and coworkers), which is very successfully working on the development of simple and application oriented methods for isotope dilution trace analysis (ref. 22).

Sample	Concentration	
	Pb (ppm)	Tl (ppm)
water-weed (lagarosiphon major)	65.4 ± 0.4	232 ± 11
water-moss (platihypnidiom ripariodes)	61.1 ± 0.3	129 ± 10
olive tree leaves (olea europaea)	29.9 ± 1.2	35 ± 4
milk powder	< 0.6	8.2 ± 1.3

Fig.11   Isotope dilution analysis of Pb and Tl in plants and food

Separation process	Blank	
	Pb (ppm)	Tl (ppm)
Ion exchange	0.8 ± 0.2	16.5 ± 1.2
Electrolytic deposition	—	5.5 ± 1.0

Fig.12   Blank values of Pb and Tl for a generally applicable analysis procedure

A separate investigation of the blanks (fig.12) has shown that the detection limit (=3 x standard-deviation of the blank) is approx. 0.6 ppm for lead (in milk powder), assuming an original sample weight of 1g. This result is obtained by using a relatively simple and, therefore, generally applicable chemical treatment.
More elaborate procedures(e.g. in a clean room laboratory) would lead to lower detection limits.
The accurate analysis of halogen traces is a very difficult task. Heumann and coworkers could show (ref.22), that the analysis of chlorine in minerals (and snow), performed by isotope dilution methods, produces better results than any other method described in the literature.
The main reason for this can be found in the principle of the isotope dilution method itself: In all other methods, the highly volatile chloride ion has to be recovered quantitatively during the disintegration process of the mineral. This is not necessary if isotope dilution is applied.
The chloride is analyzed in the negative ion mode. The negative ions are produced on a hot rhenium surface in a similar way as the positive ions of metals. The corresponding Saha-Langmuir equation, which

describes the ion yield, contains the electron affinity of the halogen (instead of the ionisation potential).

Fig. 13 shows some typical results and the separate determination of the blank. The determining source of the blank was the "suprapure" hydrofluoric acid which was used to disintegrate the samples.

Chloride traces in silicate rocks (from Black Forest)		
Sample	Spike ^{37}Cl Enrichment	Cl⁻ Concentration ppm
Granite	98 %	197 ± 3
	75 %	197 ± 2
Metablastite (porphyroblastic gneiss)	75 %	141 ± 3
Hornblende	75 %	114 ± 5
Amphibolite	75 %	9 ± 3

Blank Nr	Cl⁻ blank per analysis (μg)
1	22.4
2	19.0
3	15.0
4	16.2
5	21.5
Mean	18.8 ± 3.2

Fig.13   Chloride in minerals, analyzed by isotope dilution techniques

In a typical isotope dilution analysis only one element at a time can be analysed. However, in favourable cases, a multi-element analysis is feasible. This has been shown, for instance, with a mixture of rare earths, which can be simultaneously analysed using a composite master spike of all rare earth elements in the mixture.

ANALYSIS OF RARE EARTH ELEMENTS IN HIGH PURITY Y_2O_3

ELEMENTS	LOWEST CONCENTRATION (ppm)	STD.-DEV (ppm)
La, Pr, Tb, Ho, Tm, Lu	0.1	± 0.2
Ce, Gd, Yb	0.1	± 0.01
Nd, Sm, Eu, Dy, Er	0.01	± 0.002

Fig.14   Multi element simultaneous trace determinations on rare earths

Fig. 14 gives a summary of the results which have been obtained by a Norwegian group on a "pure" ytrium oxide sample (ref. 23). Even the elements of the first line of the table, which only have one natural isotope, could be measured within a factor of two in accuracy by comparison with other rare earth elements which have a similar behaviour in the ion source.

In summary, one can state that trace determinations, using isotope dilution techniques, in almost all cases are more precise and more accurate, compared to other methods.

There might be some realistic hope that this method will find more widespread use if reliable and low cost automated mass spectrometers become available in the future.

References

1. G. v. Hevesy and F.A. Paneth, Z. Anorg. Chem., 82 (1913) 323
2. H. Hintenberger, in M.L. Smith (Ed.), Proc.Conf.Harwell 1955, Enriched Isotopes and Mass Spectrometry, Butterworth Sci. Publ., 1956, p. 177.
3. R.K. Webster, in J. D. Waldron (Ed.), Advances in Mass Spectrometry, Vol. 1, Pergamon, 1958, p. 103.
4. R.K. Webster, in A. A. Smales and L. R. Wagner (Ed.), Methods in Geochemistry, Interscience Publ., 1960, p. 202.
5. G. Horlick, Anal. Chem., 52 (1980) 290R.
6. L.G. Hargis and J.A. Howell, Anal. Chem., 52 (1980) 306R.
7. J.A. McHard, Anal. Chem., 51 (1979) 1613.
8. A.L. Wilson, Analyst, 104 (1979) 273.
9. J.C. Van Loon, Anal. Chem., 52 (1980) 955A.
10. N.E. Holden, Pure and Appl. Chem., 52 (1980) 2349.
11. A.P. Mykytink, Anal. Chem., 52 (1980) 1281.
12. H.R. Schulten, Z. Naturforsch., 33C (1978) 484.
13. F. Tera, Proc. Apollo 11 Lunar Conf., 2 (1970) 1637.
14. D.J. Rokop, presented at the Workshop on Isotope Ratio Measurements, 28th Ann. Conf. on Mass Spectrometry and Allied Topics, New York, N.Y., May 1980.
15. D.H. Smith, Analytical Letters 12 (A7) (1979) 831.
16. K. Habfast and D. Tuttas, Appl. Note No. 43 (1980), Finnigan MAT GmbH, Bremen
17. K. Habfast, presented at the 29th Ann. Conf. on Mass Spectrometry and Allied Topics; Minneapolis, MN, May 1981.
18. H. Kanno, Bull. Chem. Soc. Japan, 44 (1971) 1808.
19. L.J. Moore, in N.R. Daly (Ed.), Adv. in Mass Spectrometry, Vol. 7A, The Institute of Petroleum, London 1978, p. 448.
20. L.J. Moore, Anal. Chem., 46 (1974) 1082.
21. K.G. Heumann, Fresenius Z. Anal. Chem. (1981) in press.
22. K.G. Heumann, Talanta 27 (1980) 567.
23. S. Haaland, Inst. for Atomenergi, Work Report No. CH-97, 1972.

STABLE ISOTOPES IN BIOMEDICAL AND ENVIRONMENTAL ANALYSIS BY FIELD DESORPTION MASS SPECTROMETRY

W.D.LEHMANN[+], H.-R.SCHULTEN[++]

[+]Abteilung Medizinische Biochemie, Institut für Physiologische Chemie, Universitäts-Krankenhaus-Eppendorf, Martinistr. 52, D-2000 Hamburg 20,
[++]Institut für Physikalische Chemie, Universität Bonn, Wegelerstr. 12, D-5300 Bonn 1, (Fed.Rep.Germany)

ABSTRACT

Applications of stable isotope abundance measurements using field desorption mass spectrometry are reviewed. The areas covered are direct isotope analysis of depleted and enriched biochemicals, quantification of endogenous and exogenous compounds in body fluids, and metal cation analysis.

INTRODUCTION

Because of its soft ionization and its simple mass spectra, field desorption (FD) mass spectrometry (MS) has a strong analytical potential in biomedical and environmental analysis (1). Consequently, efforts have been made to develop FD MS into a reproducible and quantitative method.

Although the problem of producing high temperature activated carbon FD emitters (2) of reproducible quality has been solved (3), the use of absolute ion current intensities for quantitative determinations is not recommended. The method of choice for quantifications by FD - as for all other mass spectrometric techniques - is internal standardization by stable isotope labelled analogs. Methodological investigations on the suitability of differently labelled organic compounds as internal standards in FD determinations have been focused on deuterium and ^{13}C labelled analogs (4,5). Figure 1 shows the data obtained for the desorption of a roughly equimolar mixture of D-glucose and D-glucose-1-^{13}C labelled with 90 atom % ^{13}C. The data points in Figure 1 show a uniform desorption behaviour for the cationized species of D-glucose and of its carbon-13 labelled analog. In addition, the observed and calculated intensity ratios for m/z 203 : m/z 204 agree within about 1 % showing that the FD measurement has a good accuracy. In order to avoid the elaborate evaluation of a large number of scans, signal accumulation using a multichannel analyzer or a data system has been introduced for quantitative FD analyses (6). Using this instrumentation, it has been demonstrated for a variety of biochemicals and drugs that the molecular

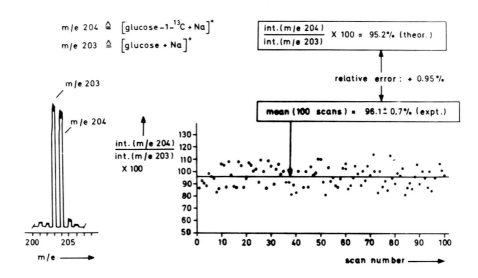

Fig. 1: FD data obtained by repetitive scanning during complete desorption of a mixture of D-glucose and D-glucose-1-C-13 (4).

ion patterns generated by FD measurements agree with those calculated on the basis of the natural abundances within about ± 0.4 % (absolute %).

On testing deuterium labelled compounds as internal standards for FD, it was found that compounds labelled with only a few deuterium atoms exhibit a desorption behaviour virtually identical to that of the non-labelled species (5). However, for deuterium labelled biochemicals containing more than about 3 or 4 deuterium atoms, differences in the desorption behaviour begin to be detectable in the FD measurements. Figure 2 shows this effect for a mixture of didansyl L-tyrosine-d_0 and didansyl L-tyrosine-d_7. Although the relative mass difference of the two compounds shown in Figure 2 is only about 1 %, a fractionated desorption can be clearly observed. As expected, the deuterated molecule shows a preferential desorption at the beginning of the measurement as the volatility of this molecule is slightly increased through the partial replacement of hydrogen by deuterium. During the measurement, the ion intensity ratio displayed in Figure 2 gradually increases as a small relative depletion with respect to the labelled material occurs on the emitter surface. The error which might be introduced into quantitative measurements by this fractionated desorption can be effectively compensated for by accumulation of the complete desorption process. This has been established by the analysis of a number of test mixtures and on the basis of calibration plots (7,8).

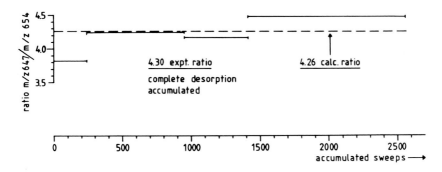

Fig. 2: Desorption behaviour of a mixture of didansyl L-tyrosine-d_0 and didansyl L-tyrosine-d_7. The complete desorption of 250 ng of the didansyl derivative was accumulated in four 2K subunits of the total 8K memory of a multichannel analyzer.

DIRECT ISOTOPE ANALYSIS OF DEPLETED AND ENRICHED BIOCHEMICALS

FD MS is capable of producing simple mass spectra of the majority of biochemicals with molecular weights up to about 2000 mass units which are characterized by molecular ions or cationized molecules as base peaks or at least as signals of high relative abundance. The isotopic patterns of these species containing the intact molecule can be determined by signal averaging techniques with a precision between 0.1-0.5 % depending on the experimental conditions. Provided only one of the incorporated elements shows an altered isotopic composition, the amount of enrichment or depletion of isotopic species of this element can be estimated by direct isotope analysis of the <u>intact</u> organic molecule. Precision and accuracy of these measurements by FD MS are sufficient to detect biosynthetic isotope effects of carbon-13 in cases where the molecule contains a large number of carbon atoms. For instance, in the quantitative evaluation of the FD mass spectrum of aquo-cyano cobyrinic acid heptamethylester, the pattern of the $[\text{Cation-H}_2\text{O}]^+$ peak at m/z 1062 clearly showed a depletion of this ion in carbon-13 compared to the average natural isotopic abundances of carbon. The two most intense isotope peaks were found to be smaller than expected from the calculated data. Figure 3 gives the FD spectrum plus the structural formula of the compound investigated and Table 1 compares the experimental and calculated abundance data (9).

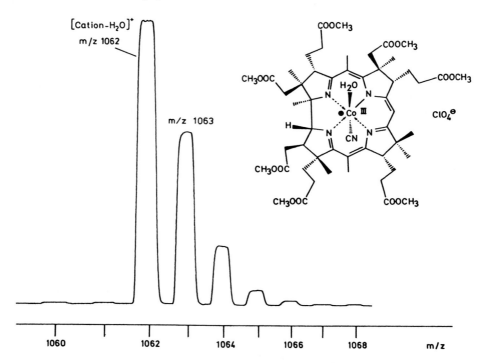

Fig. 3: Isotopic pattern of the [Cation-H_2O]$^+$ ion found in the FD mass spectrum of aquo-cyano cobyrinic acid heptamethylester perchlorate; 74 repetitive magnetic scans were accumulated (9).

TABLE 1: Experimental and calculated isotopic abundances for the organic cation displayed in Figure 3.

m/z	rel.int.calc.(%)	rel.int.exp.(%)	deviation(%)
1062	100	100	–
1063	62.94	60.67	– 2.27
1064	22.34	20.67	– 1.67
1065	5.75	5.17	– 0.58
1066	1.19	1.66	+ 0.47
1067	0.21	0.45	+ 0.24

The data given in Table 1 clearly indicate that the ion investigated is depleted in carbon-13 compared to its average natural abundance. The

FD data give a carbon-13 abundance of 1.07 ± 0.01 %, a value which agrees well with the carbon-13 abundance of 1.0696 ± 0.0002 % which was determined by isotope ratio MS on carbon dioxide after combustion of about 1 mg of the compound. These values represent about the lower end of carbon-13 abundances found in C_3 plants (10,11). The compound investigated is derived from vitamin B_{12} produced by microbiological synthesis using beet sugar molases as carbon source. Therefore, the isotope abundance found in this study will reflect mainly the carbon-13 abundance of this medium. Comparing the different techniques isotope ratio and field desorption MS for a precise carbon-13 analysis of the compound given in Figure 3, FD MS requires about three orders of magnitude less sample. However, it produces abundance data with an error about 50 times greater than the corresponding data from the isotope ratio mass spectrometer.

The estimation of the label content of isotopically labelled biochemicals is another area of application of FD MS. This has been demonstrated for a number of radiochemicals labelled with carbon-14 or hydrogen-3 (12-14) as well as for compounds labelled with the stable isotopes deuterium, carbon-13, or nitrogen-15 (6,15). The direct mass spectrometric analysis of radiochemicals allows the calculation of their specific radioactivity and, as it is possible for stable isotope labelled compounds the distribution of the species carrying different amounts of labelled atoms can be obtained in the analysis of multiply labelled substances.

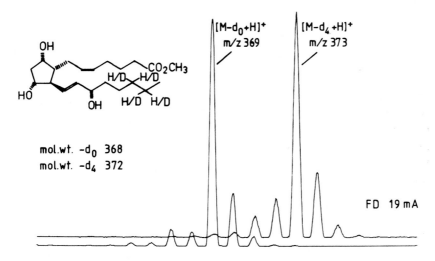

Fig. 4: Groups of protonated molecules of prostaglandin $F_{2\alpha}$ methylester and of its tetradeuterated analog measured subsequently on the same FD emitter under identical experimental conditions.

The use of FD mass spectrometry in general allows the analysis of labile and/or polar molecules without derivatization. In each case, the FD mass spectrum of the nonlabelled compound is required as a reference for a correct evaluation of the degree of labelling. This is necessary since due to surface reactions small satellite ions one and two mass units below the molecular ions or cationized molecules are observed in many cases and because for some compounds the formation of $[M]^{+\cdot}$ and $[M+H]^+$ ions are competing processes of ionization. As an example of isotopic analysis of a deuterated compound by FD, Figure 4 shows the pattern of the protonated molecules of nonlabelled and labelled prostaglandin $F_{2\alpha}$ methylester, a compound which shows extensive fragmentation under electron impact conditions. The FD mass spectrum in Figure 4 clearly shows that the main component in the deuterated prostaglandin contains 4 deuterium atoms and is accompanied by a small amount of a triply deuterated analog. A precise evaluation of the degree of labelling in this case requires the consideration of a number of superpositions and best is performed using a special calculating program.

DETERMINATION OF EXOGENOUS AND ENDOGENOUS COMPOUNDS IN BODY FLUIDS

Quantitative mass spectrometric investigations in life sciences are essential for the determination of exogenous compounds such as drugs and their metabolites. Focal points are pharmacokinetics and bioavailability studies. Also of interest are assays for endogenous substances including investigations of metabolic pathways by in vivo experiments.

As an example for the quantification of an exogenous compound, the determination of cyclophosphamide (Endoxan^R, CP) in human serum by FD MS is demonstrated (16). This drug is used in cancer chemotherapy and as immunosuppressant in the treatment of multiple sclerosis. Several studies have shown the suitability of FD for the analysis of CP and its metabolites (17-19). In quantitative studies a high degree of precision and accuracy can be obtained as has been demonstrated, for instance, in the quantification of 4-keto-cyclophosphamide from spiked human urine (20,21).

Because of the high sensitivity of FD MS for CP, sample amounts of less than 5×10^{-11} g are needed for a mass spectrometric measurement (16). Figure 5 shows the limit of detection for CP from a standard solution with a signal to noise ratio of about 20:1. This high sensitivity is of particular importance in medical investigations, where only small amounts of sample are available. Using biological material instead of pure standard solutions, the sensitivity for CP is decreased. Nevertheless, in general only 1-2 ml of a biological fluid are needed for a

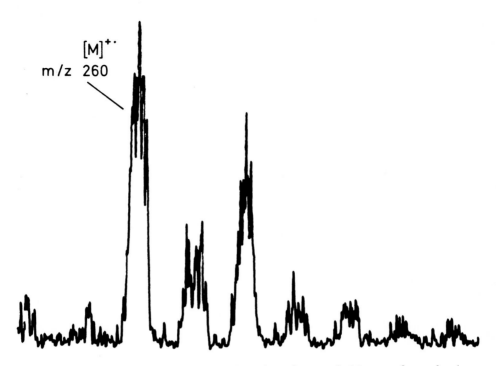

Fig. 5: Isotopic pattern of the molecular ions of 50 pg of cyclophosphamide produced by FD MS. 30 scans were accumulated on a multichannel analyzer (16).

quantification. For determination of CP in body fluids of multiple sclerosis patients using stable isotope dilution analysis, a deuterated analog of CP was used for which 1.7 % $[^2H_8]$CP, 15 % $[^2H_9]$CP, and 83.3 % $[^2H_{10}]$CP were estimated with an accuracy of 1 % (15). The molecular ion peaks of CP (m/z 260) and $[^2H_{10}]$CP (m/z 270) are used for calculating the amount of the drug in the sample. The quantification of CP from a chloroform extract of serum samples without further purification is not possible, as there are a number of FD signals of high abundance superimposed in the mass region of interest (Figure 6 a)(16). Therefore the drug was separated by high pressure liquid chromatography (HPLC) and the eluate fraction containing the drug was analyzed by FD MS (Figure 6 b). Mass spectrometric detection was necessary because the poor UV absorbance of the drug did not allow its registration using the conventional UV detector. This procedure also was employed for the determination of two CP metabolites, 4-keto CP and carboxyphosphamide, in urine samples of multiple sclerosis patients at concentrations of 0.8-23 µg/ml and using the corresponding hexadeuterated analogs as internal standards(22).

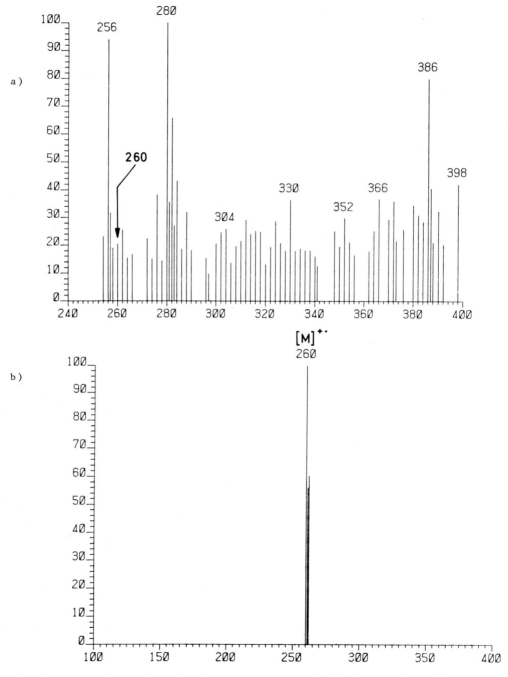

Fig. 6: Detection of CP in human serum by field desorption MS.
a) crude chloroform extract; b) chloroform extract purified by high pressure liquid chromatography (16).

In studies concerned with the identification and quantification of endogenous compounds in human body fluids, stable isotope labelled nutrients are important tools for in vivo studies in man. For instance, these can be focused on the comparison of in vitro and in vivo metabolic routes or on determinations of in vivo activities of certain enzyme systems. In connection with biomedical studies of phenylketonuria, one of the most commonly occurring inborn errors in man, a variety of tests using oral or intravenous loading with stable isotope labelled L-phenylalanine have been developed. After loading, L-phenylalanine, L-tyrosine, and a number of other metabolically related compounds have been followed in plasma and urine samples. Due to the metabolic block between L-phenylalanine and L-tyrosine in phenylketonuria, in these tests large differences have been found comparing normal healthy controls and homozygous carriers of the disease.

As an example of plasma amino acid analysis by field desorption, Figure 7 shows the analysis of a HPLC fraction of a human plasma sample containing L-leucine, L-isoleucine, and L-phenylalanine following an oral load of 25 mg/kg L-$[^2H_5]$phenylalanine in a healthy volunteer (8).

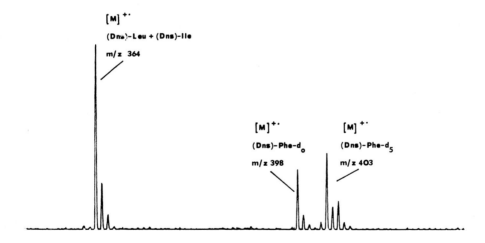

Fig. 7: FD mass spectrum of a HPLC fraction of a human plasma sample. The molecular ions of L-leucine, L-isoleucine, and L-phenylalanine are detected as dansyl derivatives. The sample amount required for this analysis corresponds to about 10 µl of plasma(8)

The FD mass spectrum given in Figure 7 shows a normal isotopic distribution for the constitutional isomers L-leucine and L-isoleucine. In contrast, the molecular ion pattern of L-phenylalanine indicates the presence of a penta- and heptadeuterated analog. These compounds represent orally ingested L-[^{2}H$_5$]phenylalanine and the internal standard L-[^{2}H$_7$]phenylalanine which has been added to the plasma sample. For this kind of plasma amino acid analysis FD MS offers a sensitivity and accuracy comparable to the established assays based on gas chromatography and mass spectrometry. However, because of the simplicity of the FD mass spectra, sample purification need not be as extensive since the chance of interfering ions being present in the mass spectrum is strongly reduced using FD.

For the analysis of organic cations such as quaternary ammonium or phosphonium compounds, FD MS offers the unique possibility of direct detection of these species. The application of chemical ionization or electron impact ionization generally requires a dealkylation of the quaternary compounds into their tertiary analogs. In contrast, the field desorption mass spectra of these compounds readily show the organic cations as base peak and only a small or negligible amount of fragment ion formation (23). FD MS has, for instance, been applied to the quantification of choline and acetylcholine in tissue samples from distinct rat brain regions (7). In view of the high sensitivity of FD for the detection of organic cations, the work up procedure of the homogenized tissue samples included only a number of extraction steps and no chromatographic separation of the organic cations. In addition, determinations of carpronium chloride (24) and of berberine chloride (25) in human urine samples have been performed by FD MS.

Summarizing the results and experiences from the area of organic analysis of biological samples by FD, it is evident that the technique provides high sensitivity and specificity, as well as a sufficient accuracy for quantitative work. The required sample work up procedures can be relatively simple and fast, although the development of these procedures may be very elaborate as these have to be newly designed in many cases and adapted to the requirements of FD MS.

TRACE AND ULTRATRACE DETERMINATIONS OF METALS

FD MS can be used as highly sensitive method for qualitative and quantitative analysis of metals (26). Alkaline, alkaline earth, and heavy metals have been determined in trace and ultratrace concentrations directly from very small amounts of biological samples using stable isotope dilution analysis. This method is limited to metals with more than one stable isotope. Cesium, for instance, has only the stable isotope ^{133}Cs, of which about 10 fg can be registered by FD MS. Therefore, it has been determined by external standardization (27,28). Lithium, occurring with two stable isotopes, ^{6}Li 7.5 %, and ^{7}Li 92.5 %, has been de-

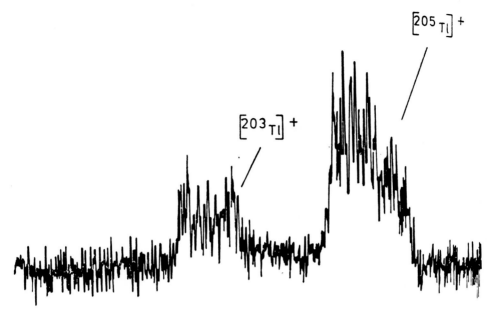

Fig. 8: Detection of 10 pg of thallium by FD from a standard solution (32).

termined by stable isotope dilution in wine, mineral water and high purity solvents (29), lithium (30) and rubidium (31) in human body fluids without pretreatment of the samples.

As can be derived from Figure 8, the limit of detection for the heavy metal thallium from a standard solution is about 10 pg. Thallium has two stable isotopes with natural abundances of 29.5 % at mass 203 and 70.5 % at mass 205. Stable isotope enriched thallium with about the reverse isotopic distribution has been used for kinetic investigations of this metal in different organs of poisoned mice. The time dependent distribution of thallium in heart, kidney, liver, and stomach of mice fed with 80-160 mg Tl/kg has been estimated (33). After addition of the isotopically enriched internal standard, the mice organs were homogenized and centrifuged. The supernatant was investigated by FD MS and the signals at m/z 203 and m/z 205 were accumulated over 50-100 scans using a multichannel analyzer. Using the same procedure, the placental transfer of thallium in pregnant mice following a dose of 8 mg Tl/kg was determined (34) and the values could be correlated with the fetal malformations observed (35). The concentration data determined in the range $10^{-3}-10^{-4}$ mol/l show a precision of $\pm$ 10 %.

In pilot studies using laser heating of the FD emitter, qualitative investigations on allmost all metals of the periodic system have been performed by FD MS (36,37). Because most metals are commercially available in a stable isotope enriched form, a wide field of applications of FD MS for metal analysis in medicine and biosciences can be expected.

ACKNOWLEDGEMENT

W.D.L. is indebted to C.O.Meese (Robert-Bosch-Krankenhaus, Stuttgart, Fed.Rep.Germany) for the sample of deuterated prostaglandin methyl ester.

REFERENCES
1) H.-R.Schulten, Int.J.Mass Spectrom.Ion Phys., 32 (1979) 97-283.
2) H.-R.Schulten, H.D.Beckey, Org.Mass Spectrom., 6 (1972) 885-895.
3) W.D.Lehmann, R.Fischer, Anal.Chem., 53 (1981) 743-747.
4) W.D.Lehmann, H.-R.Schulten, Angew.Chem.Int.Ed.Engl., 16 (1977) 184-185.
5) W.D.Lehmann, H.-R.Schulten, Biomed.Mass Spectrom., 5 (1978) 208-214.
6) W.D.Lehmann, H.-R.Schulten, H.M.Schiebel, Fresenius Z.Anal.Chem., 289 (1978) 11-16.
7) W.D.Lehmann, H.-R.Schulten, N.Schröder, Biomed.Mass Spectrom., 5 (1978) 591-595.
8) W.D.Lehmann, N.Theobald, H.C.Heinrich, Biomed.Mass Spectrom., 8 (1981) xxx.
9) H.M.Schiebel, H.-R.Schulten, Naturwissenschaften, 67 (1980) 256-257.
10) J.S.Gaffney, A.P.Irsa, L.Friedman, D.N.Slatkin, Biomed.Mass Spectrom., 5 (1978) 495-497.
11) D.A.Schoeller, P.D.Klein, J.B.Watkins, T.Heim, W.C.McLean, Jr., Am.J.Clin.Nutr., 33 (1980) 2375-2385.
12) H.-R.Schulten, R.Müller, R.E.O'Brien, N.Tzodikov, Fresenius Z.Anal.Chem., 302 (1980) 387-392.
13) L.J.Altman, R.E.O'Brien, S.K.Gupta, H.-R.Schulten, Carbohyd.Res., 87 (1980) 189-199.
14) H.-R.Schulten, W.D.Lehmann, Biomed.Mass Spectrom., 7 (1980) 468-472.
15) U.Bahr, H.-R.Schulten, J.Label.Comp.Radiopharm., 28 (1981) 571-581.
16) U.Bahr, H.-R.Schulten, O.R.Hommes, F.Aerts, Clin.Chim.Acta, 103 (1980) 183-192.
17) H.-R.Schulten, Biomed.Mass Spectrom., 1 (1974) 223-230.
18) H.-R.Schulten, Cancer Treat.Rep., 60 (1976) 501-507.

19) H.-R.Schulten, W.D.Lehmann, M.Jarman, in Quantitative Mass Spectrometry in Life Sciences, A.P.DeLeenheer and R.R.Roncucci (Eds.), Elsevier, Amsterdam, 1977, p. 187-195.
20) H.-R.Schulten, W.D.Lehmann, Mikrochim.Acta (Wien)II, (1978) 113-129.
21) W.D.Lehmann, H.-R.Schulten,
Fresenius Z.Anal.Chem., 290 (1978) 121-122.
22) U.Bahr, H.-R.Schulten, Biomed.Mass Spectrom., in press.
23) D.A.Brent, D.J.Rouse, M.C.Sammons, M.M.Bursey,
Tetrahedron Lett., (1973) 4127-4130.
24) M.Sano, K.Ohya, R.Ito, Biomed.Mass Spectrom., 6 (1979) 467-471.
25) H.Miyazaki, E.Shirai, M.Ishibashi, K.Hosoi, S.Shibata, M.Iwanaga, Biomed.Mass Spectrom., 5 (1978) 559-565.
26) U.Bahr, H.-R.Schulten, in Topics in Current Chemistry, F.L.Boschke (Ed.), Springer, Berlin 1981, p. 1-48 and references cited therein
27) W.D.Lehmann, H.-R.Schulten, Anal.Chem., 49 (1977) 1744-1746.
28) H.-R.Schulten, R.Ziskoven, W.D.Lehmann,
Z.Naturforsch., 33c (1978) 178-183.
29) H.-R.Schulten, U.Bahr, W.D.Lehmann,
Mikrochim.Acta (Wien) I, (1979) 191-198.
30) W.D.Lehmann, U.Bahr, H.-R.Schulten,
Biomed.Mass Spectrom., 5 (1978) 536-539.
31) H.-R.Schulten, B.Bohl, U.Bahr, R.Palavinskas,
Int.J.Mass Spectrom.Ion Phys., in press.
32) H.-R.Schulten, W.D.Lehmann, R.Ziskoven,
Z.Naturforsch., 33c (1978) 484-487.
33) C.Achenbach, O.Hauswirth, C.Heindrichs, R.Ziskoven, F.Köhler, U.Bahr, A.Heindrichs, H.-R.Schulten,
J.Toxicol.Environ.Health, 6 (1980) 519-528.
34) R.Ziskoven, C.Achenbach, U.Bahr, H.-R.Schulten,
Z.Naturforsch., 35c (1980) 902-906.
35) C.Achenbach, R.Ziskoven, F.Köhler, U.Bahr, H.-R.Schulten,
Angew.Chem.Int.Ed.Engl., 18 (1979) 882-883.
36) H.-R.Schulten, R.Müller, D.Haaks,
Fresenius Z.Anal.Chem., 304 (1980) 15-22.
37) H.-R.Schulten, GIT Zeitschr.Lab., 24 (1980) 916-917.

CALCIUM ABSORPTION STUDIES IN MAN BY STABLE ISOTOPE DILUTION AND
FIELD DESORPTION MASS SPECTROMETRY

W.D.LEHMANN and M.KESSLER
Abteilung Medizinische Biochemie, Institut für Physiologische Chemie,
Universitäts-Krankenhaus-Eppendorf, Martinistr. 52,
D-2000 Hamburg 20, (F.R.G.)

ABSTRACT

Field desorption mass spectrometry has been used for precise isotope abundance measurements of calcium in human plasma, urine, and saliva. Using a double isotope technique, stable isotope enriched calcium has been employed to estimate calcium absorption in man.

INTRODUCTION

The use of stable isotope enriched elements for metal bioavailability and turnover studies in man is an important area of current biochemical research, as assays based on stable isotopes allow these investigations to be performed on all groups of the population including pregnant women, children, and newborn infants. The established analytical procedures for the determination of metal isotope abundances in biological samples are thermal ionization mass spectrometry (MS) and neutron activation analysis. This study illustrates the potential of field desorption (FD) MS for stable isotope in vivo studies in man using as an example gastrointestinal calcium absorption.

EXPERIMENTAL

Mass spectrometry was performed on a VG Micromass ZAB-1F instrument equipped with a FD ion source. The source potentials were + 8 kV for the emitter and - 4 kV for the counter electrode. High temperature activated carbon emitters with an average needle length of 60 μm were used throughout. All measurements were performed in the double focusing mode. After passing through a Canberra digitizer model 6271 the electrical signals were accumulated in a multichannel analyzer type Canberra series 80. For a single isotope ratio determination 200-300 repetitive sweeps were accumulated and the peak heights were read out digitally.

TABLE 1

Isotope abundances of calcium used for the oral load (^{44}Ca enriched) and for the intravenous load (^{42}Ca enriched) compared to the natural isotope abundances of calcium

mass	^{44}Ca enriched(%)	^{42}Ca enriched(%)	Ca nat.ab.(%)
40	66.32	20.45	96.92
42	0.56	77.69	0.64
43	0.18	0.29	0.13
44	32.76	1.43	2.13
46	0.004	0.003	0.0032
48	0.19	0.15	0.179

Plasma and saliva samples were deproteinized by the addition of equal volumes of 20 % trichloroacetic acid. Urine was analyzed without pretreatment. Sample amounts between 0.5 and 2 µl of the body fluids were consumed for one analysis.

^{44}Ca enriched calcium was given orally (185.4 mg) and ^{42}Ca enriched calcium was given intravenously (27.3 mg) to a healthy adult volunteer. The corresponding isotope abundances are given in Table 1.

Isotopically enriched calcium was purchased from Rohstoff-Einfuhr GmbH, Düsseldorf, F.R.Germany. Calcium carbonate SRM 915 was supplied from the National Bureau of Standards, U.S.Department of Commerce,U.S.A.

RESULTS AND DISCUSSION

Field desorption represents a powerful analytical technique for the detection and quantification of metals in the areas of biomedical and environmental analysis [1,2]. Nevertheless, little is known about the maximal precision and accuracy of metal isotope abundance measurements

TABLE 2

Calcium isotope abundances in calcium carbonate SRM 915 determined by field desorption and thermal ionization [3] mass spectrometry

mass	rel.abundance (%) FD mass spectrometry		rel.abundance (%) thermal ionization MS	
40	96.95	± 0.02	96.941	± 0.001
42	0.642	± 0.006	0.647	± 0.001
43	0.137	± 0.006	0.135	± 0.002
44	2.063	± 0.002	2.086	± 0.004
46	0.0032	± 0.0004	0.0036	± 0.001
48	0.206	± 0.006	0.186	± 0.001

using FD MS. Therefore, as a basis for the calcium studies, we first investigated a calcium reference material. Table 2 gives the calcium isotope abundances for calcium carbonate SRM 915 as obtained by FD analysis compared to data produced by thermal ionization MS [3].

The experimental error of the FD data on average is a factor of about 5 larger than the error of the thermal ionization data. In addition, Table 2 shows a fairly good agreement between the data generated by the two different mass spectrometric techniques, only the abundance values for ^{44}Ca and ^{48}Ca show deviations in excess of the experimental error limits.

FD shows a high sensitivity for a number of metal cations, such as the alkali and alkaline earth metals. Most importantly, this high sensitivity is achieved in the analysis of crude biological fluids such as deproteinized plasma, saliva, or urine. Using these biological samples, the most abundant calcium isotope is detected at m/z 40 with a sensitivity in the order of 10^{-9} C/μg. This sensitivity allows the detection of the rare ^{46}Ca isotope in about 1 μl of human plasma, indicating a higher sensitivity of FD MS relative to a newly developed combination of a thermal ionization source with a quadrupole mass analyzer [4].

The calcium ions start to desorb from the FD emitter at heating currents exceeding about 60 mA which corresponds to dark red heat of the

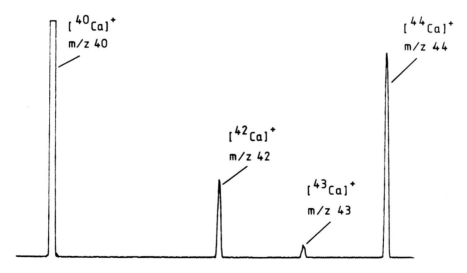

Fig. 1: Partial isotopic pattern of calcium produced by FD MS from 1 μl of human plasma. 300 repetitive sweeps were accumulated at 75-85 mA heating current. The signal at m/z 40 is registered within the linear dynamic range of the detection system.

emitter wire. At these temperatures the organic constituents of the biological samples are completely desorbed and/or pyrolytically decomposed. Therefore, virtually no interfering organic background ions are found in the FD spectra at these emitter temperatures, a situation which enables highly precise isotope abundance measurements to be performed on crude biological samples. Figure 1 demonstrates this effect showing the partial mass spectrum of Ca obtained from 1 µl of human plasma.

The absolute precision for isotope abundance measurements of ^{42}Ca and ^{44}Ca using biological fluids is about 0.02 % for the accumulation of 500 repetitive sweeps at a constant ion current.

The usefulness of FD MS for stable isotope in vivo studies in man was tested using the example of gastrointestinal calcium absorption. True absorption values for ingested calcium are difficult to measure as absorbed calcium is partially reexcreted, and as endogenous fecal calcium is lost. Using ^{45}Ca and ^{47}Ca as radiotracers a number of isotopic absorption tests have been developed and validated [5,6]. On the analogy of the double isotope method based on the subsequent administration of ^{45}Ca and ^{47}Ca [7] we attempted to use two stable isotope enriched species of calcium in combination with mass spectrometric analysis by FD. We selected ^{44}Ca enriched calcium for the oral load and ^{42}Ca enriched calcium for the intravenous load. The i.v. load was given 2 h after the oral load in order to approximate the desired condition that the two differently labelled calcium loads enter the body at the same time. Up to 76 h after the oral loading the calcium isotope abundances were followed in a number of plasma, urine, and saliva samples. Figure 2 shows the data obtained in the analysis of the plasma samples. The plasma concentration of the oral calcium shows a rapid linear increase up to maximal values between 2 and 3 h after the loading followed by an exponential decrease. The intravenous calcium exhibits a very steep decrease in the initial phase and then shows elimination kinetics gradually approaching that of the oral calcium.

Provided the oral and the intravenous calcium show an identical distribution and excretion behaviour, and provided these differently administered calcium doses have reached a complete relative equilibrium, it is possible to estimate the fractional absorption of the oral dose by a single analysis of an aliquot of a body fluid according to the following equation [7]:

$$\text{fractional absorption} = \frac{\text{oral Ca (\% of total Ca)}}{\text{i.v. Ca (\% of total Ca)}} \times \frac{\text{i.v. dose (mg)}}{\text{oral dose (mg)}}$$

For these calculations as well as for the plasma kinetic given in Figure 2 the observed calcium isotope abundances were transformed into data representing the percentage of the oral and of the i.v. dose relative to the total calcium content of the sample. Using these data, fractional absorption has been calculated from samples of plasma, urine, and saliva taken more than 24 h after the oral loading. The results are given in Table 3. The absorption values calculated from the plasma and urine samples show a good agreement and in addition even the data calculated from the saliva samples are in reasonable agreement with the plasma and urine data. The absorption values obtained by FD MS from this loading test appear to be near the upper limit of normal values determined for the absorption of a 200 mg dose of calcium in adults using radioactive isotopes of calcium [7].

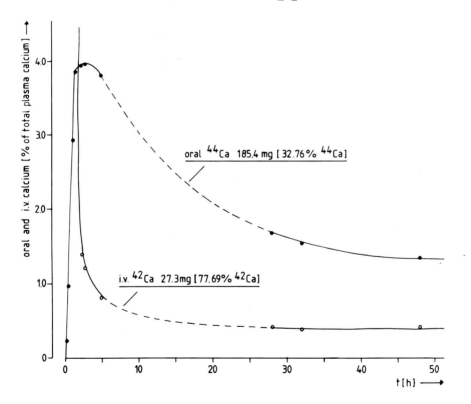

Fig. 2: Plasma kinetics of an oral and of an intravenous calcium load determined by FD MS. The complete isotopic abundances of the ^{42}Ca and of the ^{44}Ca load are given in Table 1. Three minutes after the end of the i.v. ^{42}Ca load, 23.4 % of the total calcium were found to represent the i.v. loading dose.

TABLE 3

Fractional Ca absorption values determined in different body fluids

sample	time	fractional absorption (%)	average (%)
plasma	+ 48 h	44.1	
plasma	+ 76 h	48.6	46.4 ± 3.2
urine	+ 36 h	50.9	
urine	+ 48 h	44.9	
urine	+ 56 h	49.4	
urine	+ 76 h	38.6	46.0 ± 5.5
saliva	+ 36 h	54.6	
saliva	+ 48 h	32.0	
saliva	+ 56 h	40.7	
saliva	+ 76 h	40.3	41.9 ± 9.3

CONCLUSION

It has been shown that field desorption in combination with signal accumulation is capable of following absorption and elimination of calcium in man. With respect to the large pools of exchangeable calcium in adults, loading doses of several tens of milligrams of highly enriched calcium have to be administered in these studies. However, for investigations in children and/or in trace element studies the amount of stable isotope enriched metals required for a single test is significantly reduced, a factor which markedly improves the practicability of such a stable isotope test.

It appears possible, that field desorption could develop into a widely used technique for metal analysis including stable isotope tracer studies in life sciences. This assumption is supported by the outstanding sensitivity of the technique, its sufficient accuracy, and by the fact that sample preparation steps are very simple or even unnecessary.

REFERENCES

1 W.D.Lehmann, H.-R.Schulten, Anal.Chem., 49 (1977) 1744-1746.
2 W.D.Lehmann, U.Bahr, H.-R.Schulten,
   Biomed.Mass Spectrom., 5 (1978) 536-539.
3 L.J.Moore, L.A.Machlan, Anal.Chem., 44 (1972) 2291-2296.
4 A.L.Yergey, N.E.Vieira, J.W.Hansen, Anal.Chem., 52 (1980) 1811-1814.
5 J.E.Harrison, K.G.McNeill, D.R.Wilson, D.G.Oreopulos, A.Krondl,
   J.M.Finlay, Clin.Biochem., 6 (1973) 237-245.
6 S.P.Nielsen, O.Bärenholdt, O.Munck,
   Scand.J.Clin.Lab.Invest., 35 (1975) 197-203.
7 J.A.DeGrazia, P.Ivanovich, H.Fellows, C.Rich,
   J.Lab.Clin.Med., 66 (1965) 822-829.

H.-L. Schmidt, H. Förstel and K. Heinzinger (Editors), *Stable Isotopes*
© 1982 Elsevier Scientific Publishing Company, Amsterdam — Printed in The Netherlands

AN ORGANIC PREPARATION SYSTEM FOR ^{13}C STABLE ISOTOPE RATIO ANALYSIS

A. BARRIE, M.C. CLARKE, R.A. COKAYNE
VG Isogas Ltd., Aston Way, Middlewich, Cheshire, CW10 OHT (U.K.)

ABSTRACT

Stable Isotope Ratio Analysis (SIRA) of ^{13}C in organic materials has grown in importance over the past decade. However a limiting factor has been manual conversion of samples to a suitable gas (normally CO_2) for analysis in the mass spectrometer. Traditional methods are slow and require significant operator attention and skill whilst the apparatus involved takes up considerable space. In this paper, we describe a small bench top system in which most operations are performed automatically under control of a digital sequencer. The system uses catalytic combustion at 800°C in the presence of oxygen and a helium carrier gas is used to transport reaction products through a purification section to a liquid nitrogen cooled tube where pure, dry CO_2 is collected. Typically, a CO_2 sample can be prepared from an organic solid in only 5 minutes compared with approximately 30 minutes by the traditional method. Examples of measurements are given for several foodstuffs and samples of human tissue with δ^{13}C values in the range -12 to -30°/oo relative to the PDB standard. The reference gas used was obtained by combustion of an NBS22 oil standard (-29.4 per mil).

INTRODUCTION

Recent years have seen a growth of interest in the application of ^{13}C SIRA (Stable Isotope Ratio Analysis) to life sciences. The techniques used for this were initially developed for use in geochemical research and involve the use of a special double collector, isotope ratio mass spectrometer [1,2]. For work requiring the most precise measurements, a dual inlet system is also used to allow frequent comparison with a reference. A limitation which exists is that the instrument requires a sample to be presented as a pure, dry gas whereas many organic samples of interest occur naturally as solids or liquids.

A convenient gas for measurements is CO_2 and samples are normally converted to this gas by chemical means.

Traditionally, this conversion has been carried out by manual methods based on work by Craig in 1953 {3}. This involves a glass and quartz high vacuum line which is evacuated after introducing the sample following which oxygen is admitted and combustion is carried out in a copper oxide packed furnace at 900-950°C. A mercury Toepler pump is used to recirculate the gases through the furnace to ensure complete combustion. Typical preparation times are 30 minutes or more.

The alternative method described below employs a helium carrier gas instead of a vacuum system for transporting gases and catalysts are used in the furnace to improve efficiency of combustion. In addition, the system is small enough to be contained within a bench top cabinet, is semi-automated using programmable microelectronics and solenoid valves and requires only 5 minutes preparation time per sample.

INSTRUMENTATION.

Description.

The system is shown schematically in figure 1. Sample introduction is made using a quartz ladle with a magnetic slug sealed into one end to enable it to be pushed or pulled in or out of the hot zone of the combustion tube by the action of a magnet held outside the cool part of the tube which extends outside the furnace. Low volatility solids are loaded in an open platinum boat whilst volatile samples are pre-sealed in thin walled aluminium capsules which later rupture in the furnace to release their contents. Gases are purified before admission to the combustion tube by passage through scrubber tubes containing magnesium perchlorate and Colorcarb to remove water vapour and traces of CO_2.

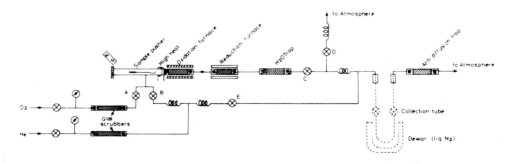

Fig. 1. Schematic of combustion system.

The oxidation furnace is maintained at 800°C around a quartz combustion tube of 11 mm i.d. with a packed length of 170 mm. Three catalysts are used, silver vanadate, silver oxide and silver tungstate, packed separately and separated by thin pads of quartz wool. The packed region is closed by 12 mm lengths of silver gauze which serve to remove sulphur compounds present as impurities.

After combustion, gases pass into the reduction tube containing a 300 mm packed length of 60-100 mesh copper powder terminated with silver gauze and maintained at 250°C. At this stage nitrogen oxides produced by combustion of nitrogen containing organic compounds are reduced to molecular nitrogen, N_2, and excess oxygen is removed. Water produced during combustion is then removed by a trap containing a 170mm length of magnesium perchlorate.

For sample collection, valve C opens and the gas mixture now containing only N_2, He and CO_2 passes to the liquid nitrogen cooled collection tube where CO_2 is collected as a solid whilst N_2 and He pass on to atmosphere through an anti diffusion trap containing silica gel which serves to minimise back diffusion of atmospheric gases.

Operation.

Process timing of valves A to E, the High Heat coil and the Pause indicator is controlled by a two part program stored in a PROM (programmable read only memory). The operating sequence is shown in figure 2. After loading the sample the operator presses a button to start the first program which flushes the system with He for 90 sec. before admitting a controlled volume of oxygen. At this point the system is ready for combustion to begin and a light comes on to indicate this condition. At this point the program stops to allow time for the operator to raise the liquid nitrogen Dewar around the collection tube and to move the sample into the hot zone of the combustion tube.

Once these manual operations are over, the sequence is restarted by a second push button and the system comes under control of the second program. Combustion proceeds for 70 sec at which point a "high heat" coil is switched on at the entrance to the hot zone to volatise any sample material which may have condensed on these cooler parts and so ensures complete combustion. After a further 10 seconds the furnace exit valve (C) opens and He is admitted (valve B) to sweep the reaction products through the system. The collection tube containing pure, dry CO_2 is then valved off and taken to the mass spectrometer for analysis.

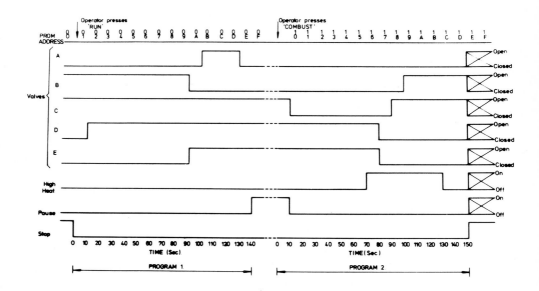

Fig. 2. Operating sequence.

RESULTS.

To test the system a number of samples were prepared with values expected to fall in the range -12 to -30°/oo with respect to PDB. A convenient reference was NBS22, an oil with a literature value of -29.4°/oo with respect to PDB{4}, and this was combusted on the system to provide a CO_2 reference. Three subsequent samples fell within ±0.3°/oo as shown in figure 3. A sample of Muscovado cane sugar labelled as produce of Guyana gave a value of -11.9°/oo, within the expected range for C4 plants and measurements on duplicate samples of wheat, honey and maple syrup gave mean values of -25.4, -24.2 and -22.5 respectively, all quite acceptable values for C3 plant origin {5,6,7}. The wheat was hand picked from a field in England, the honey was labelled "100% pure - product of more than one country" and the maple syrup was labelled as pure. The samples of human tissue gave values closely spaced around -21°/oo in keeping with a European diet mainly derived from C3 plant sources.

It is anticipated that the increase in speed and convenience brought about by this development will facilitate wider applications of ^{13}C SIRA to life sciences.

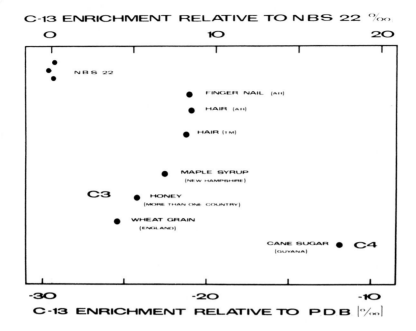

Fig. 3. $\delta^{13}C$ values obtained from combustion of various foods and samples of human tissue. C3 and C4 refer to photosynthetic pathways. NBS22 is an oil used for calibration.

REFERENCES.

1 C.R. McKinney, J.M. McCrea, S. Epstein, H.A. Allen and H.C. Urey Rev. Sci. Instrum. 21 (1950) 724-230.
2 N.I. Bridger, R.D. Craig and J.S.F. Sercombe. Adv. Mass Spectrom. 6 (1974) 365-375.
3 H. Craig, Geochim. Cosmochim. Acta 3 (1953) 53-92.
4 During this conference a comment was made by M. Schoelle that recent interlaboratory measurements indicated a revised value of -29.9°/oo for NBS22. If this is accepted, our values would require correction by -0.5°/oo.
5 J.C. Lerman and J.H. Troughton in E.R. Klein and P.D. Klein (Eds.) Proc. 2nd Int. Conf on Stable Isotopes, Oak Brook, Illinois, October 20-23, 1975, pp. 630-644.
6 J.C. Vogel, Fractionation of the Carbon Isotopes During Photosynthesis, Springer Verlag, Berlin Heidelberg 1980.
7 L.W. Doner and J.W. White Jr., Science 197 (1977) 891-892.

A NEW METHOD FOR THE DETERMINATION OF THE $^{18}O/^{16}O$ RATIO IN ORGANIC COMPOUNDS

C.A.M. BRENNINKMEIJER and W.G. MOOK
Isotope Physics Laboratory, University of Groningen, Westersingel 34, 9718 CM Groningen (The Netherlands)

ABSTRACT

A new method for the conversion of organic oxygen and water to CO_2 for $^{18}O/^{16}O$ analysis and some applications are presented. A batch of small nickel bombs containing the samples and nickel powder as catalyst is heated to 950 °C for pyrolysis. The H_2 formed diffuses away through the walls of the bombs, leaving a mixture of C, CO and CO_2. Because of the nickel catalyst, total conversion to CO_2 is obtained by leaving the bombs at 350 °C for one hour. The bombs are opened individually and the escaping CO_2 is purified and collected for mass spectrometric analysis.

The method has been applied for $^{18}O/^{16}O$ ratio determinations on cellulose extracted from peat for palaeoclimatological applications. The method works equally well for vanillin and probably for most other C, H, O organic compounds. Water samples have been converted to CO_2 by the addition of paraffin or graphite prior to pyrolysis.

INTRODUCTION

Analysis of the stable isotopic composition of carbon, hydrogen and oxygen in natural organic compounds, can yield invaluable information on a multitude of processes in nature. However, while determinations of the $^{13}C/^{12}C$ and D/H ratios are routinely performed in many laboratories, the complexity and time consuming nature of the analytical techniques available for $^{18}O/^{16}O$ analysis have limited the number of these analyses to a small fraction of the number of D/H analyses. The existing techniques, either chemical or based on pyrolyses, at least require the separate conversion of a mixture of CO and CO_2 to CO_2 for mass spectrometric analysis, a step which is prone to introduce isotope effects (refs. 1-4).

This paper reviews a new, simple method for the determination of natural $^{18}O/^{16}O$ ratios (ref. 5). This method has been applied for $^{18}O/^{16}O$ analyses of cellulose extracted from peat, demonstrating its potential for palaeoclimatological work. Many other applications are possible like tracing of foodwebs, determining the origin of food additives, further unravelling the oxygen metabolism in plants and the analysis of small water samples.

METHOD

The method is based on the total pyrolysis of organic compounds, e.g. cellulose, which results in a mixture of mainly C, CO, CO_2, H_2O and CH_4 (ref. 5). By applying nickel as sample container material during pyrolysis, H_2 is lost by diffusion, which forces several reactions to completion, resulting in C, CO and CO_2. For mass spectrometric analysis, either CO or CO_2 is required and complete conversion to either of these is necessary because of a strong oxygen isotopic fractionation.

At high pressures and low temperatures, the equilibrium $2CO \rightarrow CO_2 +$ + C shifts towards CO_2 but at very low speed. Coincidentally, nickel itself is one of the few catalysts which catalyse this reaction. The nickel container does not provide a large active surface area, and therefore nickel powder is added to the pyrolysis vessel. At a temperature of 350 ^0C complete conversion to CO_2 is obtained in one hour. A second improvement is the use of small nickel bombs for pyrolysis, which means that the conversion can be done in a batch process. In addition the sample containers will be isothermal, which excludes the formation of tar on colder parts. Therefore, the hydrogen that escapes during pyrolysis may well be used for D/H analysis (ref. 6).

Cellulose, 18 ± 2 mg, is inserted into a nickel bomb, which is machined from pure nickel rod (fig. 1). Next about 400 mg nickel powder (3 - 7 µm) is added and the bomb is provided with a nickel plug. A number of these bombs is evacuated overnight at 60 ^0C for drying and outgassing, after which helium or argon is allowed to leak into the bombs. Subsequently the plugs are argon-arc welded to the bombs. During this welding the lower parts of the bombs are clamped in copper blocks, which keeps the temperature of the contents well below 100 ^0C. Pyrolysis is done either in vacuo or in air; the former condition having the advantage that the rate of release of hydrogen can be measured. In 2 - 3 min. the temperature of the bombs is raised to 950 ^0C for pyrolysis. After 10 - 12 min. the rate of release of hydrogen becomes negligible and the bombs are cooled within 3 minutes to 350 ^0C for the conversion of CO to CO_2, which takes one hour. The bombs are opened

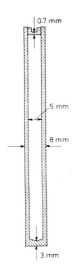

Fig. 1.
Nickel pyrolysis bomb.
The bombs are opened by piercing
the thin section in the plug.

individually and the escaping CO_2 is collected for mass spectrometric analysis. It is of importance to keep the period of pyrolysis as short as possible because the catalytic activity of the nickel powder deteriorates by exposure to CO_2 at high temperatures.

The nickel bombs can be re-used up to 20 times, gradually decreasing in lenght to 4 cm, by machining off a 2 mm wide rim for removal of the plug. After cleaning the bombs and heating them in oxygen to remove contaminants, they are reduced in hydrogen and outgassed at 1050 ^{0}C. The catalyst is a commercial nickel-powder which is reduced in hydrogen prior to use.

RESULTS

The accuracy of the method has been checked by analysing waters of known isotopic composition. About 10 mg of water is drawn into nickel capillary tubing which is subsequently sealed by spotwelding. Together with a slight excess of graphite or paraffin, the nickel capillaries are inserted into the nickel bombs for pyrolysis. The capillaries burst upon heating, allowing the water to react with the graphite or paraffin. The results are shown in table 1, indicating that only the value for the strongly depleted SLAP water shows a bias ($\delta^{18}O$ = -55.5 $^0/_{00}$). In the same table the results for repeated analysis of a commercial cellulose are shown, giving a standard deviation of 0.15 $^0/_{00}$.

Fig. 2 shows the $\delta^{18}O$ and δD results for cellulose extracted from peat sampled from a raised peat bog which has been deposited during the Subboreal-Subatlantic transition. The good correlation (r = 0.8)

TABLE 1

^{18}O determinations on cellulose and standard waters

	$\delta^{18}O$ $^0/_{00}$ (V-SMOW)	Number of determinations	Standard deviation
cellulose	26.8	28	0.15
V-SMOW	0.0	7	0.2
SLAP	-53.8	7	0.3

between δD and $\delta^{18}O$ is remarkable and is to some extend an independent proof of the good performance of the conversion process. Measurements like these show the potential of $^{18}O/^{16}O$ analyses in combination with D/H analyses for palaeoclimatological work. Basically, a shift upwards along the regression line indicates higher temperatures, whereas a shift to the right is more likely to be caused by higher evapotranspiration, i.e. dryer conditions.

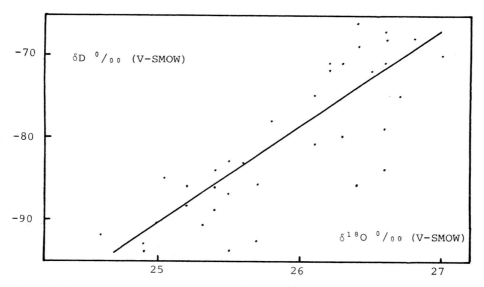

Fig. 2. The relationship between the δD and $\delta^{18}O$ of cellulose extracted from peat. The δD is determined for the carbon-bound hydrogens.

Table 2 shows the results of $\delta^{13}C$, δD and $\delta^{18}O$ determinations on vanillin samples from different origin. Vanillin is the most widely used flavoring agent. The procedure for the conversion of vanillin into CO_2 differs slightly from that of cellulose, in that respect that drying and outgassing prior to the sealing of the bombs is limited to a few hours at room temperature, because of the relatively high vapour pressure of vanillin. Duplicate measurements agree within 0.2 $^0/_{00}$ (average difference between 4 duplicate measurements). True vanillin is derived from plants belonging to the family Vanilla planifolia, which have a CAM metabolism. A substitude vanillin is usually synthesized from wood lignin (ref. 7) (C-3 metabolism) and is chemically indistinguishable from the natural product. Because of the two different carbon metabolism of C-3 and CAM plants, the $\delta^{13}C$ differs remarkably, which can be used to identify natural and synthesized vanillin (ref. 8 and 9). The difference in $\delta^{18}O$ between the two natural vanillins, NAT 1 and NAT 2, may point to a different geographical origin. The relatively large difference in δD cannot be explained easily, but it should be noted that the δD values are subject to an uncertainty because of the exchangeable hydroxyl hydrogen atom in the vanillin molecule. By considering the $\delta^{13}C$ values it can be concluded that sample SYN 1 is derived from wood lignin. Its δD seems to indicate that the origin of the wood must be from trees growing far north.

TABLE 2

Isotopic composition of different vanillin samples

Vanillin	δD $^0/_{00}$ (V-SMOW)	$\delta^{18}O$ $^0/_{00}$ (V-SMOW)	$\delta^{13}C$ $^0/_{00}$ (PDB)
NAT 1	− 55	11.8	−21.1
NAT 2	− 89	12.5	−22.4
SYN 1	−189	9.8	−27.0
SYN 2	− 50	0.3	−29.6

The $\delta^{18}O$, which differs not much from that of natural vanillin may provide information on the oxidation process used to synthesize it from lignin. Sample SYN 2 has a slightly deviating $\delta^{13}C$ value and is most likely derived from eugenol (glove oil). However, it is clearly marked by its deviating $\delta^{18}O$. This fact may be used to determine mixtures of the two synthetic vanillins.

CONCLUSIONS

The new method for the conversion of C, H, O organic compounds into CO_2 for $^{18}O/^{16}O$ analysis proves to work for cellulose and vanillin. Also the determination of the $^{18}O/^{16}O$ ratio on small water samples is possible. The small analytical error and the straightforward nature of the method opens the way to research on oxygen isotope effects in nature.

REFERENCES

1 W.E. von Doering and E.J. Dorfman, J. Am. Chem. Soc., 75(1953)5595.
2 J. Agget, C.A. Bunton, T.A. Lewis, D.R. Llewellyn, C. O'Connor and A.L. Odell, Int. J. Appl. Radiat. Isot., 16(1965)165.
3 K.G. Hardcastle and I. Friedmann, Geophys. Res. Lett., 1(1974)165.
4 P. Thompson and J. Gray, Int. J. Appl. Radiat. Isot., 28(1977)411.
5 C.A.M. Brenninkmeijer and W.G. Mook, Int. J. Appl. Radiat. Isot., 32(1981)137.
6 P. Thompson, J. Gray and S.J. Song in G.C. Jacoby (Ed.), Proc. Int. Meet. on Stable Isotopes in Tree-Ring Research, New Paltz, New York, May 22-25, 1979, NTIS/CONF.-7905180, Dec. 1980, p. 142.
7 K.V. Sarkanen and C.H. Ludwig, Lignins, Wiley-Interscience, New York, 1971, p.800.
8 J. Bricout and J. Koziet, Ann. Fals. Exp. Chim., 69(1975)245.
9 P.G. Hoffman and M. Salb, J. Agric. Food Chem., 27(1972)352.

H.-L. Schmidt, H. Förstel and K. Heinzinger (Editors), *Stable Isotopes*
© 1982 Elsevier Scientific Publishing Company, Amsterdam — Printed in The Netherlands

ANCILLARY TECHNIQUES FOR $^{13}CO_2$ GENERATION IN BIOMEDICAL STUDIES

W.W. WONG, D.A. SCHOELLER* and P.D. KLEIN
Stable Isotope Laboratory, Children's Nutrition Research Center, Department of Pediatrics, Baylor College of Medicine, Houston, Texas 77030 (USA) and
*Department of Medicine, University of Chicago, Chicago, Illinois 60637 (USA).

ABSTRACT

Vacutainers® can be used for processing CO_2 generated from the combustion of stool samples and as a reaction vessel for in vitro generation of CO_2 from urinary urea by urease for gas isotope ratio mass spectrometric measurements. Vacutainers used in conjunction with the sealed quartz tube combustion technique, a direct transfer system, and an automated purification system greatly reduce processing time for isotopic measurements on stool and urine samples.

INTRODUCTION

Vacutainers are useful particularly for collecting and transporting breath samples for mass spectrometric analysis. The samples can be introduced and retrieved easily from the Vacutainers through the rubber septum with a syringe; the Vacutainers can be shipped in large numbers without breakage; there is no isotope fractionation of breath CO_2 collected in Vacutainers (ref. 1); breath CO_2 can be stored in Vacutainers for a long period of time without changes in isotopic content (ref. 1); Vacutainers are available commercially and are reasonably economical; and most important of all, breath CO_2 in Vacutainers can be processed immediately for isotopic measurement in a gas isotope ratio mass spectrometer equipped with an automated purification inlet system (ref. 2).

The use of Vacutainers, in conjunction with an automated purification system, is one of the primary factors which has led to the growing popularity and wide scale use of $^{13}CO_2$ breath tests in clinical and nutritional studies. Breath $^{13}CO_2$ measurements alone, however, cannot provide all the information necessary for biomedical studies. Stool and urine samples also can be collected in a non-invasive manner from a test subject, however the lengthy workup process has made fecal and urinary ^{13}C analyses very laborious for use in a ^{13}C protocol.

Following is a description of the manner in which stool samples may be converted easily into CO_2 by combustion, how the CO_2 from the combusted sample can be transferred to a Vacutainer via a direct transfer system, and how Vacutainers can be used for in vitro generation of CO_2 from urinary urea by urease. These techniques enhance the measurement of ^{13}C abundance by gas isotope ratio mass spectrometry.

EXPERIMENTAL TECHNIQUES AND RESULTS

Conversion of Stool Samples to CO_2 by Combustion

In 1976, DesMarais and Hayes (ref. 3) introduced the sealed quartz tube combustion technique for conversion of organic samples to CO_2 for mass spectrometric analysis. They also introduced the use of a tube cracker inlet on the vacuum line. This technique and the tube cracker inlet enable simultaneous combustion of large numbers of samples. The collection of CO_2 from the individual sample combusted for isotope measurement, however, has remained a time-consuming process. This lengthy process can be minimized by means of a direct transfer system as shown in Fig. 1.

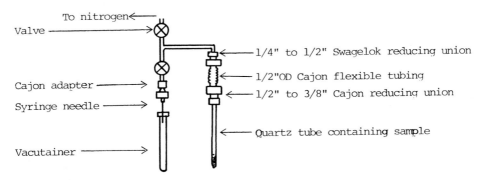

Fig. 1. Direct transfer system.

This system consists of a tube cracker system and two valves: one valve is connected to a source of pure nitrogen and the other is connected to a syringe needle via a Cajon Ultra-Torr adapter. The following procedure is used for transferring CO_2 from a combusted sample to the Vacutainer: 1) the system is purged with nitrogen; 2) after closing the valve to the syringe needle, a Vacutainer is attached to the system through the needle and then is immersed in liquid nitrogen; 3) the quartz tube containing the combusted sample is attached loosely to the system by means of the reducing union of the cracker system; 4) after purging the cracker system with N_2, the direct transfer system is isolated from the source of N_2 and the nuts of the reducing union are tightened; 5) the tube is fractured by bending the flexible tubing of the cracker system; and 6) after opening the valve to the Vacutainer for 5 min, the Vacutainer is removed from the transfer system, brought to atmospheric pressure with pure N_2, and placed directly onto the carousel of the automated purification system attached to the gas isotope ratio mass spectrometer.

Table 1 shows the isotopic carbon content of CO_2 generated from seven NBS #22 samples obtained using the sealed quartz tube combustion technique and the direct transfer system. The results demonstrate that the combustion technique and use of the transfer system did not introduce an isotope effect. The experimental mean $\delta^{13}C$ value observed was the same as the theoretical value of -29.8 ‰ for NBS #22.

TABLE 1

The $\delta^{13}C$ values of NBS #22 combusted to CO_2 in quartz tubes and transferred directly to Vacutainers

NBS #22	$\delta^{13}C$ vs PDB, ‰
1	-30.2
2	-30.2
3	-29.5
4	-29.8
5	-29.1
6	-29.0
7	-30.2
Mean $\delta^{13}C$ vs PDB =	-29.7 ± 0.5 ‰
Theoretical $\delta^{13}C$ =	-29.8 ‰

TABLE 2

The $\delta^{13}C$ values of total fecal carbon collected from premature infants, children and adults. (Condensed from Schoeller, et al. 1981)*

Subject	Diet	Feces	Mean $\delta^{13}C$ vs PDB ± SD, ‰	
Infant	milk formula	24-h	-25.7 ± 0.9 (4)[†]	
Infant	milk formula	24-h	-26.8 ± 0.4 (4)	
Infant	milk formula	24-h	-24.9 ± 1.1 (4)	
Infant	milk formula	24-h	-26.2 ± 0.5 (4)	
		Mean	-25.9 ± 0.8 ‰	
Child	solid foods	24-h	-25.5 ± 0.3 (3)	
Child	solid foods	24-h	-27.8 ± 0.5 (3)	
Child	72-h repeating diet	72-h	-19.6 ± 0.2 (3)	
Child	72-h repeating diet	72-h	-23.5 ± 0.7 (3)	
Child	72-h repeating diet	72-h	-15.4 ± 1.0 (3)	
Child	72-h repeating diet	72-h	-20.5 ± 0.1 (3)	
		Mean	-22.1 ± 4.5 ‰	
Adult	solid foods	24-h	-19.6 ± 0.4 (2)	
Adult	solid foods	24-h	-21.6 ± 0.4 (3)	
		Mean	-20.6 ± 1.4 ‰	

*Reprinted by permission of Am. J. Clin. Nutr.
[†]The number of collection runs.

Table 2 summarizes the ^{13}C data previously measured from CO_2 generated from combustion of total fecal materials collected from premature infants, children, and adults (ref. 4). With the exception of the infants, the children and adults were on uncontrolled solid diets. Even though the ^{13}C isotopic content of total fecal carbon differs by as much as 12 ‰ between individuals, the fecal ^{13}C abundance remains very constant within a single individual for as long as 9 days. Therefore, as suggested previously, changes in isotopic composition of fecal carbon can be used to monitor malabsorption of ^{13}C labeled substrates and thereby supplement the $^{13}CO_2$ breath test data.

Enzymatic liberation of CO_2 from urinary urea by urease

Table 3 summarizes the isotopic composition of CO_2 generated from urea by combustion and by urease. In the first two experiments, solid urea and a

TABLE 3

Variation in the $\delta^{13}C$ values of urea carbon generated by combustion and by urease

Form of urea	Vessel (Mode of CO_2 formation)	Mean $\delta^{13}C$ value vs PDB, ‰
Urea (solid)	Quartz tube (combustion)	-42.9 ± 0.8 (4)*
Urea (0.5M)	Quartz tube (combustion)	-42.1 ± 0.8 (4)
Urea (0.5M)	Reaction vessel (enzymatic)	-41.9 ± 0.6 (2)
Urea (0.5M)	Vacutainer (enzymatic)	-41.4 ± 0.7 (4)

*The values in parentheses are the number of individual runs.

urea solution were converted to CO_2 by combustion and the isotopic composition of the CO_2 was determined. In the last two experiments, the isotopic content of CO_2 liberated from the urea solution by urease in gas conversion reaction vessels and in Vacutainers was measured. The results from these four experiments indicate that CO_2 generated from urea by urease accurately reflects the isotopic composition of urea carbon. Thus enzymatic liberation of labeled CO_2 within the Vacutainer provides yet another source of diagnostic information from the use of substrates.

TABLE 4

The $\delta^{13}C$ values of breath CO_2 and urea carbon from six normal adults

Subject	Sex	$\delta^{13}C$ value vs PDB, ‰		Difference, ‰
		Breath CO_2	Urea carbon	(Urea - Breath)
B.W.	M	-21.6	-18.9	$+2.7$
C.I.	M	-21.3	-19.6	$+1.7$
C.L.	M	-20.9	-22.3	-1.4
P.K.	M	-22.0	-19.3	$+2.7$
W.W.	M	-20.4	-19.3	$+1.1$
L.B.	F	-21.9	-20.5	$+1.4$
	Mean	-21.4 ± 0.6	-20.0 ± 1.3	$+1.4 \pm 1.5$

Table 4 shows the $\delta^{13}C$ values of breath CO_2 and urinary urea carbon collected from six normal adults. Even though the breath CO_2 isotopic composition was constant among these individuals, the urea carbon differed by as much as 3 ⁰/₀₀. It is interesting to note that breath carbon, with the exception of subject C.L., was more depleted in ^{13}C than the corresponding urea carbon.

CONCLUSION

Vacutainers used in conjunction with a gas isotope ratio mass spectrometer equipped with an automated purification inlet system have accelerated the processing of breath samples in $^{13}CO_2$ breath tests. We have demonstrated that the same system, when combined with the sealed quartz tube combustion technique, the direct transfer system, and specific in vitro enzymatic reactions can decrease significantly the time required for ^{13}C analysis of stool and urine samples. Therefore, the workup time for stool and urine samples no longer should be a limiting factor in the design of protocols using ^{13}C labeled substrates.

ACKNOWLEDGMENTS

This work is a publication of the USDA/SEA, Children's Nutrition Research Center, Department of Pediatrics, Baylor College of Medicine and Texas Children's Hospital. The work also was supported by NIH grant #AM 28129.

REFERENCES

1 D.A. Schoeller and P.D. Klein, Biomed. Mass Spectrom., 5(1978)29-31.
2 D.A. Schoeller and P.D. Klein, Biomed. Mass Spectrom., 6(1979)350-355.
3 D.J. DesMarais and J.M. Hayes, Anal. Chem., 48(1976)1651-1652.
4 D.A. Schoeller, P.D. Klein, W.C. MacLean, Jr., J.B. Watkins and E. van Santen, J. Lab. Clin. Med., 97(1981)439-448.

A METHOD OF MASS SPECTROMETRIC ISOTOPIC ANALYSIS OF NITROGEN AND OXYGEN IN DOUBLY LABELLED NITRIC OXIDE

M.N. KERNER, K.G. ORDZHONIKIDZE, L.P. PARULAVA and G.A. TEVZADZE
Institute of Stable Isotopes, State Committee on the Uses of Atomic Energy, 380086 Tbilisi (USSR)

ABSTRACT

A method of isotopic analysis of nitrogen and oxygen directly in nitric oxide in a wide range of concentrations has been developed.

The isotopic analysis of nitrogen- and oxygen-labelled nitric oxide is limited by a number of systematic measuring errors.

As a result of the studies the sources of the errors have been found and conditions of performing the analysis that ensure the elimination or minimization of errors were established.

Equations were obtained for the determination of atomic fractions of all nitrogen and oxygen isotopes in the nitric oxide taking into account the overlapping components.

The above method allows simultaneous isotopic analyses of both oxygen and nitrogen to be performed in one and the same sample of nitric oxide labelled with oxygen and nitrogen isotopes.

In solving different problems in biology, medicine, agricultural chemistry and environmental pollution, a need arises of performing isotopic analyses of nitrogen and oxygen in nitric oxide. For these studies, the nitric oxide can be doubly labelled with the isotopes of nitrogen and oxygen.

In the separation of nitrogen and oxygen isotopes by low-temperature distillation, one of the principle methods of production of stable isotopes of oxygen and nitrogen, the product withdrawn from the separation plant is doubly labelled nitric oxide. It is evident that the nitrogen and oxygen isotope analyses can be performed to advantage directly in the nitric oxide, without any further chemical treatment. This can be done on a mass spectrometer with a resolution of not less than 15 000. We have developed an isotope analysis technique for oxygen and nitrogen in doubly labelled nitric oxide using relatively inexpensive mass spectrometers of the type available at many research laboratories.

Methods of isotopic analysis of nitrogen in nitric oxide enriched with nitrogen-15 have been developed and described in a number of works [1-5].

The studies have shown that some limitations appear during the isotopic analysis of oxygen in the nitric oxide. Nitric oxide is known to readily react with oxygen [6]; the resulting nitrogen dioxide easily dissociates in the ion source of the mass spectrometer forming nitric oxide [7]. The presence of the background ion currents of the molecular and the elemental oxygens of natural isotopic abundance leads to the dilution of the samples due to the following reactions in the ion source:

$$N^{18}O + \frac{1}{2} {}^{16}O_2 \rightarrow N^{16}O^{18}O \tag{1}$$

$$N^{16}O^{18}O + e \rightarrow N^{16}O^+ + {}^{18}O + 2e. \tag{2}$$

The degree of dilution is dependent on the composition and partial pressure of the residual gases in the instrument, on the absorption properties of the structural elements of the ion source and the analyzer chamber, and on the analysis time as well.

The investigations were carried out on two mass spectrometers, MI-1201 and MI-1305, differing in the residual gas pressures in the ion source by one and a half orders of magnitude and also in absorption properties of the ion source. The ion source in the MI-1201 is made of stainless steel, that in the MI-1305 of copper.

The results of the isotopic analyses of oxygen in the NO made on both instruments are given in table 1, together with the results for double samples of molecular oxygen obtained from the NO by dissociation in a high-voltage discharge; the latter values are supposed to be true [8].

The data in table 1 show that the MI-1305 gives incorrect values for the ^{18}O-concentration in the NO, and a considerable decrease of the concentrations is observed as the analysis proceeds. The ^{18}O-concentrations in the NO practically coincide with the actual values when the measurements are made on the MI-1201. A reduction of the values is observed only in samples 3 and 5 where the intensities of the background ion currents of water and oxygen have been intentionally increased by an order of magnitude (see table 1). These experiments lead to the conclusion that the isotopic analysis of oxygen in nitric oxide can be performed on instruments that are constructed of stainless steel when the residual gas pressure in the ion source does not exceed $5 \cdot 10^{-6}$ Pa.

TABLE 1
Isotopic analyses of oxygen in NO performed on the MI-1305 and the MI-1201 mass spectrometers

	Analyses on the MI-1305		Analyses on the MI-1201			True
	^{18}O-concentration in NO (atom-%)		^{18}O-concentration in NO (atom-%)		"$O_2^+ + O^+$" ions background currents intensities (A)	concentrations
	Initial value	1 hour later	Initial value	1 hour later		(atom-%)
1	20.5	17.8	27.2	27.2	$6 \cdot 10^{-14}$	27.2
2	25.3	22.0	32.6	32.5	$8.5 \cdot 10^{-14}$	32.6
3	30.8	27.4	39.2	37.4	$5.2 \cdot 10^{-13}$	40.3
4	40.1	36.0	50.8	50.7	$6.1 \cdot 10^{-14}$	50.8
5	52.2	47.1	68.8	66.3	$4.5 \cdot 10^{-13}$	70.4
6	64.2	58.2	79.9	79.8	$7.2 \cdot 10^{-14}$	79.8
7	69.4	62.5	84.7	84.7	$4.2 \cdot 10^{-14}$	84.7

During the isotopic analyses of nitrogen and oxygen in the doubly labelled nitric oxide, a superimposition of different isotopic forms of nitric oxide with equal mass numbers occurs. In order to distinguish the doublets on the 31 and 32 mass lines a mass spectrometer is required with a resolution of not less than 15 000 ("Class 1").

We have attempted to find a way, determining all the oxygen and nitrogen isotope concentrations in nitric oxide without resolving the doublets.

Since the probabilities of forming the different isotopic forms of nitric oxide molecules are proportional to the products of the corresponding isotope concentrations and the ionization cross sections for all the NO isotopic forms are the same, we obtain the following equations:

$$J_{30} = k \cdot X_{16_O} \cdot X_{14_N} \tag{3}$$

$$J_{31} = k \cdot (X_{17_O} \cdot X_{14_N} + X_{16_O} \cdot X_{15_N}) \tag{4}$$

$$J_{32} = k \cdot (X_{17_O} \cdot X_{15_N} + X_{18_O} \cdot X_{14_N}) \tag{5}$$

$$J_{33} = k \cdot X_{18_O} \cdot X_{15_N}, \tag{6}$$

where k is the factor of proportionality, J_{30}, J_{31}, J_{32}, J_{33} are the ion peak intensities corresponding to 30, 31, 32, 33 mass numbers and X_{16_O}, X_{17_O}, X_{18_O}, X_{15_N} the isotopic concentrations of the corresponding isotopes. We know that

$$X_{16_O} + X_{17_O} + X_{18_O} = 1 \tag{7}$$

$$X_{14_N} + X_{15_N} = 1. \tag{8}$$

Equations (3) - (8) are a system of equations with six unknowns: k, X_{16_O}, X_{17_O}, X_{18_O}, X_{14_N}, X_{15_N}.

The factor of proportionallity k can be determined by summing up equations (3) - (6) with account of equations (7) and (8):

$$k = \sum_{i=30}^{33} J_i. \tag{9}$$

We introduce the following substitutions: $a = \frac{J_{33}}{k}$, $b = \frac{J_{32}}{k}$, $c = \frac{J_{31}}{k}$, $d = \frac{J_{30}}{k}$. If we choose three equations from relations (3) - (6) (in the case of 50 atom-% N-15 and O-18 concentrations equations (4), (5) and (6) are more suitable) and taking into account equations (7) and (8), we obtain the system of three equations with three unknown concentrations, e.g., X_{18_O}, X_{17_O}, X_{15_N}, which can be reduced to a cubic equation with one unknown concentration:

$$X_{18_O}^3 - (3a+b) \cdot X_{18_O}^2 + (3a^2 + 2ab + ac) \cdot X_{18_O} - d^2 = 0. \tag{10}$$

By solving this cubic equation one can get the O-18 concentration. The remaining isotope concentrations are obtained from the equations:

$$X_{17_O} = \frac{(a+b) \cdot X_{18_O} - X_{18_O}^2}{a} \tag{11}$$

$$X_{15_N} = \frac{a}{X_{18_O}}. \tag{12}$$

The information was processed on a desk program-controlled computer.

It should be mentioned, that one more type of systematic measuring error appears during the oxygen and nitrogen isotope analysis in the doubly labelled nitric oxide: As a result of the NO dissociation with subsequent recombination of oxygen and nitrogen atoms in the ion source according to the reaction $2NO \rightarrow N_2 + O_2$, a superimposition occurs of the $^{15}N_2^+$, $^{16}O_2^+$, $^{16}O^{17}O^+$ on the ions of nitric oxide with 30, 32, 33 mass numbers. In order to resolve the resulting doublets, a Class 1 mass spectrometer is also required.

It is known [1] that by reducing the ionizing potential to 20 V it is possible to decrease the measuring error that is due to the superimposition of the $^{15}N_2^+$ on mass line 30. Our investigations showed that by means of reducing the ionizing potential to 18 V it is possible to reduce the molecular oxygen ionization to so small a value that it can be neglected; the ionization of the nitric oxide at that potential is sufficient for isotopic analysis.

In table 2 the data obtained on the MI-1201 for oxygen and nitrogen in the doubly labelled NO are given, together with the results for double samples of the mixture of the molecular oxygen and nitrogen obtained from NO by its dissociation in a high-voltage discharge. The analyses were carried out at the ionizing potential 18 V and the isotopic concentrations were calculated with equations (10), (11) and (12). The oxygen isotope concentrations in the "$O_2 + N_2$" mixture were calculated with the help of the equations presented in [9], which allowed the exact values of all oxygen concentrations in the O_2 without resolving the doublet on mass line 34 ($^{16}O^{18}O$ and $^{17}O_2^+$) to be determined.

As shown in table 2, there is agreement within the measuring error between the data for the double samples.

Thus, as a result of the studies, the task of performing the oxygen and nitrogen isotopic analysis in the doubly labelled nitric oxide has been accomplished for a wide range of nitrogen and oxygen concentrations with the help of the MI-1201 mass spectrometer.

TABLE 2

Isotopic analyses of oxygen and nitrogen in doubly labelled NO and in the "$O_2 + N_2$" mixture on the MI-1201

	Analyses in NO ionizing potential 18 V			Analyses in "$O_2 + N_2$" mixture		
	^{18}O	^{17}O	^{15}N	^{18}O	^{17}O	^{15}N
	(atom-%)	(atom-%)	(atom-%)	(atom-%)	(atom-%)	(atom-%)
1	34.3±0.1	11.5±0,1	65.5±0.1	34.2±0.1	11.4±0.1	65.5±0.1
2	41.5±0.2	13.9±0.1	68.4±0.1	41.4±0.1	13.8±0.1	68.3±0.1
3	58.1±0.2	6.96±0.07	55.7±0.2	58.0±0.1	6.94±0.07	55.8±0.1
4	78.0±0.1	2.40±0.03	19.6±0.1	78.0±0.1	2.39±0.03	19.5±0.1
5	83.6±0.1	0.80±0.01	29.6±0.1	83.6±0.1	0.78±0.01	29.5±0.1
6	83.6±0.1	0.21±0.003	16.2±0.1	83.5±0.1	0.210±0.03	16.2±0.1
7	83.8±0.1	0.901±0.01	30.2±0.1	83.7±0.1	0.89±0.01	30.2±0.1
8	84.3±0.1	0.851±0.01	29.1±0.1	84.2±0.1	0.84±0.01	29.2±0.1

REFERENCES

1 K.G. Ordzhonikidze and M.C. Mikhalashvili, Zavod. Labor., 10 (1964)1218.
2 R. Bes, G. Lacoste and J. Mahenc, Methods Phys. Anal., 6(1970) 109.
3 F. Lantelme and M. Chemla, Bull. Soc. Chim., (1968)1314.
4 L. Friedman and J. Bigeleisen, J. Chem. Phys., 18(1950)1325.
5 V.M. Akulintsev, N.M. Gorshunov and S.E. Kupriyanov, Khimia vysokikh energii, 12 6 (1978)554.
6 V.F. Volynets and M.P. Volynets, Analiticheskaya khimia azota, Nauka, Moscow, 1977, p. 25.
7 K.G. Ordzhonikidze, M.N. Kerner and L.L. Birkaya, in Proc. Techn. Comm. Meeting on Stable Isotopes in the Life Sciences, Leipzig, 14-18 Febr. 1977, IAEA, Vienna, 1977, pp. 165-170.
8 K.G. Ordzhonikidze, M.N. Kerner and L.L. Birkaya, Zavod. Labor. 8(1979)725.
9 G.Haase, Isotopenpraxis, 5(1979)140.

APPLICATION OF THE MICROWAVE PLASMA DETECTOR IN METABOLIC STUDIES
WITH THE ANTIARRHYTHMIC DRUG PROPAFENONE

H.G.HEGE and J.WEYMANN
Knoll AG, Ludwigshafen/Rhein (G.F.R.)
BASF, Untern.-Bereich Pharma, Experimentell-Medizinische Forschung
und Entwicklung

ABSTRACT

The elimination of propafenone and its metabolites in urine and bile was determined after simultaneous oral administration of ^{14}C- and ^{2}H-propafenone to dogs. The radioactivity was measured by liquid scintillation counting and the atomic emission of deuterium by the MPD.[+] The metabolites were quantified after the metabolites in urine, bile and plasma samples had been separated by thin layer chromatography, high performance liquid chromatography and gas chromatography (metabolic pattern). When compared, the results of the radioactivity measurements were very consistent with the deuterium measurements. With the MPD it is possible to monitor the excretion of propafenone and metabolites and to evaluate a metabolic pattern.

When applying the microwave plasma detector the employment of stable isotopes in metabolic studies is an interesting alternative to using radioactive isotopes.

INTRODUCTION

It has been established [1-3] that a helium microwave plasma emission source can be used as the basis for an elemental analyser for chromatography. In this paper the application of the microwave plasma detector (MPD)[+] in metabolic studies with the antiarrhythmic drug propafenone is described. The excretion of propafenone and its metabolites in urine and bile was determined after simultaneous oral administration

---

[+] Abbreviations used: MPD = microwave plasma detector (Applied Chromatography Systems Ltd., Luton, England), DSI = direct sample injector, TLC = thin layer chromatography, HPLC = high performance liquid chromatography, GC = gas chromatography, LSC = liquid scintillation counting, MBTFA = N-Methyl-bis-trifluoracetamid

of 5 mg/kg of a mixture of ^{14}C-labelled and ^{2}H-labelled propafenone to dogs. The radioactivity was measured in the eluate of a XAD-2-column by liquid scintillation counting (LSC) and the atomic emission of deuterium by the MPD via the direct sample injector (DSI). Thereafter the metabolites in urine, bile and plasma samples were separated by chromatography (TLC, HPLC or GC) and in each fraction the radioactivity and the deuterium content were determined respectively (metabolic pattern). The results of the deuterium measurements were compared with the respective radioactivity measurements.

RESULTS

Excretion

The cumulative excretion of propafenone and its metabolites in urine and bile is shown in Fig. 1.

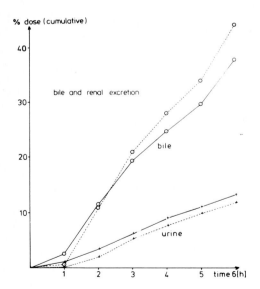

Fig. 1. Cumulative bile and renal excretion of propafenone and metabolites
——— ^{2}H measurements with the MPD via the DSI
--- ^{14}C measurements by LSC

The broken line represents the ^{14}C total radioactivity and the continuous line the excretion of total deuterium. For experimental purposes the excretion was monitored for only 6 hours. The curves are not completely superimposable but the radioactivity curve and the deuterium curve are largely parallel. As can be seen from these results, propafenone is predominantly excreted via the bile with the faeces, and about 12 % of the dose is recovered in the urine within the period indicated above. The reduced deuterium content in the bile, as compared to the radioactivity, results from the metabolic instability of the deuterium labelling (microsomal oxidation causes loss of one of the five deuterium atoms of the labelling in the phenyl ring of the propafenone molecule).

Metabolic pattern in the bile

Figure 2 shows the metabolic pattern in the bile after separation by HPLC. The eluate of the HPLC column was separated into 200 ul fractions. In a total of 150 fractions, collected within 30 minutes, the radioactivity was measured by LSC and the deuterium content with the MPD via the DSI. ^{2}H-propafenone and ^{14}C-propafenone served as external standards respectively.

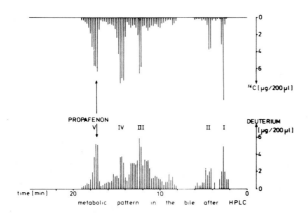

Fig. 2. Metabolic pattern in the bile after separation by HPLC.
HPLC-System: RP 18, water/methanol/buffer (ammonium formiate), linear gradient
upper part = radioactivity measurements; lower part = deuterium measurements;
ordinate scale: concentration [ug-equivalents propafenone/200 ul eluate]
abscissa: retention time [min]; the retention time of propafenone is indicated.

The figure clearly illustrates how well the two measuring methods correspond. Both plots have 5 clear-cut peaks; additional components, smaller in quantity, are recognizable.

Figure 3 shows the corresponding metabolic pattern after separation by TLC. Here, in defined zones, the silicagel is scratched off across the total width of the plate. The bands were eluated with methanol. The radioactivity content was then determined in these fractions by LSC and the deuterium content with the MPD via the DSI.

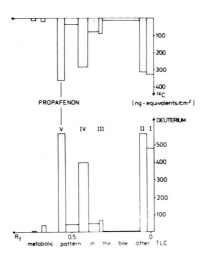

Fig. 3. Metabolic pattern in the bile after separation by TLC.
Solvents: butylacetate/propanol-2/conc. ammonia (60:35:5)
upper part = radioactivity measurements; lower part = deuterium measurements;
ordinate scale = amount [ng-equivalents propafenone/cm^2 silicagel];abscissa = Rf-value
The Rf-value of propafenone is indicated.

The radioactivity measurements and the deuterium measurements correspond well with each other. About 5 intense peaks can be recognized in the two plots along with few smaller components.

Figure 4 shows the metabolic pattern in the bile after separation by GLC. Now the chromatographic system is coupled on-line with the MPD and the radioactivity detector respectively. The figure shows a chromatogram of a bile extract which was injected into the gas chromatograph after derivatization with MBFTA. The microwave-plasma detector records the hydrogen line and the deuterium line simultaneously (Fig. 4a). On recording the hydrogen line all volatile hydrogen containing compounds are detected. The deuterium line shows specifically the deuterium containing metabolites only. The trace of the radioactivity detector (^{14}C) and the trace of the flame ionization detector (FID) are shown in Fig. 4b.

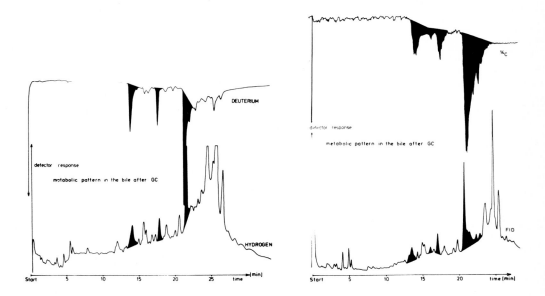

Fig. 4. Metabolic pattern in the bile after separation by GLC.
GLC-System: 2 m x 2 mm i.d. / 3 % SE 30 / chromosorb Q 100 —120 mesh/150-300°/6°/min.

Fig. 4a: upper part: MPD-deuterium line
       lower part: MPD-hydrogen line

Fig. 4b: ^{14}C-curve
       FID-curve

In comparison the radioactivity curve and the deuterium curve are very similar (in spite of the fact, that the MPD has a smaller dead volume than the radioactivity detector, which means that the MPD provides smaller peaks). Now from the retention times of the deuterium containing peaks structure elucidation of the metabolites by GC/MS can easily be performed.

CONCLUSION

With propafenone as an example it was proved that the microwave plasma detector can be used in metabolic studies. The great advantage of this detector is that it allows selective detection of deuterated compounds. Therefore, it had to be tested whether sensitivity and selectivity are sufficient to monitor the fate of propafenone in the organism. This was to be done by comparative measurements of radioactivity and deuterium.

The excretion of total deuterium and total radioactivity are comparable. With the two measuring methods the metabolic pattern in plasma, bile and urine provided identical chromatograms.

The application of radioactive labelled compounds is preferred in animal studies because of more favourable selectivity, detection limits and sample handling (the limit of detection can be reduced by increasing the specific radioactivity, the selectivity is extremely high as only the radioactive material "radiates"). However, the administration of deuterated substances and their measurement by means of the microwave plasma detector offers a true alternative for metabolic studies in humans.

REFERENCES

1 A.J.McCormack, S.C.Tong, W.D.Cooke, Anal. Chem., 37(1965)1470
2 W.R.Mclean, D.L.Stanton, G.E.Penketh, Analyst., 98(1973)432
3 K.S.Brenner, J. Chromatog., 167(1978)365

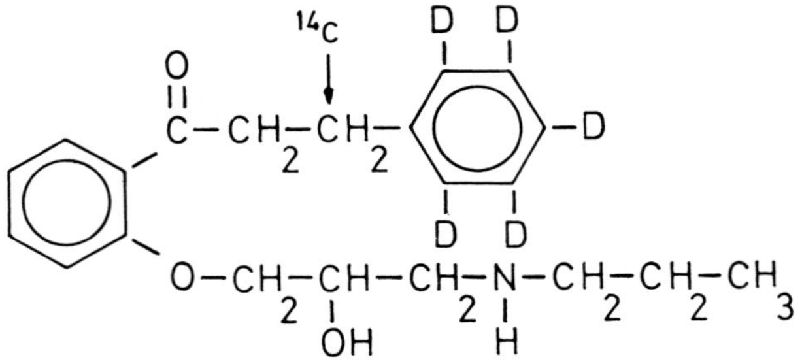

^{14}C, ^{2}H - Propafenone

AUTORADIOGRAPHIC DETERMINATION OF ^{18}O BY PROTON ACTIVATION ANALYSIS

Y. OHTA, T. INADA[+], A. MARUHASHI[+], T. KANAI[++], K. KOUCHI[++], and M. AIHARA[+++]
National Institute for Environmental Studies, Yatabemachi, Tsukuba, Ibaraki (Japan)
[+] University of Tsukuba
[++] National Institute of Radiological Sciences
[+++] Institute of Whole Body Metabolism

ABSTRACT

The determination of ^{18}O labelled compounds in biological tissue was achieved by an autoradiografic method using the nuclear reaction $^{18}O(p,n)^{18}F$. The energy of a proton beam employed for irradiation from a Van de Graaff was degraded to 2.8 Mev, from a cyclotron to about 5 Mev or 3 Mev at the sample position. The [^{18}O]-glycine has led to exposures from stomach and intestines, the intravenously administered one from the whole body. The possibility of interference from proton reactions with other elements such as ^{16}O, ^{12}C and ^{14}N was investigated. The detected sensitivity was calculated to be 1.176 µg of [^{18}O]-glycine at an optical density of 0.009.

INTRODUCTION

Oxygen is a major component of biological materials. It is difficult to observe directly the concentration of oxygen labelled compounds in biological tissue as an indicator for distribution or for a metabolic pathway. The radioisotopes of oxygen can not be applied as a tracer because of their short half-lives.
The determination of the distribution of ^{18}O-labelled compounds in biological tissue has been achieved by an autoradiographic method using the nuclear reaction $^{18}O(p,n)^{18}F$. Numerous techniques for the activation analysis of oxygen have been described before. However, there is no report of a (p,n) reaction employed in the determination of the distribution of ^{18}O in biological tissue. The photographic

emulsions were exposed to positrons and their annihilation photons from ^{18}F in a manner similar to ^3H or ^{14}C autoradiography.

Experiments were performed in several ways. ^{18}O labelled glycine (74.1 atom% enriched) was orally administered or intravenously injected to rats and mice or they were exposed to $N^{18}O_2$ (47.1 atom% enriched) and $^{18}O_2$ (99 atom% enriched). These animals were sacrificed by placing them into liquid nitrogen in order to take thin sliced samples from the whole body. The thickness of each slice was 50 μm. The slices were prepared for irradiation by using lyophilization.

The irradiation was done by a 20 Mev proton beam at its Bragg peak position in air. The beam extracted through a steel vacuum window and aluminium foils was scattered and had a flat intensity distribution of 40 cm diameter after a 250 cm flight through an air bag. The mean energy of the protons in the Bragg peak was estimated to be about 2.9 Mev being in the range of the peak in the total cross section of $^{18}O(p,n)^{18}F$ at 2.65 Mev (refs. 1,2). In addition, a 60 Mev cyclotron at maximum energy was employed for irradiation. The proton beam was scattered by a screen for dispersion and its intensity at the Bragg peak was rather flat within a radius of about 4 cm after 4 meters of flight. The average energy of the protons in the Bragg peak was estimated to be about 5 Mev (ref. 3)

RESULTS

The macroautoradiograms from the $^{18}O(p,n)^{18}F$ reaction

The samples were kept in contact with the ^3H type films for at least six hours. The typical cases are illustrated in Figs. 1-6.

Decay curves of ^{18}F radiation

The occurence of the reaction $^{18}O(p,n)^{18}F$ was confirmed by the decay curves of annihilation x-rays (0.51 Mev). The pellets were made of $Mg^{18}O$ (90 atom% enriched), ^{18}O-Isoleucine (73.5 atom% enriched), [^{18}O]-glycine administered in tissue, and non-enriched MgO. The decay curves from enriched ^{18}O compounds agreed with the known half-life (109.7 min). However, the decay curve from [^{18}O]-glycine administered in tissue showed interferences with competitive reactions.

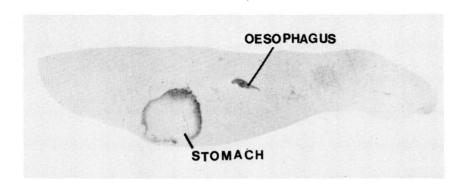

Fig. 1 The macroautoradiogram of [^{18}O]-glycine (35.75 mg/0.6 ml, ^{13}O: 17.16 mg) in stomach three minutes after oral administration. The intensity of proton irradiation was 6-9 μC at 2.8 Mev.

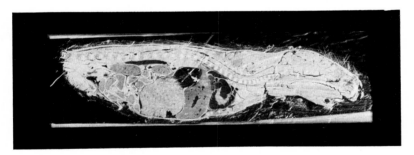

Fig. 2 The thin sliced sample of a rat corresponding to Fig. 1

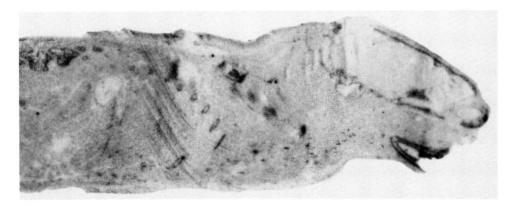

Fig. 3 The macroautoradiogram of [^{18}O]-glycine (5.38 mg/0.2 ml, ^{18}O: 2.592 mg) in rat after intravenous administration. The bone membrane, muscle, heart, and other parts except the brain can be recognized. The incident beam current was 95 nC/cm^2 at an average proton energy of 5 Mev.

Fig. 4 The thin sliced whole body sample of a rat corresponding to Fig. 3

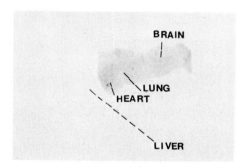

Fig. 5 The macroautoradiogram of a mouse exposed for 4 min. to ^{18}O (99 atom%). Brain, muscle, heart and lung but not liver can be recognized.

Fig. 6 The thin sliced upper body sample of a mouse corresponding to Fig. 5, set on aluminium foil for irradiation in the beam cup of a Van de Graaff.

Chemistry, Life Sciences, Biochemistry, Molecular Biology, Pharmacology, Pharmaceutical Sciences, Earth Sciences, Environmental Research

Stable Isotopes

Proceedings of the Fourth International Conference held in Julich, March 23-26, 1981

edited by H.-L. SCHMIDT, H. FÖRSTEL *and* K. HEINZINGER

ANALYTICAL CHEMISTRY SYMPOSIA SERIES, 11

1982 xvii + 758 pages
Price: US $127.75 / Dfl. 275.00
ISBN 0-444-42076-2

The 85 contributions and 15 reviews in this volume make it a comprehensive overview of the state of investigations on stable isotopes of the main bioelements as they occur in nature. The conference emphasized stable tracer applications in medicine, pharmacology, agriculture and biochemistry, and added the dimension of the theory and consequences of isotope effects, and the importance of stable isotopes in geochemistry, cosmochemistry and environmental research.

The papers compare recent results and methodology in the different disciplines obtained on the basis of isotope measurements. Interaction between the disciplines was facilitated by a common methodology which revealed that a particular problem could be approached from various angles, and stimulus given for further investigations. Examples of new possibilities are the use of NMR in biological research and the introduction of "naturally labelled" compounds in nutrition physiology. The book is essential to all those working on stable isotopes but even scientists not familiar with isotope research may benefit from the demonstration of the wide possibilities isotope application can have as a common tool in most biosciences from agriculture to zoology.

A SELECTION OF THE CONTENTS: **I. Isotope Effects, Theory and Consequences.** The Theoretical Analysis of Isotope Effects *(M. Wolfsberg).* Vapour Pressure Isotope Effects of Acetonitrile *(Gy. Jakli et al.).* Heavy-Atom Isotope Effects on Enzyme-Catalyzed Reactions *(M.H. O'Leary).* Isotope Effects on Each, C- and N-Atoms, as a Tool for the Elucidation of Enzyme Catalyzed Amide Hydrolyses *(R. Medina et al.).* **II. Geochemistry and Cosmochemistry.** Stable Isotopes

ELSEVIER SCIENTIFIC PUBLISHING COMPANY

Amsterdam *and* New York

and the Evolution of Life: An Overview *(M. Schidlowski)*. Isotope Geochemistry of Carbon *(J. Hoefs)*. Carbon Isotope Fractionation Factors of the Carbon Dioxide-Carbonate System and their Geochemical Implications *(L.E. Maxwell and Z. Sofer)*. The Isotopic Fractionation During the Oxidation of Carbon Monoxide by Hydroxylradicals and its Implication for the Atmospheric Co-Cycle *(H.G.J. Smit et al.)*. **III. Biomedical Applications. A. Pharmacology and Drug Metabolism.** Applications of Stable Isotopes in Pharmacological Research *(T.A. Baillie et al.)*. Application of ^{13}C-Labelling in the Bioavailability Assessment of Experimental Clovoxamine Formulations *(H. de Bree et al.)*. Measurement of the Pharmacokinetics of DI-([15,15,16,16-D4]-Linoleoyl)-3-sn-Glycerophosphocholine After Oral Administration to Rats *(A. Brekle et al.)*. **B. Clinical Diagnosis.** The Application of the Stable Isotopes of Oxygen to Biomedical Research and Neurology *(D. Samuel)*. In Vivo Measurement of Enzymes with Deuterated Precursors and GC/SIM *(H.-Ch. Curtius)*. Prenatal Diagnosis of Propionic and Methylmalonic Acidemia by Stable Isotope Dilution Analysis of Methylcitric and Methylmalonic Acids in Amniotic Fluids *(L. Sweetman et al.)*. Use of Water Labelled with Deuterium for Medical Applications *(E. Roth et al.)*. **C. Breath Tests and Lung Function Tests.** The Commercial Feasibility of ^{13}C Breath Tests *(P.D. Klein and E.R. Klein)*. Application of ^{13}C-Fatty Acids Breath Tests in Myocardial Metabolic Studies *(M. Suehiro et al.)*. Breath Test Using ^{13}C Phenylalanine as Substrate *(D. Glaubitt and K. Siafarikas)*. Use of the ^{13}C/^{12}C Breath Test to Study Sugar Metabolism in Animals and Men *(J. Duchesne et al.)*. **IV. Life Sciences, Agriculture, and Environmental Research.** Stable Isotopes in Agriculture *(H. Faust)*. Balance of ^{15}N Fertilizer in Soil/Plant System *(N. Sotiriou and F. Korte)*. Nitrogen Isotope Ratio Variations in Biological Material - Indicator for Metabolic Correlations *(R. Medina and H.-L. Schmidt)*. Possibilities of Stable Isotope Analysis in the Control of Food Products *(J. Bricout)*. **V. Methods. A. Analytical Developments.** Recent Applications of ^{13}C NMR Spectroscopy to Biological Systems *(N.A. Matwiyoff)*. Quantitative Mass Spectrometry with Stable Isotope Labelled Internal Standard as a Reference Technique *(I. Björkem)*. Application and Measurement of Metal Isotopes *(K. Habfast)*. Stable Isotopes in Biomedical and Environmental Analysis by Field Desorption Mass Spectrometry *(W.D. Lehmann and H.-R. Schulten)*. **B. Isotope Separation and Synthesis of Labelled Compounds.** The Production of Stable Isotopes of Oxygen *(I. Dostrovsky and M. Epstein)*. Separation of the Stable Isotopes of Chlorine, Sulfur and Calcium *(W.M. Rutherford)*. The Synthesis of Mono-and Oligosaccharides Enriched with Isotopes of Carbon, Hydrogen and Oxygen *(R. Barker et al.)*. New Approaches in the Preparation of ^{15}N Labeled Amino Acids *(Z.E. Kahana and A. Lapidot)*.

Send your order to **your bookseller** or
ELSEVIER SCIENCE PUBLISHERS
P.O. Box 211, 1000 AE Amsterdam, The Netherlands

Distributor in the U.S.A. and Canada:
ELSEVIER SCIENCE PUBLISHING CO., INC.
52 Vanderbilt Ave., New York, N.Y. 10017

Continuation orders for series are accepted.

Orders from individuals must be accompanied by a remittance, following which books will be supplied postfree.

The Dutch guilder price is definitive. US$ prices are subject to exchange rate fluctuations.
Prijzen zijn excl. B.T.W.

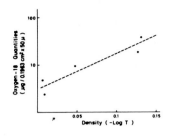

Fig. 7 Relation between ^{18}O quantities and optical density

The possibility of interferences from other proton reactions

The living body, e.g. human body, consists of about 61% hydrogen, 10% carbon, 26% oxygen, 2.5% nitrogen, and 0.5% trace elements. Therefore, the following competitive reactions can occur: ^{14}N(p,α)^{11}C, ^{16}O(p,α)^{13}N, ^{14}N(p,pn)^{13}N, ^{18}O(p,α)^{15}N and ^{27}Al(p,n)^{27}Si from aluminium foil. In our experiments interferences with the radioactive nuclides ^{11}C (20.3 min) or ^{13}N (9.96 min) were taken into account.

The detected sensitivity

The standards were prepared with a fixed ^{18}O concentration in 10% gelatin. The sliced gelatin samples were made in the same way as the sliced rat samples and irradiated at the same time. The intensity of blackness was measured by a densitometer. The relation between ^{18}O quantities and optical density is shown in Fig. 7. At 2.9 Mev 1.176 µg of ^{18}O resulted in an optical density of 0.009.

Finally, the autoradiographic method with (p,n) reactions of ^{18}O can be applied in biology, medicine, agriculture, or some engineering fields when the distribution of ^{18}O containing compounds is of interest.

ACKNOWLEDGEMENTS

The authors wish to express their gratitude to Drs. T. Ishihara and K. Kobayashi of the University of Tsukuba, for their cooperation in the operation of the accelerator "Pelletron" at the university.

REFERENCES

1 J.M. Blair and J.J. Leigh, Phys. Rev., 118(1960)495-498
2 P.M. Beard, P.B. Parks, E.G. Bilpuch and H.W. Newson, Ann. Physics, 54(1969)566-597
3 W. Fritsch, K.D. Buchs, E. Finckh, P. Pietrzyk and B. Schreiber, Z. Physik, 262(1973)65-70
4 Y. Ohta, T. Inada, T. Kanai, K. Kawachi and M. Aihara, Radiochem. Radioanal. Letters, 44(1980)419-422

V. METHODS
B. ISOTOPE SEPARATION AND SYNTHESIS OF LABELLED COMPOUNDS

THE PRODUCTION OF STABLE ISOTOPES OF OXYGEN

I. DOSTROVSKY and M. EPSTEIN

Isotopes Dept., The Weizmann Inst. of Sci., Rehovot (Israel)

ABSTRACT

The 25 years record of the production of ^{17}O and ^{18}O at the Weizmann Institute is summarized and more recent modification of the plant to increase its capacity are described.

The isotopes separation plant consists of two sections. The first is a water distillation cascade which is constructed of 36 packed fractionating columns varying in diameter from 2 cm to 15 cm and in height from 10 m to 15 m connected in a complex series-parallel arrangement. This section can produce 6 kg/a of 98-99% ^{18}O and 1.5 kg/a of 25% ^{17}O. We are currently in the process of adding stages to this cascade to raise the product concentration of ^{17}O to 40%. The second section consists of two thermal diffusion cascades fed with enriched oxygen gas made by electrolysis of the product from the first section. The output of this section is ^{17}O at a maximum concentration of 90%. Other, lower, concentration of ^{17}O can be drawn from intermediate points in the cascade.

INTRODUCTION

Ever since the discovery of ^{18}O and ^{17}O in 1929, attempts have been made to find methods of separating these isotopes from ^{16}O. These methods include distillation, diffusion, thermal diffusion, electrolysis and a wide variety of exchange reactions. Of all the processes only distillation methods are now operating for the production of ^{18}O and ^{17}O in the appreciable quantities available commercially.

The Isotopes Separation Plant of the Weizmann Institute of Science for the production of ^{18}O and ^{17}O by the fractional distillation of water, is the only one which has been working under steady state conditions for more than 25 years.

The ^{18}O cascade was described by Dostrovsky and Raviv (ref.1) and in various other publications (ref. 2 and 3) but we thought it will be useful to describe the entire plant as it is today and relate the accumulated experience with its operation.

Initiated as a research and experimental plant the plant is now producing about 6 kg/a of water enriched with 98 to 99% ^{18}O and 1500 gr/a of water enriched to 25% in ^{17}O. The capacity of the distillation cascade is now being doubled.

The production of enriched ^{17}O is much more difficult than that of enriched ^{18}O. Not only is the separation factor smaller and the natural abundance lower, but additional difficulties are created by the fact that the isotope is "sandwiched" between two more abundant ones.

In order to reach higher ^{17}O enrichment a thermal diffusion cascade follows the distillation section and concentrations of over 90% ^{17}O are obtained in this way.

GENERAL DESCRIPTION OF THE PLANT
The distillation cascade

The distillation cascade consists of packed fractionating columns connected in series-parallel area. A typical column module is shown in Fig. 1 and consists of three parts: The reboiler, the column itself and the condensing system. The columns are made of copper tubing and have diameters ranging from 15 cm to 2 cm depending on their position in the cascade. The columns are packed with Dixon rings made from 100 mesh phosphor bronze which has been treated to improve its wetting properties. The columns are thermally insulated with glass wool and external heating is provided to ensure adiabatic conditions.

An example of the operating characteristics of a 4" column packed for 14.3 meters with $1/8$" rings follows. With a pressure at the top of 260 mmHg and liquid flow rate of 120 ml/min the pressure drop across the column is 100 mmHg and the liquid hold-up is 16.5 liters. The number of theoretical plates under these conditions is 975. The flooding flow rate for such a column is 180 ml/hr. When 100 ml per day are withdrawn from the reboiler the concentrations of ^{18}O and ^{17}O in the product are 2.1% and 0.2% respectively.

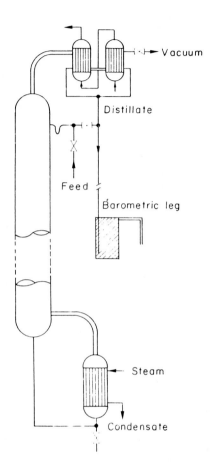

Fig. 1. Schematic description of a typical fractionating column

Two types of reboilers are used as shown in Fig. 2. Both are flash boilers. For the large diameter columns a bundle of tubes in a shell is used and for the small diameter columns a double pipe is used.

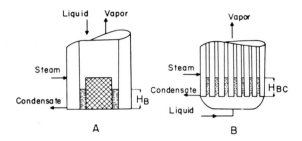

Fig. 2. Types of Reboilers. A: Double pipe type; B: Bundle of tubes in a shell.

The column system is shown in Fig. 3, with barometric leg feed and distillate lines.

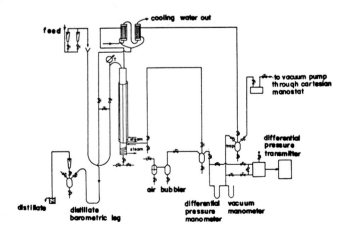

Fig. 3. A schematic description of the column system.

Before proceeding further with the enriching of ^{17}O and ^{18}O the product from the third stage is processed by a 2" "splitter" column, the function of which is to provide two streams of equal ^{17}O concentration but one high in ^{18}O (about 83%) and the other low in ^{18}O (about 14%). A separation between ^{18}O and ^{16}O is thus obtained. The bottom stream goes to the last ^{18}O enriching stage, where the ^{17}O is stripped to 0.5% and the ^{18}O is over 98%. The top stream from the splitter goes to the second part of the cascade in which only the ^{17}O is enriched. This part consists of two stages and an additional splitter. The product from the last stage which consists of two 10 m, $3/4$" diameter columns operating in parallel is about 25% ^{17}O.

The time necessary to accumulate the plant hold-up is 480 days for ^{18}O and 950 days for ^{17}O. This would be the minimum relaxation time for the plant under ideal conditions.

It is clear that with such long relaxation time very reliable and steady operation are absolutely essential. For this reason much effort was devoted to the question of optimal design and control procedures.

Flow and pressure drop dynamics

The flow dynamics is the feature on which the control of the plant is based and this is one of the main reasons for the stable operation of the plant over the years. The flow dynamics depends primarily on flow and vacuum changes. The pressure drop or pressure differential between the top and the bottom of the column serves as a very sensitive indication of its hydrodynamic state. Analysis of the pressure drop response to changes in operation conditions, gives the operator an exact picture of the status of the column.

Holdup replacement

In the event of a major fault requiring the interruption of a column operation (for example replacement of a reboiler), a procedure is available which minimizes the loss of enrichment work. It was found that if the hold-up is drained gradually and collected in many fractions the concentration profile is approximately preserved in them.

After the column is restarted with normal (or low enrichment) water and reaches stable operation, the hold-up is replaced by introducing the fractions collected in the drainage operation in the reverse order. Fig. 4 shows the concentration profile in a column before and after an interruption using the procedure described above.

A general flow sheet of the distillation cascade is shown in Fig. 5. The first part of the cascade consists of three separate stages. The first stage is the feed section. It consists of 20 columns 15 cm and 10 cm in diameter, operating in parallel, enriching water from natural abundance to about 2% ^{18}O and 0.2% ^{17}O. The product from this stage is fed into the second stage which at present consists of three pairs of columns connected in series-parallel. The product is 5 to 6% ^{18}O and 0.5% ^{17}O and the stripped stream is of natural abundance.

The third stage consists of three columns connected in series providing a product of 48% ^{18}O, 3.5% ^{17}O and stripped stream of 0.2% ^{18}O and 0.15% ^{17}O. The latter is returned to the feed of the second stage. It is impossible to design a cascade which will be simultaneously ideal for both ^{18}O and ^{17}O. Since it is more difficult to obtain ^{17}O, the cascade is designed to approximately match optimum operational conditions for separating ^{17}O and not ^{18}O. The feed points are chosen so as to avoid mixing of different ^{17}O concentration streams.

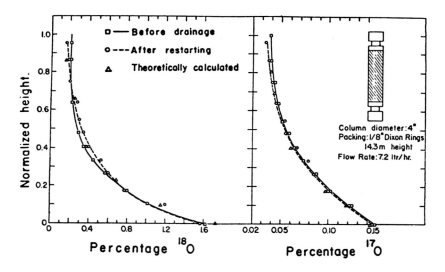

Fig. 4 Concentration profiles of ^{18}O and ^{17}O before emptying the column and after reoperation

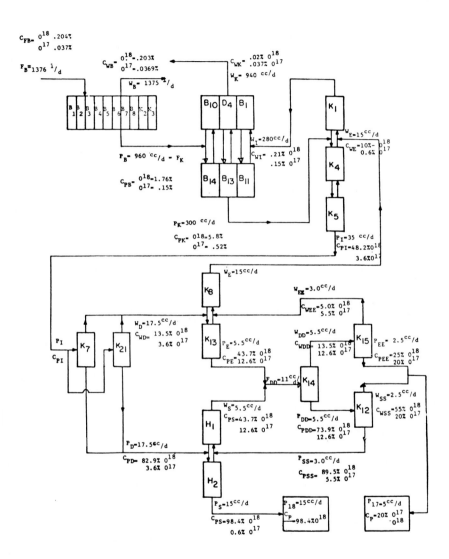

Fig. 5. Flow sheet of the distillation cascade

C – Isotopic concentrations
F – Feed flow rate
P – Product flow rate
W – Waste flow rate

The thermal diffusion cascade for enrichment of Oxygen-17

^{17}O at concentrations higher than 25% are obtained using a thermal diffusion cascade. Gaseous oxygen obtained by the electrolysis of the partially enriched ^{17}O water from the distillation cascade, is fed into thermal diffusion columns for further enrichment. A steady state condition was reached in this cascade with concentration of over 90% ^{17}O.

The thermal diffusion section consists of central-wire type columns. The circulation of the gas in each stage is achieved by thermosyphons and among the stages through thermal pumps as shown in Fig. 6.

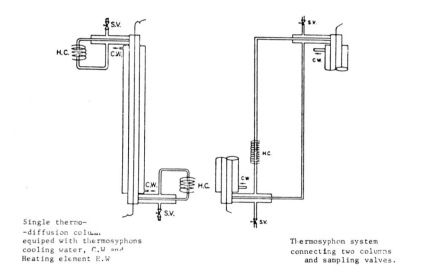

Single thermo-
-diffusion column
equiped with thermosyphons
cooling water, C.W and
Heating element H.W

Thermosyphon system
connecting two columns
and sampling valves.

Fig. 6. Single gas thermodiffusion column equiped with thermosyphon

The wire temperature is 650 to 740 °C depending on the column diameter. The calculated temperature and gas velocity profiles are shown in Fig. 7. (ref. 4.). The separation factors are 1.0066 for $^{16}O/^{17}O$ and 1.0133 for $^{16}O/^{18}O$.

The thermal diffusion cascade consists of two splitters and two stages. Each splitter is a group of 12 columns and each stage comprises 24 columns.

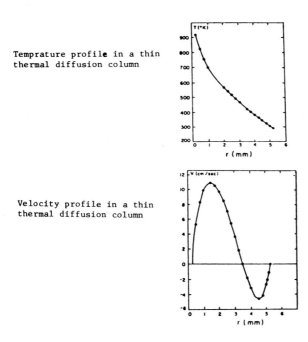

Temprature profile in a thin thermal diffusion column

Velocity profile in a thin thermal diffusion column

Fig. 7. Radial temperature and velocity profiles in a thin gas thermal diffusion column.

GENERAL COMMENTS

A new distillation cascade is now in construction in parallel to the existing one for doubling the production of ^{17}O and ^{18}O and increasing the ^{17}O enrichment to about 40%. The first stage for producing the low enrichment of about 2% ^{18}O has already been running for a year. The decision to duplicate the plant with only minor changes was made due to its high reliability and stability.

In addition to the enrichment of the isotopes of oxygen in the water, there is also enrichment of Deuterium. Although the natural abundance of the Deuterium in water is only 0.016%, the concentration in the first stage is already around 2% due to the relatively large separation factor for HDO.

The Deuterium concentration along the cascade is increased up to above 90%. For many purpose the high Deuterium concentration is undesirable. The product is therefore electrolyzed and the oxygen gas recombined with hydrogen of normal isotopic composition to give $H_2^{18}O$. Similarly the Deuterium from the electrolysis may be recombined with normal oxygen to give D_2O.

The oxygen from the electrolysis can also be used as starting material for the synthesis of many labelled compounds.

REFERENCES

1 I. Dostrovsky and A. Raviv, in J. Kistemaker, J. Bigeleisen and A.D.L Nier (Eds) Proceedings of the Symposium on Isotopes Separation, April 23-27,1957, North-Holland Publicating Comp., Amsterdam, 1958, pp. 336-348.
2 D. Wolf and H. Cohen, The Canadian Journal of Chem. Eng.,50, (1972) 621-627
3 C. Gilath, H. Cohen and D. Wolf. The Canadian Journal of Chem. Eng.,55,(1977) 168-176.
4 C. Gilath, Ph.D. Thesis, The Weizmann Instit. of Science, Israel (1973) p.95

H.-L. Schmidt, H. Förstel and K. Heinzinger (Editors), *Stable Isotopes*
© 1982 Elsevier Scientific Publishing Company, Amsterdam — Printed in The Netherlands

SEPARATION OF THE STABLE ISOTOPES OF CHLORINE, SULFUR, AND CALCIUM

W. M. RUTHERFORD

Monsanto Research Corporation, Mound Facility,* Miamisburg, Ohio 45342 (U.S.A.)

ABSTRACT

The stable isotopes of sulfur and chlorine are being separated on a practical scale by liquid phase thermal diffusion. The present capacity of a small separation system occupying 3 m^2 of laboratory floor space is 0.5 kg of chlorine isotopes per year at enrichments of 95% ^{37}Cl and 99.5% ^{35}Cl. A similar system for the separation of sulfur isotopes has a capacity of 0.3 kg/yr of ^{34}S enriched to 93%. The feasibility of separating calcium isotopes by thermal diffusion of an aqueous solution was demonstrated. It appears to be possible to enrich ^{48}Ca up to 10% in a thermal diffusion system comparable in size to those used for chlorine and sulfur isotopes.

INTRODUCTION

Research carried out over the last few years at Mound Facility has demonstrated that many isotopes of potential interest can be separated by liquid phase thermal diffusion. Within the last few years, practical processes have been developed for separating chlorine and sulfur isotopes [1] at rates of 0.3 to 0.5 kg per year in small systems requiring minimum operator attention. The development of a process to separate bromine isotopes is currently in progress. Results recently obtained indicate that a practical liquid phase thermal diffusion process can be developed for separating the isotopes of calcium [2].

LIQUID PHASE THERMAL DIFFUSION

Gas and liquid mixtures in general tend to separate slightly in the presence of a temperature gradient, with the lighter component (usually) tending to concentrate in the region of higher temperature. The elementary separation effect by itself is small and not at all useful. If the effect is combined with countercurrent thermogravitational circulation, as illustrated in Fig. 1, then it is greatly enhanced and large separations can be created along the vertical axis of

---

*Mound Facility is operated by Monsanto Research Corporation for the U.S. Department of Energy under Contract No. DE-AC04-76-DP00053.

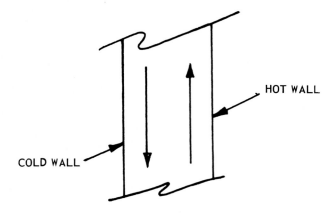

Fig. 1. Countercurrent thermogravitational circulation.

the system. It can be visualized as a process wherein the lighter isotopic compound diffuses into the stream moving upward along the hot wall, and the heavier component migrates into the stream descending along the cold wall. Under typical circumstances the spacing between the hot and cold walls is in the range from 3 to 15 mm for gases and 0.2 to 0.5 mm for liquids. The corresponding temperature differences are 400° to 800°C for gases and 50° to 200°C for liquids.

The gas phase technique has been used for many years to separate isotopes of the noble gases. Helium, neon, argon, krypton and xenon isotopes are routinely separated at Mound Facility for worldwide distribution to a wide variety of users [3].

The liquid phase process is somewhat more difficult to apply; however, it had been used with relatively low efficiency to separate uranium-235 during World War II [4]. Very little additional attention was given to the separation of isotopes by this method until 1960, when K. F. Alexander pointed out some of the advantages of liquid phase thermal diffusion for the separation of isotopes other than uranium [5]. These include simplicity and small size of the equipment, and adaptability to small-scale operations (a few kg per year or less) on a wide variety of systems. Although energy requirements are relatively large, the cost of energy for the process is relatively unimportant at the production rates required to meet currently foreseeable needs for enriched isotopes.

Reliable liquid thermal diffusion columns of simple designs have been developed at Mound Facility for use as components of multiple column separation cascades. The columns, shown schematically in Fig. 2, are concentric cylinders of stainless steel confining an annular working space containing the fluid to be separated. The inner cylinder is heated by condensing steam at 7 atm while the outer cylinder is cooled by flowing water at 15°C.

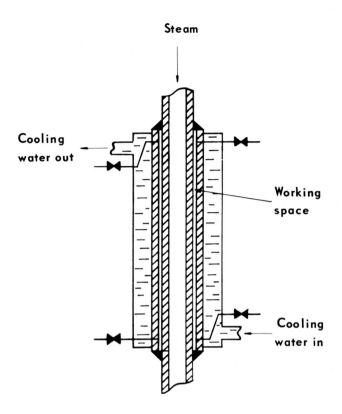

Fig. 2. Schematic diagram of thermal diffusion column for liquids.

SULFUR ISOTOPE SEPARATION

Natural sulfur comprises four stable isotopes, the relative abundances of which are given in Table 1. Our separation program is directed toward the separation of ^{34}S at enrichments greater than 90%. Carbon disulfide was chosen as the working fluid on the basis that it is a simple stable compound with known properties, readily available in high purity. It has the disadvantage of being

TABLE 1

Natural abundances of the sulfur isotopes

Isotope	Atom %
^{32}S	95.06
^{33}S	0.74
^{34}S	4.18
^{36}S	0.0136

doubly substituted in sulfur; however, the elementary isotopic thermal diffusion effect for the $C^{32}S^{34}S$ pair has been measured and has been found to be quite large [6]. The separation is readily accomplished in a 10-column system with a total length of about 12 m. The columns in the system, which is depicted schematically in Fig. 3, range in hot-to-cold wall spacing from 0.3 mm at the top of the cascade to 0.2 mm at the bottom. Isotopic exchange among the several species of CS_2 molecules does not take place at a useful rate at the operating temperature of the cascade. Some external provision must, therefore, be made for the conversion of the mixed molecule $C^{32}S^{34}S$ into the desired product molecule $C^{34}S_2$ according to the equation:

$$2\ C^{32}S^{34}S \rightleftarrows C^{32}S_2 + C^{34}S_2 \tag{1}$$

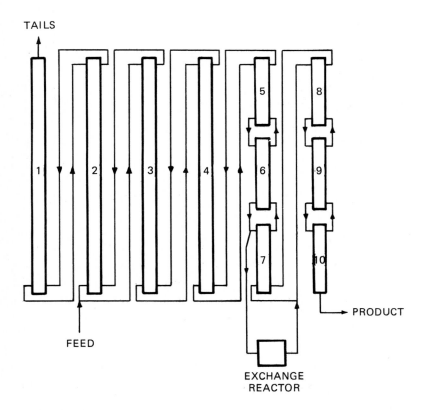

Fig. 3. A cascade of thermal diffusion columns for the separation of ^{34}S by liquid phase thermal diffusion of CS_2.

The equilibration, which takes place rapidly in the homogeneous vapor phase at 830°C, is accomplished in a side stream reactor attached to the cascade at an intermediate point.

The separation system occupies 3 m^2 of laboratory floor space, and it is capable of separating 0.2 to 0.4 kg per year of ^{34}S at enrichments exceeding 90%. The enriched carbon disulfide is converted to elemental sulfur by partial oxidation. It is anticipated that an inventory of elemental sulfur will be maintained to support distribution to users. At present Mound has no plans for offering chemical forms other than carbon disulfide and elemental sulfur.

The total investment in the sulfur separation facility is on the order of $100,000, and the total cost of the separated material amounts to about $400 per gram of ^{34}S, to which the energy contribution is less than 15%.

The separation campaign now under way is the third one that we have undertaken. The material from the first two efforts, which were primarily research projects, has already been distributed. The current campaign is expected to be a long one, and sufficient material will be separated to support usage on an expanded scale.

If needs should develop in excess of 1 kg/year, the SO_2-bisulfate chemical exchange process will become more attractive than thermal diffusion. An exchange system has been constructed at Mound [7] but it has not been operated because the demand for sulfur isotopes has not yet been sufficient.

CHLORINE ISOTOPE SEPARATION

There are two stable isotopes of chlorine which occur naturally in the proportions of 24.5% ^{37}Cl and 75.5% ^{35}Cl. Our objective has been to separate both at high enrichment by liquid phase thermal diffusion.

Three fluids were evaluated for the separation. These were 1-chloropropane, chlorobenzene and methyl chloride. 1-chloropropane was found to react slowly with the hot-wall surface producing soluble chlorides. The chlorides concentrated at the bottom of the system, precipitated, and plugged capillary connecting lines.

Chlorobenzene was a marginally satisfactory working fluid, and short campaigns were undertaken to separate chlorine-37 at 90% and chlorine-35 at 95% enrichment.

Methyl chloride, however, was the best working fluid. The critical temperature of methyl chloride is 143°C. slightly lower than the estimated hot-wall temperature of the thermal diffusion apparatus. In order to ensure single phase conditions in the separation system, therefore, the working pressure must be substantially greater than the critical pressure of 66 atm.

The chlorine isotopes are separated in a 13-column cascade similar to the one used for ^{34}S. The hot-to-cold wall spacings range from 0.3 mm for columns 1 and 2 to 0.2 mm for 12 and 13. Total length of the system is approximately 11 m, occupying about 3 m^2 of laboratory floor space. The system operates 24 hr per day, unattended, with operator intervention only to empty and refill the top and bottom reservoirs and to take samples for isotope ratio determination.

The lines interconnecting vertical stacks of columns are switched and the reservoir sizes are changed to accommodate the separation of ^{35}Cl. This can be done in a few days but would not be necessary if the system were long enough to separate both species simultaneously.

Two campaigns have now been run to separate a large quantity of the chlorine isotopes. The capacity of the system is on the order of 400 g/year of ^{37}Cl at 96% enrichment and somewhat more than 1 kg per year of ^{35}Cl at 99.5% enrichment. Mound now has accumulated 230 g of ^{37}Cl and 240 g of ^{35}Cl to support distribution of these materials to interested users. The product is routinely converted to sodium chloride for distribution. The total costs for separating the chlorine isotopes and converting them to sodium chloride amount to approximately $400/g for ^{37}Cl and $185/g for ^{35}Cl.

CALCIUM ISOTOPE SEPARATION

There are a number of attractive biomedical applications of separated calcium isotopes. At present these isotopes are available from the Oak Ridge Calutron separators in very limited quantities at very high prices ($150,000/g for ^{48}Ca).

Liquid phase thermal diffusion is a possible method for separating calcium isotopes on a useful scale. A stable, low molecular weight compound of calcium that is liquid in an appropriate temperature range does not exist; however, in principle it is possible to separate the isotopes of calcium and other elements by thermal diffusion of some solid compound dissolved in a suitable solvent. Normally, the separation of the solute from the solvent is much greater than the separation among the isotopic species of the solute; thus, nearly pure solvent tends to accumulate at the top of a thermal diffusion system and essentially all the solute concentrates at the bottom.

The solute-solvent separation can be suppressed, however, by imposing a net flow of solvent through the separation column. This concept, which was originally suggested by Korsching [8], is illustrated in Fig. 4. Solvent is injected at a controlled rate into the bottom of the apparatus and selectively removed by distillation at the top. It can be shown theoretically that the solvent flow does not affect the isotopic distribution and that selection of the proper solvent injection rate completely eliminates the solute concentration gradient.

Recent experiments have shown that this concept is effective: separation factors as high as 1.47 for the ^{40}Ca-^{48}Ca pair have been measured in work with a 0.71 m column. The effect seems to be large enough to support separation on a practical scale. Further development is expected to lead to a practical process for separating isotopes in solution and thus to the improved availability of these materials to the scientific community. It appears to be possible to

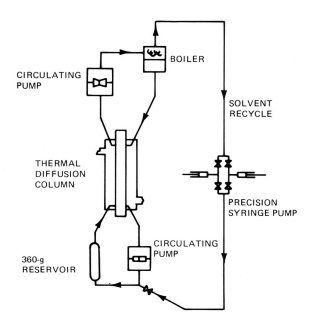

Fig. 4. The solvent counterflow process for separating calcium isotopes in aqueous solution.

enrich ^{48}Ca up to 10% in a thermal diffusion system comparable in size to those used for chlorine and sulfur isotopes. There is no apparent reason why the method cannot also be applied to potassium, magnesium, and zinc isotopes.

REFERENCES

1 W. M. Rutherford, I&EC Process Design and Development, 17(1978)77-81.
2 W. M. Rutherford and K. W. Laughlin, Science, 211(1981)1054-1056.
3 Monsanto Research Corporation, Stable Isotopes (Catalog). Mound Facility, Miamisburg, Ohio 45342, 1977.
4 P. H. Abelson, N. Rosen and J. I. Hoover, Liquid Thermal Diffusion, United States Atomic Energy Commission Report TID-5229, August 1958.
5 K. F. Alexander, Fortschr. Phys., 8(1960)1-41.
6 W. M. Rutherford, J. Chem. Phys., 59(1973)6061-6069.
7 W. M. Rutherford and W. J. Roos, in Proc. IAEA/FAO Symp. Isotope Ratios as Pollutant Source and Behavior Indicators, Vienna, November 18-22, 1974, International Atomic Energy Agency, IAEA STI/PUB/382, Vienna, 1975, pp. 295-304.
8 H. Korsching, Z. Naturforsch, 7b(1952)187-191.

ENRICHMENT OF HEAVY CALCIUM ISOTOPES BY ION EXCHANGER RESINS WITH CYCLOPOLYETHERS AS ANCHOR GROUPS

KLAUS G. HEUMANN, HANS-PETER SCHIEFER and WOLFGANG SPIESS
Institut für Anorganische Chemie der Universität, Universitätsstraße 31,
D-8400 Regensburg (F.R.G.)

ABSTRACT

Calcium isotope separations in systems using ion exchange resins with the cyclic polyether anchor groups [2_B.2.2] cryptand and dibenzo-[18]-crown-[6] are investigated. In both systems the heavier calcium isotopes are enriched in the solution phase, whereas the lighter isotopes are enriched in the resin phase. For the exchange reaction at the resin with a [2_B.2.2] anchor group the isotope separation effect ε is calculated to be in the range $\varepsilon(^{44}Ca/^{40}Ca) = (2.6-7.5) \times 10^{-3}$ and $\varepsilon(^{48}Ca/^{40}Ca) = (4.5-12.9) \times 10^{-3}$ depending on the temperature and on the water content of the eluent. These ε-values are higher by a factor 10-100 than in systems with Dowex 50. By a multi-cycle experiment 30 mg of calcium chloride were produced which showed a relative enrichment of 3.3 % for ^{48}Ca and of 1.9 % for ^{44}Ca compared with natural calcium.

INTRODUCTION

The stable heavy calcium isotopes ^{44}Ca, ^{46}Ca and ^{48}Ca are of common interest for labeling purposes, especially in medical research. Whereas the production of highly enriched calcium isotopes by physical methods, e.g. by mass separators, is expensive, a chemical preenrichment of these isotopes can reduce the expenditure. On the other hand for many labeling experiments a low isotope enrichment is sufficient, so that only chemically enriched isotopes can be used. A precondition for that is an enrichment by a factor of two or more.

Most experiments, which concern investigations for the enrichment of calcium isotopes have been done with the strongly acidic cation exchanger Dowex 50. A summary of the results of these investigations is given in Table 1. In all experiments ion exchange chromatography was used and the eluent consisted of solutions with complexing agents [1,2] as well as of solutions of acids and other electrolytes [3-6]. The isotope separation effect ε of one equilibrium stage, which is equal to the isotope separation factor α -1, was calculated from the elution curve and the measured isotope ratio in dependence on the eluted calcium amount by a method described by Glueckauf [7]. The second and third column of Table 1 show the ε-values for the isotope pairs $^{44}Ca/^{40}Ca$ and $^{48}Ca/^{40}Ca$. Positive ε-values mean an enrichment of the heavier calcium isotopes in the solution phase of the heterogeneous exchange system Dowex 50/electrolyte solution, whereas a negative ε-value shows an enrichment of the heavier isotopes in the resin phase (see equations (1) and (3)). As one can see from the results in Table 1 the separation effect per mass unit lies between $(0.3-1.6) \times 10^{-4}$ for the calcium isotopes. To employ such a chromatographic system for the chemical isotope enrichment, the ε-values should be higher by a factor of ten or more.

TABLE 1

Isotope separation effect $\varepsilon = \alpha - 1$ of calcium by experiments with the strongly acidic cation exchange resin Dowex 50

Eluent solution	$\varepsilon\ (^{44}Ca/^{40}Ca)$	$\varepsilon\ (^{48}Ca/^{40}Ca)$	References	
α-Hydroxyisobutyrate	4.7×10^{-4}	8.7×10^{-4}	Aaltonen	[1]
Citrate	-1.4×10^{-4}	-	Klinskii et al.	[2]
EDTA	-1.7×10^{-4}	-		
4 M HCl	2.1×10^{-4}	-	Russel et al.	[3]
0.1 M $BaCl_2$	-	12.7×10^{-4}	Heumann et al.	[4-6]
3 M HCl	3.7×10^{-4}	6.4×10^{-4}		
9 M HNO_3	-6.0×10^{-4}	-11.8×10^{-4}		

As it is known calcium forms stable complexes - especially in non-aqueous solutions - with crown ethers and amino-polyethers (cryptands) [8]. Jepson and DeWitt [9] were able to enrich ^{40}Ca relative to ^{44}Ca in the $CHCl_3$ phase using a liquid-liquid extraction system $H_2O/CHCl_3$ with dicyclohexyl-[18]-crown-[6]. Here they found an isotope separation effect of 10^{-3} per mass unit, which is much higher than the ε-values in the Dowex 50 systems.

Our first investigations with an exchange resin containing the cryptand [2_B.2.2] as anchor group result in relative high calcium isotope separations [10]. Therefore, the experiments reported in this paper were carried out with the commercially available resin "[2_B.2.2] polymer" (E. Merck) having [2_B.2.2] as anchor group and with a resin having dibenzo-[18]-crown-[6] as anchor group which was synthesized by a method described by Blasius et al. [11]. The two cyclic polyethers used as anchor groups have the following structure:

Cryptand [2_B.2.2]   Dibenzo-[18]-crown-[6]

Equation (1) gives the isotope exchange reaction between $^{48}Ca^{2+}$ and $^{40}Ca^{2+}$ in a system with a strongly acidic exchange resin. Contrary to that system, where the calcium ion is fixed at the resin by the anionic $R-SO_3^-$ group, calcium is complexed by the neutral polyether in case of cryptand or crown ether anchor groups. In addition, for the charge equilization an anion must be bound at the resin phase as well. Equation (2) shows the isotope exchange reaction for a resin with [2_B.2.2] cryptand anchor groups as an example. For both, equation (1) and (2) the equilibrium constant K_C given by equation (3) corresponds to the isotope separation factor α.

$$R-(SO_3)_2{}^{48}Ca + {}^{40}Ca^{2+}_{sol} \rightleftharpoons R-(SO_3)_2{}^{40}Ca + {}^{48}Ca^{2+}_{sol} \qquad (1)$$

$$R-[2_B.2.2,{}^{48}Ca]^{2+} + {}^{40}Ca^{2+}_{sol} \rightleftharpoons R-[2_B.2.2,{}^{40}Ca]^{2+} + {}^{48}Ca^{2+}_{sol} \qquad (2)$$

$$K_C = \frac{[{}^{40}Ca^{2+}_{resin}] \times [{}^{48}Ca^{2+}_{sol}]}{[{}^{48}Ca^{2+}_{resin}] \times [{}^{40}Ca^{2+}_{sol}]} \equiv \alpha = 1 + \varepsilon \qquad (3)$$

RESULTS AND DISCUSSION

Figure 1 gives the elution curve of 0.052 mmol calcium iodide in a thermostatically controled column (20°C) filled with a resin having dibenzo-[18]-crown-[6] anchor groups (2.4 g resin; column height 19 cm, diameter 0.7 cm). As eluent a mixture of 70 % by volume tetrahydrofuran and 30 % by volume methanol was used. After 10 ml of the eluent have been eluted the highest calcium concentration of nearly 100 µg/ml was achieved, which rapidly decreases to a level of 3.5 µg/ml and then continues at this level for more than 100 ml of the eluent. Up to fraction No. 25 only 35.6 % of the whole calcium amount was eluted (Table 2). The unusual elution curve can be explained by a kinetic effect. The isotope ratios $^{44}Ca/^{40}Ca$ and $^{48}Ca/^{40}Ca$ of fraction No. 3, 5 and of the combined fractions 18 to 25 are measured and presented in Table 2. In the last line of Table 2 the initial isotope ratio of the starting material is indicated, which represents values we measured for calcium of natural abundance over a long period by our thermal ionization mass spectrometer [12]. Additional details of the mass spectrometric calcium isotope determination are given elsewhere [13]. From the results in Table 2 one can see an evident enrichment of the heavier calcium isotopes in the first fractions, which decreases with increasing calcium amount eluted. At the point where 35.6 % of the calcium amount are eluted the isotope ratios are slightly higher than the values of the starting material. Because of the strong deviation of the elution curve from that with a Gaussian distribution and because of a possible kinetic effect which can also be responsible for the measured isotope separation, no ε-value can be calculated for this system by Glueckauf's plot. But a significant isotope separation could be analysed at the resin with the crown ether anchor groups.

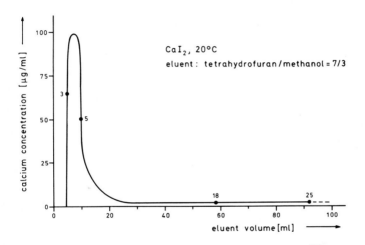

Fig. 1. Calcium elution curve using a resin with dibenzo-[18]-crown-[6] anchor groups

TABLE 2

Calcium isotope separation using a resin with dibenzo -[18]-crown-[6] anchor groups

Fraction No.	Δ m/m [%]	Isotope ratio $(^{44}Ca/^{40}Ca) \times 10^5$	$(^{48}Ca/^{40}Ca) \times 10^6$
3	1.6	2186 ± 13	1989 ± 19
5	21.7	2180 ± 9	1980 ± 22
18 - 25	35.6	2168 ± 3	1956 ± 4
starting material		2151	1937

Figure 2 shows the measured isotope ratio $^{48}Ca/^{40}Ca$ in dependence on the calcium amount Δm/m eluted. This result was obtained using 19 g commercially available resin "[2$_B$.2.2] polymer" in a column experiment with a mixture of methanol/chloroform (7/3 by volume) containing 1.65 % water by volume as eluent (column height 60 cm, diameter 1.2 cm). When measuring the $^{44}Ca/^{40}Ca$ isotope ratio a similar dependence but with smaller separation effect is found. Whereas the first part of the fractions are enriched in the heavier calcium isotopes, the second part of the eluted calcium shows a depletion of the heavier isotopes compared with the starting material. We therefore have an enrichment of the lighter calcium isotopes at the resin phase.

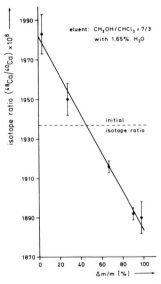

Fig. 2. Calcium isotope separation using a resin with $[2_B.2.2]$ anchor groups

For the isotope separation it is very important to point out that contrary to the results in experiments with Dowex 50 we found here a linear relationship between the isotope ratio (and therefore between the isotope separation) and the calcium amount eluted. Using the measured isotope ratio $^{44}Ca/^{40}Ca$ in the system Dowex 50/3 M HCl as an example, one can see the different shapes of the curves (Figure 3). In a system with Dowex 50 only the first and last fractions show a significant enrichment of the heavier and lighter calcium isotopes, whereas the main calcium amount shows a small isotope separation. Therefore, using a resin with $[2_B.2.2]$ anchor groups the amount of enriched calcium isotope which can be isolated is much higher than by the application of Dowex 50.

For the experiments with "$[2_B.2.2]$ polymer" the ε-values for the isotope pairs $^{44}Ca/^{40}Ca$ and $^{48}Ca/^{40}Ca$ could be determined by Glueckauf's plot [7] in dependence on the water content of the eluent and on the temperature. The results are summarized in Table 3. There is only a small difference in the ε-values using an eluent with 3.25 and 1.65 % water (temperature 20°C).
With decreasing temperature the isotope separation effect decreases. Our experiments showed that this temperature dependence, which is contrary to that found in Dowex 50 systems [4,14], is only valid when the $[2_B.2.2]$ anchor groups are (partly) protonated.

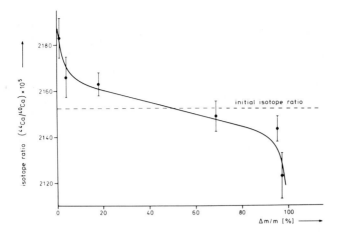

Fig. 3. Calcium isotope separation using Dowex 50-X12 (eluent: 3M HCl)

TABLE 3

Isotope separation effect ε in dependence on the temperature and the H_2O content of the eluent ([2_B.2.2] anchor groups; eluent: $CH_3OH/CHCl_3$ = 7/3)

H_2O content of eluent [Vol %]	Temperature [°C]	$\varepsilon(^{44}Ca/^{40}Ca)$	$\varepsilon(^{48}Ca/^{40}Ca)$
3.25	20	7.5 x 10^{-3}	12.9 x 10^{-3}
1.65	20	5.7 x 10^{-3}	10.4 x 10^{-3}
1.65	0	4.2 x 10^{-3}	7.7 x 10^{-3}
1.65	-21	2.6 x 10^{-3}	4.5 x 10^{-3}

The ε-values of these experiments are the highest isotope separation effects achieved for calcium isotopes in systems with ion exchangers. Compared with exchange reactions at Dowex 50 the ε-value is one to two orders of magnitude greater. Therefore, the application of resins with cryptand anchor groups seems to be a promising system for the chemical enrichment of stable calcium isotopes.

In a preliminary experiment three cycles were carried out under the same conditions (0°C; eluent $CH_3OH/CHCl_3$ = 7/3 with 1.65 % H_2O) in that column with which the monocycle experiment has been done using the resin "[2_B.2.2] polymer". After every passage of calcium through the column the first part of the calcium elution (Δ m/m = 35-69 %, see Table 4) was again given to the top of the column, whereas the second part of the calcium elution was taken out of the chromatographic system. Under the precondition that the measured dependence of the isotope ratio on the eluted calcium amount in the multi-cycle experiment is equal to that in the monocycle experiment, the lower straight line in Figure 4 is valid. After four cycles the upper line was determined. As one can see, the two lines are nearly parallel, which shows that there is indeed a linear relationship between the isotope ratio and the eluted calcium amount.

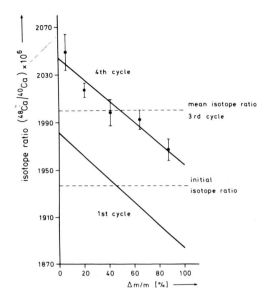

Fig. 4. Calcium isotope separation by a multi-cycle experiment using a resin with $[2_B.2.2]$ anchor groups

The mean isotope ratio $^{48}Ca/^{40}Ca$ of the selected first part of the calcium elution after three cycles is higher than the highest isotope ratio at the very beginning of the first cycle. An additional passage of this part of calcium through the column gives the indicated line for the 4th cycle. Table 4 summarizes the results of the multi-cycle experiment. Beginning with a calcium chloride amount of 210 mg of natural abundance the mean relative enrichment of cycle No. 1 and 2 is calculated for the isotope pairs $^{44}Ca/^{40}Ca$ and $^{48}Ca/^{40}Ca$ by the selected calcium amount of each cycle under the assumption that the results of Figure 2 are valid. The experimentally determined enrichment for the 3rd cycle is given in the last line of Table 4. These experimental values agree very well with the calculated values which are given in brackets. By this multi-cycle experiment it was possible to isolate 30 mg calcium chloride with a relative enrichment of 3.3 % for ^{48}Ca and of 1.9 % for ^{44}Ca compared with natural calcium.

CONCLUSION

The application of exchanger resins with cyclic polyethers as anchor groups seems to be a promising possibility for the chemical enrichment of the heavy calcium isotopes. Compared with resins from the Dowex 50 type there are the following advantages:
1. The isotope separation effect ε is higher by a factor 10-100.
2. The eluent solutions are pure solvents or mixtures of solvents without other

TABLE 4

Calcium isotope separation by a multi-cycle experiment

Cycle No.	$CaCl_2$ amount of the cycle [mg]	Mean relative enrichment [%] $^{44}Ca/^{40}Ca$	$^{48}Ca/^{40}Ca$
Starting material	210	0	0
1	144	0.42*	0.73*
2	84	0.93*	1.71*
3	30	1.87 (1.77*)	3.27 (3.32*)

* Calculated values

chemicals. Therefore, the isolation of the enriched isotopes only requires the evaporation of the solvent without additional separation from other electrolytes.
3. The regeneration of the column can be carried out very simply by water.
4. The wanted heavy calcium isotopes are enriched in the first fractions of the chromatographic separation.

The main disadvantages of the resins with cyclic polyether anchor groups are the not so fast exchange kinetics and the lower exchanger capacity compared with Dowex 50, which results in a lower number of stages in the column. But when optimizing the discussed system with "[2_B.2.2] polymer" it should be possible to produce by that system a technologically adaptable amount of heavy calcium isotopes (in the range of 10-100 mg per experiment), which is enriched e.g. in ^{48}Ca by a factor of two or more in relation to natural calcium.

REFERENCES

1 J. Aaltonen, Suomen Kemistilehti, B44(1970)1-4.
2 G.D. Klinskii, D.A. Knyazev and G.I. Vlasova, Russ. J. Phys. Chem., 48(1974)380-382.
3 W.A. Russel and D.A. Papanastassiou, Anal. Chem., 50(1978)1151-1154.
4 K.G. Heumann and H. Klöppel, Z. Anorg. Allg. Chem., 472(1981)83-88.
5 K.G. Heumann, F. Gindner and H. Klöppel, Angew. Chem. Int. Ed. Engl., 16(1977)719-720.
6 K.G. Heumann and H. Klöppel, Z. Naturforsch., 34b(1979)1044-1045.
7 E. Glueckauf, Trans. Faraday Soc., 54(1958)1203-1205.
8 J.M. Lehn and J.P. Sauvage, J. Amer. Chem. Soc., 97(1975)6700-6707.
9 B.E. Jepson and R. DeWitt, J. Inorg. Nucl. Chem., 38(1976)1175-1177.
10 K.G. Heumann and H.-P. Schiefer, Angew. Chem. Int. Ed. Engl., 19(1980)406-407.
11 E. Blasius, W. Adrian, K.-P. Janzen and G. Klautke, J. Chromatogr., 96(1974)89-97.
12 K.G. Heumann, E. Kubassek, W. Schwabenbauer and I. Stadler, Z. Anal. Chem., 297(1979)35-43.
13 K.G. Heumann, E. Kubassek and W. Schwabenbauer, Z. Anal. Chem., 287(1977)121-127.
14 D.A. Lee, J. Phys. Chem., 64(1960)187-188.

THE SYNTHESIS OF MONO- AND OLIGOSACCHARIDES ENRICHED WITH ISOTOPES OF CARBON, HYDROGEN, AND OXYGEN

R. BARKER, E.L. CLARK, H.A. NUNEZ, J. PIERCE, P.R. ROSEVEAR and A.S. SERIANNI

Division of Biological Sciences, Section of Biochemistry, Molecular and Cell Biology, Cornell University, Ithaca, N.Y. 14853 U.S.A.

ABSTRACT

A convenient, high yield synthesis of [^{13}C]-enriched monosaccharides and monosaccharide phosphates is described. The method is based on the facile condensation of ^{13}CN$^-$ with aldoses or aldose phosphates at pH 7.5 to 8.5 to yield pairs of 2-epimeric aldononitriles (cyanohydrins). These are stable at pH < 5.0 and are reduced in aqueous solution over Pd-BaSO$_4$ to imines which hydrolyze spontaneously to aldoses having one more carbon than the starting aldose and are enriched at C-1. Yields exceed 80 percent. Serial application of the synthesis permits the preparation of multiply-enriched compounds and compounds enriched at sites other than C-1. Exchange of the carbonyl oxygen of the starting aldose with [^{17}O]-or [^{18}O]-enriched water before the addition of cyanide permits the preparation of aldoses enriched with oxygen isotopes at O-2. Reduction of the aldononitriles with ^{2}H$_2$ in ^{2}H$_2$O gives [1-^{2}H]-enriched aldoses. The method, therefore, can be applied to the preparation of aldoses, which are singly or multiply enriched with C, O or H isotopes. Enriched monosaccharides are used to prepare glycosyl phosphates and nucleoside diphosphate sugars. Using glycosyl transferases as biochemical reagents, several di- and tri-saccharides related to the human Type 1 blood-group substances have been prepared. The solution conformations of the latter have been determined by NMR spectroscopy. The enzymatic preparation of several biochemically important aldose and ketose phosphates with specific sites of enrichment is described.

INTRODUCTION

The incorporation of ^{13}C at high atom-percent levels at one or more specific sites in a molecule provides a powerful probe to study chemical and biochemical phenomena [1]. For example, with [^{13}C]-enriched compounds, carbon-carbon coupling ($J_{13_C,13_C}$) can be observed readily in the carbon-13 nuclear magnetic resonance (^{13}C NMR) spectrum [2]. Since coupling through two and three bonds depends on the spatial arrangement of the coupled nuclei [2,3], valuable conformational information can be obtained. The increased sensitivity that accrues from the use of compounds with sites of [^{13}C]-enrichment greatly facilitates the examination of dilute solutions by ^{13}C NMR spectroscopy. In addition, analysis of reaction mixtures yield spectra with a single resonance for each chemical species, permitting specific evaluation of the fates of reactants and intermediates in complex chemical [4] and biochemical transformations. Carbohydrates enriched with ^{2}H at specific sites can be used to assign ^{13}C chemical shifts [5], to simplify complex ^{1}H NMR spectra [6], to evaluate isotope shifts and ^{13}C-^{2}H coupling [5,6], and to study carbohydrate-protein interactions [7]. [^{17}O]-Enriched carbohydrates have been employed to evaluate ^{17}O NMR parameters [8] and to assess solution dynamics [9].

During the past twenty years, it has been established that complex oligosaccharides confer individuality and specificity to many biological structures. As components of glycoproteins and glycolipids, they serve as recognition sites in cell-cell, antibody-antigen and other inter- and intra- cellular recognition processes. In glyco-substances, information is coded in the sequence of monosaccharide units and in the position and configuration of inter-residue linkages. The conformations of the monosaccharide rings and of the glycosidic bonds joining them determine overall molecular shape which, in turn, probably plays a key role in the inter-molecular association processes that determine recognition phenomena. Enrichment with ^{13}C at specific sites adjacent to the glycosidic bond and determination of C-C and C-H couplings across the bond can provide information on both intra- and inter-residue conformational preferences in solution, the biologically-important state.

The oligosaccharide 4 (Scheme 1) is the human Type 1 blood-group substance. Rotation about the C2 Gal - O2 Gal bond produces different magnitudes of coupling between C1 of the fucosyl residue (C1 Fuc) and C1 of the galactosyl residue (C1 Gal) (the atoms indicated by filled circles). In addition, coupling between C1 Fuc and H2 Gal (J_{C-H}) varies in a complementary fashion. These coupling constants can be evaluated by preparing 4 with [^{13}C]-enrichment at C1 Gal and/or C1 Fuc, as shown in Scheme 1.

The synthetic problem can be divided into five components:
A. Preparation of [1-^{13}C]monosaccharides
B. Preparation of [1-^{13}C]glycosyl 1-phosphates with the correct anomeric configuration (α-galactopyranosyl 1-phosphate and β-L-fucopyranosyl 1-phosphate)
C. Preparation of UDP-galactose and GDP-fucose with [^{13}C]-enrichment in the pyranosyl component
D. Purification of the glycosyl transferases: UDP-Gal transferase from bovine whey, GDP-Fuc transferase from porcine submaxillary glands

E. Preparation of [^{13}C]-enriched oligosaccharides and the determination of preferred solution conformations by NMR spectroscopy

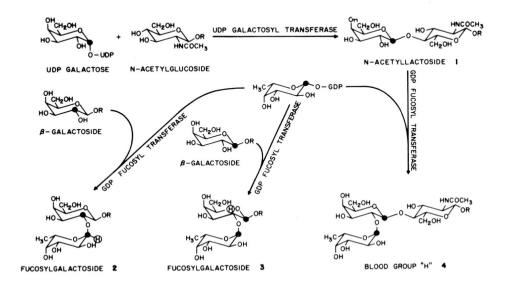

Scheme 1

A. <u>Preparation of 1-^{13}C] Monosaccharides</u>. The standard method for incorporating carbon isotopes into monosaccharides utilizes cyanide as the enriched precursor. Cyanide is condensed with a starting aldose to form 2-epimeric nitriles (cyanohydrins) which hydrolyze in alkali to aldonates. After separation by differential solubility, these are converted to lactones and reduced to the desired aldose with sodium amalgam [10] or borane [11]. The process is time-consuming and yields can be low.

It has been established [12] that α-amino-nitriles can be reduced in aqueous solution to give good yields of α-amino-aldehydes, presumably through the formation of imines which hydrolyze rapidly <u>in situ</u>. It appeared that the standard method described above could be simplified if the intermediate aldononitriles could be prevented from hydrolyzing to amides and aldonates. A study of the cyanohydrin reaction at various pH values was undertaken to explore this possibility [4]. Shown in Figure 1A is the ^{13}C NMR spectrum of a reaction mixture containing the [1-^{13}C] nitriles derived from D-erythrose and K^{13}CN and some of the intermediates formed during their decomposition to D-[1-^{13}C] arabinonate and ribonate. By comparison with standard compounds, each intermediate was identified, and the time course of the decomposition was established by acquiring spectra at intervals during the reaction (Fig. 1B). Based on this study, a simple synthesis of [1-^{13}C]-enriched aldoses was developed [13,14]. The condensation of cyanide with aldoses is rapid and complete at pH 8.0 ± 0.5 and the nitriles are completely stable at pH < 5.0. For example, D-[1-^{13}C]

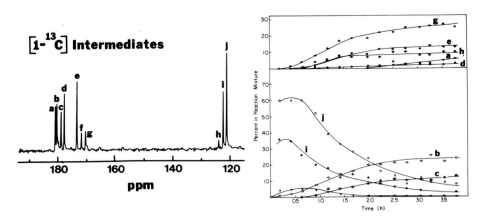

Fig. 1. ^{13}C NMR Spectrum of [1-^{13}C] Intermediates in the Reaction of K^{13}CN with D-Erythrose at pH 7.0 (left) and Reaction Profile at pH 8.5 (right). a, D-[1-^{13}C] arabinonate; b, D-[1-^{13}C] arabinonamide; c, D-[1-^{13}C] ribonamide; d, D-[1-^{13}C] arabinono-γ-lactone; e, D-[1-^{13}C] arabinonoamidine; f, D-[1-^{13}C] ribonoamidine; g,h, dimer; i, D-[1-^{13}C] arabinononitrile; j, D-[1-^{13}C] ribononitrile. Shoulder and minor resonance between peaks b and c (left) are D-[1-^{13}C] ribono-γ-lactone and D-[1-^{13}C] ribonate, respectively.

galactose is conveniently prepared from D-lyxose as shown below:

The formation of the 2-epimeric nitriles is >95 percent complete in 20 min at pH 7.8 when three molar equivalents of ^{13}CN are used. The reaction mixture is adjusted to pH 4.2 ± 0.1 with glacial acetic acid and purged with N_2 to remove the excess $H^{13}CN$. The latter is trapped in methanolic KOH and recovered essentially quantitatively. The nitriles in solution at pH 4.2 are hydrogenated over Pd-BaSO$_4$ (60 mg/mmole of nitrile) at pressures up to 60 p.s.i. H_2 to give the corresponding aldoses and small amounts of 1-amino-1-deoxyalditols. The amines are removed by treatment with Dowex 50 (H^+) and the epimeric [1-^{13}C]aldoses are separated on large columns of Dowex 50 (Ba^{2+}) using water as the eluent [15]. Yields exceed 85 percent and products of excellent purity are obtained as crystals or as dry syrups. The ratio of D-[1-^{13}C]galactose to D-[1-^{13}C]talose is approximately 1.5 to 1.

The procedure described above can be employed with minor modifications to prepare any aldose [13,14] or aldose phosphate [16] from a parent aldose or aldose phosphate, respectively, having one fewer carbon atom than the desired product. When the chain-length or the presence of a phosphate group prevents the formation of a pyranose ring in the parent aldose or aldose phosphate, the reaction with cyanide is complete with stiochiometric amounts of reactants. Nitriles with four or fewer carbons and C_3-C_5 aldononitrile phosphates are hydrogenated at pH 1.7 ± 0.1 to obtain high yields of aldose or aldose phosphates. Higher pH values, higher pressures and catalysts more active than Pd-BaSO$_4$ increase the formation of alditylamines [14].

In the preparation of aldose phosphates, separation of the 2-epimeric nitrile phosphates by anion-exchange chromatography on Dowex 1 X8 (formate) is performed prior to hydrogenation.

All of the [1-^{13}C]-enriched aldoses having 2 to 6 carbons and the terminal phosphate-substituted glyceraldehydes, tetroses and pentoses have been synthesized.

Overall yield in the nitrile hydrogenation reaction is sufficiently high to permit serial application of the synthesis. In this manner, several aldoses [14] and aldose phosphates [16] with internal [^{13}C]-enrichment have been prepared as well as several with enrichment at two sites.

The composition of the aqueous reaction mixtures can be examined easily prior to purification by ^{13}C NMR as illustrated in Figure 2. It is clear that ^{13}CN reacts completely to produce [1-^{13}C]nitriles and that hydrogenation yields only small amounts of amine by-products.

Isotopes of oxygen can be incorporated by equilibrating the starting aldose with appropriately enriched H_2O under acidic (pH 1) conditions. The incorporated isotope is trapped by addition of KCN to form the 2-epimeric 2-[O]-enriched nitriles.

[1-^{2}H]-Enriched aldoses can be produced in high yields by carrying out the reduction of nitriles in ^{2}H$_2$O using ^{2}H$_2$ [6]. Obviously, these procedures can be combined to produce multiply-enriched monosaccharides.

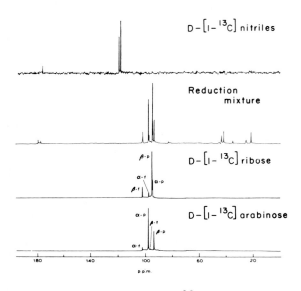

Fig 2. Preparation of D-[1-^{13}C]Arabinose and Ribose. The proton-decoupled ^{13}C NMR spectra of the intermediate [1-^{13}C]nitriles, the reduction mixture prior to purification, purified D-[1-^{13}C]ribose, and purified D-[1-^{13}C]arabinose. Resonance at ∼177 ppm in the [1-^{13}C]nitrile spectrum is due to D-[1-^{13}C]arabinono-γ-lactone. Resonances at ∼43 ppm in the spectrum of the reduction mixture are due to D-[1-^{13}C] 1-amino-1-deoxyalditols.

B. **Preparation of [^{13}C] Glycosyl 1-Phosphates.** A general synthesis of glycosyl 1-phosphates was developed by Khwaja et al. [17] and has been extended by Shibaey et al. [18] and by Behrman and coworkers [19]. Application to the synthesis of enriched compounds requires careful attention to reaction conditions, especially when labile products are involved. The reaction sequence shown below can be used to prepare either α- or β-glycosyl 1-phosphates. The glycosyl halide always contains the halide substituent in an axial orientation as predited from the anomeric effect. Hydrolysis, which is very rapid, always yields the protected product with an equatorial hydroxyl group. This compound is unstable and must be derivatized immediately if an equatorial glycosyl phosphate is desired. On the other hand, in the presence of trace amounts of moisture, an equilibrium is established which favors the axial hydroxyl arrangement and allows the preparation of axially-oriented glycosyl phosphate esters. The separation of α- and β-anomers is easily achieved by anion-exchange chromatography [20]. Alternatively, for certain glycosyl phosphates such as α-D-galactosyl 1-phosphate, reasonable yields can be obtained by the MacDonald procedure [21].

C. <u>Preparation of [^{13}C]Enriched Nucleoside Diphosphate Sugars</u>. Two procedures have been used. In both, the appropriate nucleoside monophosphate is activated to facilitate displacement by a glycosyl 1-phosphate to form the pyrophosphate bond of the nucleoside diphosphate sugar. The Moffatt procedure [22] has the advantage of an easy preparation of the activated nucleotide. A method developed by Furusawa [23] employs shorter reaction times and yields are generally higher, but the activated nucleotide is more difficult to prepare. The application of both methods to the preparation of GDP-fucose has been described [24].

D. <u>Purification of Glycosyl Transferases</u>. A general procedure for the purification of glycosyl transferases was proposed by Barker et al. [25]. The method depends on the high specificities and affinities of these enzymes for the purine or pyrimidine base of the nucleoside diphosphate sugar substrates and for the acceptor mono- or oligo- saccharides. Using affinity adsorbents with these components as ligands, the principal transferases involved in the formation of Type 1 human blood-group oligosaccharides were purified, some to homogeneity [26-28]. Each transferase catalyzes the formation of a specific glycosidic bond between the monosaccharide component of the nucleoside diphosphate sugar and the appropriate acceptor. Both the position of substitution on the acceptor and the anomeric configuration of the glycosidic bond are specified by the enzyme. Yields generally exceed 80% and 0.1 to 0.5 mmoles of product oligosaccharides can be prepared conveniently.

E. Preparation of [^{13}C] Oligosaccharides and the Determination Preferred Solution Conformations. As a first step in the synthesis of 4, the disaccharide β-D-Gal (1-4) β-D-Glc NAc-hexanolamine (Scheme 1, 1) was prepared with [^{13}C]-enrichment at Cl Gal. The necessary substrates were UDP-[1-^{13}C]Gal, prepared as described above, and β-D-GlcNAc-hexanolamine prepared chemically [29]. Galactosyl transferase isolated from bovine whey [25] was added in sufficient quantity to complete the reaction in 8 hr. The course of the reaction was followed by the appearance of UDP and the disappearance of UDP-Gal using high-pressure liquid chromatography [30]. The desired product was separated from a small amount of starting material and other by-products by ion-exchange chromatography followed by gel-filtration.

To assess the solution conformation of 1, the ^{13}C NMR spectrum at 90.5 MHz was obtained for those carbons having ^{13}C at natural abundance. This portion of the spectrum is shown in Figure 3. Resonances for C3' and C5' of the GlcNAc residue are particularly significant. These carbons are separated by three bonds from the enriched carbon at Cl of the galactosyl moiety. Depending upon the dihedral angle about the Cl Gal-Ol Gal bond, the resonances for C3' and C5' can be broadened or split due to coupling to the enriched carbon. Values of the coupling constants can range from 0 to 4 Hz as the dihedral angle changes [3]. Examination of the resonances of C3' and C5' in Figure 3 shows that both have $J_{C-C} \simeq 0$ Hz. Since the GlcNAc residue must rotate as a unit, only two orientations satisfy the condition that both dihedral angles are similar. These orientations place both C3' and C5' at either 60° or 120° with respect to Cl Gal. For steric reasons, the orientation with C3' and C5' at 60° can be eliminated, so that the result favors the preferred conformation shown in Figure 3. Additional evidence supporting this conformation was obtained from the evaluation of coupling between Cl and H4', Cl and C4', and C4' and H1.

The preferred solution conformation of 1 is similar to that observed in the crystalline state [31]. It permits the existence of a hydrogen bond between OH3' of GlcNAc and the ring oxygen of Gal. This conformation is also predicted from hard-sphere calculations [32].

Similar NMR studies with 4 containing ^{13}C at Cl Gal, Cl Fuc and/or C2 Gal permitted the evaluation of coupling constants between the enriched nucleus and the carbon and hydrogen nuclei within three bonds. From these data, the conformation shown below was deduced. It is noteworthy that the substitution of a fucosyl residue at C2 Gal does not alter the preferred orientation of Gal with respect to GlcNAc. The conformations of the constituent monosaccharide rings do not appear to change substantially from those of the simple glycosides of similar configurations.

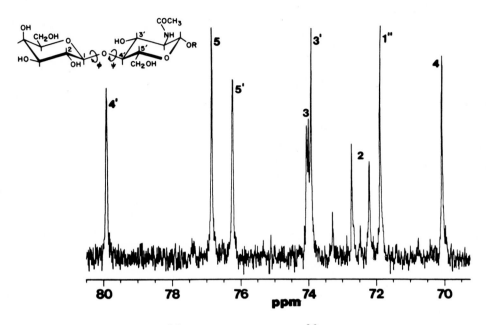

Fig. 3. Partial 90.5 MHz ^{13}C NMR Spectrum of [1-^{13}C]Gal β-(1→4)GlcNAc-β-hexanolamine in ^{2}H$_2$O. Insert shows the preferred conformation of this disaccharide determined from an interpretation of ^{13}C-^{13}C and ^{13}C-^{1}H coupling constants.

Although further studies are needed, results obtained to date demonstrate that ^{13}C labelling and NMR spectroscopy combine to permit reasonable estimates of solution conformations of biologically-important oligosaccharides. Although preferred conformations can be predicted accurately by hard-sphere calculation [32], experimental data for other systems will be needed before such predictability can be assumed. In any case, enriched compounds are essential for studies on the interaction of oligosaccharides with proteins that recognize them.

Enzymatic Synthesis of Isotopically-Enriched Aldoses and Ketose Phosphates.
Many biologically-important sugar phosphates, especially ketose phosphates, are difficult to prepare by traditional chemical methods. The chemical synthesis of such compounds with specific isotopic enrichment is particularly troublesome. Specifically [^{13}C]-enriched compounds are useful for the study of chemical and biochemical interconversions with purified enzymes [1], for the study of solution composition and rates of interconversion of tautomers [33]. We have combined chemical and biochemical reactions using commercially available enzymes to convert specifically-enriched monosaccharides to more complex biological compounds. Two examples are given: the preparation of D-[4-^{13}C]glucose and D-[2-^{13}C]ribulose 1,5-bisphosphate.

D-[4-^{13}C]Glucose was required for the preparation of di- and tri-saccharides similar to 1, with Glc substituted for GlcNac. The sequence of reactions shown below was employed. The enriched carbon was introduced as DL-[1-^{13}C]glyceraldehyde,

```
DHA  GLYCEROL KINASE    DHAP
     ATP, Mg²⁺, pH 7.2
                        +
                        H\
                          C=O    ALDOLASE    D-[4-¹³C] FRUCTOSE 1-P   ① ACID PHOSPHATASE, pH 4.5
                         /       pH 7.3              +                ② CHROMATOGRAPHY
                        CH,OH                L-[4-¹³C] SORBOSE 1-P       DOWEX 50 (Ba²⁺)
                        |
                        H₂COH

                                         D-[4-¹³C] FRUCTOSE 6-P  ← HEXOKINASE   D-[4-¹³C] FRUCTOSE
       ① ACID PHOSPHATASE, pH 4.5                                  ATP, Mg²⁺
                                              PHOSPHOGLUCOSE       pH 7.4
       ② CHROMATOGRAPHY                       ISOMERASE, pH 7.4
          DOWEX 50 (Ba²⁺)
                                         D-[4-¹³C] GLUCOSE 6-P

       → D-[4-¹³C] GLUCOSE
```

prepared by the addition of ^{13}CN to glycolaldehyde. This 3-carbon unit was condensed with dihydroxyacetone phosphate by the action of rabbit muscle aldolase to give an equimolar mixture of [4-^{13}C]-enriched L-sorbose and D-fructose 1-phosphates. Dihydroxyacetone phosphate was prepared from dihydroxyacetone and ATP by the action of glycerol kinase. The two ketose phosphates are purified by anion-exchange chromatography, dephosphorylated with acid phosphatase, and the neutral sugars separated by adsorption chromatography. To complete the conversion to D-[4-^{13}C]glucose, D-[4-^{13}C]fructose was phosphorylated at C-6 with ATP and hexokinase in the presence of phosphoglucose isomerase to given an equilibrium mixture containing ∼75 percent glucose. The mixture of 6-phosphates was purified by anion-exchange chromatography, the mixed 6-phosphates were treated with acid phosphatase, and the ketose and aldose were separated by adsorption chromatography. Overall yield of D-[4-^{13}C]glucose was approximately 30 percent based on DL-[1-^{13}C]glyceraldehyde. Yield based on recovered isotope ^{13}C is much higher, approximately 85 percent, although the label is distributed in the two ketose phosphates formed by the aldolase reaction.

D-[2-^{13}C]Ribulose 1,5-bisphosphate (RuDP) was prepared from D-[2-^{13}C]ribose 5-phosphate in 85 percent overall yield [16]. The starting [^{13}C]-enriched aldose phosphate was prepared from D-[1-^{13}C]erythrose 4-P and KCN. D-[2-^{13}C]Ribose 5-P was converted to the ketose bisphosphate by the action of phosphoriboisomerase and phosphoribulokinase in the presence of Mg^{2+}-ATP. The product D-[2-^{13}C]RuDP was purified by anion-exchange chromatography and stored as the barium salt.

SUMMARY

Monosaccharides and their derivatives have been prepared containing isotopes of carbon, hydrogen and oxygen at specific sites by a procedure based on the classical cyanohydrin reaction. Several [^{13}C]-enriched monosaccharides were chemically converted to [^{13}C]-enriched glycosyl 1-phosphates which were employed in the chemical synthesis of nucleoside diphosphate sugars. These nucleoside diphosphate sugars enriched with ^{13}C in the pyranosyl constitutents were used to prepare human blood-group Type 1 [^{13}C]oligosaccharides through the use of specific glycosyl transferases. These [^{13}C]-enriched oligosaccharides were examined by ^{13}C NMR spectroscopy to determine preferred solution conformations.

Chemical and biochemical reactions were employed to prepare several biologically-important carbohydrates. Preparations of millimole quantities of D-[4-^{13}C]glucose and D-[2-^{13}C]ribulose 1,5-bisphosphate were described. The use of specific [^{13}C]-enrichment to monitor chemical and biochemical conversions was illustrated.

REFERENCES

1. H.A. Nunez, T.E. Walker, R. Fuentes, J. O'Connor, A. Serianni and R. Barker,

1. J. Supramol. Struct. 6(1977)535.
2. T.E. Walker, R.E. London, T.W. Whaley, R. Barker and N.A. Matwiyoff, J. Amer. Chem. Soc. 98(1976)5807.
3. J.A. Schwarcz and A.S. Perlin, Can. J. Chem. 50(1972)3667. J.A. Schwarcz, N. Cyr and A.S. Perlin, Can. J. Chem. 53(1975)1872. N. Cyr, G.K. Hamer and A.S. Perlin, Can. J. Chem. 56(1978)297.
4. A.S. Serianni, H.A. Nunez and R. Barker, J. Org. Chem. 45(1980)3329.
5. P.A.J. Gorin, Can. J. Chem. 52(1974)458. P.A.J. Gorin and M. Mazurek, Can. J. Chem. 53(1975)1212. P.A.J. Gorin and M. Mazurek, Can. J. Chem. 48(1976)171.
6. A.S. Serianni and R. Barker, Can. J. Chem. 57(1979)3160.
7. L. Szilágyi, J. Harangi and L. Radics, Biophys. Chem. 6(1977)201. K.J. Neurohr, N. Lacelle, H.H. Mantsch and I.C.P. Smith, Biophys. J. 32(1980)931.
8. P.A.J. Gorin and M. Mazurek, Carbohydr. Res. 67(1978)479.
9. A. Suggett, S. Ablett and P.J. Lillford, J. Sol. Chem. 5(1976)17.
10. H.S. Isbell, N.B. Holt and H.L. Frush, Method Carbohydr. Chem. 1(1962)276.
11. S.S. Bhattacharjee, J.A. Schwarcz and A.S. Perlin, Carbohydr. Res. 42(1975) 259. P. Kohn, R. Samaritano and L. Lerner, J. Amer. Chem. Soc. 87(1965) 5475.
12. R. Kuhn and W. Kirschenlohr, Angew. Chem. 67(1955)786.
13. A.S. Serianni, H.A. Nunez and R. Barker, Carbohydr. Res. 72(1979)71.
14. A.S. Serianni, E.L. Clark and R. Barker, Carbohydr. Res. 72(1979)79.
15. J.K.N. Jones and R.A. Wall, Can. J. Chem. 38(1960)2290.
16. A.S. Serianni, J. Pierce and R. Barker, Biochemistry 18(1979)1192.
17. T.A. Khwaja, C.B. Reese and J.C.H. Stewart, J. Chem. Soc. (1970)2092.
18. V. Shibaey, Y.Y. Kusov, V.A. Pentrenko and N.K. Kochetcov, Akad. Nauk. SSSR, Ser. Khim (1974)1852.
19. H.S. Prihar and E.J. Behrman, Biochemistry 12(1973)997.
20. D.L. MacDonald, Carbohydr. Res. 6(1968)376.
21. D.L. MacDonald, J. Org. Chem. 27(1962)1107.
22. J.G. Moffatt, Methods in Enzymology 8(1966)136.
23. K. Furusawa, M. Sekine and T. Hata, J. Chem. Soc. Perkin I (1976)1711.
24. H.A. Nunez, J.V. O'Connor, P.R. Rosevear and R. Barker, Can. J. Chem., in press.
25. R. Barker, K.A. Olson, J.L. Shaper and R.L. Hill, J. Biol. Chem. 247(1972) 7135.
26. T.A. Beyer, J.I. Rearick, J.L. Paulson, J.P. Prieels, J.E. Sadler and R.L. Hill, J. Biol. Chem. 254(1979)12531.
27. T.A. Beyer, J.E. Sadler and R.L. Hill, J. Biol. Chem. 255(1980)2364.
28. J.E. Sadler, J.I. Rearick, J.L. Paulson and R.L. Hill, J. Biol. Chem. 254 (1979)4434.
29. C.K. Chiang, M. McAndrew and R. Barker, Carbohydr. Res. 70(1979)93.
30. H.A. Nunez and R. Barker, Biochemistry 19(1980)489.
31. W.J. Cook and C.E. Bugg, Acta Crystallogr., Sect. B 29(1973)907.
32. R.U. Lemieux, K. Bock, L.T.J. Delbaere, S. Koto and V.S. Rao, Can. J. Chem. 58(1980)631.
33. A.S. Serianni, J. Pierce, S. Huang and R. Barker, manuscript in preparation.

THE PREPARATION OF SOME C-DEUTERATED L-ASCORBIC ACIDS

HANS J. KOCH and RONALD S. STUART
Merck Frosst Labortories, P.O. Box 1005, Pointe Claire-Dorval, Quebec, H9R 4P8 (Canada)

ABSTRACT

2,3-O-Isopropylidene-α-L-xylo-2-hexulofuranose-6,6-d_2 and methyl-α-L-xylo-
-2-hexulopyranoside-1,1,3,4,5-d_5 were prepared from the corresponding "light" compounds by catalytic exchange in boiling deuterium oxide containing "deuterated" Raney nickel. Treatment with acidic acetone gave the correspondingly deuterated 2,3:4,6-di-O-isopropylidene-α-L-xylo-2-hexulo-
furanoses (diacetone-L-sorboses) which are intermediates in the Reichstein-Grussner synthesis of L-ascorbic acid.

INTRODUCTION

In a series of publications we have described a novel direct catalytic exchange procedure [1,2]. Non-reducing carbohydrate derivatives are readily C-deuterated, at carbon atoms bound to free hydroxyl groups, by hot deuterium oxide containing "deuterated" Raney nickel. Configuration is usually maintained during the exchange, however, prolonged treatment does lead to isomerizations. A possible mechanism has been discussed [2]. The preparation of the "deuterated" catalyst and the general experimental conditions have also been described [1,3].

Methyl glycosides are suitable substrates for the exchange, while free aldoses are reduced [4]. For example, D-glucose-2,3,4,6,6-d_5 is prepared by the Raney nickel catalyzed exchange with boiling deuterium oxide of methyl α (or β)-D-glucopyranoside followed by acid hydrolysis [2]. While O-isopropylidene groups are stable to the exchange conditions O-benzylidene groups suffer hydrogenolysis. In the case of some compounds such as alditols, furanosides, and some ketosides it is advisable to deactivate the catalyst in order to suppress isomerization. In this case the usual experimental conditions are followed, however, the "deuterated" Raney nickel catalyst is boiled under reflux with deuterium oxide for about 8 h before addition of the compound to be exchanged. The alditols of pentoses and hexoses have been fully C-deuterated by this exchange[5] as have ethylene glycol and glycerol. D-glucitol-d$_8$ is best isolated via its peracetate.

DISCUSSION AND RESULTS

For the preparation of C-deuterated L-ascorbic acids a part of the classical synthesis of T. Reichstein and A. Grussner[6] was used as the basis. In this scheme L-sorbose (1) is converted with acidic acetone into 2,3:4,6-di-O-isopropylidene-α-L-xylo-2-hexulofuranose (2) which is oxidized with potassium permanganate to the potassium salt of 2,3:4,6-di-O-isopropylidene-α-L-xylo-2-hexulofuranosonic acid (3), treatment of 3 with cold hydrochloric acid gives the free acid (4). Compound 4 is freed of its protecting groups by heating in water to afford L-xylo-2-hexulosonic acid (5), the 2-keto-L-gulonic acid of Reichstein and Grussner[6]. Treatment of 5 with diazomethane gives its methyl ester (6) which upon treatment with methanolic sodium methoxide solution followed by methanolic hydrogen chloride gives L-ascorbic acid (7). In spite of the many modifications[7] of this synthesis it is very suitable for small scale preparations and all compounds crystallize readily.

The deuterium label was introduced by Raney nickel catalyzed exchange into derivatives of L-sorbose, namely 2,3-O-isopropylidene-α-L-xylo-2-hexulofuranose (8) and methyl α-L-xylo-2-hexulopyranoside (9) both of which were converted to the correspondingly C-deuterated forms of 2,3:4,6-di-O-isopropylidene-α-L-xylo-2-hexulofuranose (2a and 2b) which were then converted to the C-deuterated L-ascorbic acids.

A fully C-deuterated L-ascorbic acid could be prepared from D-glucitol-d_8 [5] (D-sorbitol-d_8) by conversion to L-sorbose-d_7 [6,7]. This bacterial oxidation would most likely involve a considerable isotope effect [8].

Various methods [6,9] for the preparation of 2,3-O-isopropylidene-α-L-xylo-2-hexulofuranose (8) have been described. However, the following modificaton of the method of Ohle [9] makes the preparation of (8) simple. L-Sorbose (1) (100 g) and acetone (2 L) containing concentrated sulphuric acid (50 mL) were stirred for 24 hours. The mixture was neutralized with excess anhydrous sodium carbonate, filtered and the solution evaporated to a syrup which consists mainly of 2,3:4,6-di-O-isopropylidene-α-L-xylo-2-hexulofuranose (2) and 2,3-O-isopropylidene-α-L-xylo-2-hexulofuranose (8). Mild acid hydrolysis (1% aqueous H_2SO_4, 15 h, RT) followed by neutralization with excess barium carbonate, filtration and evaporation to a syrup converts 2 to 8. Compound 8 is quite stable under these conditions. Recrystallization of 8 from ethyl acetate gave the pure product. Boiling a solution of 8 (4 g) in deuterium oxide (100 mL) containing deactivated (see above) "deuterated" Raney nickel (20 mL settled volume) for 15 h gave 2,3-O-isopropylidene-α-L-xylo-2-hexulofuranose-6,6-d_2 (8a). The exchange was followed by ^{13}C.n.m.r.. Although exchange does not normally occur adjacent to an O-isopropylidene group [1,2] some deuteration was observed at position 1. However, since C-1 is converted to a carboxyl group in the Reichstein-Grussner synthesis [6] this was inconsequential. Compound 8a (3g) was converted to 2,3:4,6-di-O-isopropylidene-α-L-xylo-2-hexulofuranose-6,6-d_2 (2a) (3 g recrystallized from petroleum ether) by treatment (10 h) with acetone (100 mL) containing sulphuric acid (2 mL) at RT, followed by neutralization with excess anhydrous sodium carbonate. Compound 2a was then converted to L-ascorbic acid-6,6-d_2 by the Reichstein-Grussner method [6].

In order to introduce deuterium by exchange at positions 4 and 5 (in L-ascorbic acid), methyl α-L-xylo-2-hexulopyranoside (9) was used. The corresponding ethyl glycoside is not a suitable compound since the exchange at C-4 is very slow. This is similar to the greatly reduced rate of exchange at C-3 of ethyl α-D-glucopyranoside observed by Perlin [10]. Compound (9) is readily prepared by the method of Fischer [11]. Exchange of 9 (4.5 g) with boiling deuterium oxide (150 mL) containing deactivated "deuterated" Raney nickel (20 mL) for 15 h gave methyl α-L-xylo-2-hexulopyranoside-1,1,3,4,5-d_5 (9a), (3 g). Deuterium atoms at positions 1 and 3 are eventually lost in the

L-ascorbic acid synthesis. Treatment of (9a) (3g) at RT with acetone (100 mL) containing concentrated sulphuric acid (2 mL) for 24 h gave 1,2:4,6-di-O-isopropylidene-α-L-xylo-2-hexulofuranose-1,1,3,4,5-d_5 (2b) (2.8 g, recrystallized from petroleum ether) directly. Compound 2b can then be converted to L-ascorbic acid-4,5-d_2 by the Reichstein-Grussner proceure [6]. Attempts to exchange L-ascorbic acid directly were not successful.

REFERENCES
1 H. J. Koch and R. S. Stuart, Carbohydr. Res., 59(1977) C1.
2 H. J. Koch and R. S. Stuart, Carbohydr. Res., 67(1978)341.
3 H. J. Koch and R. S. Stuart, Carbohydr. Res., 64(1978)127.
4 J. V. Karabinos and A. T. Ballun, J. Am. Chem. Soc., 75(1953)4501.
5 H. J. Koch and R. S. Stuart, C.I.C. Conference, Vancouver, B.C., 1979.
6 T. Reichstein and A. Grussner, Helv. Chim. Acta, 17(1934)311.
7 T. C. Crawford and S.A. Crawford, Adv. in Carbohydr. Chem. and Biochem., 37(1980)79.
8 A. S. Perlin and A. Maradufu, Can. J. Chem., 49(1971)3429.
9 H. Ohle, Ber., 71(1938)562.
10 A. S. Perlin, private communication.
11 E. Fischer, Ber., 28(1895)1159.

SPECIFIC DEUTERIUM LABELLING OF 1,4-BENZODIAZEPINES

A.A. LIEBMAN, G.J. BADER, W. BURGER, J. CUPANO, C.M. DELANEY, Y-Y. LIU,
R.R. MUCCINO, C.W. PERRY AND E. THOM
Chemical Research Department, Hoffmann-La Roche Inc., Nutley, New Jersey

ABSTRACT

Specifically deuterated 1,4-benzodiazepines were prepared to assist in the rationalization of the mass spectral processes of these compounds. Aminobenzophenone precursors to the benzodiazepines were conveniently prepared from a 2,4-diphenylquinazoline which was derived from aniline and either specifically deuterated or appropriately substituted for subsequent deuteration. The preparation of 5-desmethyldiazepam and 4-diazepam derivatives is described.

INTRODUCTION

Deuterium labelled compounds have always occupied key roles in the interpretation of mass spectra. Comparison of the mass spectra of specifically labelled molecules with those of other specifically labelled or unlabelled molecules has led to satisfactory rationalizations of important mass spectral processes and has ultimately provided the basis for the development of analytical procedures for the determination of such substances in biological fluids, etc. Such a situation applies to the 5-phenyl-1,4-benzodiazepine-2-ones, a clinically important group of compounds, and reports have appeared on the mass spectral analyses of these compounds, i.e. (1). Common to the spectra of the benzodiazepines is a prominent M-1 ion and we undertook the specific deuterium labelling of these compounds to provide analogs that would be helpful in understanding the origin of this ion. An initial report on the use of these deuterated substrates has been published (2) and additional applications will be forthcoming (3).

DISCUSSION

Fentiman, Jr. and Foltz have reported (4) the synthesis of several 1,4-benzodiazepines labelled with 5 deuterium atoms (5-phenyl-d_5) and suitable for use as internal standards in the quantification of the unlabelled compounds by selected ion monitoring. Central to the preparation of these compounds is an aminobenzophenone obtained from a deuterated precursor by the use of conditions under which deuterium exchange is unlikely to occur. Therefore, a 1,4-benzodiazepine with a specific deuterium label may be prepared from the corresponding aminobenzophenone that already has the appropriate deuterium substitution. Alternatively, substituted benzodiazepines or near intermediates may be prepared along conventional lines (5) and the deuterium introduced later in the synthesis. Using these approaches, we prepared five 7-chloro-5-phenyl-1,4-benzodiazepine-2-ones specifically labelled with deuterium, as well as some N-1 CD_3-derivatives listed in the table.

The compounds prepared in this series were all syntheziseded via an intermediate aminobenzophenone. With the exceptions of compounds 17 and 21, the aminobenzophenones themselves were obtained with the necessary deuterium label and by the methylation and hydrolysis of an appropriate diphenylquinazoline. This, in turn, is assembled by reaction of the imino chloride of a benzanilide with a benzonitrile, both of which can be obtained from aniline (1), a compound readily labelled with deuterium. Thus, m-bromoaniline (2) was converted to N-benzoyl-3-bromo-4-chloroaniline (3) which yields the diphenylquinazolines (4) and (5) in a conventional reaction with benzonitrile (6). Since methylation and hydrolysis of 6-halogenated quinazolines proceeds to the corresponding N-methyl aminobenzophenones, we planned to apply these procedures here and to demethylate the resulting intermediates and carry them forward to the bromobenzodiazepines. Hydrogenolysis with deuterium gas would have provided the desired deuterated derivatives. Hydrolysis of these quinazolines, however, was not straightforward and we therefore effected hydrogenolysis at this stage to provide (6) and (7) after methylation and hydrolysis. Subsequent reactions to afford (15) and (16) were then carried out in a conventional manner (5) using the deuterated intermediates.

We obtained 9-deuterodesmethyldiazepam (17) by deuterium hydrogenolysis of a precursor, 9-iododesmethyldiazepam (20). Aminobenzophenone 14 undergoes selective iodination into the desired position and the intermediate is then converted to (20) in a conventional manner.

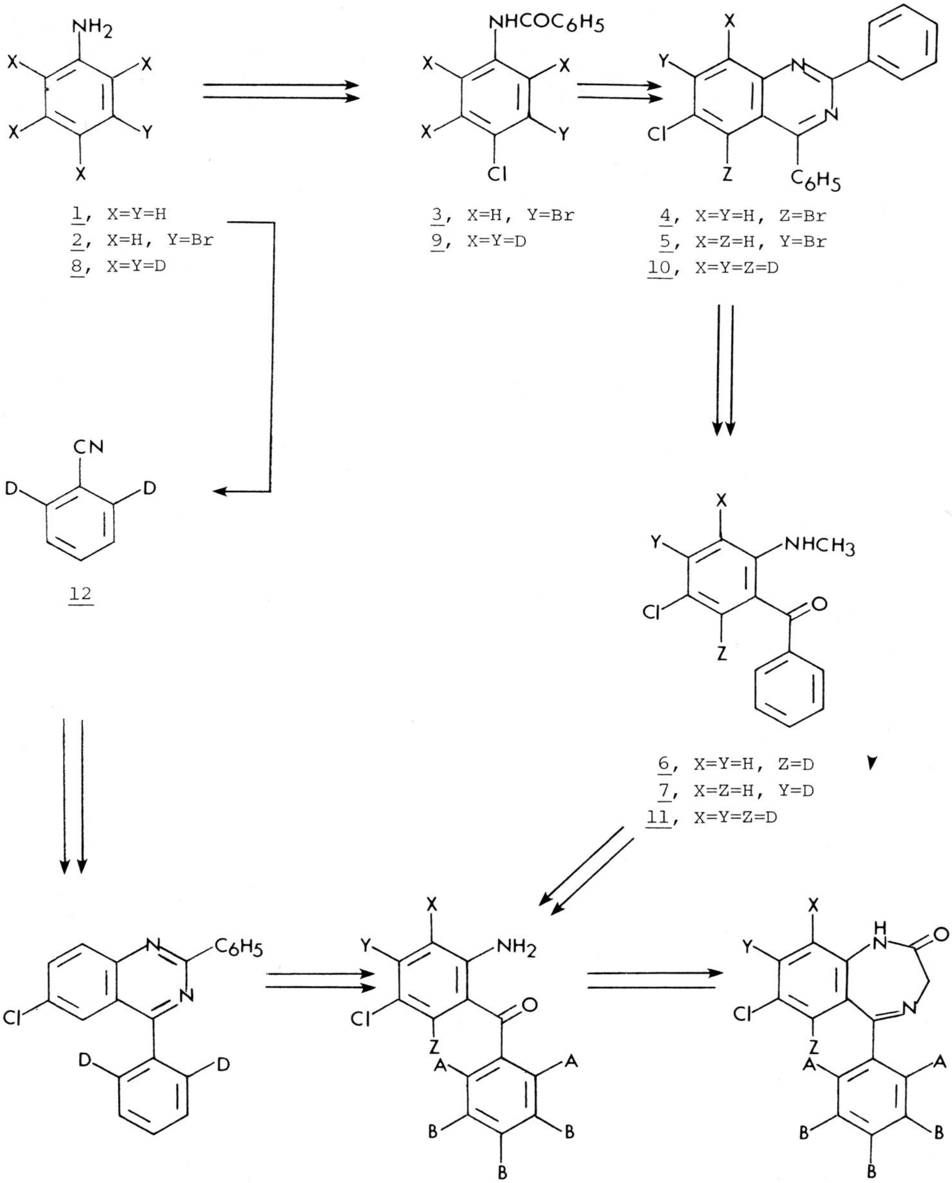

1, X=Y=H
2, X=H, Y=Br
8, X=Y=D

3, X=H, Y=Br
9, X=Y=D

4, X=Y=H, Z=Br
5, X=Z=H, Y=Br
10, X=Y=Z=D

12

6, X=Y=H, Z=D
7, X=Z=H, Y=D
11, X=Y=Z=D

13

14, X=Y=Z=A=B=H
14a, X=I, Y=Z=A=B=H

15, X=Y=A=B=H, Z=D
16, X=Z=A=B=H, Y=D
17, X=D, Y=Z=A=B=H
18, X=Y=Z=D, A=B=H
19, X=Y=Z=B=H, A=D
20, X=I, Y=Z=A=B=H

From aniline-d_5 (8), 4-chloro-2,3,5,6-tetradeuterobenzanilide (9) was prepared and converted to the quinazoline (10). Again, via an intermediate aminobenzophenone, (11), the trideuterobenzodiazepine (18) was obtained by conventional methods with 95% d_3.

For the preparation of the benzodiazepine derivative with specific deuteration in the 5-phenyl substituent, aniline was again used as the starting material. By using a series of reactions described in the literature (7), 2,6-dideuterobromobenzene was readily obtained from aniline-2,4,6-d_3. Conversion to 2,6-dideuterobenzonitrile (12) was accomplished by metal-halogen interchange, carbonation, amide formation and dehydration; all using conditions under which deuterium exchange was minimal. Reaction of 12 with the imino chloride of unlabelled benzoylaniline provided quinazoline (13) and subsequent reactions to afford the benzodiazepine (19) were then carried out as described above.

Trideuteromethyl diazepam (21), diazepam-d_8 (22), diazepam-d_{11} (23) and perdeuterodiazepam (24) were prepared by methylation of the amide nitrogen (N-1) of the appropriately labelled desmethyldiazepam with trideuteromethyl iodide and thallous ethoxide (8). Further data on these compounds are provided in the table.

Representative syntheses of the compounds prepared are described in the Experimental Section.

TABLE
Specifically deuterated benzodiazepines

Compound	R	Substitution Deuterium Position	% Overall Label (a)
15	H	6	98 (d_1)
16	H	8	95 (d_1)
17	H	9	98 (d_1)
18	H	6,8,9	95 (d_3)
19	H	2',6'	96 (d_2)
21	CD_3	N-1	97 (d_3)
22 (b)	CD_3	N-1;2',3',4',5',6'	95 (d_8)
23	CD_3	N-1;6,8,9,2',3',4',5',6'	90 (d_{11})
24 (c)	CD_3	N-1;3,3,6,8,9,2',3',4',5',6'	89 (d_{13})

(a) Determined by the mass spectrum.
(b) Prepared partly as described by Fentiman & Foltz (4).
(c) Prepared from 23 by D_2O/DMF exchange.

EXPERIMENTAL

Melting points are uncorrected. All solvents were distilled prior to use. Nuclear magnetic resonance spectra were taken on a Jeolco-C-60H or Varian HA-100 or Varian XL-100 spectrometer. The mass spectra were recorded on either a Varian-CH-5 or CEC-100 instrument at an ionizing voltage of 70eV.

3-Bromo-4-chlorobenzanilide (3)

On a 150 mmole (26 g) scale, m-bromoaniline was treated with sufficient acetic anhydride to provide 25.7 g (120 mmole) of m-bromoacetanilide. This material was dissolved in 80 ml of glacial acetic acid, stirred and treated with the dropwise addition of a solution of 8.5 g (120 mmole) chlorine in 50 ml of acetic acid to yield 8.1 g (32.6 mmole, 27%) of 3-bromo-4-chloro-acetanilide, m.p. 124-125°C. Hydrolysis of a 4.5 g portion to 3-bromo-4-chloroaniline, 3.4 g (91%) m.p. 80-81°C, was effected by refluxing the material with 60 ml of 6N HCl then by adding Na_2CO_3 to the cooled solution until alkaline reaction. A 2.5 g (12.1 mmole) portion was then dissolved in 50 ml of benzene and treated with 15.6 mmole (2.2 g) of benzoyl chloride and 20 mmole (1.8 g) of triethylamine. The resulting mixture was heated under reflux for 1 h, cooled and washed successively with 15 ml portions of 1N NaOH, 1N HCl and distilled water. The organic phase was dried and concentrated in vacuo to a residue which was crystallized from benzene to yield 2.98 g (9.6 mmole, 79%) of 3, m.p. 157-158°C, lit. (9) 128°C.

5-Bromo-6-chloro-2,4-diphenylquinazoline (4) and 7-bromo-6-chloro-2,4-diphenyl-quinazoline (5)

A 2.4 g (7 mmole) portion of 3, obtained above, and 1.56 g (7.5 mmole) of PCl_5 were stirred together then heated to 130°C for 30 min. Concentration of the mixture in vacuo removed the $POCl_3$ that was formed and the residual imino chloride, dissolved in 4 ml of o-dichlorobenzene was added, with stirring, to a mixture of 721 mg (7 mmole) benzonitrile and 935 mg (7 mmole) of aluminum chloride also in 4 ml of o-dichlorobenzene. The resulting mixture was stirred and heated to 140°C for 1 h, cooled and partitioned between $CHCl_3$ and 1N Na_2CO_3. After separation, the organic phase was dried and concentrated in vacuo and the residue crystallized from ethanol to yield 710 mg of 5, m.p. 211-212°C. Nmr (TFA) δ 8.61, 8.75 (2s, 2Ar); m/e 394 (395,396,397,398), (calcd. for $C_{20}H_{12}N_2BrCl$, 394. Anal. calcd. C, 60.7; H, 3.1; N, 7.1; Cl, 9.0; Br, 20.2. Found: C, 60.5; H, 3.0; N, 7.0; Cl, 9.0; Br, 20.2.

Evaporation of the mother liquor, followed by crystallization of the residue from benzene-ethanol provided 1 g of 4, m.p. 154-155°C. Nmr (TFA) δ 8.77 (m, 2AR); m/e 394. Anal. Found: C, 60.6; H, 2.9; N, 7.1; Cl, 8.9; Br, 20.1.

2-Methylamino-5-chloro-6-d-benzophenone (6)

A 380 mg (0.96 mmole) portion of pure 4 was dissolved in 20 ml of THF containing 0.3 mL triethylamine. The resulting mixture was treated with 70 mg of Pd/C (5%) then stirred at room temperature under an atmosphere of deuterium gas until an uptake of 1 mmole was noted. The mixture was then filtered and the filtrate concentrated in vacuo. The residue was chromatographed over 20 g of silica gel (E. Merck #7734, benzene elution) to yield, after separation, unreacted 4, some dideuteroquinazoline and the desired 6-chloro-5-deuteroquinazoline. Crystallization from ethanol afforded 42 mg of material, m.p. 190-191°C. This product and 2 ml of dimethyl sulfate were heated together at 140°C for 18 h. The mixture was then concentrated in vacuo to a residue which was dissolved in 8 ml of 5N NaOH and 8 ml ethanol and heated under reflux for 1.5 h. The reaction mixture was diluted with water, extracted with several portions of $CHCl_3$, and the organic layers combined and concentrated in vacuo. The residue was chromatographed over 20 g of silica gel (E. Merck #7734, benzene elution) and concentration of the appropriate fractions yielded 9 mg of 6. Nmr ($CDCl_3$) δ 6.70, 7.24 (2d, 2Ar), 7.55 (m, 5Ar). Demethylation was effected as described (10) with purification of the product, 4 mg, by chromatography.

7-Chloro-1,3-dihydro-5-phenyl-2H-6-d-1,4-benzodiazepine-2-one (15)

Reaction of the 4 mg of aminobenzophenone derived from 6 with bromacetyl bromide followed by treatment with ammonia to effect ring closure yielded 3 mg of 15 after purification by chromatography, m/e 271, (calcd. for $C_{15}H_{10}N_2ClDO$, 271. The desmethyldiazepam derivatives listed in the table were all prepared by these procedures. For those not requiring selective hydrogenolysis of the quinazoline intermediate, overall yields were significantly higher.

1-CD_3-5-phenyl-7-chloro-1,3-dihydro-2H-1,4-benzodiazepine-2-one (21)

Under nitrogen and with magnetic stirring, 540 mg (2 mmole) of desmethyldiazepam dissolved in 4 ml of dimethyl formamide was treated dropwise, with 500 mg (2 mmole) of thallous ethoxide. The solution was frozen, evacuated to 1 μ and 308 mg (2.12 mmole) of trideuteromethyl iodide was added by vacuum transfer. The resulting mixture was stirred at room temperature for 0.25 h. Thallous iodide was removed by filtration and the filtrate concentrated in vacuo to a residual oil which was crystallized from ethyl acetate. Recrystallization from ethyl acetate/ether afforded 276 mg (0.96 mmole, 42%) of pure 21. An additional 50% of the desired product remained in the mother liquids.

ACKNOWLEDGMENT

We thank the staff of our Physical Chemistry Department, in particular Dr. W. Benz for mass spectra, Dr. F. Scheidl for microanalyses and Dr. T. Williams for nmr spectra.

REFERENCES

1 A.J. Clatworthy, L.V. Jones and M.J. Whitehouse, Biomed. Mass Spectrom., 4(1977)248-254.
2 W. Benz, F.M. Vane and U. Rapp, Org. Mass Spectrom., 14(1979)154-159.
3 W.A. Garland and B.J. Miwa. The origin of the [M-1]$^-$ ion in the negative chemical ionization mass spectra of the 1,4-benzodiazepine-2-ones, "submitted for publication".
4 A.F. Fentiman, Jr. and R.L. Foltz, J. Labelled Compounds and Radiopharm., 13(1977)579-585.
5 L.H. Sternbach, R.I. Fryer, W. Metlesics, E. Reeder, G. Sach, G. Saucy and A. Stempel, J. Org. Chem.,27(1962)3788-3796.
6 H. Meerwein, P. Laasch, R. Mersch and J. Nentwig, Chem. Ber. 89(1956)224-238.
7 A. Streitwieser, Jr. and H.S. Klein, J. Amer. Chem. Soc.,86(1964)5170-5173.
8 E.C. Taylor, Thallous ethoxide, in Reagents for Organic Synthesis, Vol. 2, M. Fieser and L.F. Fieser, John Wiley & Sons, Inc., New York, 1969, pp.407-411.
9 N.T. Cam-Van, B.K. Diep and N.P. Buu-Hoi, Tetrahedron,20(1964)2195-2199.
10 J. von Braun, Chem. Ber.,37(1904)2812-2819.

SYNTHESIS OF ^{2}H-LABELLED PROSTAGLANDINS

C.O. MEESE and J.C. FRÖLICH
Fischer-Bosch-Institute of Clinical Pharmacology, Stuttgart (G.F.R.)

ABSTRACT

For their application in GC/MS stable isotope dilution assays, deuterium labelled prostaglandins were prepared by semi-synthetic and organic total synthetic methods (^{2}H$_4$-PGF$_{2\alpha}$, ^{2}H$_4$-PGI$_2$, ^{2}H$_4$-6-keto-PGF$_{1\alpha}$).

INTRODUCTION

Prostaglandins (PGs) represent a group of chemically closely related oxygenated and unsaturated fatty acids. Analysis of PGs in biological substrates requires application of highly specific and sensitive methods. GC/MS stable isotope dilution assays currently best fulfill these criteria. As PGs are metabolized predominantly by ß-oxidation of the upper side chain, the biochemical preparation of PG metabolites requires analogues as starting material which are labelled in the ring or in the lower side chain.
This contribution will refer to the synthesis of deuterated PGs which retain the deuterium label upon metabolism.

RESULTS

In a semi-synthetic route (Fig.1) PGE$_2$ (1) was converted into ^{2}H$_3$-PGE$_2$ [ref.1]. Reduction with NaBD$_4$ and preparative chromatography gave pure ^{2}H$_4$-PGF$_{2\alpha}$ (3). According to published procedures [ref. 2-6] the pure crystalline sodium salt of labelled prostacyclin 7 was obtained, which upon acid hydrolysis liberates the required 8,9,10,10-^{2}H$_4$-6-keto-PGF$_{1\alpha}$. Under the reaction conditions employed, only one chromatographic separation is necessary, most of the reaction steps are carried out in situ, yields are high even in 10 to 100 µmole runs, and three interesting endogenous PGs are obtained in their deuterium labelled form (PGF$_{2\alpha}$, PGI$_2$, 6-keto-PGF$_{1\alpha}$) from natural occuring PGE$_2$. 9 should be a suitable substrate for further biochemical ß-oxidation of the upper side chain. As shown by our group [ref. 7] the resultant 2,3-dinor-6-keto-PGF$_{1\alpha}$ is the major enzymatic metabolite of prostacyclin in man.

To elucidate the isotopic composition and the labelled sites, mass spectra and ^{13}C-NMR spectra of 4 and 9 were recorded. These measurements indicate that

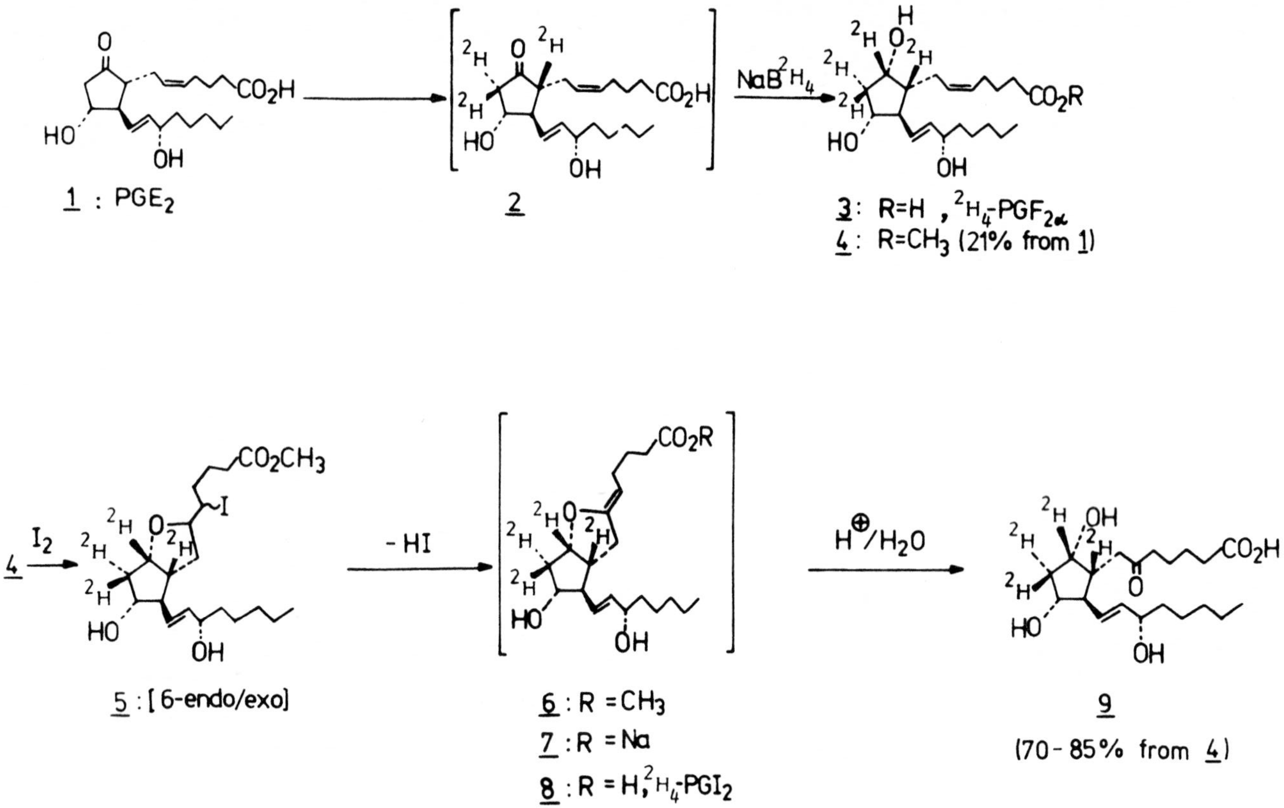

Fig. 1: Semi-synthetic route to ^{2}H-labelled prostaglandins

these compounds are highly labelled at position 10 and partially labelled at position 8 of the PG skeleton [ref.8]. Depending on the mode of preparation small amounts of unlabelled 4 or 9 are detectable (^2H_o from 0.2 to 0.6%; $^2H_3 \sim 40\%$, $^2H_4 \sim 50\%$).

In an alternative approach (Fig.2) labelled PGs were prepared utilizing the connection of the cyclic lactone 10 [ref. 9,10] with the phosphonate 11. In several steps $PGF_{2\alpha}$ (12), labelled in the lower side chain, is obtained. The requisite phosphonate 11 [ref.11] was prepared in four steps from the readily available THP-ether 13, which after C-methylation of the terminal acetylene was deuterated homogenously using Wilkinson's catalyst $((Ph_3P)_3RhCl)$. The intermediates obtained this way showed neither scrambling nor detectable amounts of unlabelled compound ($^2H_o \ll 0.1\%$).

Fig.2[+]: Synthetic preparation of ^2H-labelled $PGF_{2\alpha}$

[+]THP: 2-tetrahydropyranyl; PB: 4-phenylbenzoyl

Further transformations of 12 to other PGs, labelled in the lower side chain, have been carried out in sub-millimole scales in full analogy to the sequence described for the ring labelled PGs (Fig.1). ^{13}C-NMR spectra of 12 reveal a complete loss of signal resonance intensity of the carbon atoms 18 and 19, which is indicative of complete labelling at these positions. Within the detection limit mass spectrometric measurements on the isotopic composition showed no unlabelled material ($^{2}H_{o} \ll 0.1\%$).

REFERENCES

1   A.R.Brash, T.A.Baillie, R.A.Clare and G.H.Draffan, Biochem.Med., 16 (1976) 77-94.
2   E.J.Corey, G.E.Keck and I.Székely, J.Am.Chem. Soc., 99 (1977) 2006-2008.
3   I.Tömösközi, G.Galambos, V.Simonidesz and G.Kovács, Tetrahedron Lett. (1977) 2627-2628.
4   N.Whittaker, Tetrahedron Lett. (1977) 2805-2808.
5   R.A.Johnson, F.H.Lincoln, J.L.Thompson, E.N.Nidy, S.A.Mizsak and U. Axen, J.Am.Chem. Soc.,99(1977) 4182-4184.
6   R.A.Johnson, F.H.Lincoln, E.G.Nidy, W.P.Schneider, J.L.Thompson and U.Axen, J.Am.Chem.Soc., 100(1978) 7690-7705.
7   B.Rosenkranz, C.Fischer, K.E.Weimer and J.C.Frölich, J. Biol. Chem., 255 (1980) 10194-10198.
8   Prostanoic acid numbering: N.A.Nelson, J.Med.Chem., 17(1974) 911-918.
9   E.J.Corey, T.K.Schaaf, W.Huber, U.Koelliker and N.M.Weinshenker, J.Am. Chem. Soc., 92 (1970) 397-398.
10 N.H.Andersen, S.Imamoto and D.H.Picker, Prostaglandins 14(1977) 61-101.
11 C.O.Meese, B.Borstel and G.Beck; in preparation.

NEW APPROACHES IN THE PREPARATION OF ^{15}N LABELED AMINO ACIDS

Zvi E. KAHANA and Aviva LAPIDOT
Isotope Research, The Weizmann Institute of Science, Rehovot, ISRAEL

ABSTRACT

There is an increasing need for large quantities of ^{15}N labeled amino acids for biomedical research. Traditional methods for their preparation by organic synthesis or enzymatic biosynthesis are either long and tedious or impractical for large scale production. Microbial production has been applied for the prepartion of gram quantities of ^{15}N L-amino acids. Immobilized bacteria were used to obtain ^{15}N enriched (>95%) of L-aspartic acid, L-alanine, L-tyrosine and L-DOPA from ^{15}N ammonium chloride in high yield. Fermentation with amino acid excreting bacteria on limited ^{15}N precursors has been used for preparing of ^{15}N L-glutamic acid. Similar ways for obtaining other ^{15}N amino acids are being investigated.

INTRODUCTION

The use of biological compounds labeled with stable isotopes in clinical and in biological studies has been dramatically increasing. GCMS (gas-chromatography mass-spectrometry) and NMR (nuclear magnetic resonance) studies in biomedicine have been spurred by the greater ethical concern of using non-radioactive materials in humans. The greater availability of tracers labeled with stable isotopes further advanced this trend. Such procedures have been recently used for the determination of dynamic parameters of amino acid metabolism in humans (1,2).

Large quantities of ^{15}N labeled tracers are needed for clinical studies. Such gram quantities are not readily available at reasonable cost.

The need exists also for various ^{15}N amino acids, as branched chain amino acids, which are currently in the focal point of clinical interest in parenteral nutrition of postoperative patients and in the formulation of diets for certain cancer cases (3,4). Therefore, we have been investigating the means of producing such labeled compounds.

Traditional methods of making labeled L-amino acids are three: organic synthesis, enzymatic biosynthesis or isolation and purification from labeled protein hydrolysates. Organic synthesis, albeit well documented, is usually long and tedious. The efficiency of such methods is limited by the production of both D- and L-racemates. Therefore it has been justified for preparing only ^{15}N glycine. While this problem is overcome by enzymatic methods, which result in pure L-amino acids, the cost of the enzymes, along with their instability render this approach impractical for large scale production. For example, a linked enzymatic system for the synthesis of ^{15}N L-alanine (5) using hundreds of units of each of three purified enzymes resulted in just over one gram of product.

The separation of grams of pure amino acids from a protein hydrolysate of microorgamisms grown on precursors labeled with ^{15}N is inefficient. At most they may serve as a source for mixtures of labeled amino acids.

For these reasons we undertook a biotechnological approach for the production of ^{15}N amino acids. This methodology is available for the large scale production of amino acids (6), it can utilize the most common isotopic precursors and gram quantities of the desired labeled amino acid are excreted into the medium. Two avenues were used: synthesis by immobilized bacteria and production via fermentation. These routes were adapted for labeling several amino acids of clinical importance.

METHODS AND RESULTS

Production of ^{15}N L-amino acids by immobilized bacteria

Immobilized cell techniques (7) were adapted for the synthesis of ^{15}N L-aspartate, alanine, tyrosine and dihydroxyphenylalanine (DOPA). These are single step enzymatic reactions. The bacteria were induced for the desired enzyme and the whole cells were so used afterwards in the reaction. These cells were usually immobilized for stability and ease of handling.

For the production of ^{15}N L-aspartic acid, a 1M solution of ^{15}N ammonium fumarate (pH 8.5) was prepared from 10 g of 95% ^{15}NH$_4$Cl and 10 g fumaric acid in a trapped system. Cells of E. coli B. which had been entrapped in polyacrylamide (8), granulated and activated in 1M NaCl or in 1M ammonium fumarate for 48 hours, were carefully washed and suspended in the 1M solution of ^{15}N ammonium fumarate. The bacterial suspension was incubated with shaking at 50°C, and the appearance of aspartic acid was monitored by amino acid analysis (Dionex D 500) of timed aliquotes. Upon termination (24 - 72 hours), the beads were thoroughly washed with H$_2$O. The eluate was concentrated and the aspartate crystallized at pH 2.8. The resulting 10.8 g ^{15}N L-aspartic acid

(95 % yield based on fumarate) was recovered after ion exchange chromatography and recrystalization. Almost all remaining ^{15}N ammonium was also recovered from the supernatant. Part of this ^{15}N L-aspartate was used for the production of ^{15}N L-alanine.

For the production of ^{15}N L-alanine, Pseudomonas dacunhae (ATCC 21192) was grown and immobilized (9,10). These bacterial granules were added to 1 M ^{15}N-aspartic acid solution, brought to pH 5.5 with ammonium hydroxide, containing 10 mM pyridoxal phosphate, and incubated at 37°C with shaking.

The conversion of ^{15}N L-aspartate to ^{15}N L-alanine, as determined by amino acid analyzer, was usually completed after overnight incubation but was sometimes prolonged till 100% yield was obtained.

The ^{15}N L-alanine was further purified by ion exchange chromatography and then crystallized several times.

Amino acid analysis, NMR and mass spectrometric analyses of the products were performed. They confirmed the purity and the isotopic enrichment of 95% of the ^{15}N L-aspartic acid and the ^{15}N L-alanine. Polarimetry proved the products to be in the L- conformation.

These preparations were pyrogen free by standard rabbit body temperature measurments. Sterilized solutions for clinical studies were prepared by passage through a Millipore filter and by autoclaving.

Similar bacterial techniques (11) were adapted for preparing gram quantities of ^{15}N L-tyrosine and ^{15}N L-DOPA . ^{15}N ammonium hydroxide , pyruvate and phenol or pyrocatechol were reacted with cells of Erwinia herbicola (ATCC 21434) in acetate buffer pH 8. The biosynthesis of ^{15}N L-tyrosine was run at 28°C, while that of ^{15}N L-DOPA required antioxidation precautions as the addition of sulfite and of EDTA and was run at lower temperature for an extended period. The products which remain insoluble were recrystallized and checked for purity by all the abovementioned techniques.

Modifications of these procedures are now in progress for yield improvement.

Fermentative production of several ^{15}N L-amino acids

Methods for selection and for growth of amino acid excreting bacteria (6) were used and adapted for isotopic synthesis. Among the factors modified were the amounts of ammonium and of amino acid supplement in the medium, which were reduced to a moderate level.

For instance, Brevibacterium lactofermentum (ATCC 13689), which excretes L-glutamic acid (12) was grown in a medium where 4% ammonium sulphate were substituted with 1% ^{15}N ammonium chloride (97%), and only a tenth of the prescribed protein hydrolysate supplement was used. The pH was controlled with KOH and not with any nitrogenous base.

Up to 19 g/l ^{15}N L-glutamic acid were obtained. The yield of ^{15}N L-glutamate based on ^{15}N ammonium chloride is about 60%. Additional 10% of the ^{15}NH$_4$Cl were not used and can be recovered from the fermentation broth. In addition about 50 g/l of ^{15}N labeled bacteria were obtained. They have been used as an source for ^{15}N labeled biological compounds. For instance , hydrolysis of their proteins yielded an ^{15}N amino acids mixture used for supplementing the growth of other microorganisms producing ^{15}N labeled material.

Purification of the ^{15}N L-glutamate proceeds through precipitation at pH 3.2 and further ion exchange chromatography.

NMR, amino acid analysis and mass spectroscopy of the purified compound confirmed its purity and its ^{15}N enrichment to 97%. The ^{15}N labeled glutamate has been tested for pyrogenicity and sterilized for further clinical studies.

Mutants for the production of other ^{15}N L-amino acids were derived by selection of excretors from bacteria resistant to several amino acid analoges and from selected auxotrophes (13). They are currently tested for isotopic fermentation.

CONCLUSIONS

The biotechnological approach to the synthesis of biological compounds labeled with stable isotopes has proven effective and efficient. By this mean it is possible to make gram quantities of labeled compounds required for clinical studies, as well as to produce compounds such as antibiotics and anti-tumor drugs presently unavailable in ^{15}N form.

^{15}N precursors, as ammonium chloride, have inherently a high efficiency of labeling in contrast to ^{13}C or ^{18}O precursors which are partially expired as CO_2 . Yet, we are currently investigating the application of similar methodology to obtain efficient ^{13}C labeling. We hope the methods described in this report as well as similar approaches (14) will be a stepstone for obtaining such compounds.

ACKNOWLEDGEMENT

This work was supported by the Israeli Ministry of Commerce and Industry. We also thank the Isotope Separation Plant of the Weizmann Institute for provision of ^{15}N nitric acid.

REFERENCES

1. A. Lapidot and I. Nissim, Metabolism, 29 (1980) 230-239.
2. J. C. Waterlow , P. J. Garlic and D. J. Millward (Eds.), North Holland Publishing,Amsterdam 1978.
3. E. M. Copeland, B. V. Mavfyden Jr. and S. J. Dudrick, J. Surg. Res. 16 (1974) 241-247.
4. E. M. Copeland, J. M. Daly , D. M. Ota and S. J. Dudrick, Cancer 43 (1979) 2108-2116.
5. W. Greenway, F.R. Whatley and S. Ward, Febs Letts.,81 (1977) 286-288.
6. S. Kinoshita and K. Nakayama, in A. D. Rose (ed.), Primary Products of Metabolism, AP,London , 1978, p 209.
7. I. Chibata , T. Tosa and T. Sano. in H.J. Peppler and D. Perlman (Eds.), microbial technology vol. 2 , AP, 1979, chap. 17 ,p 433-460.
8. I. Chibata , T. Tosa and T. Sato. Appl. Microbiol. 27,(1974) 878-855.
9. I. Chibata , T. Kakomoto and J. Kato. Appl. Microbiol., 13 (1965) 638.
10. K. Yamamoto , T. Tosa and I. Chibata Biotec. Bioeng., 22 (1980) 2045-2054.
11. H. Enei , H. Nakazawa, H. Matsui and S. Okumura. Febs Letts. , 21 (1972) 39.
12. S. Kinoshita and K. Tanaka , in K. Yamada et al (Eds.) , The microbial production of amino acids , 1972 Kodansha, Tokyo, chap. 10 p 263-324.
13. K. Nakamaya , K. Araki, H. Hagino, H. Kase and H. Yoshida in K. D. MacDonald (Ed.), Second intrnational symposium on the genetics of industrial microorganisms. 1976, AP , London, p. 437-449.
14. A. Ivanof, L. Muresan, L. Quai, M. Bologa, N. Polibroda , A. Mocanu, E. Vargha and O. Barzu, Analyt. Biochem. 110 (1981) 267-269.

SYNTHESIS OF CARBON-13 AND OXYGEN-18 LABELLED COMPOUNDS

M.B. Chkhaidze, Z.N. Morchiladze, Ts.I. Obolashvili, E.D. Oziashvili
Institute of Stable Isotopes, State Committee on the Uses of
Atomic Energy, 380086 Tbilisi (USSR)

ABSTRACT

A comparative assessment of different techniques for the synthesis of some organic acids, their carbon-13 labelled derivatives and oxygen-18 labelled metal oxides has been carried out.

For the synthesis of the organic acids - succinic, malic, glycolic, acetic and their derivatives - cyanoacetic acid, benzyl cyanide, benzyl chloride - a technological scheme of synthesis based on employing potassium cyanide with carbon-13 as starting material was chosen. As a result of the studies an increase in the yield of succinic, malic, acetic, and glycolic acids by 10-15 % has been attained.

As a result of the investigation it was established that it was feasible to prepare oxygen-18 labelled ferric, calcium and titanium oxides by reacting the metal with oxygen-18 labelled nitric oxide. During the synthesis of $Fe_2{}^{18}O_3$, $Ca^{18}O$ and $Ti^{18}O_2$, a saving of oxygen-18 of 15-20 % was achieved.

INTRODUCTION

The choice of an effective method of synthesis for the production of stable isotope labelled compounds of a desired quality and with minimal isotope losses is an important aspect in the general problem of the use of stable isotopes.

It was the purpose of the present investigation to develop an efficient synthetic pathway for the preparation of carbonic acids labelled with carbon-13 in the carboxyl group, and also for the preparation of oxygen-18 metal oxides.

SYNTHESIS OF CARBON-13 COMPOUNDS

In our studies, special attention has been given to the synthesis of acetic acid, since it serves as a starting material for the production of other labelled compounds, and also because it is employed directly in biochemical research. At present, the most commonly used method for the production of acetic acid is the reaction of Grignard reagent with carbon-13 carbon dioxide, however this method has a number of disadvantages resulting in low yields (50-70 %).

Our studies show that a more feasible process for preparing acetic acid is via acetonitrile. It has been established that the yield is dependent on the completeness of the reaction of methyl iodide with KCN. However, in aqueous medium an excess of methyl iodide was needed, resulting in methanol formation which required subsequent purification of the product and caused isotope losses.

Among a number of nonaqueous solvents that have been studied in this laboratory, dimethylsulphoxide (DMSO) was found to be most advantageous. It ensured the quantitative conversion of the labelled potassium cyanide into acetonitrile, thus eliminating a purification step (reaction conditions: 1 mole $K^{13}CN$, 1 mole CH_3J in 250 ml DMSO; 1 hour, 95 °C).

Several authors [1,2] have used gaseous hydrogen chloride for the conversion of potassium acetate into acetic acid. The results obtained however, were comparable to those obtained with hydrogensulphates. Some modifications in the design of the apparatus for the conversion of sodium acetate into acetic acid ensured the effective contact of the solid phase - the sodium acetate - with the gaseous hydrogen chloride, which enable a maximum quantity of the acetic acid to be obtained. As a result of the investigations the yield of acetic acid labelled with carbon-13 was increased to 90 %.

Carbon-13 potassium cyanide was also used for the synthesis of cyanoacetic, succinic, malic acids and benzyl cyanide. For the production of glycolic acid the interaction of carbon-13 carbon monoxide with formaldehyde was chosen.

The reactions of preparing the above mentioned labelled compounds and potassium cyanide from carbon-13 monoxide are shown in Fig. 1.

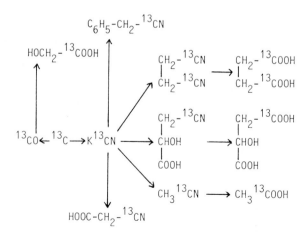

Fig. 1. Compounds prepared from carbon monoxide and potassium cyanide.

Potassium cyanide has been used for the production of succinic acid by Kushner and Weinhouse [3]. In our work however, for the preparation of the succinic dinitrile, the water-alcohol medium was replaced by dimethylsulphoxide as a solvent. It allowed an increase of 30 % in the yield, as compared with Kushner's data, without the loss of product quality (experimental conditions: 1 mole KCN + 1,5 mole $ClCH_2$-CH_2Cl in 250 ml DMSO; 3 hours, 80-100 °C).

The choice of a method for malic acid synthesis using $K^{13}CN$ as a starting material has its origin in the relative ease of the reaction. This is a two-stage process, which does not require sophisticated technology in contrast to the commonly used laboratory method for malic acid synthesis, the hydration of fumaric and maleic acids in an autoclave. The reaction between chlorolactic acid and potassium cyanide proceeds under mild conditions, and the resulting product can be isolated by continuous extraction with dimethyl ether. The yield of carbon-13 malic acid was 65-70 %.

We also studied the effects of different solvents on the completeness of the product extraction from the reaction medium. In the case of methylethylketone the ^{13}C-cyanoacetic acid output was 95-97 %.

From the work quoted in the literature the process of benzyl cyanide formation is carried out under similar conditions, i.e. the benzyl chloride solution in 95 % ethyl alcohol is added to a hot aqueous solution of potassium cyanide. Then benzyl cyanide yields were 80-85 %, although there were different approaches to the purification of the carbon-13 benzyl cyanide obtained [4-7]. The experiments we have performed showed that the product output can be increased up to 95-98 % in the case of the interaction of crystalline potassium cyanide with benzyl chloride in a 95 %-solution of ethyl alcohol (reaction conditions: 1 mole KCN + 1 mole C_6H_5-CH_2Cl in 200 ml solvent; 5 hours under refluxing).

Also of considerable interest is a method for producing glycolic acid directly from CO. According to the Patent [8], for the reaction between CO and formaldehyde, Cu_2O, CuO+Cu, and Ag_2SO_4 can be used as catalysts at a pressure between 0.1-30 atm in the temperature range 0-90 °C. The aim of our present studies was to discover new catalysts, and it was found that in the presence of selene the output could be increased up to 90 % (reaction conditions: 1 mole CO, 1.5 mole HCOOH, 0.5 g Se; 20-50°C, yield 2.5 g/hour).

PRODUCTION OF OXYGEN-18 OXIDES

For many years, the raw material for the synthesis of oxygen-18 compounds has been $H_2^{18}O$ which is obtained by rectification. Therefore, in many cases the methods for obtaining the labelled oxides were based on the direct oxidation of metals and their halides with water or water vapour. The principle drawback of the method is: considerable losses, in some cases in excess of 20-25 %.

$$Me + H_2{}^{18}O_{(v)} \longrightarrow Me^{18}O + H_2 \qquad (1)$$

$$MeCl_4 + 2H_2{}^{18}O_{(v)} \longrightarrow Me^{18}O_2 + 4HCl \qquad (2)$$

$$MeCl_4 + 2H_2{}^{18}O_{(l)} \longrightarrow Me^{18}O_2 + 4HCl \qquad (3)$$

For the last decade the most widely used process for oxygen and nitrogen isotope separation on an industrial scale has been the low temperature distillation of nitric oxide, the main starting material for the syntheses of oxygen-18 compounds. From this point of view the earlier methods of obtaining labelled metal oxides seem to be ineffective.

There is evidence in the literature indicating that metals that are oxidized in the air (Fe,Ca), are unstable with nitric oxide, especially when they are heated to red heat [9]. The data on the reaction products are however, contradictory [10,11]. Simultaneous interaction of the metal with nitrogen is highly possible and might be the cause of the contamination of the main product, or alternatively of nitrogen-15 losses.

The aim of our research was to investigate the process of the interaction of nitric oxide with some metals (Fe, Ca, Ti) with respect to the influence of starting metal properties, temperature, and pressure on the output and the chemical composition of the main product, and on gaseous reaction by-products.

For the experiments we used metal powders, both commercially available and produced in our laboratory via different methods. The commercial powders were reduced in a hydrogen stream to prevent isotopic dilution, and the calcium was distilled.

As expected, the rate of oxidation decreases, when the outer layers of metal grains are oxidized (Fig. 2), and for the completion of the oxidation an increase in pressure and long reaction times are needed. At higher temperature the rate of oxidation increases, however, some metals (Ca, Ti) react with nitrogen to form thermally stable nitrides.

The study of the rate of the concurrent processes - oxidation and the formation of nitrides - depending on temperature and the dispersion of the initial metal powder, allowed us to obtain the oxides, $Fe_2{}^{18}O_3$, $Ca^{18}O$, $Ti^{18}O_2$ with a purity comparable with that reported in the literature, ca. 98-98.5 %, and an increase in yield of 15-20 %.

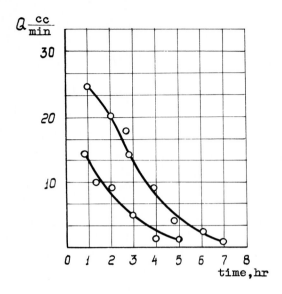

Fig. 2. Typical curves of oxidation rates for some metals.

REFERENCES
1 J.D. Cox, H.S. Turner, J.Chem.Soc., (1950)3178.
2 D.M. Hughes, R. Ostwald, J.Amer.Chem.Soc., 74(1952)2434.
3 M. Kushner, S. Weinhouse, J.Amer.Chem.Soc., 71(1949)3558.
4 V.M. Olevski (Ed.), Sintez mechenykh lekarstvennykh veshchestv, Medgiz, Moscow, 1959.
5 V.V. Bochkarev (Ed.), Methody poluchenia redioaktivnykh preparatov, Gosatomizdat, Moscow, 1962.
6 Weisgand-Hilgetag, Metody eksperimenta v organicheskoi khimii, Khimia, Moscow, 1968.
7 A.V. Isagulyants, Sinteticheskie dushistye veshchestva, 2nd edn., Ac.Sci.Arm.SSR, Erevan, 1947.
8 Patent No. 7 523 766 (1975), France.
9 Gmelins Handbuch der anorganischen Chemie, Verlag Chemie, Berlin, 59(1941)312.
10 U. Sabatier und Senderen, Compt.rend., 120(1895)1158.
11 M. Moser, Amer.Chem.J., 14(1891)324.

AUTHOR INDEX

	page		page
Abidi, S.L.	563	Cokayne, R.A.	665
Aihara, M.	685	Craig, J.C.	287
Allen, J.R.F.	529,611	Cunningham, P.T.	179
Aly, A.I.M.	475	Cupano, J.	735
Amir, J.	331	Curtius, H.-Ch.	277
Andersen, J.R.	605		
Apfelbaum, M.	337		
		Darling, W.G.	161
		Davies, D.S.	187
Baba, S.	301	Degos, F.	385
Bader, G.J.	735	Dehennin, L.	617
Baillie, T.A.	187	Delaney, C.M.	735
Barker, R.	719	Depène, E.	271
Barrie, A.	655	Dizabo, P.	55
Basset, G.	337	Dostrovsky, I.	693
Bath, A.H	161	Drifford, M.	385,549
Bazlen, O.	517	Dražďák, K.	445
Beck, Olof	611	Dubois, M.	549
Begemann, F.	93	Duchesne, J.	399
Bengsch, E.	587	Dunbar, J.	495
Benhamou, J.P.	385	Dziashvili, E.D.	753
Bertelsen, B.	605		
Bertram, H.-G.	115		
Bjorkhem, J.	593	Ebeid, M.	445
Bornhak, H.	325	Ehhalt, D.H.	147
Botter, F.	385,549	Elion, G.B.	217
Brandaenge, F.	287	Engelkemeir, A.G.	179
de Bree, H.	203	Epstein, M.	693
Brekle, A.	211		
Brenninkmeijer, C.A.M.	661		
Bricout, J.	483	Farmer, P.B.	223
Brown, P.H.	451	Faust, H.	319,325
Brunsdon, A.P.	161		421
Bryan, B.A.	451	Feldman, L.	451
Burger, W.	735	Fischer, K.	517
		Förstel, H.	173,503
			511
Catroux, G.	459	Frölich, J.C.	743
Chkhaidze, M.B.	753		
Christensen, J.E.J.	605		
Clark, E.L.	719	Galimov, E.M.	35
Clarke, M.C.	665	Germon, J.C.	459
		Glaubitt, D.	373,379

	page		page
Gold, V.	61	Kaal, J.H.M.A.	203
Graczyk, D.G.	179	Kahana, Z.E.	747
Gregersen, G.	605	Kanai, T.	685
Griffiths, J.	287	Kasuya, Y.	301
Grivet, J.-Ph.	587	Kerner, M.N.	673
Gröbner, W.	313	Kessler, M.	649
Gruenke, L.	287	Kexel, H.	83,229
Gryzewski, N.	359	Kiburis, J.H.	217
Guenot, J.	549	Klein, E.R.	347
Gupta, A.R.	29	Klein, P.D.	253,347 353,667
		Klemens, M.	271
Habfast, K.	623	Knappe, H.	147
Hachey, D.L.	235	Knick, V.C.	217
Hallaba, E.	475	Koch, H.J.	731
Harbo, H.	605	Kochen, W.	271,295
Hartig, W.	319	Kohl, Daniel H.	451
Hashimoto, Y.	247	Koritsánszki, T.	15
Hege, H.G.	679	Korte, F.	433
Helge, H.	265,359	Kouchi, K.	685
Heumann, K.G.	711	Králová, Marie	445
Hirschberg, K.	325	Kranz, C.	83
Hoefs, J.	103	Kreek, M.J.	235
Holm, J.	287	Kubát, J.	445
Holt, B.D.	179	Kumar, R.	179
Hook, C.	415	Kuster, Th.	307
Hornbeck, C.	287		
Hoskins, J.A.	223		
Hützen, H.	173,511	Lacroix, M.	393,399
Hughes, H.	187	Ladner, T.	287
		Lambe, G.	217
		Lapidot, A.	331,747
Idelson, A.	563	Larsen, E.	605
Iio, M.	367	Leclerc, A.	459
Inada, T.	685	Lehmann, W.D.	635,649
Irving, C.S.	353	Letolle, R.	459
		Liebman, A.A.	735
		Limbach, B.	229
Jákli, Gy.	15,23	Liu, Y.-Y.	735
Jakobs, C.	359	Löffler, W.	313
Janscó, G.	15		
Jarman, M.	217	Mariotti, A.	459
Jung, K.	319	Marsac, J.	337
Junghans, P.	319		

	page		page
Maruhashi, A.	685	Pallikarakis, N.	393
Matkowitz, R.	319	Paneth, P.	49
Matsuzaki, T.	439	Parulava, L.P.	673
Matwioff, N.A.	573	Perry, C.W.	735
Maxwell. L.E.	127	Picquenard, E.	55
Medina, R.	77, 313, 465	Pierce, J.	719
		Pilet, P.-E.	529,535
Meese, C.O.	743	Piiper, J.	409
Metzner, H.	517		
Meyer, M.	409,415		
Miyazaki, H.	247	Rating, D.	265,359
Möller, M.	229	Rebeaud, J.-E.	529
Moerch, L.	287	Reifsteck, A.	617
		Reimschüssel, W.	49
Mohamed, M.A.	475	Reisner, S.H.	331
Mook, W.G.	661	Rivier, L.	529,535
Morchiladze, Z.N.	753	Rosevear, P.R.	719
Morikawa, J.	367	Roth, E.	337
Morishita, T.	439	Rozanski, K.	153
Mosora, F.	393,399	Rutherford, W.M.	703
Muccino, R.R.	735		
Münnich, K.O.	153		
Muraoka, K.	439	Samuel, D.	255
		Scheid, P.	409
		Schidlowski, M.	95
Nakajima, N.	367	Schiefer, H.P.	711
Nakamura, K.	235,353	Schleser, G.H.	115
Nau, H.	265	Schmidt, H.-L.	77,83 299,465
Naylor, G.	287		
Nelson, D.J.	217	Schoeller, D.A.	353,667
Noda, S.	439	Scholler, R.	617
Noguchi, H.	439	Schulten, H.-R.	587,635
Nunez, H.A.	719	Schütze, H.	121
Nyhan, W.L.	287	Serianni, A.S.	719
		Shearer, G.	451
		Shinohara, Y.	301
Obolashvili, Ts.I.	753	Siafarikas, K.	373,379
Ohta, Y.	557,685	Smit, H.G.I.	147
Ohsawa, R.	367	Sofer, Z.	127
O'Leary, M.H.	67	Sonntag, C.	153
Olleros, T.	77	Sotiriou, N.	433
Ordzhonikidze, K.G.	673	Spieß, W.	711
		Steinhauer, H.	373
		Strauch, G.	121
		Stuart, R.S.	731
		Sturm, E.	241

	page
Suehiro, M.	367
Sutton, J.	337,385
	549
Suzuki, A.	557
Sweetman, L.	287
Tauscher, B.	271
Tevzadze, G.A.	673
Thom, E.	735
Trefz, F.K.	295
Tuttle, R.L.	217
Ueda, K.	367
van der Schoot, J.B.	203
van der Stel, D.J.K.	203
van Hook, W.A.	23
Volz, A.	147
Wakisaka, I.	557
Weber, B.	359
Wellings, S.R.	167
Wetzel, K.	135
Weymann, J.	679
White, R.H.	543
Winkler, F.J.	83
Wirtz-Peitz, F.	211
Wittfoht, W.	265
Wolfsberg, M.	3
Wong, W.W.	667
Wood, D.L.	61
Yamada, M.	557
Yoneyama, Y.	557
Zachmann, M.	307
Zagalak, B.	307
Zech, K.	241
Zierenberg, O.	211
Zöllner, N.	313

SUBJECT INDEX

Abbreviations: BT = Breath Test; CI = Chemical Ionization; detn. = Determination; GC = Gas Chromatography; IE = Isotope effect; MS = Mass Spectrometry; NA = Natural Abundance; NMR = Nuclear Magnetic Resonance; SI = Stable Isotopes; SIM = Selected Ion Monitoring. - Items occuring several times in the same contribution are indicated only once.

Abscisic acid - d_6	536
photoisomerization	539
quantification	535
synthesis	536
Abundance, natural,	
see Natural abundance	
Isotope composition	
Acarbose, effect on	
sucrose absorption	359
Acetonitrile - d_3	17
CH-stretching	20
vapour pressure, IE	16
Acetylcholine in rat brain	644
Acidemia	
methylmalonic	287,295
propionic	287
Activated Sludge Process	439
Activation analysis	685
Adrenal hyperblasia	307
Alanine-^{15}N	
tracer in neonates	331
synthesis	749
Allo-pregnanediol	307
Allopurinol	217
Alveolar gas diffusion	409
Amino acids	
biosynthesis	587
in body fluids	136
isotope composition	
see: Isotope composition NA	
isotopomers	587
metabolism	331

Amino acids-^{13}C	
intramolecular isotope distribution	587
NMR spectroscopy	587
Amino acids-^{15}N	
synthesis	747
tracer studies	331,426
Aminopyrine-^{13}C	
breath test	353,385
Ammonium-^{15}N in soil	447,449
Amniotic fluids, metabolites in	287,296
Androsterone glucuronide	301
Anharmonicity effect	8
Antarctica, SI in plants	121
Antiarrhythmic drug	679
Anticancer agent	217
Antidepressant drug	203
Antiepileptic therapy	265,271
Aquo-cyano cobyrinic acid	637
Arene oxide formation	227
Arginase reaction	
IE	77
mechanism	81
Arginine decarboxylase, IE	68
Aspartic acid-^{15}N, synthesis	748
Ascorbic acids, deuterated, syntheses	733
Atmospheric CO	152
Atmospheric moisture	
isot. compos.	154,159,513
Atmospheric pollutants	549
Atmospheric sulfate formation	179

Autoradiography, glycine-^{18}O	685
Bacteria	
immobilized	747
nitrifying	460
Bacterial overgrowth, BT	373
Benzene, CH-stretching	20
Beans (Phaseolus vulgaris)	549
1,4-Benzodiazepines, deuterated	735
Bicarbonate-^{13}C, BT	356
Bigeleisen-Mayer equation	51
Bile acids	373
Bioavailability	189
clovoxamine	203
quantitative, MS in	640
Biochemicals, isotop. labelled, field desorption MS	637
Biogenic amines, metabolism	256
Biosynthesis	
amino acids	587
dopamine	263
lipoic acid	543
prostaglandins	190
proteins in hepatitis	319
steroids	190,307
Body fluids, dtn. of	
biogenic amines	255
cholesterol	594
deuterium in	339
myo-inositol	605
endogenous a. exogenous cpds	640
^{15}N content after NO$_2$ expos.	559
nitrite after NO$_2$ expos.	560
Born-Oppenheimer approximation	3
Brain metabolism	247,255
Breath test (BT)-^{13}C	189,348
aminopyrine	353,**385**
background variations	402
pulmonary CO$_2$ output	391
clinical validation	348
commercial feasibility	347
components	348
diagnostic evaluation	387
diagnostic information	348
documentation	351
economic aspects	351
fatty acids	367
glucose metabolism	399
glycocholate	373
health care payments	347
hospitalization costs	352
in cystic fibrosis	373
in heart diseases	367
legal and medical criteria	350
legal requirements	351
maize glucose in	393,399
mathematical model	353,389
minimum precision	348
octanoate	367
outpatient	349
phenylalanine	379
reimbursement of costs	350
scores	353
sensitivity	348
sucrose	359
turnover control	401
Calcium	
absorption in man	649
analysis	649
isotopes, natural abundance	650
isotopes, separation	703,712
Calvin cycle	
carbon fractionation	97
Carbohydrate absorption	360
Carbohydrates, deuterated	733
Carbon	
isotope ratio detn.	667
equilibrium IE CO$_2$/water	521
IE see: Isotope effects	

isotope fractionation
see: Isotope fractionation

isotope geochemistry	103
isotope ratio, biomolecules	36
isotope ratio detn.	655
isotopes in soil	116
juvenile	102
sedimentary reservoirs	96,104
organic comp. in soil	115

Carbon-13

amino acids	587
aminopyrine	385
analysis	639,655,667
atmospheric CO	147

breath test
see: breath test

bicarbonate	356
clonidine	197
combustion techniques	655,667
enkephalin	582
enzymes	576
fatty acids	367

glucose

BT	394
MS	635
precursor amino acids	581
glycocholate, BT	373

NMR spectroscopy

amino acids	587
carbohydrates	719
chemical shift	720
C.utilis cells	573
nitrosamines	565
prostaglandins	743
octanoate, BT	367
phenylalanine, BT	379

syntheses

carbohydrates	719
clovoxamine	204
nitriles	753
nucleosides	723
organic acids	753
sucrose-BT	359
valproic acid	265

Carbon-14

N,N-dimethylaniline	230
nitrosamines	563
propafenone	679
testosterone	305
Carbonates	103,127,132,161,167
Carbon cycle	95,115,121

Carbon dioxide

in photosynth. O_2 evolution	518
Carbon dioxide/carbonate system	127
Carbon fixation IE	96
Carbon isotope age curves	98
Carbon monoxide	147
Carboxylation, IE	97
Carrier effect in GC	617
Catechol-O-methyltransferase, activity in vivo	247
Cellulose, oxygen isotope detn.	662
Cells, NMR	576
Central nervous system metabolism	247,255
Cerebrospinal fluid metabolites in	262

Chalk, see carbonates

Chemical ionisation (CI)	236,241,247,287
Chlorine isotopes	703
4-Chloro-benzanilide - d_4	738
7-Chloro-1,3-dihydro-5-phenyl-2H-6-d-1,4-benzodiazepine-2-one	740
Chlorophyll a	519
Chloroplasts	522
Chlorpromazine	260
Cholesterol in serum	594
Choline in rat brain	644
Chymotrypsin, nitrogen IE	72
Cirrhosis	319
Clonidine-$^{13}CO_2$, metabolism	193

Clovoxamine-^{13}C, -d, metabolism, MS	203	Didansyl-L-tyrosine-d_7, MS	636
Cluster technique	190	Dielectric effect	15
Combustion	655,667	Dilinoleoyl-glycerophosphocholine-d_4, pharmacokinetics, synthesis	211
Computer cloned patient population	353	Dimethylaminopyrine-^{13}C, BT	355
Continental drift	135	N,N-Dimethylaniline-^{15}N	230
Continental crust	138	N,N-Dimethylaniline oxide-^{18}O, -^{15}N	230
Cosmochemistry	93	N,N-Dimethylclonidine-^{13}C, -d metabolism	187
Countercurrent thermogravitational circulation	702	Dimethylnitrosamine-d, -^{15}N, metabolism	191
Cumene hydroperoxide - $^{18}O_2$	230	Diphenylhydantoin-d_5	
C. utilis, NMR	573	metabolism, pharmacokinetics	223
Cyclophosphamide	191,217,640	Distillation cascade	694
Cysteine - ^{34}S	543	Dopamine in Parkinsonism	255
Cystic fibrosis, BT	373,379	Dopamine-d_2, synthesis	248
Cytochrome P-450	195,229	Dopamine-d_4, internal standard	281
		Dopamine-^{18}O	259
Decarboxylations, IE	68,108	DOPA-^{15}N, synthesis	749
Denitrification	447,455,461,736	Drug interactions	187,191
Desmethyldiazepam-d, synthesis	736	Drug pharmacokinetics	187
Derivatizing agents in pharmacol. res.	191	Earth crust	135,138
Deuterium analysis		E.coli, biosynth. lipoic acid	543
field desorption MS	639	Electrolyte solut. structure	29
infrared, MS	339	Emission spectroscopy, ^{15}N detn.	319,424 440
Deuterium distribution in			
tropospheric water vapour	154	Endoxan, see Cyclophosphamide	
Deuterium exchange	55,189	Enzymatic reactions, IE	40,45,67,77,83
Deuterium IE, see Isotope effect, hydrogen		Enzyme activity	
		detn. in vivo	247,277
Deuterium labelled compounds, internal standards, GC/MS, syntheses: see corresp. comp.		Enzyme-substrate complex	38
		Enkephalin-^{13}C	582
		Epinephrine-d_3	247
Deuterium oxide		Equilibrium constants, IE see: Isotope effects	
enrichment	702		
medical application	187,337	Equilibrium IE, CO_2/water	521
in soil	167	Ethanol, carbon isotope ratio	489,495
Diabetes, glucose metabolism	403	hydrogen isotope ratio	492
Diamond, isotope composition	110	Evaporation, SI dynamics	123
Diazepam derivat., deuterated, synthesis	735	Evapotranspiration	159,165

Evolution, early organic	100	respiratory quotient	395,405
Excess free energy, IE on	25	Glucose inhibitor	360
External standard	681	Glutamate, ^{13}C distribution	581
Extraterrestrial material, isotope composition	111	Glutamate decarboxylase, IE	73,74
Extravascular water detn.	337	Glycine-^{15}N	319,329,331,335
		Glycine-^{18}O	685
Fatty acids-^{13}C, BT	367	Glycine pool	334
Fertilizers		Glycosyl transferases	720
denitrification	433,480	Graphite, isotope composition	109
mineral, isot. compos.	470,477	Graphitization, IE	103
natural, isot. compos.	470	Ground water	
nitrogen-15 labelled	425,433,439 445,479	hydrogen isotope ratio	163
		nitrate concentr.	168
Field desorption MS		nitrate in, isotope compos.	471,475
detn. of ^2H, ^{13}C, ^{15}N	639	oxygen isotope ratio	163,173,497,504
detn. of metal isotopes	644,649	pollution	428,475
Flavouring substances, hydrogen isotope ratio	491		
Fluvoxamine maleate, internal stand.	203	Harmonic approximation	18
Food constituents, origin of	465,483	Harmonic oscillation	52
Force constants	3,15,19	Health care delivery	347
Fractionation factors	45,61,63,459	Heart diseases, BT in	367
CO_2/carbonate	128	Hepatic function	353,385
nitrogen metabolism	465	Hepatitis	319
Frequencies	18	High CO_2 gases	127,132
Fruit beverages, authenticity	484	High density lipoprotein	212
source of water	496	Hill reaction	523
		Hippuric acid	326
GC/MS		Homovanillic acid, ^{18}O incorp.	255
see: correspond. comp.	605,617	Honey, carbon isot. ratio	657
see: isotope dilution		Human brain, pharmacol. anal.	256
GDP-fucose-^{13}C, synthesis	720	Huntington chorea patients	260
Geochemical cycles	121	Hydrogen	
Glucose, glucose-^{13}C		catalytic exchange	731
balance	398	detn. in breath	359,363
breath test	394,399	geochemical cycle	121
exogenous, endogenous	397,403	isotope composition see: Natural abundance	
in blood, postprandial	360	isotope effects see: Isotope effects, hydrogen	
in diabetics	403,406	isotope separation factor	31
MS	635	NMR spectroscopy of deuter.	
precursor for amino acids	581	pyrimidines	55

6-mercaptopurine	219
Hydrogen carbonate-$^{18}O_3$	
in photosynthesis	517,524
Hydrogen sulfide	
detn. of ^{34}S abundance	543
Hydroxylases, detn. in vivo	277
Hydroxylation (enzymatic)	
brain	256
cytochrome P-450	229
hydrogen isotope effects	224
Hydroxyl radicals	148
Hyperphenylalaninemia	280
Hypoxanthine-d_2, synthesis	219
Indole-3-acetic acid, GC/MS	529
Indole-3-acetic acid-d_5	529
Internal standard	187,596
abscisic acid-d_6	536
1,4-benzodiazepines-d	736
dopamine-d_4	281
fluvoxamine maleate	203
indole-3-acetic acid-d_5	529
methanephrine-d_6	247
methadone-d_8	235
3-methoxytyramine-d_5	247
methylcitric acid	287
methylmalonic acid	287,295
1-methyl-1,2,3,4-tetrahydro-ß-carboline-d_8	611
myo-inositol-d_6	605
nitrosamines-d	563
nitrosamines-^{15}N	563
normetanephrine-d_6	247
phenylalanine-d	279
L-phenylalanine-d_7	644
serotonin-d_4	283
steroids-d	617
testosterone-d_3	301
tyrosine-d_7	279
urapidil-d_4	241
Interstitial moisture	162

Intracellular fluid, NMR	583
Inorganic oxides-^{18}O	753
Ion exchanger	29,326,712
Ions, hydration sphere	29
Isotope abundance	
natural, and variations	
see: Natural abundance	
labelled compounds	
see: correspond. compounds	
Isotope dilution	187,303,593,645
Isotope discrimination	84
Isotope distribution	36
intramolecular	587
Isotope effect	
carbon	68,77,83,108
double	77
enzyme catal. react.	67,77,83,217,223 250,454,462
equilibrium pertubation method	69
geochemical	103
hydrogen	29,61,217,223 250,274,539
intrinsic	71
intramolecular	223
inverse	7
in vivo	189,227,274
kinetic	3,29,35,61,70,83,217 223,451,459,466
metals	637,703,711
nitrogen	77,454,462,466
methods	69
molar volume	25
oxygen	50,69,462
partition factor	70
rate determining step	70
secondary	192
solvent	61,74
theory	3,7,18,35,61,67
thermodynamic	3,15,23,74,470 521,693,703,711
transition state	70
vapour pressure	9,15

Isotope enrichment factor	461	Macroautoradiogram	686
Isotope exchange	39	Malic enzyme, IE	69
deuterium, pyrimidine derivatives	55	Manganese	
oxygen in magma products	137	in podzol	118
SO_2/water vapour	181	influence on enzymatic IE	86
Isotope fractionation		Maple syrup, ^{13}C-content	657
carbon		Marine carbonates, decarbonation	127
biologically mediated	95	Mass fragmentography	
CO_2/carbonate	127	see: Selected ion monitoring	
graphitization process	103	Membranes, NMR	575
enzymatic oxidation of pyruvate	42	6-Mercaptopurine - d, -2,8-d_2	
organic carbon/carbonate	99	metabolism, IE	217,220
photosynthesis	68	synthesis	218
plants	83	Metabolic switching	191
pressure dependence	149	Metabolic system, model	44
ribulosebisphosphate carboxylase	83	Metabolism	
in soil	119	amino acids	319,325,331,467
interpretation	35	aminopyrine	355
nitrogen		biochemical mechanisms of, investigations with stable isotopes	211
denitrification	460		
in biosphere	467	biogenic amines in brain	255
nitrate reduction	471	catecholamines	247
soybean nodules	451	cells in vivo	576
oxygen in water of plants	507,516	N,N-dimethylclonidin	187
sulfur, biologically mediated	95	diphenylhydantoin	223
Isotope fractionation factor	45,459	endogenous compounds	190
Isotopic anomalies	93	glucose	398,403,406
Isotopic shift	100	6-mercaptopurine	217
		neurotransmitter	190
Kerogen synthesis, IE	108	nitrogen compounds	319,325,426,465
4-Keto-cyclophosphamide	640	optical isomers	189
		phosphatidylcholine	215
Lead	137,143	prostaglandins	190
Lipoic acid, biosynthesis	543	proteins	319,426
Lithium	645	purines	313
Liquid phase thermal diffusion	704	steroids	190,307
Liver diseases	319	tryptophan	282
Liver perfusion, clonidine-$^{13}C_2$	198	Metal isotopes	623,635,649,703,711
Low density liver protein	212	Metanephrine deuterated	248
Lung diffusion test	409,415	Methadone, deuterated	235
Lysimeter	161,168,433		

Methemoglobin	229	carbon in	
Methionine-^{34}S	543	antarctic plants	121
3-Methoxytyramine-d_5	247	aquo-cyano cobyrinic acid	638
2-Methylamino-5-chloro-6-d-benzophenone	740	atmospheric CO	151
		biomolecules	37
Methylcitric acid-d_3	287	C_3 plants	87,487,495
1-Methyl-5-phenyl-7-chloro-1,3-dihydro-2H-1,4-benzodiazepine-2-one-d_3	740	C_4 plants	87,487,495
		CAM plants	88
Methylmalonic acid - d	287,295	carbonate minerals	103,128
Methylmalonyl-CoA mutase	287,295	diamonds	110
1-Methyl-1,2,3,4-tetrahydro-β-carboline-d	611	ethanol	489,495
		extraterrestrial material	111
Microbacterium ammoniaphilum, NMR	580	foods, relationship to origin	483
Microsomes	223,229,681	graphite	109
Microwave plasma detector	679	high CO_2 gases	132
Mid-ocean ridges	135	honey	657
Minerals		kimberlites	110
isotope composition (NA)	139	maple syrup	657
nitrogen binding capacity	445	proteins of foods	488
Monamine metabolites	255	sedimentary carbonates	103
Monosaccharides-^{13}C, NMR	719	soil	115
Monosaccharides-d	719	sugar	488
Monosaccharides-^{18}O	714	vanillin	490
Monosaccharide phosphates-^{13}C	719	wheat	657
Monothiopyrophosphates, IE	49	hydrogen in	
Multibox cloud model	155	antarctic lake water	124
Multi-element analysis	633	antarctic plants	121
Multiply enriched compounds		atmospheric water vapour	154
analysis	639	ethanol	492
syntheses	719	flavouring substances	491
Myocardial infarction, BT	368	foods	483
Myo-inositol, detn. in body fluids	605	moisture of chalk	161
GC/MS	605	precipitation	153,161
Myo-inositol-d_6		lead in minerals	143
internal standard	605	nitrogen	465
synthesis	606	nitrogen in	
		amino acids	468
Natural abundance (NA)		antarctic plants	121
calcium isotopes	650	arginine from bacteria	469
carbon, juvenile	103,110	food chains	466

N_2-fixing plants	454
natural fertilizers	470
nitrates	471,475
soybeans	451
urea	470
oxygen in	
atmospheric sulfate	179
ground water	497,504
hydrosphere	138
moisture of chalk	161
plants	503
precipitation	161,173,503
rocks	138
tap water	505,512
water in wine	495
silicon in minerals	139
strontium	140,143
Neurotransmitters	190,255,283
NIH-shift	192
Nitrapyrine	460
Nitrate	
fertilizer, ^{15}N content	477
ground water	168,471,475
surface water, ^{15}N content	475
reduction in vivo	
isotope fractionation	471
Nitrate-^{15}N in soil	447
Nitrate respiration	455
Nitric oxide, isotope analysis	673
Nitrifying bacteria	460
Nitriles-^{13}C, synthesis	753
Nitrite in plasma	560
Nitrite, reduction	460
Nitrogen	
geochemical cycle	121
immobilization in soil	446
isotope discrimination	466
isotope effects	
see: Isotope effects, nitrogen	
isotope fractionation	
see: Isotope fractionat. nitrogen	
metabolism	
fractionation factors	465
mineralization	446
partitioning of isotopes in plants	452
recycling of N atoms	465,475
Nitrogen-15	
amino acids synthesis	747
amino acids, tracer	331,467
detn. by	
automatic analyzer	325
emission spectroscopy	321,424
field desorption MS	639
GC/MS	424
MS with nitric oxide	673
NMR	565
excretion kinetics	319
excretion pool	329
fractionation	459
glycine-^{15}N	319,329
labelled compounds	
isolation	424
renal excretion	313,329
metabolic pool	320,329
$^{15}N_2$ preparation	424
natural abundance (NA)	425
nitrate in ground water	475
nitrogen dioxide	459,557
nitrosamines	563
proteins	325
syntheses, see: compounds	
tracer in	
Activated Sludge Process	439
animal nutrition	421,426
animal tissues	427,558
agriculture	421
N-fixation	426
urea	433
uric acid	313
Nitrogen-15 in	
fertilizers	425,433
food chains	466

pharmacological research	188
soil	445,475
soybean	451
Nitrogen dioxide	557
Nitrogen fixing plants	455
Nitrogen turnover	325
Nitrosamines, detn.	563
Nitrosamines-d, -^{15}N	563
NMR spectroscopy	573,587
see also: Carbon-13 Hydrogen Nitrogen-15	
Non-protein nitrogen	326,426
Norepinephrin - d_3	247,256
Nucleoside diphosphate sugars-^{13}C	723
Nutrients, labelled	643
Oceanic crust	136
Octanoate-^{13}C, BT	367
Oil recovery	127
Oligosaccharides-d, -^{13}C, -^{18}O	719,728
Organic acids-^{13}C	753
Osmotic coefficients, IE	23
Oxidation	
enzymatic	195,229
sulfur dioxide	180
Oxides, inorganic - ^{18}O	753
Oxygen	
isotope abundance see: Natural abundance, oxygen	
isotope effects see: Isotope effects, oxygen	
isotope geochemistry	137
photosynthetic	517
stable isotopes, production	693
Oxygen-17	
enrichment	700
Oxygen-18	
atmospheric sulfate	179
breathing apparatus for $^{18}O_2$	259
detn. by gamma activation	339

$H_2^{18}O$, tracer	229,338,511
labelled compounds, back exchange	189
$^{18}O_2$, in vivo tracer	285
photosynthetic O_2	517
Oxygen-18 in	
biomedical research	255
enzymatic oxidation	192,229
pharmacological research	187
photosynthesis	517
Oxygen isotope analysis	661,673
Parkinson patients	260
Partition factor	70
Partition function ratio, reduced	5,17
Pentaaquochromium (III) ions	61
Peroxydicarbonic acid	526
Perturbation theory	11
Pharmacokinetic protocol	355
Pharmacokinetic studies	
dilineoyl-glycerophosphocholine	211
diphenylhydantoin	288
quantitative MS in	640
urapidil	241
valproic acid	265
Phenylalanine-d	279,643
Phenylalanine-^{13}C, BT	379
Phenylalanine-4-hydroxylase	277
Phenylalaninemia	278
Phenylketonuria	280,643
Phosphatidylcholine	211,584
Postprandial blood glucose	360
Phosphoenolpyruvate carboxylase, IE	68
Phosphorus-32 labelled sludge	439
Photoisomerization abscisic acid	539
Photosynthesis	
carbon fixation	96
carbon isotope fractionation	68,83
isotope ratio of O_2	517
oxygen isotope discrimination	522
Plant hormones	529,535

Plants
  antarctic, isotopes NA   121
  C_3/C_4-, CAM-plants, isotopes in   68,85
                                          122,487,495
  carbon isotope fractionation   83
  N-nutrition   433
  oxygen isotopes in water   503
  pollutants uptake   549
  transpiration   512
Plate tectonics   135
Podzol, carbon isotopes   115
Potential energy surface   3
Precipitation
  antarctic, origin of   124
  continental effect   157
  hydrogen isotope ratio   153,163,484
  oxygen isotope ratio in   153,163
                                       173,179
Pregnanediol, MS   311
Pregnenediol   307
Pregnenolone-d   307
Prenatal diagnosis   287,300
Primordial heterogeneities   93
Propafenone -d_4, -^{14}C, metabolism   679
Propionyl-CoA carboxylase deficiency   287
Prostaglandins, -d   190,743
Proteins
  carbon isotope composition   488
  half life time   468
  metabolism   426
  separation   327
  turnover in hepatitis   319
Proteins-^{15}N, isolation   325
Proton activation analysis   685
-Proton irradiation   686
Proton-NMR (PNMR)
  see: Hydrogen
Proton transfer   62
Pseudohermaphrodism   308
Pulmonary diffusion capacity   409
Purines, metabolism   313

Pyrimidines, deuterium exchange   55
Pyrolysis of organic compounds   661

Quinazolines, syntheses   739

Radiochemicals, MS analysis   639
Rare earth elements   633
Rate constants   5,84
Rate determining step   70
Rayleigh distillation   29
Respiratory MS   411
Respiratory quotient   395,405
Rhizobium japonicum in denitrific.   455
Ribulose-1,5-bisphosphate   83
  carboxylase, IE
Rocks, isotope compos.   138
Rubidium, isot. dilut.   645

Schizophrenia   285
Sea floor spreading   135
Selected ion monitoring
  abscisic acid   538
  in enzyme activ. detn.   247,260,277
  in gas chromatography   617
  myoinositol   605
  nitrosamines   563
  pregnenolone metabolites   307
  testosterone   301
  tetrahydro-β-carbolines   611
  urapidil   244
  valproic acid   266
Separation cascade   704
Separation factor in ion exchange   30
Serotonin   257,280
Serotonin-d_4 283
Sewage treatment   439
Silicon, isotope geochemistry   137,139
Sludge, -^{15}N, -^{32}P labelled   439
Soil
  ammonium exchange   445
  ammonium fixation   449

C/N ratio	118
carbon isotope ratio	115
denitrification in	460
exchangeable nitrogen	436
hydraulic conductivity	167
interstitial water	168
nitrogen pool	434
nitrogen-15 content	475
nitrogen transformation	445
organic carbon compounds	115
water isotope content	505
movement	161,167
potential	168
Soil/plant system	
^{15}N labelled fertilizers	433
Solar system, origin of	93
Solutions (aqueous)	
electrolyte structure	23
urea, thermodyn. properties	23
tetramethylurea, thermodyn. prop.	23
Solvent activity, IE	25
Stable isotopes	
enrichment	714
separation	693,703,711
Stearic acid-d, GC/MS	213
Stereochemical marker	189
Steroids, biosynthesis, metabol.	190,307
Steroids-d	617
Stretching vibrations, CH	15
Strontium, isotope geochem.	137
Subduction zones	135
Sucrose, malabsorption, BT	359
Sugar, carbon isotope NA	488
Sulfane sulfur, ^{34}S	543
Sulfate	
bacterial reduction	96
in precipitation, oxygen isot. ratio	179
Sulfate-^{34}S in biosynthesis of sulfur compounds	543
Sulfur	
elemental	97
isotope geochemistry	98,464
terrestrial cycle	95
Sulfur-34, tracer in biosynthesis	543
detn.	543,551
sulfur dioxide in plants	549
non enzymatic oxid.	181
separation of	706
Surface water	
oxygen isotope composition	173
nitrate, isotope composition	475
Sustained release formulation	203
Symmetry number	4,36,51
Tap water, isotope content	485,505,512
Testosterone, -d_3, -^{14}C	301
biological effects	305
metabolism	301
Tetrahydrocarbolines	
origin in biological material	611
Tetramethylurea, aqueous sol.	23
Thallium, field desorption MS	645
Thermal diffusion cascade for ^{17}O	700
Thermal ionization MS	625
Thiocycteine-^{34}S, tracer	547
Thylakoids	522
Tissues	
in vivo metabolism by NMR	577
Tissue pools	
equilibrium with plasma pools	355
Tobacco mosaic virus	584
Tomography	256
Total body water detn.	337
Trace metals in biological material	631
Transition state	5,70
Transition state structures	49
Transpiration	165,503,512
Tryptamine-d_4, intern. stand.	611
Tryptophan, metabolism	282

Tryptophan-d_5 for detn. of enzyme activ.	277
Tryptophan-5-hydroxylase in vivo	283
Tryptophan pyrrolase, detn. in vivo	285
L-Tyrosine, MS analysis	636
Tyrosine-d_2, detn. of tyrosin hydroxylase in vivo	277
L-Tyrosine-d_7, MS analysis	636
L-Tyrosine-^{15}N, synthesis	749
UDP-Galactose-1-^{13}C, synthesis	720
Urapidil pharmacokinetics	241
Urapidil-d_4, synthesis	242
Urea	
CO_2 generation technique	667
nitrogen NS	
see: Natural abundance	
separation by ion exchanger	326
thermodyn. propert. of aqueous solutions	23
Urea-^{15}N, fertilizer	433
Urease, IE, mechanism	77
Uric acid - ^{15}N	
endogenous production	316
pool	315
renal clearance	313
separation by ion exchanger	326
Urine	
see: Body fluids	
Vacutainer	667
Valproic acid	
metabolism	266
antiepileptic therapy	271
urinary elimination	268,275
Valproic acid-^{13}C, pharmacokinetics, BT	265
Valproic acid-d, -d_7, pharmacokin.	275
Vanillin	
carbon isotope ratio	490
oxygen isotope ratio	661

Vapour pressure	
acetonitrile	15
differences	16,25
IE	9,25
osmotic coefficient IE	24
Vibrational frequencies	19,59
Virial coefficients	17
Vitamin B_{12}	639
Water	
extravascular detn.	337
hydrogen isotope composition	153,484
oxygen isotope composition	173,484 503,512
oxygen isotope fractionation in plants	507
total body water detn.	337
Water in	
fruits, NA	495
wine, NA	495
wine, detection	495
soil, movement	167
Water-^{18}O, tracer	229,511
Water cycle	173,503
Water vapour	
hydrogen isotope composition	153
oxygen isotope composition	505
Wheat, carbon isotope ratio	657
Wind tunnel	549
Wine	
carbon isotope ratio	489,498
detn. of sugar addition	495
detn. of water addition	495
oxygen isotope ratio	497
Xanthine-d_8, synthesis	219
Xanthine oxidase, IE	219
Zero point energy	6